TRENDS IN HIGH PRESSURE BIOSCIENCE AND BIOTECHNOLOGY

Progress in Biotechnology

Volume 1   New Approaches to Research on Cereal Carbohydrates (Hill and Munck, Editors)

Volume 2   Biology of Anaerobic Bacteria (Dubourguier et al., Editors)

Volume 3   Modifications and Applications of Industrial Polysaccherides (Yalpani, Editor)

Volume 4   Interbiotech '87. Enzyme Technologies (Blazej and Zemek, Editors)

Volume 5   In Vitro Immunization in Hybridoma Technology (Borrebaeck, Editor)

Volume 6   Interbiotech '89. Mathemathical Modelling in Biotechnology (Blazej and Ottová Editors)

Volume 7   Xylans and Xylanases (Visser et al. Editors)

Volume 8   Biocatalysis in Non-Conventional Media (Tramper et al. Editors)

Volume 9   ECB6: Proceedings of the 6th European Congress on Biotechnology (Alberghina et al., Editors)

Volume 10   Carbohydrate Bioengineering (Petersen et al., Editors)

Volume 11   Immobilized Cells: Basics and Applications (Wijffels et al., Editors)

Volume 12   Enzymes for Carbohydrate Engineering (Kwan-Hwa Park et al., Editors)

Volume 13   High Pressure Bioscience and Biotechnology (Hayashi and Balny, Editor

Volume 14   Pectins and Pectinases (Visser and Voragen, Editors)

Volume 15   Stability and Stabilization of Biocatalysts (Ballesteros et al., Editors)

Volume 16   Bioseparation Engineering (Endo et al., Editors)

Volume 17   Food Biotechnology (Bielecki et al., Editors)

Volume 18   Molecular Breeding of Woody Plants (Morohoshi and Komamine, Editors)

Volume 19   Trends in High Pressure Bioscience and Biotechnology (Hayashi, Editor)

The illustration on the cover is of an ear of ripened rice and was drawn by Miss Haruko Hikita

Progress in Biotechnology 19
TRENDS IN
HIGH PRESSURE BIOSCIENCE
AND BIOTECHNOLOGY

Proceedings First International Conference on High Pressure Bioscience and Biotechnology (HPBB- 2000), 26-30 November 2000, Kyoto, Japan

Edited by

Rikimaru Hayashi
Kyoto University, Graduate School of Agriculture,
Oiwake-cho, Kitashirakawa, Sakyo-ku, Kyoto 606-8502, Japan

2002

ELSEVIER
Amsterdam - London - New York - Oxford - Paris - Shannon - Tokyo

ELSEVIER SCIENCE B.V.
Sara Burgerhartstraat 25
P.O. Box 211, 1000 AE Amsterdam, The Netherlands

First edition 2002

Library of Congress Cataloging in Publication Data
A catalog record from the Library of Congress has been applied for.

ISBN: 0-444-50996-8

♾ The paper used in this publication meets the requirements of ANSI/NISO Z39.48-1992 (Permanence of Paper).
Printed in The Netherlands.

PREFACE

The papers in this book comprise the proceedings of the First International Conference on High Pressure Bioscience and Biotechnology (HPBB-2000), organized by R. Hayashi, C. Balny, J.C. Cheftel, K. Heremans, S. Kaneshina, C. Kanno, C. Kato, D. Knorr, S. Kunugi, H. Ludwig, A. Suzuki, J. Frank, K. Gekko, H. Iwahashi, J. Jonas, N. Klyachko, P. Masson, A. Noguchi, M. Osumi, J.L. Silva and K. Yamamoto, which was held during Nov. 26 to 30, 2000 in Kyoto, Japan.

A number of conferences devoted to various aspects of high pressure in the field of biological science have been organized and industrial applications of high pressure techniques in food processing has been developed in Japan over the past ten years: Domestic symposia, organized by the Japanese High Pressure Group, on biologically related fields have been held each year since 1988; The European High Pressure Research Group (EHPRG) has a long tradition in bringing together European scientists from various fields; The concept of having joint Japanese and European meetings on High Pressure Bioscience was elaborated by a small group of scientists and the first meeting was held in France (1992) and subsequent meetings have been held in Japan (1995), Belgium (1996) and Germany (1998). During the development of this area of science and technology, it became necessary to organize a formal international association, because of the world wide interest and the establishment of such a group was proposed at the last joint meeting in Heidelberg, Germany. HPBB-2000 in Kyoto represented the first conference sponsored by the Association and it was a great pleasure to welcome about 200 participants from more than 20 countries in Asia, Australia, Europe, North and South America, in addition to large delegations from France and Germany.

Interest in high pressure has become widespread in the field of bioscience including food sciences, pharmacy and medical fields. Some high pressure techniques are used in the production of industrial products. Moreover, high pressure is a valuable tool for the study of natural macromolecules including biomembranes which are composed, primarily, of lipid and protein. Many intermediate processes in the pressure-induced protein unfolding have been discovered, as a result. This book covers the entire range of current high pressure bioscience and its possible applications.

We wish to thank the authors for their prompt submission of manuscripts, and all participants of the Conference for creating a cordial atmosphere and for then frank and open-minded discussion.

We hope that this publication will stimulate further developments in basic and applied research as well as a new area of scientific-industrial cooperation.

September 2001, Kyoto

Editor

Rikimaru Hayashi

INTERNATIONAL ORGANIZING COMMITTEE OF HP CONFERENCE-2000

(alphabetical order)

Chairman	R. Hayashi	Kyoto Univ., Japan
Executive Committee	C. Balny	INSERM U128, France
	J.C. Cheftel	Montpellier Univ., France
	K. Heremans	Katholieke Univ., Belgium
	S. Kaneshina	Tokushima Univ., Japan
	C. Kanno	Utsunomiya Univ., Japan
	C. Kato	Japan Marine Sci. & Tech. Center, Japan
	D. Knorr	Berlin Univ. of Technology, Germany
	S. Kunugi	Kyoto Inst. of Technology, Japan
	H. Ludwig	Heidelberg Univ., Germany
	A. Suzuki	Niigata Univ., Japan
Steering Committee	J. Frank	Delft Univ., the Netherlands
	K. Gekko	Hiroshima Univ., Japan
	H. Iwahashi	Natl. Inst. of Biosci. & Human Tech., Japan
	J. Jonas	Beckman Inst., USA
	N. Klyachko	Moscow State Univ., Russia
	P. Masson	CRSSA, France
	A. Noguchi	The Ministry of Agriculture, Japan
	M. Osumi	Japan Women's Univ., Japan
	J. Silva	Rio de Janeiro Federal Univ., Brazil
	K. Yamamoto	Rakuno Gakuen Univ., Japan
Scientific Advisors	M. Fujimaki	Emeritus Prof. of Tokyo Univ.
	K. Horikoshi	Director of Japan Marine Sci. & Tech. Center
	S. Kimura	Former Director of Natl. Food Research Inst.
	T. Ooi	Emeritus Prof. of Kyoto Univ.
	K. Suzuki	Emeritus Prof. of Ritsumeikan Univ.

THE CONFERENCE HAS BEEN SUPPORTED BY:

Ajinomoto Co., Inc.

Asahi Breweries

Daiichi Giken Co., Ltd.

Echigo Seika Co., Ltd.

Ehime-ken Fruit Growers Co-operative Association

Fuji Oil Co., Ltd.

Harumi Handa

Hokurei Co., Ltd.

Kirin Brewery Co., Ltd.

Kiyoshi Hayakawa

Koma International Inc.

Kobe Steel, Ltd.

M&S Instruments Trading Inc.

Meiji Kenko Ham Co., Ltd.

Meiji Milk Products Co., Ltd.

Meiji Seika Kaisha, Ltd.

Meidi-Ya Food Factory Co., Ltd.

Morinaga Milk Industry Co., Ltd.

Nestle Science Promotion Committee

Nihonshokken Co., Ltd.

The Naito Foundation

Nikken-kasei Co.

Nippon Meat Packers, Inc.

Nisshin Food Products Co., Ltd.

Nisshin Seifun Co., Ltd.

Nitto Best Corporation

Oriental Yeast Co., Ltd.

Otsuka Chemical Co., Ltd.

Riken Chemical Industries, Ltd.

Rokko Butter Co., Ltd.

Shimadzu Corporation

Snow Brand Milk Products Co., Ltd.

Takara Shuzo Co., Ltd.

Tamotsu Shoji

Teramecs Co., Ltd.

Toppan Printing Co., Ltd.

Yamamoto Hydraulic Machinery Works, Ltd.

Yamazaki Baking Co., Ltd.

Yasushi Ooba

Yotsuba Milk Products Co., Ltd.

European High Pressure Research Group (EHPRG)

Institute National de la Santé et de la Recherche Médicale, France

Japan Society for Bioscience, Biotechnology, and Agrochemistry

Japanese Research Group of High Pressure Bioscience & Food Science

Ministry of Education, Culture, Sports, Science and Technology

The Japanese Biochemical Society

ACKNOWLEDGEMENTS:

We wish to express our gratitude and special appreciation to Ms. Setsuko Yasui, Conference secretary and to Dr. Michiko Kato and Mr. Joji Mima for their assistance in preparing these conference proceedings.

CONTENTS

Trends in High Pressure Bioscience and Biotechnology
R. Hayashi (editor)

High pressure bioscience and biotechnology:
A century and a decade perspective

K. Heremans
Department of Chemistry, Katholieke Universiteit Leuven
B-3001 Leuven, Belgium

In the last decade of the past century we have noticed an explosion in activities in the field of high pressure bioscience and biotechnology. It is a fascinating history of new experimental approaches, new concepts and new applications that has his roots in many branches of the history of scientists, science and technology.

1. INTRODUCTION

It all started a century ago with the innovative technological developments of Bridgman which allowed him to reach pressures far beyond those available at that time and his famous "observation of possible biological interest" on the white of an egg. But people were more interested in the "hot ice" that he made! During the time that Bridgman continued his physical studies, Basset and Macheboeuf laid the foundation for a long tradition of biological high pressure studies in France. Suzuki is at the origin of the "Kyoto Hot Spot". The number of laboratories in Kyoto in the field of high pressure bioscience is indeed remarkable.

During the last decade we have noticed interesting applications of high pressure technologies of the food industry in Japan, followed by the USA. Europe issued Novel Foods Regulations because, as seen in other issues, it is more hesitant to start new adventures.

This First International Conference on HP Bioscience & Biotechnology (HPBB-2000) is the outcome of a long process that started in Japan from the Japanese High Pressure Group in the Biologically Related Fields, a group that was and still is the brainchild of Prof. R. Hayashi. As we gather in this beautiful Kyoto during the last few weeks of the present century, it is appropriate to express our thanks to Prof. R. Hayashi and his team of staff and students under the silent guidance of Miss Setsuko Yasui. They will take care of us in this beautiful part of the world. When walking through Kyoto, I always keep in mind the advertisement that I discovered in the subway: "If you do not love your own garden, you cannot appreciate the beauty of the garden of your neighbour". Indeed, Kyoto is the not only the historical and cultural center of Japan, it is also an intellectual and spiritual center of the world.

2

2. A CENTURY PERSPECTIVE: THE FASCINATING HISTORY

If we extend the scope of our perspective over more than a century, it becomes clear that the development of the concept of pressure gradually developed into the simple idea of a force on a surface.

Whereas the contributions of Archimedes of Syracuse (c287-212 BC), Blaise Pascal (1623-1662) and Evangelista Toricelli (1608-1647) are well known, it was Simon Stevin van Brugghe (1548-1620) who stated and explained the "hydrostatic paradox", the fact that pressure exerted by a fluid is determined by the vertical height and not by the quantity of the fluid [1]. It was Toricelli, one of the last students of Galilei who observed that the "The height of the mercury column varied from day to day". It was Pascal who extended the analysis of Stevin and supplemented the work of all his predecessors by combining the statics of liquid and gases. His famous experiment on the Puy de Dôme showed that the mercury column was held up by the pressure of the air and not by the abhorrence of nature for vacuum as has been assumed since the time of Aristotle. Further details can be found in the monograph of Dijksterhuis [2].

The development of high pressure bioscience is not documented in a single volume. Several books contain chapters that discuss some topics related to bioscience. Bridgman himself always kept a lively interest in the field as may be judged from the last chapter in his book "The physics of high pressure"[3] and in the many reviews that are collected in his "Collected experimental papers"[4]. The most extensive source on the older literature is probably to be found in "The kinetic basis of molecular biology" by Johnson et al. [5] or the newer edition "The theory of rate processes in biology and medicine" [6]. Books that concentrate only on pressure effects are: "High pressure effects on cellular processes" by Zimmerman [7] and "The effects of pressure on organisms" a symposium volume edited by Sleigh and Macdonald [8].

Over a period of about a century the emphasis in our understanding of the effect of pressure on organisms shifted from the question of how organisms can survive under those conditions to the recent discovery that some organisms actually like these extreme conditions. This contrast can be illustrated from two quotations from the end of the 19th century and the end of the 20th century. In 1884 Regnard wrote: "Le physiologiste n'a donc plus (qu') à rechercher si, sous ces pressions énormes de plus de 600 atmosphères que supporte le fond de l'Océan, la vie existe; il doit maintenant essayer de déterminer dans quelles conditions elle peut se produire". This phrase was written with the reporting of the fascinating observations made during the "Talisman" Atlantic Deep-Sea Expedition in 1883 [9].

The recent volume on Extremophiles by Horikoshi and Grant [10] shows how well life is adapted to the extreme conditions of temperature, pressure, cold, salinity, pH, etc. An important point that is becoming clear is that under such extreme conditions one encounters almost exclusively prokaryotes, many of which belong to the archaea, an evolutionary line of microorganisms quite distinct from the other prokaryotes. The book of Gross, "Life on the edge" nicely illustrates the biotechnological developments that have resulted from the research in that field [11].

My personal view, which is strongly biased because of my interest in the behavior of molecules, is that the story really started with the observations of Bridgman. In 1914 P.W. Bridgman – he obtained the Nobel Prize for Physics in 1946 - reported "a fact of possible biological interest": "If the white of an egg is subjected to hydrostatic pressure at room temperature, it becomes coagulated, presenting an appearance much like that of a hard boiled egg". Being interested in phase transitions of compounds as a function of pressure and temperature, he notes: "The effect of temperature, which is not large, seems to be such that the ease of coagulation increases at low temperatures, contrary to what one might expect" [12].

It may be noted that nobody was interested in this observation for immediate exploitation. To the surprise of Bridgman, there was some interest from the pressmedia in the "hot ice" that he made! The complete story can be found in Bridgman's biography [13]. With a strategy developed to search for new pressure-induced transitions, Bridgman observed new phases of water among them the so-called "hot ice". He reports: "It is well known that under ordinary conditions water is abnormal in many respects. The effect of high pressure is to wipe out this abnormality…the modification of ice stable at high pressures giving indications of being the last form, corresponding to the completely normal liquid" [14].

Bridgman continued his studies on the physical properties of liquids and solids. Every new discovery stimulated him to develop new pressure equipment to reach higher pressures. These new technical developments were of great help to other scientists who, at about the same time, tried to develop procedures to sterilise milk with high hydrostatic pressure. It is of particular interest to read the reports of Hite [15] were the constant concern for technical problems with the pressure equipment hindered progress in the field. As may be judged from his reviews, Bridgman took an interest in the biological effects of pressure. The lifelong interest of Bridgman in the challenge for the making of diamonds is also well documented [16].

In France, Basset and Macheboeuf reported their results on the effect of high pressure on the inactivation of enzymes, viruses, antigens and antibodies, some of them being quite pressure stable, microorganisms, tissues and cells. In the USA, Johnson and coworkers at Princeton, studied various biosystems at low pressures in order to get insight in the physicochemical mechanisms [5, 6].

In Japan, the research concentrated on the denaturation of proteins. Here we see the first report on the pressure and temperature effects on ovalbumin and hemoglobin with the data reported as a phase diagram [17]. At low temperatures and high pressures, negative activation energies are reported and interpreted as the pressure-induced penetration of water into the protein as the first step in the denaturation process. High pressure computer simulations support this proposal [18]. The developments since then have been reviewed elsewhere [19]. It is now possible to make a unified phenomenological description of the cold, heat and pressure denaturation of proteins.

3. A CENTURY PERSPECTIVE: NEW CONCEPTS

This conference takes place almost exactly a century after Max Planck held his famous lecture, to be precise on December 14, 1900, on the interpretation of the black-body radiation. This lecture was considered as the birthday of Quantum Physics. It was the time of the development of chemical thermodynamics especially by Josiah Willard Gibbs. But since his papers were rather mathematical, he was understood by only a few. The contributions of van't Hoff, Helmholtz and Planck were more accessible for chemists. In 1884 van't Hoff derived an equation for the effect of temperature on a chemical equilibrium ($\delta \ln K/\delta T = \Delta H/RT^2$) based on the Clapeyron-Clausius equation (1834). Incidentally, he noticed that the heat absorbed or released is not the driving force for a chemical equilibrium as proposed by Bertholet. A few months later, Le Chatelier, stated the *Le Chatelier principle* based on the equation proposed by van't Hoff. Van't Hoff also derived an equation for the effect of temperature on the reaction

4

rate, but the equation became known as the Arrhenius (1889) equation ($\delta \ln k/\delta T = Ea/RT^2$). More details on the development of chemical thermodynamics can be found in Laidler [20].

The derivation of the equations describing the effect of pressure on the equilibrium was then a trivial matter except for the fact that it is much more difficult to verify experimentally. In 1887 Planck proposed the equilibrium equation ($\delta \ln K/\delta p = - \Delta V/RT$), followed by the kinetic equation ($\delta \ln k/\delta p = - Va/RT$) proposed by van 't Hoff (1901) and Evans and Polanyi (1935). This last equation is often found in high pressure articles or books as it gives for the first time a molecular interpretation of the activation volume in terms of intrinsic volume effects and solvent effects.

The combined effects of pressure and temperature on physical equilibria was first discussed by Bridgman in 1915. The application of these equations was used in the early seventies by Brandts and Hawley. The treatment makes use of the change of the Gibbs free energy change as a function of pressure and temperature:

$$d(\Delta G) = (\Delta V)\, dp - (\Delta S)dT$$

With the relation between ΔG and the equilibrium constant K, an expression is obtained, assuming that ΔV and ΔS are pressure and temperature dependent, that is used to analyze thermodynamic as well as kinetic data for the denaturation of proteins, the inactivation of enzymes and microorganisms, and many other biochemical and microbiological phenomena. The most salient feature of the equation is that it described the cold, the heat as well as the pressure denaturation of proteins.

A molecular interpretation of kinetic and thermodynamic data is usually based on data obtained from low molecular weight model systems. This is done in terms of intrinsic volume and solvent effects as first proposed by Evans and Polanyi. But the separation of these effects can only be based on molecular models. These models, again, start from some basic assumptions about the pressure and temperature behavior of matter. Occasionally one can hear very lively discussions among chemists at conferences up to the point that a lonely physicist in the audience asks the question "How real are these activation volumes?". The audience starts laughing, but during the coffee break, the reaction was different: People realize that there is thermodynamics and kinetics on the one hand and the molecular reality on the other hand and that it is not at all obvious how to bring them together even in the post-Boltzmann era.

For proteins and other biopolymers there is the additional factor of uncertainty: the packing of the atoms which is not always perfect. In order to take this effect into account Kauzmann (1959) proposed to consider the volume of a protein in solution as being composed of three contributions: The atoms, the cavities and the hydration. It is clear that pressure as well as temperature effects give rise to changes in the cavities as well as the hydration. Independent data on the cavities and the hydration are hard to get and most investigators are forced to make assumptions on the contributions of these factors. One possible technique that could be used to probe the role of cavities in proteins is positron annihilation lifetime spectroscopy. It is a powerful technique for determining the size of cavities in synthetic polymers.

4. A DECADE PERSPECTIVE: NEW APPLICATIONS

Somewhat a decade ago, the use of high pressure found new applications and industrial realizations, following the suggestions made by Hayashi and coworkers. Although Hite in the USA, Macheboeuf and coworkers in France, had studied the possible applications of high pressure for the treatment of milk and vaccines respectively, it was in Japan that the first products were put on the market. This was followed by an explosion of new papers exploring various possibilities mostly in food science but also in medical and pharmaceutical applications. The challenge for the industry was to develop new high pressure machines that could handle the vast amount of material that is usually treated in food preservation technologies.

This was not an easy job, both technically and from a commercial point of view. The developments made in Japan are recorded in the volumes (in Japanese with one page abstracts in English) edited by R. Hayashi under the title: High Pressure Bioscience Conferences in Japan [21]. A total of eight volumes has been published up to now.

The first conference on High Pressure Bioscience & Biotechnology in Europe was organized by C. Balny in 1992 in Montpellier (France). The publication of the proceedings proved to be a commercial success [22]. Other conferences were held, both in Japan and Europe, the proceedings of which are available [23]. There is also the long NATO Advanced Study Institute tradition, the proceedings of the most recent one contains a number of chapters on bioscience [24]. Planned high pressure conferences can always be found on the website of the European High Pressure Research Group [25].

The European Community supported the first research project on the application of high pressure in food science in 1992 under the direction of D. Knorr (TU Berlin). Other projects followed on dairy products under the direction of B.E. Brooker (IFR, Reading) and on the Kinetic aspects of high pressure treatment of food components, directed by M. Hendrickx (K.U. Leuven).

5. THE FUTURE: THE NEVER-ENDING HISTORY

Should we look for more molecular details? Or should we try another approach, e.g. the "polymer science" approach? Should we pay more attention to the dynamics of the glassy state, to water dynamics? Or should we look for a fresh instrument? "A fresh instrument serves the same purpose as foreign travel; it shows things in unusual combinations". This quotation by A.N. Whitehead dates from 1925 but is still valid today!

From molecules to macromolecules, from macromolecules to genes, from genes to cells and from cells to organisms. These are the levels for the conceptual interpretation of our experiments. When we try to relate these levels, we must be careful in our reasoning. Or should we make more errors in order to stimulate progress in our knowledge? We certainly must open the doors and windows of our mind and look for "unusual combinations"! Not only in theory, but also in practice. Many of our established industrial processes are not very well understood at the molecular level. In this repect there is still a lot to be learned from Mother Nature! We look forward seeing all of you in Dortmund for HPBB-2002 reporting on your new findings!

REFERENCES

1. J. Frank & K. Heremans in High Pressure Research in the Biosciences and Biotechnology, K. Heremans (Ed.), Leuven University Press (1997) 9-12.
2. E.J. Dijksterhuis, The mechanization of the world picture, O.U.P. (1961).
3. P.W. Bridgman, The Physics of High Pressure, Dover Publications (1970).
4. P.W. Bridgman, Collected Experimental Papers, Harvard University Press (1964)
5. F.H. Johnson, H. Eyring & M.J. Polissar, The Kinetic Basis of Molecular Biology, Wiley (1954).
6. F.H. Johnson, H. Eyring & B.J. Stover, The Theory of rate Processes in Biology and Medicine, Wiley (1974).
7. A.M. Zimmerman, High Pressure Effects on Cellular Processes, Acad. Press. (1970)
8. M.A. Sleigh & A.G. Macdonald, The effects of Pressure on Organisms, C.U.P. (1972).
9. In ref 5.
10. K. Horikoshi and W.D. Grant, Extremophiles. Microbial Life in Extreme Environments, Wiley-Liss (1998).
11. M. Gross, Life on the Edge, Plenum Trade (1998).
12. P.W. Bridgman, J. Biol. Chem. 19, 511-512 (1914).
13. M.L. Walter, Science and Cultural Crisis. An Intellectual Biography of Percy Williams Bridgman, Stanford University Press (1990).
14. P.W. Bridgman, Proc. Am. Acad. Arts Sci. 48, 309-362 (1911).
15. B.H. Hite, West. Va. Univ. Agr. Expt. Sta. Bull. 58, 16-35 (1899).
16. R.M. Hazen, The Diamond Makers, C.U.P. (1999).
17. K. Suzuki, Rev. Phys. Chem. Japan. 29, 91-97 (1960).
18. B. Wroblowski, J.F. Diaz, K. Heremans and Y. Engelborghs, Proteins: Structure, Function and Genetics, 25: 446-455 (1996).
19. K. Heremans & L. Smeller, Biochim. Biophys. Acta, 1386, 353-370 (1998).
20. K.J. Laidler, The World of Physical Chemistry, O.U.P. (1993).
21. R. Hayashi, Use of Pressure in Food (1989) and subsequent volumes.
22. C. Balny, R. Hayashi, K. Heremans & P. Masson (Eds.) (1992) High Pressure and Biotechnology, INSERM/Libbey.
23. a) High Pressure Bioscience and Biotechnology, R. Hayashi & C. Balny (Eds), Elsevier (1996); b) High Pressure Research in the Biosciences and Biotechnology, K. Heremans (Ed.), Leuven University Press (1997); c) Advances in High Pressure Bioscience and Biotechnology, H. Ludwig, Ed., Springer, Heidelberg (1999).
24. R. Winter & J. Jonas, High Pressure Molecular Science, Kluwer Academic (1999).
25. http://www.kuleuven.ac.be/ehprg

Trends in High Pressure Bioscience and Biotechnology
R. Hayashi (editor)

Structural features and dynamics of protein unfolding

R. Lange[a], E. Mombelli[b], J. Torrent[a], J. Connelly[a], M. Afshar[c], C. Balny[a]

[a]INSERM U128, IFR 24, 1919 Route de Mende, 34293 Montpellier, France

[b]Mathematical Biology Division, National Institute for Medical Research
The Ridgeway, Mill Hill, NW7 1AA London, United Kingdom

[c]RiboTargets, Granta Park, Abington, Cambridge CB1 6GB, United Kingdom

The heat and pressure induced unfolding process was compared for 3 model proteins: Sso7d from the thermophilic archaebacterion Sulfolobus solfataricus, Ribonuclease A, and trypsin. The methods of choice were fluorescence and UV absorbance (4th derivative mode) under equilibrium conditions, unfolding kinetics after a sudden change of the equilibrium by a pressure jump, and by molecular dynamics simulation. Protein stability appears to depend strongly on hydrophobic interactions, such as stacking between aromatic residues and van der Waals interactions. The mechanism of stabilization appears to be protein specific, and it is yet too early to draw general conclusions. However, evidence for an energetic equivalency of unfolded states could be found. Furthermore, the dynamics of protein unfolding can be interpreted in terms of the so-called "new view". The p-jump technique appears to be promising for further kinetic investigations.

1. INTRODUCTION

Proteins are built up of typically many hundred amino acids. Each amino acid residue interacts with all others to a different degree by electrostatic and hydrophobic interactions or by hydrogen bonding. Furthermore, proteins interact strongly with their solvation shell (usually water), and sometimes also between each other. Their folding and unfolding processes (which may be induced by pressure, temperature or by a chemical denaturant) may therefore be expected to be complex: multiple pathways comprising many intermediates are theoretically possible. Indeed, some first studies with single molecules suggest stepwise folding/unfolding pathways [1]. Nevertheless, the most frequently observed re-sult is a two-state transition between a native and a denatured protein structure. Sometimes, an intermediate state (it is fashion to call it «molten globule») is detected.

8

The two- or three-state transition is generally interpreted as resulting from a high cooperativity of the various elementary folding/unfolding steps. This is of course in contradiction with the stepwise unfolding observed for single molecules.

For a better understanding of the principles underlying protein unfolding mechanisms, more experimental as well as theoretical work is needed. Our approach is a thorough comparison of temperature and pressure induced protein unfolding by fluorescence and 4th derivative UV absorbance spectroscopy, as well as by a kinetic analysis of pressure jump induced protein folding and unfolding reactions. Furthermore, we used molecular dynamics simulation in order to get an insight into the very first protein unfolding events.

We used 3 very different protein models. Sso7d is a very small (7 kDa) ribonuclease and DNA binding protein from the hyperthermophilic archaebac-terion Sulfolobus solfataricus, an organism which grows optimally at 89 °C in the acid hot springs of Southern Italy [2]. The protein, which has an SH3 like structure of orthogonally disposed β-sheets, in addition to a small external α-helix, is characterized by a small hydrophobic core and a positively charged envelop. Its binding to DNA has been shown to considerably increase the DNA melting temperature. For this study, we used wild-type Sso7d, as well as its F31A mutant [3].

Another model was Ribonuclease A from bovine pancreas. This model was chosen because it is one of the best known proteins in the field of protein unfolding. We used the wild-type protein, but also 14 mutant forms, each differing in a single point mutation of a chain folding initiation site. All muta-tions were conservative. They concerned the aliphatic chain of several selected hydrophobic amino acid residues [4].

The last model was trypsin, which is a relatively large globular protein. This protein was chosen because of its hysteresis like structural response as a function of pressure [5]. The 2 other models mentioned above, do not show such a phenomenon.

2. MATERIAL AND METHODS

Proteins

Bovine pancreatic ribonuclease A was from Sigma. It was further purified as described [6]. Mutant and wild type genes were expressed in Escherichia coli and isolated as reported previously, using an FPLC Mono S HR 5/5 cation-exchange column[4]. By this procedure the following RNase A variants were prepared: I106A/L/V, I107A/L/V, V108A/G, A109G, Y115W, V116A/G, and V118A/G.

Sso7d from the archaebacterium S. solfataricus was expressed in E. Coli and purified as described by Fusi et al. [2]. The F31A mutant form was obtained by site-directed mutagenesis as described [3].

Bovine pancreatic trypsin was purchased from Sigma. The fluorescent probe 8-anilinonaphthalene-1-sulfonate (ANS) was from Molecular Probes Co.

Methods

The UV spectra were recorded with a Cary 3E (Varian) spectrophotometer. The monochromator proceeded in steps of 0.1 nm with a data acquisition time of 0.3 s per step and a bandwidth of 1 nm. Each spectrum was the result of 5 accumulations. The 4th derivatives of the UV spectra were evaluated as described [7,8], using a mean derivation window of 2.6 nm, optimal for tyrosine. The resulting derivative spectra reflect the averaged polarities of the tyrosines. Fluorescence spectra were taken with an Aminco Bowman Series 2 luminescence spectrometer with excitation and emission slits of 4 nm.

Spectral (absorbance and fluorescence) measurements were performed using a thermostated high pressure cell which was placed into the sample compartment of the spectrometers. The high pressure cell was of Marval X12 steel, the windows of sapphire, and the sample cuvette (quartz) was closed by a Dura-Seal polyethylene stretch film. Pressure jumps were performed by opening a pneumatic valve between the cell containing the sample and another cell which had been pressurized differently. This procedure allowed an equilibration of pressure between the two cells within 5 ms. The equipment permitted to undertake positive and negative pressure jumps within the range of atmospheric pressure up to 700 MPa. Typically, pressure jumps of 50 or 100 MPa were done. The accompanying small change of temperature could be neglected.

For RNase A and Sso7d, the buffer was MES from Sigma. Experiments with trypsin were performed in 1 mM HCl at 1 °C [5].

3. RESULTS

Ribonuclease A

Under high pressure, as well as at high temperature, the UV absorbance spectrum of RNase A was blue-shifted. This spectral change appears very clearly in the 4th derivative mode, as shown in Figure 1. This blue-shift reflects the increase in polarity in the tyrosine environment due to hydration as the protein unfolds. The spectral changes are characterized by clear isosbestic points, suggesting that the protein unfolding is a two-state transition. The transition was found to be completely reversible. From the change of the signal amplitude as a function of temperature and pressure, the thermodynamic parameters ΔG, ΔH, ΔS, and ΔV were determined for each mutant.

Surprisingly, we obtained virtually identical free energy values for pressure and temperature induced unfolding. This fact is reflected in Figure 2: clearly, the ΔG_u values determined from pressure and temperature effects are strongly correlated. This indicates that the unfolded states of RNase A are degenerate: temperature and pressure induced denaturation lead to equivalent energy levels. This is surprising, as temperature and pressure are expected to induce different protein conformations [9].

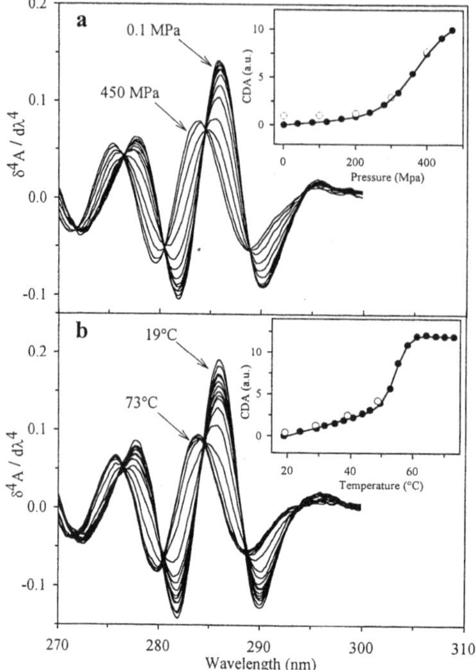

Figure 1. Effects of pressure and temperature on the 4th derivative spectrum of RNase A V116A (1 mg /ml, at pH 5).
a) effect of pressure at 40 °C,
b) effect of temperature at atmospheric pressure.
The inset shows the temperature unfolding curve for this variant. The open circles show the folding reaction. The cumulative difference amplitude (CDA) corresponds to the total signal change. The solid line is the nonlinear least-squares fit of the data based on a two-state model.

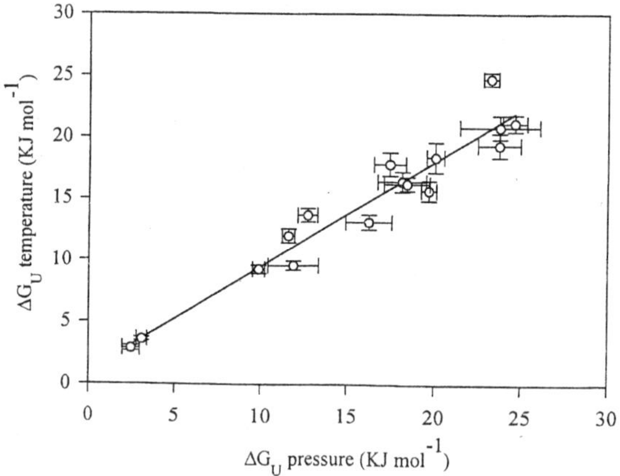

Figure 2. Correlation between the unfolding energies of the different RNase A variants determined from pressure and temperature denaturation curves at 40°C and 0.1 MPa. The standard errors are indicated.

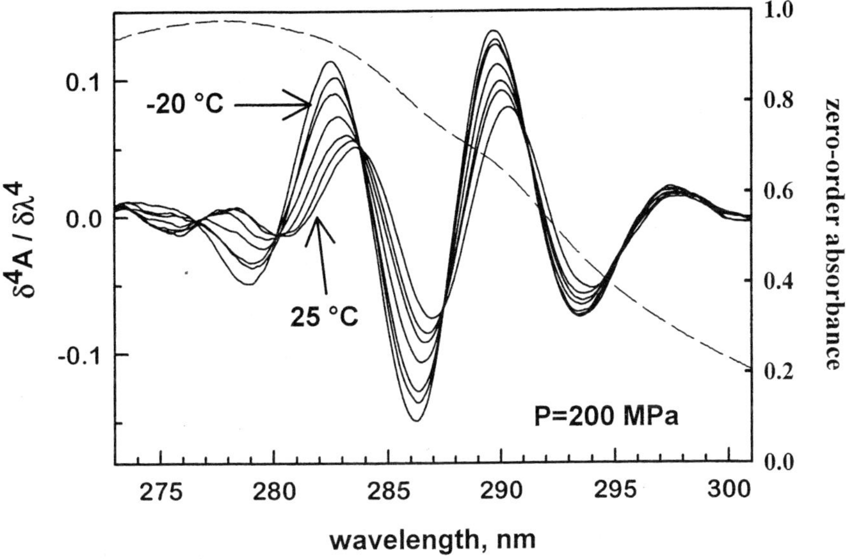

Figure 3. Fourth derivative UV spectra of Sso7d F31A as a function of temperature. Cold denaturation between +25 and –20 °C; phosphate buffer, 50 mM, pH 7.4.

Sso7d

The power of the 4th derivative method is immediately evident from Figure 3. Whereas the absolute UV spectrum was rather broad, a significantly better resolution was obtained in the 4th derivative mode. The Figure shows the result obtained by a cold denaturation experiment. That is, first pressure was raised to 200 MPa, and then temperature was decreased from +25 to –20 °C. Again, clear isosbestic points were obtained, indicating a two-state transition from the folded to the unfolded protein. The results obtained with heat, as well as with high pressure, yielded very similar results: each time a pronounced blue-shift of the tyrosine bands was observed, and the transitions were reversible. In order to obtain a further mechanistic insight, we investigated the kinetics of the unfolding / folding process. For that, we used both, experimental and theoretical approaches: pressure jumps, and molecular dynamics simulations.

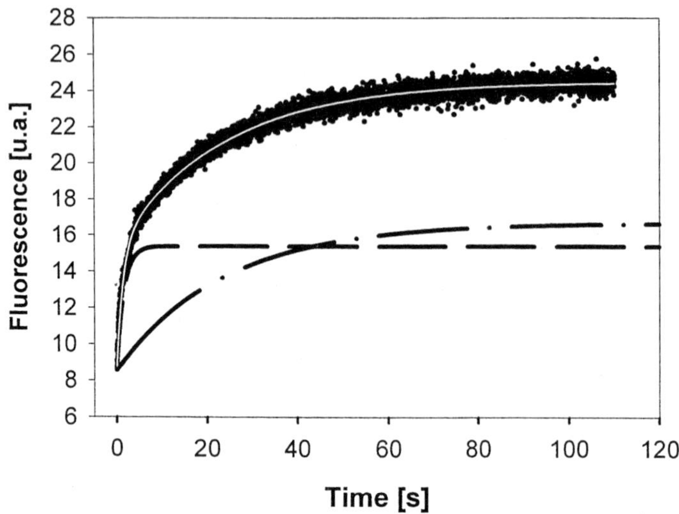

Figure 4. Kinetics of Sso7d F31A after a pressure jump from 300 to 350 MPa in the presence of ANS. The broken lines represent the fitted fast and slow phases.

- Pressure jumps. We used ANS to probe the exposition of hydrophobic domains in course of the protein unfolding. Indeed, when structured hydrophobic protein domains become accessible to ANS, the ANS fluorescence increases strongly (normally it is quenched by the polar water molecules). As shown in Figure 4, a positive pressure jump induced a strong enhancement of ANS fluorescence. The kinetics of this fluorescence increase reflect the protein unfolding kinetics. Clearly, 2 kinetic phases can be distinguished. This suggests, that protein unfolding of Sso7d occurs within at least two steps.

- Molecular dynamics simulation. We simulated the trajectory of the unfolding events after a sudden heating of the molecule to 400 K. The simulations suggested a complex unfolding mechanism within the first nanosecond. The most intriguing feature was that the small C-terminal α-helix which lies upon the surface of the protein, starts suddenly to unwind, and to swing around like a whip. After that, other parts of the protein outer layer began to change conformation. However, the very solid hydrophobic core in the center remained in its original configuration [10].

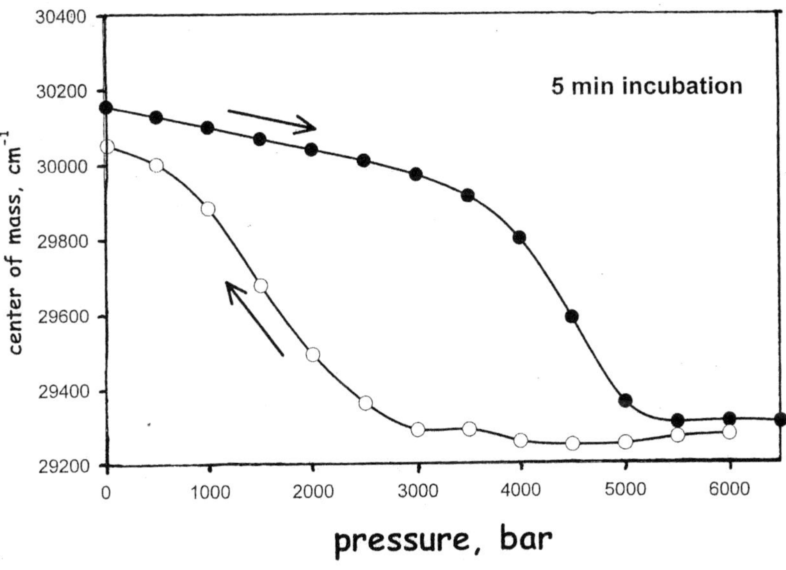

Figure 5. Pressure induced unfolding of trypsin at pH3 and 0 °C. Center of mass of the fluorescence spectra as a function of increasing and decreasing pressure.

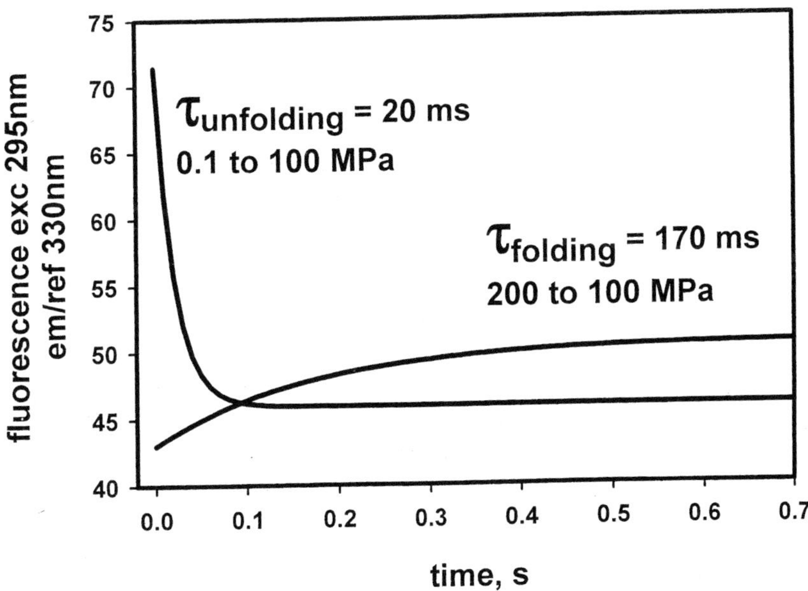

Figure 6. Kinetics of pressure induced folding and unfolding of trypsin. The experimental conditions were those of Figure 5. The final pressure was 100 MPa.

Trypsin

At 0°C and pH3, the tryptophan fluorescence emission spectrum of trypsin was strongly red-shifted. This effect is best viewed by plotting the center of the spectral mass (csm) as a function of pressure (csm is the intensity weighted averaged wave-number of the spectrum [5]). As shown in Figure 5, again a two-state transition between the folded and the unfolded state appeared. However, a strong hysteresis effect was observed: for increasing and decreasing pressures, the values of $p_{1/2}$ were 450 and 150 MPa respectively. This argues for a more complex folding/unfolding mechanism, and we effectuated therefore a kinetic analysis.

Figure 6 shows the kinetics of the change in fluorescence intensity after a pressure jump. We compared positive and negative pressure jumps, the final pressure being 100 MPa in both cases. Now, from a thermodynamic point of view, it should not matter, whether the equilibrium is reached from a lower or from a higher pressure. However, our experiments showed that the relaxation times for the positive and the negative p-jump were significantly different (20 and 170 ms respectively). These experimental findings are therefore in contradiction with the classical thermodynamic theory.

4. DISCUSSION

In the light of several diseases, such as Creutzfeld-Jacob, Huntington, etc., which imply protein conformational changes, there is actually a real need for a better understanding of protein folding/unfolding mechanism. Indeed, classical thermodynamic theory does not seem to be appropriate to describe macromolecular conformational changes, and we have therefore to collect experimental information from as many proteins as possible. Here we compared the thermal and pressure stability, as well as the mechanism of unfolding of three very different proteins: RNase A as the classical folding model, Sso7d as the representative of proteins from thermophilic organisms, and trypsin, a mammalian protein of common properties.

Their common feature for structural stabilization is the importance of hydrophobic interactions [11, 12]. This appears very clearly for RNase A, where the mutants differing in the chain-length or the configuration of hydrophobic residues are considerably less stable. For Sso7d, it is the stacking of aromatic amino acids in the hydrophobic core, which determines the stability. And finally, for trypsin, the cold denaturation results point again in favor of stabilizing hydrophobic interactions.

As to the mechanism of the folding/unfolding processes, three results appear to be important:

a) In each case, spectral measurements at equilibrium can be interpreted by simple two-states transitions. This does not mean that the reactions occur without intermediates (for example, the molecular dynamics simulation with Sso7d suggests the occurrence of intermediates). However, it indicates that if there is an intermediate, it will be a kinetic and not a thermodynamic intermediate.

b) The results obtained with RNase A indicate that the energy levels of the unfolded states are equivalent. This is in opposition of our expectation of different structural effects of pressure and temperature. A way to overcome this problem is the interpretation by the so-called "new view" [13]. This theory describes the energetic landscape of a protein as a funnel. The most simple situation would be a smooth surface of the funnel. This would explain both the two-state transitions and the energetic equivalency of the unfolded states.

c) The real situation appears to be more complex: molecular dynamics simulations as well as experimental pressure jump kinetics indicate the presence of intermediates in the course of the unfolding process. Furthermore, as shown by the example of trypsin, the kinetics are path-dependent, i.e., the same equilibrium is obtained by different mechanisms depending on the initial conditions. This observation is in clear contradiction with classical thermodynamics. Interestingly, this does not appear to be an isolated observation: similar, path-dependent unfolding kinetics have been reported recently for the major heat shock protein of E. coli [14]. A possible interpretation of such "non-classical" kinetics can be given again by the concept of the 'new view'. Indeed, a tentative explanation would be that the funnel-shape energetic landscape of the protein folding / unfolding reaction is not smooth, but rugged.

5. CONCLUSION

Clearly, our understanding of protein folding/unfolding mechanisms is still at the beginning. Whereas much is known now about equilibrium conditions, relatively little information is yet available to explain the dynamics of these processes [15]. It appears therefore evident, that much more experimental kinetic data, as well as information from dynamics simulation studies are needed in order to understand the principles underlying the mechanism of protein folding and unfolding. Experimentally, pressure-jump kinetics may reveal as a very useful tool. By this method, irreversible side-reactions can be avoided. Furthermore, this method is suitable for both positive and negative jumps, and it allows also to conduct kinetic experiments in highly viscous media. The technical problems of this method are now resolved, and suitable instrumental set-ups for both fluorescence and absorbance measurements are available [15].

ACKNOWLEDGEMENTS

E. Mombelli, J. Torrent and J. Connelly are grateful to INSERM for long term fellowships (Postes Verts). The authors thank C. Valentin for excellent technical assistance.

REFERENCES

1. G. Zocchi, Proc. Natl. Acad. Sci. USA, 94 (1997) 10647.
2. P. Fusi, G. Tedeschi, A. Aliverti, S. Ronchi, P. Tortora and A. Guerritore, Eur. J. Biochem., 211 (1993) 305.
3. E. Mombelli, M. Afshar, P. Fusi, M. Mariani, P. Tortora, J.P. Connelly, R. Lange, Biochemistry, 36 (1997) 8733.
4. J. Torrent, J.P. Connelly, M.G. Coll, M. Ribo, R. Lange, M. Vilanova, Biochemistry, 38 (1999) 15952.
5. K. Ruan, R. Lange, F. Meersman, K. Heremans and C. Balny, Eur. J. Biochem., 265 (1999) 79.
6. J. Alonso, M.V. Nogués and C.M. Cuchillo, Arch. Biochem. Biophys., 246 (1986) 681.
7. R. Lange, J. Frank, J.L. Saldana and C. Balny, Eur. Biophys. J., 24 (1996) 277.
8. R. Lange, N. Bec, V.V. Mozhaev and J. Frank, Eur. Biophys. J., 24 (1996) 284.
9. J. Zhang, X. Peng, A. Jonas and J. Jonas, Biochemistry 34 (1995) 8631.
10. R. Lange, E. Mombelli, J. Connelly, M. Afshar, P. Fusi, C. Balny, P. Tortora Adv. High Press. Biosc. Biotech., H. Ludwig, ed., (1999) Springer, 211.
11. V.V. Mozhaev, K. Heremans, J. Frank, P. Masson and C. Balny, Proteins, 24 (1996) 81.
12. K.A. Dill, Biochemistry, 29 (1990) 7133.
13. H.S. Chan and K.A. Dill, Proteins, 30 (1998) 2.
14. D.A. Leeson, F. Gai, H.M. Rodriguez, L.M. Gregoret and R.B. Dyer, Proc. Natl. Acad. Sci. USA, 97 (2000) 2527.
15. R. Lange, E. Mombelli and C. Balny, High Pressure Research, 19 (2000) 297.

Trends in High Pressure Bioscience and Biotechnology
R. Hayashi (editor)
© 2002 Elsevier Science B.V. All rights reserved.

A discussion of the physical basis for the pressure unfolding of proteins

Catherine A. Royer

Centre de Biochimie Structurale, INSERM U554, CNRS UMR 5048, 29 rue de Navacelles
34090 Montpellier Cedex, France

The application of pressures in the range below 8 kbar leads to the disruption of the native structures of most proteins in solution. This occurs because the volume of the system (protein + solvent) is smaller when the protein adopts an unfolded conformation. The underlying contributions to this decrease in volume are discussed. In addition for several protein systems, the kinetic basis for the destabilization has been determined. The results are analyzed in terms of the volumes of activation for the folding and unfolding reactions. These values in turn are interpreted in light of our limited understanding of the volume change, in terms of the position of the transition state along the reaction coordinate.

1. INTRODUCTION

Due to a decrease in system volume upon the unfolding of proteins in solution, the application of relatively low hydrostatic pressures leads to a decrease in the relative stability of the folded *vs.* unfolded state. This effect of pressure has been known since the 1940's, but continues to be perceived as counterintuitive, when it is not altogether ignored by the protein folding community. Despite this general lack of understanding and interest, the effects of pressure should be considered as fundamental (in the same right as the effects of temperature which are relatively well understood) to the study of protein stability and the specificity of the folded state. The characterization of the phase diagram of any material must be considered as the very least one should do to begin to comprehend its properties. To date, however, the phase diagrams of only a few protein systems have been reported. Below we discuss the various contributing factors to the pressure/temperature phase behavior of proteins. Secondly, in order to understand, and eventually modulate for a number of practical purposes any chemical reaction, the characterization of the kinetics of the reaction and the response of the kinetic parameters to changes in temperature and pressure must be undertaken. Interestingly, while a large number of studies have appeared describing the temperature behavior of protein folding and unfolding reaction kinetics, very few reports have appeared in which the pressure dependence of these properties has been investigated. We resume here the results from a few systems that we have studied and draw conclusions concerning the properties of the transition state of the reaction. We note that we consider here only systems in which the protein is in a simple equilibrium between a folded state and an unfolded ensemble. Pressure has also been used to characterize reaction intermediates, but that topic will not be addressed herein.

2. CONTRIBUTIONS TO THE VOLUME CHANGE OF UNFOLDING

Regardless of the atomic mechanisms underlying the decrease in system volume upon the unfolding of proteins, it is clear in general, that the effect arises from a change in the interaction of the protein with the solvent. Moreover, this decrease in volume upon unfolding is quite small in magnitude, ~-100 ml/mol, which for small single domain proteins corresponds to approximately 1% of the total protein volume. The task of identifying the underlying contributing factors remains relatively daunting. The volume of the solution is the sum of the volume of the atoms and covalent bonds, and the free volume. The covalent bonds of the protein do not change in length over this pressure range, and of course nor does the volume associated with the atoms of the solvent and the protein. Thus, we can consider that the volume change upon the unfolding of proteins arises due to changes in the free volume of the solution of protein and water. Changes in the free volume arise from changes in the non-covalent interactions between the atoms, and these changes occur due to the exposure of amino acid residues and the peptide backbone, that in the folded form are buried in the native structure, and inaccessible to solvent.

2.1 Electrostriction

Consider the disruption of an ion pair that is buried in the native structure and its subsequent exposure to the surrounding aqueous environment. Water molecules will interact more strongly (and thus at a shorter distance) with these exposed charged groups than with other water molecules, and a phenomenon termed electrostriction will occur. For a single ion pair such effects can be quite significant, resulting in a decrease in volume of nearly 20 ml/mol ion pair. This effect is responsible for the pressure dependence of the pKa of ionisable buffers, for example. The magnitude of the contribution of electrostriction to the total decrease in volume upon protein unfolding will depend upon the individual protein.

2.1 Exposure of polar moieties

The exposure of polar, uncharged moieties may also contribute to the decrease in volume, depending upon relation between the packing density of these groups in the protein interior, the specific volume of the bulk water and their specific volume in interaction with the new hydration layer. While the magnitude of this effect per mole exposed polar moiety, is surely much smaller than for an exposed disrupted ion pair, in general, the number of exposed polar moieties, including the backbone, is much larger than the number of exposed ion pairs. Hence, this contribution could indeed prove significant to the total decrease in volume.

2.3 Hydrophobic hydration

Proteins, in their native structure bury a significant proportion of their hydrophobic surface, which upon unfolding also becomes exposed to the solvent. Based on transfer studies of hydrophobic compounds from neat liquid to water (Table 1), it has been assumed by many that the transfer of hydrophobic groups from the interior of proteins to aqueous solution would lead to a very large decrease in volume. A few investigators have questioned this assumption (1-4). In fact, the density of neat liquids of hydrophobic compounds is much smaller than that of hydrophobic groups in proteins. In fact, proteins are very tightly packed. A methyl group in a protein interior has a specific volume which is approximately 5 times smaller than neat methane. Thus the large decrease in specific volume observed in transfer studies of hydrophobic compounds to water is largely due to their low density in the pure form. Interestingly, it has been noted in a transfer study of water into solutions of hydrophobic compounds that its specific volume increased (5) (Table 2). Thus, water molecules hydrating hydrophobic moieties that are exposed upon unfolding may well lead to an overall increase, rather than a decrease in the system volume.

Table 1.
Specific volume of hydrocarbons in various solvents (ml/mol) (5).

Solvent	Methane	Ethanol
Per-fluoro-n-heptane	68.4	82.9
n-heptane	60.0	69.3
Carbon tetrachloride	51.7	66.0
Water	37.3	51.2

Table 2. Specific volume of water in various organic solvents (ml/mol) (6).

Solvent	Methane
1,2 dichloroethane	20.1 ± 1
benzene	22.1
1,1,1-trichloroethane	22.3
Carbon tetrachloride	31.6
Water	18.0

2.2 Loss of cavities and voids

As noted above, protein interiors are very well-packed. But as for all substances, according to scaled particle theory, there exist voids in the interior of proteins between the atoms. In addition, to the voids due to the spherical nature of the atoms, actual cavities exist in the protein structure, which are often small enough or inaccessible enough as to exclude solvent. Upon unfolding of the protein, the interaction of the exposed surface with the relatively small water molecules leads to a decrease in system volume akin to that which is achieved by adding small spheres to a box containing larger spheres. The smaller spheres can take up excluded volume in between the larger spheres. Moreover, the unfolding of the protein also eliminates the free volume due to actual cavities in the structure, and this space is occupies by water molecules. Thus, the elimination of packing defects and scaled particle free volume necessarily leads to a decrease in system volume upon unfolding. Actual cavities

20

in proteins (defined as occupying approximately the volume of one water molecule) can constitute up to 2% of the total protein volume (7).

2.5 A word on compressibility changes upon unfolding

The volume changes which have been reported for protein folding reactions represent the results of fits of pressure denaturation profiles obtained by measuring some observable (fluorescence, UV absorbance, chemical shift, enzyme activity, etc.) as a function of increasing pressure. Assuming a two state equilibrium between a folded, native structure and an unfolded, denatured state, and taking into account whatever mass action considerations apply (monomer unfolding or unfolding/dissociation of oligomers), then one can fit the profile taking into account only the first virial coefficient, that is the change in volume between the two states, ΔVu.   Alternatively, it has been proposed that the change in isothermal compressibility, $\Delta\beta u$, between the two states is significant, and thus that this parameter should also be taken into account. Obviously the value of the volume change upon unfolding obtained from analysis of the unfolding profiles is not the same depending upon which equation is used.   First of all, one must be careful in comparing the volume change values found in the literature since these have not necessarily been obtained using the same method of data analysis. Secondly, and more importantly, the correctness of one or other of these approaches merits discussion. Use of the $\Delta\beta u$ term in the analysis of the unfolding profiles we have obtained for staphylococcal nuclease, *trp* repressor and P13[MTCP1] by fluorescence, or other methods does not improve the fits, and therefore we have not used this approach.   Eftink (8) has shown that global analysis of the temperature and pressure dependence of the stability of staphylococcal nuclease necessitates inclusion of a heat capacity change, ΔCp ,and a change in the coefficient of thermal expansion, $\Delta\alpha u$, but does not require a $\Delta\beta u$ term.   Finally, we have recently measured directly the specific volume of staphylococcal nuclease as a function of both temperature and pressure using a high pressure densitometer (9).   We find no significant difference in the compressibility of the folded and unfolded states of the protein. Moreover, the magnitude of the decrease in volume upon unfolding by pressure at 40°C observed directly by this method is in good agreement with that obtained from simple fits (not taking into account a $\Delta\beta u$ term) of the data at the same temperature.   These findings, taken together, suggest that the difference in compressibility between the folded and unfolded states may be negligible.   We note that the accuracy of any of these methods is no better than 10%, such that we cannot conclusively state that this is true. However, in the absence of more accurate information, the rule of thumb should be to fit the data with the simplest model which accurately describes the experimental results, within the statistical error of the measurements.   This should include the uncertainty arising from the correlation between the parameters of any fit, and requires rigorous confidence limit testing procedures.

3. THE KINETIC BASIS FOR PRESSURE-INDUCED UNFOLDING OF PROTEINS

If the application of pressure leads to the destabilization of the folded structure of proteins, relative to their unfolded state, this necessarily arises because pressure effects the rate constants of folding, unfolding or both, since the equilibrium constant is simply the ratio between these two.   Using pressure jump relaxation methods, we have characterized the pressure dependence of the rate constants for the folding and unfolding of three proteins, staphylococcal nuclease, *trp* repressor, and P13[MTCP1].   We note that although the first two proteins exhibit complex unfolding/refolding kinetics at atmospheric pressure using chemical denaturants, under pressure, the relaxation profiles obtained from a number of observables (including those used at atmospheric pressure) exhibit pure single exponential relaxation. The third exhibits simple two-state behavior under all conditions. Using Eyring transition state theory, the effect of pressure on the rate constants can be ascribed to the sign and magnitude of the volumes of activation. For example, for the folding rate constant, we can write:

$$kfp = kfo(exp(-p\Delta V^{\ddagger}/RT)) \qquad\qquad (1).$$

Thus, if the volume of activation for folding is positive then the folding rate constant will decrease as a function of pressure, the forward reaction will therefore slow down. A negative activation volume results in an increase in the reaction rate. Determination of the magnitude and sign of the activation volume reveals information about the nature of the transition state in the folding/unfolding reaction. A negative activation volume for unfolding informs us that the system volume of the protein in the transition state is smaller than in the folded state. A decrease in volume between the folded and the transition state is indicative of disruption of structure and increased hydration of the transition state. A positive activation volume for folding reveals that the transition state volume is larger than that of the unfolded state. This means that the rate-limiting step n folding is accompanied by significant dehydration. The relative magnitudes help to place the transition state along the reaction coordinate.

3.1 Staphylococcal nuclease

The pressure dependence of the stability of staphylococcal nuclease is arguably the best characterized to date. Studies monitoring the pressure dependence of the fluorescence emission of WT and in some cases mutant staphylococcal nuclease at equilibrium and in many instances the kinetics of folding as a function of pH, temperature, concentration of an osmolyte stabilizer and of the denaturant guanidine hydrochloride (8, 10-16) have been carried out. Moreover, the pressure dependence at equilibrium and the pressure dependent kinetics of the FTIR spectrum (15), the 1-D hisitidine $^{1}\varepsilon$ proton NMR spectrum (10,17) and the small angle X-ray scattering signals (15,16) have been characterized. Recently, we reported the direct measurement of the specific volume of staphylococcal nuclease as a function of temperature and pressure using high pressure, variable temperature densitometry (9). What can be learned from this large body of data is that under pressure the two state model for nuclease unfolding holds very well both at equilibrium and in the kinetics. The lack of effect of the additives on the apparent volume change of unfolding is also interesting, as is the observation that the effect of osmolyte is primarily to increase the rate constant for folding. This latter observation is consistent with a general destabilization of the unfolded state by osmolyte. In terms of the pressure dependence of the kinetics of folding and unfolding, these studies have demonstrated that the formation of tertiary structure (as observed by fluorescence and histidine $^{1}\varepsilon$ proton NMR), the formation of secondary structure (as observed by FTIR) and collapse (as monitored by SAXS) all exhibit the same rate kinetics and pressure dependence, and thus all depend upon the same rate limiting step. Most interestingly, it was found that the activation volume for folding (~+80-90 ml/mol) is nearly equal to the total apparent volume change for folding (~+80 ml/mol), indicating that the hydration and packing of the transition state are nearly identical to that of the folded state (Figure 1).

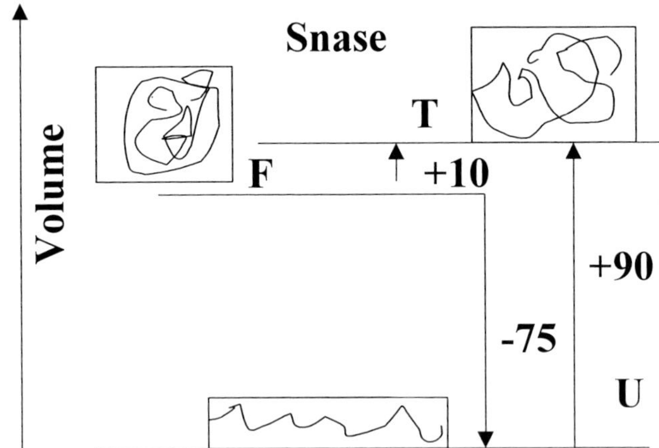

Figure 1.Volume diagram for the unfolding of Staphylococcal nuclease at room temperature.

3.2 *trp* Repressor

The *trp* repressor is a dimeric protein of 26000 MW for the dimer. The interface between the two monomers is quite extensive and consists of the intertwining of three helices from each monomer. In pressure-jump relaxation studies of the fluorescence signal from its intrinsic tryptophan residues, the unfolding/refolding was found to be single exponential, unlike the kinetics at atmospheric pressure, indicating that the intermediates and/or unfolded state heterogeneity are disfavored by pressure (18). In this case the activation volume for unfolding was found to be negative, whereas that for folding, as in the case of nuclease was positive. The magnitude of the activation volume for folding corresponds to approximately 2/3 of the total volume change upon folding, whereas that for unfolding accounts for the

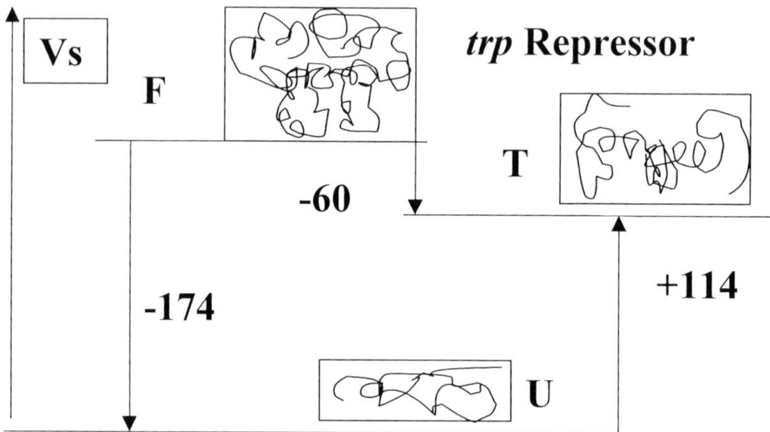

Figure 2. Volume diagram of the *trp* repressor

remaining 1/3. Moreover, the concentration dependence of the folding rates demonstrates that the rate-limiting step in the folding of this protein involves dimerization. Thus, taken together with the values and signs of the activation volumes for folding and unfolding, these results indicate that in the transition state, the dimer interface involving helices A-D and F, is structured, whereas the DNA reading heads involving helices D and E remain disordered.

3.3 P13[MTCP1]

Recently, we have undertaken the study of the folding properties of a small oncogene product the structure of which folds into a canonical filled β-barrel, with a novel and unique topology (19). Since topology has been suggested as one of the primary determinig factors in controlling folding rates, and in may cases stability (20), it is of interest to characterize the folding of novel topologies. In studies of the guanidine hydrochloride, temperature and pressure dependence of the folding of P13MTCP1 we have found that it is the slowest folding protein of β-structure studied to date (manuscripts in preparation). Moreover, due to the large positive activation volume for folding, the folding reaction is slowed even more under pressure such that at 3 kbar, the relaxation occurs over several hours.

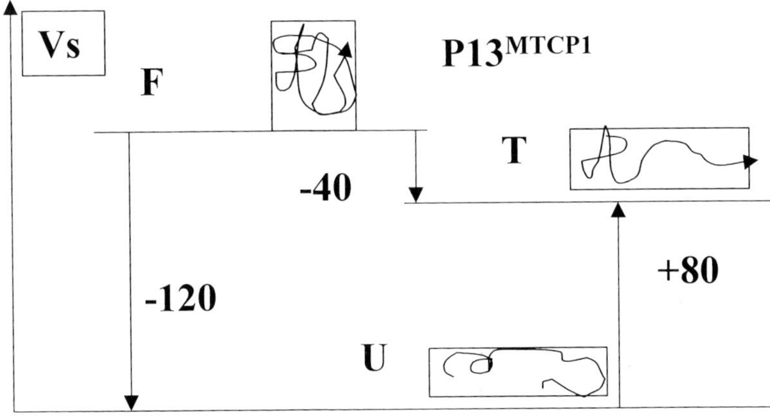

Figure 3. Volume diagram of P13[MTCP1].

4. CONCLUSIONS

In addition to the systems mentioned above, the protein tendamistat (21) also exhibits a positive activation volume for folding that places the transition state at approximately 60% folded, much as in the case of *trp* repressor and P13[MTCP1]. However, the number of thorough studies of the kinetics of pressure induced unfolding is relatively small. Thus, any generalizations must await the constitution of a much larger database. In addition, it is important to stress that the field needs a much more profound understanding of the fundamental basis of the volume change upon unfolding. Because this volume change is likely to represent the small difference that arises from the superposition of large values of opposite sign (not uncommon in protein thermodynamics in general), designing experiments to determine the sign and magnitude of the various contributing factors to the overall value of the volume change likely represents the most challenging aspect of high pressure protein folding.

REFERENCES

1. Hvidt, A. (1975) *J. theor. Biol.* **50**, 245.

2. Klapper, M. H. (1971) *Biochim. Biophys. Acta* **229**, 557.

3. Harpaz, Y., Gerstein, M. & Chothia, C. (1994) *Structure* **2**, 641.

4. Prehoda, K E, & Markley, J L (1996), *In* High-Pressure Effects in Molecular Biophysics and Enzymology. J L Markley, D B Northrop, and C A. Royer, eds. Oxford University Press, New York. 33.

5. Masterton, W L (1954) *J. Chem. Phys.* **22**, 1830.

6. Masterton, W L & Seiler, H (1968) *J. Phys. Chem.* **72**, 4257.

7. Rashin, A A, Iofin, M & Honig, B (1986) *Biochemistry* **25**, 3619.

8. Eftink, M R &Ramsay, G D (1996) *In* High-Pressure Effects in Molecular Biophysics and Enzymology. J. L. Markley, D. B. Northrop, and C. A. Royer, editors. Oxford University Press, New York. 62.

9. Seeman, H., Royer, C.A. & Winter, R. (2001) *J. Mol. Biol. (in press).*

10. Royer, C A, Hinck, A P, Loh, S N, Prehoda, K E, Peng, X, Jonas, J & Markley, J L (1993), *Biochemistry*, **32**, 5222.

11. Vidugiris, G J A, Markley, J L & Royer, C A (1995) *Biochemistry,* **34**, 4909.

12. Frye, K J & Royer, C A (1997) *Protein. Sci.,* **6**, 789.

13. Vidugiris, G J A, Truckses, D M, Markley, J L & Royer, C A (1996) *Biochemistry,* **35**, 3857.

14. Frye, K.J. & Royer, C.A. (1998) *Protein Sci 7*, 2217.

15. Panick, G, Malessa, R, Winter, R, Rapp, G, Frye, K J & Royer, C A (1998) *J. Mol. Biol.,* **275**, 389.

16. Panick, G., Vidugiris, G.J.A., Malessa, R., Rapp, G., Winter, R. & Royer, C. (1999). *Biochemistry*, **38**, 4157.

17. Prehoda, K.E., Mooberry, E.S. & Markley, J.L. (1998) *in* Protein Dynamics, Function and Design, J.-F.Lefèvre & R.E. Holbrook, eds., Plenum, New York, p.59.

18. Desai, G., Panick, G., Zein, M., Winter, R. & Royer, C. A. (1999) *J. Mol. Biol*, **288**, 461.

19. Guignard, L., Padilla, A., Mispelter, J., Yang, Y.-S, Stern., M.-H., Lhoste, J.M. & Roumestand C. (2000). *J. Biomol. NMR* **17**, 215.

20. Alm, E. & Baker, D. (1999) *Curr. Opin. Struct. Biol.* **9**, 189.

21. Pappenberger G, Saudan C, Becker M, Merbach AE, Kiefhaber T. (2000) *Proc. Natl. Acad. Sci. U S A.*, **97**, 17.

Trends in High Pressure Bioscience and Biotechnology
R. Hayashi (editor)
© 2002 Elsevier Science B.V. All rights reserved.

Thermal and pressure stability of Phe46 mutants of ribonuclease A

E. Chatani[a], R. Hayashi[a], R. Lange[b] and C. Balny[b]

[a]Division of Applied Life Science, Graduate School of Agriculture, Kyoto University, Sakyo, Kyoto 606-8502, Japan

[b]INSERM U128, IFR 24, 1919 Route de Mende (CNRS), F-34293 Montpellier, Cedex 5, France

To investigate roles of Phe46, which is located in the hydrophobic core, of the protein bovine pancreatic RNase A, the thermal and pressure denaturation of wild-type and three Phe46 mutant RNase A (F46V, F46E, and F46K) were analyzed by means of fourth derivative UV absorbance spectroscopy under pressures of 0.1-400 MPa or temperatures of 4-70 °C. The red-shift of λ_N on the fourth derivative UV spectrum and the change in far-UV CD spectra of F46E and F46K mutant RNase A suggest that the hydrophobicity of Phe46 is important in maintaining the original RNase A conformation. All mutant enzymes, as well as the wild-type enzyme showed a two-state transition curve during both thermal and pressure denaturation, and the T_m value was decreased significantly by the mutation of Phe46. Moreover, the Gibbs free energy, as calculated from the thermal transition (ΔG_t) was significantly larger than that calculated from the pressure transition (ΔG_p) in F46V, F46E, and F46K, while no difference between ΔG_t and ΔG_p was observed in the case of the wild-type RNase A. These results led to the conclusion that the pressure-unfolded state in the mutant enzymes contains some folded structures.

1. INTRODUCTION

The structure of many globular proteins is typically comprised of a hydrophobic core, which is generally thought to be important for folding and stability (1). Bovine pancreatic ribonuclease A (RNase A) is typical of this type of protein and has been studied as a model protein in this respect. Some hydrophobic residues in RNase A serve as a chain-folding initiation site (CFIS) in folding (2, 3), as a structural determinant (4), or for structural stability (2, 3, 5).

Phe46 is located in the center of the hydrophobic core (Fig. 1), and is well conserved in mammalian ribonucleases. To study the role of this residue on protein folding, structure construction, and stability, three mutant RNase A, F46V, F46E, and F46K were produced and their thermal and pressure denaturation were monitored by a fourth derivative UV spectroscopic method (6, 7). The use of pressure as a protein-structural perturbant provides useful information, *i.e.*, the volume change in unfolding, which reflects changes in the structure and hydration of the native and/or denatured proteins. The goal of this study is clarify not, only the importance of Phe46 in maintaining the original conformation and stability of RNase A, but also to examine the difference in thermal and pressure effect on the unfolding of Phe46 mutants.

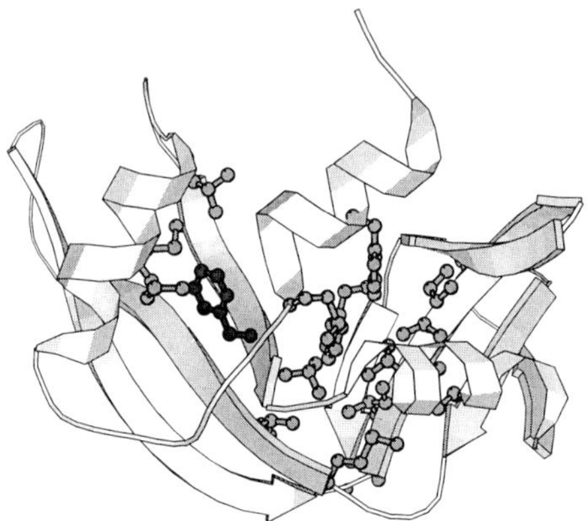

Fig. 1. Ribbon model of wild-type RNase A. Ball-and-stick models show the amino acid residues which comprise the hydrophobic core. Phe46, located in the center of the hydrophobic core, is colored by dark gray.

2. MATERIALS and METHODS

2.1. Preparation of wild-type and mutant enzymes

Commercial RNase A (type III-A) from Sigma (Missouri, USA) was used as a wild-type enzyme. F46V, F46E, and F46K mutant RNase A were produced using the *Escherichia coli* expression system which was constructed according to the method of delCardayre *et al*. with minor modifications (8, 9). The expressed protein was purified using the same method as previously described (10).

2.2. Thermal or pressure denaturation monitored by UV Spectroscopy

In a typical experiment, the protein was dissolved in 50 mM MES buffer, pH 5.5 to a concentration of 73 μM.   After increasing either temperature or pressure, the protein solution was incubated for 3 min, which is sufficient for equilibration to be attained, and its absorption spectrum was measured, in the range of 250-310 nm.

The fourth derivation of the measured spectra was carried out according to a previously described method (6, 7).   The spectra measured under high pressure were corrected for pressure-dependent changes in volume before calculating the fourth derivative.

2.3. CD spectroscopy

In a typical experiment, the protein was dissolved into 50 mM MES buffer, pH 5.5 at a concentration of 10 μM.   The CD spectrum from 190 to 260 nm of the protein solution was measured with a 0.5 mm-long optical path at 6.8°C, at which conditions, the existence of thermally-denatured protein is negligible for all RNase A enzymes.

3. RESULTS and DISCUSSION

3.1. Contribution of Phe46 to determining RNase A conformation

The wavelength at the highest peak of the fourth derivative curve in the native state, λ_N (see Fig. 2) of wild-type and F46V were nearly 260.0 nm, but the comparable peaks for F46E and F46K were red-shifted to 285.1 and 284.3 nm, respectively.   The CD spectra and the estimation of secondary structure from the amino acid sequence using a protein sequence analysis (PSA) server also suggested some conformational destruction around position 46 in F46E and F46K.   Thus, the hydrophobic nature of the amino acid residue at the position 46 appears to be necessary for retaining the correct conformation of RNase A.   A positive charge may destroy the conformation more significantly than a negative charge, which suggests that some interaction between the π electron of Phe46 and the surrounding charged groups in the wild-type enzyme is important in maintaining protein structure.

3.2. Contribution of Phe46 to conformational stability, and difference between thermal- and pressure-induced unfolded states observed by the replacement of Phe46

When the amplitude of $d^4\varepsilon/d\lambda^4$ at λ_N (Fig. 2) was plotted against pressure or temperature, both the thermal and pressure denaturation plots of all mutant enzymes, as well as the wild-type RNase A, gave a two-state sigmoidal curve which could be easily fitted to the following equations:

$$A = \frac{(A_n - mT) - (A_d - qT)}{1 + e^{-[(\Delta H - T\Delta S)/RT]}} + (A_d - qT) \qquad \text{(thermal denaturation)}$$

$$A = \frac{A_n - A_d}{1 + e^{-[(\Delta G_p + P\Delta V)/RT]}} + A_d \qquad \text{(pressure denaturation)}$$

where ΔH, ΔS, ΔG, and ΔV are enthalpy change at T_m, entropy change at T_m, the Gibbs free energy change at T (K) and 0.1 MPa, and the volume change by pressurization at T (11), respectively. The m and q terms in the equation for thermal denaturation were introduced, in order to take into consideration the change in the intrinsic tyrosine spectrum by increasing the temperature (3). We conclude that both pressure and temperature cooperatively induce the unfolding of a tertiary structure, as evaluated by the tyrosine environment, in both the wild-type and mutant enzymes.

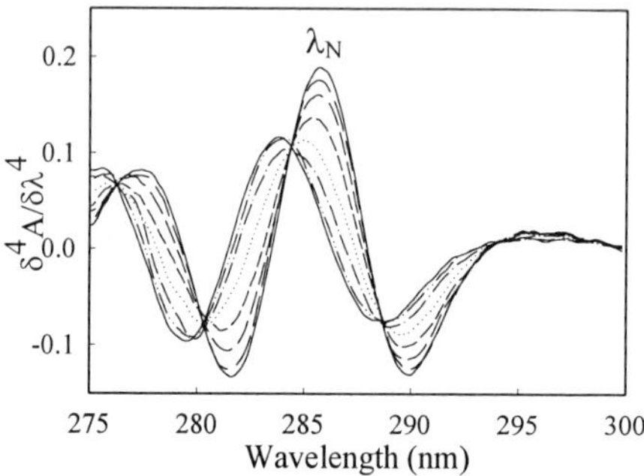

Fig. 2. A typical fourth derivative spectra showing the pressure transition of F46V RNase A monitored at 40.0°C. λ_N shows the wavelength at the highest peak in this figure.

The T_m values of wild-type and F46V, F46E, and F46K RNase A were 58.3±0.5, 46.7±0.5, 28.1±0.5, and 24.6±0.8°C, respectively. The destabilization effect, resulting from the replacement of Phe46 with the smaller hydrophobic residue, valine, was more dramatic than that by the replacement of Phe120 (10), suggesting that Phe46 is more important in protein stability than Phe120. In addition, the very drastic conformational destabilization of F46E and F46K suggests that the introduction of a charged group changes the conformation of RNase A and decreases its conformational stability as well.

To compare the difference in unfolded states by temperature and by pressure, attempts were made to calculate the ΔG of thermal denaturation (ΔG_t) and of pressure denaturation (ΔG_p) at constant conditions of temperature and pressure. However, while ΔG_t at 0.1 MPa and some T (K) can easily be obtained by the equation, $\Delta G_t = \Delta H - T\Delta S$, it was impossible to measure pressure denaturation of all enzymes at the same temperature because the T_m of the wild-type and mutant enzymes varied from 24 to 59°C (see above). To overcome this problem, pressure denaturation curves were obtained at various temperature under which the existence of thermal-denatured proteins can be ignored, each ΔG_p was plotted against temperature and pressure, and the ΔG_p at 0.1 MPa, 0°C was then obtained by fitting the 3D plot to the equation:

$$\Delta G_p = \Delta\alpha \cdot P(T-273)+\Delta V \cdot P-\Delta S(T-273)+\Delta G_{p \text{ at } 0.1 \text{ MPa, } 0°C}$$

where $\Delta\alpha$, ΔV, and ΔS are volume expansibility, volume change at 0.1 MPa and 0°C, and entropy change at 0.1 MPa and 0°C, respectively.   The values of ΔG_p at 0.1 MPa and 0°C, obtained by the curve fitting was used for the comparison with ΔG_t at 0.1 MPa and 0°C .

As a result, while no energetic differences between pressure- and temperature-unfolded states of wild-type RNase A were evident, which is consistent with the previous report by Mombelli et al. (3), significant differences were detected among F46V, F46E, and F46K in which ΔG_p was obviously smaller than ΔG_t (Fig. 3).   The ΔV values for F46V, F46E, and F46K at 0°C and 0.1 MPa were −89.7±8.3, -83.3±6.2, and 79.4±8.1 ml/mol, respectively, and these values were clearly smaller than the ΔV of wild-type, -114.8±11.5 ml/mol.   The smaller values of ΔG_p and ΔV for the mutant enzymes suggest that the pressure-induced unfolded states of F46V, F46E, and F46K retain some partially-folded structures.

Table 1.   Comparison of ΔG_t and ΔG_p at 0°C and 0.1 MPa

Enzyme	ΔG_t (kJ/mol)			ΔG_p (kJ/mol)		
Wild type	92.5	±	30.2	106.9	±	4.6
F46V	57.2	±	4.5	31.6	±	3.4
F46E	23.9	±	2.9	15.9	±	0.8
F46K	16.5	±	2.9	9.3	±	2.0

See the text for details of the calculation of ΔG_t and ΔG_p.

The results of DSC measurements provide support for the conclusion that conformational cooperativity is weakened in the case of F46E and F46K.

The Phe46 is positioned inside the hydrophobic core and interacts with both the right and left parts of the core to make up one compact domain (Fig. 1).   The replacement of Phe46 with other amino acid residues, especially charged amino acid residues, might allow the two parts to separate.   Pressure might induces the step-wise unfolding of the two parts and, as a result, conformational cooperativity is lost.

32

REFERENCES

1. M. Vlassi, G. Cesareni, and M. Kokkinidis, J. Mol. Biol., 285 (1999) 817.
2. M.G. Coll, I. I. Protasevich, J. Torrent, M. Ribo, V. M. Lobachov, A. A. Makarov, and M. Vilanova, Biochmical and Biophysical Research Communications, 265 (1999) 356.
3. J. Torrent, J.P. Connelly, M.G. Coll, M. Ribo, R. Lange, and M. Vilanova, Biochemistry, 38 (1999) 15952.
4. B. Gutte, J. Biol. Chem., 252 (1977) 663.
5. N. Tanimizu, H. Ueno, and R. Hayashi, J. Biochem., 124 (1998) 410.
6. R. Lange, J. Frank, J.L. Saldana, and C. Balny, Eur. Biophys. J., 24 (1996) 277.
7. R. Lange, N. Bec, V.V. Mozhaev, and J. Frank, Eur. Biophys. J., 24 (1996) 284.
8. S.B. delCardayre, M. Ribo, E.M. Yokel, D.J. Quirk, W.J. Rutter, and R.T. Raines, Protein Eng., 8 (1995) 261.
9. R.W. Dodge, and H.A. Scheraga, Biochemistry, 35 (1996) 1548.
10. E. Chatani, N. Tanimizu, H. Ueno, and R. Hayashi, J. Biochem., 129 (2001) 917.
11. E. Mombelli, M. Afshar, P. Fusi, M. Mariani, P. Tortora, J.P. Connelly, and R. Lange, Biochemistry, 36 (1997) 8733.

Trends in High Pressure Bioscience and Biotechnology
R. Hayashi (editor)

Pressure and temperature-induced denaturation of carboxypeptidase Y and procarboxypeptidase Y

Michiko Kato, Rikimaru Hayashi, Reinhard Lange[*], and Claude Balny[*]

Laboratory of Biomacromolecules, Division of Applied Life Sciences, Graduate School of Agriculture, Kyoto University, Sakyo-ku, Kyoto 606-8502, Japan
[*] INSERM U-128, IFR 24, 1919 Route de Mende (CNRS), F-34293 Montpellier, Cedex 5, France

Pressure-induced denaturation of carboxypeptidase Y (CPY) and procarboxypeptidase Y (proCPY) was investigated by measuring the intrinsic fluorescence in the temperature range of 20-80°C or the pressure range of 0.1-700 MPa.   The pressure-induced denaturation of CPY showed at least a three-state transition, while temperature-induced denaturation of CPY showed a two-state transition.   The pressure and temperature-induced denaturation of proCPY showed a simple two-state transition.

1. INTRODUCTION

Carboxypeptidase Y (CPY), a vacuolar enzyme from *Saccharomyces cerevisiae* belongs to the family of serine proteases (1).   Procarboxypeptidase Y (proCPY) is a precursor of CPY which consists of a mature region (421 amino acid residues) and a propeptide of 91 amino acid residues at the N-terminus.   The propeptide is an essential for folding, as has been shown by both in vivo and in vitro studies (2-4), and for maintaining CPY in an inactive state.

The intrinsic protein fluorescence which is mainly due to tryptophan and tyrosine residues (5) reflects the conformation of proteins: the wavelength of the maximum fluorescence, the center of the spectral mass, changes depending on the polarity of the environment of these residues (6). Since CPY contains 10 tryptophan and 24 tyrosine residues, which are evenly distributed throughout the entire protein molecule, the denaturation of CPY can be measured by changes in its intrinsic fluorescence (7).

In this report, differences between pressure-induced denaturation and temperature-induced denaturation of CPY and proCPY were observed by measuring fluorescence changes and the conformation of the two proteins is discussed.

2. EXPERIMENTAL PROCEDURES

Materials: Carboxypeptidase Y (CPY) was obtained from Oriental Yeast Co. (Lot 21003805) (Osaka, Japan). ProCPY was purified from yeast strain BJ2168 transformed by pTSY3 in our laboratory. 3-(N-Morpholino) propanesulphonic acid, Mops, was obtained from Nacalai Tesque (Kyoto, Japan).

Fluorescence spectroscopy: Fluorescence measurements of CPY and proCPY were made using an Aminco-Bowmann Series 2 luminescence spectrometer (SLM), which was modified to accommodate a pressure cell. Protein concentrations of 0.1mg/ml in 50 mM Mops buffer (pH 7.0) were used for the pressure and temperature measurements, and measurements were made using a 5 mm diameter quartz cuvette. The pressure or temperature was increased, typically in steps of about 50 MPa or 5 °C, respectively and the fluorescence of the protein was observed after a 5 min pause to allow conformational equilibrium to be attained. Intrinsic fluorescence was measured by exciting at 280 nm (4 nm slit) and emission was recorded between 310-410 nm (4 nm slit, 1 nm step size). The fluorescence spectra of the sum of the tryptophan and tyrosine contributions were quantified by specifying the center of the spectral mass, $\langle \upsilon \rangle$, as defined and used by Weber and coworkers as follows (8):

$$\langle \upsilon \rangle = \Sigma \upsilon_i \cdot F_i / \Sigma F_i$$

where υ_i is the wavenumber and F_i is the fluorescence intensity at υ_i.

T_m and P_m are temperature and pressure, respectively, which is a half of the center of the spectral mass.

Table 1

Effects of pressure and temperature on CPY and proCPY

	In situ experiments		*Ex situ* experiments	
	CPY	proCPY	CPY	proCPY
P_m (MPa) determined by fluorescence	340 [b]	253		
Activity (%) after pressure treatment [a]			90.8	48.1
T_m (°C) determined by fluorescence	58.6	54.5		
T_m (°C) determined by CD at 222 nm [b]	57	57		
Activity (%) after heat treatment [c]			35.8	35.6

[a] Activity after treatment at 400 MPa and 25 °C for 10 min (10).
[b] Taken from reference (9).
[c] Activity after treatment at 50 °C for 1 hr (10).

3. RESULTS

3.1. *In situ* temperature-induced changes of CPY and proCPY

The intrinsic fluorescence spectra of CPY and proCPY were measured against temperature up to 80 °C at 0.1 MPa. The unfolding curves of CPY and proCPY showed a simple two-state transition with a T_m of 58.6 and 54.5 °C, respectively (Table 1). The T_m of proCPY was found to be about 4 °C lower than that of CPY, indicating that proCPY was less stable than CPY to temperature. By decreasing the temperature from the highest temperature tested, changes in $<\upsilon>$ of CPY and proCPY were partially reversible.

Based on measurement of the circular dichroism (CD) at 222 nm with an increase in temperature at a rate of 0.5 °C/min, CPY and proCPY also showed the same T_m at 57 °C (9).

3.2. *In situ* pressure-induced changes of CPY and proCPY

The intrinsic fluorescence spectra of CPY and proCPY were measured against pressure up to 700 MPa at 25 °C. The $<\upsilon>$ of CPY was red-shifted with increasing pressure, indicating exposure of the tryptophan and tyrosine residues to the solvent. The unfolding of CPY showed at least a three-state transition, and did not follow a simple two-state transition. The complete unfolding of CPY was not observed at pressures up to 700 MPa at 25 °C. Based on the clear sigmoidal shape of the second transition, the P_{m2} was calculated as to be

340 MPa (Table 1).   The P_m of the first transition, P_{m1}, was estimated at low pressures of below 50 MPa and the P_m of the third transition, P_{m3}, was observed at 540 MPa or higher.

The $<\upsilon>$ of proCPY was red-shifted with an increase in pressure and showed a simple two-state transition with a P_m of 253 MPa (Table 1).   After the release of pressure from the highest pressure tested, changes in $<\upsilon>$ of CPY and proCPY were incompletely reversible.

3.3. *Ex situ* temperature-and pressure-induced changes of CPY and proCPY

After temperature treatment at 45 °C for 1hr, the activities of CPY and proCPY remained at 90.6% and 83.2%, respectively (data not shown).   After temperature treatment at 50 °C for 1hr, activities of CPY and proCPY were determined to be 35.8% and 35.6%, respectively (Table 1)(10).   These data indicate that there are no significant differences between CPY and proCPY with respect to temperature treatment.

After pressure treatment at 400 MPa and 25 °C for 10 min, the activity of CPY remained at 90.8%, but that of proCPY decreased to 48.1%, activity of which was measured after activation using proteinase K (10).   These data indicate that CPY is more stable than proCPY with respect to pressure treatment.

4. DISCUSSION

The temperature-induced denaturation curves of CPY and proCPY showed a simple two-state transition with similar T_m values.   These results indicate that temperature induces the cooperative denaturation of proteins.

The pressure-induced denaturation of CPY showed at least a three-state transition with increases in pressure up to 700 MPa.   Relatively low pressures (below 150 MPa) induced a small change in $<\upsilon>$, indicating only small conformational changes in CPY.   This finding is consistent with a previous report (7) in which it was shown that no hydrophobic core was exposed, as evidenced by no increase in ANS-binding fluorescence and by only a small loss in catalytic activity.   Higher pressure, 150 to 500 MPa, induced a large decrease in $<\upsilon>$, indicating a large conformational change.   At these pressures, it has been demonstrated that ANS-binding fluorescence increased with a decreased activity (7).   This result supports the view that the pressure-denatured CPY molecule at 150-500 MPa exists in a molten globule-like state (11).   A multiple transition *via* the molten-globule has also been reported for other proteins (12, 13). Higher pressures in excess of 500 MPa induced a larger red-shift in $\langle \upsilon \rangle$, although the protein appeared to be incompletely denatured. The three-state transition seemingly reflects a sequential unfolding of at least two structural domains (7). That is, two domains, one of which is the β-sheet-rich central core and another of which is the helix-rich external part of CPY molecule, unfold independently.

On the other hand, the pressure-induced denaturation of proCPY showed a simple two-state transition. The difference between the pressure denaturation of CPY and proCPY may be explained as follows. The propeptide sterically covers the active site, and thereby prevents substrate binding, as is the case for aspartic proteinase (14) and subtilisin (14, 15). Although the X-ray crystal structure of proCPY has not yet been solved, the charge-relay system of the mature CPY seems to be incompletely formed by some interference of the propeptide (16). This suggests that the cleft of the active site, which is located at the interface of the two domains is covered with the propeptide and, as a result, proCPY behaves as a one domain, resulting in cooperative unfolding.

In conclusion, CPY is composed of two structural domains, which show different sensitivities to pressure. Therefore pressure induces a multi step conformational change. On the other hand, proCPY denatures via a two-state transition as the propeptide combines the two structural domains so as to form one domain.

REFERENCES

1. Hayashi, R. (1976) Methods Enzymol. 45, 568-587.
2. Ramos, C., Winther, J. R., and Kielland-Brandt, M. C. (1994) J. Biol. Chem. 269, 7006-7012.
3. Winther, J. R., and Sorensen, P. (1991) Proc. Natl. Acad. Sci. U.S.A. 88, 9330-9334.
4. Winther, J. R., Sorensen, P., and Kielland-Brandt, M. C. (1994) J. Biol. Chem. 269, 22007-22013.
5. Hamaguchi, K. (1992) in The Protein Molecule, Conformation, Stability and Folding, Hamaguchi, K. ed., Japan Scientific Soc. Press, Tokyo, pp. 1-19.
6. Ruan, K., Lange, R., Bec, N., and Balny, C. (1997) Biochem. Biophys. Res. Commun. 239, 150-154.

7.Dumoulin, M., Ueno, H., Hayashi, R., and Balny, C. (1999) Eur. J. Biochem. 262, 1-10.

8.Silva, J., Miles, E., and Weber, G. (1986) Biochemistry 25, 5780-5786.

9.Haruta, N. (1997) A master's thesis in Applied Life Sciences, Graduate School of Agriculture, Kyoto University.

10.Koyama, T. (2000) A master's thesis in Applied Life Sciences, Graduate School of Agriculture, Kyoto University.

11.Kunugi, S., Yanagi, Y., Kitayaki, M., Tanaka, N., and Uehara-Kunugi, Y. (1997) Bull. Chem. Soc. Jpn. 70, 1459-1463.

12.Masson, P., and Clèry, C. (1996) in High Pressure Bioscience and Biotechnology, Hayashi, R. and Balny, C. eds., Elsevier Science B. V., The Netherlands, pp. 117-126.

13. Ptitsyn, O. B. (1995) Adv. Prot. Chem. 47, 83-229.

14. Khan, A. R., and James, M. N. G. (1998) Protein Sci. 7, 815-836.

15. Bryan, P., Wang, L., Hoskins, J., Ruvinov, S., Strausberg, S., Alexander, P., Almog, O., Gilliland, G., and Gallagher, T. (1995) Biochemistry 34, 10310-10318.

16. Sørensen, S. O., and Winther, J. R. (1994) Biochim. Biophys. Acta 1205, 289-293.

Trends in High Pressure Bioscience and Biotechnology
R. Hayashi (editor)
© 2002 Elsevier Science B.V. All rights reserved.

Compression and expansion of biomatter: predicting the unpredictable ?

K. Heremans[a], F. Meersman[a], H. Pfeiffer[a], P. Rubens[a] and L. Smeller[b]

[a]Department of Chemistry, Katholieke Universiteit Leuven, B-3001 Leuven, Belgium

[b]Department of Biophysics and Radiation Biology, Semmelweis University, Budapest H-1444, Hungary

In this paper the stability phase diagram for proteins and starch is discussed and the role of hydration and the cavities is emphasized. Examples of experimental approaches show that infrared spectroscopy is particularly suited for the study of protein unfolding and aggregation. The temperature and pressure effects on water and the relation to the stability diagram of biopolymers are considered together with recently reported results on synthetic polymers in the absence of any solvent.

1. INTRODUCTION

New directions for the interpretation of the observed effects in water-soluble biopolymers, proteins and starch, come from the observation of pressure-induced amorphization in inorganic substances, liquid crystals (but not in lipids) and synthetic polymers. Pressure-induced disordering is contrary to the behaviour that one expects for most materials. However, it has recently been observed in a number of one component systems such as H_2O, SiO_2, some other inorganic systems [1] and synthetic polymers [2]. For water this can be interpreted as a struggle between the desire for compactness and the inclination for optimal hydrogen bonding between the water molecules. Similar mechanisms may play a role in other systems. For two component systems it has been observed in water soluble synthetic polymers [3], proteins [4, 5] and starch [6, 7].

Suzuki [4] made systematic observations on the kinetics of denaturation of ovalbumin and carbonylhemoglobin. He plotted his results in a P, T, k-diagram and connected the points with the same rate constant (k). He noted a changeover in sign for the activation energy ($\Delta H^{\#}$) and volume ($\Delta V^{\#}$) from positive at high temperature and low pressure to negative at high pressure and low temperature. Suzuki interpreted the *negative activation energy* by assuming the following mechanism:

$$P + n\,W \leftrightarrow P(W)n \rightarrow Pd \qquad mPd \rightarrow (Pd)m \qquad (1)$$

In the first step water is pressed into the protein which then unfolds in a second step. In the third step the unfolded protein aggregates. High pressure computer simulations have

40

supported the water penetration model [8, 9]. More recently this has been discussed by Hummer et al. [10], Heremans and Smeller [11] and Vidiguris and Royer [12].

It is, however, noteworthy that the presence of water is not a prerequisite for the pressure-induced amorphization per se as may be seen from the observed effects with dry synthetic polymers and inorganic substances [1, 2].

Differences in pressure and temperature unfolding result in the formation of gels with different properties. The characteristics of pressure-induced gels depend on the type of protein, the concentration, the pressure level and holding time and the solvent conditions such as pH and salt. Of special interest is the observation that, in the case of proteins, a pressure pretreatment at room temperature may also have profound effects on the texture of the gels. In general pressure-induced gels are *softer* than temperature-induced gels, and contain non-incorporated water. This creates perspectives for the use of pressure in the development of unique textures [13].

2. THE PRESSURE AND TEMPERATURE BEHAVIOUR

The three common possible physical ways to denature a protein are the heat, pressure, and cold denaturation. In the thermodynamic description the denaturation is a reversible process. Given the equilibrium assumption in the transition state theory, the thermodynamic theory can be extended to kinetic data. The advantage of the thermodynamic model is that it gives the basis for a simple mathematical description.

The change in Gibbs free energy as a function of pressure and temperature is given by:

$$d\,(\Delta G) = - (\Delta S)\,dT + (\Delta V)\,dP \tag{2}$$

where ΔG gives the difference in partial molar Gibbs free energy between the denatured and the native state. Integration of the equation with the assumption that both ΔS and ΔV are temperature as well as pressure dependent leads to the following expression [11]

$$\Delta G = \Delta G^0 - \Delta S^0 (T - T_0) - \frac{\Delta C_p}{2T_0}(T - T_0)^2 + \Delta V^0 (P - P_0) + \frac{\Delta \beta}{2}(P - P_0)^2 \\ + \Delta \alpha (P - P_0)(T - T_0) \tag{3}$$

with the definitions of ΔC_p, $\Delta \beta$ and $\Delta \alpha$ as the change in heat capacity, compressibility and thermal expansion respectively.

These equations provide a consistent thermodynamic description of the heat, pressure and cold denaturation. The cross term in the equation (3) finds its origin in the pressure and temperature dependence of the volume changes:

$$\Delta V = \Delta V^0 + \Delta \alpha \left(T - T_0\right) - \Delta \beta \left(P - P_0\right) \tag{4}$$

The limits of the equations are that only differences in the thermal expansion factor and compressibility factor and heat capacity between the denatured and the native state are obtained. For absolute quantities one needs additional data from densitometry, calorimetry and ultrasonic velocimetry. Since the equation gives a relation between partial molar quantities, the usual difficulties arise for a molecular interpretation of these quantities. It is the common practice to rely on low molecular mass model systems. However, in many model systems the cavity term needs no consideration. In macromolecules this term cannot be neglected.

3. PROTEIN CAVITIES AND/OR HYDRATION?

The search for a correlation between hydration, compressibility, thermal expansion and heat capacity will always depend on a specific model that has to be used for the interpretation of thermodynamic data. But even when detailed structural information is available it is difficult to rationalize thermodynamic data [14]. The same applies to volume changes [15]. A comparison between the simplest model for the partial molar volume of a small compound and that for a protein:

$$V_i = V_{atoms} + V_{cavities} + \Delta V_{hydration} \tag{5}$$

suggests that apart from electrostatic interactions, hydrogen bonding and hydrophobic effects, cavities might play a crucial role in the volumetric properties of proteins. Their role has certainly been neglected when compared to the more popular hydrophobic effect. However, the fact that an elliptic phase diagram is observed for protein denaturation and starch gelation points to other factors than hydrophobic interactions as is usually assumed for proteins.

It is well known that the directionality of the hydrogen bond gives rise to the formation of cavities in ice I. The role of cavities in the volumetric properties of proteins was probably first suggested by Silva and Weber [16]. They suggested a common cause for the dissociation of oligomeric proteins and the unfolding of monomeric proteins: the imperfect packing of the atoms at the intersubunit surfaces and the restrictive effect of the covalent bond architecture of the protein.

Further examples as to the effect of cavities on the pressure behaviour may be found in the studies of the F31A mutant of the 7 kDa protein Sso7d from *S. solfataricus* where the creation of a cavity drastically reduces the pressure stability [17]. Along the same lines one expects the volume change for the unfolding of apomyoglobin to be larger than for myoglobin because of the elimination of the large void volume that occurs from the elimination of the heme group [12].

One possible technique that could be used to probe the role of cavities in proteins is *positron annihilation lifetime spectroscopy*. It is a powerful technique for determining the size of cavities and pores in materials. An estimate of the compressibility of the cavities in proteins should also be possible with this technique. For an epoxy polymer, the free volume reduces exponentially with pressure with a reduction to half of the free volume at about 200 MPa [18]. The compressibility of the cavities is ca. 250 Mbar^{-1} as estimated from the experimental data [19]. These numbers are much higher than the ones characteristic for proteins and suggest that there may be a considerable contribution of the reduction in cavity size to the compressibility of a protein.

4. UNFOLDING, PHASE SEPARATION, AGGREGATION

As pointed out before, proteins may be unfolded or denatured by heat, pressure or cold treatment. In recent years several spectroscopic techniques have been used to obtain a deeper insight in the difference between temperature and pressure-induced unfolding. These include NMR [20], UV spectroscopy [21], fluorescence and infrared spectroscopy combined with small angle X-ray scattering [22]. In combination with FTIR spectroscopy the diamond anvil cell may be used to follow *in-situ* the structural changes of the biomacromolecules during the compression and decompression phase.

FTIR spectroscopy can easily be used to study aggregation phenomena since it is insensitive to scattering, in contrast to fluorescence or circular dichroism techniques. Furthermore, certain types of aggregation display specific IR-bands.

4.1 Phase separation

In the previous discussions we neglected an important aspect of macromolecule behaviour: phase separation and its relation to unfolding, aggregation and gel formation. High pressure treatment of protein-polysaccharide mixtures has led to interesting new possibilities for the creation of new textures that show some specific features when compared to those obtained by heat treatment.

Aqueous solutions of biopolymers can undergo phase separation by two possible mechanisms: complex coacervation and thermodynamic incompatibility. In the case of coacervation the system separates into a solvent rich and a solvent poor phase. The solvent poor phase contains the precipitate of the polymer complex. In the case of the thermodynamic incompatibility, the two phases are solvent rich and one phase is enriched in protein whereas the other is enriched in polysaccharide. The conditions for complete mixing, phase separation and coacervation can be rationalised by the Flory-Huggins equation developed for synthetic polymers. The main difference between synthetic and biopolymers is the monodispersity of the latter. In addition there is the highly specific structure of proteins in contrast with random coil polymers. Also, in real food systems there is the fact that we usually deal with, at least, a ternary system and the complexity of the medium composition which has an important effect on the behaviour of the macromolecules.

An interesting overview of these concepts and their application in the dairy systems has been given by Syrbe et al. [23]. Recent comparisons between temperature and pressure-induced gels are on ovalbumin and sulphated polysaccharides by Galazka et al. [24] and on β-lactoglobulin with alginate and pectins by Dumay et al. [25]. Dumoulin and Hayashi [13] have reviewed several strategies that demonstrate the usefulness of high-pressure as a unique tool to texturize food and provide products with novel properties.

4.2 Protein aggregation

In many *in vitro* studies of protein folding the formation of aggregates is usually considered as an undesirable side effect which obscures the folding process as such. However, because of its important role in a number of molecular diseases as well as its role in the formation of gels, the mechanism of the formation of aggregates, and the possible role of folding intermediates, deserves closer attention. The general picture that is emerging from a number of studies is that under partial denaturation conditions proteins may acquire intermediate conformations that have a strong tendency to aggregate [26, 27]. San Biagio et al. suggested that apart from the formation of an intermediate conformation aggregation requires two other processes, being a mesoscopic phase separation and protein cross-linking, simultaneously to take place [28, 29].

The pressure-temperature phase diagram does not give any information about the mechanism of the unfolding and the aggregation of the unfolded state. Nor does it give information on the possible role of intermediates in the unfolding process which have been observed in a number of temperature and high pressure studies.

Recent results for metmyoglobin suggest that pressure induces partially unfolded states that may play a very important role in the aggregation of proteins. The high pressure unfolding of horse heart metmyoglobin results in an intermediate form that shows a strong tendency to aggregate when heated after pressure release. These aggregates are similar to those that are usually observed upon temperature denaturation [30]. Thus a pressure pre-treatment may have profound effects on the temperature-induced gel formation in food materials. It remains to be investigated whether the effects observed in myoglobin and lipoxygenase are a general property of proteins.

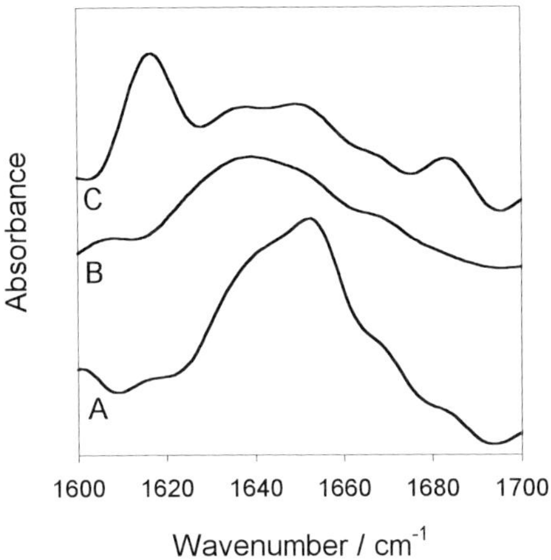

Figure 1: Reduced lysozyme at ambient conditions (A), at 10kbar (B) and at ambient conditions after pressure treatment (C). The presence of the two bands at 1618 and 1683 cm^{-1} in (C) is characteristic for antiparallel β-sheet aggregation.

The influence of the presence of disulphide bonds on the stability and aggregation behaviour was investigated. Lysozyme, with four disulphide bonds, was taken as a model. It is found that the reduced protein readily aggregates *after a pressure treatment* at ambient temperature conditions. This is shown in Figure 1, where the IR-bands typical of intermolecular antiparallel β-sheet aggregation at 1618 and 1683 cm^{-1} can be seen. A similar result has been obtained for horseradish peroxidase [31]. In contrast to the reduced protein such aggregation is not observed for the native lysozyme, as is the case for most proteins. Denisov et al. [32] suggested that cleavage of the disulphide bonds leads to a denatured state that has a higher flexibility rather than a changed structure. Our results seem to confirm this finding.

The temperature behaviour of the native and the reduced lysozyme was also investigated. The reduced lysozyme readily aggregates upon heating, whereas the native protein does not. However, when cooled the native protein also aggregates at temperatures below 75°C. The aggregation is the result of a kinetic competition between refolding and aggregation. Interestingly, it is found that when the temperature is raised again above 75°C the aggregation bands at 1618 and 1683 cm^{-1} disappear. This phenomenon of temperature-induced

44

disaggregation has been observed before [33, 34]. We suggest that it is a property of all proteins.

5. POLYSACCHARIDES: STARCH

Starch granules are semicrystalline particles mainly composed of high molecular mass polymers: the linear amylose and highly branched amylopectin. Recently, many studies focused on the ultrastructure of the starch granule to get a deeper insight in the stability and the gelatinization mechanism.

Thevelein et al. [35] showed that the application of 200 MPa induces an upward shift of 5°C for the gelatinization temperature. Hayashi and Hayashida [36] showed that pressure treated starches show an increased amylase digestibility at room temperature. Pressure-induced gelatinization has recently regained attention from *in-situ* studies [6, 7]. A gelatinization mechanism similar to the temperature-induced mechanism can be proposed. High pressure induces hydration of the amorphous phase followed by a distortion of the crystalline region leading to destruction of the granular structure.

Starch shows absorption bands in the infrared region between 1200-950 cm^{-1} that are sensitive to polymer conformation and hydration and can be used to follow *in-situ* the gelatinization of starch granules as a function of temperature and pressure. Temperature as well as pressure induces cooperative changes in the infrared spectra of starch granules that are related to the mechanism of gelatinization.

The shape of the gelatinization diagram of starch [7] is very similar to the stability diagram for the denaturation of proteins and the cloud point of water soluble synthetic polymers [3]. This is quite remarkable given the major differences in composition between these polymers. A change in conformation of the polymer from coil in solution to a more compact form is proposed in the phase separated form [3].

For starch it seems that the imperfect packing of amylose and amylopectin chains in the starch granule plays an important role in the observed changes. Both components when examined separately do not show cooperative changes upon heating or pressure increase. The combination of these components may give rise to specific interactions that determine the particular behavior of starch. In this respect it is of interest to note that the phase diagram of starch is unusual in comparison with other polysaccharides [37].

6. CONCLUSIONS

Can the compression and expansion of biomatter be predicted? It seems that a simple thermodynamic model that assumes a temperature and pressure dependence of the reaction volume can give a good phenomenological description of the stability diagram. It would be highly desirable to measure directly the changes in heat capacity, thermal expansion and compressibility. Water is a reaction partner in the pressure-induced *disordering* of biomatter. The contribution of the cavities should be obtained in a separate experimental approach to obtain a better understanding of the role of water. It is suggested that this may be possible with positron annihilation lifetime spectroscopy.

An important aspect, unique for proteins, is the aggregation that takes place upon unfolding. A further characterization of the intermediates in these processes is highly desirable.

REFERENCES

1. Katayama, Y. et al., 1999 A first-order liquid-liquid phase transition in Phosphorous, *Nature*, 403, 170-173.

2. Rastogi, S., G.W.H. Höhne and A. Keller. 1999 Unusual pressure-induced phase behavior in crystalline poly(4-methylpentene-1): Calorimetric and spectroscopic results and further implications, *Macromolecules*, 32: 8897-8909.

3. Kunugi, S., K. Takano, N. Tanaka, K. Suwa and M. Akashi. 1997. Effects of pressure on the behavior of the thermoresponsive polymer poly(N-vinylisobutyramide), *Macromolecules*, 30: 4499-4501

4. Suzuki, K. 1960. Studies on the kinetics of protein denaturation under high pressure, *Rev. Phys. Chem. Japan*, 29: 91-98.

5. Hawley, S.A. 1971. Reversible pressure-temperature denaturation of chymotrypsinogen, *Biochemistry*, 10: 2436-2442

6. Douzals Douzals, J.P., J.M. Perrier-Cornet, P. Gervais and J.C. Coquille. 1998. High-pressure gelatinization of wheat starch and properties of pressure-induced gels, *J. Agric. Food Chem.* 46: 4824-4829

7. Rubens, P. and K. Heremans 2000 Pressure-temperature gelatinization phase diagram of starch: an in situ Fourier Transform infrared study, *Biopolymers*, 524-530.

8. Kitchen, D.B., L.H. Reed and R.M. Levy. 1992. Molecular dynamics simulation of solvated protein at high pressure, *Biochemistry*, 31: 10083-10093.

9. Wroblowski, B., J.F. Diaz, K. Heremans and Y. Engelborghs. 1996. Molecular mechanisms of pressure induced conformational changes in BPTI, *Proteins: Structure, Function and Genetics*, 25: 446-455

10. Hummer G. et al. 1998. The Pressure Dependence of Hydrophobic Interactions Is Consistent With The Observed Pressure Denaturation of Proteins, *Proc. Natl. Acad. Sci. USA*, 95: 1552-1555.

11. Heremans, K. and L. Smeller. 1998. Protein structure and dynamics at high pressure, *Biochim. Biophys. Acta*, 1386: 353-370

12. Vidugiris, G.J.A. and C.A. Royer. 1998. Determination of the volume changes for pressure-induced transitions of apomyoglobin between the native, molten globule, and unfolded states, *Biophys. J.*, 75: 463-470.

13. Dumoulin, M. and R. Hayashi. 1998. High Pressure, a unique tool for food texturization, *Food Sci. Technol. Int. Tokyo*, 4: 99-113.

14. Cooper, A. 1999. Thermodynamic analysis of biomolecular interactions, *Curr. Opin. Chem. Biology*, 3: 557-563

15. Chalikian, T. and K.J. Breslauer. 1998. Thermodynamic analysis of biomolecules: a volumetric approach, *Curr. Opin. Struct. Biol.*, 8: 657-664

16. Silva, J.L. and G. Weber. 1993. Pressure stability of proteins, *Ann. Rev. Phys. Chem.* 44: 89-113.

17. Fusi, P. et al. 1997. The extremely heat- and pressure resistant 7-kDa protein P2 from the archaeon *Sulfolobus solfataricus* is dramatically destabilized by a single aminoacid substitution, *Proteins: Structure, Function and Genetics*, 29: 381-390.

18. Deng, Q. and Jean, Y.C. (1993) Free volume distributions of an epoxy polymer probed by positron annihilation: pressure dependence, *Macromolecules* **26**, 30-34.

19. Deng, Q., Sundar, C.S. and Jean, Y.C. (1992) Pressure dependence of free-volume hole properties in an epoxy polymer, *J. Phys. Chem.* **96**, 492-495.

20. Zhang, J., X. Peng, A. Jonas and J. Jonas. 1995. NMR study of the cold, heat and pressure unfolding of ribonuclease A, *Biochemistry*, 34: 8631-8641.

21. Mombelli, E., M. Afshar, P. Fusi, M. Mariani, P. Tortora, P., J.P. Connelly and R. Lange. 1997. The role of phenylalanine 31 in maintaining the conformational stability of ribonuclease P2 from *Sulfolobus solfataricus* under extreme conditions of temperature and pressure, *Biochemistry*, 36, 8733-8742.

22. Panick, G., G. Vidugiris, R. Malessa, G. Rapp, R. Winter and C.A. Royer. 1999. Exploring the temperature-pressure phase diagram of staphylococcal nuclease, *Biochemistry*, 38: 4157-4164.

23. Syrbe, A., W.J. Bauer and H. Klostermeyer. 1998. Polymer science concepts in dairy systems – An overview of milk protein and food hydrocolloid interaction *Int. Dairy Journal*, 8: 179-193.

24. Galazka, V.B., D. Smith, D.A. Ledward and E. Dickinson. 1999. Interactions of ovalbumin with sulphated polysaccharides: effects of pH, ionic strength, heat and pressure treatment, *Food hydrocolloids*, 13: 81-88.

25. Dumay, E., A. Laligant, D. Zasypkin and J.C. Cheftel. 1999. Pressure- and heat-induced gelation of mixed β-lactoglobulin/polysaccharide solutions: scanning electron microscopy of gels, *Food hydrocolloids*, 13: 339-351.

26. Booth, D.R. et al. 1997. Instability, unfolding and aggregation of human lysozyme variants underlying amyloid fibrillogenesis, *Nature*, 385: 787-793.

27. Fink, A.L. 1998. Protein aggregation: folding aggregates, inclusion bodies and amyloid, *Folding and design*, 3: R9-R23

28. San Biagio, P.L. et al. 1999. Interacting Processes in Protein Coagulation, *Proteins: Structure, Function, and Genetics*, 37:116-120.

29. San Biagio, P.L. et al. 1996. Self-Assembly of Biopolymeric Structures below the Threshold of Random Cross-Link Percolation, *Biophys. J.*, 70: 494-499.

30. Smeller, L., P. Rubens and K. Heremans. 1999. Pressure effect on the temperature induced unfolding and tendency to aggregate of myoglobin, *Biochemistry*, 38: 3816-3820.

31. Smeller et al. in this volume.

32. Denisov, V.P., Jonsson, B-H. and Halle, B. 1999. Hydration of Denatured and Molten Globule Proteins, *Nature Struct. Biol.*, 6: 253-260.

33. Holzbaur, I.E., English, A.M. and Ismail, A.A. 1996. FT-IR study of the thermal denaturation of horseradish and cytochrome c peroxidases in D2O. Biochemistry 35: 5488-5494.

34. Dubois, J., Ismail, A.A., Chan, S.L. and Ali-Khan, Z. 1999. Fourier transform infrared spectroscopic investigation of temperature- and pressure-induced disaggregation of amyloid A. *Scand. J. Immunol.* 49: 376-380.

35. Thevelein, J., J.A. Van Assche, K. Heremans and S.Y. Gerlsma. 1981. Gelatinisation temperature of starch as influenced by high pressure, *Carbohydrate Res.*, 93: 304-307

36. Hayashi, R. and Hayashida, A. 1989. Increased amylase digestibility of pressure-treated starch. *Agric. Biol. Chem.*, 53, 2543-2544.

37. Gekko, K. 1992. Effects of pressure on the sol-gel transition of food macromolecules. in High Pressure and Biotechnology; Balny, C., Hayashi, R, Heremans, K., Masson, P. Eds.John Libbey Eurotext: Montrouge, 1992, 105-113.

Trends in High Pressure Bioscience and Biotechnology
R. Hayashi (editor)
© 2002 Elsevier Science B.V. All rights reserved.

Fluctuation of apomyoglobin monitored from
H/D exchange and proteolysis under high pressure

Naoki Tanaka[a], Chizuko Ikeda[a], Kenji Kanaori[b], Kazumi Hiraga[b],
Takashi Konno[c] and Shigeru Kunugi[a]

[a]Department of Polymer Science & Engineering and [b]Department of Applied
Biology, Kyoto Institute of Technology,
Matsugasaki, Sakyo, Kyoto 606-8585, Japan.

[c]Department of Physiology, Fukui Medical University,
Yoshida, Fukui, 910-1193, Japan.

We investigated fluctuation of apomyoglobin (apoMb) by monitoring H/D exchange and
limited proteolysis under high pressure. The intrinsic fluorescence spectrum showed that a
large fraction of apoMb molecules is in the native conformation in the pressure range from
0.1 to 150 MPa. In this pressure range, apoMb was shown to be not locally unfolded by
limited proteolysis experiments. To investigate the pressure effect on the fluctuation of
apoMb in this condition, H/D exchanges of the backbone amino acids was monitored by   2D
NMR. The protection factors of H/D exchange (P) for several residues decreased with the
increase of the pressure, and most of the residues are located on the B and E helices which are
around the heme pocket. This result suggests that pressure-induced enlargement of
fluctuation of apoMb is due to the progressive hydration to the heme pocket region.

1. INTRODUCTION

To characterize the dynamics of proteins is important for understanding their physical
properties (1) and functions (2). High pressure is a unique intensive parameter that can give
volumetric information, and has been combined with NMR, H/D exchange and
phosphorescence to study protein dynamics. These studies have generated conflicting

48

pictures of pressure effects on the protein dynamics. The dominant effect of applied pressure is to slow the dynamics, whereas the dynamics of some proteins are enhanced by an increase in pressure. This inconsistency has been explained by the fact that the pressure increase induces a volume reduction of the protein solution by two opposing factors: a decrease in the size of the internal cavities and an increase in hydration (3). The pressure-induced reduction of the size of the cavities will suppress the fluctuation whereas progressive hydration under high pressure will enlarge it.

In this study we characterized the dynamics of apomyoglobin (apoMb) by using a combination of fluorescence, limited proteolysis and H/D exchange. The picture of dynamics of apoMb has oscillated between two extreme views. Results from mass spectroscopic study of apoMb (4) and conformational analysis of its single amino acid mutants (5) suggested that apoMb has many of the same characteristics of a molten globule. On the other hand, a result from NMR has shown that the majority of the apoMb conformation adopts the structure similar to that of holoMb (6). We investigated the pressure effect on the conformational state of apoMb to increase our knowledge about its dynamic properties.

2. EXPERIMENTAL PROCEDURE

Sperm Whale Mb was purchased from Biozyme (Gwent, UK). Fluorescence spectra at elevated pressures were obtained with a high pressure cell, having three optical windows manufactured by Teramecs Co. (Kyoto, Japan), combined with a Shimadzu RF5000 spectro-fluorimeter (Kyoto, Japan). The aqueous solution of apoMb at pH 6.0 was diluted into D_2O at pH 6.0 to initiate the H/D exchange reaction under high pressure. H/D exchange was quenched by adding the heme to reconstitute holoMb. COSY NMR measurement was performed by a Brucker 500ARX spectrometer.

Limited proteolysis was performed by incubating apoMb solutions with protease in polypropylene tubes at pH 6.0 at an elevated pressure. SDS-PAGE was performed using a slab gel with a concentration of 16.5 % and a tricine buffer system. The N-terminal sequence analysis was performed on the peptide samples isolated by blotting from the gel utilizing Applied Biosystems (Foster City, CA) protein sequencer (model 476A) equipped with an on-line analyzer (model 610A) of phenylthiohydrantoin-derivatives of amino acids.

3. RESULTS AND DISCUSSION

3.1. Intrinsic fluorescence of apoMb under high pressure.

Figure 1 shows the average emission wavelength of apoMb at pH 6.0 and 293 K in the pressure range from 0.1 to 400 MPa. A pressure increase caused a red-shift in the maximum wavelength of the fluorescence. The fluorescence excited at 295 nm is due to the contribution of Trp7 and Trp14 which are located at the A helix in the N terminal region. Thus, the fluorescence changes in this

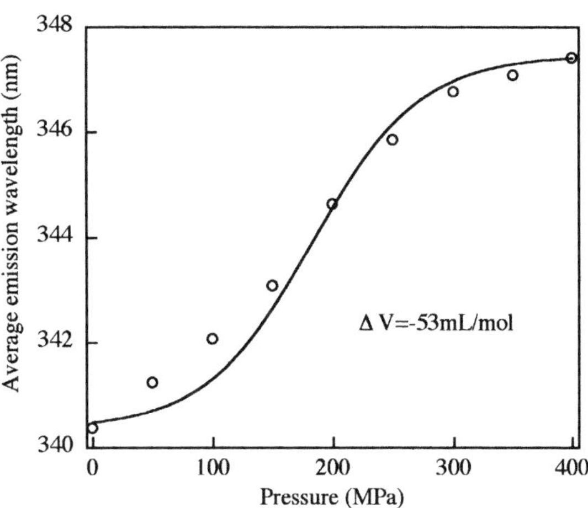

Figure 1. The averaged wavelength of the fluoresence of apoMb in the pressure range from 0.1 to 400 MPa.

figure reflect the change in the local environment around the A helix. The average wavelength of the fluorescence of apoMb at pH 6.0 and 400 MPa was 13 nm lower than that at pH 2.2, indicating that apoMb is not fully unfolded in this condition. These observations indicate that the conformational transition from the native to the intermediate state of apoMb takes place in the pressure range between 150 and 200 MPa at pH 6.0. We analyzed the transition curve in the averaged wavelength of fluorescence versus pressure by fitting the data in terms of a two-state transition, and obtained a volume change as -53 mL/mol. Results consistent with ours have been reported by other groups studying pressure-induced denaturation of apoMb (7, 8). Vidugiris and Royer (8) have investigated the effects of pressure on horse heart apoMb, and have reported that a conformational transition occurs in the pressure range from 50 to 150 MPa with a volume change of -70 mL/mol. The transition pressure and the volume change obtained from our result of sperm whale apoMb were slightly different from what Vidugiris and Royer obtained, which could be explained by the difference in apoMb species and the

50

solution conditions.

3.2. Limited proteolysis of apoMb under high pressure.

We have investigated the conformational state of apoMb by limited proteolysis for further insights into the conformational state under high pressure. The apoMb solution, in the presence of subtilisin, was incubated at an elevated pressure, and the pattern of protein degradation was monitored with SDS-PAGE. The pattern of proteolysis at 0.1 MPa showed two major fragments of approximate molecular masses of 7 and 10 KDa. Blotting and a N-terminal protein sequencing of the fragment of 7 KDa gave amino acid sequences of Ala-Gln-Ser-His and Ala-Thr-Lys-His which correspond to the sequences of 90-93 and 94-97 of myoglobin. The initial nicking of apoMb occurred at Leu89-Ala90 and His93-Ala94 peptide bonds at the F helix region, indicating that the local structure of the F helix region of apoMb is disordered.

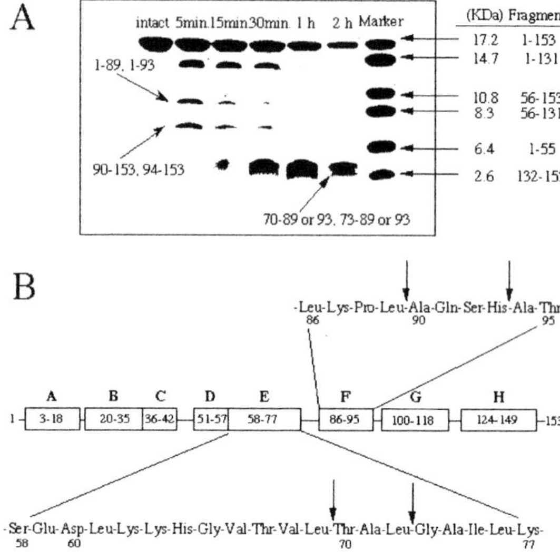

Figure 2. A, Time course of the limited proteolysis of apoMb with subtilisin at 150 MPa monitored by SDS-PAGE; B, The cleaved site on the amino acid sequence of apoMb.

The band for the intact apoMb became intense by an increase of pressure from 0.1 MPa to 150 MPa, indicating that the rate of proteolysis with subtilisin is reduced. This is consistent with the previous result, which states that protease activity of subtilisin was reduced by an increase of pressure (9). On the other hand, the SDS-PAGE pattern above 150 MPa showed that the rate of proteolysis is accelerated by an increase in pressure. The result from fluorescence has shown that a pressure-induced transition to the intermediate state occurs above 150 MPa, and that the acceleration of proteolysis originates from the increase of the conformational fluctuation of the substrate protein.

Figure 2A shows the time course for the digestion of apoMb by subtilisin at 150 MPa. The digestion of apoMb by subtilisin gave the short fragments of 2-3 KDa, in addition to the two major fragments from the cleavage of the F helix. Blotting and a protein sequencing of these short fragments gave a mixture of the sequences. The comparison of all the possible combinations of the obtained sequences to the original sequence of myoglobin allows the identification of the sequences of these short fragments. We have found that the combined sequences of Tyr-Ala-Leu-Gly and Gly-Ala-Ile-Leu correspond to the sequences of 70-73 and 73-76 on the E helix, respectively (Figure 2B). This result indicates that Leu69-Tyr70 and Leu72-Gly73 on the E helix were cleaved at 150 MPa, which may lead to a conclusion that local unfolding of the E helix occurs under high pressure. However, it is also possible that the F helix region of intact apoMb is cleaved first and a sequential cleavage of the fragments occurs in the proteolysis process of apoMb under high pressure. To confirm the validity of this process, the fragments from the cleavage at the F helix produced by proteolysis at 0.1 MPa were treated with subtilisin under high pressure. The time course of the pattern of SDS-PAGE showed the rapid production of the short fragments (2-3 KDa) which is similar in size to those shown in Figure 2A. Thus, these short fragments are produced by the cleavage of the E helix sequentially after the cleavage of the F helix region. This result suggests that a pressure increase from 0.1 to 150 MPa does not induce local unfolding of the E helix region of the intact apoMb. This is consistent with the result that the rate of proteolysis was decreased by a pressure increase from 0.1 to 150 MPa, since the rate of the proteolysis should be enhanced by local unfolding.

3.3. H/D exchange of apoMb under high pressure.

H/D exchange in apoMb at an elevated pressure was monitored by NMR spectroscopy. After various exchange-out periods in the range from 20 minutes to 2 hours at elevated pressures, the exchange of the backbone amide proton was monitored by the volume change of the individual cross-peaks of NH and CH hydrogen in the COSY spectra. The effect of pressure on the H/D exchange rate of each amino acid residue is related to the change in protection against the exchange. This is expressed by the protection factor, $P = k_{rc}/k_{obs}$, where k_{obs} is the H/D exchange rate constant obtained experimentally, and k_{rc} is that of each backbone amide proton in the unfolded state calculated from the theoretically predicted rates of the random coil polypeptide, correcting for the value at high pressure.   As shown in Table 1, P value decreased by an increase of pressure for every observed backbone amide proton of apoMb, indicating that H/D exchange rate was enhanced.

The present results from limited proteolysis suggest that small conformational fluctuations of apoMb provide pathways for H/D exchange at low pressures. More details concerning the H/D exchange mechanisms of apoMb under high pressure can be inferred from the result in Table 1. The H/D exchange process   through the native sate is explained by the local unfolding model or the penetration model (10). In the former model, the segments on the secondary structure undergo local unfolding, and the decrease in the P value should have a

Table 1
Amide proton protection factors of apoMb at various pressures

Residue	Helix position	Pressure			
		0.1 MPa[11]	60 MPa	100MPa	150MPa
Val 10	A8	100,000	900	400	100
Leu 11	A9	<100,000	9,000	2,000	400
Ile 28	B9	4,000	1,000	200	<1
Leu 29	B10	20,000	3,000	<1	<1
Ile 30	B11	70,000	4,000	600	100
Arg 31	B12	5,000	7,000	<1	<1
Phe 33	B14	6,000	5,000	900	<1
Lys 34	B15	4,000	6,000	<1	<1
Val 66	E9	3,000	<1	<1	<1
Leu 69	E12	5,000	6,000	<1	<1
Thr 70	E13	4,000	<1	<1	<1
Leu 72	E15	100	500	<1	<1
Gly 73	E16	3,000	<1	<1	<1
Ile 75	E18	2,000	1,000	300	100
Ile 112	G13	60,000	10,000	1,300	300
Val 114	G15	200,000	20,000	5,000	900
Leu 115	G16	60,000	4,000	1,000	100
Lys 133	H10	6,000	14,000	<1	<1

similar magnitude on the same secondary structure unit. In the penetration model, on the other hand, solvent molecules diffuse into the protein matrix mediated by a small rapid fluctuation of interior atoms on the order of tenth to several angstrom. In this model, the amplitudes of the decrease in the P value do not necessarily correlate to the secondary structure unit, because there may be several steps required for the exchange of a particular site. The relative decrease in the H/D exchange rate of apoMb in Table 1 varied in amplitude among the amide protons on the same helix. Therefore, H/D exchange in apoMb is explained best by the penetration model in which H/D exchange is mostly mediated by a small rapid fluctuation. The positions of amino acid residues on which the P value significantly decreased are on the B and E helices around the heme pocket that is composed of hydrophobic side chains. Since hydrophobic hydration produces a negative volume change, hydration at the heme pocket will be promoted by applied pressure to increase the contact area to water. Furthermore, this part of the molecule should be flexible because of the removal of the heme, and may allow water penetration to occur relatively easily under high pressure.

CONCLUSION

The present results indicate that the fluctuations at the heme pocket of apoMb are enhanced by an increase of pressure from 0.1 to 150 MPa. Since the hydrophobic hydration mediates the diffusion of the solvent into the protein matrix to gain the negative volume change, the pressure-induced enhancement of fluctuation is explained well by the progressive hydration on the heme pocket. The rate of H/D exchange in protein has been suggested to be correlated to its compressibility (10), and a detailed analysis of H/D exchange in proteins under high pressure would provide useful information for understanding the effect of pressure on compressibility of protein.

54

REFERENCE

1. H. Frauenfelder, F. Parak and R. D. Young, Ann. Rev. Biophys. Biophys. Chem., 17 (1988) 451.

2. S. Kunugi, S. Fujiwara, S. Kidokoro, K. Endo and S. Hanzawa, FEBS Letter, 462 (1999) 231.

3. P. Cioni and G. B. Strambini, J. Mol. Biol., 291 (1999), 955.

4. F. Wang and X-j. Tang, Biochemistry 35 (1996) 4069.

5. L. Lin , R. J. Pinker, K. Forde, G. D. Rose, and N. R. Kallenbach, Nature Struct. Biol., 1 (1994) 447.

6. D. Eliezer, and P. E. Wright, J. Mol. Biol., 263 (1996) 531.

7. E. Bismuto, G. Irace, I. Sirangelo and E. Gratton, Biochemistry, 35 (1996) 1173.

8. G. Vidugiris and C. A. Royer, Biophys. J., 75 (1998) 463.

9. S. Kunugi, I. Kobayashi, K. Takano and Y. Murakami, 69 (1996) Bull. Chem. Soc. Jpn., 3375.

10. C. Woodward, I. Simon and E. Tuchsen, Molecular and Cellular Biochemistry, 48 (1982) 135.

11. F. M. Hughson, P. E. Wright and R. L. Baldwin, Science, 249 (1990) 1544.

Trends in High Pressure Bioscience and Biotechnology
R. Hayashi (editor)
                    55

EFFECT OF PRESSURE TREATMENT ON HYDROPHOBICITY AND SH GROUPS INTERACTIONS OF MYOFIBRILLAR PROTEINS

N. Chapleau, S. Delépine, M. de Lamballerie - Anton

ENITIAA, GEPEA, BP 44322, NANTES Cedex 3, FRANCE

e-mail : chapleau@enitiaa-nantes.fr, anton@enitiaa-nantes.fr

ABSTRACT

Response surface methodology was used to study the simultaneous effect of two independent variables : **Pressure** [0 – 600 MPa] and **Time** [0 – 1800 s] according to a central composite design in stars. The significant responses were the surface hydrophobicity, the ratio -SH_{free} / -SH_{total} and the flowing properties. Surface hydrophobicity was determined using 1-anilinonaphtalene-8 sulfonate (ANS) as hydrophobic fluorescence probe. The ratio -SH_{free} / -SH_{total} was determined using the 5-5'-Dithiobis-2-nitrobenzoic acid. Flowing properties (flow index and viscosity coefficient) of solutions were stemmed from modelised flow curves performed with a rheometer AR1000.

Results show that high pressure treatment induces an increase of the surface hydrophobicity of myofibrillar proteins (x 3) from 0 MPa to 440 MPa, and from 450 MPa to 600 MPa the surface hydrophobicity is stable. The same trend are obtained with the sulfhydryl interactions and the rheological index n. The viscosity coefficient K decreases with the increase of pressure : the viscosity decreases in relation to a change of structure of the proteins. The statistical analysis of responses show that the most important significant effects are the pressure and its quadratic effect compared with the effect of time.

Keywords : myofibrillar proteins, high pressure, functional properties, hydrophobicity, sulfhydryl, rheological behaviour

INTRODUCTION

High pressure treatment is a potential tool for the modification of food functional properties and could be used apart from its bacteriological effect to create new products or new textures. Proteins modifications are mainly involved in high pressure functional properties modifications [1], thus gelling [2], emulsifying [3] or foaming properties change under the effect of high pressure.

Protein denaturation under pressure is a complex phenomenon induced by the disruption of both hydrophobic bonds, sulfhydryl interactions and salt bridges. High pressure act by altering the balance of intramolecular and solvent-protein interaction. The impact of pressure induced denaturation of protein depends on factors such as temperature of treatment, pH, ionic strength solvent, pressure and time level.

In this study we focused on hydrophobicity and sulfhydryl interactions of myofibrillar proteins. Bovine myofibrillar proteins were chosen because of the important role of their functional properties in meat products. Myofibrillar proteins were extracted from meat and solubilised in a high ionic strength buffer (0.6 KCl). High pressure treatment was performed on the extract.

1. MATERIALS & METHODS

1.1. Sample source

Bovine muscles (*Biceps femoris*) obtained from a commercial abattoir (SOVIBA, le Lion d'Angers France) one day *PM* were stored at + 4°C before the extraction of myofibrillar proteins.

1.2. Extraction of myofibrillar proteins

Myofibrillar proteins were extracted according to the method of Offer *et al.* (1973) [4]. Protein samples at 10 g/l in 0,6 M KCl 40 mM phosphate buffer (pH 6.0) were vacuum conditioned in a polyethylene bag before pressure treatment.

1.3. High pressure treatment

Protein samples were pressurised at 10°C (+/- 7°C) in a reactor of 1.5 litres from ACB Pressure System (Nantes, France).

1.4. Rheological measurements

One hour after high pressure treatment, viscosity was measured at 20°C with a rheometer AR 1000 (TA Instrument, New Castle, U.K.) equipped with a plate/cone system (6 cm diameter cone, angle 2°). Flow curves were determined using an increasing shear rate : 0.03 to 1000 s^{-1} in 2 minutes. The experimental flow curves were modelled using the Power law model :

$$\tau = k \, \dot{\gamma}^{\,n} \tag{1}$$

where k = viscosity coefficient (Pa.s), $\dot{\gamma}$ = shear rate (s^{-1}) and n = flow index (n<1 corresponds to a shear thinning behaviour, n>1 corresponds to a shear thickening behaviour, and n=1 corresponds to a Newtonian behaviour). The choice of the model was got by the AR 1000 software after calculation of the validation domain of the results.

1.5. Measurement of SH

For the determination of SH Ellman [5] procedure was found most suitable. The reagent, 5,5'-dithiobis-2-nitrobenzoic acid (DTNB) 10 mM was added to solution of myofibrillar proteins (phosphate buffer 40 mM pH 6.0 and KCl 0,6 M) for measurement of $-SH_{free}$. For $-SH_{total}$ the same phosphate buffer was used with 6 M urea. The optical density were measured against appropriate blanks at 412 nm and calculations were carried out by using the extinction coefficient 1.36×10^4 given by Ellman [5].

1.6. Proteins surface hydrophobicity

Determination of myofibrillar proteins hydrophobicity was determined by 8-anilino-1-naphtalene sulphonic acid (ANS) according to the method described by Kato and Nakai (1980) [6]. 10 µl of a solution of ANS 8 mM was added to a dilution of myofibrillar protein (phosphate buffer 40 mM pH 6.0 and KCl 0,6 M) from 0 to 0.2 g/l (6 spots). The ANS-myofibrillar proteins conjugate was then excited at 380 nm and the relative fluorescence intensity was measured and recorded at 490 nm. The relative fluorescence was compared to the control for each sample. The result was expressed in percent of hydrophobicity compared with the control.

1.7. Experimental design

Analysis of variance was carried out using the Statgraphics computer programme. Level of significance was set for $p \leq 0.05$. Response surface methodology was used to study the simultaneous effect of two independent variables (Pressure [0 – 600 MPa] and Time [0 – 1800 s]) according to a central composite design in stars. Assessment of error was based on three replications of the treatment of the central conditions. Only the significant factors for hydrophobicity, ratio $-SH_{free}$ / $-SH_{total}$ and rheological indexes were retained to determine the influence of the treatment. Experimental design was replicated two times with two samples of myofibrillar proteins.

1.8. Characterisation of protein fraction

SDS polyacrylamide gel electrophoresis (SDS-PAGE) 12.5% was carried out on samples of myofibrillar proteins after high pressure treatment to determine proteins patterns.

2. RESULTS & DISCUSSION

2.1. Modification of surface hydrophobicity under high pressure treatment

All the results of experimental design are presented in table and figure 1.

Variable	R^2	Significant effects[1]
Hydrophobicity	92.91 %	P^{**} - PxP^{**} - TxT^{**} - T^{**}
Ratio $-SH_{free}$ / $-SH_{total}$	94.26 %	P^{**} - PxP^{**} - TxT^{*} - T^{*}
Flow index	94.80 %	P^{**} - PxP^{**} - TxT^{**} - T^{**}
Viscosity coefficient	95.41 %	P^{**} - PxP^{**} - TxT^{**} - T^{**} - PxT^{*}

R^2 : Representation of the model

[1] graded by decreasing-shaped order for * $p \leq 0.05$ or ** $p \leq 0.001$

Linear effects : P Pressure and T Time

Quadratic effects : PxP Pressure and TxT Time

Interaction effect : PxT Pressure and Time

Table 1 : Results of the experimental design

58

Results show that in our buffer conditions, pressurisation of myofibrillar proteins modifies the surface hydrophobicity of proteins. Treatment above 400 MPa during 800 seconds leads to the maximum surface hydrophobicity : three times more than control. The hydrophobicity begins to change only for short treatment higher than 150 MPa, but in the range of 150 to 400 MPa the increase is more important.

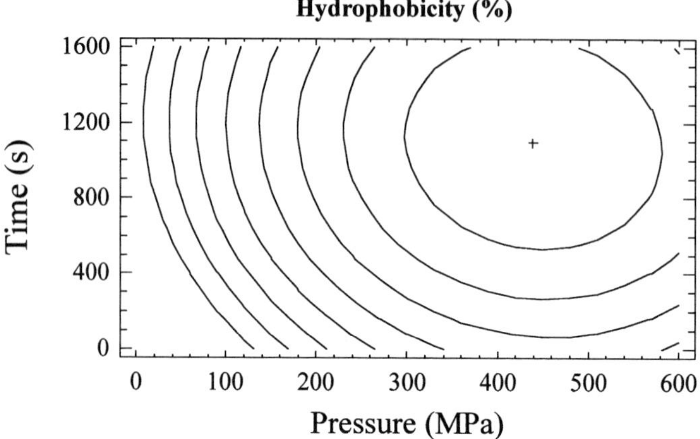

Figure 1 : Surface curves showing the effect of pressure and time on hydrophobicity.

The significant effects (table 1) show a good resolution of the quadratique model and show the main effect of pressure (linear and quadratic) in front of the effect of time, the interaction pressure was not often present. A block effect between the two proteins samples is explained by the different sensibility of extracted proteins, thus the level of hydrophobicity was not the same but the effect of pressure gave the same result.

This result confirms the impact of pressure on solubilised proteins, and the hypothesis of depolymerisation and irreversible modification of the structure, this depolymerisation increasing the surface reactivity of proteins.

2.2. Modification of rheological flow behaviour induced by high pressure treatment

Results of rheological flow were divided in two parts, in the first part we discussed from the impact of pressure on the rheological behaviour and especially on the flow index n and in the last part we argued from the effect of pressure on the viscosity coefficient K which imparted the consistence of the myofibrillar proteins solutions.

Effect on flow index n : High pressure induced a decrease of the flow index n, *i.e.* a loss of the rheofluidifying behaviour (figure 2). The behaviour of the myofibrillar solution begins to change as soon as the pressure increases, and a Newtonian behaviour appears for treatment higher than 400 MPa 800 seconds.

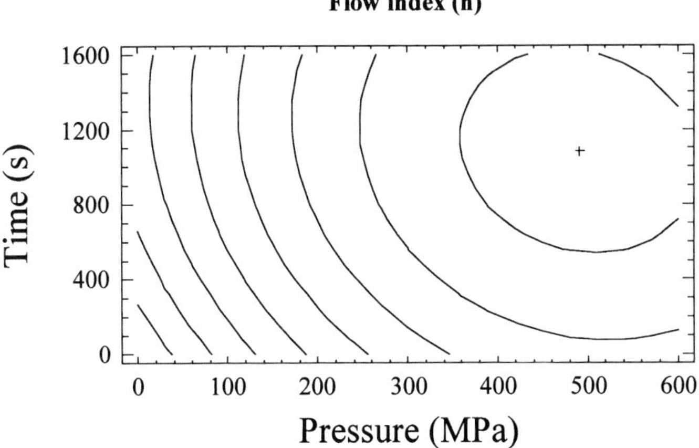

Figure 2 : Surface curves showing the effect of pressure and time on flow index n.

Effect on viscosity coefficient K : High pressure induces a change of the consistency of the myofibrillar proteins solution (figure 3). Even if pressure level is low : after treatment at 100 MPa during 800 seconds consistency is reduce by half. Visual change can be observed after treatment at 300 MPa 900 seconds, but changes could be induced before this scale. The maximum of loss was obtained after treatment up to 460 MPa during 1015 seconds, this value corresponds to a loss of 1 log cycle of K (65 to 6.39 mPa.s). This value corresponds to 6 times the viscosity of water.

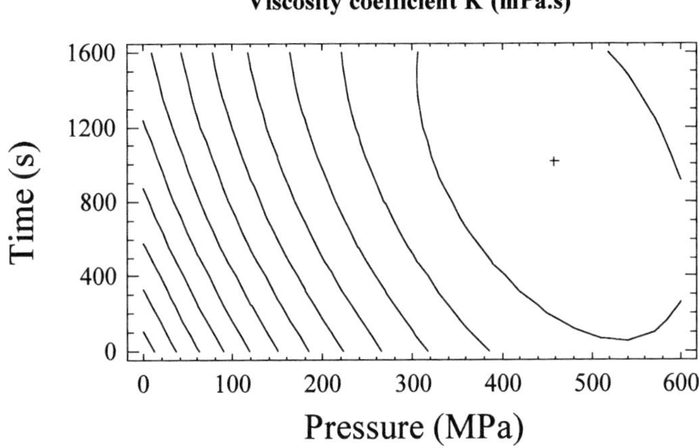

Figure 3 : Surface curves showing the effect of pressure and time on viscosity coefficient K (mPa.s).

These results characterize the rheological properties of myofibrillar proteins solution after high pressure treatment. They show a loss of rheofluidifying properties. For the most important treatment (600 MPa during 900 seconds) the rheofluidifying properties were transformed in a Newtonian behaviour. This result show that a modification of proteins properties may be induced by a change of their structures. Myosin and actin were changed under the effect of pressure many authors shew that pressure induced a dissociation of the actomyosin complex and could produce the formation or the dissociation of new aggregates [7], [8]. The dissociation of the actomyosin complex could produce a change of the solution flow by the formation of a new layout between actin and myosin. The hypothesis of depolymerisation of proteins strengthened the results obtained with hydrophobicity because these phenomena went with an increasing of the surface of proteins by the destruction of their 3D structures.

2.3. Increase of accessibility of reagent sulfhydryls (-SH$_{free}$) after pressure treatment

The number of free SH grew with the increasing pressure and time of treatment, and the most significant effect is due to the level of pressure (figure 4). For the control we observed only 37 % of free SH, this value doubles after treatment at 450 MPa for 1200 seconds. This result confirms the hypothesis observed with surface hydrophobicity : high pressure treatment induces a change of molecules conformation, thus the properties of solution changed.

These modifications increase the surface of molecule accessible by the solvent and the number of accessible reagent sulfhydryles.

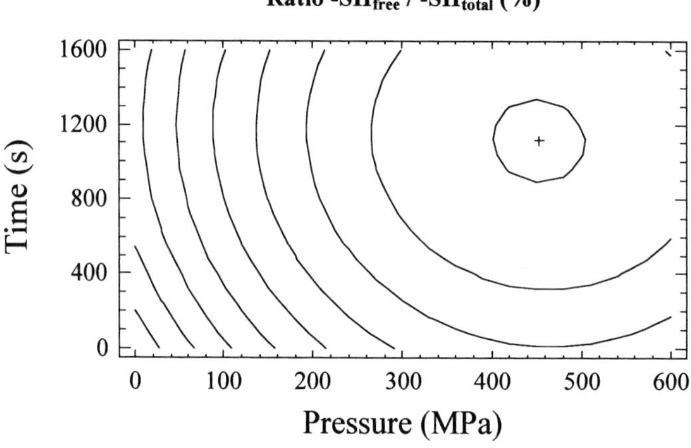

Figure 4 : Surface curves showing the effect of pressure and time on the Ratio -SH$_{free}$ / -SH$_{total}$ (%).

2.4. Characterisation of protein fraction

Figure 5 presents the SDS-PAGE of the samples, all the samples showed the pattern characteristic of myofibrillar proteins : myosin heavy chain, C protein, actin and myosin light chains. After pressure treatment no change was distinguishable by SDS-PAGE.

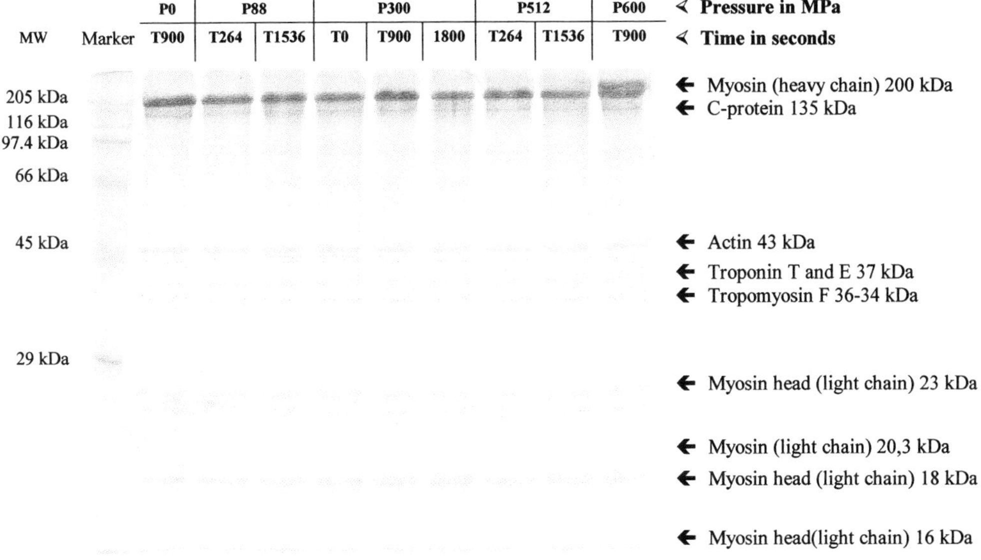

Figure 5 : SDS-PAGE patterns of myofibrillar proteins fractions.

CONCLUSION

This study shows that high pressure treatment of myofibrillar proteins in buffer solution induces modifications of surface hydrophobicity and sulfhydryl interactions and changes of the rheological behaviour and specially modification of flow. These results let think that pressure treatment produces changes in the structure of proteins that induce the results obtained. If we superpose our results we show a concomitance between the results : the hypothesis of depolymerisation and irreversible modification of the structure was put forward. The experiment shows that the pressure level is more important than the duration of the experiment.

More work is necessary to obtain detailed information concerning the denaturation stage and the interactions between myofibrillar proteins. This knowledge would allow to understand the new properties of these molecules which represent a new potential of food functional properties.

ACKNOWLEDGEMENT

This work was supported by the Region Pays de la Loire. We gratefully acknowledge INRA - LEIMA laboratory for the material assistance.

REFERENCES

[1] R. Hayashi, Y, Kawamura, T. Nakasa and O. Okinaka, Agricultural biological chemistry, 53 (11) (1989) 2935-2939.

[2] N. Chapleau, S. Jung and M. de Lamballerie – Anton, High pressure research, 19 (2000) 61-68.

[3] M. Anton, N. Chapleau, V. Beaumal, S. Delépine and M. de Lamballerie – Anton, Innovative Food Science & Emerging Technologies (IFSET) (2001) in press.

[4] G. Offer and C. Moos, J. Mol. Bio., 74 (1973) 653-676.

[5] L.G. Ellman, Archives of Biochemistry and Biophysics, 74 (1958) 443-450.

[6] A. Kato and S. Nakai, Biochemica et biophysica, 634 (1980) 13-20.

[7] T. Sano, S. Noguchi, J. Matsumoto and T. Tsuchiya, Journal of Food Science, 54 (1989) 800-804.

[8] I. Hayakawa, Y. Linko and P. Linko, Lebensm.-Wiss. u.-Technol., 29 (1996) 756-762.

Trends in High Pressure Bioscience and Biotechnology
R. Hayashi (editor)

Effects of mutation and ligand binding on the compressibility of a protein

Kunihiko Gekko, Tadashi Kamiyama, Eiji Ohmae, and Katsuo Katayanagi

Department of Mathematical and Life Sciences, Graduate School of Science, Hiroshima University, Higashi-Hiroshima 739-8526, Japan

The partial specific volume (v^0) and adiabatic compressibility (β_s^0) of *E. coli* dihydrofolate reductase (DHFR) are largely influenced by binding ligands (DHF, THF, NADPH, NADP$^+$, and methotrexate) and single amino acid substitutions at Gly67, Gly121, and Ala145 located in three flexible loops (18 mutants). The variations in v^0 and β_s^0 are mainly ascribed to the amount of cavity with a minor contribution of hydration (solvent accessible surface area). The β_s^0 of DHFR changes alternatively by binding or releasing the coenzyme and substrate. A mutant with a large β_s^0 value shows high enzymatic activity. These results demonstrate that a small alteration in the local structure due to ligand binding and mutation is dramatically magnified in the overall protein dynamics to affect the function, possibly *via* modification of cavity or long range interactions.

1. INTRODUCTION

It is well known that the adiabatic compressibility (β_s^0) and partial specific volume (v^0) sensitively reflect the structural characteristics of native protein and the conformational change due to denaturation although they are macroscopic quantities involving two contributions, surface hydration and internal cavity [1,2]. However, only limited data have been reported on the compressibility and volume changes due to ligand binding and amino acid substitution although such volumetric study is essential for understanding the role of the structural flexibility of protein in the function. To address the structure-flexibility-function relationship of protein, we have examined the effects of ligand binding and mutation on β_s^0 and v^0 of a model protein, *E. coli* dihydrofolate reductase (DHFR) [3,4]. This enzyme is a monomeric protein of 159 amino acids with no disulfide bond and catalyzes the reduction of dihydrofolate (DHF) to tetrahydrofolate

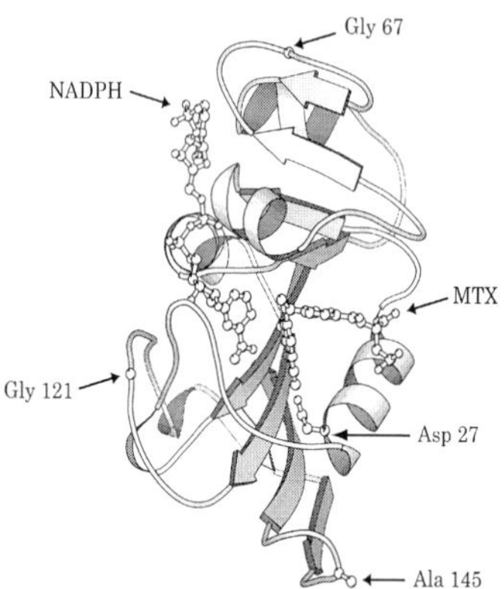

Figure 1. The structure of a DHFR·NADPH·MTX ternary complex (PDB-ID: 1rx3) taken from Sawaya and Kraut [6]. Positions Gly67, Gly121, Ala145 (mutation sites) and Asp27 (catalytic site) are indicated by arrow.

(THF) with the aid of coenzyme, NADPH [5]. The enzymatic function is strongly repressed with an inhibitor, methotrexate (MTX). The X-ray structure of a DHFR ·NADPH·MTX ternary complex is shown in Fig.1 [6].

2. METHODOLOGY

2.1. Sample preparation

The wild-type and all mutant DHFRs were obtained from *E. coli* strain HB101 and purified on a MTX-agarose affinity column. The oligonucleotides used for site-directed mutagenesis and the detailed mutant construction are described in previous papers [7,8]. The enzyme was fully dialyzed against 10 mM phosphate buffer (pH 7.0) containing 0.1 mM EDTA and 0.1 mM dithiothreitol, and centrifuged at 14,000 rpm for 20 min to remove the aggregates. The concentrations of all DHFRs were determined by absorption measurements.

For the ligand binding experiments, the sample solutions were prepared by mixing a given weight of stock solutions of enzyme and ligand. The final concentration of enzyme was adjusted to be 3 mg·cm^{-3}. Each ligand was added in the minimal excess to make a 1:1 complex with enzyme, taking into consideration its dissociation constant.

2.2. Sound velocity and density measurements

The partial specific volume and adiabatic compressibility were determined by sound velocity and density measurements. The sound velocity in protein solution, u, was measured by means of a "sing-around pulse method" at 5.9 MHz with an accuracy of 1 cm·s⁻¹. The solution density, d, was measured with a precision density meter DMA-02C (Anton Paar, Gratz), with an accuracy of 1×10^{-6} g·cm⁻³. The apparatus and detailed experimental procedures are described in previous papers [1-4].

The adiabatic compressibilities of the sample solution (β) and solvent (β_0) were calculated with the Laplace equation, $\beta = 1/du^2$, using the sound velocity and density data set for the sample solution and solvent. The partial specific volume ($v^{\,0}$) and adiabatic compressibility ($\beta_s^{\,0}$) of protein at infinite dilution were calculated using the following equations

$$v^0 = \lim_{c \to 0} (1/c)[1 - (d - c)/d_0] \tag{1}$$

$$\beta_s^0 = -(1/v^0)(\partial v^0/\partial P) = (\beta_0/v^0)\lim_{c \to 0}(1/c)[(\beta/\beta_0) - (d - c)/d_0] \tag{2}$$

where c is the concentration of the protein in grams per milliliter of solution. For DHFR-ligand complexes, the v and β_s values measured at a DHFR concentration (3 mg·cm⁻³) were regarded as those at infinite dilution since the protein concentration dependence of v and β_s is negligibly small.

The partial specific volume of a protein in water is expressed as the sum of three contributions: constitutive atomic volume, v_c, the internal cavity due to imperfect atomic packing, v_{cav}, and the volume change due to hydration, Δv_{sol}.

$$v^0 = v_c + v_{cav} + \Delta v_{sol} \tag{3}$$

Since the volume of constitutive atoms is assumed as incompressible, the adiabatic compressibility is mainly determined by cavity and hydration as follows,

$$\beta_s^0 = -(1/v^0)[\partial v_{cav}/\partial P + \partial \Delta v_{sol}/\partial P] \tag{4}$$

The cavity contributes positively and hydration negatively to v^0 and β_s^0, then β_s^0 can be positive or negative depending on the magnitude of both terms.

2.3. Calculation of accessible surface area and cavity

To estimate the contributions of hydration and cavity to v° and β_s°, the solvent accessible surface area (ASA) and the total cavity volume (V_{cav}) were calculated from the X-ray structures reported by Sawaya and Kraut [6] for the wild-type DHFR and its ligand complexes, and those for the DHFR mutants determined in our laboratory (K. Katayanagi *et al.,* to be published). The ASA and V_{cav} were calculated with a probe radius of 1.4 and 0.9 Å, respectively, using a program GRASP.

3. RESULTS AND DISCUSSION

3.1. Effects of ligand binding

Table 1 shows the v° and β_s° values of DHFR-ligand complexes at 30℃ [3]. Evidently, binding of these ligands (binary and ternary complexes) brings about large changes of v° (0.734 − 0.754 $cm^3 \cdot g^{-1}$) and β_s° (6.6 − 9.8 $Mbar^{-1}$).

In general, the enzyme-ligand complex formation proceeds in three states: formation of net bonds between enzyme and ligand, changes in hydration of the interacting species, and conformational changes of the enzyme. The solute-solute interaction in water will accompany dehydration of each solute molecule, then the v° and β_s° should increase on ligand binding. However, the observed changes in both parameters are not necessarily positive. As shown in the forth and fifth columns of Table 1, the V_{cav} changes positively or negatively ($\pm 40\%$) depending

Table 1

The values of v°, β_s°, ASA, and V_{cav} of DHFR-ligand complexes at 30℃ [3].

Ligand	v° ($cm^3 \cdot g^{-1}$)	β_s° ($Mbar^{-1}$)	ASA ($Å^2$)	V_{cav} ($Å^3$)
Apo-DHFR	0.741 ± 0.001	8.1 ± 0.7	8350	250
NADPH	0.741 ± 0.001	7.3 ± 0.2	8397	224
NADP$^+$	0.743 ± 0.005	8.0 ± 0.7	7982	283
DHF	0.754 ± 0.002	8.7 ± 0.2	8105	262
THF	0.736 ± 0.002	8.0 ± 0.3	8138	152
THF·NADPH	0.741 ± 0.002	8.3 ± 0.4	8180	326
THF·NADP	0.733 ± 0.004	6.6 ± 0.7	7871	264
MTX·NADPH	0.754 ± 0.001	9.8 ± 0.9	7979	341

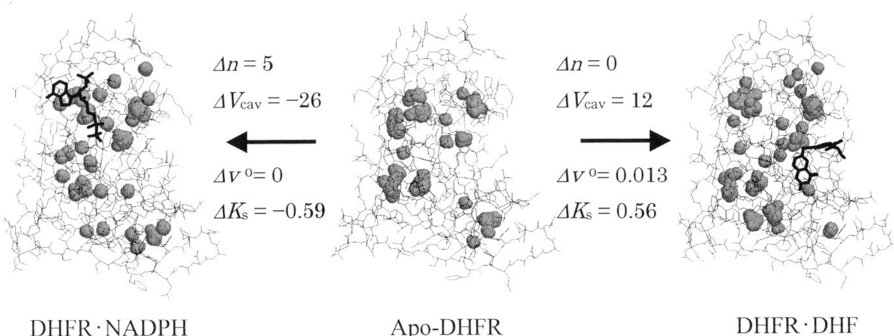

Figure 2.   The cavity distribution in some kinetic intermediates of DHFR.

on the ligands and ASA decreases (at most 6%) for any ligand except for NADPH. Interestingly, the β_s^0 increases with increasing V_{cav} ($r = 0.486$). These results suggest that a large modification of internal cavity occurs to compensate for hydration effect and the cavity plays a dominant role in ligand-induced compressibility change of DHFR. It is highly possible that only a small change of cavity brings about a large influence on the compressibility because the apparent compressibility of cavity is very large [1]. Fig. 2 illustrates the cavity distribution in the DHFR molecules with and without ligands (NADPH and DHF), together with the changes in partial specific volume (Δv^0), compressibility ($\Delta K_s = \Delta(v^0 \beta_s^0)$), number of cavities ($\Delta n$), and total cavity volume (ΔV_{cav}). Ligand binding affects not only the total cavity volume but also the size and distribution of

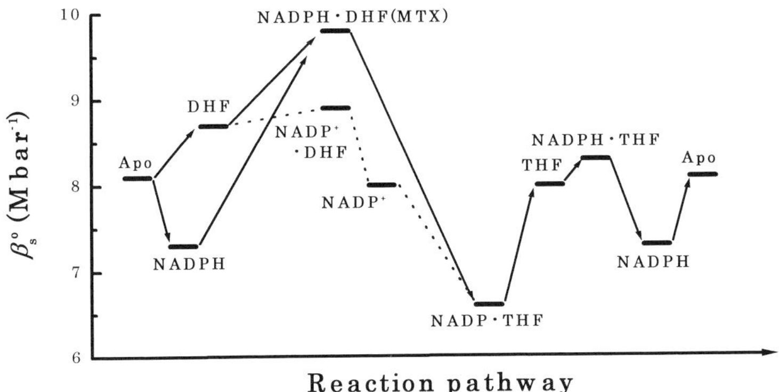

Figure 3. The β_s^0 level of the kinetic intermediates of DHFR as a function of the reaction pathway [3].

cavities. Our recent high pressure NMR study of folate-bound DHFR reveals that major differences in conformation occur in the NADPH binding site and the flexible loops [9].

DHFR catalyzes the NADPH-linked reduction of DHF to THF in the kinetic reactions cycling through five intermediates [5]. The $\beta_s{}^o$ of these intermediates would be helpful for understanding how the enzyme reaction is linked to the flexibility of the enzyme structure. Fig. 3 shows the $\beta_s{}^o$ level of the kinetic intermediates as a function of the reaction coordinate, in which we assumed a complex DHFR·NADPH·MTX for the transient state DHFR·NADPH·DHF since the X-ray structures of both complexes are quite similar [6]. Evidently, the conformational flexibility of DHFR changes alternatively by binding or releasing the coenzyme and substrate. It is expected that the transient state DHFR·NADPH·DHF is most flexible in the intermediates and changes to the most rigid state DHFR·NADP⁺·THF after the hydride transfer from NADPH to DHF.

3.2. Effects of mutation

As shown in Fig. 1, DHFR has several flexible loops, the residues 9-24, 64-72, 117-131, and 142-149. The B-factors of these loops are remarkably affected by binding MTX, suggesting the participation of the loops in dynamics and function of this enzyme. Table 2 shows the v^o and $\beta_s{}^o$ values for some of DHFR mutants at

Table 2
The v^o, $\beta_s{}^o$ (15°C) and kinetic parameters (25°C) of DHFR mutants [4].

DHFRs	v^o (cm³·g⁻¹)	$\beta_s{}^o$ (Mbar⁻¹)	K_m (μM)	k_{cat} (s⁻¹)	k_{cat}/K_m (μM⁻¹·s⁻¹)
wild	0.723 ± 0.001	1.7 ± 0.3	1.3	24.6	18.9
G67D	0.724 ± 0.003	3.0 ± 0.7	1.2	20.6	17.2
G67V	0.721 ± 0.003	1.1 ± 0.5	1.3	19.1	14.7
G67C	0.724 ± 0.001	1.7 ± 0.3	1.2	24.6	20.5
G121C	0.725 ± 0.001	5.5 ± 0.5	1.5	8.4	5.6
G121S	0.721 ± 0.002	0.7 ± 1.0	1.8	5.2	2.9
G121Y	0.724 ± 0.001	−0.7 ± 0.6	2.5	4.4	1.8
A145G	0.726 ± 0.001	3.1 ± 0.2	1.2	21.2	17.6
A145H	0.728 ± 0.001	3.8 ± 0.6	0.8	18.8	26.9
A145R	0.725 ± 0.002	0.8 ± 0.1	1.4	19.9	14.2

Gly67 (7 mutants), Gly121 (6 mutants), and Ala145 (5 mutants), which are located at the center of three loops [6]. Evidently, single amino acid substitutions bring about large variations in v° (0.710 – 0.733 $cm^3 \cdot g^{-1}$) and β_s° (–1.8 – 5.5 $Mbar^{-1}$), from the corresponding values for the wild-type DHFR. The β_s° value increases with increasing v°, as found statistically for other globular proteins [1]. This is evidence that the cavity contributes positively and hydration negatively to both v° and β_s°. Recently, we have carried out the X-ray crystallographic analyses of DHFR mutants at sites 67 and 121 and found that the B-factors of main chain atoms and the cavities at positions far from the mutation sites are influenced by the mutation without any significant change in ASA (K. Katayanagi *et al.*, to be published), and the V_{cav} of mutants increases with increasing β_s° (correlation coefficient, $r = 0.510$). These results suggest that the β_s° changes due to mutation are dominantly ascribed to modification of internal cavities.

The steady-state kinetic parameters, K_m and k_{cat}, for the enzyme reaction of DHFRs are listed in Table 2 [7,8,10]. The K_m is only slightly influenced by the mutation at three positions but k_{cat} significantly changes, especially at position 121, leading to a large modification of enzyme activity (k_{cat}/K_m). This is very interesting because it is unlikely that the mutation sites directly participate in the enzyme reaction: α-carbons of Gly67, Gly121 and Ala145 are separated respectively by 29.3, 19.0 and 14.4 Å from the catalytic site Asp27, and at

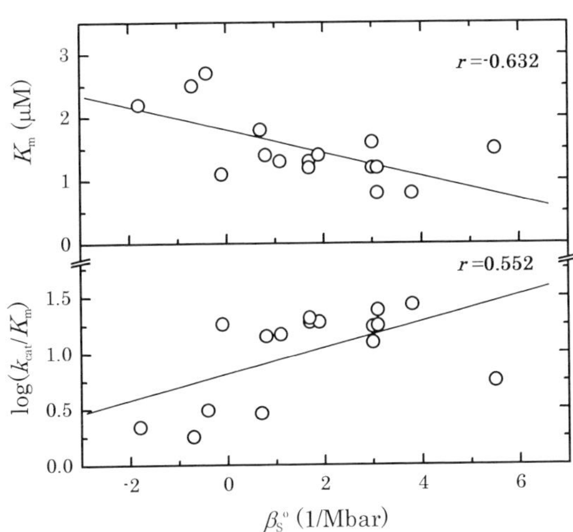

Figure 4. Plots of K_m and $\log(k_{cat}/K_m)$ against β_s° of DHFRs [4].

shortest by 8.5, 10.6 and 29.2 Å from the NADPH molecule [6]. As shown in Fig. 4, the K_m decreases with increasing $\beta_s{}^0$ ($r = -0.632$), on the other hand, the log (k_{cat}/K_m) and k_{cat} increase ($r = 0.552$, $r = 0.410$). These correlations suggest that the structural flexibility contributes favorably for the function of this enzyme through the increased affinity for the substrate (DHF) and the enhanced catalytic reaction rate. It is important that the flexibility-mediated enhancement of enzyme function can be brought about by mutations in the loop regions far from the catalytic site, because this gives a new standpoint for the molecular design of functional proteins.

As revealed in this study, the compressibility of DHFR is sensitively influenced by ligand binding and single amino acid substitutions in the flexible loops, due to large modifications of the internal cavity compared with surface hydration. These findings demonstrate that the conformational flexibility is an important factor to regulate the enzyme function and that a small alteration of local structure is dramatically magnified in the overall protein dynamics, possibly *via* "stepwise atomic reconstruction or long-range cooperative effect". Further insight into the molecular mechanism of compressibility-structure-function relationship of proteins should be obtained through the combination study with X-ray and high-pressure NMR analyses.

REFERENCES

1. K. Gekko & Y. Hasegawa, Biochemistry, 25 (1986) 6563.
2. Y. Tamura & K. Gekko, Biochemistry, 34 (1995) 1878.
3. T. Kamiyama & K. Gekko, Biochim. Biophys. Acta, 1478 (2000) 257.
4. K. Gekko, T. Kamiyama, E. Ohmae, & K. Katayanagi, J. Biochem., 128 (2000) 21.
5. C.A. Fierke, K.A. Johnson, & S.J. Benkovic, Biochemistry, 26 (1987) 4085.
6. M.R. Sawaya & J. Kraut, Biochemistry, 36 (1997) 586.
7. K. Gekko, Y. Kunori, H. Takeuchi, S. Ichihara, & M. Kodama, J. Biochem., 116 (1994) 34.
8. E. Ohmae, K. Iriyama, S. Ichihara, & K. Gekko, J. Biochem., 119 (1996) 703.
9. R. Kitahara, S. Sareth, H. Yamada, E. Ohmae, K. Gekko, & K. Akasaka, Biochemistry, 39 (2000) 12789.
10. E. Ohmae, K. Ishimura, M. Iwakura, & K. Gekko, J. Biochem., 123 (1998) 839.

Trends in High Pressure Bioscience and Biotechnology
R. Hayashi (editor)
© 2002 Elsevier Science B.V. All rights reserved.

Effect of glycosylation on the mechanism of renaturation of carboxypeptidase Y

M. Dumoulin[a,†], S. Matsukawa[a], H. Ueno[a,‡], C. Cléry[b], P. Masson[b] and R. Hayashi[a]

[a]Division of Applied Life Sciences, Graduate School of Agriculture, Kyoto University, 606-8502 Kyoto, Japan

[b]Centre de Recherches du Service de Santé des Armées, Unité d'enzymologie, 38702 La Tronche, France

A comparative study of the denaturation of the glycosylated and unglycosylated forms of carboxypeptidase Y (CPY) at high pressure and low temperature was conducted, in order to better understand the role of the carbohydrate moiety in the structural stability of this enzyme. Our findings show that, in the absence of its carbohydrate moiety, the structural fluctuations of CPY are increased, resulting in a lower kinetic stability of unglycosylated CPY against pressure-assisted cold denaturation compared to the fully glycosylated form.

1. INTRODUCTION

Carboxypeptidase Y (EC 3.4.16.1; CPY) from *Saccharomyces cerevisiae* is a 61 kDa vacuolar monomeric glycoprotein. The central question of how the carbohydrate moiety, which accounts for about 16 % of the molecular mass, is involved in the conformational stability and function of this enzyme is not entirely clear at the present time. In the present approach, high pressure combined to low temperature was used, in order to investigate the stability of glycosylated (WtCPY) and unglycosylated CPY (ΔglyCPY, obtained by site-directed mutagenesis) and thus, to elucidate the structural role of the carbohydrate moiety.

In a preceding study, it was shown that the *in-situ* pressure-induced denaturation of both WtCPY and ΔglyCPY involves at least three transitions and the formation of two molten globule states (Scheme 1, [1]). For ΔglyCPY, the first two transitions clearly occurred at pressures lower than the pressures required for inducing the denaturation of glycosylated CPY, suggesting that the carbohydrate moiety protects CPY against pressure-induced denaturation.

$$N \rightleftharpoons N' \rightleftharpoons MG_1 \rightleftharpoons MG_2 \qquad\qquad \text{Scheme 1}$$

The goal of the present study was to develop a better understanding of the mechanism of involved in the pressure-induced denaturation of CPY, and to determine whether the carbohydrate moiety affects this mechanism. The kinetics of renaturation were monitored by activity measurements after pressure release.

Present address : [†] Centre for Protein Engineering, Institute of Chemistry B6, University of Liège, B-4000 Liege, Belgium; [‡]Department of Food Science, Nara Women's University, Nara 630-8506, Japan.

This work was supported by a grant from the Japanese Ministry of Education, Culture and Sport.

The structural properties of the irreversibly unfolded state were investigated by: intrinsic fluorescence, ANS-bound fluorescence, CD measurements and non-denaturing PAGE. In parallel, pressure-induced unfolding equilibrium was monitored by electrophoresis by means of capillary gels under pressure.

2. MATERIALS & METHODS

WtCPY and ΔglyCPY were produced and purified as described previously [1].

2.1. Capillary gel electrophoresis under high pressure

Electrophoreses in capillary non denaturing gels were performed under high pressure in a thermostated high pressure vessel with electrical connections, essentially as previously described [2]. 3.5 µg of each enzyme, diluted in running buffer which also contained the tracking dye were loaded onto each gel. Three gels loaded with WtCPY and three gels loaded with ΔglyCPY were run simultaneously. The microelectrophoresis cell was placed in the high pressure vessel which had been pre-equilibrated at the desired temperature (from 2 to 25 °C). Pressure (up to 250 MPa) was then applied, and the system was allowed to equilibrate for a period of 10 minutes. To stack the protein sample, a pre-electrophoresis was carried out at a constant intensity of 0.17 mA/gel for 5 minutes. Electrophoresis was then run at a constant intensity of 0.3 mA/gel for 20 minutes. A 8.26 mM Tris/0.1 M glycine buffer, pH 8.3 was used as the running buffer. After depressurization, the gels were removed from the capillary tubes. For each enzyme sample, one gel was stained for activity and two gels for protein content. Activity staining was carried out essentially as described previously [3]. The staining solution was prepared immediately prior to use by dissolving 2.5 mg of N-acetyl DL-phenylalanine β-napthyl ester in 1.5 ml of acetone, followed by the addition of 10 ml of 50 mM Tris/HCl pH 7.5 and 2.5 mg of Fast Blue BB salt. Protein was stained with Coomassie Brilliant Blue. The stained gels were analyzed using a video densitometer (Vilber Lourmat, Marne-la-Vallée, France) to determine protein migration distances.

2.2. *Ex situ* measurements

The protein concentration was 0.6 mg/ml in 10 mM phosphate buffer pH 7. Pressurization was achieved as described previously [4]. About 3-5 min interval existed between the release of pressure and the first measurement of pressurized samples.

Enzymatic activity was measured spectrophotometrically using a Shimadzu UV-210A digital double beam spectrophotometer which was equipped with a temperature water bath set at 25 °C. Both sample and reference cuvettes containing 995 µl of 1 mM ATEE in 50 mM sodium phosphate buffer pH 7.5 were incubated for 5 minutes in the cell holder. 5 µl of the enzyme solution were then added to the sample cuvette and the change in absorbance at 237 nm was recorded for 5-10 minutes. The activity was calculated from the linear part of the curve using the Sigma plot program (Version 4.0). The activity of a non-pressurized enzyme sample, which had been stored on ice was taken as 100 %.

Fluorescence emission spectra were measured using a Shimadzu RF 5300 PC spectrofluorimeter in a 0.5 mm path length quartz cuvette. The excitation wavelength was 280 and 350 nm for intrinsic and ANS-bound fluorescence, respectively. Spectra were recorded from 300 to 450 nm and from 420 to 600 nm for intrinsic and ANS-bound fluorescence, respectively. The protein was diluted to 0.05 mg/ml in 10 mM phosphate buffer pH 7.

For CD measurements, the enzyme concentration was adjusted to approximately 0.2 mg/ml in 10 mM sodium phosphate buffer pH 7 and, CD measurements were carried out at room temperature immediately after the release of pressure. Spectra were recorded on a JASCO J-720W spectropolarimeter (JASCO Co., Tokyo, Japan), in the far UV range (190-250 nm) in a 0.5 mm path length quartz cuvette. The final spectrum was the result of ten accumulations.

3. RESULTS

3.1. Electrophoresis in capillary gels under high pressure

Figure 1A shows the electrophoresis patterns for WtCPY, run at 25 °C, under 0.1, 150 and 250 MPa. The gels were stained for protein. At atmospheric pressure a single band (N) corresponding to the native CPY was observed, as expected, which exhibited activity. The same pattern was observed under a pressure of 150 MPa. The fact that Bromophenol blue and the protein band migrated slower under pressure than at atmospheric pressure is due to the effect of pressure on the buffer conductivity. Thus, it is unlikely that the protein mobility retardation was due to structural changes. When electrophoresis was performed under a pressure of 250 MPa, a new band corresponding to a protein state having a higher hydrodynamic volume appeared. This band also exhibited enzyme activity. Repetition of the experiment over a range of temperatures (2 to 25 °C) showed that the pressure-temperature region where the high hydrodynamic volume form appeared, corresponds closely to the region where the first molten globule (MG_1) is formed, as was observed by fluorescence measurement under pressure [1]. Thus, this form is likely MG_1 [1].

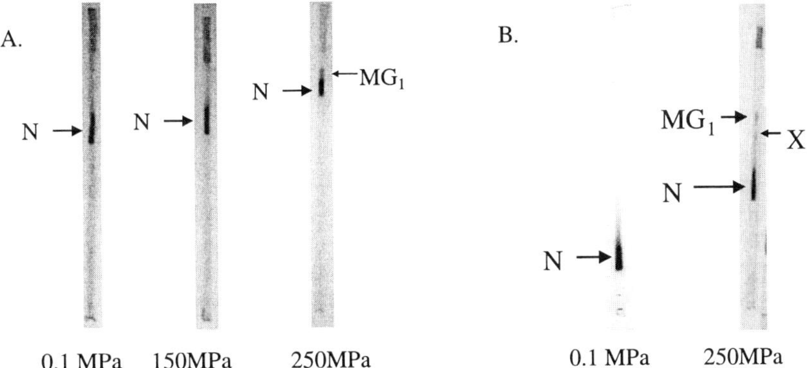

Figure 1. Polyacrylamide gel electrophoresis under different pressures at 25 °C. Electrophoresis were performed into capillary rods of 12% acrylamide under pressure in 8.26 mM Tris/0.1 M glycine, pH 8.3. Gels were stained with Coomassie Brilliant Blue. A) WtCPY, B) AglyCPY.

74

The results relative to ΔglyCPY were somewhat different (Fig. 1B). At atmospheric pressure and 25 °C, a single band (N), corresponding to the native protein was observed. At 250 MPa and 25°C, two new bands with a higher hydrodynamic volume were visible. Electrophoresis runs under pressure were conducted at different temperatures (from 2 to 25°C). The results indicated that the band having the highest hydrodynamic volume was formed under milder pressure/temperature conditions than that having the intermediate hydrodynamic volume. The pressure-temperature region where the highest hydrodynamic volume form appeared, corresponds closely to the region where the first molten globule (MG$_1$) is formed, as observed by fluorescence measurements under pressure. By analogy with the WtCPY, it is assumed that the form having the highest hydrodynamic volume corresponds to the MG$_1$ observed by *in situ* fluorescence measurements [1]. The significance of the form having an intermediate hydrodynamic volume is presently unclear: this form appeared in significant amounts under conditions where only traces of MG$_1$ were formed. From spectroscopic measurements, the transition from MG$_1$ to MG$_2$ occurs at a much higher pressure. Thus, it is unlikely that this band corresponds to MG$_2$. Close inspection of the densitograms showed that the intermediate band was much broader than the band having the highest hydrodynamic volume. This suggests that the intermediate band may correspond to a continuum of intermediate states, which occur between MG$_1$ and MG$_2$. These multiple transitions were undetectable by fluorescence measurements.

3.2. *Ex situ* measurements

Figure 2 summarizes the results for WtCPY under pressure. The activity was measured: a) under pressure (dash-dot-dot line); b) immediately after pressure release (dashed line) and, c) 1 hr after pressure release (solid line). Pressurization was carried out at 25°C and from 100 to 400 MPa for 30 minutes. A comparison of the dash-dot-dot line and dashed line indicates that,

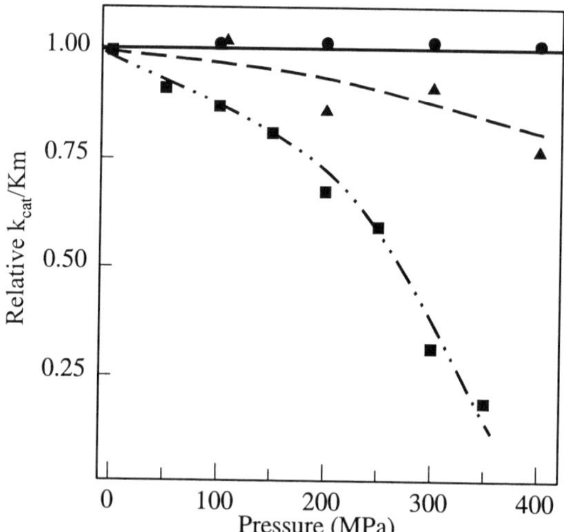

Figure 2. Effect of pressure on CPY activity. Dash-dot-dot line, under pressure (from [1]); dashed line, immediately after pressure release; solid, 1 hr after pressure release. [Enz] = 0.6 mg/ml in 10 mM sodium phosphate buffer pH 7; duration of pressurization: 30 min, temperature of pressurization: 25°C.

after treatment at a pressure up to 100 MPa, the WtCPY recovered all of its activity on pressure release. When the applied pressure was higher than 100 MPa, the activity was not immediately regained on pressure release, however the recovered activity reached 100% at 1 hr after the pressure release. Immediately after pressure treatment at 400 MPa and 25°C, WtCPY exhibited about 75% of its initial activity. *In situ* at 25 °C, WtCPY was almost completely inactivated by treatment of 400 MPa ([1] and Fig. 2). A rapid but partial renaturation of WtCPY then occurred, on pressure release. This rapid recovery, which occurs on pressure release is referred to as "the burst". The remaining 25% of the activity was slowly recovered on storage on ice. ΔglyCPY exhibited a behavior which is virtually identical to that of WtCPY (data not shown).

Figure 3 shows the amplitude of the burst for WtCPY and ΔglyCPY after various pressure/temperature treatments. WtCPY and ΔglyCPY were pressurized at given temperatures in the range of -22 to 25 °C and pressure, in the range of 0.1 to 400 MPa for 30 minutes. The remaining activity was then measured immediately after the pressure release (*e.g*, within 3-5 minutes) and as a function of time after pressure release, with the samples being kept on ice (Fig. 4).

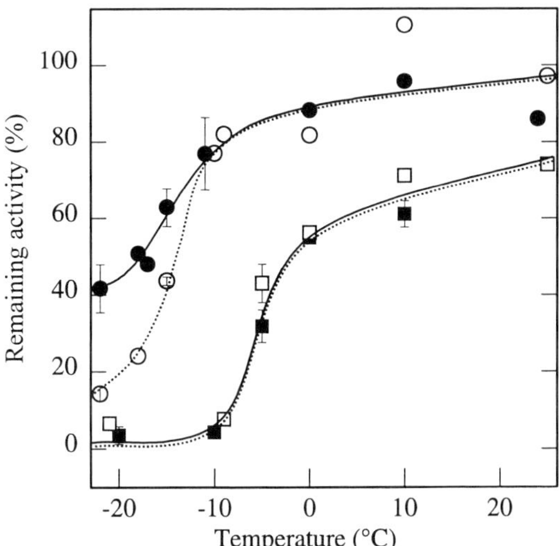

Figure 3. Pressure–temperature diagram for the activity of WtCPY and ΔglyCPY measured immediately after pressure release. [Enz]= 0.6 mg/ml in 10 mM sodium phosphate buffer pH 7; duration of pressurization: 30 min. (● and ■), WtCPY; (○ and □), ΔglyCPY. (○ and ●), 200 MPa; and (□ and ■), 400 MPa.

Immediately after ceasing the treatment at P= 400 MPa and T= 25°C, WtCPY exhibited 75 % of its initial activity (■) The remaining activity only slightly decreased as the temperature of pressurization was further lowered to 0 °C, *i.e.*, the activity of CPY was about

55% of its initial activity after a treatment at 400 MPa and 0 °C. In turn, the remaining activity abruptly decreased when the temperature of pressurization was further lowered to -10 °C. This suggests that the reversibility of the denaturation decreases with the temperature of pressurization. The curve for inactivation after a pressure treatment at 200 MPa (●) and different temperatures, exhibited a profile similar to that obtained after a treatment at 400 MPa, but the residual activity was shifted to higher values. After a treatment at 200 MPa and -22 °C, WtCPY had about 40% of its initial activity.

The open symbols in Fig. 3 show results obtained with ΔglyCPY. After treatment at 400 MPa the remaining activity of ΔglyCPY was similar to that of WtCPY at all temperatures of pressurization investigated. After pressurization at 200 MPa and temperatures higher than -10 °C, the two enzymes also displayed the same behavior (O and ●). After a treatment at this pressure, ΔglyCPY was inactivated to a greater extent than WtCPY as the temperature of pressurization was decreased further, *i.e.*, under -22°C, the remaining activity was 15% *versus* 40 % for WtCPY. The *in situ* inactivation is correlated to the second unfolding transition (corresponding to the formation of the first molten globule). Pm_2, the pressure at mid-transition from the native like state to the first molten globule state, decreased with decreasing temperature and, from our previous work, Pm_2 at -20 °C was estimated to be about 100 MPa [1]. Therefore, it can be concluded that under -20~-22°C and 200 MPa both enzymes were nearly fully inactivated *in situ*. Thus, the higher activity exhibited by WtCPY may result from a higher reversibility compared to ΔglyCPY.

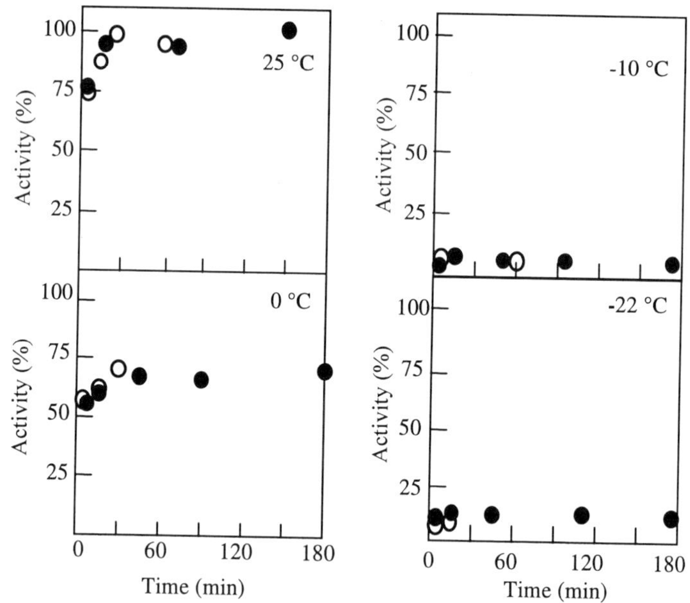

Figure 4. Kinetics of activity recovery upon storage on ice after different P/T treatments. [Enz]= 0.6 mg/ml in 10 mM sodium phosphate buffer pH 7; duration of pressurization: 30 min. (●), WtCPY; (o), ΔglyCPY).

Figure 4 shows the time course for activity recovery after a pressure treatment at 400 MPa, and at different temperatures. After treatment at 25 or 10 °C, WtCPY regained its full activity after about 1 hr of storage on ice (■). Following a treatment at 0°C, the magnitude of the burst was 50% and the recovery of activity was small. After treatment at 400 MPa and -10 or -20°C, WtCPY was nearly completely inactivated and no activity was recovered up to 3 hours after decompression. Storing the pressurized enzymes at 25°C had no effect on the recovery of activity (data not shown). The same pattern of renaturation was obtained for ΔglyCPY (Fig. 4, ○). The pressure-temperature phase diagram of water shows that, water would have been in the liquid state in all conditions investigated except at 400 MPa and -20 °C. Therefore, formation of ice cannot be responsible for the lower degree of reversibility of pressure-induced cold denaturation.

The low renaturation on refolding is often due to the aggregation of partially folded intermediates [5]. Thus, to determine whether the lowest reversibility observed for both enzymes after a treatment at subzero temperature was due to aggregation, a sample was pressurized for 30 minutes at -20°C and 200 MPa and then subjected to electrophoresis under non denaturing conditions. Under these conditions, a second band having a higher hydrodynamic volume (slow-migrating band) was observed on the gel for both for WtCPY and ΔglyCPY. For both enzymes, the increase in hydrodynamic volume was estimated to be 45% by gel filtration chromatography (data not shown). Therefore, the slow migrating band does not correspond to a dimer, and its lower mobility, compared to N, reflects a change in its charge/mass ratio.

The structural properties of the irreversibly denatured state were also investigated. In this form, aromatic and hydrophobic residues of the core appear to be more exposed to the solvent as revealed by intrinsic and ANS-bound fluorescence. A significant amount of secondary structure remained and this state has a higher hydrodynamic volume than the native state. Consequently, the irreversibly denatured state has some of the properties of the molten globule-like state. The fact that pressure cannot cause full unfolding of WtCPY could be due to protection by the glycan chains. However, the irreversibly denatured state of ΔglyCPY was virtually identical to that of WtCPY (data not shown).

The difference in reversibility, observed after treatment at a pressure of 200 MPa at -20 °C, may be due to a difference in kinetic stability (rate of irreversible denaturation). The half-time of inactivation after a treatment under -20 and 200 MPa was 9 and 20 minutes for WtCPY and ΔglyCPY, respectively. Thus, under these conditions of pressurization, the glycosylated CPY was found to be two fold more kinetically stable than the unglycosylated CPY. This means that a higher energy input is necessary to unfold WtCPY compared to ΔglyCPY. This suggests that, under the specific conditions of pressurization, the carbohydrate moiety significantly reduced the structural fluctuations of the protein. Such an increase in kinetic stability for glycosylated enzymes *versus* their glycan-free counterparts has been reported for other enzymes as well [6]. The absence of a continuum for intermediates between MG_1 and MG_2 for WtCPY is also consistent with a decrease in structural fluctuations due to the presence of a carbohydrate moiety. Transitions leading to these intermediates were not observed by intrinsic fluorescence or by ANS-bound fluorescence. This suggests that these intermediates share similar structural properties with one another and with MG_1. It is likely that they result from discrete packing rearrangements of some amino acids whose conformational fluctuations would be larger without the stabilizing effect of the carbohydrate moiety.

4. CONCLUSION

Results show that the absence of a carbohydrate moiety has no effect on the structure of the pressure-induced irreversibly denatured state. However, in the absence of the carbohydrate moiety, the structural fluctuations of CPY are increased. This lowers the kinetic stability of unglycosylated CPY against pressure-assisted cold denaturation.

REFERENCES

1. M. Dumoulin, H. Ueno, R. Hayashi, and C. Balny. Eur. J. Biochem., 262 (1999) 475.
2. P. Masson. In: W. B. Holzapfel and N. S. Isaacs (Eds), High-Pressure Techniques in Chemistry and Physics. A practical Approach. Oxford University Press, Oxford (1997) 353
3. M. Pacaud, and J. Uriel. Eur. J. Biochem., 23, (1971) 435.
4. M. Dumoulin, S. Ozawa and R. Hayashi. J. Food Sci., 63 (1998) 92.
5. N. Schülke, and F. X. Schmid. J. Biol. Chem, 263, (1988), 8832.
6. J. W. Tams and K. G. Welinder. FEBS Lett., 421, (1998) 234.

Trends in High Pressure Bioscience and Biotechnology
R. Hayashi (editor)
© 2002 Elsevier Science B.V. All rights reserved.

Pressure Studies on Protein Folding, Misfolding, Protein-DNA Interactions and Amyloidogenesis.

D. Ishimaru, L. M. T. R. Lima, A. Ferrão-Gonzales, P. A. Quesado, L. M. Maiolino, J. L. Silva and D. Foguel.

Centro Nacional de Ressonância Magnética Nuclear, Departamento de Bioquímica Médica, Instituto de Ciências Biomédicas, Universidade Federal do Rio de Janeiro.

ABSTRACT

Hydrostatic pressure is a useful tool for dissecting macromolecular interactions at the molecular level. Nonpolar interactions are determining factors in protein folding, protein aggregation and protein-nucleic acid recognition. Because nonpolar interactions are entropic and compressible, they are more sensitive to pressure and low temperatures. We have studied problems of macromolecular recognition using hydrostatic pressure as the primary tool and employing several spectroscopic techniques, especially fluorescence, circular dichroism and high-resolution nuclear magnetic resonance. High pressure has the unique property of stabilizing partially folded states of a protein which degree of dissimilarity from the native state may range from drifted conformations to molten-globule states. The competition between correct folding and misfolding, which in many proteins leads to formation of insoluble aggregates is an important problem in the biotechnology industry and in human diseases such as amyloidosis, Alzheimer's, prion and tumor diseases. Because of its ability to sequester folding intermediates, pressure has been used to direct the folding in one direction or the other and to explore intermediates, which are at the junction of the routes for folding and aggregation.

1. INTRODUCTION

Proteins play the major functions in cells either as isolated molecules or forming macromolecular complexes held together by noncovalent interactions. A unique feature of the 'biological' world, as compared to the 'inorganic' world, is the large number of interactions among its components that involve relatively weak energies (less than 10 kcal/mol). The energetic coupling among these interactions is likely the basis of the high specificity of macromolecular recognition. Life depends on macromolecular recognition, especially at the level of protein folding and protein-nucleic acid (NA) interactions. The mechanism by which unstructured proteins spontaneously fold to their native functional form is still not completely understood (1). To this end, it is necessary to isolate and describe the intermediate structures that occur during folding. Here, it is reviewed that in several cases these intermediates have been trapped under pressure and their dynamics and structure have been characterized by fluorescence, light scattering, NMR and hydrodynamic methods. Pressure affects the equilibrium between denatured or dissociated and native forms in the direction of the form that occupies a smaller volume (2-5). The structural region of the protein that is most sensitive to pressure is the hydrophobic core, especially when cavities are present (5, 6). It is described below the unique characteristic of pressure to isolate molten globules or to favor the state that appears to have a segment of the protein in a partially folded state ("partial" molten globule).

The isolation of folding intermediates is crucial to the understanding of protein misfolding and protein aggregation. In the last decade, several diseases, such as Alzheimer's disease and other amyloidogenic diseases, spongiform encephalopathies (caused by prions), inherited emphysema, cystic fibrosis and likely many cancers are caused by protein misfolding (7). Biotechnology companies also face many problems with proteins aggregating into inclusion bodies when they are expressed in

80

bacteria. Figure 1 shows how functional proteins and macromolecular assemblies may divert from the folding pathway to a misfolding and aggregation dead end.

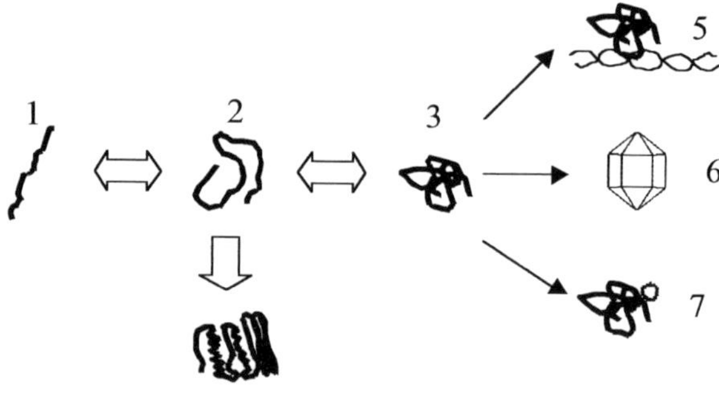

Figure 1. Schematic representation of the folding and assembly of protein and multimolecular complexes. (1) Unfolded protein; (2) Folding intermediate; (3) Tertiary structure; (4) Protein misfolding - Aggregate; (5) Protein-DNA interaction; (6) Virus; (7) Protein-ligand interaction.

To control misfolding and aggregation, we have to understand the energetics, the dynamics and the mechanisms of correct assembly. As described below, pressure has been useful not only to investigate the correct pathways (leading to 3, 5, 6 and 7 products – Figure 1), but also to the incorrect, misfolded form (product 4).

1.1. Protein folding and protein-protein interactions: The volume cavity model for pressure-induced dissociation and unfolding.

The mechanisms by which one dimension information (linear sequence of amino acid residues) is transferred to four dimensions (3D-structure and dynamics) remains one of the major challenges in Biology. Although most of the thermodynamic studies have been done with conventional perturbing agents, such as high temperature, urea, guanidine, etc, they offer limited information, especially because they cause drastic changes in protein structure, not likely to occur at physiological conditions. Folding of a protein, or association of one protein with another, is accompanied by an increase in volume because of the additive effects of the formation of solvent-excluding cavities and the release of bound solvent. Water is released as nonpolar amino acid residues are buried, as well as when salt linkages are formed (2-5). The Gibbs free energy (and the equilibrium constant) for an interprotein or intraprotein interaction will depend on the standard volume change (ΔV) of the reaction:

$$\Delta G(p) = \Delta G(o) + p\Delta V \tag{1}$$

$$\ln (\alpha_p^{\,n}/(1 - \alpha_p)) = p(\Delta V/RT) + \ln (K_{do}/n^n C^{(n-1)}) \tag{2}$$

where $\Delta G(p)$ and $\Delta G(o)$ are the free energies for association/folding at pressure p and at atmospheric pressure respectively; K_{do} is the equilibrium constant for dissociation or denaturation at atmospheric pressure; ΔV is the volume change, α_p is the extent of reaction at pressure p, and n is the number of dissociating subunits.

The packing among the different components play a major role in stabilizing proteins. Because of the huge number of atoms involved, packing defects cannot be avoided, which lead to formation of cavities (8). Cavities are also related to the metastability of some proteins and macromolecular assemblages (9-12). Recent studies show that hydrostatic pressure is a useful tool to investigate packing and cavities (6, 10-12).

Coiled-coil proteins are excellent models to evaluate the balance between packing and hydrophobic interactions. Studies with the large two-stranded coiled-coil protein tropomyosin show that, under pressure, this protein is substantially denatured (13), but that enough residual coiled-coil region remains to maintain a 'denatured dimer'. On the other hand, the short coiled-coil dimer, formed by the 31 amino acid polypeptide chains of the leucine zipper GCN4-p1, dissociated completely under pressure to unfolded monomers, with the decrease in volume reflecting the rupture of the hydrophobic interactions (14). The volume change of unfolding for the leucine zipper dimer corresponds to a decrease in volume of 141 $Å^3$ per molecule of dimer; whereas ΔV for a tetrameric mutant corresponds to a decrease in volume of 315 $Å^3$ per tetramer. The larger volume change for the tetramer reflects a cavity in the middle of each leucine and isoleucine layer as seen by X-ray diffraction data (15). Royer and coworkers have obtained strong evidence (6) for the hypothesis that the internal cavities in the protein's structure contribute to the magnitude of the observed volume change. Experimental and theoretical approaches indicate that the underlying mechanism of pressure unfolding is the penetration of water into the protein matrix (16, 17).

1.2. Protein folding intermediates isolated under pressure
It has been generally accepted that to unveil protein folding it is necessary to isolate and describe the intermediate structures. Some of these intermediates have been trapped under pressure and their dynamics and structure have been characterized by different methods. Several model proteins have been used for studies of protein folding and dimerization, including the E2 DNA-binding domain (E2-DBDD) from human papillomavirus (18, 19), LexA repressor (20) and Arc repressor (16, 21-23). Arc repressor dimer reversibly dissociates into partially folded subunits under pressure (21) and the structure of the pressure-denatured monomer was partially determined by two-dimensional ^{1}H NMR (24, 25). The pressure-induced population of partially folded intermediates has also been found for lysozyme (26), ribonuclease A (27), apomyoglobin (28), E2-DBD (19) the Ras binding domain of RalGEF (29) and DHFR (30) as determined by high-resolution NMR. The partially folded conformation of pressure-denatured proteins binds a substantial amount of water (16) and pressure denaturation does not proceed when water is withdrawn. This latter result has been corroborated by a theoretical approach that postulates the infiltration of water into the protein matrix (17, 31).

Small monomeric proteins are the best models for studying protein folding but they are often too stable for the pressures normally attainable in standard equipment. Recently, we made use of the lower stability of a non-covalent complex formed by two complementary fragments of the chymotrypsin inhibitor-2 to study the folding equilibrium (32) and kinetics (33). The pressure-denaturation of the complex was completely independent of protein concentration, indicating the formation of a denatured heterodimer. The structural characteristics of the pressure denatured state of the CI-2 complex can be explained by a molten globule conformation, similar to those obtained for E2-DBD (18, 19) and Arc repressor (21-24). Further kinetic studies under pressure confirmed a denatured CI-2 complex with residual persistent structure (33).

2. PROTEIN FOLDING AND PROTEIN-DNA INTERACTIONS:

Genes are turned on and off by the binding of proteins to specific DNA sequences (34). In several cases, the formation of a specific protein-DNA complex is followed by an increase in nonpolar interactions and release of solvent, which explain the entropy-driven character of the specific binding, while solvent is not displaced in nonspecific complexes (23). On the other hand, eukaryotic transcription factors seems to be less sensitive to hydration effects and cooperative interactions among the proteins (homo or hetero interactions) play a dominant role in the binding to the specific DNA sites (19).

2.1. The role of specific DNA in tightening protein-protein interactions in Arc and LexA repressor:

DNA recognition by Arc repressor is tightly coupled to the competence of the given sequence of base pairs in stabilizing the native dimeric state of Arc repressor (22). The Arc repressor-operator DNA complex was cold-denatured at sub-zero temperatures under pressure, indicating that the formation of the specific complex is followed by an increase in nonpolar interactions and release of solvent, which would explain its entropy-driven character, while solvent would not be displaced in nonspecific complexes (23). With LexA repressor, we found that the protein is a dimer at nanomolar concentrations and that the dissociation constant lies in the picomolar range (20). Whereas non-specific DNA has no stabilizing effects, specific DNA induces tightening of the dimer and a 750-fold decrease in the K_d. Accordingly, the LexA dimer only loses its ability to recognize a specific DNA sequence by RecA-induced autoproteolysis.

2.2. E2 DNA-Binding Domain (E2c) Protein of Human Papillomavirus 16.

E2 proteins from papillomaviruses are involved in viral DNA replication and regulation of transcription. The papillomavirus E2 protein is comprised of an N-terminal transactivation domain (E2n) separated by a flexible hinge region from the C-terminal DNA binding and dimerization domain (E2c). This dimeric domain, from a high-risk human papillomavirus (HPV-16), can be pressure-dissociated to a monomeric state that presents substantial residual structure, through a highly reversible transition (18). Further unfolding of this high-pressure state can be achieved by low concentrations of urea. NMR studies have confirmed that the pressure-dissociated monomer of E2c is highly folded (19).

The dissociation of E2c-DNA complexes (with specific and non-specific sequences) cannot be obtained using pressure alone (19). Taking into account that pressure and urea studies on the stability of E2c provide equivalent thermodynamic parameters, we obtained pressure isotherms in different urea concentration for both specific and non-specific DNAs. A plot of ΔG_o versus pressure (Figure 2.A) is linear for both the E2c complexes of E2c dimer. Both specific and non-specific DNA sequences promoted a large stabilization of E2c dimer when compared to the protein alone, with a higher change in free energy for the specific DNA sequence ($\Delta\Delta Go = 6.7$ kcal/mol) than for the non-specific DNA ($\Delta\Delta Go = 4.5$ kcal/mol). The difference between the effects produced by specific and nonspecific DNA is likely the basis for the sequence discrimination.

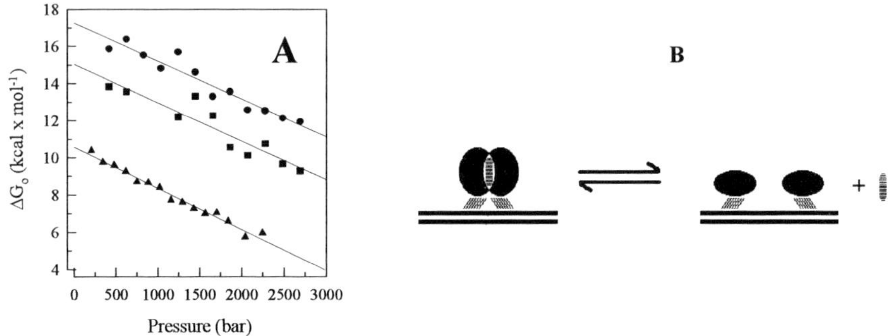

Figure 2. A) Plot of ΔG_o versus pressure for (▲) 1.0 μM E2c alone, (■) 250 nM [E2c:*poly*AT] complex (18 mer) and (●) 250 nM [E2c:E2DBS] (18 mer); B) Diagram of the monomer-dimer equilibrium of E2c bound to DNA. Volume change upon protein-protein and protein-DNA interaction correspond to hatched symbols.

The free energy change increases dramatically on DNA binding with no effect in the volume change, which indicates that no extra surface area is involved when E2c dissociates on the DNA. These results indicate that E2c dimers dissociate into monomers that are still bound to DNA (Figure 2.B). The retention of a large amount of tertiary structure by the dissociated monomer as evidenced by fluorescence (18) and NMR (19) data corroborate the hypothesis of a functional monomeric unit of E2c. These studies open new avenues to the development of drugs targeted to the monomer that will make it possible to prevent or treat human papillomavirus infection.

3. INCORRECT FOLDING, OFF-PATHWAY AGGREGATION AND AMYLOIDOGENESIS

The competition between correct folding and misfolding, which in many proteins leads to formation of insoluble aggregates, is an important problem in the biotechnology industry and in human diseases. The off-pathway aggregation of proteins often occurs *in vivo* with heterologous proteins that are over-expressed in *Escherichia coli*, resulting in the formation of inclusion bodies or amorphous aggregates within the cell. Pressure may affect the two pathways (aggregation/misfolding and correct folding) as recently demonstrated in the tailspike protein of bacteriophage P22 (35), rhodanese (36), myoglobin (37) and in the amyloidogenic protein transthyretin (38).

3.1. P22 Tailspike
High pressure (2.4 kbar) treatment dissociated tailspike aggregates and resulted in formation of monomers and native folded trimers, without changes in buffer, dilution of reagents, or modification of reaction conditions (35). Our results also indicate that aggregation intermediates have similar specific chain recognition events as those that are involved in proper folding events, and suggest that an increased understanding of this specificity will lead improved folding methodologies. In the practical side, the use of high pressure appears to be an efficient tool for combating aggregation in a variety of research and industrial settings (35, 39).

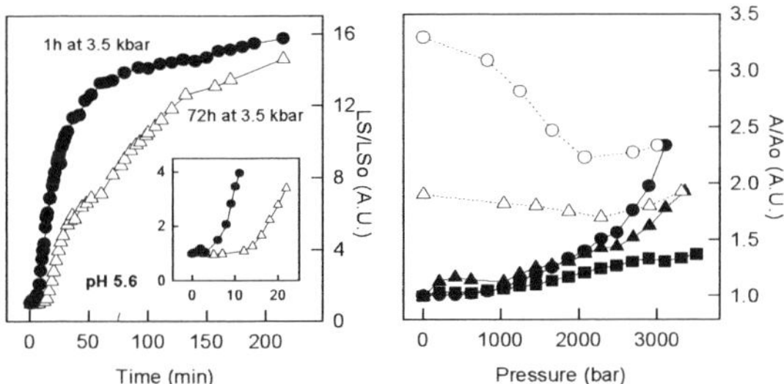

Figure 3. Left panel: Aggregation of TTR after a cycle of compression-decompression as followed by light scattering (LS) increase. (●) At pH 5.6, 3.5 μM TTR were compressed at 3.5 kbar during 1 h or 72 h (△). Then pressure was release and the LS was followed and normalized by the initial value (LS₀). Right panel: Bis-ANS binding as a function of pH. (■) pH 7.5; (▲) pH 5.6 and (●) pH 5. The hollowed symbols represent the decompression pathways. [TTT]= 1μM and [bis-ANS]= 10 μM.

3.2. Transthyretin (TTR)
Transthyretin (TTR) is a tetrameric human plasma protein responsible for familial amyloid polyneuropathies (FAP) and senile systemic amyloidosis (SSA) (40). In vitro studies have shown that

84

at acidic pH TTR dissociates into partially folded monomers and refolding from this denatured state lead to aggregation (41). The main hypothesis is that aggregation occurs after dissociation into unfolded or partially folded monomers. Hydrostatic pressure produced reversible denaturation of TTR to a state that maintained residual structure and still associated as tetramers (Figure 3) (38). Among other properties, partially folded TTR induced by pressure binds bis-ANS, similar to the acid-denatured protein that undergoes aggregation into amyloid fibrils (Figure 3). Under extreme high pressure, there was no aggregation, but at intermediate pressures, after compression and decompression, the protein aggregated into amyloid fibrils (38). These data show that pressure can be used as a controlled tool to explore intermediates, which are at the junction of the routes for folding and misfolding/aggregation.

3.3. Tumor Suppressor Protein P53

The tumor suppressor protein p53 is a 393 amino acid residue-transcriptional factor with an important role at the cell cycle control, specially after cellular stresses such as genotoxic damage, cytokines, hypoxia and alterations of ribonucleotide pools (42). Most human cancers ($\approx$50%) result from mutations in the p53 protein, mainly at its DNA binding domain (p53/DBD), affecting the DNA binding ability and/or the protein stability. Our results demonstrate that p53/DBD undergoes denaturation when submitted to hydrostatic pressures at 37°C, leading to a misfolded and aggregated state. However, when we investigated the pressure effects at 4°C, the protein denaturation was displaced to higher pressures and aggregation was non-cooperative with denaturation (Figure 4).

Under pressure and subzero temperatures, we observed the formation of a partially folded state of the tumor suppressor protein with no aggregation. Taken together, these results suggest that, by carefully tuning high pressure and temperature, we might achieve stable intermediates of the protein folding, providing unprecedented targets for the development of antagonists capable of blocking protein misfolding and aggregation, potential drugs against tumoral diseases. The pressure denaturation curves have a large hysteresis, a characteristic of a metastable state.

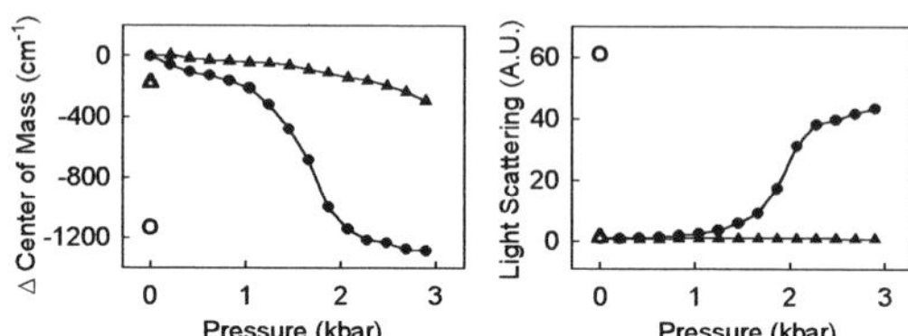

Figure 4. Pressure-induced denaturation of the tumor suppressor protein p53 up to 3.1 kbar at 37°C (squares) or at 4°C (circles). Closed symbols correspond to the compression cycle and open symbols represent values after return to atmospheric pressure. (A) Center of spectral mass based on the tryptophan emission fluorescence.. (B) Light scattering data.

Figure 5 shows a scheme that can be applied to transthyretin and to p53/DBD. In both cases, pressure populates a partially folded, pre-aggregated state. Depending on the temperature and applied pressure, the decompression may lead to a metastable state or to an aggregate. The folding "half-funnel" represented in Figure 5 shows clearly how pressure can allow us to fully explore intermediates and determine their structures, potential targets against tumor and amyloidogenic diseases.

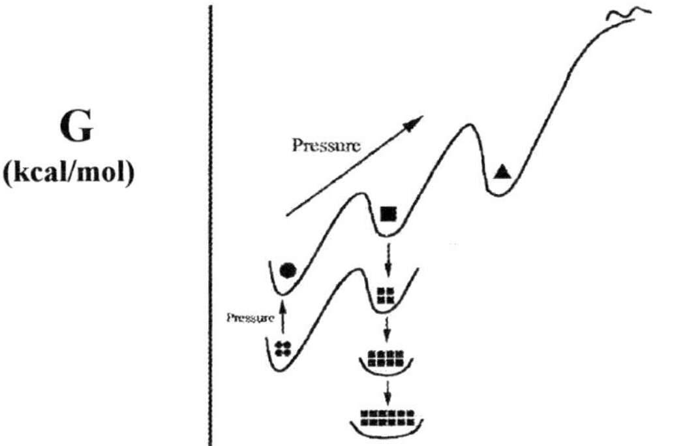

Figure 5. Free-energy diagram for dissociation, denaturation and aggregation of proteins. The native oligomer and the monomer are represented in circles, the altered, monomer, oligomer and aggregates in squares, the monomer with residual structure in triangle and unfolded protein as a line. Distance along vertical axis indicates differences in Gibbs free energy among different protein states.

4. CONCLUSIONS AND PERSPECTIVES:

The increasing use of structural methods, such as NMR, to characterize the pressure-denatured state has placed pressure as one of the more promising techniques to unravel the missing links in the "Protein Folding Problem". We have learned that nonpolar interactions are determining factors in protein folding, protein-nucleic acid recognition and protein misfolding/aggregation. Pressure has permitted characterization of the high degree of plasticity of proteins. Recent detailed studies on the kinetics of pressure unfolding and folding (43) together with several theoretical approaches have increased our understanding of the pressure effects. Furthermore, the isolation of folding intermediates is crucial to the understanding of protein misfolding and protein aggregation, which have been recently tackled by pressure. As outlined in the free-energy diagram for protein folding and misfoding/aggregation (Figure 5), the population of an amyloidogenic intermediate without proceeding to aggregation is a unique property of pressure, which opens the prospect to characterize the structure of the amyloidogenic form.

REFERENCES

1. S.E. Radford, Trends Biochem. Sci., 25 (2000) 611.
2. J.L. Silva and G. Weber, Annu. Rev. Phys. Chem., 44 (1993) 89.
3. J. Jonas and A. Jonas, Annu. Rev. Biophys. Biomol. Struct., 23 (1994) 287.
4. V.V. Mozhaev, K. Heremans, J. Frank, P. Masson and C. Balny, Proteins, 24 (1996) 81.
5. J.L. Silva, D. Foguel, A.T. Da Poian and P.E. Prevelige, Curr. Opin. Struct. Biol., 6 (1996) 166.
6. K.J. Frye and C.A. Royer, Protein Sci., 10 (1998) 2217.
7. E.H. Koo, P.T. Lansbury Jr. and J.W. Kelly, Proc. Natl. Acad. Sci. USA, 96 (1999) 9989.
8. A.E. Eriksson, W.A. Baase, X.-J. Zhang, D.W. Heinz, M. Blaber, E.P. Baldwin and B.W. Matthews, Science, 255 (1992) 178.
9. C. Lee, S.H. Park, M.Y. Lee and M.H. Yu, Proc. Natl. Acad. Sci. USA, 97 (2000) 7727.
10. D. Foguel, C.M. Teschke, P.E. Prevelige and J.L. Silva, Biochemistry, 34 (1995) 120.
11. P.C. De Souza, R. Tuma, P.E. Prevelige, J.L. Silva and D. Foguel, J. Mol. Biol., 287 (1999) 527.

12. L.P. Gaspar, A.F. Terezan, A.S. Pinheiro, D. Foguel, M.A. Rebello, J.L. Silva, J. Biol. Chem., 276 (2001) 7415.
13. M.C. Suarez, S.S. Lehrer and J.L. Silva, Biochemistry, 40 (2001) 1300.
14. J. L. Silva and J. R. Cortines. Submitted (2001).
15. E.K. O'Shea, J.D. Klemm, P.S. Kim and T. Alber, Science, 254 (1991) 539.
16. A.C. Oliveira, L.P. Gaspar, A.T. Da Poian and J.L. Silva, J. Mol. Biol., 240 (1994) 184.
17. G. Hummer, S. Garde, A.E. Garcia, M.E. Paulaitis and L.R. Pratt, Proc. Natl. Acad. Sci. USA, 95 (1998) 1552.
18. D. Foguel, J.L. Silva and G. Prat-Gay, J. Biol. Chem., 273 (1998) 9050.
19. L.M.T.R. Lima, D. Foguel and J.L. Silva, Proc. Natl. Acad. Sci. USA, 97 (2000) 14289.
20. R. Mohana-Borges, A.B. Pacheco, F.J. Sousa, D. Foguel, D.F. Almeida and J.L. Silva, J. Biol. Chem., 275 (2000) 4708.
21. J.L. Silva, C.F. Silveira, A. Correia and L. Pontes, J. Mol. Biol., 223 (1992) 45.
22. J.L. Silva and C.F. Silveira, Protein Science, 2 (1993) 945.
23. D. Foguel and J.L. Silva, Proc. Natl. Acad. Sci. USA, 91 (1994) 8244.
24. X. Peng, J. Jonas and J.L. Silva, Proc. Natl. Acad. Sci. USA, 90 (1993) 1776.
25. X.D. Peng, J. Jonas J and J.L. Silva, Biochemistry, 33 (1994) 8323.
26. D.P. Nash and J. Jonas, Biochemistry, 36 (1997) 14375.
27. J. Zhang, X. Peng, A. Jonas and J. Jonas, Biochemistry, 34 (1995) 8631.
28. S. E. Bondos, S. Sligar and J. Jonas, Biochim. Biophys. Acta, 1480 (2000) 353.
29. K. Inoue, H. Yamada, K. Akasaka, C. Herrmann, W. Kremer, T. Maurer, R. Doker and H.R. Kalbitzer, Nat. Struct. Biol., 7 (2000) 547.
30. R. Kitahara, S. Sareth, H. Yamada, E. Ohmae, K. Gekko and K. Akasaka, Biochemistry, 39 (2000) 12789.
31. N. Hillson, J.N. Onuchic and A.E. Garcia, Proc. Natl. Acad. Sci. USA, 96 (1999) 14848.
32. R. Mohana-Borges, J.L. Silva and G. de Prat-Gay, J. Biol. Chem., 274 (1999) 7732.
33. R. Mohana-Borges, J.L. Silva, J. Ruiz-Sanz and G. de Prat-Gay, Proc. Natl. Acad. Sci. USA, 96 (1999) 7888.
34. C.O. Pabo and R.T. Sauer, Annu. Ver. Biochem., 61 (1992) 1053.
35. D. Foguel, C.R. Robinson, P.C. de Sousa Jr, J.L. Silva and A.S. Robinson, Biotechnol. Bioeng., 63 (1999) 552.
36. B.M. Gorovits and P.M. Horowitz, Biochemistry, 37 (1998) 6132.
37. L. Smeller, P. Rubens and K. Heremans, Biochemistry, 38 (1999) 3816.
38. A.D. Ferrao-Gonzales, S.O. Souto, J.L. Silva and D. Foguel, Proc. Natl. Acad. Sci. USA, 97 (2000) 6445.
39. R.J. St John, J.F. Carpenter and T.W. Randolph, Proc. Natl. Acad. Sci. USA, 96 (1999) 13029.
40. J.W. Kelly, Curr. Opin. Struct. Biol., 8 (1998) 101.
41. Z. Lai, W. Colon and J.W. Kelly, Biochemistry, 35 (1996) 6470.
42. P.A. Hall and D.P. Lane, Curr. Biol., 7 (1997) R144.
43. C. A. Royer, *in* High Pressure Molecular Science, J. Jonas and R. Winter (eds.), Kluwer Academic Publishers, The Netherlands, (1999) p473-496.

Trends in High Pressure Bioscience and Biotechnology
R. Hayashi (editor)
                    87

High Pressure Gel Mobility Shift Analysis and Molecular Dynamics: Investigating Specific Protein-Nucleic Acid Recognition

T.W. Lynch[a,b], M.A. McLean[a,b], D. Kosztin[a], K. Schulten[a,c], S.G. Sligar[a,b,*]

[a]Beckman Institute for Advanced Science and Technology, [b]Department of Biochemistry, and [c]Department of Physics, University of Illinois, Urbana-Champaign, 405 N. Mathews Avenue, Urbana, Illinois, 61801, USA

In this report, we describe the application of electrophoresis for monitoring protein-DNA binding equilibria at high pressure utilizing the gel mobility shift assay. The first protein-DNA recognition complex studied using this methodology is the restriction endonuclease *Bam*HI binding the cognate DNA recognition sequence. The application of hydrostatic pressure to the specific recognition complex of *Bam*HI-DNA favors dissociation, which is apparent due to the increase in the equilibrium dissociation constant (K_d) at elevated pressures. From the dependence of K_d on pressure, the volume change (ΔV) of dissociation was determined. Molecular Dynamic (MD) simulations on the *Bam*HI-DNA complex at both ambient and elevated pressures were performed to identify the structural origins of the observed experimental results. The simulation trajectories have identified important protein-DNA recognition elements that are disrupted with pressure. The trajectories also illustrate an increased hydration of the *Bam*HI-DNA interface at elevated pressure. Both of these calculated pressure effects would favor dissociation of the complex. The combination of MD simulations and high pressure gel shift analysis proved useful in identifying these factors for maintaining *Bam*HI-DNA complex stability.

1. INTRODUCTION

Previously described experiments using high pressure electrophoresis have demonstrated the usefulness of this technique in biochemical research. Hydrostatic pressure is a convenient and efficient method for perturbing macromolecular equilibrium and the conformation of biological systems. Polyacrylamide gel electrophoresis is an excellent method for separating biomolecules according to size, charge, and conformation. Combining pressure and electrophoresis allows the determination of the stoichiometry of dissociation and the thermodynamic parameters ΔV and K_d. Previous investigations have incorporated high pressure electrophoresis towards understanding oligomeric protein assemblies (1), monomeric

---

* Corresponding author. This work was supported by National Institutes of Health Grants GM31756 and GM33775 (to S.G.S), and NIH grant RR05969, National Science Foundation grants BIR-9318159 and BIR-23827 EQ, the MCA 93S028P computer time grant at Pittsburgh Supercomputing Center, and the Roy J. Carver Charitable Trust (to K.S.).

protein conformation (2), and macromolecular association (3). The method of gel mobility shift analysis at high pressure to study protein-nucleic acid interactions however was never utilized until now.

The gel electrophoresis mobility shift assay is a useful and popular tool for studying protein-nucleic acid interactions. (4,5) The mobility shift assay consists of adding a DNA binding protein to a solution containing a radiolabeled DNA oligomer, harboring a recognition sequence, which can be tightly bound by protein. Following equilibration, the protein-DNA complex is separated from free DNA and/or free protein by electrophoresis. The resolved species are then visualized and their stoichiometries are determined. A flatbed electrophoresis vessel, capable of maintaining hydrostatic pressures up to 2 kbar, was constructed for performing high pressure gel mobility shift assays to monitor the effects of pressure on protein-nucleic acid complexes.

The *Bam*HI endonuclease binds the DNA recognition sequence 5'-GGATCC-3' with remarkable specificity. The enzyme catalyzes double strand hydrolysis in the presence of divalent cations (Mg^{2+}) after the first 5'-G, leaving staggered four base pair overhangs. (6,7) Numerous x-ray structures of the *Bam*HI-DNA complex have been determined that illustrate the interactions involved in site-specific recognition. (8) The factors that contribute to the stability of this complex cannot be identified with structural analysis alone. Therefore, it is necessary to combine structural information with thermodynamic analysis to understand the physical basis for sequence specificity and protein-nucleic acid complex stability. (9) To probe the pressure stability of the specific *Bam*HI-DNA complex we have employed the use of high pressure binding assays.

Molecular Dynamic (MD) simulations are an important theoretical tool and have been used to model the detailed behavior of protein-nucleic acid complexes. (10,11) These calculations offer detailed insight towards understanding the thermodynamic contributions associated with site-specific recognition. To explore the effects of pressure on the cognate *Bam*HI-DNA complex, MD simulations of the complex at both ambient and elevated pressure were performed. The structural origin of the observed pressure effects on the specific *Bam*HI-DNA complex was identified using this analysis and discussed below.

1.1 The specific *Bam*HI-DNA complex

The main feature of *Bam*HI structure is a large six-stranded mixed β-sheet surrounded by α-helices. The x-ray structure of *Bam*HI bound to a DNA oligonucleotide containing the cognate recognition sequence 5'-GGATCC-3' provides insight into site-specific recognition. This complex illustrates a wide range of protein-DNA interactions including side chain atoms, main chain atoms, and bound water molecules. While the DNA does not significantly change form, *Bam*HI undergoes a sequence of conformational changes on complex formation. The most notable structural change is the unraveling of the C-terminal α^7 helices, which are referred to as arms. The arm from the L monomer is positioned along the DNA phosphate backbone towards the *Bam*HI core, while the arm from the R monomer makes specific contacts in the DNA minor groove. The R monomer arm makes direct contacts with DNA through three protein residues, Asp[196], Gly[197], and Met[198]. The unraveled arm, which inserts in the minor groove is pictured in Figure 1.

The (R monomer) arm is an important recognition element, as demonstrated by an investigation by Aggarwal and coworkers that intended to dissect the mechanism of DNA strand hydrolysis. (13) In the study, the cognate *Bam*HI-DNA crystal was soaked with Mn^{2+}, to initiate DNA hydrolysis, and cleavage only occurred in the DNA strand bound by the R monomer with the arm situated in the DNA minor groove. (13)

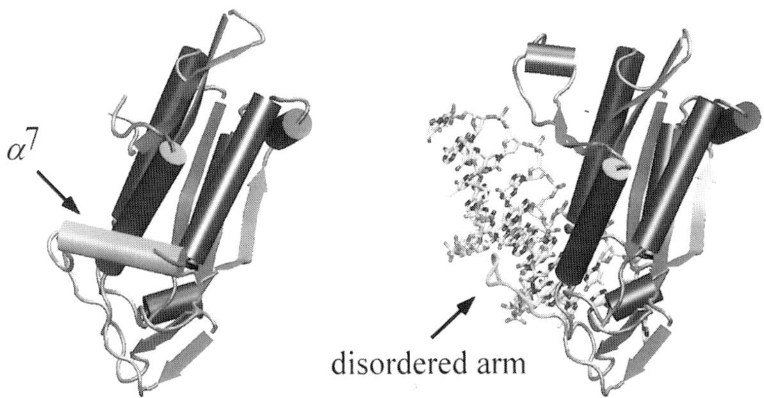

Figure 1. The free (left) and specific bound (right) forms of *Bam*HI. In the free form the C-terminal end exists as helix α^7, when bound to specific DNA this terminus becomes disordered and inserts into the minor groove. The R monomer in which the arm inserts into the minor groove is pictured for clarity.

2. EXPERIMENTAL PROCEDURES

2.1 High pressure gel shift assays

DNA oligonucleotides were purchased from Integrated DNA Technologies (Coralville, IA) and purified separately by polyacrylamide gel electrophoresis. Two complementary sequences (5'-CTCGTATAAT*GGATCC*GCAGTAAGCT-3') containing the *BamHI cognate recognition sequence* were used for binding analysis. The complementary oligonucleotides were 5' end-labeled with [γ-^{32}P]ATP (Amersham) by T$_4$ polynucleotide kinase (Life Technologies) as previously described (14). The labeled DNA strands were extracted with phenol-chloroform-isoamyl alcohol (24:25:1) and passed through a Bio-Rad P6 spin column to separate the labeled strands from the unincorporated nucleotides. Complementary oligonucleotides were annealed by mixing equimolar concentrations of the DNA strands, heating the mixture to 90° C, and allowing to cool slowly to room temperature. Equilibrium binding assays were performed in the buffer consisting of 10 mM Tris-HCl, 150 mM NaCl, 100 μg/mL bovine serum albumin, 0.1 mM EDTA (pH 8.0), 1 mM dithiothreitol, pH 7.9. Various concentrations of *Bam*HI were incubated at 25° C with the ^{32}P-labeled DNA (1 pM). The samples were loaded onto 15% polyacrylamide gels (37.5:1) in 0.5X TBE (45 mM Tris-borate, 1 mM EDTA) buffer, the electrophoresis vessel was pressurized, and the gels were run at 9 V/cm for 1.5 hours. The assays were run at four pressures: 1) ambient, 2) 150 bar, 3) 300 bar, and 4) 500 bar. Each binding assay was performed at least four times for each pressure tested. Gels were fixed, dried, and exposed using a Molecular Dynamics Phosphorimaging screen. Band intensities of the complexed and uncomplexed DNA were quantified using ImageQuant software with the volume measurement utility.

The equilibrium dissociation constant (K_d) was determined as described previously (15), using the relationship:

$$\Theta^{-1} = 1 + (K_d/[E_t]) \qquad (1)$$

where Θ is the fraction of bound DNA, E_t equals total enzyme concentration, and K_d is the equilibrium dissociation constant. Each binding assay was performed at least four times for each pressure tested. The volume change of BamHI-DNA dissociation was determined using the relationship:

$$\delta \ln K_d/\delta P = -\Delta V/RT \qquad (2)$$

where P equals pressure, V is the volume change, R is the universal gas constant, and T is the temperature.

2.2 Molecular Dynamic simulations

The protein-DNA system was constructed using the crystal structure of the cognate BamHI-DNA complex (Protein Data Bank entry 1bhm). (8) The DNA contains 11 base pairs with the overhanging 5' Thymine base removed and missing protein residue atoms were modeled using X-PLOR. (16) The protein-DNA system was energy minimized using the Powell algorithm to remove unfavorable contacts and reduce the strain in the system. A pre-equilibrated cube of water molecules (dimensions 88Å x 88Å x 88Å) was superimposed on the protein-DNA system and water molecules closer than 2.4 Å to the protein-DNA complex were removed. Thirty-two sodium ions were added to the system, by replacing water molecules, to bring the resulting protein-DNA-solvent system to charge neutrality. The system contains approximately 65,300 atoms. After equilibration, the system was simulated under ambient and elevated pressure for one nanosecond (ns) at 297 K. The MD simulations were performed using the program NAMD2 (17), with v.26 of the CHARMM force field (18). All hydrogen bonds were constrained using SHAKE and a time step of 2 femtoseconds (fs) was used. The system was simulated using Langevin dynamics in an NpT ensemble with periodic boundary conditions and full electrostatics were calculated using the particle mesh Ewald (PME) method. (19) For the simulation at elevated pressure, a gradient of 50 bar/100 picoseconds (ps) was used until a pressure of 400 bar was attained. Constant pressure was maintained using the Langevin piston method (20) with a piston period of 200 fs and damping time constant of 100 fs. Interfacial water molecules are defined as those molecules that are 3 Å from both the protein and DNA throughout the simulation. Simulation snapshots were created using the molecular graphics program VMD (http://www.ks.uiuc.edu).

3. RESULTS ANS DISCUSSION

3.1 High pressure gel mobility shift analysis

The system we have chosen for our initial study is the specific BamHI-DNA complex. The specificity and stability of this complex has previously been shown to be sensitive to pressure

(21), making it an ideal system for the first high pressure gel shift analysis. The application of hydrostatic pressure during the gel shift analysis of *Bam*HI causes a shift in the equilibrium dissociation constant for the cognate site. After measurement of the dissociation constant at several elevated pressures the volume change (ΔV) associated with *Bam*HI dissociation was also determined using. (Figure 2)

Gel mobility shift analysis of *Bam*HI binding to a specific DNA oligonucleotide, pictured in Figure 2, clearly illustrates that elevated hydrostatic pressure shifts the binding equilibria towards a lower affinity binding state.

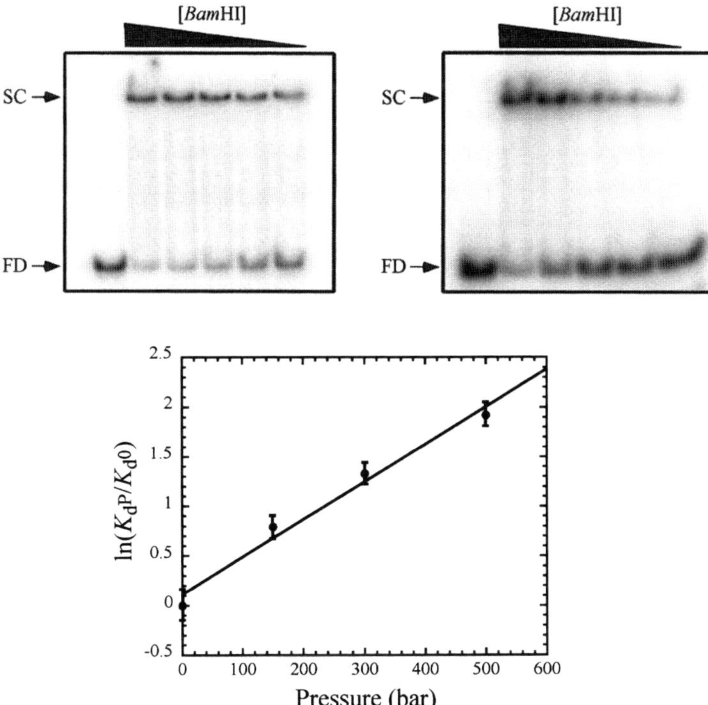

Figure 2. Gel mobility shift analysis at ambient pressure (top left) and 300 bar elevated pressure (top right). An identical protein titration was used for both illustrated binding assays. **SC** – specific *Bam*HI-DNA complex, **FD** – free DNA probe. The dependence of dissociation constant (K_d) on hydrostatic pressure (bottom).

The measured equilibrium dissociation constants (K_d), over the range of pressures tested, are 0.7 ± 0.1 nM (ambient), 1.5 ± 0.1 nM (150 bar), 2.5 ± 0.3 nM (300 bar), 4.6 ± 0.4 nM (500 bar). The dependence of dissociation constant on pressure, plotted in Figure 2, revealed a volume change (ΔV) of -92 ± 8 mL/mol for the pressure induced dissociation of the specific *Bam*HI-DNA complex. The loss of binding may be due to a structural change that decreases the overall volume of the system that favors dissociation, or an increased hydration of the protein-DNA interface that could disrupt specific protein-DNA interactions.

3.2 Molecular Dynamic simulation trajectories

The MD simulation calculations were performed to provide insight in to the observed high pressure effects during the binding analysis. The first trajectory calculated was the cognate *Bam*HI-DNA-solvent complex at ambient pressure; the second was the complex being exposed to a pressure gradient. We have compared the two simulations both visually and through interaction energy analysis to determine the effects of pressure at the molecular level. The previously identified direct and water-mediated protein-DNA contacts in the cognate *Bam*HI-DNA complex (21) were monitored during both trajectories. The measurement of protein-DNA interaction energies offers detailed analysis that may have been overlooked during the initial screening of the trajectories.

In viewing the trajectories of both systems, the most noticeable effects of pressure are manifested in the removal of the (R monomer) arm from the DNA minor groove. Interaction energy analysis between the arm residues (Asp[196], Gly[197], and Met[198]) and DNA throughout both simulations confirms that the application of pressure disrupts these specific interactions. The direct and water-mediated contacts are maintained during ambient pressure simulation. Figure 3 illustrates the dissociated arm observed in the elevated pressure simulation versus the intact arm, which remains sequestered in the minor groove during the ambient pressure simulation.

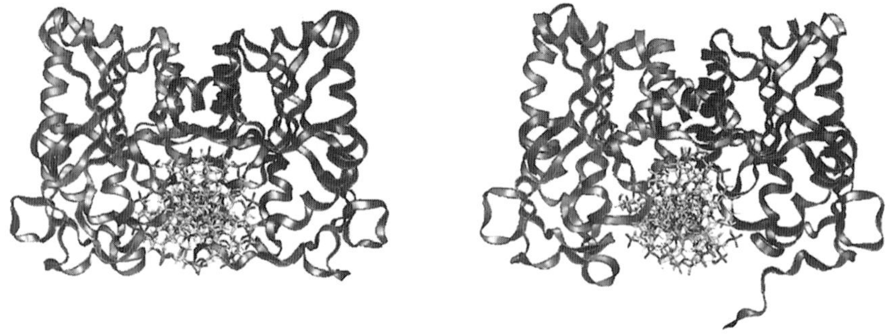

Figure 3. MD simulation snapshots of the equilibrated specific *Bam*HI-DNA complex (left) and the complex at an elevated pressure of 400 bar (right). At elevated pressure, the arm becomes dissociated from the DNA minor groove (right) and contacts near the protein-DNA interface are apparently disrupted.

The detailed interaction energy analysis reveals that the contact between Lys[193] and the DNA phosphate backbone is lost at elevated pressure. This loss of interaction was not identified in the initial trajectory screening phase. The simulation at elevated pressure illustrates that Lys193 is no longer within hydrogen bonding distance in the pressure equilibrated state of the *Bam*HI-DNA complex.

Endonuclease function relies on both enzyme specificity and the ability to catalyze DNA strand hydrolysis. A part of the interaction energy analysis focused on the effects of pressure on the active site residues Glu^{77}, Asp^{94}, Glu^{111}, and Glu^{113}. *Bam*HI has been shown previously to maintain catalytic activity at pressure up to 1.2 kbar. (22) The interaction energy analysis confirms that pressure does not affect the interactions between DNA and the catalytic residues, which is consistent with the aforementioned experimental observations.

The effect of pressure-induced hydration was probed by determining the number of water molecules associated with the *Bam*HI-DNA interface during both simulations. The simulations reveal that a number of water molecules enter the protein-DNA interface at elevated pressure, which are absent at ambient pressure. The water molecules, illustrated in Figure 4, are those that reside at the *Bam*HI-DNA interface in the equilibrated states of the complex at ambient and elevated pressure (400 bar).

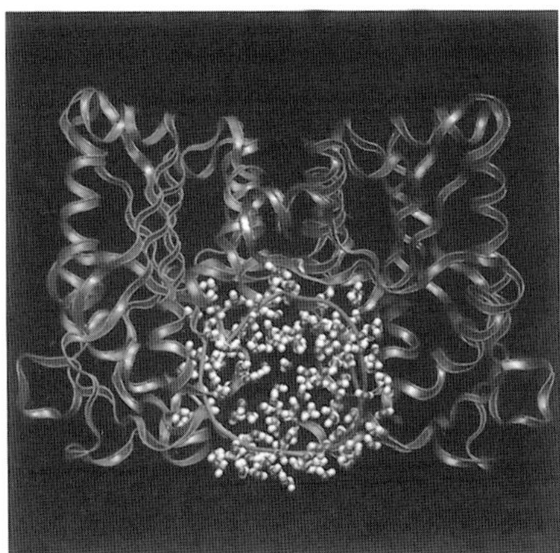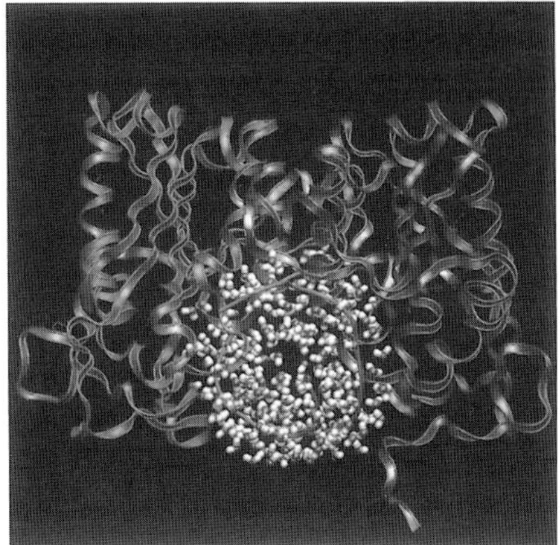

Figure 4. MD simulation snapshots of all water molecules within 3 Å of both the protein and DNA at both ambient pressure (left) and 400 bar (right). Water molecules are highlighted in white.

5. CONCLUSIONS

In this report, we have examined the pressure dependence of the interaction between the restriction endonuclease *Bam*HI and its cognate DNA recognition sequence. Application of hydrostatic pressure favors the dissociation of the specific complex indicating that the unliganded state of *Bam*HI displays a smaller molecular volume than that of the bound state (ΔV = -92 ± 8 mL/mol). To examine the possible structural origins of the observed volume change we have performed MD simulations on the complex at elevated pressure. The MD simulation trajectories illustrate two key features that give rise to the loss of binding affinity. First, the recognition arm is displaced from the minor groove and second, there is an observed increase in hydration at the protein-nucleic acid interface. Both of these structural changes are consistent with the observation of a negative volume change. Removal of the arm exposes the charges of the DNA that would result in reordering of the solvent in the minor groove and

94

around the exposed recognition arm. These results demonstrate the importance of combining theory and experiment to gain insight into thermodynamic equilibria and identify key factors involved in specific protein-nucleic acid recognition.

REFERENCES

1. A.A. Paladini, G. Weber, L. Erijman, Anal. Biochem., 218 (1994) 364.
2. L. Erijman, A.A. Paladini, G.H. Lorimer, G. Weber, J. Biol. Chem., 268 (1993) 25914.
3. P. Masson and J. Reyboud, Electrophoresis, 9 (1988) 157.
4. M.G. Fried, Electrophoresis, 10 (1981) 366.
5. J. Carey, Methods Enzymol., 208 (1991) 103.
6. G.A. Wilson and F.E. Young, J. Mol. Biol., 97 (1975) 123.
7. A. K. Aggarwal, Curr. Opin. Struct. Biol., 5 (1995) 11.
8. M. Newman, T. Strzelecka, L.F. Dorner, I. Schildkraut, A.K. Aggarwal, Science, 269 (1995) 656.
9. L.A. Marky and K.J. Breslauer, Proc. Natl. Acad. Sci. USA, 84 (1987) 4359.
10. D. Kosztin, T.C. Bishop, K. Schulten, Biophys. J., 73 (1997) 557.
11. S. Sen and L. Nilsson, Biophys. J., 77 (1999) 1782.
12. M. Newman, T. Strzelecka, L.F. Dorner, I. Schildkraut, A.K. Aggarwal, Structure, 2 (1994) 439.
13. H. Viadiu and A.K. Aggarwal, Nature Struct. Biol., 5 (1998) 910.
14. B.M. Brown and R.T. Sauer, Biochemistry, 32 (1993) 1354.
15. T.W. Lynch and S.G. Sligar, J. Biol. Chem., 275 (2000) 30561.
16. A.T. Brünger, X-PLOR, Version 3.1: A system for X-ray crystallography and NMR, The Howard Hughes Medical Institute and Department of Molecular Biophysics and Biochemistry, Yale University (1992).
17. M. Nelson, W. Humphrey, A. Gursoy, A. Dalke, L. Kalè, R.D. Skeel, K. Schulten, Int. J. Supercomput. Appl. High Perform. Comput., 10 (1996) 251.
18. A.D. MacKerrel, D. Bashford, M. Bellott, R.L. Dunbrack, J. Evanseck, M.J. Fried, S. Fischer, J. Gao, H. Guo, S. Ha, D. Joseph, L. Kuchnir, K. Kuczera, F.T.K. Lau, C. Mattos, S. Michnick, T. Ngo, D.T. Nguyen, B. Prodhom, B. Roux, M. Schlenkrich, J. Smith, R. Stote, J. Straub, M. Watanabe, J. Wiorkiewicz-Kuczera, D. Yin, M. Karplus, FASEB J., 6 (1992) A143.
19. B.A. Luty, M.E. Davis, I.G. Tironi, W.F. Vangunsteren, Mol. Sim., 14 (1994) 11.
20. S.E. Feller, Y.H. Zheng, R.W. Pastor, B.R. Brooks, J. Comp. Phys., 103 (1995) 4613.
21. H. Viadu and A.K. Aggarwal, J. Mol. Biol., 5 (2000) 889.
22. C.R. Robinson and S.G. Sligar, Proc. Natl. Acad. Sci. USA, 92 (1995) 3444.

Trends in High Pressure Bioscience and Biotechnology
R. Hayashi (editor)

Aggregation and gel formation of proteins after combined pressure-temperature treatment

L. Smeller[a*], F. Meersman[b], J. Fidy[a], K. Heremans[b]

[a] Institute of Biophysics and Radiation Biology, Semmelweis University,
H-1444 Budapest, Puskin u. 9. PF. 263 Hungary

[b] Department of Chemistry, Katholieke Universiteit Leuven, Celestijnenlaan 200D, B-3001
Leuven, Belgium

1. INTRODUCTION

It is well known that proteins can be denatured not only by high temperature, but also by high pressure or low temperature [1,2,3]. The unfolded states formed after these treatments were investigated by several methods and they were found to be different [4,5]. FTIR measurements proved that the main difference between the temperature unfolded and pressure unfolded proteins can be found in the intermolecular interactions formed after the denaturation. The infrared spectrum of the temperature unfolded protein shows specific bands characteristic for the intermolecular antiparallel beta sheet hydrogen bonding [6]. These bands are absent in the case of pressure or cold unfolded proteins [5].

Hawley gave a consistent theoretical description of the protein unfolding caused by temperature and/or pressure changes [3]. According to this theory, proteins have their native structure inside an elliptic boundary. Crossing this elliptic phase boundary either by increasing the temperature or the pressure, or lowering the temperature one unfolds the protein. This phenomenological theory cannot describe the differences between the proteins denatured by different methods. The theory also omits the intermolecular interactions between the protein molecules.

In order to unravel the details of this aggregation process, combined pressure and temperature treatments were used to proteins. Our previous studies on myoglobin proved that pressure unfolding induces an intermediate structure, which has an increased tendency for aggregation after the pressure release [7]. In the present study we investigated horseradish peroxidase. This is a heme protein, similar to myoglobin, but its tendency for aggregation at high temperature was reported to be very low [8].

* Corresponding author. *e-mail smeller@puskin.sote.hu

2. EXPERIMENTAL

The pressure and temperature cycle was carried out in a thermostated diamond anvil cell (DAC) mounted in the sample container of the FTIR spectrometer. Barium sulfate was used as an internal pressure standard in all cases [9]. The infrared spectra were obtained with a BRUKER IFS66 FTIR spectrometer equipped with a liquid nitrogen cooled broad band MCT solid state detector. 256 interferograms were co-added after registration at a resolution of 2 cm^{-1}. The secondary structure was determined by fitting the resolution enhanced amide I' band of the spectrum [10,11]. The overlapping components of the amide I' band were narrowed by the Fourier self-deconvolution developed by Kauppinen et al. [12]. The optimal parameters were determined from the observation of the power spectrum [13]. The deconvoluted spectra were then fitted with Gaussian functions.

Horseradish peroxidase (HRP) was purchased from Sigma (RZ=3.0) To purify the basic form (isoenzyme C) of this protein, column chromatography was used [14]. The protein was applied to a CM Sepharose (Pharmacia) in a column equilibrated with 5mM Tris buffer at pH 7.4 and eluted with 0.01 M NaCl. The RZ values of the fractions were checked by a Cary 4E (Varian) absorption spectrophotometer, and the fraction RZ=3.4 was used. The native heme group was removed from the enzyme by acid methyl ethyl ketone [15] to produce apoHRP. Substituted HRP was prepared from apoHRP and free base (metal free) mesoporphyrin (MP). The apoprotein was reconstituted with aliquots of MP dissolved in ethanol. Ca^{++}-free enzyme was obtained from the native one. Ca^{++} ions were removed by adding EDTA to the protein solution. Molar concentration of EDTA was one order of magnitude greater than that of the protein. To reduce the disulfide bridges, merchaptoethanol was used (HRPred.).

Samples for FTIR spectroscopy were prepared at a concentration of 75mg/ml in a deuterated buffer. The solutions were stored overnight to ensure sufficient H/^{2}H-exchange.

3. RESULTS AND DISCUSSIONS

3.1. Pressure experiments

Since the native HRP is very pressure stable [16], and also does not show aggregation after heat unfolding [8], we studied several modified forms of the enzyme. There were three main points where we changed the protein, the heme pocket, the Ca^{++} binding sites and the disulfide bridges. The latter ones were believed to be responsible for the extreme stability of the secondary structure. The changes at the heme pocket resulted in the apoHRP, MP-HRP (where the heme is substituted with a metal-free mesoporphyrin). Substrate bound protein was also studied, using benzohydroxamic acid (BHA) as substrate.

Fig 1. shows the amide I region of the FTIR spectra of native HRP at atmospheric pressure and at 13 kbar. This is a conformation sensitive vibration containing mainly C=O stretching. The oxygen atom participates in hydrogen bonding in the folded proteins, thus the position of this band indicates the types of secondary structures occurring in the protein. The complete secondary structure is, however, hard to determine, because of the composite character of the band. On the other hand, global structural changes like unfolding can simply be followed by tracking the maximum position of this band. Intermolecular interactions are also reflected in the amide I range (1600-1700 cm^{-1}) of the spectrum. A specific pair of "side bands" appears at

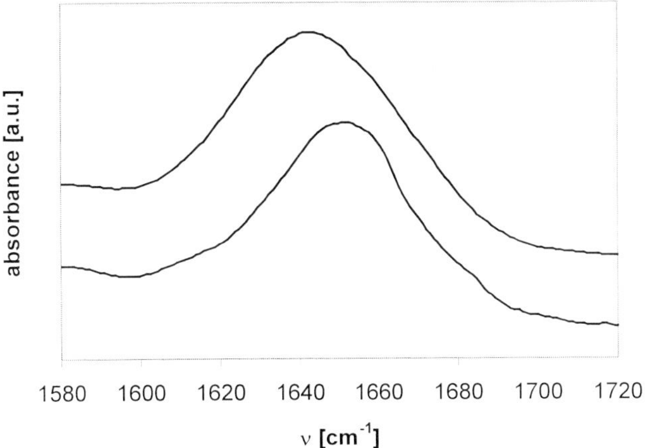

Fig. 1. Amide I region of the infrared spectrum of horseradish peroxidase at ambient pressure (bottom curve) and at 13 kbar (top curve).

around 1615 cm^{-1} and 1685 cm^{-1} indicating the hydrogen bond pattern of the intermolecular antiparallel beta sheet.

For the native HRP the maximum position of the amide I band can be found at 1650 cm^{-1}, indicating that the dominant secondary structure is helical. This is in agreement with the X-ray crystallographic results [17]. The broadness of the spectrum taken at high pressure is characteristic of the unordered secondary structure, indicating the unfolded protein. When the maximum position of the amide I band was plotted against the pressure, a sigmoid curve was found. The midpoint of this transition ($p_{1/2}$) was used to characterize the pressure stability.

The pressure stability was reduced in almost all modified cases, i.e. the midpoint of the pressure unfolding was lower than that of the native protein. The pressure stability increases in the following order:

HRPred. < apoHRP < HRP Ca^{++} free < MP-HRP <≈ HRP+BHA <≈ HRP native

As it can be seen from this sequence, the most prominent stabilizing effect on the secondary structure was provided by the presence of the disulfide bonds. Removal of the heme group also reduces the unfolding pressure, but this effect is not as drastic as it was observed in the case of myoglobin. For myoglobin the drop of $p_{1/2}$ to 25% was reported upon removal of the heme [16]. In contrast, the present experiments showed that the $p_{1/2}$ of apoHRP was more than 80% of the native one.

Substitution of the heme or substrate binding did not decrease the $p_{1/2}$ significantly. These results show that the interaction between the heme and the polypeptide chain does not contribute significantly to the overall stability of the protein. The absence of the Ca^{++} ions

98

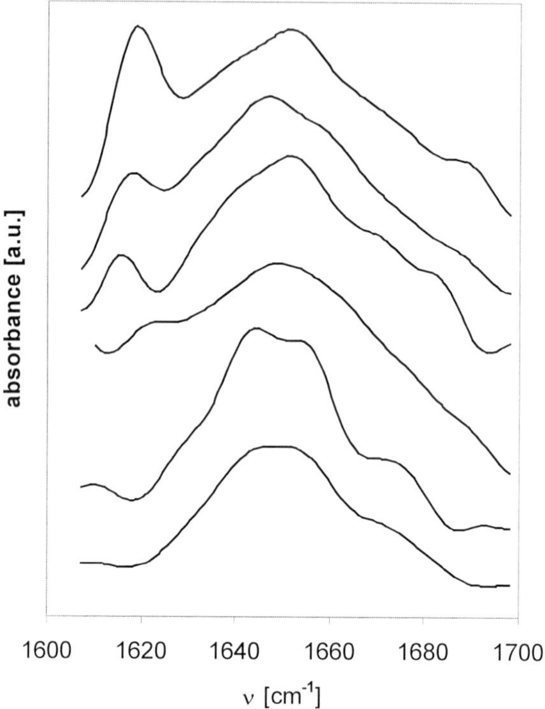

Fig. 2. Infrared spectra of differently modified HRPs at temperatures of
ca. 10°C below the denaturation temperature, recorded after a pressure
treatment of 13 kbar at 25 °C. From top to bottom: HRPred, apoHRP,
HRP Ca^{++} free, MP-HRP, HRP native, HRP+BHA. The spectra were
resolution enhanced as described in the experimental section.

also reduces the stability. This is in agreement with the earlier studies on HRP where the
reduction of enzymatic activity was reported in the absence of Ca^{++} [18]

3.2. Combined pressure-temperature treatment.

It is known that the aggregation tendency of the proteins increases if the heating is
preceded by a pressure treatment [7]. Therefore we studied the effect of combined pressure-
temperature treatment in the case of the above mentioned forms of HRP. In this experiment,
the pressure unfolded proteins were heated after the release of the pressure. Fig 2. shows the
FTIR spectra of the different forms of HRP after a pressure treatment of 13 kbar, recorded at
temperatures slightly below the unfolding temperature. The temperature of denaturation is 85

°C for the native HRP. It is reduced by 10 °C for the apo form, and by 20 °C for the MP-HRP and HRPred.

The spectra on Fig. 2. show clearly that the removal of the disulfide bridges makes the protein the most prone for forming intermolecular interactions. In this case the specific side bands characteristic for the intermolecular aggregation are already visible after the release of the pressure, and they remain almost unchanged during the heating cycle. This means, that the disulfide-bridges play a very important role in the pressure stability of the protein, and reducing the disulfide-bridges not only prevents the protein from refolding into a native-like conformation, but aggregation occurs immediately.

The removal of the Ca^{++} ion has a similar effect. This agrees with the results of NMR experiments [19], concluding that at least one Ca^{++} ion is necessary for maintaining the structure of the heme pocket. Our experiments show that without Ca^{++} the protein adopts a structure after the pressure treatment, which is less thermostable than the native (Ca^{++} containing) protein. This form again has an increased tendency for aggregation.

Changes made in the heme pocket have less effect on the aggregation tendency of the pressurized protein. Except for the complete removal of the heme (apoHRP) there is no aggregation tendency before the thermal denaturation. ApoHRP behaves similarly to the HRPred and the Ca^{++} free enzyme. This, however, can be foreseen from the IR spectrum of the apoHRP at ambient conditions. The spectrum indicates the increased amount of unordered state, compared to the native protein. The pressure stability of the remaining structure is, however, about the same as in the holoprotein, the apo form seems to form intermolecular antiparallel beta structure already at room temperature.

If the heme is substituted by a metal free mesoporphyrin (MPHH), the side bands characteristic for the intermolecular interactions appear around 65 °C, but their intensity is considerable less than for the reduced, or Ca^{++}-free form. The side bands however increase above the denaturation temperature, which is similar to the case reported earlier for myoglobin and lipoxygenase [7,20].

Binding the substrate BHA does not influence the aggregation tendency of the pressure-treated protein. Fidy *et al.* [21] reported the softening of the heme pocket in MP-HRP upon binding a similar substrate. According to the unchanged pressure stability and aggregation tendency, this softening must have been caused by a rearrangement restricted to the heme pocket.

Summarizing, we can say that the three stabilizing factors (disulfide-bridges, Ca^{++}, heme) contribute to the 3D structure of the HRP to a different extent. On the basis of the aggregation tendency of the pressure treated protein, the most important stabilization is provided by the disulfide-bridges and the Ca^{++} ions. Presence of the heme or similar prosthetic groups is also important for avoiding the aggregation. The substrate binding is a local effect, it does not increase the aggregation tendency.

REFERENCES

1   K. Heremans, and L. Smeller Biochim. Biophys. Acta, 1386 (1998) 353.
2   P. L. Privalov, Yu. V. Griko, and S. Yu. Venyaminov J. Mol. Biol., 190 (1986) 487.
3   S. A. Hawley, Biochemistry, 10 (1971) 2436.
4   J. Zhang, X. Peng, A. Jonas, and J. Jonas. Biochemistry, 34 (1995) 8631.
5   F. Meersman, L. Smeller and K. Heremans, High Pressure Research, 19 (2000) 263.

6   A. A. Ismail, H. H. Mantsch, and P. T. T. Wong, Biochim. Biophys. Acta, 1121 (1992) 183.

7   L. Smeller, P. Rubens and K. Heremans, Biochemistry, 38 (1999) 3816.

8   I. E. Holzbaur, A. M. English, and A. A. Ismail, Biochemistry, 35 (1996) 5488.

9   P. T. T. Wong, and D. J. Moffat Appl. Spectr., 43 (1989) 1279.

10   D.M. Byler, and H. Susi, Biopolymers, 25 (1986) 469.

11   L. Smeller, K. Goossens and K. Heremans, Vib. Spectr., 8 (1995) 199.

12   J. K.Kauppinen, D. J. Moffat, H. H. Mantsch, and D.G. Cameron, Appl. Spectrosc., 35 (1981) 271.

13   L. Smeller, K. Goossens and K. Heremans, Appl. Spectr. 49 (1995) 1538.

14   K, G. Paul, Acta Chem. Scand., 12 (1958) 1312.

15   F. W. Teale, J. Biochim. Biophys. Acta, 35 (1959) 543.

16   L. Smeller, K. Goossens, and K. Heremans. 1996. *In* High Pressure Science and Technology. Ed: W.A. Trzeciakowsky,. World Scientific Publishing Co. Pte. Ltd. pp863.

17   M. Gajhede, D. J. Schuller, A. Henriksen, A. T. Smith and T. L. Poulos, Nature Structural Biology, 4 (1997) 1032.

18   I. Morishima, M. Kurono, Y. Shiro, J. Biol. Chem., 261 (1986) 9391.

19   S. Ogawa, Y. Shiro, I. Morishima, Biochim. Biophys. Res. Commun. 90 (1979) 674.

20   L. Smeller, P. Rubens, K. Heremans, High Pressure Research 19 (2000) 323.

21   J.Fidy, J. M. Vanderkooi, J.Zollfrank and J.Friedrich, Biophys. J., 63 (1992) 1605.

Trends in High Pressure Bioscience and Biotechnology
R. Hayashi (editor)
                                                     101

BEHAVIOR OF ACTIN UNDER HIGH PRESSURE

Y. Ikeuchi[a], A. Suzuki[b], T. Oota[b], K. Hagiwara[b] and C. Balny[c]

[a]Department of Bioscience and Biotechnology, Graduate School of Agriculture,
Kyushu University, Fukuoka, 812-8581, Japan
[b]Department of Applied Biological Chemistry, Faculty of Agriculture, Niigata
University, Niigata, 950-2181, Japan
[c]INSERM, Montpellier, Cedex 1. France

ABSTRACT

The pressure effects on the behavior of actin were studied, *in situ*, by using ε ATP. For studies of the collapse process of the actin structure, spectroscopic measurements such as NMR and biochemical assays were also achieved after pressure release.

The dissociation rate constants of nucleotides from actin molecule (the decay curve of the fluorescence intensity of ε ATP-G-Actin) followed first order kinetics. The disappearance of the characterized ^{1}H NMR signal at 2.055 ppm, which is considered to be specific of the methyl proton of methionine (Met 43 and 46) in actin, and the loss in biochemical activities (DNaseI inhibition capacity, polymerizability) at 300 MPa occurred nearly simultaneously. This suggested a rapid collapse of the steric structure around the upper region called "pointed end" of actin molecule following the dissociation of the nucleotide bounding (ATP).

1. INTRODUCTION

Ikkai and Ooi [1] did a thorough study on the pressure effects on G- and F-actin   giving the following conclusions: 1. Without ATP, actin was irreversibly denatured above 150 MPa, a denaturation achieved above 250 MPa in the presence of ATP. The amount of protein denatured by pressure was dependent on the initial protein concentration. 2. ATP prevented actin from pressure-induced denaturation. 3. In the presence of ATP under pressure, the reversible F-G transformation liberated both ADP and Pi. 4. From the pressure denaturation curve, the volume change was estimated to be - 82 mL/mol of monomer. However, all these measurements were performed after pressure-release. Therefore it was interesting to perform measurements under pressure in order to get further detailed thermodynamic information on the pressure-induced denaturation and to observe possible pressure-reversible phenomenon.

In the present experiments, a Hitachi F2000 fluorospectrophotometer equipped with a high-pressure vessel allowing HP *in situ* observations was used. Furthermore, spectroscopic measurements such as NMR, CD and biochemical assays were performed after pressure-release.

2. MATERIALS AND METHODS

2.1 Protein preparations

Actin preparations from rabbit skeletal muscle were obtained from acetone dried powder according to the procedure of Pardee and Spudich [2]. 1:N6-Ethenoadenosine 5'-Triphosphate (ε ATP) was synthesized from ATP (Sigma Co., Mo. USA) according to the method of Secrist III et al. [3]. ε ATP-labeled G-actin was prepared as described by Waecher and Engel [4].

2.2 High pressure apparatus

Two different types of high-pressure devices were used for this study. One was a device consisting of a thermostated high-pressure vessel equipped with sapphire windows and a pump capable to reach a pressure up to 400 MPa (Teramecs Co., Ltd., Kyoto). The other was basically the same apparatus as a cold isostatic pressing device (CIP) used conventionally for the molding of ceramics (Nikkiso Co. Ltd., Tokyo).

2.3 Fluorescence spectroscopy

Fluorescence measurements were made using a Hitachi F2000 fluorospectrophotometer, equipped with the high-pressure vessel. For the ε ATP or ε ADP bound to actin monitoring under pressure, the excitation and emission wavelengths were 360 nm and 410 nm, respectively.

2.4 NMR measurement

All 1D NMR spectra were recorded on a Bruker DPX-400 (400 MHz) spectrometer at 298 K using a 5 mm ^{1}H probe (Sigemi Co. Japan). 3-(Trimethylsilyl)-1-propanesulfonic acid (DSS) was used as an internal reference. The presaturation procedure was adopted to suppress the solvent signals. 1D spectra were recorded with data points of 64K, and 512 scans. Heavy water (99.9 atomic % D) was obtained from Aldrich Chemical Co. (Milwaukee, WI, USA).

3. RESULTS AND DISCUSSION

Under high-pressure, ε ATP-bound to actin emitted a strong fluorescence at 410 nm. The fluorescence intensity at 410 nm of the ε ATP-G-actin was higher than the ε ADP-F-actin. The fluorescence of both G- and F-actin and ε ATP buffer increased when they were exposed to a pressure of 250 MPa. However, the fluorescence intensity increasing of the ε ATP buffer itself was smaller than the ε ATP-bound to G-actin. Therefore, this fluorescence increasing seems to be mainly due to a conformational change of actin under pressure. When the pressure was elevated from 0.1 MPa to 400 MPa, the relative intensity of F-actin increased with a rise to around 230 MPa, then reaching a plateau up to 250 MPa. On a further rise in pressure, it began to decrease. At 400 MPa it dropped almost to the same level as the ε ATP buffer did. Thus, the decrease in fluorescence intensity evidently corresponds to the dissociation of ε ADP-bound to F-actin. In G-actin a similar pattern as F-actin was obtained except that the intensity had already begun to decrease when the pressure reached 230 MPa. This indicates that F-actin is somewhat more resistant to pressure than G-actin.

The time dependence of the relative fluorescence intensity of ε ADP-F-actin at several pressure values was investigated. The fluorescence intensity increased with pressure and, depending the pressure reached, increased more as a function of time when the pressure has been stabilized. The extent of this increasing was dependent of the applied pressure. This may be attributable to the increase of the amount of the depolymerized actin because the ε ATP-bound to G-actin generates a stronger fluorescence than the ε ADP-F-actin does. The time dependence of the change in intensity decreased at pressure values higher that 250 MPa, following first-order kinetics (see Figure 1). Assuming that the dissociation rate constant of ε ADP from F-actin corresponds to its denaturation rate, the volume change for the denaturation was estimated to be - 67 mL/mol (Figure 2), a value similar to those obtained for the G-actin (- 72 mL/mol).

We examined the *in situ* exchange of ε ATP-bound to G-actin by external ε ATP or ATP at 100 MPa where G-actin is not denatured. When the pressure was maintained constant (100 MPa) the fluorescence intensity did not change in the presence of ε ATP, whereas it exponentially decreased in the presence of ATP. Both G- and F-actin exposed to a pressure of 100 MPa for 5 min showed the same biochemical activity (DNase I inhibition capacity) after pressure-release. This implied that the decrease in the fluorescence intensity in the presence of ATP was not attributable to the denaturation of G-actin. These data suggested a rapid exchange between the bound and the external nucleotides at relatively low pressure (such as 100 MPa). ε ATP bound to F-actin is not easily exchanged with the external nucleotides at the atmospheric pressure unless any external forces applied [5]. Hence, to determine whether ε ADP-bound to F-actin is capable of exchanging the nucleotides under pressure, similar experiments as in the case of ε ATP-G-actin were conducted. The results

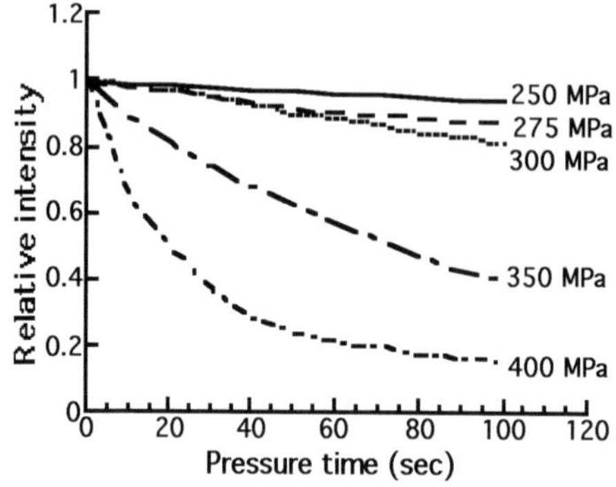

Figure 1. Time course of the release of ε ADP bound to F-actin under pressure.

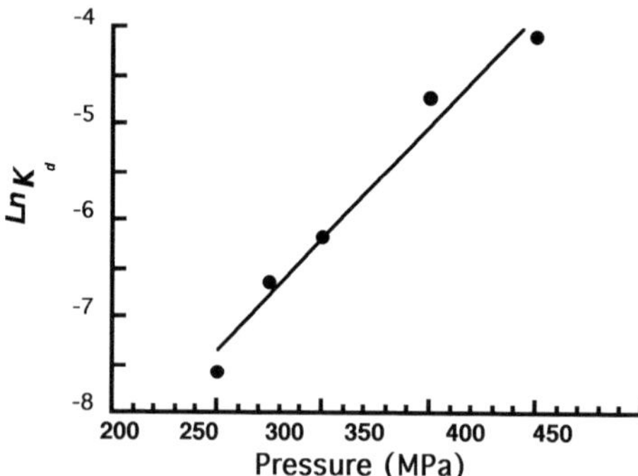

Figure 2. Logarithmic plot of the dissociation rate constant of ε ADP bound to F-actin as a function pressure.

indicated that the ε ADP-bound to F-actin could be replaced by the external ATP in the pressure range (0.1-230 MPa) where the irreversible denaturation does not take place. F-actin is not denatured even in the presence of EDTA, while EDTA deprives G-actin of the divalent cation, leading to a quick irreversible denaturation. Then, the fluorescence measurements of ε ADP-F-actin were made either in the absence or in the presence of both EDTA and ATP to confirm the pressure dissociation-association equilibrium of actin. No change in the intensity was observed even at a constant pressure of 100 MPa whatever the presence of EDTA. This result could be interpreted as the following sequential reactions: ε ADP-F-actin was first depolymerized to ε ADP-G-actin, quickly exchanging its with external ε ATP, and then polymerized again accompanying the liberation of phosphate from the ε ATP-bound to G-actin. This shows that the cycling F->G->F transformation (F-G being in equilibrium under a certain pressure values) is occurring without pressure-denaturation.

We have recorded the ^{1}H NMR spectra of G-actin with ATP immediately after pressure release. The signal intensity at 2.055 ppm began to decrease above 250 MPa, disappearing completely at 400 MPa. The sharp NMR signal at 2.055 ppm is considered as a probe from the methyl proton of methionine (Met44 and Met47) in G-actin [6]. These methionine groups are located in a highly flexible region in G-actin and are a part of the DNaseI binding loop. In Figure 3, the relative integral ratio (R=2.055 ppm/8.3 ppm) of G-actin exposed to 300 MPa was plotted as a function of time on the basis of spectra illustrated in the inset of Figure 3. The disappearance rate of the signal at 2.055 ppm is clearly first order, resembling to the behavior of the DNaseI inhibitory activity (data not shown). These facts suggest that the pressure-induced denaturation of G-actin with ATP is characterized by a large conformational change of the protein near the DNaseI binding site. The reason for disappearance of the specific signal at 2.055 ppm at 400 MPa is not completely understood.   One possibility could regions are deeply buried into the molecule, or a refolding process occurs after decompression.

In conclusion, the pressure-dissociation rates of nucleotides from the actin molecule (i. e., the decay curves of the fluorescence intensity of ε ATP-G-actin) nicely followed first order kinetics. From their rate constants, the calculated volume change was closed to the value obtained by Ikkai and Ooi [1] after pressure-release. In addition, the pressure-denaturation of G-actin is coupled with a loss in the exchangeability of bound ATP against external ATP. The disappearance of the characterized ^{1}H NMR signal at 2.055 ppm, which is considered as

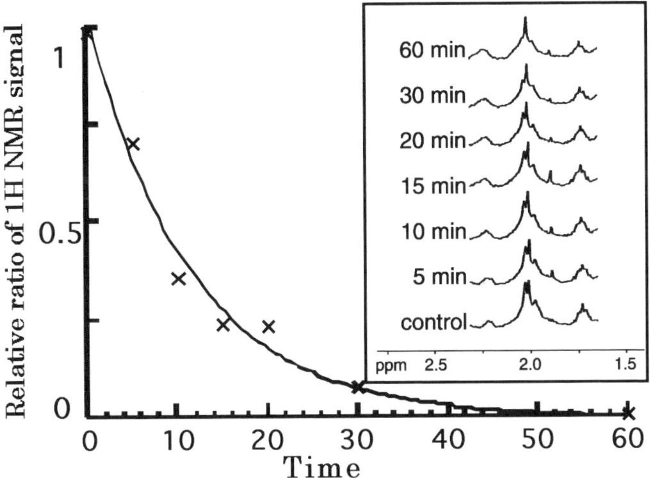

Figure 3. Time course of the relative ratio of ^{1}H NMR signals (2.055 ppm / 8.3 ppm) of G-actin based on the ^{1}H NMR spectra of G-actin after release of pressure (Inset). G-actin with ATP was pressurized at 300 MPa for the designated times.

coming from the methyl proton of methionine in the vicinity of DNaseI binding site in actin, and the loss in the DNase I inhibition capacity were almost identical.

The DNaseI binding site is located on the surface of the actin molecule. Taking these facts in consideration, we will infer that the rapid collapse of the three dimensional structure around the upper region called "pointed end" (e.g. burying into the inside of the molecule) is caused by the dissociation of the bound nucleotide (ATP).

4. ACKNOWLEGEMENT

This study was supported in part by a Grant-in-Aid for Scientific Research from the Ministry of Education, Science, Sports and Culture of Japan (No. 10460118)

REFERENCES

1. T. Ikkai & T. Ooi, Biochemistry, 5 (1966) 1551.

2. J. D. Pardee & J. A. Spudich, in Methods in Cell Biology   (Wilson, L. ed.) Vol. 24, pp. 271-289, Academic Press, New York , 1982.

3. J. A. Secrist III, R. Barrio, J. Leonard & G.Weber, Biochemistry, 11 (1972) 3499.

4. F. Waechter & J. Engel, Eur. J. Biochem., 57 (1975) 453.

5. F. Oosawa & M. Kasai, in Subunits in Biological Systems (Timasheff, S.N. & Fasman, G.D., eds) part A, pp.261-322, Dekker, New York, 1971.

6. D. Heintz, H. Kany &H. R. Kalbitzer, Biochemistry, 35 (1996) 12686.

Trends in High Pressure Bioscience and Biotechnology
R. Hayashi (editor)

Effect of pressure and pressure-denaturation on fast molecular motions of solvated myoglobin

W. Doster, M. Diehl, H. Schober, W. Petry and J. Wiedersich

Technische Universität München, Physikdepartment E13, D - 85748 Garching, Germany

The influence of pressure in the range up to 8 kbar on the pico-second dynamics of myoglobin and its hydration shell has been studied by neutron time-of-flight spectroscopy. The quasi-elastic scattering decreases with pressure due to a reduction in the rate and amplitude of hydrogen bond fluctuations. Irreversible denaturation occurs around 4 kbar. The hysteresis observed on completing the pressure cycle suggests a structural contribution to the dynamic effects of less than 10 %: Water interacts more strongly with the unfolded state.

1. Background

The increase in packing density with pressure should enhance both, polar and non-polar interactions and thus protein stability. The onset of pressure denaturation of globular proteins is presumably triggered by a breakdown of intra-molecular voids which are penentrated by water [1]. The destabilising action of water on the native state has been clarified in the context of cold-denaturation of proteins [2]: Water molecules tend to assume well defined positions around protein residues at lower temperatures. The resulting decrease in enthalpy favours the unfolded state. A similar ordering effect can be achieved by application of pressure. The common features of cold- and pressure denaturation and their differences have not been studied systematically. In this contribution we investigate molecular motions as a monitor of protein-water interactions. The synergetic effects of temperature and pressure variations on molecular mobility may provide new insight to the folding - unfolding problem. Inelastic neutron scattering allows to record the trajectories of hydrogens, attached either to water molecules or to protein residues. By hydrogen/deuterium exchange one can modify the scattering cross-section without affecting chemical properties significantly. This allows to focus either on protein dynamics or water motions. With this method it has been shown previously, that a decrease in the temperature mainly decreases the amplitude of fast structural motions [3,4]. High pressure leads to a similar reduction in molecular displacements.

2. Experimental Details

The neutron time-of-flight spectrometer IN6 at the Institut Laue-Langevin in Grenoble was operated at the flux maximum, implying a wavelength of 5.1 Å. An instrumental resolution of 100 μeV, (FWHM) was used. The corresponding time window ranges from 0.1 ps to 15 ps. The pressure cell consists of a spindle-operated piston in a cylinder, which exerts pressure on two small aluminum bars which were wedge-shaped on one end such that they could fit together. The sample was compressed between the wedges. An expansion strip was employed to determine the pressure. The pressure could be reproducibly adjusted between 1 and 8 kbar to an accuracy of 0.2 kbar. The strong scattering by the pressure cell was corrected for.

108

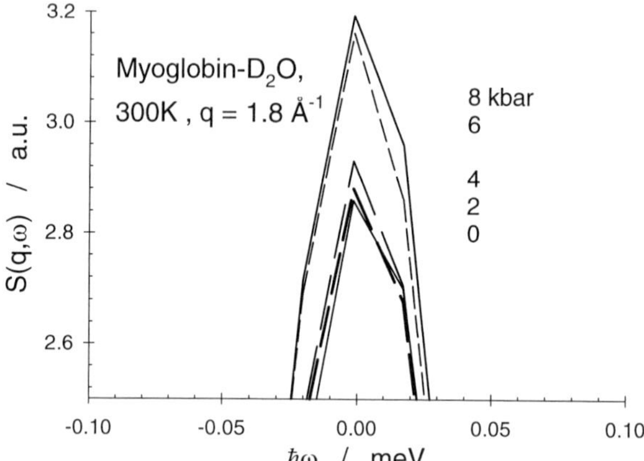

Fig. 1: Central peak of the scattering function of myoglobin in concentrated D$_2$O solution (0.7 g/g protein). The scattering vector was at q = 1.8 Å^{-1}.

3. Results

Fig. 1 shows the central portion of the scattering function of myoglobin in D$_2$O. The deuterated solvent is chosen to enhance the relative signal of the non-exchangeable protein hydrogens. The elastic intensity, at ω =0, reflects those hydrogens which do not move within the instrumental time window of 15 ps. The rigid fraction smoothly increases with pressure, but the large gap in intensity between 4 and 6 kbar points to a structural change. According to the infrared spectrum in the amide region taken after the experiment, the protein was irreversibly denatured. This result suggests to assign the transition observed at 4 kbar to pressure denaturation. It follows that the denatured, less ordered state exhibits a lower flexibility than the compact, native structure of myoglobin.

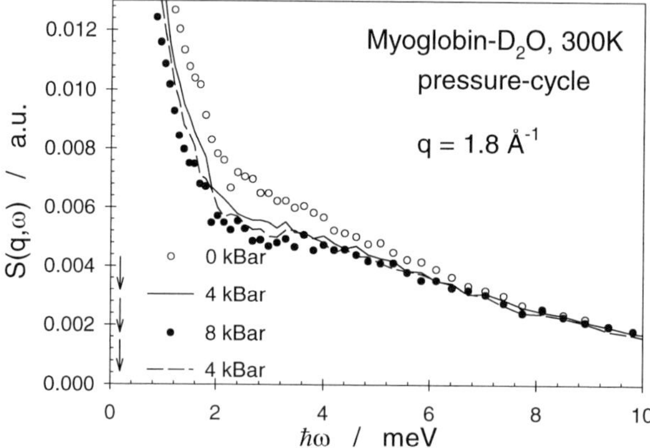

Fig. 2: Scattering function myoglobin in concentrated D$_2$O-solution (0.7 g/g) along the pressure cycle. The scattering vector was at Q = 1.8 Å^{-1}. 1 meV corresponds to 8 cm^{-1}.

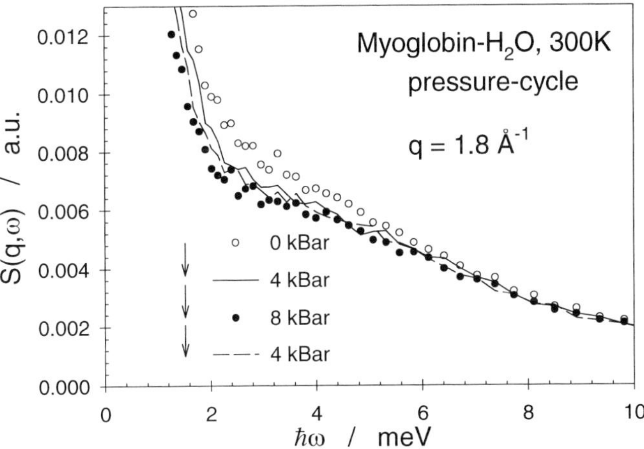

Fig. 3: Scattering function of myoglobin in H_2O solution (0.7 g/g) along the pressure cycle.

In fig. 2 we focus on a quasi-elastic region of the scattering spectrum. The inelastic intensity in the low frequency region decreases with pressure, while excitations above 6 meV are less affected. More precisely, the wings of the central peak decrease implying, that activated structural fluctuations occur less frequently at high pressure. As an alternative explanation one could invoke that fluctuations occur with the same frequency but on a smaller spatial scale. The spectra taken at 4 kbar are nearly independent of pressure history. Thus most of the pressure-induced dynamic effects are reversible. Fig. 3 shows the results of the same experiment performed however with a protonated solvent, myoglobin in H_2O. In this case the cross-section of water dominates the scattering function. The effects of pressure on water mobility are somewhat smaller but otherwise similar to those observed with the protein. The spectra above 100 µeV reflect fast librational flips of water molecules interacting with the protein surface [5].

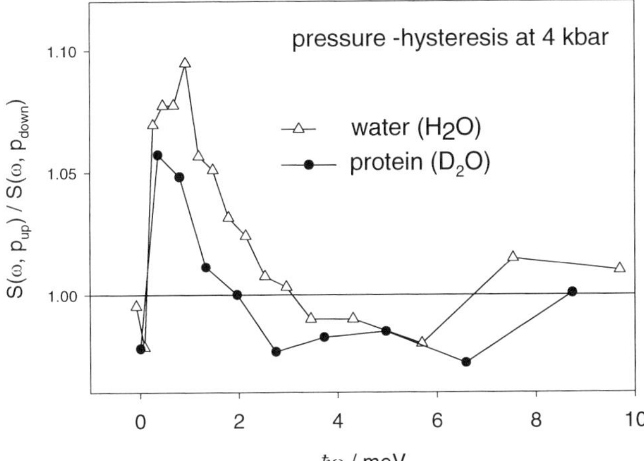

Fig. 4: Ratio of the scattering functions before / after pressure treatment at 8 kbar.

Diffusion along the surface occurs on a much longer time scale. High pressure generally enhances the strength of intermolecular hydrogen bonds. Thus, the changes observed in fig. 3 imply most likely a slowing down of surface water librations.

The effects of pressure on molecular motions have to be distinguished from those which occur as a result of structural changes. In fig. 4 we display the ratio of the scattering functions at 4 kbar before and after exposure of the sample to a pressure of 8 kbar. A deviation of this ratio from unity indicates that the spectrum of molecular motions has been modified by irreversible denaturation. Both the water and the protein spectrum exhibit a ratio of less than unity at zero frequency, while it increases up to 1.1 and 1.05 respectively at 1 meV. A ratio above unity at $\omega > 0$ indicates larger quasi-elastic scattering in the native state. Possibly, water in the unfolded state interacts more strongly with the now exposed protein residues which decreases its mobility. The similar spectral shape in fig. 4 suggests that structure-related changes of water and protein mobility are correlated.

4. Discussion and Conclusion

We present, to our knowledge, the first neutron scattering spectra of a protein at high pressure. The method of isotopic exchange is used to discriminate between protein fluctuations on pico-second time scale and those in the hydration shell. Pressure decreases the amplitude and rate of molecular motions. Further dynamic effects occur by pressure-induced structural changes. Pressure-unfolding affects protein and water mobility in parallel resulting in enhanced rigidity, presumably because of a lower free volume. Heat denaturation in contrast increases the structural flexibility. Hummer et al. [1] propose that water is penetrating into the voids and in between secondary structure at high pressure. According to their theoretical analysis hydrophobic aggregates dissolve above pressures of a few kbars. A closely related phenomenon is the weakening of the hydrophobic effect with decrease in the temperature. Privalov [2] attributes cold-denaturation to an ordering of water molecules around hydrophobic groups which lowers their enthalpy, destabilising the native state. Similarly, high pressure will force water molecules to assume more specific positions around protein residues. At 4 kbar, myoglobin has reached a stability limit at our conditions. As shown in figs. 2 and 3, the quasi-elastic spectra at 4 kbar are closer to spectra taken at 8 kbar than to those taken at ambient pressure. This suggests that most of the voids in the hydration shell have been consumed at 4 kbar, implying a decrease in compressibility. The gain in enthalpy due to enhanced order may reach a critical limit at this pressure and the free energy of the native state becomes positive.

The project is supported by a grant of the Deutsche Forschungsgemeinschaft, Forschergruppe Hochdruckbehandlung von Lebensmitteln.

References

1. G. Hummer, S. Garde, A. Garcia and M. Paulaitis, Proc. Natl. Acad. Sci (USA) 95 (1998) 1552
2. P.L. Privalov, Biofizika 32 (1987) 742
3. W. Doster, S. Cusack and W. Petry, Nature 337 (1989) 554.
4. M. Diehl, W. Doster, W. Petry and H. Schober, Biophys.J. 73 (1997) 2726
5. M. Settles and W. Doster, Roy. Chem. Soc. Faraday Discussion 103 (1996) 269.

Trends in High Pressure Bioscience and Biotechnology
R. Hayashi (editor)
© 2002 Elsevier Science B.V. All rights reserved.

Structural changes in chicken myosin subfragment-1 induced by high hydrostatic pressure

T. Iwasaki and K. Yamamoto

Department of Food Science, Rakuno Gakuen University,
Ebetsu, Hokkaido 069-8501, Japan

ATP-induced fluorescence increment of pressurized S1 was almost the same as unpressurized one at least up to 150 MPa, whereas it decreased above 200 MPa. The binding of ε-ADP to S1 decreased at 250–300 MPa. S1 pressurized at 100–250 MPa and unpressurized S1 bound to F-actin similarly, although binding of S1 to actin decreased when the pressure treatment was done above 250 MPa. S1 was easily cleaved by tryptic digestion into three domains. Tryptic fragments of S1 digested after pressure treatment were essentially the same as those of unpressurized one. On the other hand, additional fragments appeared when the digestion was performed under pressure at 300 MPa. It is concluded that pressure-induced structural changes of S1 begin to occur about 150 MPa, and ATPase and actin binding sites lose those intrinsic structures at 250–300 MPa.

1. INTRODUCTION

Subfragment-1 (S1) is a head portion of myosin molecule, and it contains ATPase and actin binding sites. Balint *et al.* [1] showed that trypsin cleaved the heavy chain of S1 into three fragments, which were 27, 50 and 20 kDs. The binding sites of ATP and actin are located at the junctions of 25-50 kD and 50-20 kD, respectively [2–5]. Morphological and physicochemical studies of the effect of hydrostatic pressure on myosin have been reported [6–8]. Turbidity of myosin solution after pressure treatment markedly increased within 5 minutes at 210 MPa, and the shape of myosin aggregate looked like daisy wheel [6]. Hydrophobicity of myosin after pressure application up to 250 MPa increased, but it reached a plateau above 300 MPa at pH 7.0, and the helix content decreased at 250–300 MPa [7]. Myosin filaments form a gel by pressure treatment with head-to-head interaction among filaments [8]. Yamamoto *et al.* [6–8] indicated that the head (S1) was the most pressure-sensitivity portion in myosin molecule. The pressure-induced changes in actin binding and ATP binding sites are not well understood. In this paper, we present high pressure effect on the structure of S1 using biochemical and photochemical methods.

2. MATERIALS AND METHODS

2.1. Preparation of proteins. Myosin was prepared from chicken pectoral muscle according to Offer *et al* [9]. Subfragment-1 (S1) was prepared by a modification of the method of Weeds and Pope [11]. Chymotryptic digestion of myosin was performed at 23 °C for 7 min. S1 was separated from the digested myosin by ammonium sulfate fractionation and purified by Whatman DE-52 chromatography using a NaCl gradient from 0 to 0.2 M. Actin was prepared from rabbit skeletal muscles according to Spudich and Watt [10].

2.2. Application of hydrostatic pressure. Two milliliters of S1 (0.2–3 mg/ml) solution in 0.05–0.2 M NaCl (pH 7.0) was put into a plastic tube. Application of pressure was performed at 20 °C. After pressure release, the S1 solution was ultracentrifuged at $194,000 \times g$ for 60 min to remove aggregates. S1 in the supernatant was used in the present study.

2.3. Fluorescence measurement. The tryptophan fluorescence of S1 was measured in 0.2 M NaCl and 20 mM Na phosphate (pH 7.0) at 20 °C according to Werber *et al* [12]. Increment of ATP-induced fluorescence (ΔF) was expressed as follows :

$$\Delta F\ (\%) = [(F_{+ATP} - F_{-ATP})\ \text{pressurized}] / [(F_{+ATP} - F_{-ATP})\ \text{unpressurized}] \times 100$$

Here, F_{-ATP} and F_{+ATP} are fluorescence intensities at 334 nm before and after addition of ATP, respectively.

The fluorescence of ε-ADP was measured at 20 °C using an excitation wavelength of 320 nm and an emission wavelength of 410 nm. The interaction of ε-ADP with S1 was studied by quenching the fluorescence of the free nucleotide with acrylamide as described by Ando *et al* [13]. In control experiment, the fluorescence of ε-ADP was measured under the same condition without S1. The ATP binding ability of pressurized S1 was estimated from the difference of fluorescence, which obtained from unpressurized S1.

2.4. Binding experiment. The binding of pressurized S1 to F-actin was measured as described by Chalovitch and Eisenberg [14], and Yamamoto *et al* [15].

2.5. Acto-S1 ATPase measurement. Actin activated S1 ATPase activity was measured with a pH-stat at pH 7.0 and 25 °C in a reaction mixture containing 0.02 M NaCl, 0.5 mM ATP and 2 mM $MgCl_2$. The protein concentrations of S1 and F-actin were 0.38 and 1.6 µM, respectively.

2.6. Electron microscopy. A drop of acto-S1 (0.1 to 0.2 mg/ml) was applied to a carbon-coated grid, which had been glow-discharged just before use. The grid was negatively stained with 2% aqueous uranyl acetate. The specimens were observed in a Hitachi H-800 electron microscope at an accelerating voltage of 75 kV. Images were recorded at a magnification of 40,000.

2.7. SDS-PAGE of tryptic digested fragments. S1 was digested with 1/400 (w/w) of TPCK-trypsin. Electrophoresis was performed on a 7.5 % polyacrylamide gel [15].

3. RESULTS AND DISCUSSION

It is known that the structural change of S1 is induced by binding of ATP [12]. We measured fluorescence spectrum of S1. The fluorescence intensity of unpressurized S1 with ATP was higher than that without ATP. There was no difference in fluorescence spectrum of pressurized S1 before and after addition of ATP. ATP-induced fluorescence increment (ΔF) by pressure treatment is shown in Figure 1. The degree of increasing of fluorescence did not change up to 150 MPa, although they were decreased above 200 MPa. Since the fluorescence intensity of S1 after addition of ATP was lower than that of S1 before addition ATP at 300 MPa, the differential spectra of ATP-induced fluorescence intensity was negative value. The result in Figure 1 indicates a global structural change of S1 by pressure.

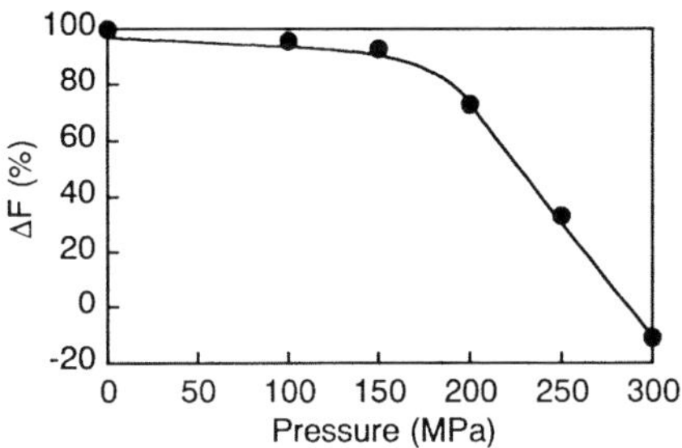

Figure 1. The ATP-induced fluorescence increment of S1 after pressure treatment. One µM of S1 (MW: 120,000) in 0.2 M NaCl (pH 7.0) was pressurized at 0.1–300 MPa. Excitation wavelength was 295 nm.

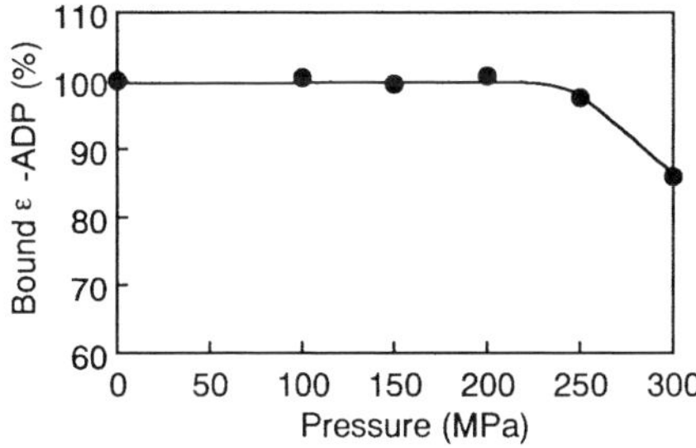

Figure 2. Fluorescence spectrum of S1 with or without ε-ADP. S1 (1 µM) was dissolved in 0.1 M NaCl, 200 mM acrylamide, 2 mM MgCl$_2$, and 20 mM Na phosphate (pH 7.0) with or without ε-ADP (8 µM). The bound ε-ADP was calculated from the fluorescence intensity at 410 nm before and after addition ε-ADP.

114

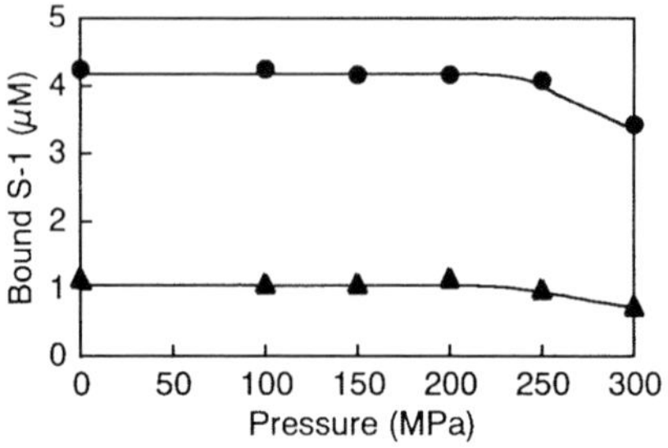

Figure 3. Effect of pressure treatment on the binding of S1 to F-actin. F-actin was mixed with pressurized S1 in 50 mM NaCl, 5 mM Na phosphate (pH 7.0), and 1 mM MaCl$_2$. The mixture was incubated for about 20 min at room temperature, then centrifuged at 135,000×g for 10 min. Protein concentration of the supernatant (unbound S1) was measured by UV absorbance at 280 nm. Actin (MW: 42,000) concentration was 5 μM. ●, 5 μM S1; ▲, 2μM S1.

We also investigated pressure effect on ATP binding site of S1. Ando *et al.* [13] developed a method to measure nucleotide binding to S1. They found that the fluorescence of free ε-ADP or ε-ATP was quenched by acrylamide, but the quenching was prevented when ε-nucleotides bound to S1. We titrated pressure treated S1 with ε-ADP in the presence of acrylamide using this method (Figure 2). The intensity of the acrylamide nonquenchable fluorescence, i.e., the amount of ε-ADP bound to S1, decreased with elevating pressure. The amount of ε-ADP bound to S1 did not change up to 200 MPa, and decreased above 250 MPa. This result indicates that ε-ADP is hardly bind to pressurized S1 at 300 MPa. The effect of pressure treatment on the actin binding to S1 was studied (Figure 3). Bound S1 was calculated by subtracting unbound S1 from total S1. The binding ability did not change up to 200 MPa; however, it decreased above 300 MPa. This result shows that the structure of actin binding sites of S1 is changed by pressure above 300 MPa.

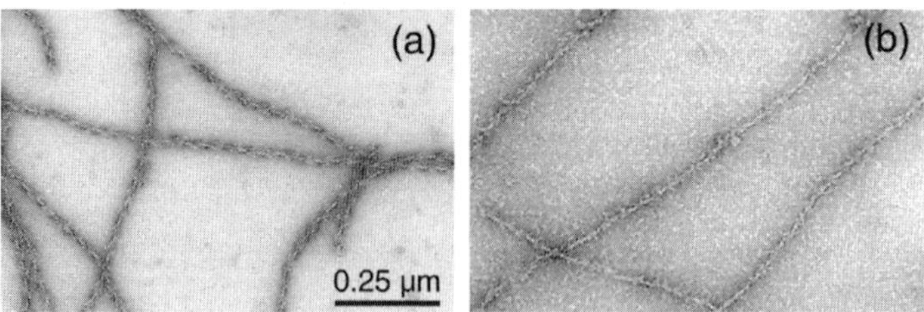

Figure 4. Transmission electron micrographs of acto-S1. (a), control; (b), acto-S1 formed from pressurized S1 at 300 MPa for 10 min and unpressurized F-actin.

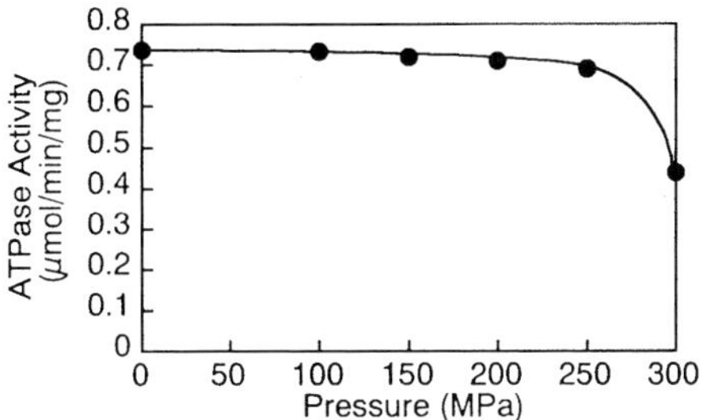

Figure 5. Effect of pressure treatment of S1 on acto-S1 ATPase activity. Acto-S1 ATPase activity was assayed in 20 mM NaCl, 0.5 mM ATP, and 2 mM $MgCl_2$ with 0.38 μM S1 and 1.6 μM actin at pH 7.0 and 25 °C.

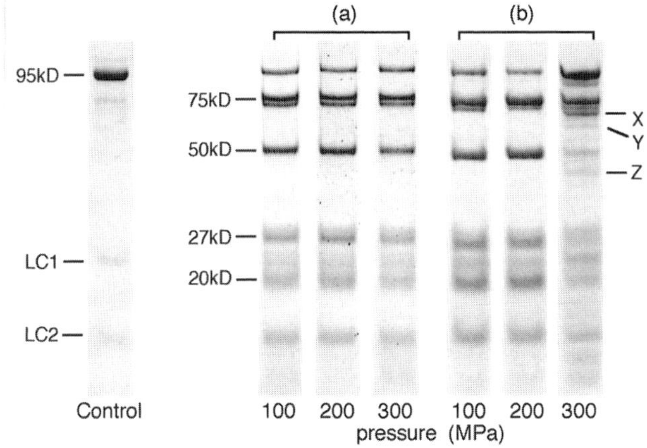

Figure 6. SDS-PAGE of S1 digested with trypsin. S1 in 0.1 M NaCl and 10 mM Tris-HCl (pH 7.0) was digested with 1/400 (w/w) of trypsin at 20 °C for 10 min. Control denotes undigested S1. Digestion was done after (a) and under (b) pressure treatment. X, Y and Z designate unknown fragments.

Figure 4 shows a morphology of acto-S1 prepared from pressure treated S1 at 300 MPa and unpressurized actin. A typical arrow head structure was observed in the control acto-S1 complex (a). After exposure to 300 MPa, binding of S1 to F-actin occurred, however acto-S1 seemed irregular and no typical arrow head structure was observed (b).

We measured actin activated S1 ATPase activity to ensure the interaction between F-actin and pressurized S1 took place. Figure 5 shows pressure effect of S1 on acto-S1 ATPase activity. The activity remained at almost the same level as an unpressurized S1 at least up to

250 MPa, but the activity was decreased above 300 MPa. This result agrees with decreasing actin binding ability of pressurized S1 shown in Figure 3, though pressure treated S1 at 300 MPa still retains ATPase activity about 0.43 μmol/min/mg. The decreasing actin binding ability of S1 is due to structural alteration of actin binding site in S1. Decreasing actin binding and nucleotide binding occurred at almost same pressure as shown in Figures 2 and 3.

Figure 6 shows the pattern of tryptic fragments of S1. S1 is composed of three domains, i.e., 27 kD (N-terminal), 50 kD, and 20 kD (C-terminal) fragments. In the present study, additional fragments appeared in SDS-polyacrylamide gel. The tryptic fragment of S1, in which pressure-induced S1 aggregates were removed, did not change from 0.1 MPa to 300 MPa (a). However, those produced under pressure at 300 MPa showed additional bands (b). The quantity of 27, 50, and 20 kD fragments were decreased, while the band of S1 heavy chain (95 kD) was denser than that in 200 MPa, and three novel fragments appeared between 75 kD and 50 kD (band X, Y), and 50 kD and 27 kD fragments (band Z). It seemed that appearance of unknown tryptic fragments of S1 is correlated with turbidity, in which S1 solution was pressurized up to 300 MPa (data not shown), increased.

The present study indicates; (1) pressure application up to 200–250 MPa induces global deformation of S1 without losing its biological activities such as ATPase and actin binding, (2) ATPase and actin binding sites lose those intrinsic structures with higher pressure above 250 MPa.

REFERENCES

1. M. Balint, I. Wolf, A. Tarcsafalvi, J. Gergely and F.A. Sreter, Arch. Biochem. Biophys., 190 (1978) 793.
2. A. Muhlrad and T. Hozumi, Proc. Natl. Acad. Sci. USA, 79 (1982) 958.
3. T. Hozumi and A. Muhlrad, Biochemistry, 20 (1981) 2945.
4. D. Mornet, P. Pantel, E. Audemard and R. Kassab, Biochem. Biophys. Res. Commun., 89 (1979) 925.
5. K. Yamamoto and T. Sekine, J. Biochem., 86 (1979) 1869.
6. K. Yamamoto, S. Hayashi and T. Yasui, Biosci. Biotech. Biochem., 57 (1993) 383.
7. K. Yamamoto, Y. Yoshida, J. Morita and T. Yasui, J. Biochem., 116 (1994) 215.
8. K. Yamamoto, T. Miura and T. Yasui, Food Structure, 9 (1990) 269.
9. G. Offer, C. Moos and R. Starr, J. Mol. Biol., 74 (1973) 653.
10. J.A. Spudich and S. Watt, J. Biol. Chem., 246 (1971) 4866.
11. A.G. Weeds and B. Pope, J. Mol. Biol., 111 (1977) 129.
12. M.M. Werber, A.G. Szent-Gyögyi and G. Fasman, Biochemistry, 11 (1972) 2872.
13. T. Ando, J.A. Duke, Y. Tonomura and M.F. Morales, Biochem. Biophys. Res. Commun., 109 (1982) 1.
14. J.M. Chalovich and E. Eisenberg, J. Biol. Chem., 257 (1982) 2432.
15. K. Yamamoto and C. Moos, J. Biol. Chem., 258 (1983) 8395.

Trends in High Pressure Bioscience and Biotechnology
R. Hayashi (editor)

In situ measurements of the solubility of protein crystals under high pressure

Yoshihisa Suzuki[a], Tsutomu Sawada[b], Satoru Miyashita[c], Gen Sazaki[d], Toshitaka Nakada[e], Hiroshi Komatsu[f], Toshiaki Arao[a], Katsuhiro Tamura[a]

[a]Department of Chemical Science and Technology, Faculty of Engineering, The University of Tokushima, 2-1 Minamijosanjima, Tokushima 770-8506, Japan

[b]National Institute for Research in Inorganic Materials, 1-1 Namiki, Tsukuba 305-0044, Japan

[c]Physics, Liberal Arts and Sciences, Toyama Medical and Pharmaceutical University, 2630 Sugitani, Toyama 930-0194, Japan

[d]Institute for Materials Research, Tohoku University, 2-1-1 Katahira, Aoba-ku, Sendai 980-8577, Japan

[e]Department of Physical Sciences, Faculty of Science and Engineering, Ritsumeikan University, 1-1-1 Nojihigashi, Kusatsu, Shiga 525-8577, Japan

[f]Faculty of Policy Studies, Iwate Prefectural University, 152-52 Takizawa-aza-sugo, Takizawa, Iwate 020-0173, Japan

Using a two-beam interferometer and a high-pressure cell with transparent windows, we measured the solubility of lysozyme crystals under high pressure *in situ*. The change in the concentration with time during equilibration was measured accurately and continuously starting from a supersaturated state (growth relaxation) and an undersaturated state (dissolution relaxation). The concentration for the dissolution relaxation reached to a constant value in a week, which did not coincide with a concentration for the growth relaxation in a comparable period. From a theoretical point of view, we regarded the asymptotic concentration for the dissolution relaxation as the solubility.

The molar enthalpy of thermal denaturation of pressurized (1hour at 100 MPa) sample (lysozyme: 5mg/ml) was also measured with a differential scanning microcalorimeter (DSC). The enthalpy was not different from that of non-pressurized sample. Thus, the effect of the irreversible pressure denaturation of lysozyme up to 100 MPa on the results of our experiments was negligible.

1. INTRODUCTION

The effects of pressure on protein crystallization have received much attention since Visuri *et al.* [1] had reported that the growth rate of crystals of glucose isomerase was significantly enhanced with increasing pressure. So far, several crystallization studies under high pressure have used hen egg-white lysozyme as a model [2-8]. To understand crystal growth under high pressure, a basic property is the pressure dependence of the solubility. Here, the solubility is defined as the equilibrium concentration for the solution coexistent with crystals. Several groups [2, 4, 6, 7] have commonly reported an increase in the solubility of lysozyme with increasing pressure, but there are large differences in the value of the increase among them. In some studies [2, 6], the solubility of lysozyme under high pressure was measured only from the supersaturated state by *ex situ* methods. In such a case, the solubility is always uncertain from 0 mg/ml to the asymptotic concentration in those studies. Takano *et al.* [7] have measured the equilibrium pressure *in situ* from supersaturation and undersaturation. But, they did not measure the concentration itself, and the error of the equilibrium pressure was still large. Needless to say, the solubility is indispensable to determine the supersaturation, which is a principal parameter to discuss the mechanism of crystal growth. Thus, the high accuracy is required for the solubility.

We have developed a novel *in situ* method to measure the solubility accurately under high pressure using an optical interferometer [9]. With this method, the change in the concentration with time during equilibration under high pressure can be followed continuously for the same sample to a high precision starting from a supersaturated state and an undersaturated state, and we can discuss whether the system comes to the equilibrium or not. Here, we report the pressure dependence of the solubility of tetragonal hen egg-white lysozyme in water [10]. We also checked whether the irreversible pressure denaturation occurred or not during the measurements. The denaturation should affect our results.

2. EXPERIMENTAL PROCEDURE AND MATERIALS

An oil pressure cell with transparent sapphire optical windows which is combined with a light interferometer of a Mach-Zehnder type was used [9]. The change in concentration of the solution under high pressure is measured directly by the *in situ* observation of the shift of the interference fringes. The pressure in the cell was well controlled automatically (accuracy of pressure: $\pm$ 0.5 MPa) by a feedback system with a pressure sensor and an electric screw pump (NOVA Swiss). A Cu jacket with a Peltier element was attached to the pressure cell for temperature control ($\pm$ 0.2 ℃). The more detailed description on the experimental instruments is given in our previous report [9].

In the high-pressure cell, two cylindrical glass capillaries were set parallel to each other and perpendicular to the optical path [9]. One capillary (sample capillary) was filled with lysozyme solution to be measured together with a few crystallites, and the other only with the buffer solution as a reference. One end of the capillaries was closed with glue and the opposite end remained open to the pressure transmitting oil (Durasyn 166, Amoco Chemicals Co. Ltd.: The components of the oil are polymers of C_{10} alpha olefin [11]) so that pressure could be transmitted smoothly into the capillaries. Here, we assumed that the effect of the oil on the solubility is negligible, since the solubility at 0.1 MPa for the solution in contact with the oil was not different from that without the oil.

The relative change in the concentration, ΔC, is defined as $\Delta C \equiv C - C_e(0.1\mathrm{MPa})$, where C is the concentration of the solution and C_e (0.1MPa) is the solubility at 0.1 MPa. ΔC is obtained by measuring the shifts of the interference fringes as we reported before [9]. Here, the precision in ΔC is determined as ± 0.05 mg/ml by the precision of the fringe shift [9] and the concentration dependence of the refractive index of the solution, which is measured as

$$n=(1.34097\pm0.00009)+(16\pm1)\times10^{-5}C. \tag{1}$$

Here, n is the refractive index of the solution, and C [mg/ml] is the concentration of lysozyme. This was measured at 20℃ with an Abbe refractometer and a light source giving the D line of Na (λ = 589.3nm), and the coefficient of C does not depend on the wavelength so much. When equilibrium is reached under high pressure, we write ΔC as ΔC_e.

The preparation of samples and the experimental conditions of this work are as follows. The six times recrystallized lysozyme (Seikagaku Kogyo Co. Ltd.) was used without further purification. All other chemicals were of reagent grade. Lysozyme solution was prepared by mixing two solutions. One was made by dissolving lysozyme in 0.02 M (M means mol/l hereafter) sodium acetate buffer (pH 4.66). The other was also made by dissolving sodium chloride in the same 0.02 M acetate buffer. Crystallites in the sample capillary were made by using a supersaturated solution (30 mg/ml lysozyme and 0.83 M NaCl in 0.02 M acetate buffer (pH 4.66)): The supersaturated solution was transferred into the sample capillary and incubated at 15℃. After obtaining a few crystallites, several non-equilibrium states were produced as initial states of equilibration processes by changing the temperature or the pressure. Temperature was kept constant at 20.0 $\pm$ 0.2 ℃ during the equilibration processes. Under this condition, crystals with the tetragonal structure are known to grow [2], which was confirmed with the observation of the crystal shape in our experiment.

The possibility of the irreversible pressure denaturation was checked with the DSC (CSC4100; Calorimetry Science Corporation). To check the pressure denaturation, the following two samples (lysozyme: 5mg/ml, NaCl: 0.83M, total amount: 0.8 ml) were prepared. One was left at 0.1 MPa (non-pressurized sample). The other was pressurized up to 100 MPa, kept at that pressure for 1 hour at 20℃, and then, the pressure was released (pressurized sample). The molar enthalpy of thermal denaturation was measured for the two samples with the DSC. If the pressure denaturation had already occurred in the pressurized sample, the enthalpy of the pressurized sample should be smaller than that of non-pressurized sample.

3. RESULTS AND DISCUSSIONS

To confirm whether the sample solution was in equilibrium with crystals or not, equilibration processes were initiated by intentionally producing non-equilibrium states. When experiments were started from the undersaturated state (dissolution relaxation), the concentration relaxed to a constant value within about a week (Fig. 1). But, when they were started from the supersaturated state (growth relaxation), the concentration did not reach the same value as the asymptotic concentration for the dissolution relaxation in a comparable period.

To determine which process is more suitable for the solubility measurement, let us make a theoretical consideration. Since lysozyme crystals have faceted faces, the surface of the crystal is considered to be molecularly smooth, and new growth centers have to be supplied by the special mechanisms such as two dimensional nucleation or the dislocation mechanism [12]. In such mechanisms, the density of the growth centers rapidly decreases when supersaturation decreases, and the change in the concentration for the growth relaxation can be significantly retarded at low supersaturations.

120

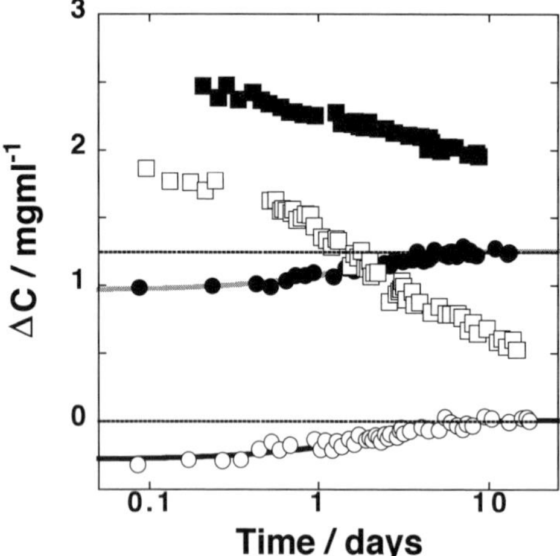

Fig. 1. Transitional change in concentration with time. ΔC is the relative increase in concentration from the solubility at 0.1 MPa. Closed and open marks indicate the data taken at 100 MPa and 0.1 MPa, respectively. Circles and squares indicate the data for the dissolution and the growth relaxation, respectively.

However, in the process of dissolution, edges and apexes of the crystal are always active as the dissolution centers independently of the degree of undersaturation. In such a case, the concentration may approach the equilibrium value more easily. Therefore, the dissolution relaxation is considered to be more suitable for the solubility measurement than the growth relaxation.

On the analogy of the adhesive growth mechanism [13] where growth centers exist on the crystal surface independently of the degree of supersaturation, the dissolution rate may be assumed to be proportional to the degree of undersaturation. If the concentration of the solution is regarded as uniform, as it was achieved in the present measurements probably because of convection in the solution, the dissolution relaxation is described as the following simple exponential function.

$$\Delta C = \Delta C_e + (\Delta C_0 - \Delta C_e)\exp(-t/\tau) \tag{2}$$

Here, ΔC_0 is the value of ΔC at t = 0, and τ is the relaxation time, which is proportional to the volume of the solution. The results of fitting to the data of the dissolution relaxation with eq. (2) are also shown in Fig. 1 by solid lines; the relaxation phenomena are well described by eq. (2) (τ = 2.4 days for 0.1 MPa, τ = 2.2 days for 100 MPa). Thus, we regarded the asymptotic concentration for the dissolution relaxation as the equilibrium concentration. On the other hand, it was confirmed that the data in the growth relaxation can not be fitted well with eq. (2). The data in the growth relaxation contained the information concerning the mechanisms of crystal growth. However, it seems to be difficult to discuss the growth mechanism on the sole basis of our data, because growth phenomena of the faceted crystal strongly depend on the orientation of the crystal and the growth mechanism of the faces [12]. In order to do this, contributions from different crystals and different faces must be separated.

ΔC_e at high pressures which were determined from the dissolution relaxation are shown in Fig. 2. The fit to the data by a quadratic function (solid line in Fig. 2) gives

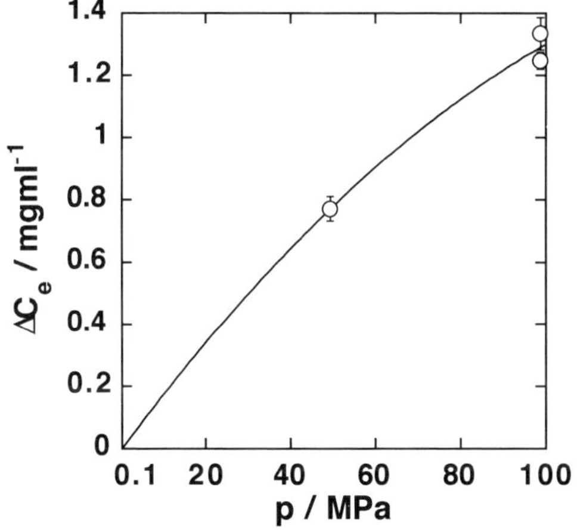

Fig. 2. The relative increase in the solubility, ΔC_e, of lysozyme with pressure. The errors are calculated from the standard deviations of ΔC (p) and ΔC (0.1 MPa) after the 7th day. The eq. (3) is expressed as the solid line.

$$\Delta C_e\,(p) = (1.8\pm0.2)\mathrm{X}10^{-2}(p\text{-}0.1) - (5.7\pm1.4)\mathrm{X}10^{-5}(p\text{-}0.1)^2 \text{ [mg/ml]}, \tag{3}$$

where, p is the pressure in MPa.

The absolute value of the solubility can be obtained by adding the solubility at 0.1 MPa to ΔC_e. The solubility at 0.1 MPa and 20℃ was determined from the measurement of the refractive index of the solution in equilibrium with the crystals, and found after combining with eq. (1) as

$$C_e\,(0.1 \text{ MPa}) = 4.7\pm0.2 \text{ [mg/ml]}. \tag{4}$$

Thus, the pressure dependence of the absolute value of the solubility are expressed as

$$C_e\,(p) = (4.7\pm0.2) + (1.8\pm0.2)\mathrm{X}10^{-2}(p\text{-}0.1) - (5.7\pm1.4)\mathrm{X}10^{-5}(p\text{-}0.1)^2 \text{ [mg/ml]}. \tag{5}$$

Pressure dependence of the solubility of hen egg-white lysozyme was measured *in situ* by continuous interferometric measurements of the concentration of the solution. The supersaturation under high pressure will be determined accurately from the pressure dependence of the solubility. It will allow us to quantitatively study the effect of pressure on the crystal growth kinetics [14].

As mentioned above, pressure made an important role to control protein crystallization. But, if an irreversible pressure denaturation occurred, our result would become less reliable and less precise. Thus, we checked the possibility of the irreversible pressure denaturation during our experiments. To check the possibility, the thermal denaturation temperature, T_d, and the molar enthalpy of denaturation, ΔH_d, were measured for the pressurized and non-pressurized lysozyme solution (lysozyme: 5mg/ml). Calorimetric measurements were made by the DSC at a 0.5℃/min. heating rate (from 30℃ to 100℃). The results are listed in Table in comparison with data by Kechnashvili *et al.* [15]. From Table, ΔH_d of pressurized sample is not different from that of non-pressurized one.

Table. The thermal denaturation temperature T_d [°C] and the molar enthalpy of the denaturation ΔH_d [kJmol^{-1}] compared to data by Kechnashvili, *et al.* [15].

	pH	Salt	lysozyme [mg/ml]	T_d [°C]	ΔH_d [kJmol^{-1}]
<This work>					
Non-pressurized	4.66	0.83 M NaCl	5	76.6 ± 0.2	430 ± 50
Pressurized	4.66	0.83 M NaCl	5	76.6 ± 0.2	430 ± 40
<Kechinashvili et al.>	2.0	None	1~5	56*	440*
	2.5	None	1~5	66*	490*
	4.0	None	1~5	78*	590*

* Estimated from graph shown in the reference

Thus, the irreversible pressure denaturation did not occur during the experiments. If the reversible denaturation occurred during the experiments, denatured proteins would precipitate because of high NaCl concentration (0.83 M). But such kind of precipitate could not be observed during the experiments.

REFERENCES

1. K. Visuri, K. Kaipainen, J. Kivimaki, H. Niemi, M. Leisola, and S. Palosaari, Bio/Technol. 8 (1990) 547.
2. M. Gross and R. Jaenicke, FEBS Lett. 284 (1991) 284.
3. C. A. Schall, J. M. Wiencek, M. Yarmush, and E. Arnold, J. Crystal Growth 135 (1994) 548.
4. Y. Suzuki, S. Miyashita, H. Komatsu, K. Sato, and T. Yagi, Jpn. J. Appl. Phys. 33 (1994) L1568.
5. M. V. Saikumar, C. E. Glatz, M. A. Larson, J. Crystal Growth 151 (1995) 173.
6. B. Lorber, G. Jenner, and R. Gieg*, J. Crystal Growth 158 (1996) 103.
7. K. J. Takano, H. Harigae, Y. Kawamura, M. Ataka, J. Crystal Growth 171 (1997) 554.
8. M. V. Saikumar, C. E. Glatz, M. A. Larson, J. Crystal Growth 187 (1998) 277.
9. Y. Suzuki, T. Sawada, S. Miyashita and H. Komatsu, Rev. Sci. Instrum. 69 (1998) 2720.
10. Y. Suzuki, T. Sawada, S. Miyashita, H. Komatsu, G. Sazaki, T. Nakada, J. Crystal Growth 209 (2000) 1018.
11. Products data of Amoco Chemicals Co. Ltd.
12. A. A. Chernov, Modern Crystallography III, Crystal Growth (Springer, Berlin, 1984) ch. 3.3.
13. A. A. Chernov, Modern Crystallography III, Crystal Growth (Springer, Berlin, 1984) ch. 3.1.
14. Y. Suzuki, S. Miyashita, G. Sazaki, T. Nakada, T. Sawada, H. Komatsu, J. Crystal Growth 208 (2000) 638.
15. N. N. Khechinashivili, P. L. Privalov and E. I. Tiktopulo, FEBS Lett. 30 (1973) 57.

Trends in High Pressure Bioscience and Biotechnology
R. Hayashi (editor)

Effects of pressure on growth kinetics of protein crystals

Yoshihisa Suzuki[a], Satoru Miyashita[b], Tsutomu Sawada[c], Gen Sazaki[d], Toshitaka Nakada[e], Hiroshi Komatsu[f], Toshiaki Arao[a], Katsuhiro Tamura[a]

[a]Department of Chemical Science and Technology, Faculty of Engineering, The University of Tokushima, 2-1 Minamijosanjima, Tokushima 770-8506, Japan

[b]Physics, Liberal Arts and Sciences, Toyama Medical and Pharmaceutical University, 2630 Sugitani, Toyama 930-0194, Japan

[c]National Institute for Research in Inorganic Materials, 1-1 Namiki, Tsukuba 305-0044, Japan

[d]Institute for Materials Research, Tohoku University, 2-1-1 Katahira, Aoba-ku, Sendai 980-8577, Japan

[e]Department of Physical Sciences, Faculty of Science and Engineering, Ritsumeikan University, 1-1-1 Nojihigashi, Kusatsu, Shiga 525-8577, Japan

[f]Faculty of Policy Studies, Iwate Prefectural University, 152-52 Takizawa-aza-sugo, Takizawa, Iwate 020-0173, Japan

Normal growth rates of {110} and {101} faces of tetragonal hen egg-white lysozyme crystals were measured *in situ* under 0.1, 50, and 100 MPa. The solubilities under high pressure enabled us to evaluate supersaturation under high pressure. We found that both growth rates decreased with pressure under the same supersaturation. This means that the surface growth kinetics depends on pressure significantly. The *birth and spread* model was applied to understand the pressure effects on the growth kinetics. It was found that the increase in the average ledge surface energy of the two-dimensional nuclei with pressure explained the decrease in the growth rate.

The molar enthalpy of thermal denaturation of pressurized (1hour at 100 MPa) sample (lysozyme: 5mg/ml) was also measured with a differential scanning microcalorimeter (DSC). The enthalpy was not different from that of non-pressurized sample. Thus, the effect of the irreversible pressure denaturation of lysozyme up to 100 MPa on the results of our experiments was negligible.

1. INTRODUCTION

Growth of large and high-quality protein single crystals is a bottleneck for the analysis of their three-dimensional structures. Pressure is expected to be one of the important parameters to control the growth process [1-2], since hydrostatic pressure affects the whole system uniformly and can be changed very rapidly. Unlike the other experimental parameters, pressure easily influences not only the crystal structure but also the structure of water (solution). Thus, a study of pressure effects should lead to a deeper understanding of the mechanisms for protein crystal growth. Many reports have shown that pressure influences the protein crystallization process from solutions [3-12]. Several studies have shown that *in situ* measurements are the most effective in analyzing the protein crystallization [6, 9, 11, 12]. Suzuki *et al.* [6] reported for the first time that the solubility of lysozyme increased, and the nucleation and the growth rates decreased as pressure increased. The pressure effect on the solubility was analyzed, but the effect on the growth kinetics was left unexplained. Recently, Takano *et al.* [9], Suzuki *et al.* [11] and Sazaki *et al.* [12] obtained the phase diagrams of lysozyme under high pressure using *in situ* techniques. They reported that the solubility of the tetragonal crystals increased with pressure. Nobody, however, has reported pressure effects on the growth kinetics of protein crystals before our work [13]. With a use of the solubility data, the growth rate can be expressed as a function of supersaturation σ under certain pressure, and we can throw light on the pressure effects on the growth kinetics.

In this study, we built up an *in situ* observation system under high pressure, and measured the growth rate of the tetragonal lysozyme crystals [13]. With the solubility data, the pressure effects of growth kinetics were investigated. We also checked whether the irreversible pressure denaturation occurred or not during the measurements. The denaturation should affect our results.

2. EXPERIMENTAL PROCEDURE

A high-pressure vessel with transparent sapphire windows for the *in situ* observation was used [13]. The vessel was filled with the hydraulic oil. The hydrostatic pressure (up to 120 MPa) was applied to the vessel with a hand pressure pump through the oil. The pressure was monitored by Bourdon type pressure gauge (Heise) within an accuracy of $\pm$ 0.1 MPa. We could maintain the vessel pressure manually within $\pm$ 0.2 MPa. The temperature of the vessel was measured by a thermocouple. The vessel was placed in an air conditioned box and the temperature was kept constant within $\pm$ 0.1 °C. The inner cell for crystal growth was made of glass slides and set in the vessel [13]. The crystals and the solution were put in the cell. The solution and the oil were separated by soft silicone tubes sealed with glue. The pressure was applied to the solution via these tubes. Crystallization process was observed by an optical microscope.

Hen egg-white lysozyme was used as a model protein, since we have the solubility data under high pressure [11]. The six times recrystallized lysozyme (Seikagaku Kogyo Co. Ltd.) was used without further purification. The buffer solution was 0.02 M sodium acetate (pH=4.66), and the precipitant was sodium chloride (0.83M). The temperature was kept constant at T = 20.0 $\pm$ 0.1°C. Under this crystallization condition, tetragonal form crystals bounded by {110} and {101} faces were obtained.

Supersaturated lysozyme solution of 50 mg/ml in concentration was incubated under atmospheric pressure (0.1 MPa) for about 10 hours to get seed crystals of 200-300 μm in size. The solution in the inner cell was replaced with the fresh solution of a given lysozyme concentration through the silicone tubes with a peristaltic pump. Then the inner cell was inserted into the high pressure vessel, and pressure was applied. It took no more than 10 minutes to change the solution and to apply pressure. The normal growth rates of {110} and {101} faces were measured by taking micrographs [14].

The solubilities of lysozyme at T = 20.0 $\pm$ 0.1 °C under pressure p [MPa], $C_e(p)$ [mg/ml] were expressed as [11],

$$C_e(p) = (4.7 \pm 0.2) + (1.8 \pm 0.2) \times 10^{-2} p - (5.7 \pm 1.4) \times 10^{-5} p^2 \ . \tag{1}$$

The supersaturation (σ) defined by $\ln C / C_e$ was estimated by using eq. (1) under each pressure, where C [mg/ml] is the lysozyme concentration.

The possibility of the irreversible pressure denaturation was checked with the DSC (CSC4100; Calorimetry Science Corporation). To check the pressure denaturation, the following two samples (lysozyme: 5mg/ml, NaCl: 0.83M, total amount: 0.8 ml) were prepared. One was left at 0.1 MPa (non-pressurized sample). The other was pressurized up to 100 MPa, kept at that pressure for 1 hour at 20℃, and then, the pressure was released (pressurized sample). The molar enthalpy of thermal denaturation was measured for the two samples with the DSC. If the pressure denaturation had occurred in the pressurized sample, the enthalpy of pressurized sample should be smaller than that of non-pressurized sample.

3. RESULTS AND DISCUSSION

Figures 1 and 2 show the normal growth rates of {110} and {101} faces with σ, respectively. Here, σ at each pressure was calculated by eq. (1). The vertical error bars indicate the standard deviations of the growth rates measured by 2 or 3 crystals. The horizontal error bars are estimated by the error of the solubility measurements. Figures 1 and 2 show that: (1) As pressure increased, the normal growth rates of the {110} faces, G_{110}, and those of the {101} faces, G_{101}, decreased. (2) The growth rates reached almost zero for the small value of σ. No growth was observed till the supersaturation reached the critical value, σ_c, which suggests that the crystals grew after two dimensional nuclei formed. (3) The growth rates had maxima at σ=2.3~2.5. This phenomenon was observed under 0.1 MPa and explained by the cluster formation [15] or the impurity incorporation [16]. Here, we focused on (1) and (2), to elucidate the pressure effect on the growth kinetics.

To confirm the growth mode, we observed the surface topography on the (110) face by an atomic force microscope (AFM) under 0.1 MPa [13], and concluded that, in the range of σ from 0.9 to 2.4, crystals grew by the *birth and spread* mechanism [17], which is expressed as [18],

$$G = k_1 (C / C_e)^{2/3} ((C - C_e) / C_e)^{2/3} \sigma^{1/6} \exp(-k_2 / \sigma) \tag{2}$$

where k_1 and k_2 are expressed as

$$k_1 = (\pi / 3)^{1/3} a^2 h^{7/6} v \lambda_0^{-2/3} C_e^{4/3} \exp((\varepsilon_{ad} - \varepsilon - 2(E - \phi)) / 3kT) \ , \tag{3}$$

$$k_2 = \pi \Omega \alpha^2 h / (3k^2 T^2) \tag{4}$$

126

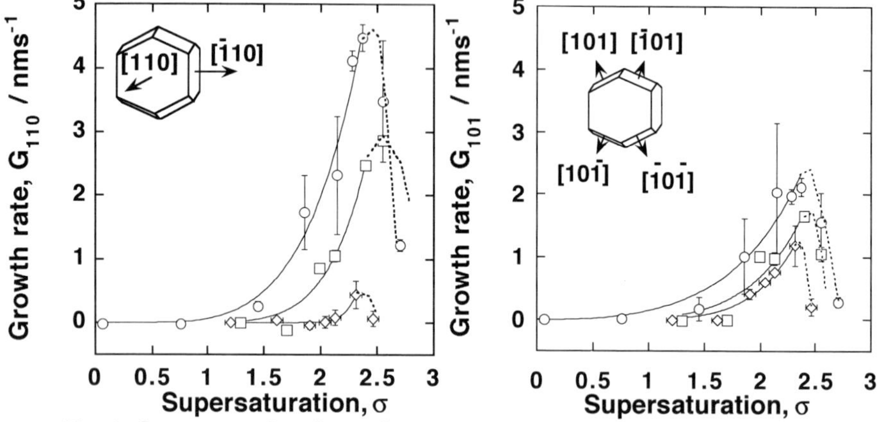

Fig. 1. Supersaturation dependence of the growth rates of the {110} face of a tetragonal lysozyme crystal [13]. Pressure: 0.1 MPa (○), 50 MPa (□), and 100 MPa (◇). The solid lines are fittings by the birth and spread model. The broken lines are guides for eyes of each plot.

Fig. 2. Supersaturation dependence of the growth rates of the {101} face of a tetragonal lysozyme crystal [13]. Pressure: 0.1 MPa (○), 50 MPa (□), and 100 MPa (◇). The solid lines are fittings by the birth and spread model. The broken lines are guides for eyes of each plot.

In eqs. (3) and (4), the following symbols are used; a: the distance between the molecules in the crystal (= $\Omega^{1/3}$, Ω: the molecular volume in the crystal), h: the height of the step (h=5.7nm for {110} face and h=3.4nm for {101} face measured by AFM), v: the thermal frequency of a solute, λ_0: the average distance between the kinks on a step, ε_{ad}: the adhesive energy of a solute to the crystal surface, ε: the activation energy for a solute to be incorporated to a critical nucleus, E-ϕ: the activation energy for dehydration of a solute, α: the average surface energy on the step ledge, k: Boltzmann constant.

We fitted the obtained data by eq. (2) with the least mean square method. The solid curves in Figs. 1 and 2 represent the best fits. They well reproduce the experimental data within the errors. The obtained values of k_1 and k_2 are listed in Table 1. On the both faces, the values of k_1 and k_2 increased with pressure. Increase in k_1 means increase in the growth rate while increase in k_2 leads to decrease in the growth rate. The results shown in Figs. 1 and 2 indicate that the change in k_2 dominated the pressure dependence of the growth rate. Equation (4) shows that the increase in k_2 corresponds to the increase in α. However, the value of k_1 of the {110} face at 100 MPa for the best fit (Table 1) is extraordinarily large and this value has less physical meanings. This is owed to the lack of data and the error of the growth rate. As in equation (3), there are too many factors to determine which dominate the increase in k_1 with pressure. To avoid such uncertainties, we fixed k_1 for each fitting. Since it is difficult to estimate a k_1 value under high pressure, we used the k_1 at 0.1 MPa as the fixed value. These fitting results are also listed in Table 1, and k_2 of both faces also increased with pressure.

Table 1. List of the fitted values of k_1 and k_2 according to eq. (2) [13].

Coefficients	Pressure (MPa)		
	0.1	50	100
{110} face			
k_1 (nm/s)	0.84	3.5	7.1×10^6
k_2	3.6	8.7	45
k_2 (k_1=0.84 (nm/s) fix)	3.6	5.4	9.6
{101} face			
k_1 (nm/s)	0.14	0.33	0.86
k_2	1.0	3.8	6.5
k_2 (k_1=0.14 (nm/s) fix)	1.0	2.0	2.6

These results also support that α increases with pressure.

If we regard the growth mechanism as the *birth and spread* type, the other possibility which reduces the growth rate with pressure is the decrease in k_1 (eq. (3)). But for the best fit, k_1 always increases with pressure. If we unreasonably fix k_2 for fitting, k_1 decreases with pressure, but the error of the fitting becomes significantly large. Thus, the most suitable reason that the growth rate decreases with pressure is the increase in α.

The evaluated values of the average surface energy on the step ledge, α, with eq. (4) are plotted against pressure in Fig. 3. The values of α under 0.1 MPa are in the same order of those reported previously [14, 18]. The values of α of the {110} face are larger than those of the {101} under each pressure. The values of α of both faces increased with pressure. Takano *et al.* reported the same pressure dependence of the surface free energy of p-xylene crystals grown from p- and m- xylene mixture solutions (6:4) [19]. In this case, we think that the decrease in the lattice constants (the decrease in the surface area) with pressure probably cause the increase in the surface free energy. Kundrot *et al.*, however, reported that the lattice constants, a and c, of tetragonal lysozyme crystals under 100 MPa were by 0.6% less and 0.1% more than those under 0.1 MPa, respectively [20]. As shown in Fig. 3, the values of α of {110} face for the best fit increased more than two times for instance. This large changes in α could not be explained only by the changes in the surface area with pressure. The other evaluation of the change in α was also made. When the pressure is applied, the stress is applied to the crystal. The mechanical work by this stress is equal to the change in Helmholtz free energy of the crystal. We can estimate the mechanical work of compression per unit area of crystal surface, since Young modulus of tetragonal lysozyme crystal has already been reported [21]. However, even if we use the dynamic Young modulus (1.1 GPa; it is larger than the static Young modulus (= 0.2 GPa)), the works with increasing pressure up to 100 MPa (0.1 mJ/m^2 for the {110} face, 1.6×10^{-4} mJ/m^2 for the {101} face) are still much less than the increases in α of this study (Fig. 3). At present, it is not clear why the values of α increased so much with pressure. Further studies, especially on hydration and dehydration, binding state between the molecules and so on, are needed.

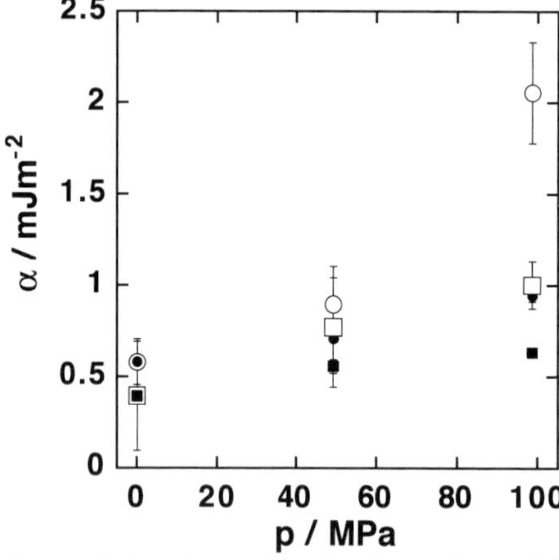

Fig. 3. Pressure dependence of the average surface energy of the step ledge, α, of two dimensional islands on the {110} surface ($\bigcirc$, $\bullet$) and the {101} surface ($\square$, $\blacksquare$) [13]. Open marks express the results of the best fits, and closed marks express those of the fit when the value of k_1 of eq. (3) is fixed. The vertical error bar was estimated by the standard deviation of the fitting.

Figure 3 also shows that the change in α of the {110} face with pressure is larger than that of {101} face. This larger increase in α of {110} face leads to larger decrease in the normal growth rate of {110} face. The change in the growth rates of each face results in morphological change of the crystal shape with pressure. Thus, pressure as well as supersaturation is found to be one of key parameters to control the crystal shape.

As mentioned above, pressure made an important role to control protein crystallization. But, if an irreversible pressure denaturation occurred, our result would become less reliable and less precise. Thus, we checked the possibility of the irreversible pressure denaturation during our experiments. To check the possibility, the thermal denaturation temperature, T_d, and the molar enthalpy of denaturation, ΔH_d, were measured for the pressurized and non-pressurized lysozyme solution (lysozyme: 5mg/ml). Calorimetric measurements were made by the DSC at a 0.5℃/min. heating rate (from 30℃ to 100℃). The data of the apparent excess molar heat capacity, C_p, with temperature are shown in Fig. 4. All thermal denaturation processes in this study were irreversible. Using the peak temperature and the area from the baseline, T_d and ΔH_d are calculated. The results are listed in Table 2 in comparison with data by Kechnashvili *et al.* [22]. From Table 2, ΔH_d of pressurized sample is not different from that of non-pressurized one. Thus, the irreversible pressure denaturation did not occur during the experiments. If the reversible denaturation occurred during the experiments, denatured proteins would precipitate because of high NaCl concentration (0.83 M). But such kind of precipitate could not be observed during the experiments.

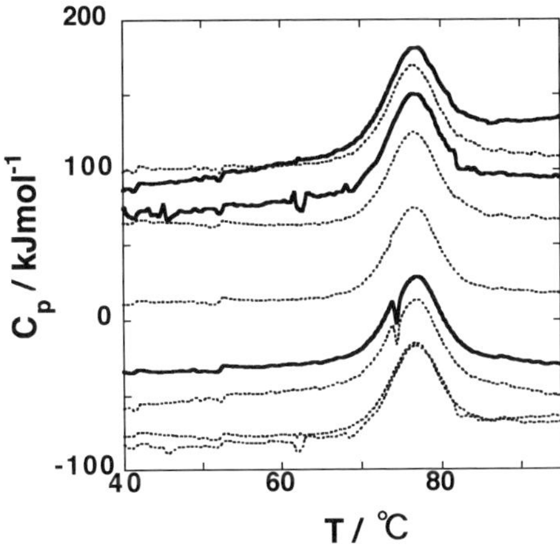

Fig. 4. Apparent excess molar heat capacity of lysozyme solution (lysozyme: 5mg/ml, NaCl: 0.83 M, pH=4.66). Solid and dotted lines indicate the data of pressurized and non-pressurized samples, respectively.

Table 2. The thermal denaturation temperature T_d [℃] and the molar enthalpy of the denaturation ΔH_d [kJmol^{-1}] compared to data by Kechnashvili, *et al.* [22].

	pH	Salt	lysozyme [mg/ml]	T_d [℃]	ΔH_d [kJmol^{-1}]
<This work>					
Non-pressurized	4.66	0.83 M NaCl	5	**76.6 ± 0.2**	**430 ± 50**
Pressurized	4.66	0.83 M NaCl	5	**76.6 ± 0.2**	**430 ± 40**
<Kechinashvili et al.>	2.0	None	1~5	56[*]	440[*]
	2.5	None	1~5	66[*]	490[*]
	4.0	None	1~5	78[*]	590[*]

* Estimated from graph shown in the reference

REFERENCES

1. T. Sawada, K. Takemura, K. Shigematsu, S. Yoda and K. Kawasaki, Phys. Rev. E 51 (1995) R3834.
2. T. Sawada, K. Takemura, K. Shigematsu, S. Yoda and K. Kawasaki, J. Crystal Growth 158 (1996) 328.
3. K. Visuri, E. Kaipainen, J. Kivimaki, H. Niemi, M. Leisola and S. Palosaari, Bio/Technology 8 (1990) 547.
4. M. Gross and R. Jaenicke, FEBS Lett. 284 (1991) 87.
5. C. A. Schall, J. M. Wiencek, M. Yarmush, and E. Arnold, J. Crystal Growth 135 (1994) 548.

6. Y. Suzuki, S. Miyashita, H. Komatsu, K. Sato, and T. Yagi, Jpn. J. Appl. Phys. 33 (1994) L1568.

7. M. V. Saikumar, C. E. Glatz and M. A. Larson, J. Crystal Growth 151 (1995) 173.

8. B. Lorber, G. Jenner and R. Giege, J. Crystal Growth 158 (1996) 103.

9. K. J. Takano, H. Harigae, Y. Kawamura and M. Ataka, J. Crystal Growth 171 (1997) 554.

10. M. V. Saikumar, C. E. Glatz, M. A. Larson, J. Crystal Growth 187 (1998) 277.

11. Y. Suzuki, T. Sawada, S. Miyashita, H. Komatsu, G. Sazaki, T. Nakada, J. Crystal Growth 209 (2000) 1018.

12. G. Sazaki, Y. Nagatoshi, Y. Suzuki, S. D. Durbin, S. Miyashita, T. Nakada, H. Komatsu, J. Crystal Growth 196 (1999) 204.

13. Y. Suzuki, S. Miyashita, G. Sazaki, T. Nakada, T. Sawada, H. Komatsu, J. Crystal Growth 208 (2000) 638.

14. S. D. Durbin and G. Feher, J. Crystal Growth 76 (1986) 583.

15. A. Nadarajah, E. L. Forsythe, M. L. Pusey, J. Crystal Growth 151 (1995) 163.

16. F. Rosenberger, P. G. Vekilov,   M. Muschol, B. R. Thomas, J. Crystal Growth 168 (1996) 1.

17. A. I. Malkin, A. A. Chernov and I. V. Alexeev, J. Crystal Growth 97 (1989) 765.

18. K. Kurihara, S. Miyashita, G. Sazaki, T. Nakada, Y. Suzuki, H. Komatsu, J. Crystal Growth 166 (1996) 904.

19. K. J. Takano, M. Wakatsuki, J. Crystal Growth 171 (1997) 591.

20. C. E. Kundrot and F. M. Richards, J. Mol. Biol. 193 (1987) 157.

21. V. N. Morozov and T. Ya. Morozova, J. Biomolecular Structure & Dynamics 11 (1993) 459.

22. N. N. Khechinashivili, P. L. Privalov and E. I. Tiktopulo, FEBS Lett. 30 (1973) 57.

Trends in High Pressure Bioscience and Biotechnology
R. Hayashi (editor)

Pressure effects on the structure and phase behavior of phospholipid-gramicidin bilayer membranes

J. Eisenblätter, M. Zein and R.Winter

University of Dortmund, Physical Chemistry I, Otto-Hahn-Straße 6, D-44227 Dortmund, Germany

We investigated the effect of incorporation of gramicidin D (GD) on the structure and phase behavior of aqueous dispersions of DPPC phospholipid bilayers. SAXS, DSC, ^{2}H-NMR and FTIR spectroscopy were used to detect topological and conformational changes upon incorporation of GD into the lipid bilayer. The data show that, depending on the GD concentration, the structure of the temperature- and pressure-dependent lipid phases is significantly altered by the insertion of the polypeptide, but also the lipid matrix has the ability to modulate the conformation of the inserted polypeptide. No pressure-induced unfolding of the polypeptide is observed up to pressures of 10 kbar.

1. INTRODUCTION

Understanding the molecular basis of interactions between phospholipids and membrane proteins in cell membranes is an important aim of structural biology. Here, we report on studies of protein-lipid interactions using differential scanning calorimetry (DSC), small-angle x-ray scattering (SAXS), Fourier-transform infrared (FTIR) and ^{2}H-NMR spectroscopy. We investigated the effect of incorporation of 2-5 mol% gramicidin D (GD) on the structure and phase behavior of aqueous dispersions of fully hydrated DPPC bilayers in the temperature range 0-80 °C at pressures up to several kbar. The primary sequence of gramicidin consists of only 15 amino acid residues which have alternate L and D chirality, and all side-chains are nonpolar. As a consequence, the peptide is able to adopt conformations of β-helices, which would be unacceptable for an all-L-amino acid peptide. The helices can be right- or left-handed, and they can differ in the number of amino acid residues per turn and therefore in length and diameter [2-5]. Individual molecules can fold into single-stranded helices, which are stabilized by intramolecular hydrogen bonds, and can associate to helical dimers (HD) in which two single-stranded helices are joined end-to-end (Fig. 1), such as the ss $\beta^{6.3}$ helical dimer which has an internal pore diameter of 3-4 Å and the total length of the dimer is ~26 Å [5]. Also, double-stranded helices (DH) can be formed (Fig. 1) in which the two strands run either parallel or antiparallel, such as the ds $\beta^{5.6}$ double helix which has a length of ~36 Å and a maximum lumen of 3 Å.

Pressure was applied in this study so as to be able to fine tune the lipid chain-lengths and conformation and to select specific lipid lamellar phases. For example, the phospholipid bilayer thickness increases by ~1 Å/kbar in the liquid-crystalline phase, and up to six gel

132

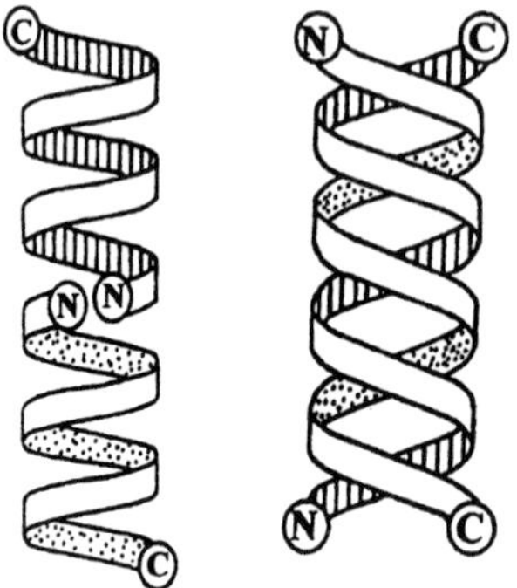
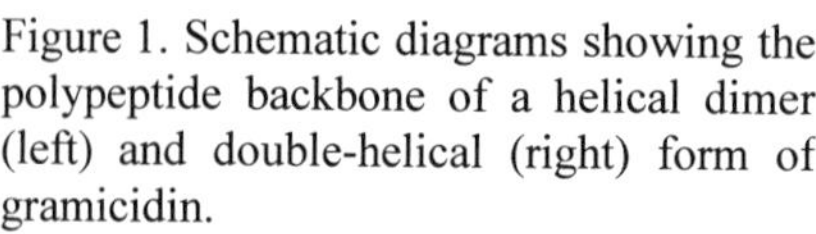

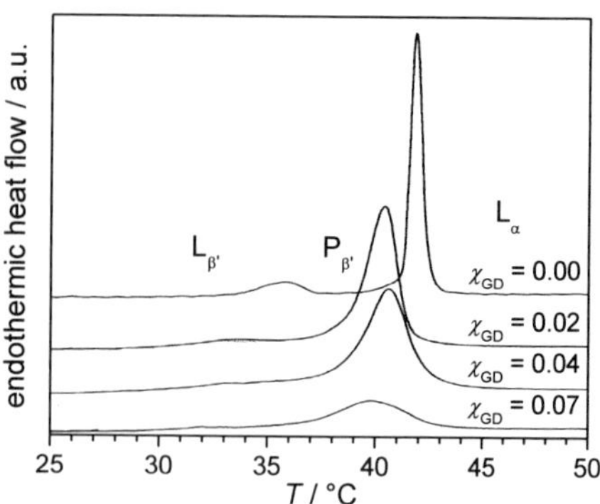

Figure 1. Schematic diagrams showing the polypeptide backbone of a helical dimer (left) and double-helical (right) form of gramicidin.

Figure 2. Differential scanning calorimetry of DPPC dispersions with different molar ratios of GD.

phases have been found in fully hydrated DPPC dispersions [1]. IR spectral parameters obtained by FTIR spectroscopy, such as frequencies, intensities, and bandwidths, and the ^{2}H-NMR spectral parameters were used to detect structural and dynamic changes upon incorporation of GD into the lipid bilayers. Analysis of the infrared amide I band shape allowed us to obtain also information on the corresponding peptide conformation adopted in the lipid environment. The SAXS and DSC data yielded information about changes in the lipid bilayer topology and phase behavior, respectively.

2. EXPERIMENTAL

1,2-Dipalmitoyl-*sn*-glycero-3-phosphatidylcholine (DPPC) was obtained from Sygena Lipids (Liestal, Switzerland), chain-perdeuterated DPPC-d$_{62}$ was purchased from Avanti Polar Lipids (Alabaster, AL, USA). Gramicidin D (GD) was purchased from Sigma Chemical (Deisenhofen, Germany). DPPC-GD mixtures were prepared by co-dissolving the appropriate amounts in chloroform and vortex mixing the solution. The solvent was then removed under vacuum using a Speed Vac Sc 110 (Savant, Farmingdale, NY, USA) for at least 16 h. Fully hydrated dispersions were then prepared by adding doubly distilled water, heating the hydrated mixtures above the main phase transition temperature of the lipid, vortex mixing and freezing the samples in liquid nitrogen. At least five *freeze-thaw-vortex* cycles were performed until homogenous lipid dispersions were obtained.

Small-angle x-ray scattering (SAXS) experiments were performed at beam line A2 of the HASYLAB at DESY. The reciprocal spacings $s = (2/\lambda)\sin\theta$ (λ wavelength of radiation, 2θ scattering angle) were calibrated by the diffraction pattern of rat-tail collagen. For the pressure studies a pressure cell with flat diamond windows was used, suitable for studies up to pressures of 8 kbar at temperatures ranging up to 150 °C [6]. The lamellar lattice constant d of the lipid bilayer stacking can be calculated from the diffraction patterns using Bragg's equation.

For the high pressure ^{2}H-NMR experiments a probe made of a titanium alloy (Ti-6Al-4V) was used [7] which fits into the room-temperature bore of a wide-bore superconducting magnet (diameter 72 mm). The samples were placed in a glass tube, fitted to the solenoid rf-coil, and closed with a flexible polymer foil to separate the sample from the pressurizing fluid (methylcyclohexane). The experimental setup allows studies at pressures up to 3 kbar and temperatures ranging from 5 to 80 °C. A Bruker MSL 400 spectrometer was used operating at a resonance frequency of 61.4 MHz. All deuteron NMR experiments were performed using the quadrupolar echo pulse sequence $\pi/2_x$-τ_1-$\pi/2_y$-τ_2-acquisition. In the liquid-crystalline phase perdeuterated lipids show ^{2}H-NMR spectra, which are superpositions of axially symmetric quadrupolar powder patterns of all C-D bonds. From the sharp edges the quadrupolar splittings

$$\Delta\omega_Q = \frac{3}{4}\frac{e^2 qQ}{h} S_{CD} \tag{1}$$

can be obtained, where $e^2 qQ/h$ (the static quadrupole coupling constant) for ^{2}H(D) in C-D bonds is about 167 kHz [8], and the C-D bond order parameter S_{CD} for distinguishable deuterons can be calculated. It is defined by:

$$S_{CD} = \frac{1}{2}\left\langle 3\cos^2\theta - 1\right\rangle \tag{2}$$

where θ is the angle between the C-D bond vector and the bilayer normal. Furthermore, in all phases studied the first spectral moment M_1 of the ^{2}H-NMR spectra can be calculated, which is proportional to the average chain order parameter of the lipid:

$$M_1 = \frac{\int_0^\infty \omega \cdot f(\omega)\,d\omega}{\int_0^\infty f(\omega)\,d\omega} = \frac{4\pi}{3\sqrt{3}}\left\langle \Delta\omega_Q\right\rangle \propto \left\langle S_{CD}\right\rangle \tag{3}$$

For the temperature dependent FTIR studies, lipid dispersions were filled into a 25 μm thick infrared cell with CaF_2 windows. For the pressure dependent studies, small amounts of the homogenous dispersions were placed, together with powdered α-quartz, in the 0.5 mm diameter hole of a 0.05 mm thick stainless steel gasket mounted on a diamond anvil cell (Type IIa diamonds from Diamond Optics, Tucson, AZ, USA), which was thermostated by an external water thermostat. Pressure was determined from the shift of the 695 cm^{-1} phonon band of α-quartz [9]. The pressure can be determined with an accuracy of ±100 bar using this method. The infrared spectra were collected on a Nicolet Magna 550 Fourier-transform infrared spectrometer. The amide I mode of gramicidin appears in the spectral range 1700 - 1600 cm^{-1} and may result of overlapping absorption bands which belong to a particular type of secondary structure of gramicidin [10]. The fractional intensities of the secondary elements were calculated from band-fitting procedures assuming a Gaussian-Lorentzian lineshape function. The bands associated with particular types of secondary structure elements were determined by Naik and Krimm [10] using normal mode calculations. We analysed spectral changes in the amide I band region following the absorption bands occuring at about 1631 and 1646-1648 cm^{-1}, which are expected to originate from the single-stranded helical dimers $\beta^{4.4}$ [HD(1)] and $\beta^{6.3}$ [HD(2)], respectively, and at about 1636, 1656 and 1666-1669 cm^{-1}, which originate from double-stranded antiparallel and parallel $\beta^{5.6}$ helices (DH). In several papers, a single band around 1631 cm^{-1} has been assigned to a $\beta^{6.3}$ helical dimer structure. This assignment is based on the corresponding CD spectra, however, and a single-stranded, right-handed channel conformation of gramicidin A with 6.5 residues per turn forming a 4 Å diameter pore in fluid DMPC bilayers has been determined using solid-state NMR spectroscopy [11-13]. Owing to these uncertainties in conformer assignment we denote the two helical dimer states as HD(1) and HD(2), respectively.

3. RESULTS AND DISCUSSION

As can be seen from the DSC traces shown in Figure 2, the $L_{\beta'}-P_{\beta'}$ pretransition of pure DPPC occurs at ~34 °C, and the $P_{\beta'}-L_\alpha$ gel to liquid-crystalline (main) phase transition at ~41 °C. Adding GD concentrations up to about 2 mol% leads to a downward shift and broadening of the transition peaks. The DSC data indicate that the pretransition vanishes and broad gel-fluid coexistence regions are formed at higher GD concentrations.

In the 2800-3000 cm^{-1} region appear infrared absorption bands due to antisymmetric and symmetric stretching modes of the methylene chains of DPPC, at about 2920 and 2850 cm^{-1}, respectively. The wavenumbers of these bands are conformation sensitive and for this reason respond to temperature- and pressure-induced changes of the *trans/gauche* ratio in acyl chains [14]. The position of the CH$_2$ stretching vibration is a qualitative measure of the number of *gauche* conformers in the acyl chains. Formation of *gauche* conformers results in a shift to higher frequencies. At the main phase transition of phospholipids, which results in melting of the acyl chains, the band shifts by about 2 cm^{-1}. Figure 3 exhibits the temperature dependence of the wavenumber of the symmetric CH$_2$ stretching absorption band of pure DPPC and the DPPC-GD (5 mol%) mixture. The main phase transition of the pure DPPC dispersion is clearly seen at about 41 °C by a sharp increase in the wavenumber. The gel to fluid phase transformation of the DPPC-GD (5 mol%) mixture seems to take place over a temperature range from about 33 to 46 °C.

Figure 4 shows the temperature dependent infrared spectra of the DPPC-GD (5 mol%)

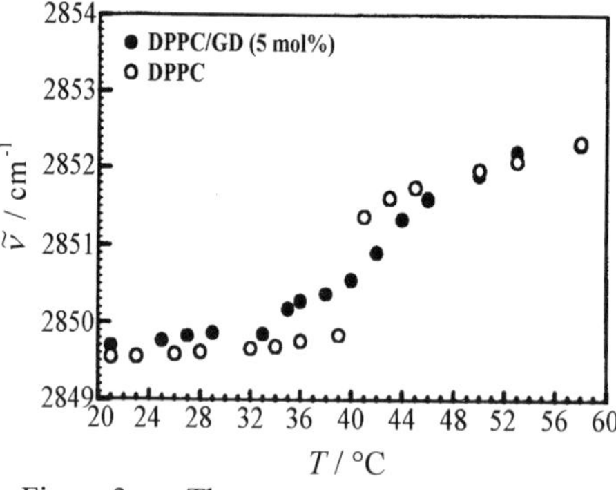

Figure 3. The temperature dependent wavenumber of the symmetric CH$_2$ stretching mode of DPPC and DPPC-GD (5 mol%).

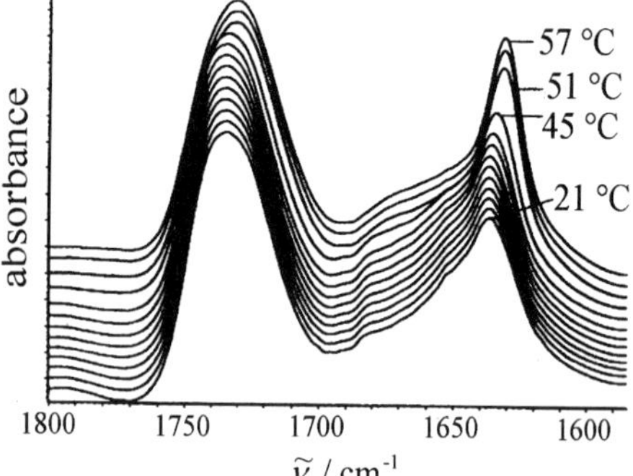

Figure 4. Temperature dependent IR spectra of DPPC-GD (5 mol%) in the IR amide I region.

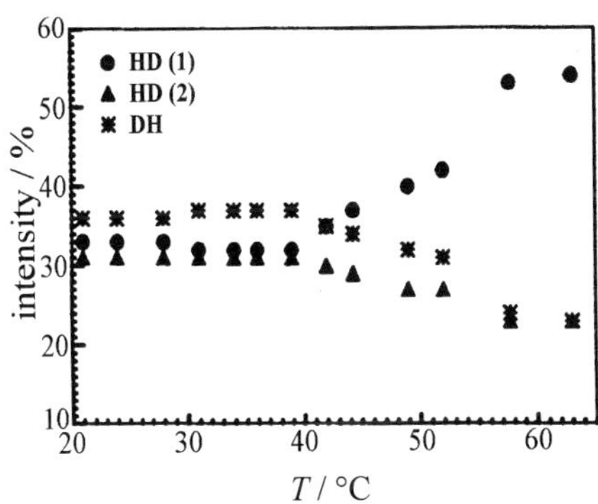

Figure 5. Temperature dependence of the fractional band intensities of gramicidin conformations in DPPC-GD (5 mol%).

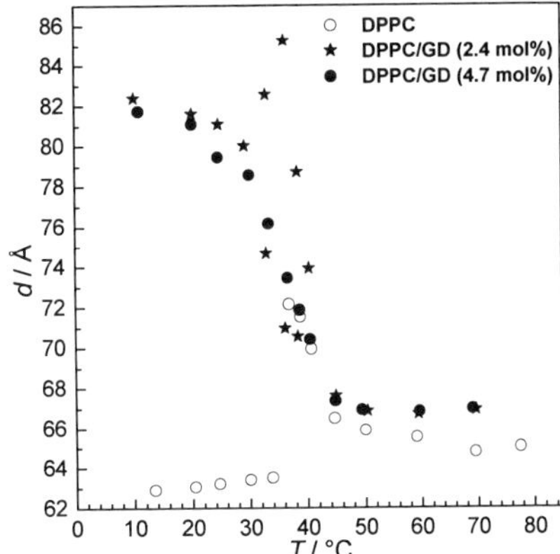

Figure 6. Temperature dependence of the lamellar lattice constant d of DPPC and the DPPC-GD mixtures.

system in the amide I region. We recognize that the maximum of the amide I band is shifted to higher frequencies at temperatures below about 42 °C, i.e., below the main transition temperature of the pure lipid component (T_m = 41.5 °C), indicating that significant structural rearrangements occur with respect to the peptide conformation. The fractional intensities of the secondary structural elements are shown in Figure 5. At temperatures above 43 °C the fractional intensity of the HD(1) conformation increases substantially with temperature, whereas the fractional intensities of the other conformers decrease.

The lamellar lattice constants d of the DPPC-GD samples studied (Fig. 6) show that incorporation of GD leads to an extraordinary large lamellar lattice constant in their gel

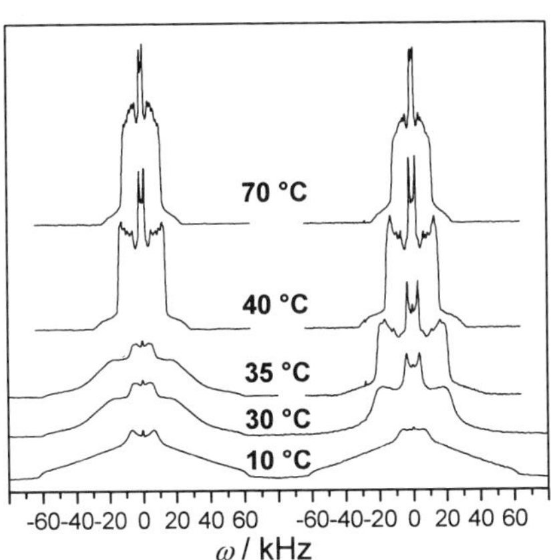

Figure 7. ^{2}H-NMR spectra of DPPC-d$_{62}$ (left) and DPPC-d$_{62}$-GD (4.7 mol%) (right) at selected temperatures.

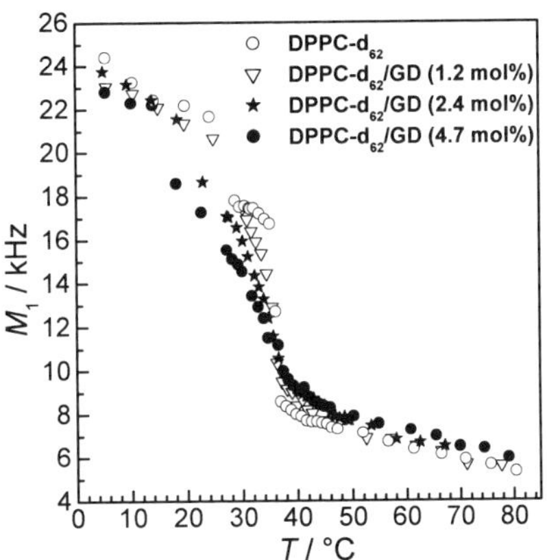

Figure 8. Temperature dependence of the first spectral moment of DPPC-d$_{62}$ and the DPPC-d$_{62}$-GD mixtures.

phases. This swelling might be induced by the formation of the longer double helical form of gramicidin. Figure 6 shows, that in the broad fluid-gel co-existence region between 30 and 45 °C the lattice constant of the DPPC-GD (4.7 mol%) mixture is continuously decreasing until a homogeneous fluid phase is formed.

The influence of gramicidin insertion on the acyl chain segmental order could be determined by the analysis of the ^{2}H-NMR spectra of perdeuterated lipid samples. The experiments show the well-known downward shift of ~4 °C in transition temperature of perdeuterated lipids relative to the transition temperature of the normal lipid. The spectra of the pure DPPC-d$_{62}$ bilayers exhibit a drastic change in lineshape at the main phase transition temperature (Fig. 7). Incorporation of GD broadens the phase transition, so that in the fluid-gel co-existence region the NMR lineshapes show characteristics of both, motionally averaged and rigid lattice type spectra. This is also obvious in the temperature dependence of the first moment M_1, shown in Figure 8. The M_1 data show that incorporation of GD leads to a significant decrease of the molecular order of the acyl chains in the gel phase, whereas in the fluid phase the mean quadrupolar splitting, and therefore the mean order parameter, increases slightly. The calculated C-D bond order parameter profiles (data not shown) in the liquid-crystalline phase show, that the methylene bond order parameter increases about 30% in the DPPC-d$_{62}$-GD (4.7 mol%) system, independent of the position in the acyl chain. The methyl group order parameter shows a slight increase, only.

To analyse the effect of pressure on the structure and phase behavior of the DPPC-GD mixtures, we studied the pressure dependence of the CH$_2$ symmetric stretching absorption band of DPPC and the amide I absorption band of gramicidin. In pure DPPC dispersions for example at 63 °C, the CH$_2$ vibrational mode is shifted to lower wavenumbers when the pressure passes 1.2 kbar, owing to the pressure-induced L$_\alpha$ to P$_{\beta'}$ gel phase transition. Further increase in pressure leads to the pressure-induced interdigitated gel phase, L$_{\beta i}$, around 2 kbar, and to two further gel phases, Gel III and Gel IV, at around 4 and 5.7 kbar, respectively [1]. The pressure dependence of the wavenumber of the symmetric CH$_2$ stretching absorption band of the DPPC-GD (5 mol%) mixture shows a shift to lower wavenumbers around 1 kbar, when the gel state is reached. The subsequent non-linearity of $\tilde{v}(p)$ indicates that further phase transformations occur. In comparison with the pure lipid system, they appear at different pressures, however, and broad phase co-existence regions have to be envisaged.

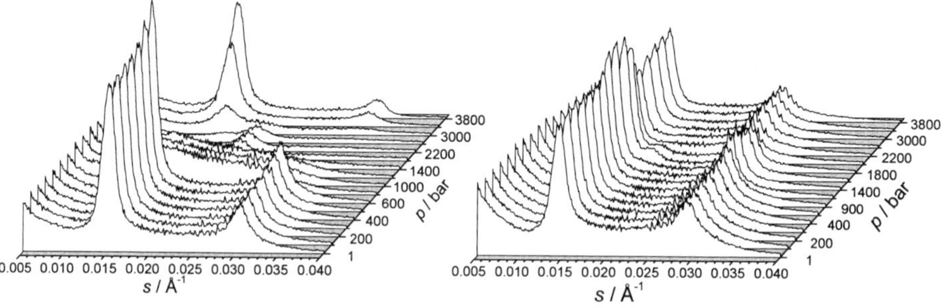

Figure 9. Pressure dependent SAXS data of DPPC at 55 °C.

Figure 10. Pressure dependent SAXS data of DPPC-GD (4.7 mol%) at 55 °C.

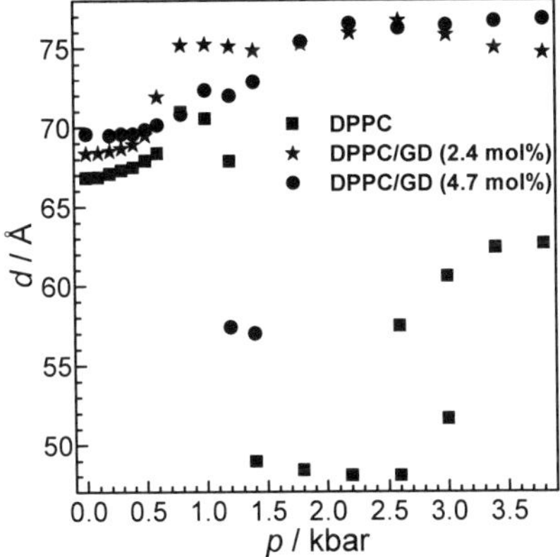

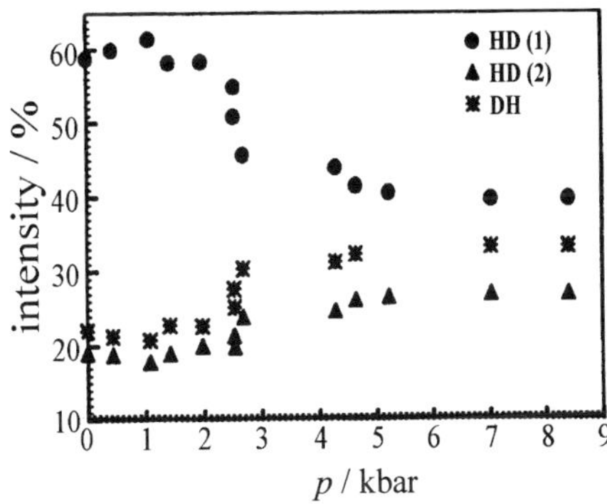

Figure 11. Pressure dependence of the lamellar lattice constant d of DPPC and the DPPC-GD mixtures at 55 °C.

Figure 12. Pressure dependence of the fractional band intensities of GD conformations in DPPC-GD (5 mol%) at 63 °C.

These changes are clearly visible in the SAXS patterns of DPPC and DPPC-GD (4.7 mol%) shown in Figs. 9 and 10. Owing to the hydrophobic mismatch in the DPPC-GD sample the interdigitated gel phase is not formed any more (Fig. 11). The lamellar lattice constant of the DPPC-GD sample increases continuously, and at intermediate pressures two distinguishable peaks are observed in the diffraction patterns which are due to the formation of a two-phase coexistence region. Like in the ambient pressure data the d-spacings in the gel phases are found to be much larger than in pure DPPC.

The ^{2}H-NMR lineshapes of the DPPC-d_{62}-GD (4.7 mol%) mixture show significant changes in the high pressure gel phases at 55 °C and are also indicative of the absence of the $L_{\beta i}$ phase. From the M_1 data we conclude that the molecular order in the liquid-crystalline phase is increased and in the high pressure gel phases is decreased.

The IR spectrum in the amide I region of the DPPC-GD (5 mol%) system exhibits significant changes over the pressure-range covered, and the spectral shapes are similar to those observed in the temperature dependent studies. Figure 12 shows the pressure dependence of the fractional intensities of the secondary structures. Only a small change in conformer proportion is observed when the fluid-gel phase region is entered at around 1 kbar. A significant change is observed at around 2.5 kbar when the fractional intensity of the HD(1) component decreases and the fractional intensities of the other conformers increase correspondingly. In comparison with the low temperature gel phase data at ambient pressure, in the high-pressure gel state a higher percentage of HD(1) states is observed, however.

4. CONCLUSIONS

The gramicidin conformers differ in the number of amino acid residues per turn and therefore in length. The lengths have been found to be ~24 Å for the dimeric single-stranded

138

right-handed $\beta^{6.3}$-helix and ~36 Å for the left-handed antiparallel double-stranded $\beta^{5.6}$-helix. For the $\beta^{4.4}$-helix no experimental data are available so far, its length will be greater than that of the dimeric $\beta^{6.3}$-helix. For comparison, the hydrophobic fluid bilayer thickness is about 30 Å for DPPC bilayers [15]. The hydrophobic thicknesses of the gel phases are 4-5 Å larger.

In DPPC lipid bilayers, the DH form is prominent in the gel state, and the HD is abundant at high temperature. Owing to the formation of broad fluid-gel phase co-existence regions, a continuous change in the population ratio is observed for the intermediate temperature regions. Destabilization of the HD conformation in the thicker DPPC gel phase membranes is probably related to the hydrophobic mismatch between lipid length and polypeptide hydrophobic surface. The changes observed may be attributed, at least partially, to the ability of the double helical conformation to tolerate more hydrophobic mismatch than the helical dimer. By formation of gel-fluid co-existence regions, the system has an efficient means of avoiding large hydrophobic mismatch. In these phase-separated regions, a molecular sorting mechanism, i.e., a selective accumulation of lipid species that hydrophobically match the protein surface, seems to be very likely. In the different pressure-induced gel phases, a similar tendency is observed. The larger the mismatch, the higher is the population of the double-helical structures. Interestingly, under high mismatch conditions, such as those in the pressure-induced gel phases, no further changes in polypeptide conformation take place. No pressure-induced unfolding of the polypeptide GD in phospholipid bilayers is observed at pressures up to 10 kbar. We might conclude that the membrane matrix has a strong stabilizing effect on the polypeptide structure.

Acknowledgements

We gratefully acknowledge financial support by the Deutsche Forschungsgemeinschaft (DFG) and the Fonds der Chemischen Industrie.

REFERENCES

1. C. Czeslik, O. Reis, R. Winter and G. Rapp, Chem. Phys. Lipids, 91 (1998) 135.
2. B. A. Wallace, Annu. Rev. Biophys. Chem., 19 (1990) 127.
3. R. E. Koeppe, II, Annu. Rev. Biophys. Biomol. Struct., 25 (1996) 231.
4. D. J. Chadwick and G. Cardew (eds.), Gramicidin and Related Ion Channel-forming Peptides, Novartis Foundation Symposium 225, John Wiley & Sons, New York, 1999.
5. B. A. Wallace, J. Struct. Biol., 121 (1998) 123.
6. R. Winter and C. Czeslik, Z. Kristallogr., 215 (2000) 454.
7. J. Eisenblätter, A. Zenerino and R. Winter, Magn. Reson. Chem., 38 (2000) 662.
8. J. H. Davis, Biochim. Biophys. Acta, 737 (1983) 117.
9. P. T. T. Wong, D. J. Moffat and F. L. Baudais, Appl. Spectros., 39 (1985) 733.
10. V. M. Naik and S. Krimm, Biophys. J., 49 (1986) 1131 and 1147.
11. M. Bouchard and M. Auger, Biophys. J., 65 (1993) 2484.
12. S. V. Sychev, L. I. Barsukov and V. T. Ivanov, Eur. Biophys. J., 22 (1993) 279.
13. F. Kovacs, J. Quine and T. A. Cross, Proc. Natl. Acad. Sci. USA, 96 (1999) 7910.
14. O. Reis, R. Winter and T. W. Zerda, Biochim. Biophys. Acta, 1279 (1996) 5.
15. T. Gil, J. H. Ipsen, O. G. Mouritsen, M. C. Sabra, M. M. Sperotto and M. J. Zuckermann, Biochim. Biophys. Acta, 1376 (1998) 245.

Trends in High Pressure Bioscience and Biotechnology
R. Hayashi (editor)
© 2002 Elsevier Science B.V. All rights reserved.

Effect of pressure on the bilayer phase transition of diacylphosphatidylethanolamine

Shoji Kaneshina[a], Shigeru Endo[a], Hitoshi Matsuki[a] and Hayato Ichimori[b]

[a]Department of Biological Science and Technology, Faculty of Engineering,
The University of Tokushima, Minamijosanjima, Tokushima 770-8506, Japan
[b]Anan National College of Technology, Minobayashi, Anan, Tokushima 774-0017,
Japan

All the temperatures of the main transition between lamellar gel (L_β) and liquid crystaline (L_α) phases for dilauroyl- (DLPE), dimyristoyl- (DMPE), dipalmitoyl- (DPPE), and distearoyl-phosphatidylethanolamine (DSPE) bilayer membranes were elevated by pressure. The temperature (T) – pressure (p) phase boundary curve was best described by a smooth curve and not a linear function. The values of slope (dT/dp) at ambient pressure were 0.230, 0.251, 0.264 and 0.275 K MPa^{-1} for DLPE, DMPE, DPPE and DSPE, respectively. The volume change associated with the main transition from L_β to L_α phase, which was estimated from the Clapeyron-Clausius equation, was 13.0, 19.5, 27.2 and 35.3 cm^3 mol^{-1} for DLPE, DMPE, DPPE and DSPE, respectively, which increased linearly with an increase in the acyl chain length of lipids. Thermodynamic properties for the main transition of phosphatidylethanolamine (PE) bilayer membranes were compared with those for phosphatidylcholine (PC) bilayer membranes. Regarding the enthalpy and volume changes, the values for PEs and PCs with the same acyl chain are similar in magnitude to each other. However, the main-transition temperatures for PE bilayer membranes are siginificantly higher than those for PC. The difference in the main-transition temperature between PE and PC bilayer membranes may be attributable to the head group interaction in the bilayers.

1. INTRODUCTION

Effect of pressure on lipid bilayer membranes is of particular interest to the studies of pressure reversal of anesthesia [1], pressure adaptation in deep sea organisms [2], and high-pressure sterilization in food processing [3-6]. Lipid bilayer membranes composed of phosphatidylcholines containing two identical linear saturated fatty acyl chains have been most thoroughly studied. We have shown the pressure effect on the bilayer phase transition of a series of diacylphosphatidyl-cholines [7-10], and extend these measurements and thermodynamic analysis to another major phospholipid group, i. e. phosphatidylethanolamines (PEs) bilayer membrane. So far, few studies have been reported on the effect of pressure on the bilayer phase transition of PEs [11,12]. Russell and Collings [11] determined the gel-liquid crystal coexistence curve for three members of PE by the method of high-pressure adiabatic compression. We observed two kinds of phase transitions for dimyristoylphosphatidylethanolamine bilayer membrane under high pressure [12]; one is the high-enthalpy transition from the subgel phase to the liquid crystalline phase and the other is the main transition between gel and liquid crystalline phases. The present study demonstrates the pressure effect on the phase transition from the lamellar gel phase to the liquid crystalline phase of dilauroyl- (DLPE), dimyristoyl- (DMPE), dipalmitoyl- (DPPE), and distearoyl-phosphatidylethanol-amine (DSPE) bilayer membranes. And then, we compare the present results for PEs with the previous results for phosphatidylcholines (PCs).

2. EXPERIMENTAL

Synthetic PEs used in this study were abbreviated as follows.
>   DLPE; 1,2-didodecanoyl-*sn*-glycero-3-phosphoethanolamine,
>   DMPE; 1,2-ditetradecanoyl-*sn*-glycero-3-phosphoethanolamine,
>   DPPE; 1,2-dihexadecanoyl-*sn*-glycero-3-phosphoethanolamine,
>   DSPE; 1,2-dioctadecanoyl-*sn*-glycero-3-phosphoethanolamine.

All of the phospholipids were purchased from Sigma, and used without further purification. The phospholipid multilamellar vesicles were prepared by suspending each phospholipid in water at 1 mmol kg^{-1}. Water was distilled twice from dilute alkaline permanganate solution. The phase transition under high pressure was observed by two kinds of optical methods. One is the observation of isothermal barotropic phase transition and the other is the isobaric thermotropic phase transition. The general arrangement of the high-pressure apparatus has been described previously [13]. The sudden change in transmittance accompanying the phase transition of lipid bilayer membranes was followed at 560 nm.

3. RESULTS AND DISCUSSION

3.1. Bilayer phase transition under high pressure

It is well known that the saturated diacylphosphatidylethanolamines having fatty acyl chains of 12 - 18 carbons undergo a high-enthalpy transition upon first heating [14]. This transition has been identified as a conversion from a dehydrated, lamellar crystalline (L_c) phase to a liquid crystalline (L_α) phase. The transition is not reversible. Rather, what happens upon cooling is that the L_α phase converts to a lamellar gel (L_β) phase with hydrocarbon chains orientated perpendicular to the plane of the bilayer. Subsequent heating-cooling cycles show a reversible L_β/L_α transition. In the present study, the L_β/L_α transition, that is, the main transition was observed at various pressures.

The pressure effect on the main-transition temperatures of PE and PC bilayer membranes is shown in Fig. 1. The temperature (T) – pressure (p) phase boundaries between L_β and L_α phases for a series of PEs are drawn in Fig. 1(a). Both results of the isobaric and isothermal phase transition give a simple line on a $T - p$ diagram

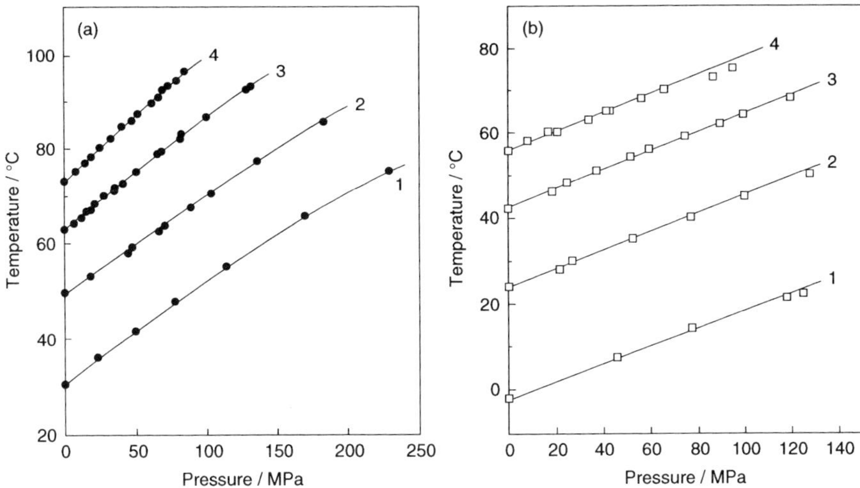

Fig. 1. The effect of pressure on the main-transition temperatures of lipid bilayer membranes. (a) the L_β/L_α transition for phosphatidylethanolanine (PE), (1) DLPE, (2) DMPE, (3) DPPE, (4) DSPE. (b) the P_β'/L_α transition for phosphatidylcholine (PC), (1) DLPC, (2) DMPC, (3) DPPC, (4) DSPC.

for each lipid. The temperatures of the main transition at ambient pressure were in good agreement with those in the literatures [14,15]. As is seen from Fig. 1(a), the $T - p$ curves for the main transition of PEs were best described by a smooth curve and not a linear function. On the other hand, the $T - p$ phase boundaries between ripple gel (P_β') and L_α phases for a series of PCs are shown in Fig. 1(b), which were taken from the previous results [7]. The main-transition temperatures for PC bilayer membranes increased almost linearly with an increase in pressure. The values of slope , dT/dp, at ambient pressure for the main transition of PEs were in the range of $0.230 - 0.275$ K MPa^{-1} depending on the acyl chain-length, which are summarized in Table 1 together with other thermodynamic quantities for the main transition. Russell and Collings [11] reported that the values of dT/dp for the main transition were 0.226 K MPa^{-1} for DLPE, 0.257 KMPa^{-1} for DMPE and 0.293 K MPa^{-1} for DPPE, respectively, which are comparable with the present results. Most significant difference in both series of PE and PC is the main-transition temperature of PE to be higher than that of PC with corresponding acyl chains. Actually there exists the temperature difference of about $20 - 30$ °C.

3.2. Thermodynamic properties of phase transition

The enthalpy change (ΔH) associated with the main transition has been determined by the differential scanning calorimetry. A review of bilayer phase

Table 1
Thermodynamic properties of main transition for bilayer membranes of diacylphosphatidylethanolamines and diacylphosphatidylcholines

Lipid	Transition temperature		dT/dp	ΔH	ΔS	ΔV
	(°C)	(K)	(K MPa⁻¹)	(kJ mol⁻¹)	(J K⁻¹ mol⁻¹)	(cm³ mol⁻¹)
DLPE	30.6	303.8	0.230	17.2	57	13.0
DMPE	49.8	323.0	0.251	25.1	78	19.5
DPPE	63.1	336.3	0.264	34.7	103	27.2
DSPE	73.1	346.3	0.275	44.4	128	35.3
DLPC	-2.1	271.1	0.200	7.5	28	5.5
DMPC	23.9	297.1	0.212	24.7	83	17.6
DPPC	42.0	315.2	0.220	36.4	115	25.4
DSPC	55.6	328.8	0.230	45.2	137	31.6

transitions by Koynova and Caffrey [14] referring the hydrated PEs has appeared. Most reliable data for a series of PE are taken from the literature [15] and shown in Table 1. The chain-length dependence of ΔH in a series of PE is shown to be linear (Fig. 2), indicating that the bulk of the heat of the transition derives from a chain order-to-disorder change. On the other hand, in a series of PC the enthalpy change of main (P_β'/L_α) transition increased with increasing acyl-chain length and the data have been best described by a smooth curve and not a linear function [7]. From the comparison between PE and PC series, however, it may be concluded that the enthalpy changes of the main transition derived from a chain order-to-disorder change are almost the same magnitude for both PE and PC with the same acyl chains except for the dilauroyl-compounds.

The values of entropy change ($\Delta S = \Delta H/T$) associated with the main transition were also shown in Table 1.

The volume change (ΔV) associated with the main transition was calcu-

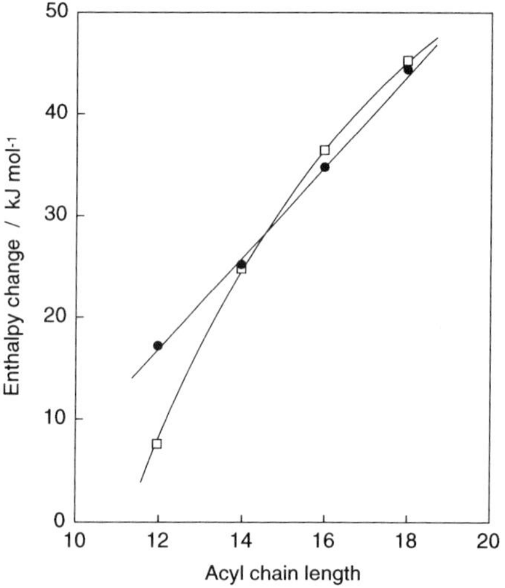

Fig. 2. The enthalpy change for the main transition of PE and PC bilayer membranes as a function of acyl-chain length: (●) PE, (□) PC.

lated from the Clapeyron-Clausius equation

$$\mathrm{d}T/\mathrm{d}p = \Delta V/\Delta S \tag{1}$$

and is also summarized in Table 1. The values of ΔV estimated by Eq. (1) increased linearly with an increase in the acyl-chain length as is shown in Fig. 3(a). Direct measurements of the volume change associated with the main (L_β/L_α) transition of PE bilayers have been reported by the methods of dilatometry [16,17] and differential scanning densitometry [18]. Koynova and Hinz [18] determined the volume change of the main transition (L_β/L_α) for PE series, which showed the volume change increasing with the length of the acyl-chains in the range of $18.8 - 29.9$ cm^3 mol^{-1}. Wilkinson and Nagle [16,17] reported the volume changes of the main transition for DLPE and DMPE bilayer membranes, which showed smaller values than those by Koynova and Hinz. These reported values of ΔV, which are shown in Fig. 3(a), are comparable with the present results estimated from the Clapeyron-Clausius equation. The transition volumes for PE series are compared with those for PC series in Fig. 3(b). The values of ΔV for PCs were taken from our previous data [7]. As is seen from Fig. 3(b), the volume change of the main transition for PE bilayer membranes is slightly larger than that for PC bilayer membranes. The interlipid hydrogen bond formation between PE head groups in the bilayer gel phase, which is mentioned below, may be responsible for this larger volume change associated with the main transition of PE bilayers.

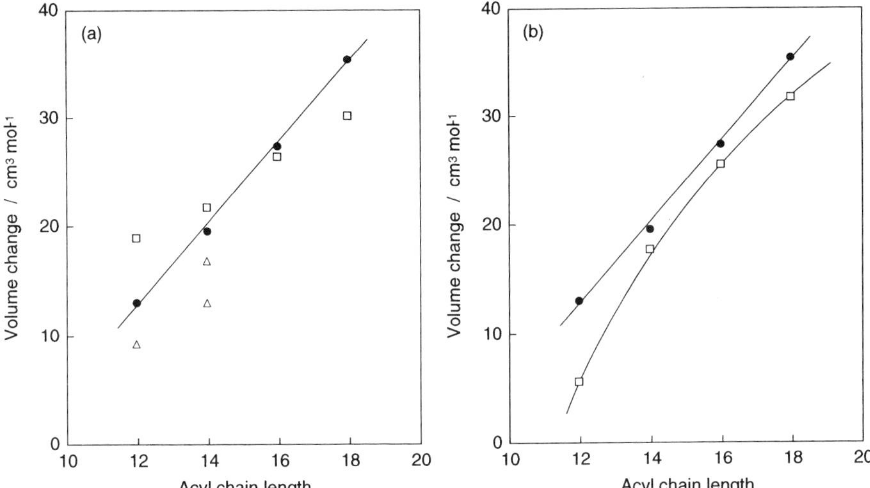

Fig. 3. The volume change for the main transition of PE and PC bilayer membranes as a function of acyl-chain length. (a) PE bilayer membranes; (●) present results, (□) densitometry, (△) dilatometry. (b) comparison between (●) PE and (□) PC.

3.3. Comparison between PE and PC bilayers

Thermodynamic properties for the main transition of a series of PC bilayer membranes [7] are also included in Table 1 for comparison. The enthalpy and entropy changes of the main transition for PEs and PCs with the same acyl chain are similar in magnitude to each other. The volume change of the main transition for PE bilayer membranes is slightly larger than that for PC bilayer membranes. Since the main transition brings about the chain melting in bilayer membranes, a resemblance in thermodynamic quantities of the main transition between PE and PC with the same acyl chain seems to be reasonable. However, the most surprisingly difference between PEs and PCs lies in the temperature of the main transition. The main-transition temperatures for PE membranes are significantly higher than those for PC with the corresponding acyl chain. The difference in the main-transition temperature between PE and PC bilayer membranes amounts to $17.5 - 32.7\ °C$ depending upon the acyl chain-length, and may be attributable to the head group interaction in the bilayer membranes. Figure 4 shows schematic drawing of the difference in the size and electrostatic interaction between PE and PC head-groups in the gel phase of bilayer membranes. Because the choline head-group is bulky, the acyl chains are tilted away from perpendicular. The larger PC head-group prevents chains from coming close enough together to minimize their van der Waals interactions. However, as a results of tilting, the distance between the all-*trans* chains is reduced and the van der Waals interactions are minimized. The chain tilting is responsible for polymorphism in the gel phase of PC bilayer membranes. On the other hand, in the L_β phase, the chains of PE molecules are oriented perpendicular to the plane of the bilayer. One likely explanation for the difference of $17.5 - 32.7\ °C$ in the main-

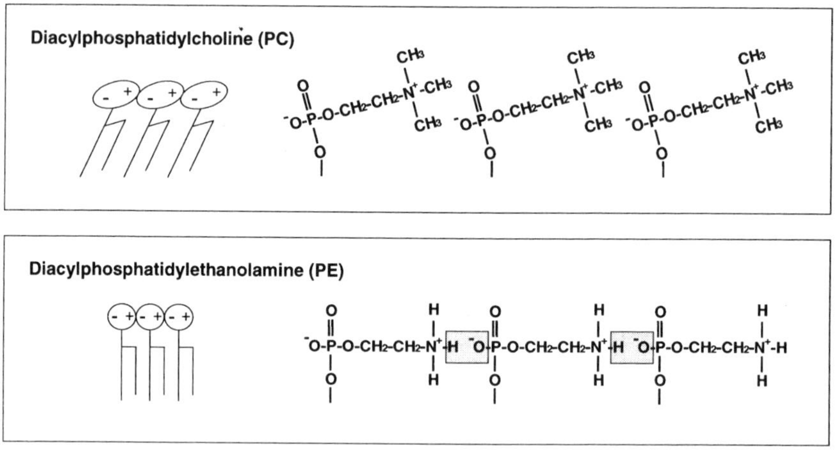

Fig. 4. Schematic drawing for head-group packing of phospholipid bilayer membranes.

transition temperature is the presence transient hydrogen bonding between the protons of the quarternary nitrogen of the PE head group and the phosphate group on adjacent PE head group. Such bonds would inhibit lateral expansion of the bilayer thus raising the temperature of main transition.

Alternative explanation is the hydrogen bond formation between the quaternary nitrogen of the PE head group and the (sn-2 chain) carbonyl group on adjacent PE head group. Lewis and McElhaney [15] measured the carbonyl-group infrared stretching band intensities in the $1650 - 1800$ cm^{-1} region, comparing results for DMPE and DMPC in both the liquid crystalline and gel phases. They found that, while one could account for the DMPC spectrum with two overlapping bands, an understanding of the DMPE band required an additional, third, band centered at $1705 - 1710$ cm^{-1}. Since the carbonyl groups of both PE and PC can take part in hydrogen bonding with water, but only the PE can form interlipid hydrogen bonds involving the carbonyl groups, the third band can be interpreted as a C = O stretch when the carbonyl is involved in hydrogen bonding with a PE moiety. Pink et al. [19] have modeled hydrogen bond formation in phospholipid bilayers formed from lipids with PE head groups. They calculated the fraction of C = O groups taking part in interlipid hydrogen bonding, and showed the satisfactory agreement with experiments carried by Lewis and McElhaney.

4. ACKNOWLEDGMENTS

This study was supported in part by a Grant-in-Aid for Scientific Research (B) (2) (11440206) from Japan Society for the Promotion of Science.

REFERENCES

1. K. T. Wann and A. G. Macdonald, Prog. Neurobiol., 30 (1988) 271.
2. A. R. Cossins and A. G. Macdonald, J. Bioenerg. Biomembr., 21 (1989) 115.
3. C. Balny, R. Hayashi, K. Heremans and P. Masson (eds.), High Pressure and Biotechnology, John Libbey Eurotext Ltd., France, 1992.
4. R. Hayashi and C. Balny (eds.), High Pressure Bioscience and Biotechnology, Elsevier, Amsterdam, 1996.
5. K. Heremans (ed.), High pressure Research in the Biosciences and Biotechnology, Leuven University Press, Belgium, 1997.
6. H. Ludwig (ed.), Advances in High Pressure Bioscience and Biotechnology, Springer, Heidelberg, 1999.\
7. H. Ichimori, T. Hata, T. Yoshioka, H. Matsuki and S. Kaneshina, Chem. Phys. Lipids, 89 (1997) 97.
8. H. Ichimori, H. Matsuki and S. Kaneshina, Chem. Lett. (1998) 75.
9. S. Kaneshina, H. Ichimori, T. Hata and H. Matsuki, Rev. High Pressure Sci. Technol., 7 (1998) 1277.
10. H. Ichimori, T. Hata, H. Matsuki and S. Kaneshina, Biochim. Biophys. Acta, 1414 (1998) 165.
11. N. D. Russell and P. J. Collings, J. Chem. Phys., 77 (1982) 5766.

12. H. Ichimori, S. Endo, T. Hata, H. Matsuki and S. Kaneshina, in: H. Ludwig (ed.) , Advances in High Pressure Bioscience and Biotechnology, Springer, Heiderberg, 1999, p. 149.
13. S. Maruyama, H. Matsuki, H. Ichimori and S. Kaneshina, Chem. Phys. Lipids, 82 (1996) 125.
14. R. Koynova and M. Caffrey, Chem. Phys. Lipids, 69 (1994) 1.
15. R. N. A. H. Lewis and R. N. McElhaney, Biophys. J., 64 (1993) 1081.
16. D. A. Wilkinson and J. F. Nagle, Biochemistry, 20 (1981) 187.
17. D. A. Wilkinson and J. F. Nagle, Biochemistry, 23 (1984) 1538.
18. R. Koynava and H. J. Hintz, Chem. Phys. Lipids, 54 (1990) 67.
19. D. A. Pink, S. McNeil, B. Quinn and M. J. Zuckermann, Biochim. Biophys. Acta, 1368 (1998) 289.

Trends in High Pressure Bioscience and Biotechnology
R. Hayashi (editor)

Effect of deuterium oxide on the phase transitions of phospholipid bilayer membranes under high pressure

H. Ichimori[a], F. Sakano[b], H. Matsuki[b] and S. Kaneshina[b]

[a]Anan National College of Technology, Minobayashi, Anan, Tokushima 774-0017, Japan
[b]Department of Biological Science & Technology, Faculty of Engineering,
The University of Tokushima, Minamijosanjima, Tokushima 770-8506, Japan

The present study demonstrates the substitution effect of hydrogen oxide (H_2O) by deuterium oxide (D_2O) on the phase transitions of dipalmitoylphosphatidylcholine (DPPC) and distearoylphosphatidylcholine (DSPC) bilayer membranes under high pressure. Substitution effect of H_2O by D_2O was as follows. With respect to the main transition from the ripple gel (P_β') phase to the liquid crystalline (L_α) phase for DPPC and DSPC bilayer membranes, there was no significant difference in the main transition temperature in H_2O and D_2O. On the other hand, the pretransition temperature between lamellar gel (L_β') phase and P_β' phase for both lipids was significantly raised by the substitution of H_2O by D_2O. In H_2O, a pressure-induced interdigitated gel ($L_\beta I$) phase has been observed at pressures above 100 MPa for DPPC and 70 MPa for DSPC, respectively. It has been known that the $L_\beta I$ phase of DPPC bilayer in D_2O is observed at higher pressure beyond 150 MPa by the method of neutron diffraction. In the present study, the temperature-pressure phase diagrams for DPPC and DSPC bilayers showed the transition temperature between P_β' and $L_\beta I$ phases to be lowered in D_2O, on the other hand, the transition temperature between L_β' and $L_\beta I$ phases to be raised in D_2O under high pressure. In other words, the pressure for appearance of the $L_\beta I$ phase was elevated by the substitution of H_2O by D_2O. The difference in transition temperatures by the water substitution is discussed by the difference of hydrophobic interaction in H_2O and in D_2O.

1. INTRODUCTION

The phase behavior of diacylphosphatidylcholines bilayer membranes has been most extensively studied by various physical techniques. We have shown the pressure effect on the bilayer transitions of a series of diacylphosphatidylcholines containing linear saturated acyl chains [1,2]. It is well known that the bilayer membrane of diacylphosphatidylcholines having saturated fatty acyl chains of 14 – 18 carbons undergoes the pretransition from the lamellar gel (L_β') phase to the ripple gel (P_β') phase and sequentially the main transition from the P_β' phase to the liquid crystal (L_α) phase. In addition, the pressure-induced interdigitated gel ($L_\beta I$)

phase is known to be observed under high pressures. Braganza and Worcester [3] and Winter and Pilgrim [4] reported the observation of the pressure-induced $L_\beta I$ phase of dipalmitoylphosphatidylcholine (DPPC) bilayer membranes in deutrerium oxide (D2O) by the method of neutron diffraction at high pressure beyound 150 MPa and 160 MPa, respectively, which seems to be a somewhat higher pressure than our results in hydrogen oxide (H2O). In order to make clear the difference in solvent effect between H2O and D2O, we now extend previous measurements and thermodynamic analysis for the bilayer phase-transitions to the systems including D2O solvent. The present study demonstrates the substitution effect of H2O by D2O on the phase transitions of DPPC and DSPC, i.e., distearoylphosphatidyl-choline, bilayer membranes under high pressure.

2. MATERIALS AND METHODS

Synthetic DPPC, 1,2-dipalmitoyl-*sn*-glycero-3-phosphocholine, and DSPC, 1,2-distearoyl-*sn*-glycero-3-phosphocholine, were obtained from Sigma. Multilamellar vesicles were prepared by suspending DPPC or DSPC in the water at 2 mmol kg^{-1} using a Branson model 185 Sonifier at a temperature several degrees above the phase-transition temperature. The phase transitions under high pressure were observed by two kinds of optical methods, i.e., the isothermal barotropic phase transitions and the isobaric thermotropic phase transitions. The general arrangement of the high pressure apparatus and the experimental procedures have been described in detail previously [1,2].

3. RESULTS AND DISCUSSION

A typical measurement of isothermal barotropic phase transition for DSPC bilayer membrane is shown in Fig. 1. The light transmittance of vesicle suspensions in the range of 540 – 580 nm was measured at 65 °C and various pressures (Fig. 1A). The transmittance at 560 nm was depicted in Fig. 1B as a function of pressure. The transmittance-pressure curve shows the sudden decrease of the transmittance accompanying the pressure-induced transition. The phase transition pressure was determined as a function of temperature from the transmittance-pressure profiles at various temperatures.

The temperature (T) – pressure (P) phase diagram of the DPPC bilayer membrane in D2O is shown in Fig. 2. The temperatures of main transition and pretransition at ambient pressure were 42.0 and 36.0 °C, respectively. The temperatures for both transitions increase with increasing pressure. The values of dT/dP for the main transition and pretransition were 0.220 and 0.13 K MPa^{-1}, respectively. With respect to the main transition of DPPC bilayer membrane in D2O, Winter and Pilgrim [4] and Braganza and Worcester [3] reported the value of dT/dP to be 0.22 and 0.23 K MPa^{-1}, respectively, which are in good agreement with the present result. However, they could not observe the pretransition because of the small-angle neutron scattering method under high pressure. As is seen from Fig. 2, we observed the interdigitation of DPPC bilayer membrane in D2O; a triple point on the phase diagram among $P_\beta{}'$, $L_\beta{}'$ and $L_\beta I$ phases was found at 160 MPa and 56 °C. The observation of the $L_\beta I$ phase by Winter and Pilgrim [4] is also shown in Fig. 2, which is in good agreement with the present results. Braganza and Worcester [3] reported also the observation of the $L_\beta I$ phase in D_2O at high pressure beyound 150 MPa.

Fig. 3 shows the $T-P$ phase diagram of DPPC bilayer membrane in H_2O, which has been reported in our previous paper [5]. The temperatures of main transition and pretransition at ambient pressure were 42.0 and 34.3 °C, respectively. Regarding the main transition of DPPC bilayer membrane, there was no significant difference in the temperature of transition and the value of dT/dP in between two solvents, H_2O and D_2O. On the other hand, the pretransition temperature was

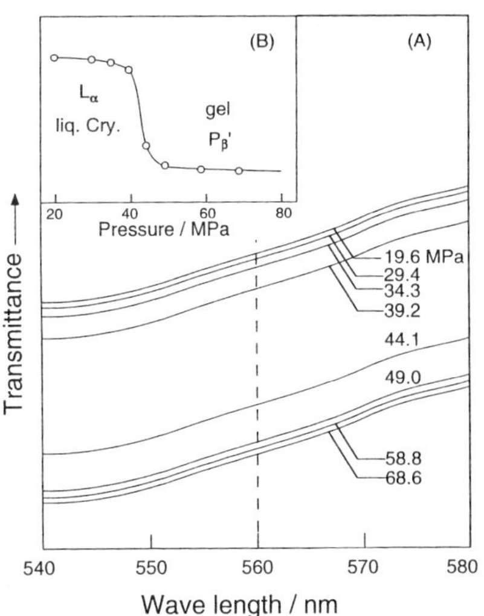

Fig. 1. Isothermal barotropic phase transition of DSPC bilayer membrane in H_2O at 65 °C.

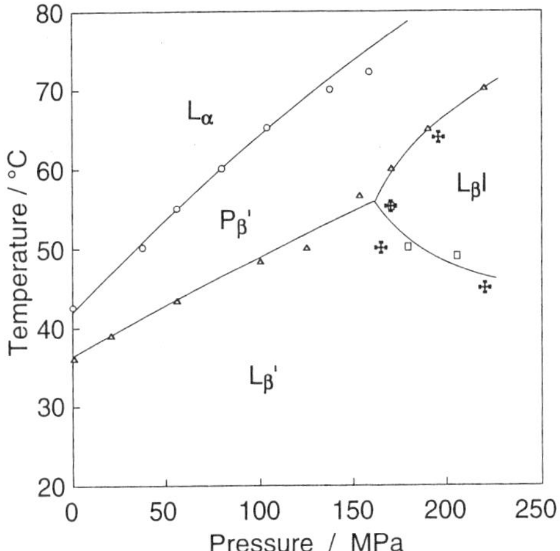

Fig. 2. Phase diagram of DPPC bilayer membranes in D_2O. Date by Winter and Pilgrim are also shown (✠).

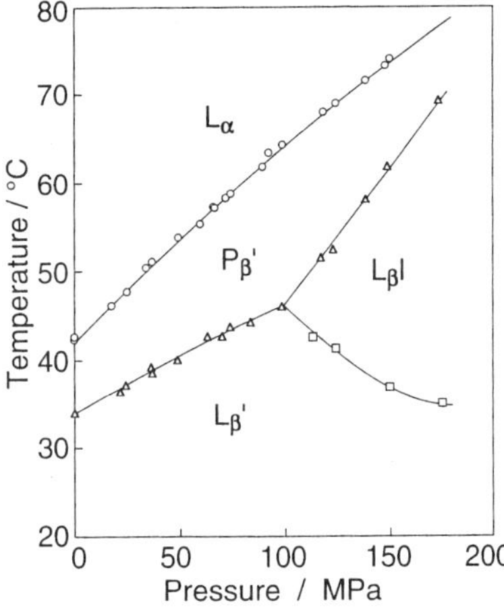

Fig. 3. Phase diagram of DPPC bilayer membranes in H_2O.

150

significantly raised by the substitution of H_2O by D_2O. The interdigitation in H_2O has been observed at pressure above 100 MPa; a triple point on the phase diagram among P_β', L_β' and $L_\beta I$ phases was found at 100 MPa and 45 °C. It is known from the comparison between Fig. 2 and Fig. 3 that the pressure-induced interdigitation of DPPC bilayer membrane is observed at higher pressure in D_2O than in H_2O.

The $T-P$ phase diagrams of the DSPC bilayer membrane in D_2O and in H_2O are shown in Figs. 4 and 5, respectively. We recognized no difference of the main-transition temperature (55.6 °C at ambient pressure) in both solvents, H_2O and D_2O. On the other hand, the pretransition temperatures in D_2O and in H_2O were 52.8 °C and 50.9 °C, respectively. The substitution of H_2O by D_2O brought about an obvious raise in temperature of the pretransition. The triple point on the phase diagram among P_β', L_β' and $L_\beta I$ phases was found at 90 MPa and 65 °C in D_2O, and at 70 MPa and 59 °C in H_2O. Braganza and Worcester [3] reported the appearance of the $L_\beta I$ phase beyond 90 MPa by the neutron diffraction method, which are comparable with the present result. It may be concluded that the pressure, at which the pressure-induced $L_\beta I$ phase is observed, is elevated by the solvent substitution of H_2O by D_2O.

Consequently, the substitution of H_2O by D_2O does not affect the main-transition temperature of DPPC and DSPC bilayer membranes. Contrarily, the temperatures of phase transition between gel phases, i.e., L_β'/P_β', $L_\beta'/L_\beta I$ and $L_\beta I/P_\beta'$, were significantly affected by the substitution of H_2O by D_2O. Judging from the present

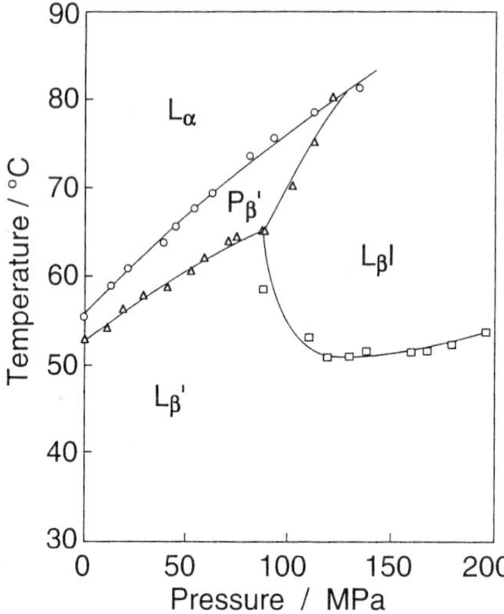

Fig. 4. Phase diagram of DSPC bilayer membranes in D_2O.

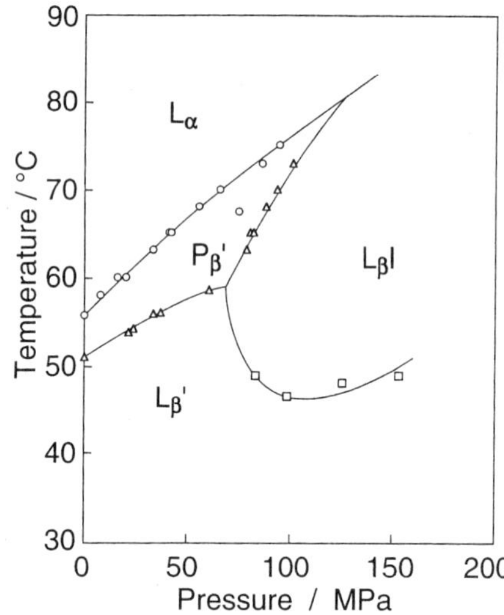

Fig. 5. Phase diagram of DSPC bilayer membranes in H_2O.

situation, we may say that the water substitution is effective at the bilayer-water interface rather than at the hydrocarbon-chain domain.

The hydrophobic interaction among the phospholipid molecules, which is the driving force of membrane formation, is primary based on hydrogen bonds of water molecules. The energy required to break one mole of hydrogen bonds in the liquid has been estimated to be 5.52 kJ mol^{-1} and that of deuterium bonds has been found to be 6.53 kJ mol^{-1}. The difference in the transition temperatures of lipid bilayers by the water substitution is discussed by the difference of hydrophobic interaction in H_2O and in D_2O, in other words, by the difference in hydrogen bonding between H_2O and D_2O. Now we consider the interactions among the head groups, which include some effects of aqueous medium due to hydration and hydrophobic interactions. Interaction free energy between lipid molecules in bilayer is principally composed of an attractive interaction arising from hydrophobic force or interfacial tension and a repulsive interaction arising from electrostatic head-group repulsion and steric repulsions. Thus the interfacial free energy (μ) per molecule in bilayer is represented by

$$\mu = \gamma \cdot A + C/A \tag{1}$$

where γ is an interfacial free energy per unit area characteristic of a liquid hydrocarbon-water interface, A is the molecular area of lipid in bilayer and C is free energy constant [6]. The minimum free energy is given when the molecular area is

$$A = A_0 = \sqrt{C/\gamma} \tag{2}$$

The value of γ is greater in D_2O than in H_2O, which indicates that the optimal surface area occupied by a phosphatidylcholine molecule on the membrane interface is smaller in D_2O than in H_2O. Molecular area of lipid in the bilayer gel phases increases in the order of $L_\beta' < P_\beta' < L_\beta I$. Since greater value of γ in D_2O brings about the decrease of A_0, we may explain qualitatively that the substitution of H_2O by D_2O brings about the elevation of temperature of the $L_\beta' \rightarrow P_\beta'$ and $L_\beta' \rightarrow L_\beta I$ transitions and the depression of temperature of the $L_\beta I \rightarrow P_\beta'$ transition. Consequently, the region of $L_\beta I$ phase on the phase diagram is shifted to the direction of higher pressure. Ohki [7] reported the thermotropic phase transition of DPPC and dihexadecylphosphatidylcholine bilayers in H_2O and D_2O by the method of DSC. His data, which are restricted to measurements at ambient pressure, are in good agreement with the present results.

4. ACKNOWLEDGMENTS

This study was supported in part by a Grant-in-Aid for Scientific Research (B) (2) (11440206) from Japan Society for the Promotion of Science.

REFERENCES

1. H. Ichimori, T. Hata, T. Yoshioka, H. Matsuki and S. Kaneshina, Chem. Phys. Lipids 89 (1997) 97.
2. H. Ichimori, T. Hata, H. Matsuki and S. Kaneshina, Biochim. Biophys. Acta 1414 (1998) 165.
3. L. F. Braganza and D. L. Worcester, Biochemistry 25 (1986) 2591.
4. R. Winter and W. C. Pilgrim, Ber. Bunsenges. Phys. Chem. 93 (1989) 708.
5. S. Maruyama, H. Matsuki, H. Ichimori and S. Kaneshina, Chem. Phys. Lipids 82 (1996) 125.
6. J. N. Israelachivili, D. J. Mitchell and B. W. Ninham, Biochim. Biophys. Acta 470 (1977) 185.
7. K. Ohki, Biochem. Biophys. Res. Commun. 174 (1991) 102.

Trends in High Pressure Bioscience and Biotechnology
R. Hayashi (editor)
                                        153

Effect of pressure on the bilayer phase transitions of N-methylated dipalmitoylphosphatidylethanolamines

Hitoshi Matsuki[a], Masataka Kusube[a], Hayato Ichimori[b] and Shoji Kaneshina[a]

[a]Department of Biological Science and Technology, Faculty of Engineering,
The University of Tokushima, Minamijosanjima, Tokushima 770-8506, Japan
[b]Anan National College of Technology, Minobayashi, Anan, Tokushima 774-0017,
Japan

The main-transition temperatures between gel and liquid crystalline phases for dipalmitoylphosphatidylmonomethylethanolamine (DPMePE) and dipalmitoylphosphatidyldimethylethanolamine (DPMe$_2$PE) under high pressure were determined from differential scanning calorimetry (DSC) and optical methods. The thermodynamic quantities of main transition (enthalpy, entropy, and volume) were calculated and the results were compared with those of dipalmitoylphosphatidylethanolamine (DPPE) and dipalmitoylphosphatidylcholine (DPPC) to clarify the effect of N-methylation of DPPE on the main transition. The main-transition temperatures of N-methylated DPPEs increased linearly by applying pressure while they decreased with increasing the size of head group by stepwise N-methylation. On the other hand, there was no significant difference in thermodynamic quantities between the phospholipids. The difference in hydrophilic interactions between the different-sized polar head groups produces a great difference on the main-transition temperature and the packing arrangement of bilayer membrane in the gel phase.

1. INTRODUCTION

There are many kinds of phospholipids by various combinations of hydrophobic aliphatic groups and hydrophilic polar head groups which link with glycerol skeleton. The difference in polar head groups among phospholipids greatly affects the geometrical structures of bilayer membranes. It is known in an erythrocyte membrane that outer membranes are more abundant in phospholipids with a large head group such as diacylphosphatidylcholine (PC) while inner membranes with a small head group such as diacylphosphatidylethanolamine (PE). The phase behavior and membrane properties of PC and PE bilayers have been examined and compared by several researchers [1-8]. In our recent studies, the pressure effect on the bilayer phase transitions of dipalmitoylphosphatidylcholine (DPPC) and dipalmitoylphosphatidylethanolamine (DPPE) multilamellar vesicles was elucidated [9-11]. The present study concerned with the main transitions from the gel phase to the liquid crystalline phase of N-methylated DPPEs, dipalmitoylphosphatidylmonomethylethanolamine (DPMePE) and dipalmitoylphosphatidyldimethylethanolamine (DPMe$_2$PE), under high pressure. By comparing the results with those of DPPC and DPPE, the influence of different-sized polar head groups among N-methylated DPPEs on the main transition was considered.

2. EXPERIMENTAL

DPMePE (1,2-dipalmitoyl-sn-glycero-3-phospho-N-methylethanolamine) and DPMe$_2$PE (1,2-dipalmitoyl-sn-glycero-3-phospho-N,N-dimethylethanolamine) were purchased from Avanti Polar Lipids and used as received. Water was distilled twice from dilute alkaline permanganate solution. The phospholipid multilamellar vesicles were prepared by suspending each phospholipid in water at 1 mmol kg^{-1} using a vortex mixer. The suspension was sonicated for a couple of times at a temperature several degrees above the main transition. The phase transitions of lipid bilayer membranes under atmospheric pressure were observed by a Micro Cal MCS high-sensitivity differential scanning calorimeter with heating rate of 0.33 K min^{-1}. The phase transitions under high pressure were observed by two methods of high-pressure light transmittance: isothermal barotropic and isobaric thermotropic phase-transition methods. The high-pressure measurements were explained in detail elsewhere [9,12]. The transmittance changes accompanying the phase transitions were pursued at 560 nm.

3. RESULTS AND DISCUSSION

We have shown in the previous study [9-11] that both DPPC and DPPE bilayer membranes undergo main transitions at temperatures of 42.0 and 63.1 °C under atmospheric pressure, respectively. At the main-transition temperature, the membrane state of DPPC varies from the ripple gel (P$_\beta$') phase to the liquid crystalline (L$_\alpha$) phase while that of DPPE does from the lamellar gel (L$_\beta$) phase, which is metastable, to the L$_\alpha$ phase. Figure 1 demonstrates the heating DSC thermograms of DPMePE and DPMe$_2$PE bilayer membranes in the temperature range near the main-transition temperatures of DPPC and DPPE under atmospheric pressure. In the figure are also included the results of DPPE and DPPC. The thermograms of DPMePE and DPMe$_2$PE bilayer membranes exhibited similar high-enthalpy transitions to DPPC and DPPE at 58.6 and 50.0 °C, respectively. The transition temperatures were in good agreement with those reported by Casal and Mantsch [13]. They identified from the DSC and FT-IR study that the transitions corresponded to the main transition from the gel phase to the liquid crystalline phase. Comparing the main-transition temperatures among N-methylated DPPEs, it is understandable that the transition temperature decreased in the order of head-group size: DPPE > DPMePE > DPMe$_2$PE > DPPC. The enthalpy changes of main transition (ΔH) were obtained from the peak areas of DSC thermograms in Fig. 1 and the corresponding entropy changes (ΔS) were calculated by the equation

$$\Delta S = \Delta H/T \tag{1}$$

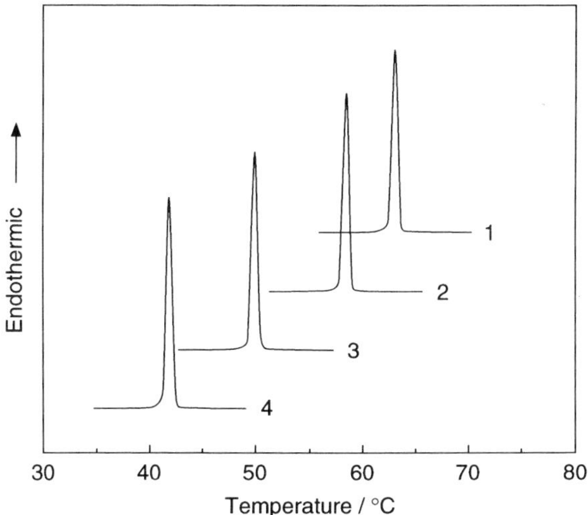

Fig. 1. DSC thermograms of *N*-methylated DPPE bilayer membranes: (1) DPPE, (2) DPMePE, (3) DPMe$_2$PE, (4) DPPC.

The values of ΔH and ΔS for *N*-methylated DPPEs are summarized in Table 1. In contrast to the transition temperature, the ΔH and ΔS values for *N*-methylated DPPEs had almost the same values as each other.

The main-transition temperatures of DPMePE and DPMe$_2$PE bilayer membranes under high pressure were investigated by two light-transmittance measurements. A typical example of isothermal barotropic measurements for the DPMe$_2$PE solution is depicted in Fig. 2. The measured transmittance in the wavelength range from 540 nm to 580 nm at constant temperature decreased with increasing pressure. The change of transmittance with applying pressure was pursed from the

Table 1

Thermodynamic properties of main transition for bilayer membranes of *N*-methylated dipalmitoylphosphatidylethanolamines

Lipid	Transition temp.		dT/dp	ΔH	ΔS	ΔV
	(°C)	(K)	(K MPa^{-1})	(kJ mol^{-1})	(J K^{-1} mol^{-1})	(cm^3 mol^{-1})
DPPE	63.1	336.3	0.264	34.6	103	27.2
DPMePE	58.6	331.8	0.246	35.5	107	26.3
DPMe$_2$PE	50.0	323.2	0.230	35.5	110	25.3
DPPC	42.0	315.2	0.220	36.4	115	25.3

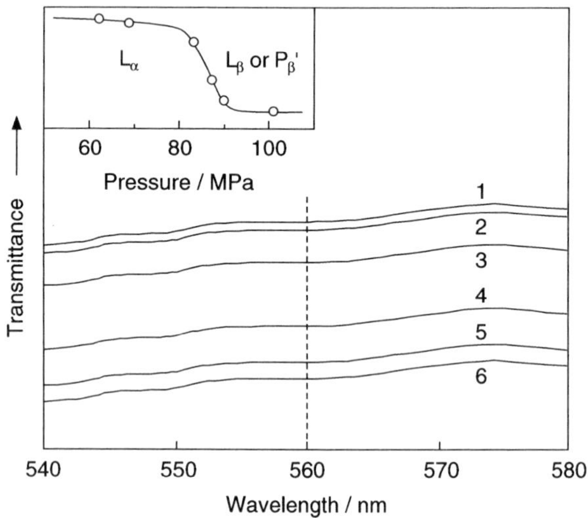

Fig. 2. Transmittance vs. wavelength curves of DPMe$_2$PE bilayer membrane at 70 °C under various pressures: (1) P = 63 MPa, (2) 69, (3) 84, (4) 87, (5) 90, (6) 101. Inset in the figure indicates the isothermal barotropic phase transition curve of DPMe$_2$PE bilayer membrane at 560 nm and 70 °C.

transmittance vs. pressure curve which was constructed by reading the transmittance values at 560 nm. The transmittance abruptly decreased at a certain pressure as seen in inset in Fig. 2. We determined the pressure as the main-transition pressure at constant temperature. Alternatively, the transition temperature at constant pressure was determined from the isobaric thermotropic measurements. The pressure dependence of main-transition temperatures, i. e., temperature (T) – pressure (p) phase boundaries, for DPMePE and DPMe$_2$PE bilayer membranes together with that for DPPC and DPPE is shown in Fig. 3. The transition temperatures of all N-methylated DPPEs increased with an increase in pressure and the variation of the $T - p$ curves was nearly linear. The slopes of the curves (dT/dp) under atmospheric pressure are also given in Table 1. The volume change of main transition (ΔV) for DPMePE and DPMe$_2$PE bilayer membranes can be evaluated by substituting T, ΔH, dT/dp values into the Clapeyron-Clausius equation

$$\mathrm{d}T/\mathrm{d}p = T\Delta V/\Delta H \tag{2}$$

The ΔV values are compared with those of DPPC and DPPE in Table 1. The ΔV values for N-methylated DPPEs were not appreciably different from each other. The effect of head-group size on ΔV was very small as seen in the ΔH and ΔS values.

The striking similarity in thermodynamic quantities of the main transition among

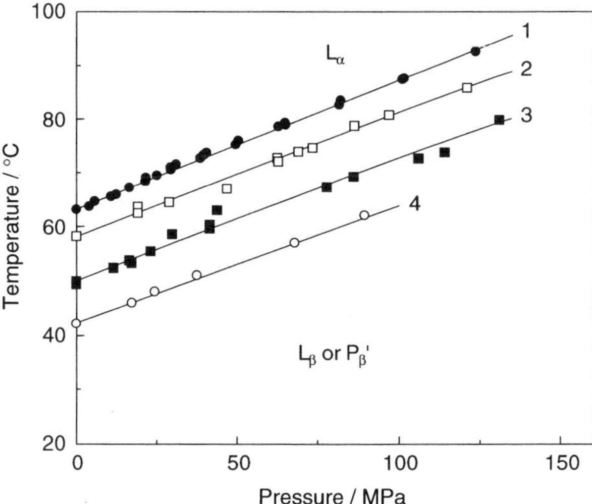

Fig. 3. Pressure dependence on the main-transition temperature of N-methylated DPPE bilayer membranes: (1) DPPE, (2) DPMePE, (3) DPMe$_2$PE, (4) DPPC.

N-methylated DPPEs seems to be attributable to the same hydrophobic part in the molecules. Because the main transition occurs from the *trans-gauche* conformation change of hydrocarbon chain due to the chain melting in bilayer membranes, it is expected that the N-methylated DPPE molecules with the same dipalmitoyl-chain mutually occur similar chain melting in the membrane. Accordingly, they show similar thermodynamic behavior of the main transition. On the other hand, there exists a significant difference in the main-transition temperatures among N-methylated DPPEs. It was suggested from the previous study [11,14] that the difference in the main-transition temperatures between PC and PE bilayer membranes might be caused by that in the polar head group interaction in the gel phases. The acyl chains of PC bilayer membranes in the P$_\beta$' phase are tilted away from the perpendicular because the larger choline head group interferes with close packing pattern with respect to bilayer normal. Whereas, the acyl chains of PE bilayer membranes in the L$_\beta$ phase are oriented perpendicular to the bilayer plane. It is conceivable that transient hydrogen bonding between the protons of the quaternary nitrogen of the PE head-group and the phosphate group on adjacent PE head group imposes a greater constraint in the bilayer compared with the PC bilayer [11]. The constraint raises main-transition temperature of PE bilayer membranes pronouncedly. In the case of DPMePE and DPMe$_2$PE, they have intermediate transition temperatures between DPPC and DPPE and the transition temperatures decrease stepwise with increasing the size of head group by N-methylation. With increasing the number of methyl groups in the head group, the cross-sectional area of the molecules increases and the interchain distances expand. The interactions between head groups in the polar region of the bilayer may be influenced by the different hydrogen bonding capabilities of N-methylated groups. Such hydrogen bonding between head groups in the gel phase may be weakened with N-methylation of head group. Further, the interactions between acyl chains are also influenced and the chains tend toward tilting with N-methylation. Although the gel phases of both phospholipids have not yet been assigned either the $P_{\beta'}$ or the L_{β} ones, Casal and Mantsch [13] described that the temperature dependence of the frequencies and bandwidths of the CH_2 scissoring modes suggests the possibilities of acyl-chain tilting in DPMePE and $DPMe_2PE$ bilayer membranes. The present results evidently show that the different-sized polar head groups of phospholipids produce a greater influence on the main-transition temperature and the packing arrangement of bilayer membranes, rather than the thermodynamic quantities of the transition.

4. ACKNOWLEDGMENTS

This study was supported in part by a Grant-in-Aid for Scientific Research (B) (2) (11440206) from Japan Society for the Promotion of Science.

REFERENCES

1. T. J. McIntosh, Biophys. J., 29 (1980) 237.
2. D. A. Wilkinson and J. F. Nagle, Biochemistry, 20 (1981) 187.
3. H. H. Mantsch, S. C. His, K. W. Butler and D. G. Cameron, Biochim. Biophys. Acta, 728 (1983) 325.
4. R. Koynova and H. J. Hinz, Chem. Phys. Lipids, 54 (1990) 67.
5. R. Koynova, A. Koumanov and B. Tenchov, Biochim. Biophys. Acta, 1285 (1996) 101.
6. M. Kodama and T. Miyata, Colloids Surfaces A: Physicochem. Eng. Aspect, 109 (1996) 283.
7. H. Aoki and M. Kodama, Thermochim. Acta, 308 (1998) 77.
8. M. Kodama and H. Aoki, Netsu Sokutei, 27 (2000) 19.
9. H. Ichimori, T. Hata, T. Yoshioka, H. Matsuki and S. Kaneshina, Chem. Phys. Lipids, 89 (1997) 97.
10. H. Ichimori, T. Hata, H. Matsuki and S. Kaneshina, Biochim. Biophys. Acta, 1414 (1998) 165.
11. S. Kaneshina, S. Endo, H. Matsuki and H. Ichimori, in : R. Hayashi (ed.), Trends in High Pressure Bioscience and Biotechnology, Elsevier, Amsterdam, 2001, in press.
12. S. Maruyama, H. Matsuki, H. Ichimori and S. Kaneshina, Chem. Phys. Lipids, 82 (1996) 125.
13. H. L. Casal and H. H. Mantsch, Biochim. Biophys. Acta, 735 (1983) 387.
14. H. Ichimori, S. Endo, T. Hata, H. Matsuki and S. Kaneshina, in: H. Ludwig (ed.), Advances in High Pressure Bioscience and Biotechnology, Springer, Heiderberg, 1999, p. 149.

Trends in High Pressure Bioscience and Biotechnology
R. Hayashi (editor)
© 2002 Elsevier Science B.V. All rights reserved.

Enzymes in membrane-like surfactant-based media: perspectives for pressure regulation

N.L. Klyachko[1*], P.A. Levashov[1], R. Köhling[3], J. Woenckhaus[3], C. Balny[2], R. Winter[3], and A.V. Levashov[1]

[1]Department of Chemical Enzymology, Faculty of Chemistry, Moscow State University, 119899 Moscow, Russia
[2]INSERM U 128, IFR 24, 1919, route de Mende, 34293 Montpellier, France
[3]University of Dortmund, Department of Chemistry, Otto-Hahn-Str. 6, D-44221 Dortmund, Germany

Many enzymes *in vivo* function in a tight contact with biological membranes. Their catalytic activity depends on the nature of membrane microenvironment. One of the factors that can play a key role is the lipid structure and phase transitions occuring at different conditions. The ternary surfactant-water-organic solvent systems were proposed as a medium for modeling the membrane environment of enzymes. The surfactant molecules in these systems (including membrane lipids) can form aggregates of different structure. Enzymes of various origins were solubilized in such systems and were found to function in different mesophases: reverse micellar, lamellar, reverse hexagonal, cubic, etc. Phase transitions in the ternary system resulting from the change of the concentration ratio can cause dramatic changes in the enzyme catalytic activity. Pressure was found to be one of the powerful factors in the enzyme structure and activity regulation. The paper discusses perspectives of using pressure as a new factor in enzyme regulation because of its participation in structural rearrangements in the ternary surfactant-based system.

1. INTRODUCTION

It has become clear now that biological membranes provides not only the barrier function but play a key role in all the biochemical processes being in many cases a versatile regulator of biocatalytic processes in the cell by influencing properties of intracellular enzymes (for reviews see, for example [1,2]). Many enzymes, being separated from biomembranes, partially or fully lose their catalytic activity, as well as substrate specificity and stability. But even those enzymes that usually regarded, as unbound (freely diffusing) cannot avoid contacts with the "water/organic medium" interface acting mostly on or near such interface [1,2]. Many concepts of structural models of biological membranes including the most fruitful one, formulated by Singer

* This work was supported in part by COST Chemistry Program, Projects D10/0003/98 and D10/0004/98

160

and Nicolson in their fluid mosaic model, based on the bilayer structural organization of lipids [3]. However, numerous lipids found in biomembranes have molecular configurations which favor the formation of nonbilayer patterns, and their unique property to form aggregates of various structures (lipid polymorphism) plays an important role in biological membranes [4,5]. Local structural transitions between lamellar and hexagonal lipid phases were first discovered long ago [6]. It was shown that in membranes lipids can form cylinders packed in a hexagonal pattern, in which polar head groups of lipid molecules line the walls of narrow water channels. Such structural rearrangements were found to influence the catalytic activity of some enzymes entrapped in proteoliposomes [7,8]. In recent years, other types of nonbilayer lipid structures have been discovered, such as the cubic phase and lipidic particles representing lipid reverse micelles sandwiched between monolayers of the lipid bilayer (see, for example [4]). To study the effect of the lipid polymorphism on enzymes *in vivo* is difficult because the structure of the living matter is a very complex one with numerous interrelated processes occurring simultaneously. Thus, to create systems modeling the conditions of the enzyme microenvironment in biological membranes is the necessity in the enzyme study.

2. TERNARY SYSTEMS: SURFACTANT-WATER-ORGANIC SOLVENT

One of the promising approaches is based on the solubilization of enzymes in microheterogeneous media of the surfactant-water-organic solvent type. At present, hundreds of phase diagrams for such systems have been described; two widely used ones are presented in Fig. 1. As seen, depending on concentrations of the components, various types of structures can be generated in such systems, including normal (L_1) and reverse (L_2) micelles, as well as different liquid crystalline structures, such as lamellar (D), hexagonal (F), cubic (I'_2) phases. The systems differ from sonicated proteolipid suspensions in water (proteoliposomes), widely employed for modeling membrane processes.

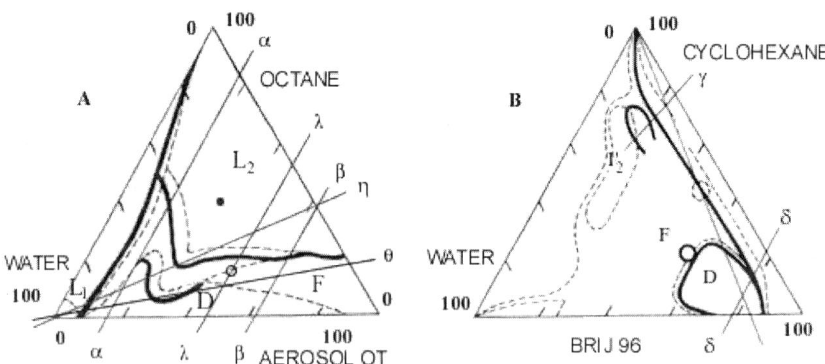

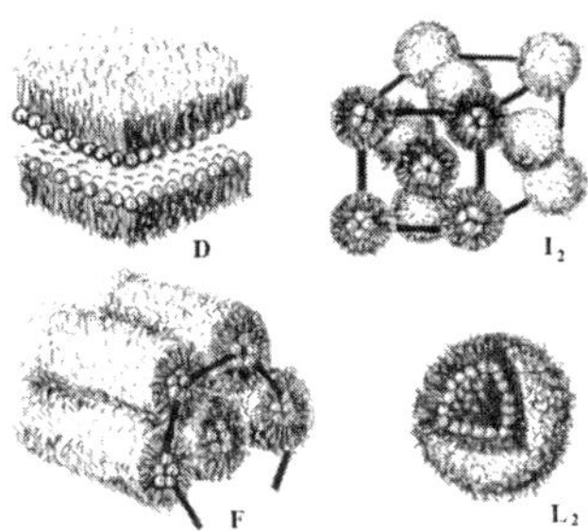

Fig. 1. Phase diagrams of ternary systems Aerosol-OT-water-octane (A) and Brij 96-water-cyclohexane (B), and schematic representation of reverse micelles (L_2), lamellar (D), reverse hexagonal (F) and cubic (I_2) phases [10].

The most important features of such systems: they form spontaneously (no sonication is necessary) by mixing the components; they are thermodynamically stable systems (equilibrium is easy to reach, hence, the phase transitions are reversible); they are optically transparent (convenient spectroscopy can be used). Phase boundaries can be easily found from the change in conductivity, viscosity and/or optical isotropy. As an example, the dependence of the conductivity on water/Brij 96 molar ratio is shown in Fig. 2 (the bottom part). As seen, the transition from isotropic dynamic reverse micelles to viscous liquid crystalline lamellar or reverse hexagonal phases is accompanied by a sharp increase in conductivity.

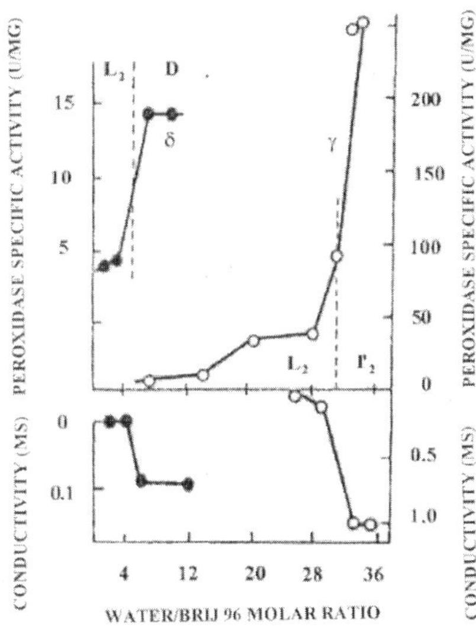

Fig. 2. The effect of phase transitions in the system Brij 96-water-cyclohexane on the peroxidase catalytic activity (top) and the conductivity (bottom). The changes in the system follow cross-sections γ ($L_2 \to I_2$) and δ ($L_2 \to D$) in Fig.1b.

3. ENZYME REGULATION BY MATRIX STRUCTURAL REARRANGEMENTS

Transitions from one phase to another can be followed at fixed concentrations of the surfactant (along the cross-sections á, â, ë or ä in the phase diagram in Fig.1), at fixed surfactant/organic solvent ratio (cross-sections ã, ç, è) or at fixed surfactant/water ratio (cross-section å). These transitions were shown to be a significant regulatory factor for many enzymes studied in such systems (for a review see, for example [10]). The observed changes in the catalytic activity of horseradish peroxidase and horse liver alcohol dehydrogenase are presented in Figs. 2 and 3. As seen, in some cases the difference in activity upon phase transition can exceed 10 times. Moreover, since the size of the surfactant aggregate can be in the same range, as the size of the protein molecule, the geometric factor can play an important role in the enzyme regulation. The special attention should be paid to non-spherical enzymes. Regulation of the catalytic activity by structural rearrangements is demonstrated by the example of alcohol dehydrogenase (Fig. 3.). Its molecule represents an ellipsoid with dimensions 45x55x110 Å, and the maximal activity can be observed in reverse hexagonal phase under conditions when the diameter of inner aqueous channels of cylinders formed by Aerosol OT molecules is equal to the middle axis of the enzyme molecule (55 Å). In this case, the coincidence between the surfactant matrix and the enzyme molecule is realized to the fullest extent.

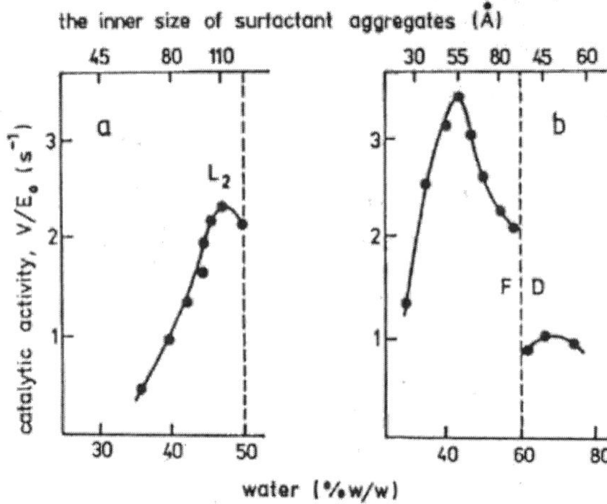

Fig. 3. Regulation of the horse liver alcohol dehydrogenase catalytic activity in the system Aerosol OT-water-octane with micellar (L2), lamellar (D) and reverse hexagonal (F) packing of surfactant molecules; (a) - cross-section η in Fig. 1; (b) - cross-section θ in Fig. 1.

4. THE ROLE OF PRESSURE IN THE ENZYME REGULATION IN TERNARY SURFACTANT-WATER-ORGANIC SOLVENT SYSTEMS

It is clear now that pressure can be a powerful factor in enzyme regulation (for reviews see, for example, [11,12] and refs. therein). It has been demonstrated recently that the application of hydrostatic pressure to enzymes placed in nano-containers or restricted compartments, such as reverse micelles, can bring an additional advantages for both increasing the enzyme stability and modulating the enzyme activity. The importance of protein-micelle or/and protein-protein contacts was clearly demonstrated, using α-chymotrypsin, peroxidase, malic dehydrogenase and some other enzymes, no such an effect was revealed in aqueous solution (see review [12] and refs. therein). An example of the pressure effect on a functioning bi-enzymic complex in reverse micelles of Aerosol OT in octane, tetrameric rabbit muscle lactic dehydrogenase (LDH) and D-glyceraldehyde-3-phosphate dehydrogenase (GPDH), is presented in Fig. 4. A pronounced effect of GPDH activation and stabilization upon pressure application can be observed only in the presence of LDH when the complex between GPDH dimer and LDH tetramer is formed in reverse micelles. Indeed, both enzymes are involved in the system of glycolysis possessing a natural potential in a complex formation. As seen from Fig. 4, the catalytic activity of GPDH in the presence of LDH enhances with increasing pressure, being always higher than that in the absence of LDH. Whereas at 400 bar the GPDH activity decreases dramatically, the GPDH-LDH complex remains active. The difference between two activities becomes 10 times.

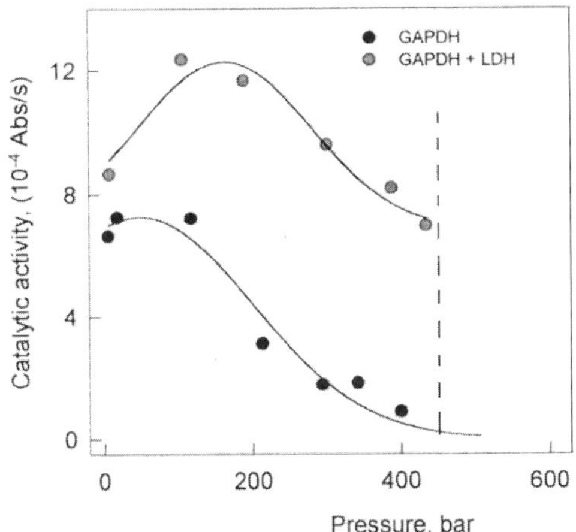

Fig. 4. Pressure dependence of the catalytic activity of glyceraldehyde-3-phosphate dehydrogenase (GPDH) in the presence (shadowed circles) and absence (black circles) of lactic dehydrogenase (LDH) in reverse micelles of 0.1 M Aerosol OT in octane at water/surfactant molar ratio equal to 36.
The vertical dashed line shows the pressure limit after which a phase transition occurs in the system.

One more interesting feature has to be pointed out. The experiment with the GPDH-LDH bi-enzymic system was carried out at hydration degree 36 [13], which is far enough from the phase boundaries, as it can be seen from Fig. 1. However, application of very moderate pressure causes a phase transition in the system followed by the turbidity increasing (vertical dashed line in Fig. 4). The structural rearrangements in the Aerosol OT-water-octane system under pressure were also registered at different surfactant concentrations and water contents by using FTIR spectroscopy or even visually in light microscope [14].

164

A particular study of the phenomenon observed was carried out using the small-angle neutron scattering (SANS) technique [15]. The SANS profiles were obtained along different cross-sections in Fig. 1a following the phase diagram of the Aerosol OT-water-octane system. By applying pressure to the system we found that pressure can induce the appearance of a new phase. As an example, Fig. 5 shows the SANS profiles for two of the samples (black and white circles in Fig. 1a) at different pressures. A peak at lower Q corresponds to the dense water-in-oil droplet structure (micellar phase L_2) and the peak at higher Q represents the lamellar structure. As seen, the peak position at lower Q shifted with increasing pressure and the peak height decreased; subsequently, the peak at higher Q grew above 1.5 kbar (Fig. 5a) and 1 kbar (Fig. 5b), respectively. At some points of the phase diagram the transition occurs at much lower pressures (less than 1 kbar). Thus, the experiment reveals that pressure, inducing the phase transitions in the membrane-like surfactant based system, could be a promising additional factor in enzyme regulation.

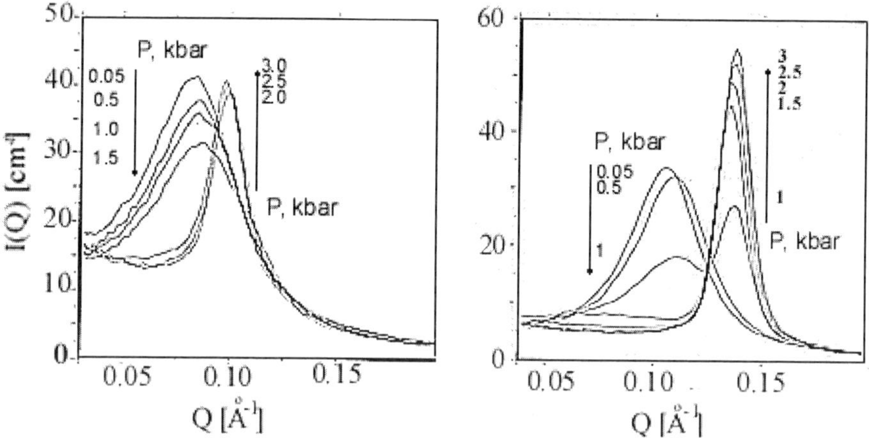

Fig. 5. SANS profiles for the samples shown in black (a) and white (b) circles in Fig. 1a at different pressures. Curves are shifting in the direction of pressure changes (shown as arrows).

ACKNOWLEDGMENTS

N.L. Klyachko would like to express her gratitude to Mr. S. Shipovskov for the help in the preparation of the figures.

REFERENCES

1. A.H. Maddy, ed., Biochemical Analysis of Membranes, Chapman and Hall, New York (1976).
2. C.J.Masters, CRC Crit. Rev. Biochem., 11 (1981) 105-143.
3. S.J. Singer, G.L. Nicholson, Science, 175 (1972) 720-731.
4. B. de Kruijff, Nature, 329 (1987) 587-588.
5. H. Goldfine, N.C.Jonston, J. Mattai, G.G.Shipley, Biochemistry, 26 (1987) 2814-2822.
6. V. Luzatti, in: Biological Membranes, vol. 2 (D. Chapman and D.F.Wallach, eds.), Acad. Press, London (1973) 1-89.
7. S.W. Hui, T.P. Stewart, P.L.Yeagle, A.D. Albert, Arch. Biochem. Biophys., 207 (1981) 227-240.
8. R.M.C. Dawson, J. Amer. Oil Chem. Soc., 59 (1982) 401-406.
9. N.L. Klyachko, A.V. Levashov, A.V. Pshezhetsky, N.G. Bogdanova, I.V. Berezin, K. Martinek, Eur. J. Biochem. 161 (1986) 149-154.
10. N.L. Klyachko, A.V. Levashov, Kabanov, Yu.L. Khmelnitsky, K. Martinek, in: Kinetics and Catalysis in Microheterogeneous Systems (Surfactant Sci., Ser., v.38) (M.Gratzel, K.Kalyanasundaram, eds.), Marcel Dekker, New York, Basel (1991) 135-181.
11. V.V. Mozhaev, K. Heremans, J. Frank, P. Masson, C. Balny, Proteins: Struct. Func. Gene., 24 (1996) 81-91.
12. C. Balny, N.L. Klyachko, in High Pressure Molecular Science (R. Winter and J. Jonas, eds.), Kluwer Acad. Publ., The Netherlands (1999) 423-436.
13. N.L. Klyachko, P.A. Levashov, A.V. Levashov, C. Balny, in: Advances in High Pressure Bioscience and Biotechnology (H. Ludwig, ed.), Springer-Verlag, Berlin, Heidelberg (1999) 283-286.
14. N.L. Klyachko, S.V. Shipovskov, F. Meersman, K. Heremans, in: Trends in High Pressure Bioscience and Biotechnology (R. Hayashi, ed.), Elsevier Sci. B.V., Amsterdam (paper in this volume).
15. R. Köhling, J. Woenckhaus, N.L. Klyachko, R. Winter, manuscript in preparation.

Trends in High Pressure Bioscience and Biotechnology
R. Hayashi (editor)

High-pressure-induced hemolysis is characterized by release of membrane vesicles from human erythrocytes

T. Yamaguchi and S. Terada

Department of Chemistry, Faculty of Science, Fukuoka University, Jonan-ku, Fukuoka 814-0180, Japan

Upon exposure of human erythrocytes to high pressures (0.1-200 MPa), hemolysis and vesiculation start to occur at about 140 MPa. At 200 MPa, the value of hemolysis is about 45%, and the diameter of released vesicles is about 460 nm. Furthermore, when transmembrane proteins are cross-linked with cytoskeletal ones using diamide, high-pressure-induced hemolysis is greatly suppressed. In this case, the diameter of vesicles is about 230 nm. These results suggest that high-pressure-induced hemolysis may be associated with the size of released vesicles.

1. INTRODUCTION

The biological membranes are mainly composed of proteins and phospholipids. The association between these components may be controlled by hydrophobic, ionic, and van der Waals interactions [1]. The membrane structure of human erythrocytes has been studied as a prototype of such biological membranes. The stability and deformability of the erythrocyte membrane are controlled by the interactions between transmembrane proteins and cytoskeletal ones [2]. The transmembrane proteins such as band 3 and glycophorin C associate with linking proteins such as ankyrin and protein 4.1 [2]. These linking proteins attach the cytoskeleton, which consists of spectrin, protein 4.1, and actin, to the red cell plasma membrane. Thus, membrane protein-protein interactions play an important role in the structure and function of the erythrocyte membrane. The default in these interactions causes the destruction of the membrane, i.e., hemolysis. Therefore, it is expected that the study of the hemolysis provides the useful information on the interactions among membrane components.

It is well known that pressure affects the hydrophobic, ionic, and van der Waals interactions. Thus, it seems likely that the membrane structure of erythrocytes is affected by pressure. In fact, high pressure induces the hemolysis [3]. In this paper, we describe that high-pressure-induced hemolysis is associated with the size of released vesicles.

2. MATERIALS AND METHODS

2.1. Materials
Diazinedicarboxylic acid bis-(N,N'-dimethylamide) (diamide) was purchased from Sigma. All other chemicals were of reagent grade.

2.2. Chemical modification of erythrocytes
Human blood was obtained from the Fukuoka Red Cross Blood Center. The blood was centrifuged at 750 g for 10 min at 4°C. The plasma and buffy coat were carefully removed.

The erythrocytes were washed three times with phosphate buffered saline (PBS; 10 mM sodium phosphate, 150 mM NaCl, pH 7.4). The erythrocytes in PBS (at 20% hematocrit) were treated with diamide (0.5 mM) at 100 MPa for 30 min at 37°C. The erythrocytes were washed three times in PBS and used for the experiments of hemolysis and vesiculation.

2.3. Hemolysis

The hemolysis of the erythrocytes under high pressure was measured as follows. The erythrocytes suspended at 0.3% hematocrit in PBS were put into a syringe-type cell with a piston. The sample cell was placed in a pressure bomb made of stainless steel. The pressure was generated by means of a hand-type pump and monitored with a Heise pressure gauge. The mixture of ligroin and kerosine (v/v 1:1) was used as the pressure transmitting fluid. The erythrocyte suspension was compressed at a rate of 20 MPa/min, incubated at 37°C for 30 min at various pressures (110-200 MPa). Then, the samples were decompressed up to atmospheric pressure at a rate of 40 MPa/min. The suspension was centrifuged at 750 g for 10 min at 35°C. Hemoglobin release into the supernatant was measured at 542 nm. One hundred percent of hemolysis was performed by adding Triton X-100 (0.02%) into the suspension.

2.4. Scanning electron microscopy

Erythrocyte suspensions (5 ml at 0.3% hematocrit) were subjected to a pressure of 200 MPa for 30 min at 37°C and centrifuged for 10 min at 750 g. The pellets were washed once with 10 volumes of PBS and then suspended in 3 ml of the same buffer containing 0.5% bovine serum albumin. Glutaraldehyde was added to a final concentration of 1% and the suspension was incubated at 22°C for 2 h. The cells were washed twice with 40 volumes of the same buffer without bovine serum albumin. Fixed erythrocytes were dehydrated with solutions of increasing acetone concentration (50, 80, 90, 95, and 100% acetone) and then suspended in 2 ml of amyl acetate. The samples were dried with a Hitachi critical point dryer (model HCP-1) and then coated with Pt-Pd using an Eiko sputtering outfit (model IB-3). A Hitachi S-430 scanning electron microscope was used.

2.5. Light scattering

The sizes of membrane vesicles were measured at 25°C by using a submicron particle sizer (model 370, NICOMP, Calif. USA) with laser wavelength at 488 nm.

3. RESULTS

3.1. High-pressure-induced hemolysis

When human erythrocytes were subjected to various pressures (0.1-200 MPa) for 30 min at 37°C, the hemolysis began to occur at a pressure of about 140 MPa. At higher pressures, the degree of hemolysis increased (Fig. 1). For instance, the value of hemolysis at 200 MPa was about 45%. To cross-link membrane proteins, the erythrocytes were treated with 0.5 mM diamide at 100 MPa. When diamide-treated erythrocytes were exposed to a pressure of 200 MPa, the value of hemolysis was about 3%.

3.2. High-pressure-induced vesiculation

To examine the shape of high-pressure-treated erythrocytes, the erythrocytes were exposed to a pressure of 200 MPa for 30 min at 37°C, decompressed, and fixed with glutaraldehyde. The scanning electron micrography of 200 MPa-treated erythrocytes shows the membrane vesicles formed on the surface of red cells (Fig. 2). We can observe various size of membrane vesicles on the cell surface. These vesicles may be released into the medium from the membrane surface. If so, the released vesicles are readily separated from mother cells by centrifugation. The membrane vesicles in the supernatant are detected by measuring the

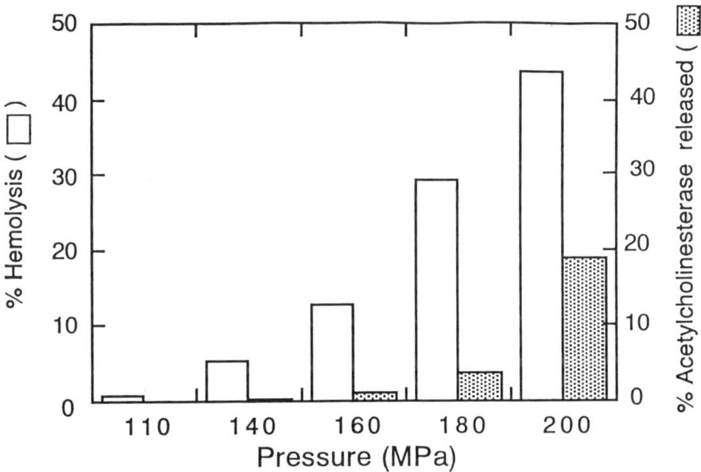

Figure 1. Hemolysis and vesiculation as a function of pressure in human erythrocytes

activity of acetylcholinesterase. As with hemolysis, the activity was also observed above 140 MPa and increased with increasing pressure (Fig. 1).

3.3. Size of high-pressure-induced vesicles
The activity of acetylcholinesterase suggests the release of membrane vesicles from the membrane surface of pressure-treated erythrocytes. So, we attempted to examine the size of vesicles by using a light scattering technique. Using this method, it was found that the diameter of membrane vesicles released from 200 MPa-treated erythrocytes is about 480 nm. On the other hand, when the diamide-treated erythrocytes, in which transmembrane proteins are cross-linked with cytoskeletal ones, were exposed to a pressure of 200 MPa, the diameter of released vesicles was about 230 nm.

4. DISCUSSION

When human erythrocytes are exposed to high pressures, hemolysis starts to occur at pressures of 130 ~140 MPa. In 200 MPa-treated cells, spectrin molecules are partially detached from the membrane [3]. Spectrin is a rod-like heterodimer molecule comprising two large subunits, α (240 kDa) and β (220 kDa) [4]. The spectrin dimers form tetramers upon head-to-head association. In general, oligomeric proteins dissociate under high pressure. Therefore, it seems likely that the association of spectrin molecules and the interactions of spectrin with ankyrin are perturbed by a pressure of 200 MPa. Thus, the cytoskeletal network is destroyed by high pressure.

When 200 MPa-treated erythrocytes are incubated at 37°C and atmospheric pressure, hemoglobin release from the membrane is suppressed [3]. Upon the shift from 37°C to 0°C, however, hemoglobin within the membrane is released again [3]. This indicates that the size of the membrane holes induced by pressure is dependent on temperature, i.e., resealed membranes reopen at 0°C [3].

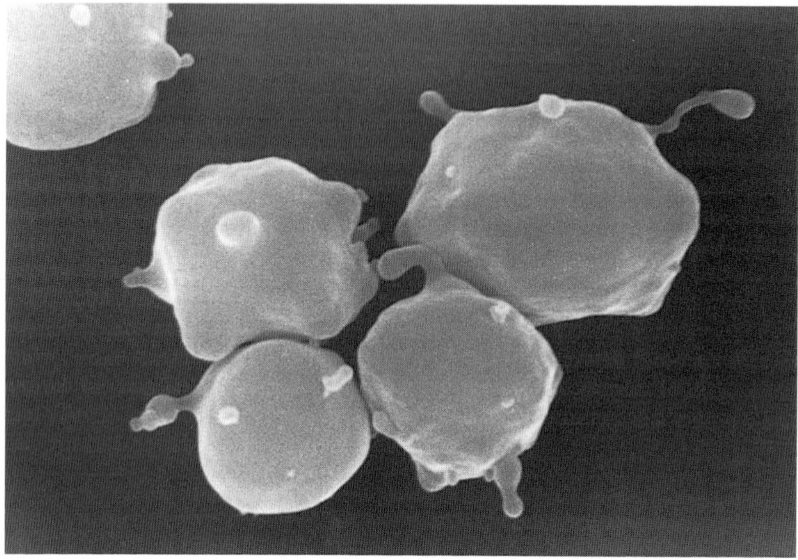

Fig. 2 Scanning electron micrograph of 200 MPa-treated erythrocytes.

The vesiculation of human erythrocytes is performed by various methods. For instance, membrane vesicles induced by ATP depletion [5], Ca^{2+}-loading [6], or incubation with dimyristoylphosphatidylcholine (DMPC) liposomes [7] are spectrin-free and band 3-rich. On the other hand, membrane vesicles containing cytoskeletal proteins such as spectrin and actin in addition to band 3 are released upon the exposure of erythrocytes to high pressure (200 MPa) [8]. The average size of such vesicles is about 480 nm in diameter. In addition, the membrane protein composition of the released vesicles is modulated by cross-linking of membrane proteins in erythrocytes. In diamide-treated erythrocytes, protein 4.1-rich vesicles are released by high pressure [9]. In this case, the average size of released vesicles is about 230 nm and pressure-induced hemolysis is greatly suppressed (about few percents). These results suggest that high-pressure-induced hemolysis is associated with the size of released membrane vesicles.

REFERENCES

1. C. Tanford, The Hydrophobic Effect; Formation of Micelles and Biological Membranes,Weley, New York, 1973.
2. V. Bennett, Biochim. Biophys. Acta 988 (1989) 107.
3. T. Yamaguchi, H. Kawamura, E. Kimoto,and M. Tanaka, J. Biochem., 106 (1989) 1080.
4. E. Ungewickell and W. Gratzer, Eur. J. Biochem., 88 (1978) 379.
5. H.U. Lutz, S.-C. Liu, and J. Palek, J. Cell Biol., 73 (1977) 548.
6. D. Allan, M.M. Billah, J.B. Finean, and R.H. Michell, Nature 261 (1976) 58.
7. P. Ott, M.J. Hope, A.J. Verkleij, B. Roelofsen, U. Brodbeck, and .L.L.M.Van Deenen, 641(1981) 79.
8 T. Yamaguchi, T. Kajikawa, and E. Kimoto, J. Biochem., 110 (1991) 355.
9. T.Yamaguchi, T. Saeki, and E. Kimoto, J. Biochem., 1147 (1993) 1.

Trends in High Pressure Bioscience and Biotechnology
R. Hayashi (editor)
© 2002 Elsevier Science B.V. All rights reserved.

Protein aggregation in the system "Aerosol-OT-water-octane" and its regulation by pressure application

N.L.Klyachko[1], S.V.Shipovskov[1], F.Meersman[2], K.Heremans[2]

[1]Department of Chemical Enzymology, Faculty of Chemistry, Moscow State University, 119899 Moscow, Russia[*]

[2]Department of Chemistry, Katholieke Universiteit Leuven, Belgium

Changes in secondary structure of α-chymotrypsin were studied in the AOT-water-octane system by Fourier Transform InfraRed spectroscopy in conjunction with pressure. Phase transition in the system, protein structure changes and octane crystallization were observed at different pressure ranges. It was demonstrated that pressure-induced changes in protein structure (formation of partially unfolded states) were responsible for its aggregation when pressure released. This aggregation was accompanied by appearance of special bands (1622 and 1685 cm-1) in the amide I region of the infrared spectra which usually observed upon temperature denaturation. The possibility to regulate the conditions of the formation of protein aggregates is discussed.

1. INTRODUCTION

Protein aggregation plays an important role both in positive and negative ways. For food technology, gel formation of proteins can improve the texture and stability of food [1]. In protein engineering, protein aggregation is one of the factors that can accompany the process of protein folding producing inclusion bodies [2,3]. In pharmacy, protein aggregation is one of the problems in long-term storage or delivery of protein drugs [4,5]. And the most intriguing phenomena involving protein aggregation are a variety of, so called, molecular or conformational diseases, such as amyloidoses, prion encephalopathies, Alzheimer's dementia, etc. (for reviews see [3,6,7].

To follow the structural changes in the protein molecule it is convenient to use Fourier Transform InfraRed (FTIR) spectroscopy elucidating the secondary structure by the analysis of the shape of the amide I band[8]. It was found for several proteins that protein inactivation associated with protein aggregation caused by temperature or temperature with pressure led to appearance of specific bands at 1615-1618 and 1685 cm^{-1} assigned to an intermolecular antiparallel β-sheet structure stabilized by hydrogen bonding (see [9] and refs.). It would be interesting to find out the conditions

---

[*] The work was supported in part by COST Chemistry Program, Projects D10/0003/98 and D10/0004/98

where the appearance of such aggregates can be regulated and studied in more detail.

Ternary systems "Surfactant-water-organic solvent" was found to be convenient to study the behavior of proteins and enzymes at ambient and elevated pressures and temperatures (for reviews see, for example, [10,11,12]). Such systems can be easily prepared, they formed spontaneously being an optically transparent (pseudo-homogeneous) allowing thus, to use different techniques. For reverse micelles, for example, it was found that the parameters of the system and the size of an inner polar cavity can be varied in wide range by changing water content, and such nano-containers of a protein size can solubilize biomolecules and other substances of different nature. Depending on the conditions, the situation can be realized when only one protein molecule can be introduced into one micelle. If the size of a micelle is big enough to entrap two or more protein molecules, this can also be realized. The aim of this work is to study the possibility of obtaining the special protein aggregates mentioned above in the AOT-water-octane system under pressure application.

2. MATERIALS AND METHODS

N-octane, α-chymotrypsin were purchased from Sigma (USA), Bis(2-ethylhexyl)sodium-sulfosuccinate (AOT) was from Fluka (Switzerland), D_2O was from Cambridge Isotope Laboratories, Inc. (USA).

The system AOT-water-octane was obtained by dissolving AOT in octane (0.1–1 M) followed by addition of an enzyme solution and/or D_2O to obtain a certain molar ratio of water and surfactant, w_0 which was varied from 5 to 40. α-Chymotrypsin was dissolved in 10 mM deutereted Tris-HCl (pD=7.6) using a protein concentration of 100-200 mg/ml. The solution was centrifuged for a few minutes and stored overnight to allow the H - D exchange of the amide group protons.

High pressure was generated in a diamond anvil cell (DAC) (DIACELL products, UK) where the pressure was built up by means of a screw mechanism. For the pressure experiments, the solution was mounted in a stainless steel gasket (0.05 mm thickness) of DAC, and the optical pathlength was 0.05 mm. Barium sulfate was used as an internal pressure standard in all cases [13].

The infrared spectra were recorded by a Bruker IFS66 FTIR spectrometer equipped with a liquid nitrogen cooled broad band MCT solid state detector. 250 interferograms were co-added after registration at a resolution of 2 cm^{-1}. The temperature kept constant at 25^0C.

3. RESULTS AND DISCUSSION

Before to analyze protein-containing AOT-water-octane system we studied phase behavior of the components of the system at different pressures. First of all, the crystallization of pure octane occurred when pressure reached 8 kbar. This phenomenon was observed by FTIR (Fig.1) and visualized also by using light microscope (Fig.2). Fig. 1 shows the changes in the band position of the infrared spectra responsible for the CH_2-, CH_3-group vibrations as a dependence of applied pressures. As seen, the significant changes occur when pressure is increased up to about 8 kbar. As seen from Fig.2, these changes can be attributed to the appearance of a new phase in the octane solution under high pressure.

In our next step, we studied how pressure can affect infrared spectra of the AOT-water-octane system. The band position of the ester bond vibrations in the AOT molecule versus pressure is given in Fig.3. As seen, the sharp shift of this band occurred at pressures about 2 kbar. This experiment was carried out at 1 M AOT and w_0=10. The same behavior of the system was found at lower surfactant concentrations down to 0.7 M. There were no changes in the system at 0.6 M AOT and down. The phenomenon observed can be attributed to the phase transition occurred in the system. Indeed, the phase transition in AOT (1 M)-water-octane system at very moderate pressures have been recently observed by using small-angle neutron scattering (SANS) [14].

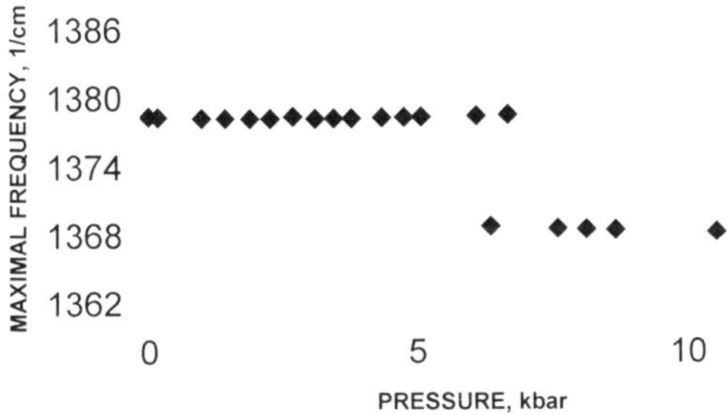

Fig.1. Pressure dependence of symmetrical CH$_3$-vibrations band in pure octane.

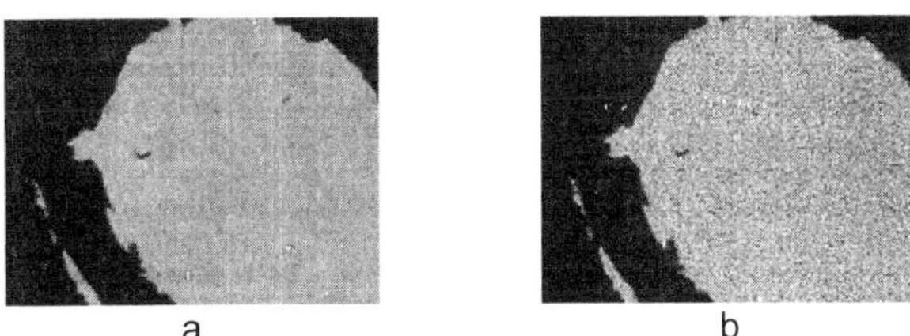

Fig.2. Structural changes in pure octane seen in light microscope: a - 1 bar, b - 10 kbar.

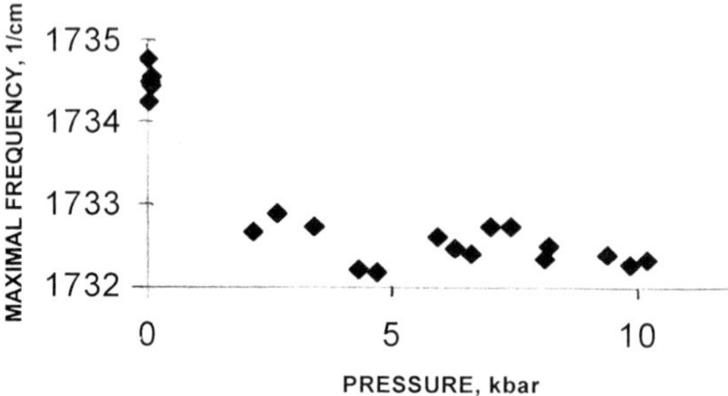

Fig.3. Pressure dependence of AOT ester bond vibrations band in the system AOT-water-octane. w_0=10, 1 M AOT.

The α-chymotrypsin structural changes under pressure in the AOT-water-octane system was studied by recording infrared spectra in the range 1600-1700 cm^{-1} characterizing the protein amide bond vibrations (amide I region). Fig.4 shows the changes in the peak area of amide I band of α-chymotrypsin in 0.6 M AOT in octane at w_0=40 upon pressure is increasing. At ambient pressure, the strongest maximum in α-chymotrypsin spectrum appears at 1638 cm^{-1} that represents β-structures in accordance with [15]. As seen, an essential shift of the maximum is observed in the pressure range 6-8 kbar. The spectral changes were found to be irreversible; histeresis was observed in all cases when pressure was going down, as shown in Fig.4 (shaded squires). The shift to higher frequency 1650 cm^{-1} can be assigned to the structural changes in the protein molecule, probably the appearance of more disordered fragments [8], indicating that protein unfolds (at least partially) under pressure. Such unfolding is possible at high hydration degrees where the size of the inner cavity of a micelle is much larger than that of a protein entrapped. This was realized in our case; we have chosen hydration degree (w_0) 25 and 40. At such hydration degrees we could expect the formation of the protein aggregates because the size of the micelle is big enough to include two and more α-chymotrypsin molecules [10]. The infrared spectra of α-chymotrypsin in the AOT-water-octane system at w_0 25 and different pressures presented in Fig.5. As seen, new bands at 1622 and 1685 cm^{-1} appeared in the spectrum when pressure released. Appearance of these bands is characteristic of intermolecular antiparallel β-structures stabilized by hydrogen bonding which were attributed to the aggregation of unfolded or partially unfolded protein [9,16]. These aggregation specific bands were observed in the case of temperature unfolded proteins [9,16], not observed in the pressure case (pressure caused this phenomenon only with temperature pretreated samples [9]). These results would indicate the process of protein aggregation occurred with pressure treated protein in the surfactant-water-organic solvent system. However, it was shown that experimental conditions can be found where no aggregation observed after pressure released. Thus, the system AOT-water-octane can be a promising one to study the process of protein aggregation at high pressures, to regulate the parameters and conditions in order to favor or prevent such aggregation and understand the mechanism of the process in more detail.

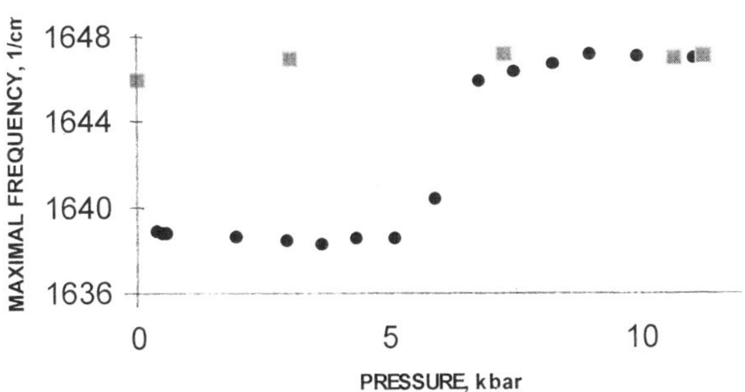

Fig.4. Pressure dependence of maximal frequency in α-chymotrypsin infrared spectrum. Black circles shows pressure increasing, shaded squires shows pressure decreasing.

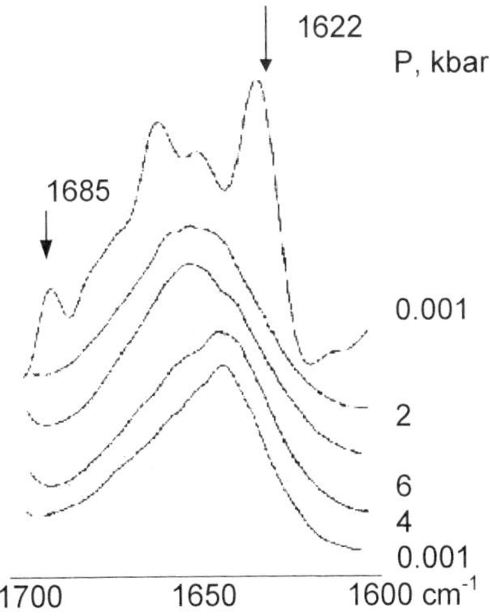

Fig.5. Infrared spectra of α-chymotrypsin in the AOT-water-octane system at different pressures. Spectra presented in chronological order from bottom to top.

176

ACKNOWLEDGMENTS

Authors express their gratitude to Prof. A.V. Levashov for fruitful critics and helpful discussion.

REFERENCES

1. M.G.Semenova, Curr.Opin. Colloid Interface Sci., 3 (1998) 627-632.
2. A.Mitaki, J.King, Biotechnology (N.Y.), 7 (1989) 690-697.
3. A.L.Fink, Folding and Design, 3 (1998) R9-R23.
4. T.Arakawa, S.J.Prestrelski, W.C.Kenney, J.F.Carpenter, Adv. Drug Deliv. Rev., 10 (1993) 1-28.
5. H.R.Costantino, S.P.Schwendeman, R.Langer, A.M.Klibanov, Biokhimiya, 63 (1998) 22-30.
6. J.W.Kelly, Curr. Opin. Struct. Biol., 8 (1998) 101-106.
7. R.W.Carrell, B.Gooptu, Curr. Opin. Struct. Biol., 8 (1998) 799-809.
8. D.M.Byler, H.Susi, Biopolymers, 25 (1986) 469-487.
9. L.Smeller, P.Rubens, K.Heremans, Biochemistry, 36 (1999) 3816-3820.
10. K.Martinek, N.L.Klyachko, A.V.Kabanov, Yu.L.Khmelnitsky, A.V.Levashov, Biochim. Biophys. Acta, 981 (1989) 161-172.
11. P.L.Luisi, M.Giomini, M.P.Pileni, B.H.Robinson, Biochim. Biophys. Acta, 947 (1988) 209-246.
12. C.Balny, N.L.Klyachko. In: High Pressure Molecular Science (R.Winter, J.Jonas, eds.), Kluwer Acad. Publishers, Dordrecht, Boston, London. 1999. Pp. 423-436.
13. P.T.T.Wong, D.J.Moffat, Appl. Spectrosc., 43 (1989) 1279.
14. J.Woenckhaus, R.Kuhling, N.L.Klyachko, R.Winter, Manuscript in preparation.
15. P.T.T.Wong, K.Heremans, Biochim. Biophys. Acta, 956 (1988) 1-9.
16. A.A.Ismail, H.H.Mantsch, P.T.T.Wong, Biochim. Biophys. Acta, 1121 (1992) 183-188.

Trends in High Pressure Bioscience and Biotechnology
R. Hayashi (editor)

Hydration and conformation changes during enzyme catalysis: from molecular enzymology to enzyme engineering and biotechnology

Patrick Masson

Centre de Recherches du Service de Santé des Armées,
Department of Toxicology, Enzymology Unit, BP 87, 38702 La Tronche cedex, France

The study of the effects of pressure on enzymes, has contributed to a better understanding of the mechanisms of enzymatic reactions, protein unfolding processes, and the molecular determinants of extreme protein stability. These advances have led to the emergence of a physical engineering of enzymes and high-pressure enzymology applied to food science, medicine, the pharmaceutical industry and environmental biotechnology.

1. INTRODUCTION

The growing interest in high pressure bioscience and biotechnology is the fruit of some 120 years of research on the effects of hydrostatic pressure on biological systems. Particularly, during the past ten years there has been a very strong renewal of interest in high pressure biology. There are several reasons: the use of high pressure for food processing [1], the development of powerful high pressure tools for the kinetic and structural analysis of macromolecules [2-4], and the potential applications of extremozymes [5].

Pressure is an important parameter: 60% of the biosphere is deeper than 1000 meters, i.e., supporting a pressure of 0.1 kbar (1 bar = 10^5 Pa). The average of pressure on the ocean floor is close to 0.4 kbar and, the pressure at the deepest abyssal trench is 1.1 kbar [6]. Thus, the existence of rich and diversified life thriving under high hydrostatic pressure is one aspect of the fascinating problem of the adaptation of life to extreme environments. For example, two extreme biotopes on Earth combine the effects of pressure (0.3-0.4 kbar), heat and subzero temperatures: deep-sea hydrothermal vents [7,8] and deep Antarctic ice close to the surface of the subglacial Lake Vostok [9]. Moreover, water, pressure and temperature are thought to have played a crucial role in prebiotical chemistry and in the origin and evolution of terrestrial life [10]. The study of extremophiles has been encouraging to astrobiologists looking for extraterrestrial life [11]. In this respect, Europa – one of the galilean moons of Jupiter – possesses an ice-covered ocean of liquid water which might harbor life [12].

2. PRESSURE EFFECTS ON ENZYME STRUCTURE

Early works in the thirties by Macheboeuf and Basset [13] on pressure-inactivation of enzymes led to the idea that high pressures denature proteins. The folded conformation of proteins results from a delicate balance between numerous stabilizing and destabilizing interactions. The net stability of native protein conformations is between 20 and 60 kJ/mol, which is only equivalent to the free energy of a small number of weak interactions [14]. Because pressure alters interatomic distances, it perturbs weak interactions, thus inducing conformational and volume changes. Complete unfolding is the extreme conformational transition. At this point it is important to remind that the partial molar volume of proteins, V_0, is the sum of two contributions: the intrinsic volume (V_{int}) and the hydration volume (ΔV_w). The intrinsic volume is composed of two terms: the total volume of atoms (ΣV_{at}) which is incompressible and the volume of voids or cavities due to imperfect packing (ΣV_v); the hydration volume is proportional to the solvent-accessible surface area [15]. Therefore, the volume change of proteins in action is

$$\Delta V_0 = \Delta V_v + \Delta\Delta V_w \tag{1}$$

In the case of pressure-induced protein denaturation, the contribution of internal cavities to negative volume changes on unfolding has been clearly demonstrated [16]. The contribution of hydration change is more difficult to assess [17,18]. Compressibility, $\beta = -V^{-1}(\partial\Delta V/\partial P)$, that is directly linked to the volume fluctuations is a flexibility parameter. The adiabatic (isoentropic) compressibility (β_s) is mainly determined by internal cavities and surface hydration.

$$\beta_s = -V^{-1}(\partial\Delta V_v/\partial P + \partial\Delta V_w/\partial P) \tag{2}$$

Compression of voids contributes positively to β_s whereas hydration contributes negatively.

The combined action of pressure (P), temperature (T), pH, solvents and various chemicals may have additive, synergistic or antagonistic effects on enzyme stability and activity. For example, pressure can stabilize proteins/enzymes at high temperature [19] or facilitate cold denaturation [20]. Elliptic phase diagrams constructed for constant free energy difference between native and denatured state on the pressure-temperature plane describe the combined action of P and T on protein conformational stability [4]. Similarly, isokinetic phase diagrams describe the inactivation of enzymes by pressure and temperature [21,22]. These diagrams are very useful for controling the operational stability/activity of enzymes under nonconventional conditions and at elevated pressure.

3. VOLUME CHANGES IN CHEMICAL REACTIONS

Because pressure (P) affects the rate (k) and equilibrium constants (K) of reactions according to the Le Chatelier's principle, it therefore has been used for investigating mechanisms [23] and modulating reactivity and selectivity in chemical processes [24]. Using the formalism of the transition–state theory, activation volumes, $\Delta V^{\neq}$, can be determined from the pressure dependence of rate constants (eqn. 3). A similar equation can be derived for volume changes, ΔV, associated with thermodynamic equilibria (eqn. 4):

$$(\partial Lnk/\partial P)_T = -\Delta V^{\neq}/RT \qquad \text{and} \qquad (\partial LnK/\partial P)_T = -\Delta V/RT \tag{3 and 4}$$

The experimental values of ΔV (or $\Delta V^{\ddagger}$) are the sum of various contributions: interactions between chemical groups (ΔV_{int}), formation and breakage of bonds (ΔV_{int}), conformational terms (ΔV_{conf}), changes in solvation (ΔV_w) of the system's various components, changes in the solvent structure:

$$\Delta V = \Sigma \Delta V_{int} + \Delta V_{conf} + \Delta V_w \tag{5}$$

Solvent effects have been carefully studied, and it has long been known that there is a prominent contribution of solvation-related terms to activation volumes. Water molecules stabilize polar transition states. Electrostriction of solvent around charged groups is accompanied by negative volume changes [25]; it depends on the solvent polarity: the lower the dielectric constant of solvent the more negative the value of $\Delta V^{\ddagger}$. Thus, ionogenic reactions such as the Menshutkin reactions are favored by pressure in less polar solvents [26]. The dielectric constant, viscosity and internal pressure of solvents change with P. The solubility of reactants and products also changes with P. Thus, pressure-tuning of the physical properties of solvent has important implications for applications of pressure in synthesis [27]. The contribution of solvation to $\Delta V^{\ddagger}$ and ΔV can be quantified in the case of reactions between small molecules [27,28].

4. ENZYME CATALYSIS UNDER PRESSURE

The theoretical and experimental aspects of the pressure effects on enzyme-catalyzed reactions have been extensively investigated [29-32]. Environmental conditions and intrinsic factors modulate the action of pressure on enzyme catalysis. Due to the size of enzymes and their flexible three-dimensional and nonhomogenous structure, temperature, solvents, cosolvents, salts, pH, and other physicochemical parameters may affect the affinity and reactivity of active sites, the overall conformation and stability of enzymes and the quaternary structures. Moreover, the dependence of volume changes on substrate concentration, changes in compressibility, and a complex interdependence of variables give rise to additional kinetic complications. Thus, the resulting effects on ligand/substrate binding and catalysis can be hard to grasp and interpret. It is sometimes difficult to assess the importance of the different elementary contributions to volume changes. However, as we will see, valuable mechanistic information can be extracted from the phenomenological analysis of experimental data.

4. 1. Dependence of $\Delta V^{\ddagger}_{obs}$ on substrate concentration

$\Delta V^{\ddagger}_{obs}$ depends on substrate concentration ([S]). In the case of the simple Michaelis-Menten model, differenciation of the rate equation with respect to pressure gives [29]:

$$\Delta V^{\ne}_{obs} = \Delta V^{\ne}_{cat} - \frac{\Delta V_b}{1 + [S]/Km} \tag{6}$$

In this equation, $\Delta V^{\ddagger}_{cat}$ is the activation volume of the catalytic step and $\Delta V_b = -\Delta V_{Km}$ is the volume change upon substrate binding. Keeping in mind that binding constants (e.g., K_s, K_m, K_i) and catalytic constants are composite parameters, any change in $\Delta V^{\ddagger}_{cat}$ and ΔV_b with pressure may reflect a pressure effect on a particular elementary step. At low [S] $\ll$ Km, $\Delta V^{\ddagger}_{obs} = \Delta V^{\ddagger}_{cat} - \Delta V_b$ and, at high [S], $\Delta V^{\ddagger}_{obs}$ is dominated by the contribution of $\Delta V^{\ddagger}_{cat}$. Fig. 1 illustrates this substrate concentration dependence.

Now let us consider a reaction in which a metastable intermediate is formed, e.g., an acyl enzyme (EA) as for serine hydrolases (Scheme 1), the activation volume is that of the rate-limiting step:

$$E\ +\ S\ \underset{P_1}{\overset{k_2}{\rightleftharpoons}}\ ES\ \overset{k_3(H_2O)}{\rightarrow}\ EA\ \underset{P_2}{\rightarrow}\ E \qquad\qquad \text{(Scheme 1)}$$

180

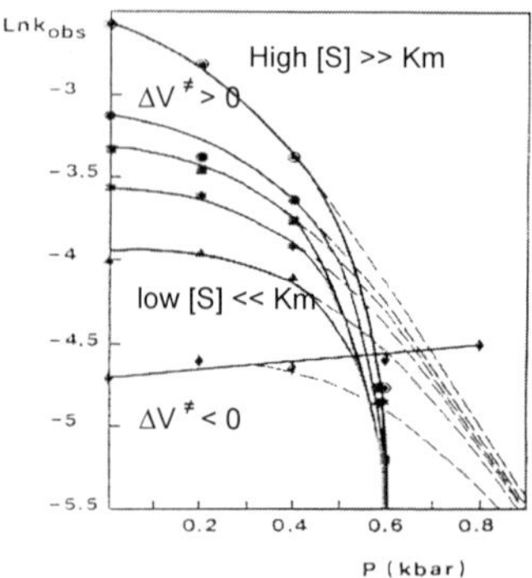

Fig. 1. Carbamylation of human butyrylcholinesterase (BuChE) by N-methyl-(7-dimethylcarbamoxy)quinolinium under single-turnover conditions in 50 mM phosphate pH 7.4 at 35°C (adapted from [33]).

For such a system, the catalytic constant and the activation volume are respectively:

$$k_{cat} = \frac{k_2 k_3}{k_2 + k_3} \tag{7}$$

$$\Delta V^{\neq}_{cat} = \frac{k_3 \Delta V^{\neq}_2 + k_2 \Delta V^{\neq}_3}{k_2 + k_3} \tag{8}$$

If acylation is the rate-determining step (i.e., $k_2 \ll k_3$), it follows that $\Delta V^{\neq}_{cat} \approx \Delta V^{\neq}_2$. On the contrary, if deacylation is rate-determining ($k_3 \ll k_2$), $\Delta V^{\neq}_{cat} \approx \Delta V^{\neq}_3$. Thus, change in $\Delta V^{\neq}_{cat}$ with pressure may reflect change in the rate-limiting step (cf. Fig.3).

4.2. Enzyme-catalyzed reaction versus uncatalytic reaction.

Comparison of catalytic and uncatalytic reaction, provides information on enzyme hydration and conformation changes during the catalytic cycle. As seen in Fig. 2 and Table 1, spontaneous and enzyme-catalyzed hydrolysis of N-methyl indoxyl acetate (NMIA) shows opposite volume profiles. BuChE-catalyzed reaction at high [S] is accompanied by $\Delta V^{\neq}_{cat} > 0$, while the uncatalytic reaction shows $\Delta V^{\neq}_{unc} < 0$ as expected for chemical hydrolysis of an ester ($\Delta \Delta V^{\ddagger}_w \triangleleft \Delta V^{\ddagger}_{int}| < 0$). Unlike chemical hydrolysis, hydrolysis of NMIA by BuChE involves several steps (Scheme 1) that could be partly rate-limiting. Moreover, enzyme catalysis is delocalized over the whole active site and, thus involves numerous elementary contributions. Comparison of wild-type and mutated enzymes, kinetic studies in various solvents and compressibility measurements may shed light on the effects of hydration and cavities on enzyme dynamics [34,35]. Here, data on the D70G mutant of BuChE (Table1) suggest that residue D70, located at the entrance of the enzyme active site gorge at 12Å far from the active serine, plays a role in substrate binding (formation of the first encounter complex) and catalytic steps. Independent studies showed that residue D70 controls the motion of water molecules in the active site gorge, flexibility of the gorge and conformational stability of the enzyme [36]. Therefore, we should point out that, whatever the enzyme system, the exact contribution of hydration changes and conformation change to the values of ΔV cannot be simply determined.

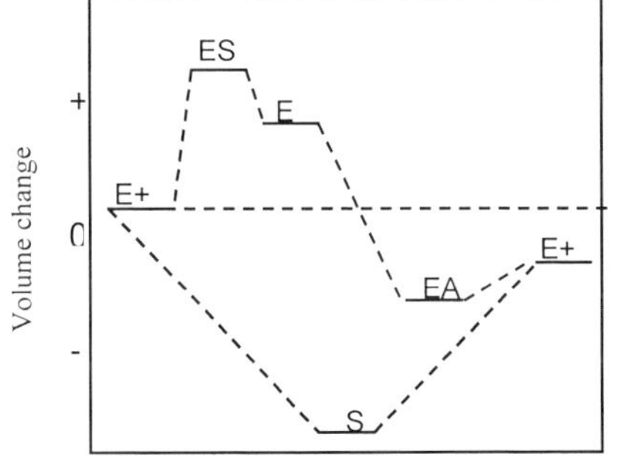

	uncatalytic	wt	D70G
$\Delta V^{\neq}$ (ml/mol)	-26.40±2.5	+10.6±1.0	+2.8±1.1
ΔV_b (ml/mol)	-	+2.9±1.0	+11.5±1.1

Fig. 2. Volume profile for spontaneous hydrolysis ($S + H_2O \xrightarrow{S^{\neq}} P_1 + P_2$) of NMIA and hydrolysis by wild-type human butyrylcholinesterase in 10mM bisTris /HCl pH 7.0 at 25°C (unpublished).

Table I. Activation volumes ($\Delta V^{\neq}_{ES}$) and binding volumes for hydrolysis of NMIA by wild-type BuChE and its D70G mutant compared to $\Delta V^{\neq}$ for uncatalytic (spontaneous) hydrolysis (unpublished).

4. 3. Pressure dependence of volume changes

Pressure dependence of volume changes causes non-linear pressure dependence of catalytic constants. Three phenomena account for nonlinear plots: compressibility changes, rate-determining change, and unfolding (dislocation of the architecture of binding/active sites). Compressibility changes give parabolic plots. Fig. 3 shows the pressure dependence of a rate constant, k_{cat}, that fits the quadratic equation:

$$Lnk = Lnk_0 - \frac{\Delta V^{\neq}}{RT}P + \frac{\Delta \beta^{\neq}}{2RT}P^2 \tag{9}$$

where $\Delta \beta^{\ddagger}$ is the absolute activation compressibility (-$\partial \Delta V^{\ddagger}/\partial P$).

However, changes in a rate-determining step with pressure can produce similar plots. Indeed, for a multistep mechanism as in Scheme 1, according to eqn 7, $\Delta V^{\neq} = 0$ when $k_2 \Delta V^{\neq}_3 = - k_3 \Delta V^{\neq}_2$; i.e., at P = 1 kbar in Fig.3. Thus, below 1 kbar, the rate-limiting step is thought to be deacylation and, acylation becomes rate-limiting beyond this pressure. Thus, without further information care must be taken in ascribing compressibility changes to nonlinear pressure-dependent behaviors. Pressure-induced unfolding (disruption of binding/active sites) is much easier to probe :a) loss of enzyme activity precedes loss of substrate/ligand binding ; b) decrease in rate constant or affinity constant occurs in a narrow pressure range ; c) the pressure effect can be modulated by environmental conditions. As seen in Fig. 4, D₂O shifts the pressure denaturing threshold by stabilizing the binding site of human butyrylcholinesterase.

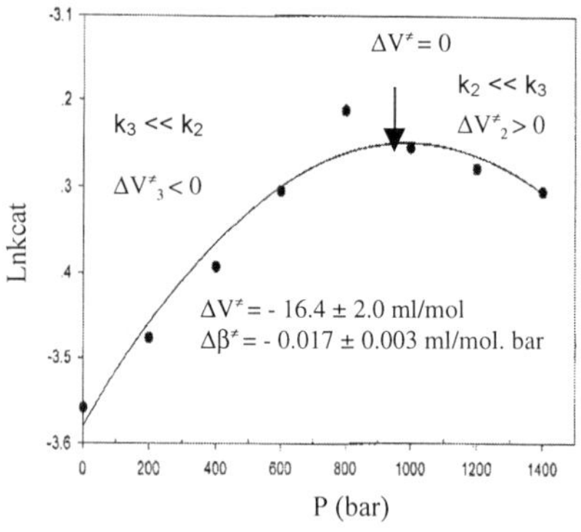

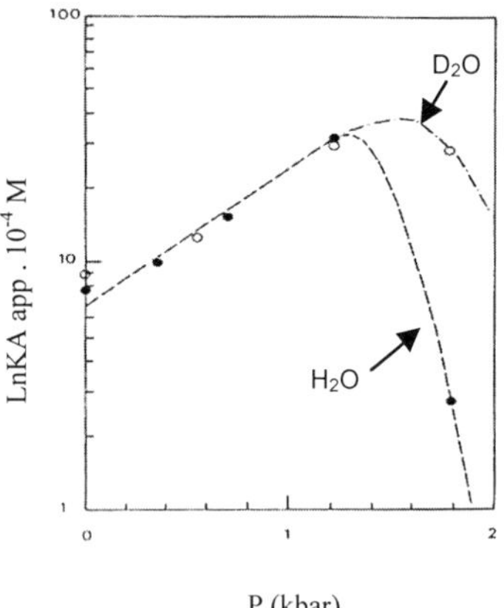

Fig. 3. Hydrolysis of benzoylcholine chloride by wild-type butyrylcholinesterase (k_{cat}) in 10 mM Tris/HCl pH 7.4 at 25°C as a function of pressure (unpublished).

Fig. 4. Binding of phenyltrimethyl-ammonium iodide to wild-type BuChE (K_A) in Tris/glycine buffer made in water (pH = 8.4) and 99 % heavy water (pD = 8.8) at 35°C (adapted from [37]).

4. 4. Multiple dependence of volume changes: effects of solvent, cosolvent, additives and temperature on stability and activity of enzymes working under pressure

The stability and activity of enzymes depend on the solvent. In Fig. 4, the isotopic effect of heavy water on stability of butyrylcholinesterase is clear. Because D-bonds are stronger than H-bonds, heavy water increases the structuration of the hydration shell. Addition of additives or cosolvents to the medium can exert a protective effect on enzymes against pressure inactivation.

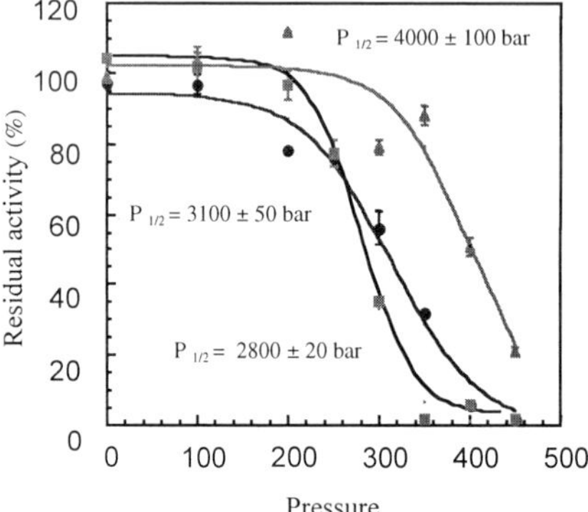

Fig. 5. Effects of additives on the pressure sensitivity of human butyrylcholinesterase in 10 mM Tris/HCl buffer pH 7.4 at 25°C:●, no additive; ■, 20 % (v/v) methoxyethanol; ▲, 20 % (w/v) trehalose. $P_{1/2}$ refers to the pressure causing 50% enzyme inactivation (unpublished).

The protective effect of sugar, polyols, glycerol, salts and amino acids is due to preferential hydration of the protein, i.e., preferential exclusion of the protectant molecules from the protein surface [38]. It was suggested that glycerol reduces the specific volume and adiabatic compressibility of proteins due to release of "lubricant" water molecules from cavities, which in turn decreases the amplitude of protein thermal motions [38,40], but the effect of stabilizer polyols of higher molecular weight appears to be more complex [41,42]. On the contrary, additives that bind preferentially cause protein destabilization [38]. These latter, e.g., urea, organic solvents such as propylene carbonate strengthen sensitivity of enzymes to pressure [43]. As seen in Fig. 5, trehalose protects BuChE against inactivation whereas methoxyethanol favors pressure inactivation of the enzyme.

It is well known that viscosity of solvents can affect the association rate constants of substrates/ligands to the active centers of enzymes, particularly in the case of diffusion-controlled reactions [44]. The dielectric properties and structure of the solvent also control the rates of polar reactions [45,46]. In addition to its action on elementary catalytic steps, solvent affects the compressibility and flexibility of enzymes under pressure. These solvent effects can be modulated by temperature. Thus, different values of volume changes for an enzyme-catalyzed bimolecular reaction can be determined; they reflect conformational changes and reorganization of the solvation shell as a function of the experimental conditions [47]. However, due to complex interdependence of variables, the individual contributions of solvent and temperature to volume changes can be difficult to estimate.

4. 5. Effect of osmotic pressure

Hydrostatic pressure shifts macromolecular equilibria towards the hydrated states. Therefore, hydrostatic pressure and osmotic pressure have antagonistic effects in macromolecular processes. The application of osmotic pressure (π) - generated by osmolytes (polyols, sugars, dextrans) added to the medium - to proteins has been used to study changes in hydration during enzyme reactions and conformational changes [48]. "Osmotic volumes" can be determined [49,50].

$$\Delta V^{\neq}_{osm} = - RT\,(\partial Lnk\,/\,\partial\pi)_T \tag{10}$$

According to the osmotic stress theory as developed by Parsegian, Rand and Rau [51], these volume changes allow quantitative measurements of hydration changes, i.e., the numbers of water molecules (n) in the process:

$$Pr\text{-}nH_2O \; \leftrightharpoons \; Pr^* + nH_2O \tag{Scheme 3}$$

$$n_w = \Delta V^{\neq}_{osm}\,/\,v_w \tag{11}$$

$v_w = 30\ \text{Å}^3$ is the molecular volume of a water molecule. However, only rarely can the actual changes in hydration accompanying enzymatic reactions be determined. Apparent osmotic volumes are not directly linked to the effects of solutes on catalytic reactions and equilibria. Rather, according to Timasheff [52], n_w is solely a measure of the mutual perturbations of the chemical potentials of the cosolvent and the protein. In this view, osmotic stress is simply a restricted case of preferential interactions. Despites these restrictions, the osmotic pressure approach has proven to be a useful tool for probing the participation of water in enzyme reactions [36,53,54] and quantifying changes in water-accessible surface area from preferential interaction coefficients [55].

5. APPLICATIONS

5.1. Enzyme inactivation

Among biotechnological applications, high-pressure inactivation of enzymes has been primarily used for food preservation [56]. This process can be facilitated by operating at low temperatures. Knowledge of the P/T phase diagram is a prerequisite for optimizing conditions. This technology is of great interest in cosmetology and pharmacy.

5.2. Enzyme catalysis

Because pressure controls the enzyme specificity (k_{cat}/K_m), chemoselectivity, regioselectivity and enantioselectivity of organic reactions, biosynthesis and biotransformations by enzymes operating under high pressure is a new and promising aspect of enzyme engineering [57,58]. In particular, the enzymatic synthesis of pharmaceuticals under mild conditions has proven to be possible., e.g., the lipase-catalyzed enantioselective synthesis of esters such as ibuprofen [59] and the synthesis of pharmacological peptides by thermolysine [60]. Enzyme synthesis of polyols and sweeteners also seems to be interesting under high pressure. These reactions are performed in bioreactors at pressures, generally, lower than 2 kbar, i.e. pressures that do not alter the stability and functionality of enzymes. However, the instability of pressure-synthesized chemicals can be a technical limitation, e.g., aspartame was found to be susceptible to degradation under pressure at neutral pH [61]. Partial proteolysis of natural proteins (e.g. from milk or rice) under pressure has been used to reduce their antigenicity and to increase their digestibility [62].

Pressure can be used to promote reactions catalyzed by naturally-occurring enzymes in plant cells. For instance, pressure was found to favor the hydrolysis of glucosinolates of cruciferous plants, leading to bioactive isothiocyanates. Glucoraphanin from broccoli juice can be converted to sulforaphane, which displays antitumor activity, anticarcinogenic properties and nematocidal properties [63]. Such reactions can be used for food functionalization and for producing "nutraceuticals" to be used in complementary therapy.

The ability of enzymes to catalyze reactions in nonconventional media has also been investigated under pressure. Biocatalysis in supercritical fluids (e.g., CO_2), in neat organic solvents, in reversed micelles and in water-restricted media ($a_w<1$) under hydrostatic pressure is becoming a fruitful field of research [64,65]. As for organic reactions, the dielectric constant of solvents controls the enzyme catalytic power [66]. Thus, the use of nonpolar solvents is expected to increase the stability and efficiency of enzymes at elevated pressure for performing reactions thermodynamically unfavorable in water. Lastly, because pressure controls the transition state geometry, enantioselective synthesis of chiral compounds by enzymes is one of the most striking potential applications.

5.3. Optimizing barophily

By operating in the presence of stabilizing cosolvents [67,68], highly concentrated lyotropic salts, or in organic media [69,70], it can be possible to increase the thermobarostability of industrial enzymes. Chemical modifications (N-terminal capping, glycosylation, PEGylation), cross-linking and immobilization also increase the resistance of proteins to pressure and heat. However, the use of barothermozymes is probably the simplest and the most efficient way. Lessons from extremozymes should help to make mutants of mesophilic enzymes with improved activity, new specificity and enhanced stability at high pressure and temperature [71], and in organic media [72]. This can be achieved through rational design (site-directed mutagenesis) or random mutagenesis. However, it should be remembered that increased stability reduces conformational plasticity, which in turn lowers the catalytic activity. Thus, a compromise must be found between stability and flexibility.

6. CONCLUSION

The high pressure approach has proven to be a powerful tool for dissecting enzyme mechanisms and investigating the molecular bases of protein stability. There are numerous potential applications of the high pressure technology in enzymology. Analytical and industrial applications in the different fields of biosciences, including medicine and pharmaceutical science [73] require extended experimental work for optimizing the stability, activity and selectivity of enzymes. In this respect, extremozymes are good models for significant advances in the use of high pressure enzymology. However, several strategies can be implemented for optimizing the use of enzymes under pressure.

ACKNOWLEDGEMENTS

This paper was written within the framework of the European COST action D10. The author wishes to thank M-T. Froment, N. Bec and F. Ribes for their contribution to unpublished works and C. Balny for valuable discussions and encouragements. Author's unpublished works were supported by DGA/DSP/STTC-99CO029.

REFERENCES

1. M. Dumoulin and R. Hayashi, Food Sci. Technol. Int. Tokyo, 4 (1998) 99-113.
2. J. Jonas, L. Ballard and D. Nash, Biophys. J., 75 (1998) 445-452.
3. G. Panick, R. Malessa, R. Winter, G. Rapp, K.J. Frye and C.A. Royer, J. Mol. Biol., 275 (1998) 389-402.
4. K. Heremans and L. Smeller, Biochim. Biophys. Acta, 1386 (1998) 353-370.
5. F. Niehaus, C. Bertoldo, M. Kähler and G. Antranikian, Appl. Microbiol. Biotechnol., 51 (1999) 711-729.
6 K. Horikoshi, Curr. Opin. Microbiol., 1 (1998) 291-295.
7. E. Bloch, R. Rachel, S. Burggraf, D. Hafenbradl, H.W. Jannasch and K.O. Stetter, Extremophiles. 1 (1997) 114-121.
8. D. Prieur, TIBTECH, 15 (1997) 242-244.
9. B. Price, Proc. Natl. Acad. Sci. USA, 97 (2000) 1247-1251.

186

10. D.K. Alargov, S. Deguchi, K. Tsujii and K. Horikoshi, in this volume.

11. D. Koerner and S. LeVay, The scientific quest for extraterrestrial life, Oxford Univ. Press, New York, 2000.

12. K. K. Khurana, M.G. Kivelson, D.J. Stevenson, G. Schubert, C.T. Russell, R.J. Walker and C. Polanskey, Nature, 395 (1998) 777-780.

13. M.A. Macheboeuf,. J. Basset, and G. Levy, Ann. Physiol. Physicochim. Biol. 9 (1933) 713-722.

14. N. Pace, TIBS, 15 (1990) 14-17.

15. T.V. Chalikian and K.J. Breslauer, Biopolymers, 39 (1996) 619-626.

16. K. J. Frye and C.A. Royer, Prot. Sci., 7 (1998) 2217-2222.

17. L.R. Murphy, N. Matubayasi, V.A. Payne and R.M. Levy, Folding and Design, 3 (1998) 105-118.

18. T.V. Chalikian and K.J. Breslauer, Curr. Opin. Struct. Biol., 8 (1998) 657-664.

19. F.T. Robb and D.S. Clark, J. Mol. Microbiol. Biotechnol., 1 (1999) 101-105.

20. S. Kunugi, H. Yamamoto, M. Makino, T. Tada and Y. Uehara-Kunigi, Bull. Chem. Soc. Jpn., 72 (1999) 2803-2806.

21. A. Weingand-Ziadé, F. Renault and P. Masson, Biochim. Biophys. Acta, 1340 (1997) 245-252.

22. Indrawati, A.M. Van Loey, L.R. Ludikhuyze, and M.E. Hendricks, J. Agric. Food Chem., 47 (1999) 2468-2474.

23. R. van Eldik, Angew. Chem. Int. Ed., 25 (1986) 673-682.

24. G. Jenner, Tetrahedron, 53 (1997) 2669-2695.

25. S.D. Hamann, Rev. Phys. Chem. Jpn., 50 (1980) 147-168.

26. N.S. Isaacs, Tetrahedron, 47 (1991) 8463-8497.

27. G. Kohnstam. In: Progress in Reaction Kinetics, G. Porter, Ed., Pergamon Press, Oxford, 1970, vol. 5, pp. 335-408

28. R. Van Eldik, T. Asano and W.J. Le Noble, Chem. Rev., 89 (1989) 549-688.

29. E. Morild, Adv. Prot. Chem., 34 (1981) 93-166.

30. Y. Taniguchi and S. Makimoto, J.Mol. Catal., 47 (1988) 323-334.

31. P.C. Michels, D. Hei and D.S. Clark, Adv. Prot. Chem. 48 (1996) 341-376.

32. V.V. Mozhaev, K. Heremans, J. Frank, P. Masson and C. Balny, Proteins: Struct. Funct & Genet., 24 (1996) 81-91.

33. P. Masson and C . Balny, Biochim. Biophys. Acta 954 (1988) 208-215.

34. T. Kamiyama and K. Gekko, Biochim. Biophys. Acta, 1478 (2000) 257-266.

35. K . Gekko, T. Kamiyama, E. Ohmae and K. Katayanagi, J. Biochem., 128 (2000) 21-27.

36. P. Masson, C. Cléry, P. Guerra, A. Redslob, C. Albaret and P-L. Fortier, Biochem. J., 343 (1999) 361-369.

37. P. Masson and C. Balny, Biochim. Biophys. Acta 1041 (1990) 223-231.

38. S.N. Timasheff, Adv. Prot. Chem., 51 (1998) 355-432.

39. A. Priev, A. Almagor, S. Yedgar and B. Gavish, Biochemistry, 35 (1996) 2061-2066.

40. M.M.C. Sun, N. Tolliday, C. Vetriani, F.T. Robb and D.S. Clark, Prot. Sci., 8 (1999) 1056-1063.

41. A. Almagor, A. Priev, G. Barshtein, B. Gavish, S. Yedgar, Biochim. Biophys. Acta, 1382 (1998) 151-156.

42. S.Fujii, K. Obuchi, H. Iwahashi, T. Fujii and Y. Komatsu, Biosci. Biotech. Biochem, 60 (1996) 476-478.

43. P. Masson, P. Gouet and C. Cléry, J. Mol. Biol., 238 (1994) 466-478.

44. B.B. Hasinoff, Biochim. Biophys. Acta 704 (1982) 52-58.

45. E.L. Mertz and L.I. Krishtalik, Proc. Natl. Acad. Sci. USA, 97 (2000) 2081-2086.

46. T.C. Bruice and S.J. Benkovic, Biochemistry 39 (2000) 6267-6274.

47. C. Balny and F. Travers, Biophys. Chem., 33 (1989) 237-244.

48. C. Reid and R.P. Rand, Biophys. J., 72 (1997) 1022-1030.

49. V.A. Parsegian, R.P. Rand and D.C. Rau, Meth. Enzymol., 259 (1995) 43-94.

50. V.J. LiCata and N.M. Allewell, Meth. Enzymol., 295 (1998) 42-61.

51. V.A. Parsegian, R.P. Rand and D.C. Rau, Proc. Natl. Acad. Sci. USA, 97 (2000) 3987-3992.

52. S.N. Timasheff, Proc. Natl. Acad. Sci. USA, 95 (1998) 7363-7367.

53. C.D. Di Primo, E. Deprez, G. Hui Bon Hoa and P. Douzou, Biophys. J., 68 (1995) 2056-2061.

54. V.J. LiCatta and N.M. Allewell, Biochim.Biophys. Acta 1384 (1998) 306-314.

55. E.S. Courtenay, M.W. Capp, C.F. Anderson and M.T. Record Jr., Biochemistry, 39 (2000) 4455-4471.

56 . M. Hendrickx, L. Ludikhuyze, I. Van den Broeck and C. Weemaes, Trends in Food Sci. Technol., 9 (1998) 197-203.

57. S. Kunugi, Annu. N.Y. Acad. Sci., 672 (1992) 293-304.

58. K. Nakamura, Trends Biotechnol., 8 (1990) 288-292.

59. A. Overmeyer, S. Schrader-Lipplelt, V.Kasche, and G. Brunner, Biotechnol. Lett., 21 (1999) 65-69.

60. S. Kunugi, M. Kitayaki, Y. Yanagi, N. Tanaka, R. Lange, and C. Balny, Eur. J. Biochem., 248 (1997) 567-574.

61. P. Butz, A. Fernandez, H. Fister, and B. J. Tauscher, Agric. Food. Chem., 45 (1997) 302-303.

62. F. Maynard, A. Weingand, J. Hau and R. Jost, Int. Dairy J., 8 (1998) 125-133.

63. B. Tauscher, High Pressure Res,19 (2000) 11-18.

64. S.V.Kamat, E.J. Beckman and A.J. Russell, Crit. Rev. Biotechnol., 15 (1995) 41-71.

65. J.S. Dordick, Y.L. Khmelnitsky and M.V. Sergeeva, Curr. Opin. Microbiol., 1 (1998) 311-318.

66. J. Lim and J.S. Dordick, Biotechnol. Bioeng., 42 (1993) 772-776.

67. V.V. Mozhaev, R. Lange, E.V. Kudryashova, and C. Balny, Biotechnol. Bioengn., 52 (1996) 320-331.

68. V. Athès and D. Combes Enz. Microb. Technol., 22 (1998) 532-537.

69. R.V. Rariy, N. Bec, N.Klyachko, A.V. Levashov, and C.Balny, Biotechnol. Bioengn., 57 (1998) 552-556.

70. N. Fontes, E. Nogueiro, M. Elvas, T. Corrêa de Sampaio, and S. Barreiros, Biochim. Biophys. Acta, 1383 (1998) 165-174.

71. D. W. Hough and M.J. Danson, Curr. Opin. Chem. Biol., 3 (1999) 39-46.

72.G.E. Sellek and J.B. Chaudhuri, Enz. Microb. Technol., 25 (1999) 471-482.

73. P. Masson, C. Tonello and C. Balny, J. Biomed. Biotechnol., 1 (2001) 1-4.

Trends in High Pressure Bioscience and Biotechnology
R. Hayashi (editor)

Enzyme-substrate specific interactions: *in situ* assessments under pressure

A. Fernández García, P. Butz, R. Lindauer & B. Tauscher

Institute for Chemistry and Biology, Federal Research Centre for Nutrition,
Haid-und-Neustr. 9, 76131 Karlsruhe, Germany

Peroxidase (POD) enzyme was irreversibly inactivated (>90%) after 2 minutes at 500 MPa/25 °C in the presence of guaiacol. POD reaction with ferulic acid or with catechol monoethyl ether as substrates is not influenced by pressure up to 500MPa (25 °C). POD reaction with guaiacol or with vanillin as substrates is strongly inhibited (reversibly) by pressures above 300MPa (25 °C). Polyphenoloxidase (PPO) reaction with chlorogenic and caffeic acid as substrates is not influenced by pressure up to 500MPa (25 °C). PPO reaction with pyrocatechol as substrate is strongly inhibited (reversibly) by pressures above 300MPa (25 and 40 °C).

1. INTRODUCTION

To avoid quality loss during and after high pressure treatment of fruits and vegetables, if possible, enzyme reactions should be inhibited during pressure build-up and enzymes should be irreversibly inactivated before storage [1, 2]. This is however often not feasible with pressure alone and high temperatures (not desirable for food treatment) must be applied additionally.

Would it be possible to selectively and irreversibly inactivate undesired enzymatic activity at lower temperatures and pressures within minutes?

Assessment of two of the most temperature- and high pressure- resistant enzymes -peroxidase (POD) and polyphenoloxidase (PPO)- measuring activities with a variety of substrates in situ using a high pressure optical cell has been conducted.

2. MATERIALS AND METHODS

Horseradish peroxidase (POD) [Boehringer Mannheim] and tyrosinase (PPO) [Sigma] were tested in excess of substrates.

In situ measurements took place in an optical cell U102 [High Pressure Research Centre, Polish Academy of Sciences] for 700 MPa placed in a spectrophotometer [Phamacia Biochorm 4060]. Opposite shapphire windows allowed the light transmission from 200 to 800 nm. A high pressure intensifier hand pump external to the system was coupled to the optical cell with a high pressure capillary tube. Stable temperature was reached with a thermostat [Haake F3] connected to the system. Post-pressure measurements were performed in the multiple cell sample holder of the [Pharmacia] spectrophotometer after treatments in a high pressure device as described [3].

3. RESULTS AND DISCUSSION

The development of a product of the reaction from PPO and pyrocatechol was followed at 300

190

nm (Fig. 1). Pressures of 500MPa have nearly stopped the reaction, but after pressure release normal rate was reestablished.

Other PPO substrates tested were chlorogenic and caffeic acid. No remarkable differences were observed *in situ* under pressure (results not shown).

Different substrates with a similar structure to guaiacol - catechol-monoalkyl-ether - were tested (Fig. 2). So far, a reduction in the activity of POD when pressurised with vanillin as substrate was observed, but after pressure release, the normal rate was restored. So the common structure of these substrates might be determinant on the variation of K_m under high pressure.

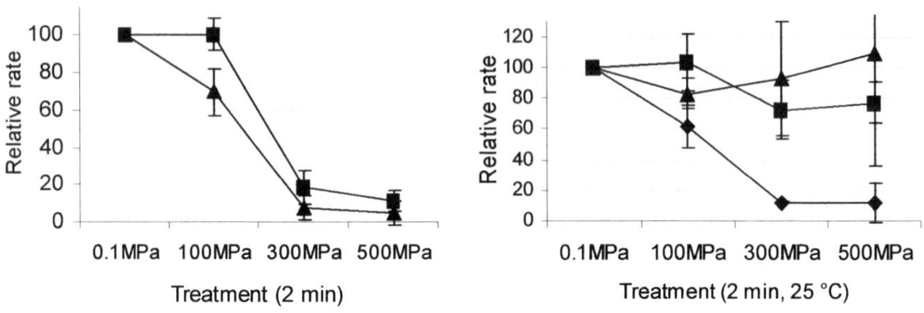

Fig. 1: Relative rate of the reaction of polyphenoloxidase with pyrocatechol.
▲ 25 °C
■ 40 °C

Fig. 2: Relative rate of the reaction of peroxidase with different substrates.
▲ ferulic acid
■ catechol monoethyl ether
◆ vanillin

When POD was pressurised with guaiacol as substrate, pressure effects on the reaction were rapidly detected, and as soon as pressure was built up, a deceleration was evident (Fig. 3).

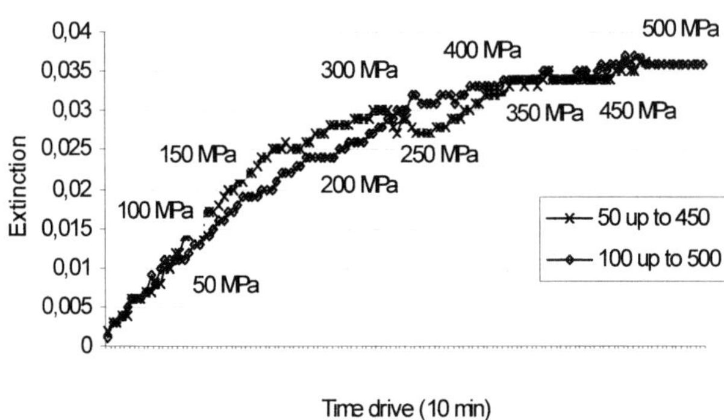

Fig. 3: Development of product at 435 nm of the reaction of peroxidase and guaiacol in dependence of high pressure intensity.

And measurements of activity after pressure release show that the inactivation of POD was dependent on the concentration of substrate, in this case, on guaiacol concentration in the pressurised mixture, as reported in Fig. 4.

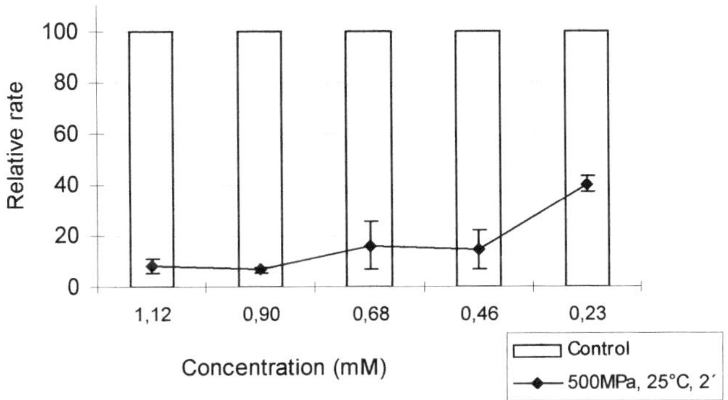

Fig. 4: Relative rate of the reaction of guaiacol and peroxidase after pressure release in dependence of substrate (guaiacol) concentration.

Following experiments evidenced that inactivation was only achieved when POD was pressurised in the presence of H_2O_2 and guaiacol. Fig. 5 shows the results after pressure was released: effects of addition of different substrates indicate that the enzyme was inactivated but H_2O_2 was not limiting the reaction. POD itself was not inactivated at 25 °C and 500MPa, and the rate was normal after pressure when guaiacol was used for testing enzyme activity (results not shown, but confirmed by the literature).

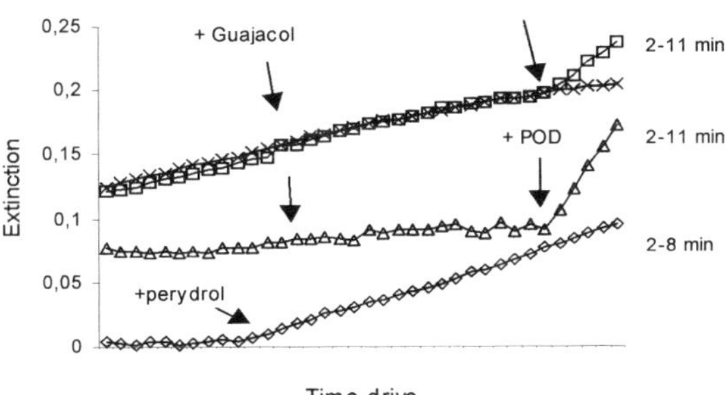

Fig. 5: Development of product at 435 nm of the reaction of peroxidase and guaiacol after pressure release when adding different components, but not to the control (a cross in the diagramm).

× control; □ control/plus guaiacol, plus fresh POD; △ (POD, perydrol and guaiacol) 500MPa, 25 °C, 2 min/plus guaiacol, plus fresh POD; ◇ (POD and guaiacol) 500MPa, 25 °C, 2 min/plus guaiacol perydrol.

192

Effects of guaiacol on the structure of peroxidase under pressure are under investigation.
Further experiments are required to determine the possibility of inactivating POD in a mixture of substrates. Existence of other substrates showing a similar effect is being also examined.

4. CONCLUSIONS

The influence of pressure on the reaction rate during pressure build-up and at end pressure is different for different substrates: substrate specific pressure dependence of K_m and V_{max}.
Peroxidase (POD), one of the highly pressure resistant enzymes, was irreversibly inactivated (>90 %) after 2 minutes at 500MPa/25 °C in the presence of guaiacol.

5. ACKNOWLEDGEMENTS

FAIR project 98-5031 made possible this investigation.

REFERENCES

1. C. Balny and P. Masson, Food Reviews International, 9 (1993) 611.
2. M. Hendrickx, L. Ludikhuyze, I. Van den Broeck and C. Weemaes, Trends Food Sci. Technol. 9 (1998) 197.
3. P. Butz, W.-D. Koller, B. Tauscher and S. Wolf, Lebensm-Wiss u Technol. 27 (1994) 463.

Trends in High Pressure Bioscience and Biotechnology
R. Hayashi (editor)
© 2002 Elsevier Science B.V. All rights reserved.

193

High pressure enhancement of cellulase activities

Takashi Yamanobe[a] and Kaoru Obuchi[b]

[a]Bioresources Laboratory and [b]Molecular and Cell Biology Laboratory,
National Institute of Advanced Industrial Science and Technology,
Higashi Tsukuba Ibaraki 305-8566, Japan

High pressure application of 150MPa did not affect CMCase activity but enhanced β-glucosidase and Avicelase activities of a cellulase system secreted from *Acremonium cellulolyticus* (Yamanobe et al., *Agric. Biol. Chem.* 1987; **51**: 65-74.) by 1.5 and 1.2 folds, respectively, at the optimum temperature under ambient pressure, 50°C. Addition of another cellulase system from *Aspergillus niger* made CMCase activity pressure sensitive, and the high pressure application doubled the activity of the mixture. The high pressure application did not enhanced filter papaer saccharifying (FPS) activity of *Acremonium* system in early stage of the reaction, but enhanced by 1.5 folds during prolonged reaction. With respect to the enzyme mixture, the high pressure application doubled this activity in the early stage, but the factor of the enhancement decreased to 1.5folds later. These time courses suggested that the rate-determining activity of FPS is CMCase activity in the early stage, and turns to be β-glucosidase activity later.

The high pressure application increased the optimum temperature of FPS reaction by the enzyme mixture by 15°C, and at 150MPa, the activity at 65°C was three times as high as that at 50°C. Therefore, the high pressure application enhanced FPS activity at the optimum temperature by 7.8 folds.

1. Introduction

Cellulolytic enzyme system comprises three major components, *i.e.* 1,4-β-D-glucan glucanohydrolase (CMCase; EC 3.2.1.4), 1,4-β-D-glucan cellobiohydrolase (Avicelase; EC 3.2.1.91) and β-glucosidase (cellobiase; EC 3.2.1.21). These saccharify cellulose synergistically, and well balanced activity composition is required to achieve effective saccharification. *Acremonium cellulolyticus* produces cellulolytic enzyme system with the activities appreciably balanced and several mutants of excellent productivity were isolated[1].

194

Cellulase from *Aspergillus niger* is deactivated at much higher pressure than 100MPa[2]. On the other hand, moderately high pressure accelerates enzymatic reactions, if the activation volume, $\Delta V^{\neq}$, is of negative value. $\Delta V^{\neq}$ is defined as difference in volumes in transition state and ground state and given by partial differentiation of activation free energy with respect to pressure, P, or:

$$\Delta V^{\neq} = -RT\left(\frac{\partial \ln k}{\partial P}\right)_T \qquad \text{(Eq. 1)}$$

where k, T, and R denote rate constant, absolute temperature, and gas constant, respectively[2]. Activation volume of any cellulase reaction has not been reported yet presumably due to the complexity of the reaction. This equation is also valid for thermal denaturation of enzymes. If activation volume of denaturation is of positive value, folded structure is kinetically stabilized under high pressure. It is expected that such enzymes can work at higher temperature under high pressure with higher activity than the optimum temperature at ambient pressure. This possibility may provides practical strategy to develop useful cellulolytic enzyme system for green chemistry. In this work, we investigated pressure dependence of cellulolytic activities and stability of the enzyme system from *Acremonium cellulolyticus*, and found eight fold enhancement of FPS activity of a mixture of enzyme systems from *Acremonium cellulolyticus* and *Aspergillus* sp. by high pressure application of 150MPa.

2. Materilas and methods

2.1 Enzymes and substrates

Cellulolytic enzyme preparations from a mutant of *Acremonium cellulolyticus* (Acremozyme™) were donated by Meiji Seika Co Ltd . (Tokyo) and Novo Nordisk Bioindustry Co. Ltd. (Denmark), respectively. Avicel-SF™, CMC (n≒500), and salicin were purchased from Asahi Kasei Co. Ltd. (Osaka), Tokyo Kasei Co. Ltd. (Tokyo), and Nakarai Tesque Co. Ltd.(Kyoto), respectively. Filter paper (No. 1) was purchased from Whatman International Ltd. (UK).

2.2 Activity determination

Catalytic reactions were expressed with amount of reducing sugars liberated in the following reaction mixture, *Acremonium* system of 0.00432 Avicelase international unit (IU) with 0.25% Avicel-SF for Avicelase activity; *Acremonium* system of 0.00923 β-glucosidase IU with 0.5% salicin for β-glucosidase activity; the systems from *Acremonium cellulolyticus* (0.00701 CMCase IU) and *Aspergillus* sp. (0.00520 CMCase IU) with 0.25% CMC and 0.25% CMC for CMCase activity, respectively; *Acremonium* system of 0.00701 CMCase IU or mixture of *Acremonium* system of

0.0154 CMCase IU and *Aspergillus* system of 0.0094 CMCase IU with a sheet of filter paper (FP, 6 x 3cm, 157.4±1.2mg) for FPS activity. The composition of the mixed enzyme systems is expressed with CMCase activity, because CMCase activity is the only significant activity of *Aspergillus* system. The reaction in Britton-Robinson's buffers (1/25M) pH of which was adjusted so that pH values become 4.5 at each pressure. The pressure-mediated pH-shift was estimated by the method of Neuman and Kauzman[3]. The reaction mixture (2ml) was loaded in a screw capped vial tube made of low-density polyethylene, which were inserted to a SUS pressurizing vessel thermostated at 50°C in a water bath, and pressurized using a TP-500 hand pump (Teramecs Co., Kyoto). Water was used as pressurizing medium in this system.

After the reaction duration of 60min, aliquot of the mixture (1ml) was taken, and liberated reducing sugars were quantified as increased absorbance at 500nm using the Somogyi-Nelson method[4]. Mean value and standard deviation were obtained from quadruplicate experiments.

2.3 High pressure DSC

An MC-2 microcalorimeter was equipped with high pressure applicator (Teramecs Co. Kyoto), which comprises two high pressure capillaries, capillary holder, a TP200MC high pressure pump driven by a stepping motor, a PG200D digital pressure gauge and SUS tubes connecting amoung the elements, and controlled by a desk-top PC. The pressurizing medium was water. The capillaries were made of SUS, one end of which was silver blazed. Outer diameter and inner volume of the capillary were 1.0mm and about 100μl, respectively.Sample protein solution and reference distilled water were loaded in the capillaries and inserted into the sample and reference cells of the calorimeter, respectively[5].

After pressure of the system reached designated value at the 5°C, temperature of the DSC cells were equilibrated at the temperature, and then increased to 95°C at the rate of 1°C/min. When the temperature reached 95°C, the cells were cooled to decrease to 5°C and equilibrated again, and the temperature scan was repeated. Trace of the second scan was subtracted from the first scan in order to eliminate the system dependent curvature. For thus obtained profile, cubic polynomial baseline was generated and corrected denaturation profile was obtained by subtraction of the baseline again.

3. Results and discussion

3.1 Pressure dependence of cellulase activities

Enzymatic activities were estimated as reducing sugar productivity in one hour at the optimum temperature of *Acremonium* cellulolytic enzyme system at ambient pressure, 50°C, and shown as the

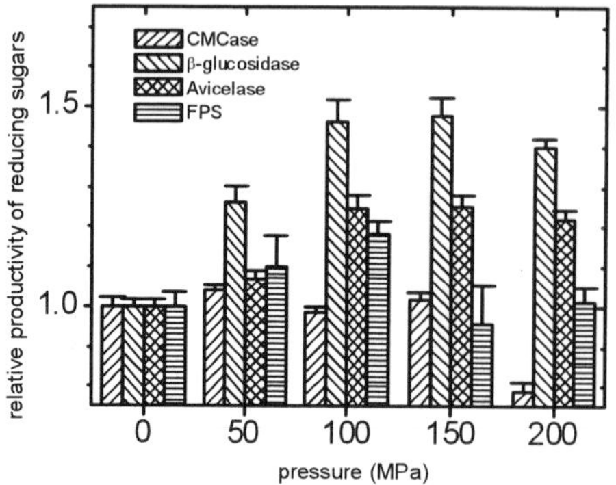

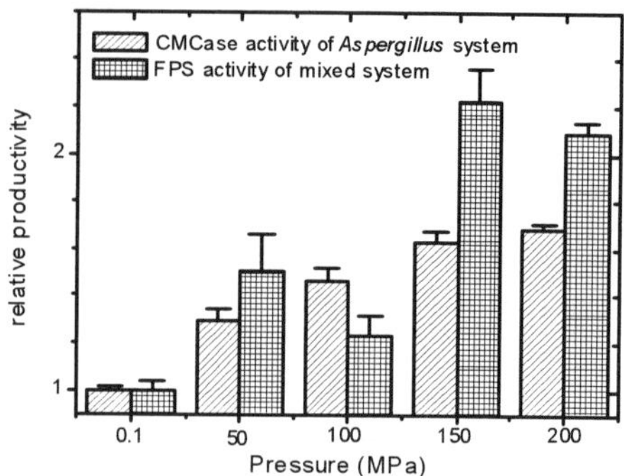

Fig. 1 Pressure dependence of acitivities of *Acremonium* **system.** Concentration of reducing sugar liberated in 48hr was normalized by the value at ambient pressure. Error bars denote standard deviation of the values from quadraplicate experiments.

Fig. 2 Pressure dependence of acitivities of *Aspergillus* **system and mixed system.** Concentration of reducing sugar liberated in 48hr was normalized by the value at ambient pressure. Error bars denote standard deviation of the values from quadraplicate experiments.

relative values normalized by the value at 0.1MPa in Fig. 1. Avicelase and β-glucosidase activities increased with increasing pressure up to 150MPa, and then, decreased at 200MPa.. The pressure application over the range of 0.1 to

150MPa did not affect CMCase activity, which declined at 200MPa. Pressure application of 150MPa did not increase FPS activity so significantly. Enhancement of β-glucosidase activity did not contribute to the FPS activity.

The cellulase system from *Aspergillus* sp. essentially comprises CMCase activity. Relative CMCase activity of this system increased with increasing pressure (Fig. 2). This activity was more baro-stable than the activity of *Acremonium* system, because the relative activity declined at 200MPa in less extent than that of *Acremonium* system. Significant enhancement of FPS activity of the mixed system was observed (Fig. 2). The extent of activity enhancement by the high pressure (150MPa) application appeared to result from synergistic effect of the enhancement of *Acremonium* Avicelase, *Acremonium* β-glucosidase, and CMCase of the mixed enzyme.

The pressure dependence illustrated in Figs. 1 and 2 imply that values of activation volume are negative for Avicelase and β-glucosidase activities of *Acremonium* system and CMCase activity of *Aspergillus* system.

3.2 Time course of FPS activity

With progress of FPS reaction, amount of reducing sugar liberated by *Acremonium* system increased almost constantly for 48hours at both 0.1 and 150MPa, implying that FPS activity of

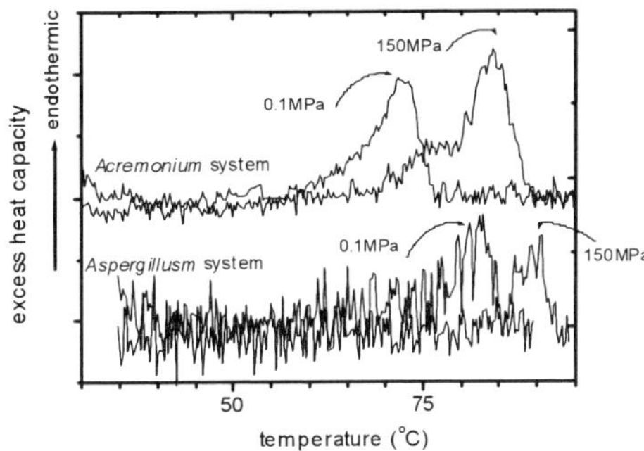

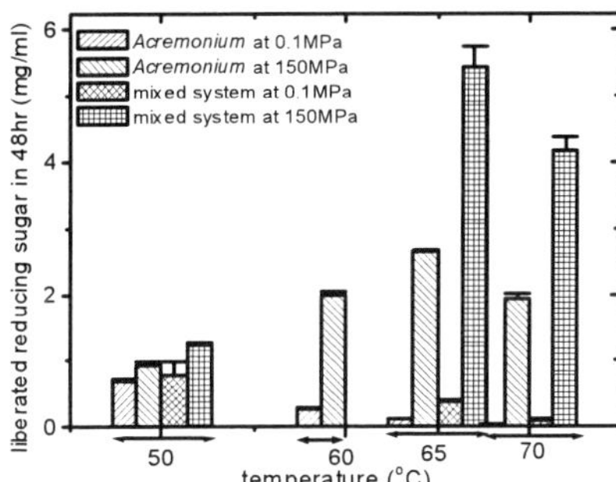

Fig. 3 Corrected DSC profiles of cellulase system. Scans were conducted using the high pressure DSC system described in the experimenyal section.

Fig. 4 Thermal stability of productivity. Error bars denote standard deviations obtained from quadraplicate experiments.

Acremonium system is stable even at 150MPa. Relative FPS activity at 150MPa, normalized by the value at 0.1MPa, increased up to 1.387±0.004 in the first 24 hours and held steady in the next 24 hours (1.33±0.09 at 48hr). The steady relative activity fell to the value between the values of β-glucosidase (1.47±0.04 at 150MPa) and Avicelase (1.25±0.02 at 150MPa), suggesting that the rate determining activity turns to be either β-glucosidase activity or Avicelase activity with the progress of FP saccharification. Relative FPS activity of the mixed enzyme decreased to 1.48±0.04 in the first 24 hours, and then was steady in the next 24 hours (1.56±0.04 at 48hr). These values are comparative to the value of β-glucosidase activity, suggesting that the rate determining activity turns to be β-glucosidase activity. Although mixing with *Aspergillus* system improved high pressure enhancement of FPS activity only in early stage, the mixing is more effective to increase FPS activity than increasing of the amount of *Acremonium* system: the pressure application increased the hydrolysis degree at 48hr with the mixed enzyme in more extent (1.56 fold) in comparison with the *Acremonium* system (1.33 fold).

3.3 High pressure stabilization of molecular structure and catalytic activity

When thermal denaturation of the system from *Acremonium* was analyzed at 0.1 and 150MPa by high pressure DSC, the traces of rescans were without significant endotherms. Thus the denaturation observed in the initial scan was irreversible. The high pressure application of 150MPa increased T_m values by about 10 and 15°C for *Aspergillus* and *Acremonium* systems, respectively (Fig. 3). These up-shifts of denaturation endotherms show that the high pressure application decreases denaturation rates of the systems at particular temperature[6] and stabilizes molecular structures. Consequently, activation volume of thermal denaturation of the rate determinig activity for the systems are of positive values.

It was expected that the catalytic activities are also stabilized under high pressure. FPS activity of *Acremonium* system at ambient pressure increased with increasing temperature up to 50°C, while the maximum activity at 150MPa was observed at 65°C (Fig 4), implying that the high pressure application of 150MPa stabilized the rate determining catalytic activity, as well as the strusture, of the system by 15°C . The productivity at 150MPa 65°C was three times as high as that at 0.1MPa 50°C. Better stabilization was observed for FPS activity of the mixed enzyme: the productivity at 150MPa 65°C was 7.8 times as high as that at 0.1MPa 50°C.

It is concluded that the high pressure application to FPS reaction enhanced the catalytic activities by two factors synergistically:
(1) negative activation volume of catalytic reaction enhances the activity according to Eq. 1;
(2) positive activation volume of thermal denaturation enables increase in optimum temperature at 150MPa according to Eq1 1, and this increase enhances the activity according to the Arrhenius equation.

REFERENCES

1. T. Ymanobe, Y. Mitsuishi, and Y. Takasaki. Agric. Biol. Chem. 51, (1987) 65.

2. R. Jaenicke. Ann. Rev. Biophys. Bioeng. 10, (1981) 1.

3. RCJr. Neuman, W. Kauzman, and A. Zipp. J. Phys. Chem. 77, (1973) 2687.

4. MJ. Somogyi. J. Biol. Chem. 195, (1952) 19.

5. K. Obuchi, H. Iwahashi and Y. Komatsu. Acquired barotolerance and thermal stability of cellular components. in Advances in high pressure bioscience and biotechnology, Springer-Verlag, Berlin, 1998.

6. K. Obuchi, H. Iwahashi, J.R. Lepock and Y. Komatsu. Yeast. 16, (2000) 111.

7. K. Obuchi, and T. Yamanobe. This book.

Trends in High Pressure Bioscience and Biotechnology
R. Hayashi (editor)
© 2002 Elsevier Science B.V. All rights reserved.

Kinetics of lipoxygenase inactivation in soybean and green beans

Indrawati, A.M. Van Loey, L.R. Ludikhuyze, M.E. Hendrickx

Laboratory of Food Technology, Department of Food and Microbial Technology, Katholieke Universiteit Leuven, Kasteelpark Arenberg 22, B-3001 Heverlee, Belgium.

Lipoxygenase (LOX) inactivation due to a combined pressure (up to 700 MPa) and temperature (-10 to 70°C) treatment was studied in a model system i.e. commercially purified soybean LOX (0.4 mg/mL) in Tris HCl buffer (10mM;pH9). LOX inactivation in green bean juice and intact green beans (*in situ* study) were investigated to study the influence of natural intrinsic food complexity on the LOX inactivation kinetics. It was noticed that LOX could be irreversibly inactivated by a high pressure-temperature treatment both in soybean and in green beans. The inactivation followed a first order reaction. Studies on green beans showed that the pressure needed to obtain the same rate of LOX inactivation at a certain temperature was higher in green bean juice than intact green beans. Based on the kinetic data obtained, a mathematical model describing pressure temperature dependence of D values for LOX inactivation and iso-D-value contour diagrams for LOX inactivation either in the model system or in the real food product as a function of pressure and temperature were constructed.

1. INTRODUCTION

Lipoxygenase (LOX) is widespread in plant tissues, especially in legumes. Its enzymatic reaction, resulting in some undesirable effects (e.g. destruction of essential fatty acids, development of off-flavors, color degradation), can be prevented by performing a blanching treatment prior to further processing. Thermal treatment is an effective way to inactivate LOX, however it also causes other food quality degradations. Therefore, high pressure processing is proposed as an alternative to replace this treatment. The objective of this research was to study the effect of high pressure (up to 700 MPa) combined with temperature (-10 to 70°C) on enzyme inactivation particularly on lipoxygenase. The inactivation study was carried out on a kinetic basis in different systems, i.e. a model system (commercially purified enzyme diluted in buffer solution) and a real food system (i.e. green bean juice and intact green beans – *in situ*).

2. MATERIALS AND METHODS

2.1. Model system and LOX activity measurement

Lipoxygenase type 1 B from soybean (Sigma) 0.4 mg/mL in Tris HCl buffer (10mM;pH9) was used as a model system. The enzyme activity was measured spectrophotometrically at 234 nm and 25°C using linoleic acid as a substrate and sodium borate buffer (12.5mM; pH9).

2.2. Real food system and LOX activity measurement

Green beans (*Phaseolus vulgaris* L.) were used as a real food system. Different levels of food complexity, i.e. LOX in green bean juice and in intact green beans (*in situ*) were studied. The juice was obtained by squeezing the beans and subsequently centrifuging (25900g; 4°C; 15 min.). The juice and green beans (for *in situ* study) were stored under liquid nitrogen until use. The enzyme activity either in juice or *in situ* was measured polarographically at 25°C using linoleic acid as substrate and phosphate buffer (10mM;pH6).

2.3. LOX extraction for *in situ* study

The (un)treated beans were cut into small slices (3-4 mm) and frozen in liquid nitrogen. The frozen beans were ground and the resulting bean powder was extracted using phosphate buffer (10mM;pH6) (containing 0.5% Tween 20 and 0.5% Triton X-100). The homogenized mixture was filtered through a folded paper filter (Schleicher & Schuell 595½ ø 185 mm) at 5°C and the resulting filtrate was used for LOX activity measurement.

2.4. High pressure treatment

2.4.1. *Combined pressure-temperature treatment of soybean LOX and green bean juice*

Pressure temperature treatments for LOX inactivation were performed in a thermostated multivessel high pressure equipment (8 vessels, Resato, Roden, Holland). This HP equipment is especially designed for simultaneous experiments at the same pressure and temperature combinations during different preset times. The enzyme solution (i.e. model system) and green bean extract were filled in flexible microcups (Elkay, 0.375 mL) without creating air bubbles and closed with parafilm. For pressure experiments at subzero temperature, the samples were wrapped in vacuumed polyethylene pockets to protect the extract solution from pressure medium contamination. Pressure was built up at a constant pressurization rate of 100-125 MPa/min. After an equilibrium time of 5 minutes to ensure achievement of isobaric isothermal conditions, vessels were decompressed at predetermined time intervals.

2.4.2. *Combined pressure-temperature treatment of green beans in situ*

Pressure-temperature treatments were carried out in a one vessel pilot scale high pressure apparatus (Engineered Pressure Systems Int., Belgium). The beans were vacuum-packed in polyethylene plastic bags. The pressure was built up and decompressed automatically. The beans were treated individually for a certain preset time at the same pressure and temperature combination.

2.5. Data analysis

In the whole P/T area studied, LOX inactivation obeys a first order kinetic model. Hereto, the Thermal Death Time model could be used to describe the P/T inactivation rate as D value (decimal reduction time, 2.303/k, time needed to reduce the initial activity with 90%, equation 1). To describe the pressure dependence of D values, z_p values were calculated according to equation 2.

$$log(A) = log(A_o) - \frac{t}{D} \tag{1}$$

$$D = D_{ref} * 10^{\frac{(P_{ref} - P)}{z_p}} \tag{2}$$

2.6. Formulation of a mathematical model

An empirical mathematical model describing the combined pressure and temperature dependence of the LOX inactivation rates (i.e. D values) was constructed to fit the entire kinetic inactivation data set. As a measure for the ability of a model to fit all experimental data (i) a visual inspection of residual plots was performed; (ii) the corrected r^2 (equation 3) and (iii) the model standard deviation (equation 4) were calculated.

$$\text{corrected } r^2 = \left[1 - \frac{(m-1)(1 - \frac{SSQ_{regression}}{SSQ_{total}})}{(m-j)} \right] \tag{3}$$

$$SD = \sqrt{\frac{SSQ_{residual}}{(m-j)}} \tag{4}$$

where m = number of observations and j = number of model parameters.

3. RESULTS AND DISCUSSIONS

3.1. Kinetics of LOX inactivation due to combined pressure and temperature treatment

Under all experimental conditions studied, it was noticed that combined pressure temperature inactivation of LOX either in soybean or in green beans was irreversible. The inactivation could be described as a first order reaction. Therefore, the LOX inactivation rate was estimated as D values. At constant temperature, the irreversible LOX inactivation could be enhanced by increasing pressure. The calculated z_p values at various constant temperatures are given in Table 1.

Table 1
Pressure dependence of D values for LOX inactivation in different model systems at different constant temperatures

Temperature (°C)	z_p value (MPa)		
	Soybean LOX in Tris HCl buffer (10mM;pH9)	LOX in green bean juice	LOX in intact green beans
-10	97.09 (r^2=0.887)	116.28 (r^2=0.966)	/
0	78.74 (r^2=0.981)	123.46 (r^2=0.975)	/
10	75.19 (r^2=0.998)	144.93 (r^2=0.966)	200.00 (r^2=0.995)
20	68.97 (r^2=0.983)	169.49 (r^2=0.968)	/
30	126.58 (r^2=0.999)	400.00 (r^2=0.941)	147.06 (r^2=0.976)
40	212.77 (r^2=0.966)	357.14 (r^2=0.935)	212.77 (r^2=0.971)
50	/	161.29 (r^2=0.962)	555.56 (r^2=0.986)

At elevated pressure, the D value was maximal around 30°C (for soybean) and 10°C (for green beans), indicating the highest pressure stability of LOX at that temperature. The D value could be decreased (i.e. higher LOX inactivation rate) either by a temperature increase as well as by a temperature decrease respectively above and below that temperature. An antagonistic effect of pressure and low/subzero temperatures on LOX inactivation of soybean as well as in green beans was observed.

3.2. Effect of food complexity on pressure inactivation of green bean LOX

Thermal inactivation of LOX either in green bean juice or *in situ* was irreversible. Under isothermal conditions, LOX inactivation in the juice showed a biphasic behavior, i.e. a succession of two first order inactivation reactions, namely of a heat labile and a heat stable fraction [1]. In contrast to thermal inactivation, a single phase (i.e. a first order reaction) in the pressure temperature inactivation curves of LOX in green bean juice or *in situ* was observed. It could be caused by a partial LOX inactivation during pressure built up, which is accompanied by a temperature increase due to adiabatic heating of the pressure medium and during the equilibrium time of 5 minutes (e.g. pressurization at 300 MPa and 60°C resulting in 12% of inactivation while at least 50% of LOX inactivation occurred for pressurization above 500 MPa at 30°C). Based on the calculated D values, it was noticed that the pressure needed to obtain the same D value of LOX inactivation at a certain temperature was less *in situ* than in green bean juice. From Table 1, it can be seen that the evolution of the z_p value as a function of temperature for LOX inactivation in green bean juice and *in situ* is not similar.

3.3. Mathematical modeling to describe combined pressure temperature dependence of D values

In literature, no general applicable kinetic model that describes pressure temperature enzyme inactivation data is available. In the present study, the modeling was approached starting from a basic thermodynamic equation governing the behavior of a system during a pressure and a temperature change as previously described by Hawley [2] and Morild [3]. It is often used to understand a systems reversible response towards pressure and temperature. This thermodynamic model can be converted into a kinetic model (equation 5) through the transition state theory of Eyring, suggesting that enzyme inactivation is accompanied by a formation of a metastable/transition activated state ($^{\neq}$) which exists in equilibrium with the native enzyme.

$$ln(D_{obs}) = ln(D_0) + \frac{\Delta V_0^{\neq}}{R_T T}(P - P_o) - \frac{\Delta S_0^{\neq}}{R_T T}(T - T_o) + \frac{1}{2} \frac{\Delta \kappa^{\neq}}{R_T T}(P - P_o)^2 \tag{5}$$

$$+ \frac{\Delta \zeta^{\neq}}{R_T T}(P - P_o)(T - T_o) - \frac{\Delta C_p^{\neq}}{R_T T}\left[T(ln \frac{T}{T_o} - 1) + T_o \right]$$

The model parameters are estimated using a non linear regression analysis, involving an iterative numerical procedure based on the minimal sum of squares. No trend in residuals was noticed as a function of temperature, pressure or experimental D values. The estimated values of the model parameters are summarized in Table 2 and a good agreement between the natural logarithm (*ln*) of D values predicted by equation (5) and the natural logarithm of the experimentally determined D values was observed for all systems (Figure 1).

Table 2
Predicted model parameters for LOX inactivation of soybean and in green beans based on equation (5) at reference pressure 500 MPa and reference temperature 298K

Parameter	Soybean LOX in Tris HCl buffer (10mM;pH9)	LOX in green bean juice	LOX in intact green beans
D_o (min)	179.92 ± 15.46^a	171.87 ± 20.52^a	93.24 ± 12.46^a
$\Delta V_o^{\neq}$ (cm³.mol⁻¹)	-34.20 ± 2.22	-30.81 ± 2.42	-35.67 ± 4.45
$\Delta S_o^{\neq}$ (J.mol⁻¹.K⁻¹)	-20.65 ± 7.20	90.63 ± 12.29	139.41 ± 23.95
$\Delta \kappa^{\neq} (*10^{-2} cm^6.J^{-1}.mol^{-1})$	-1.40 ± 1.06	-0.54 ± 1.55	-7.34 ± 2.62
$\Delta \zeta^{\neq}$ (cm³.mol⁻¹.K⁻¹)	0.64 ± 0.05	0.44 ± 0.10	0.02 ± 0.16
$\Delta C_p^{\neq}$ (J.mol⁻¹.K⁻¹)	3046.55 ± 207.16	2466.71 ± 323.41	2811.82 ± 394.92
		Quality of fitting	
corrected r^2	0.985	0.988	0.986
SD	0.45	0.40	0.44

[a] : asymptotic standard error

By inserting all estimated model parameters into equation (5), pressure temperature combinations resulting in a specific preset D values for LOX inactivation of soybean and in green beans were predicted and illustrated in Figure 2.

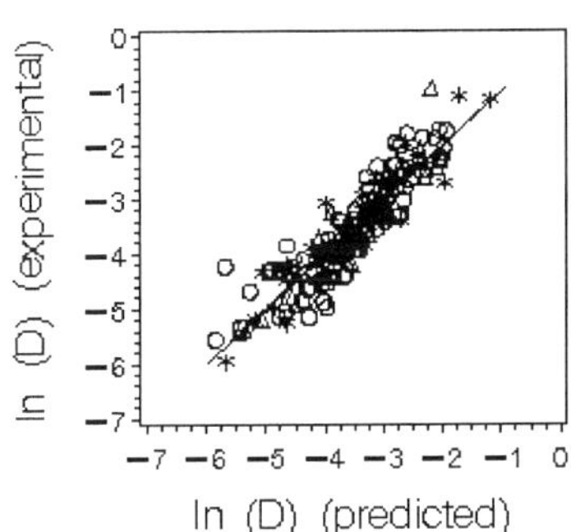

Figure 1. Relation between the *ln* of experimentally determined D values of combined P/T inactivation of soybean LOX (o); LOX in green bean juice (Δ); in intact green beans (*) and the *ln* of predicted D values according to equation (5)

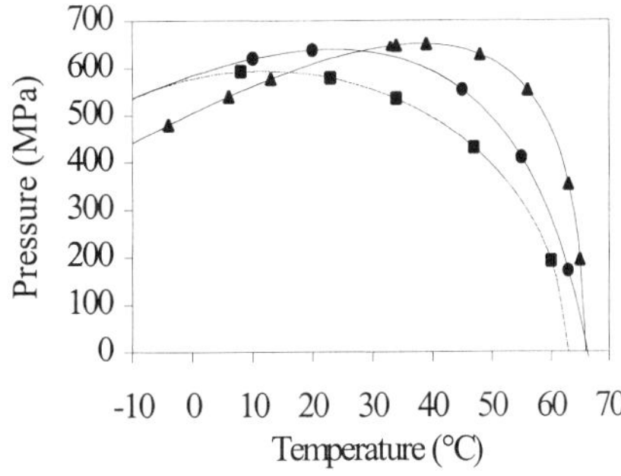

Figure 2. Pressure temperature kinetic diagram of LOX inactivation in different sources, i.e. soybean (▲); green bean juice (●) and green beans *in situ* (■) for D value of 10 min.

It was observed that the predicted pressure temperature levels for LOX inactivation either in green beans or of soybean was very similar in the high temperature ($\pm$ 65°C) and low pressure (< 200 MPa) area. However, at lower temperatures (30 - 60°C) and higher pressure, significant differences became noticeable, the stability being reduced from soybean over green bean juice to intact green beans. At low (<30°C) and subzero temperatures, the ranking became reverse. In the latter case, soybean LOX could be inactivated at lower pressure and temperature as compared to LOX in green beans. In this study, a kinetic approach to describe pressure temperature dependence of D values for LOX inactivation has been successfully transferred and applied from model system, i.e. soybean LOX to a real food product, i.e. green beans.

4. CONCLUSIONS

Similarities between the kinetic behavior of LOX inactivation in a model system (purified soybean LOX 0.4 mg/mL in 10 mM Tris HCl buffer pH 9) and a real food product (i.e. green beans) are (i) the applicability of the first order inactivation model; (ii) the applicability of the same mathematical model describing the pressure temperature dependence of LOX inactivation rate constants; (iii) the existence of an antagonistic effect of pressure at low/subzero temperature and (iv) the evidence of the highest pressure stability of LOX inactivation at room temperature. Studies on LOX inactivation in the crude extract are representative for intact vegetable especially in terms of the global inactivation behavior of LOX in a specific vegetable. Nevertheless, the level of pressure and temperature needed for LOX inactivation is dependent on the food complexity.

REFERENCES

1. Indrawati et al, Biotechnol. Prog., 15(2) (1999) 273.
2. S.A. Hawley, Biochemistry, 10 (1971) 2436.
3. E. Morild. Adv. Prot. Chem., 34 (1981) 93.

Acknowledgements
Our acknowledgements are addressed to *Katholieke Universiteit Leuven* (Research Council)-F.W.O-Vlaanderen and the European Commission (project FAIR-CT96-1175).

Trends in High Pressure Bioscience and Biotechnology
R. Hayashi (editor)

Effects of high pressure treatment on rabbit muscle proteasome

S. Yamamoto[a], Y. Otsuka[a], K. Masuda[b], Y. Ikeuchi[c], T. Nishiumi[b] and
A. Suzuki[b]

[a]Graduate School of Science and Technology, Niigata University, and
Department of Applied Biological Chemistry, Faculty of Agriculture,
Niigata University, Ikarashi, Niigata 950-2181, Japan
[c]Division of Food Animal Science, Graduate School of Agricultural Science,
Kyushu University, Hakozaki, Fukuoka 821-8581, Japan

ABSTRACT

This paper describes the effects of high hydrostatic pressure treatment on the
structure and activities of rabbit skeletal muscle proteasome.

It was cleared that the high pressure treatment up to 50~100 MPa caused the
activation of the proteasome as observed in the mild denaturing treatment with
SDS, heat etc. This activation is probably due to the increase of the interaction of
the substrate and active site of the proteasome unfolded by high pressure
treatment.

1. INTRODUCTION

Proteasome (multicatalytic enzyme complex: MCP) isolated from tissues in a
latent form can be activated various chemicals and treatment such as polylysine
[1,2], SDS [1, 3, 4] and fatty acids [3], and heat treatment [5,6] and dialysis
against water [7]. Recently, the activation of the proteasome by high pressure
treatment has been reported [8,9].

High hydrostatic pressurization is one of the new technologies for tenderizing
meat or accelerating meat conditioning [10-14]. The activation of proteasome
induced by high pressure may be involved in the causes of the pressure-induced
modification or breakdown of muscle proteins. However it is not clear why the
proteasome is activated by high hydrostatic pressurization.

This paper describes the effects of high hydrostatic pressure treatment on the
structure and activities of rabbit skeletal muscle proteasome.

2. MATERIALS AND METHODS

2.1. Purification of proteasome
Proteasome was purified from rabbit skeletal muscle by the method of Otsuka
et al. [8].

2.2. Determination of proteasome activities under high pressure

Proteasome activities were assayed by fluorometric measurement of the release of 7-amini-4-methylcoumarin (-NH-Mec; AMC) after incubation with synthetic substrates according to the procedure described by Otsuka *et al.* [8]. The substrates, Suc-Leu-Leu-Val-Tyr-NH-Mec (LLVY), Boc-Leu-Ser-Thr-Arg-NH-Mec (LSTR) and Ala-Ala-Phe-NH-Mec (AAF), were chosen to assay distinct proteolytic activities. The activities under high pressure (0.1-300 MPa) were measured by HITACHI F-2000 spectrofluorimeter equipped with TP-500 high hydrostatic pump (TERAMECS Co., Ltd.).

2.3. Determination of FITC-labelled casein hydrolyzing activities

The mixture of proteasome and FITC-labelled casein in 100 mM Tris-HCl, 1mM DTT (pH 8.0), was vacuum-sealed in a small polyethylene bag and pressurized at 0.1~400 MPa for 90 min at 37 ℃ by the method of Homma *et al.* [15]. Then, the released products from the FITC-casein were measured by the method of Lonergam *et al.* [16].

2.4. Measurement of fluorescence spectra and center of spectral mass of proteasome under high pressure

The changes in fluorescence spectra of the proteasome were measured by HITACHI F-2000 spectrofluorimeter equipped with TP-500 high hydrostatic pump (TERAMECS Co., Ltd.). Proteasome solution (0.03mg/ml) in 40 mM Tris-HCl, 10 mM EDTA, 10 mM 2-mercaptoethanol and 100 mM NaCl (pH 7.5), was exposed to high pressure of 0.1-400 MPa for 5 min. The changes in center of spectral mass were calculated by the method of Ruan *et al.* [17].

2.5. Measurement of the secondary structure of proteasome

After high hydrostatic pressure treatment (0.1-400MPa, 10℃, 5min), CD spectra of proteasome solution (0.1mg/ml) in 40 mM Tris-HCl, 10 mM EDTA, 10 mM 2-mercaptoethanol and 100 mM NaCl (pH 7.5) was recorded on a JASCO J-725 spectropolarimeter at 20℃. The secondary structure of proteasome was analyzed by the program of Yang *et al.* [18].

3. RESULTS AND DISCUSSION

3.1. Effects of high pressure on synthetic peptides and FITC-labelled casein hydrolyzing activities

Changes in LLVY hydrolyzing activity under high pressure are shown in Fig.1. During the reaction of 30 min, the degradation of LLVY linearly increased with the pressure applied up to 150 MPa and was higher than that of the unpressurized (0.1 MPa). When the pressure of 200 MPa or more applied, the progress of the degradation was not observed after first reaction of 2 min. As shown in the Fig.1, the optimal pressure of the degradation was 50 MPa. Ostuka *et al.* [8] showed that the optimal pressure inducing the highest LLVY

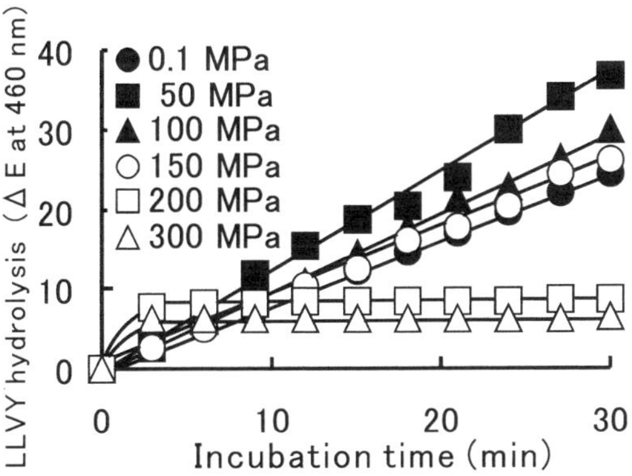

Fig.1. Effect on LLVY hydrolyzing activities

hydrolyzing activity of the pressure treated-proteasome was about 100 MPa. This difference may be due to the difference in the analytical methods. The former was measured directly under high pressure, and the latter was measured after pressure removed. The increase of proteasome activity is probably due to the increase of the interaction of the substrate and active site of the proteasome unfolded by high pressure treatment.

Km and Vmax values of proteasome for synthetic peptides under high pressure are shown in Table 1. The calculated Km and Vmax values for LLVY and AAF increased under high pressure up to 150 and 100 MPa, respectively, whereas the Km and Vmax values for LSTR were not influenced.

Effects of pressurization on casein degrading activity are shown in Fig.2. During the reaction of 90 min, the degradation of FITC-labelled casein linearly increased with the increase of the pressure applied up to 150 MPa. When the pressure of 400 MPa applied, the degradation of the casein at first 30 min of

Table 1 *Km and Vmax* values of proteasome under high pressure

Pressure	LLVY		LSTR		AAF	
	Km	*Vmax*	*Km*	*Vmax*	*Km*	*Vmax*
(MPa)	(µM)	(pmol/min/ml)	(µM)	(pmol/min/ml)	(µM)	(pmol/min/ml)
0.1	30.63	397.08	16.97	139.63	12.36	173.13
50	41.95	685.32	17.18	195.49	16.06	270.36
100	34.09	509.25	17.14	167.19	15.33	209.10
150	31.14	453.80	17.13	160.84	11.41	161.19
200	8.02	67.95	15.01	117.65	5.01	54.05
300	7.89	61.98	10.20	83.60	3.23	27.18

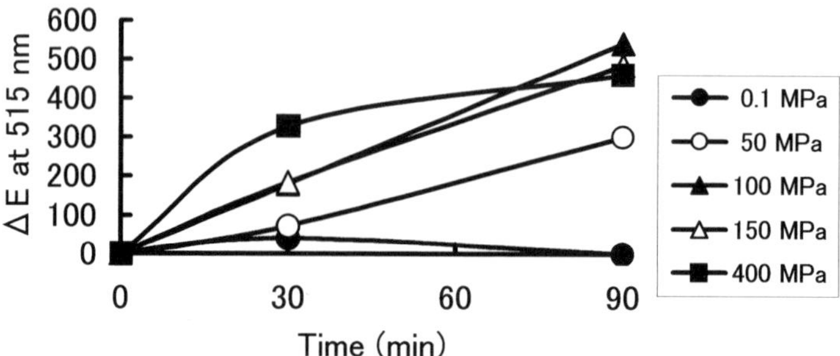

Fig.2. Effect on FITC-labelled casein hydrolyzing activities

reaction was highest. After 30 min of the reaction, the progress of the degradation was slower than that of the proteasome pressurized at more lower pressures, and finally at the 90 min reaction time, the degradation was less than that of the proteasomes pressurized at 100 and 150 MPa. These activation are probably due to the increase of the interaction of the substrate and active site of the proteasome unfolded by high pressure treatment. The difference in optimal pressure for synthetic peptides and FITC labelled-casein hydrolyzing activities may be due to the difference of active site of proteasome for substrates.

3.2. Changes in fluorescence spectra and center of spectral mass of proteasome under high pressure

The red shift and the decrease of the fluorescence of the proteasome gradually progressed with the increase of high pressure applied.

From the calculation of the center of mass (Fig.3), the pressure causing the

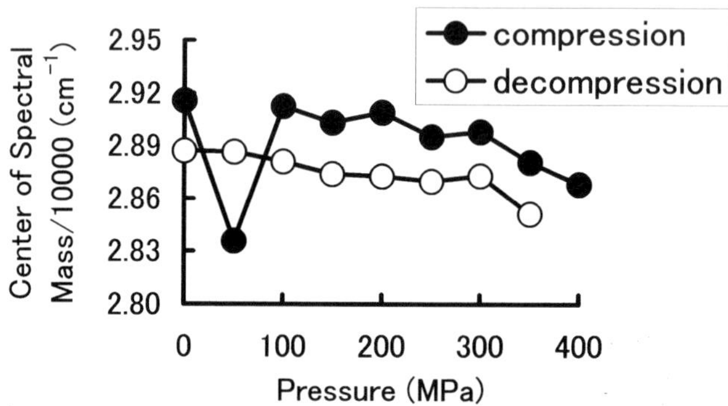

Fig.3. Changes in center of spectral mass of proteasome

Table 2 Effect of pressure on the secondary structure of proteasome

Pressure (MPa)	Helix (%)	Beta (%)	Turn (%)	Random (%)	Total (%)
0.1	54.0 ± 5.9	18.3 ± 6.5	7.1 ± 4.3	20.7 ± 3.7	100.0
50	53.3 ± 3.6	21.3 ± 1.4	8.9 ± 1.8	16.4 ± 3.9	100.0
100	48.8 ± 3.3	21.3 ± 3.9	9.5 ± 0.3	24.4 ± 0.3	100.0
200	52.2 ± 3.0	22.4 ± 0.1	5.6 ± 1.3	19.9 ± 4.3	100.0
300	45.6 ± 1.9	27.1 ± 2.4	2.2 ± 2.2	25.3 ± 1.8	100.0
400	38.7 ± 10.1	24.9 ± 1.6	2.9 ± 2.9	25.4 ± 3.2	100.0

denaturation of the proteasome was estimated to be 50 MPa.

3.3. Changes in the secondary structure of proteasome

Changes in the secondary structure of proteasome induced by high pressures are shown in Table 2. On the basis of the CD spectra, the decrease of the α-helix content of the proteasome was observed first at 100 MPa and then further decreased with the increase of the pressure applied.

ACKNOWLEDGEMENTS

This work was supported in part by Grant- in- Aid for Scientific Research (10460118) from the Ministry of Education, Science and Culture of Japan.

REFERENCES

1. K.Tanaka, K.Ii, A.Ichihara, L.Waxman and A.L.Goldberg, J.Biol.Chem., 261 (1986), 15197.
2. R.L.Mellgren, Biochim. Biophys. Acta, 995 (1990), 181.
3. S.Wilk and M.Orlowski, J. Neurochem., 40 (1983), 842.
4. B.Dahlmann, M.Rutschmann, L. Kuehn and H.Reinauer, Biochem.J., 228 (1985), 171.
5. D.L.Mykels, J.Exp.Zool., 250 (1989), 244.
6. D.L.Mykels, Biochem.Biophys. Acta, 274 (1989), 216.
7. M.J.McGuire, M.L.McCullough, D.E.Croall and G.M.DeMartino, Biochim. Biophys.Acta, 995 (1989), 181.
8. Y.Otsuka, N.Homma, K.Shiga, J.Ushiki, Y.Ikeuchi and A.Suzuki, Meat Sci., 49 (1998), 365.
9. F.Gardrat, B.Fraigneau, V.Montel, J.Raymond and J.L.Azanza, Eur.J.Biochem., 262 (1999), 900.
10. P.E.Bouton, A.L.Ford, P.V.Harris, J.J.Macfarlane and J.M.O'Shea, J.Food Sci., 42 (1977), 132.

11. R.H.Locker and D.J.C.Wild, Meat Sci., 10 (1984), 207.
12. J.J.Macfarlane, Developments in Meat Science-3 (R.A.Lawrie, ed.), Elsevier Applied Science Publisher, Essex, 1985.
13. A.Suzuki, K.Kim, N.Homma, Y.Ikeuchi and M.Saito, High Pressure and Biotechnology (C.Balny, R.Hayashi, K.Heremans and P.Masson, eds.), Colloques INSERM/John Libbey Eurotext, Montrouge, 1992.
14. A.Suzuki, N.Homma, A.Fukuda, K.Hirao, T.Uryu and Y.Ikeuchi, Meat Sci., 37 (1994), 369.
15. N.Homma, Y.Ikeuchi and A.Suzuki, Meat Sci., 41 (1995), 251.
16. S.M.Lonergan, M.H.Johnson and C.R.Calkins, J.Food Sci., 60 (1995), 72.
17. K.Ruan, R.Lange, Y.Zhou and C.Balny, Biochem.Biophys.Res.Commun., 249, (1998) 844.
18. J.T.Yang, C.-S.C.Wu and H.M.Martinez, Methods Enzymol., 130 (1986), 208.

Trends in High Pressure Bioscience and Biotechnology
R. Hayashi (editor)
                              211

Molecular mechanisms of pressure-regulation at transcription level in piezophilic bacteria

C. Kato, K. Nakasone, A. Ikegami, H. Kawano, R. Usami, and K. Horikoshi

The DEEP STAR Group, Japan Marine Science and Technology Center, 2-15 Natsushima-cho, Yokosuka, Kanagawa 237-0061, Japan

Deep-sea bacteria have unique systems for gene and protein expression controlled by hydrostatic pressure.   One of the σ factors, σ^{54}, was found to play an important role on the pressure-regulated transcription in a deep-sea piezophilic bacterium, *Shewanella violacea*.   A glutamine synthetase gene (*glnA*) has been targeted as a model for the pressure-regulated promoter to investigate the transcriptional regulation by the σ^{54} factor.   Recognition sites for σ^{54} and σ^{70} factors were observed at an upstream region of the *glnA* and also NtrC-binding sites were identified at this same region.   Primer extension analyses revealed that the transcription initiation sites of both promoters were determined and that the transcription from the σ^{54} site was regulated by elevated pressure.   The σ^{54} promoter is known to be activated by a two components signal transduction system, NtrB-NtrC phospholylated relay.   Our results suggested that this system might be regulated by deep-sea conditions and that the gene expression controlled by the σ^{54} promoter was actually regulated by pressure.   We proposed a possible model of the molecular mechanisms for pressure-regulated transcription.

1. INTRODUCTION

The psychrophilic, moderately piezophilic bacterium *Shewanella violacea* DSS12 is a deep-sea isolate from a mud sample collected at the Ryukyu Trench (depth: 5110 m), which grows optimally at 30 MPa and 8°C, but also grows at atmospheric pressure (0.1 MPa) and 8°C (1, 2).   We have targeted this strain to elucidate the molecular basis of gene regulation in piezophilic bacteria at different pressure conditions, because it is useful as a model bacterium for comparing the various features of bacterial physiology under pressure conditions (3). Isolation and characterization of several pressure-regulated *cis*-acting elements have been conducted (4-6).   Through the sequence analyses of these *cis*-elements in pressure-regulated genes, the consensus sequence for RNA polymerase sigma factor σ^{54} was found upstream of the genes and the *S. violacea* σ^{54} was shown to bind to this region (5-7).   Thus we have focused on the molecular mechanisms of σ^{54}-dependent transcription under different pressure conditions.   The σ^{54}-containing RNA polymerase has been shown to be responsible for the transcription of several genes, e.g. nitrogen metabolic genes such as *glnA* operon (8).   Our approach to understand the basis of gene expression under defined conditions is by detailed characterization of the components of the transcriptional machinery and the accessory factors involved.   In previous studies, this regulation has been shown to be mediated by one of the σ

factors, σ^{54} and a two components regulatory system composed of the bacterial signal-transducing protein NtrB and the bacterial enhancer-binding protein NtrC (6, 7, 9).

This review summarizes our recent work in transcriptional regulation under pressure conditions by the σ^{54} factor in *Shewanella violacea*. We describe the upstream *cis*-elements isolated from this strain, which are involved in transcriptional regulation under high-pressure conditions. Characterization of several *trans*-acting factors such as σ factors and two component transcriptional regulators contributing to this pressure-regulation in this piezophilic bacterium, is also discussed.

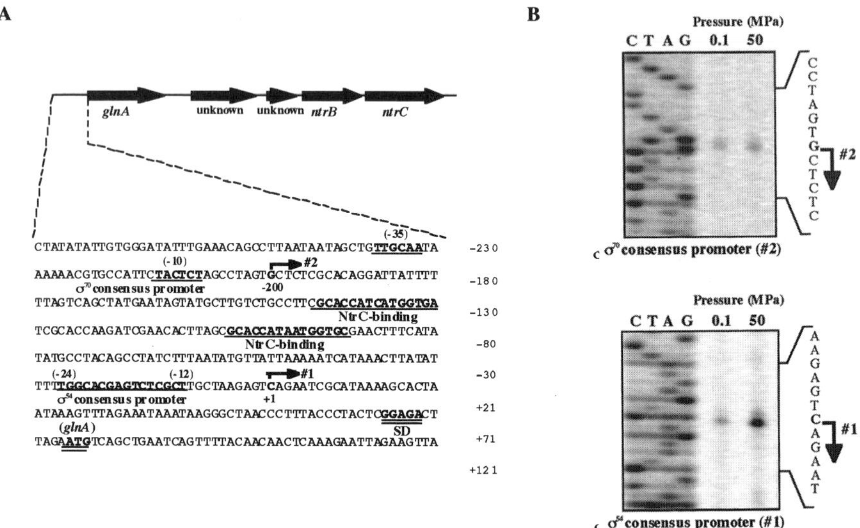

Fig. 1. Gene organization around the *glnA* gene, nucleotide sequence of the *glnA* promoter region and primer extension analysis of the *glnA* gene under different pressure conditions. (A) The coding regions and their direction of transcription are indicated by arrows. The gene names are shown under the corresponding coding regions. Transcription initiation sites (#1 and #2) are marked by arrows above the nucleotide sequence, and the coordinate +1 is defined as the first nucleotide of the transcript #1. The consensus promoter sequences (σ^{70} type: -35 and -10, σ^{54} type: -24 and -12) and the consensus sequences of the NtrC-binding site are underlined. The ATG start codon and potential ribosome binding site (SD) are indicated by double-underlined. (B) Primer extension analysis of the *glnA* gene in *S. violacea* under different pressure conditions. Total cellular RNA was prepared from cells grown at 0.1 and 50 MPa. The transcripts (#1 and #2) are shown by arrows.

2. GENE EXPRESSION OF GLUTAMINE SYNTHETASE AT ELEVATED HYDROSTATIC PRESSURE

We have reported on the isolation and characterization of a glutamine synthetase gene (*glnA*) from *S. violacea* (6). Fig. 1A shows the gene organization around *glnA* operon composed of five ORFs and nucleotide sequence of the *glnA* promoter region. In order to determine the transcription initiation sites and the effect of cultivation under elevated hydrostatic pressure on the *glnA* transcription, primer extension analysis was conducted. Through analysis of the 5'-ends of the mRNA, two distinct transcription products were identified as shown in Fig. 1B. The σ^{70} consensus sequence (10), -10 [TTGACA] and -35

[TATAAT], was identified upstream of the region corresponding to transcript #2, and the σ^{54} consensus sequence (11), [TGGYAYR-N4-TTGCA] was found upstream of the region corresponding to transcript #1, as shown in Fig. 1A. Therefore, it is evident that there are two tandem promoters of *glnA* in *S. violacea*, as seen in the case of *glnA* in *E. coli* (12, 13). The level of transcript #1 was elevated at 50 MPa, however the level of transcript #2 was not enhanced at this pressure (Fig. 1B). Transcripts #1 and #2 were expressed at a relatively low consistent level under atmospheric pressure conditions. In contrast, the expression of transcript #1 from the σ^{54}–dependent promoter was highly induced under high-pressure conditions (50 MPa) as compared to the expression of transcript #2 from the σ^{70}–dependent promoter. The σ^{54}–dependent promoter appears to play an important role in the transcription of *glnA* under high-pressure conditions in *S. violacea*, just as in the case of the pressure-regulated operon (5). In other bacteria, transcription from the σ^{54}–dependent promoter, such as in the case of *glnAp2*, is regulated by an enhancer-binding protein NtrC (14). The NtrC-binding sites [consensus: TGCACCA-N3-TGGTGCA] are essential for regulation of transcription by the σ^{54}–containing RNA polymerase (12). Two NtrC binding sites were also found from the promoter region of *glnA* in *S. violacea* (Fig. 1A).

3. ISOLATION AND CHARACTERIZATION OF THE RNA POLYMERASE σ FACTOR, σ^{54}

In addition to the results for the characterization of the *glnA* operon discussed above, we have characterized other genes, including the pressure-regulated operon cloned from *S. violacea*, that is controlled by elevated pressures at the level of transcription (5). In this report, we have analyzed a σ^{54}–like factor that recognizes a DNA element, designated as region A, upstream of the pressure-regulated operon. The RNA polymerase σ factor, σ^{54}, has been shown to be responsible for transcription of nitrogen-regulated genes such as glutamine synthetase and nitrogen fixation genes (8). Furthermore, in a number of cases, σ^{54} is required for the expression of genes that are not subject to nitrogen control, e.g., hydrogenase genes in *E. coli* and xylene degradation genes in *Pseudomonas putida* (15, 16). In *S. violacea*, transcription of these pressure-responsible genes may be dependent on the σ^{54}–containing RNA polymerase and its related transcription factors. From this background, we have isolated the *rpoN* gene encoding σ^{54} from the piezophilic bacterium *S. violacea* and also investigated the expression of the *rpoN* gene product under different pressure conditions. The structure of this fragment, containing eight ORFs, is shown in Fig. 2A. This *rpoN* gene, consisting of 1,476 bp, was found to encode a putative protein consisting of 492 amino acid residues with a predicted molecular mass of 55,359 Da (7). The nucleotide sequence of the upstream region is also shown in Fig. 2A.

To determine the transcriptional initiation site and the effect of cultivation under elevated hydrostatic pressure on *rpoN* transcription, primer extension analysis was conducted. As shown in Fig. 2B, the single major product showed as "A" residue at the transcription initiation site under both atmospheric and high-pressure conditions. Putative core promoter sequences, [-35; TAGCCT] and [-10; GAGAAG], which might be recognized by the typical σ^{70}–containing RNA polymerase, were observed (Fig. 2A). The analysis also showed that each transcript was detected with the same prominence under both atmospheric and high-

214

pressure conditions, indicating the *rpoN* gene is expressed at a consistent level under both pressure conditions (7).

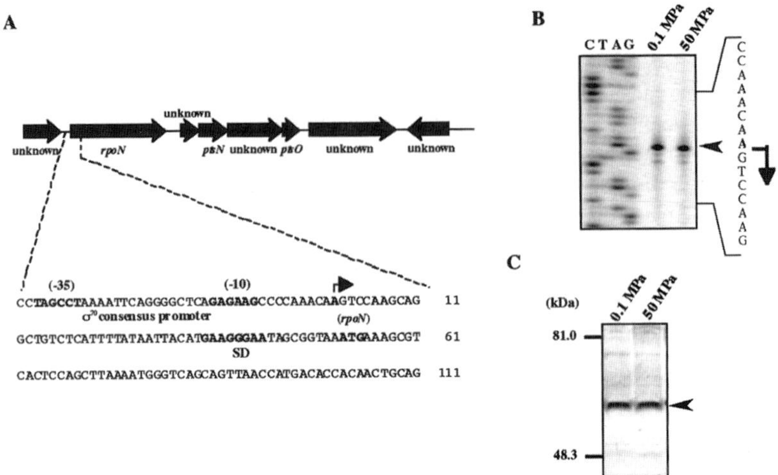

Fig. 2. Gene organization around the *rpoN* gene, nucleotide sequence of the *rpoN* promoter region, primer extension analysis of the *rpoN* gene and Western blot analysis of σ^{54} under different pressure conditions. (A) The coding regions and their direction of transcription are indicated by arrows. Transcription initiation site is marked by arrows above the nucleotide sequence and double-underlined and the consensus promoter sequence is underlined. The start codon and potential ribosome binding site (SD) are double-underlined. (B) Primer extension analysis of the transcription initiation site of the *rpoN* gene in *S. violacea*. Total cellular RNA was prepared from cells grown at 0.1 and 50 MPa. The arrow shows the transcription start site. (C) Western blot analysis of the σ^{54} protein under different growth conditions. The arrow indicates the position of the σ^{54} protein with a mass of 55 kDa. The sizes of the molecular mass standards are indicated in kDa.

Western blot analysis was also performed to examine the expression of σ^{54} in *S. violacea*. The cells were cultured at a pressure of 0.1 or 50 MPa, and cell lysates were prepared and fractionated by SDS-PAGE. The gel was then incubated with antiserum against *P. putida* σ^{54} (7). As shown in Fig. 2C, bands of equivalent intensity were detected in the case of cells grown at both 0.1 and 50 MPa. These bands each corresponded to a mass of 55 kDa consistent with the predicted molecular mass of *S. violacea* σ^{54}. The results of primer extension and Western blot analyses together indicated that σ^{54} was expressed at a relatively constant level in *S. violacea* under both pressure conditions, 0.1 and 50 MPa, at both the transcriptional and translational levels. These observations are consistent with the finding that the intracellular concentration of σ^{54} in *E. coli* remained consistent under various growth conditions (17). Our findings suggest the possibility that the level of functional σ^{54} molecules may be controlled by the availability of certain regulatory factors such as NtrC. The role of this activator is to catalyze the isomerization of closed complexes between the σ^{54}-containing holoenzyme and the promoter, and to open the complexes (18). In *S. violacea*, transcription by the σ^{54}-containing RNA polymerase may be very dependent on such transcription factors.

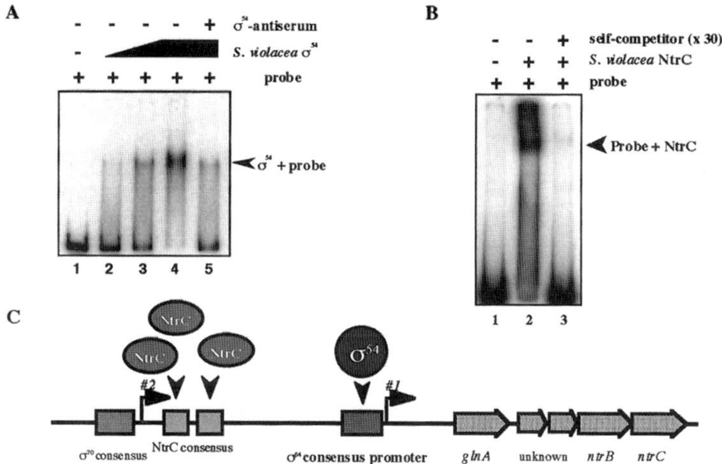

Fig. 3. Interaction of the σ^{54} and the NtrC factors to the promoter region of the *glnA–ntrBC* operon. (A) An EMSA of the interaction of the σ^{54} protein from *S. violacea* with the promoter region of the *glnA* gene and the effect of antiserum containing anti-σ^{54} polyclonal antibodies. The labeled fragment was incubated with increasing concentrations of *S. violacea* σ54 protein: 0, 30, 60, 150 and 150 ng, (in lane 1, 2, 3, 4 and 5, respectively). Antiserum containing σ54 polyclonal antibodies was simultaneously added to the reaction mixture (lane 5). (B) EMSA of the interaction of the NtrC protein from *S. violacea* with the promoter region of the *glnA* gene. Lane 1, free probe (0.1 pmol); lanes 2 and 3, the probe incubated with the σ^{54} protein (120 ng); lane 3, the same mixture as in lane 2 with self-competitor (x 30). (C) Cartoon model for interaction of the σ^{54} and the NtrC factors to promoter region of *glnA–ntrBC* operon.

4. RECOGNITION OF THE σ^{54} FACTOR AND NTRC PROTEIN TO THE *glnA* PROMOTER REGION

As mentioned above (Fig. 1A), we have identified potential regulatory elements, consensus sequences for σ^{70}, σ^{54} and NtrC, upstream of the *glnA* operon (6). Furthermore, an electrophoretic mobility shift assay (EMSA) was performed to confirm whether or not purified recombinant *S. violacea* σ^{54} recognized the DNA fragment containing the σ^{54} consensus promoter sequence of the *glnA* operon. A [γ-^{32}P] DNA fragment containing the σ^{54} consensus sequence was incubated with various concentrations of purified *S. violacea* σ^{54} protein in the presence or absence of antiserum against σ^{54} (6). As shown in Fig. 3A, an increase in the retarded band in proportion to the amount of protein was observed (lanes 2, 3 and 4 in Fig. 3A) and a decrease in the retarded band as a result of treatment with the antiserum against σ^{54} was also detected (lane 5 in Fig. 3A). These results indicated that a specific binding of the *S. violacea* σ^{54} to the promoter region of *glnA* occurred, as observed previously in the case of the pressure-regulated operon (7). In this bacterium, the intracellular concentrations of σ^{54} were similar under several pressure conditions described above. Transcription from σ^{54}–dependent promoters is known to be regulated by the enhancer-binding protein NtrC, and the initiation of transcription is responsible for the conversion of NtrC to active NtrC-phosphate by the protein kinase NtrB (14). Pressure-regulated transcription from the σ^{54}–dependent promoter in *S. violacea* may be controlled by several transcription factors, as in the case of other nitrogen regulation systems (19). In addition, we performed an EMSA, to confirm whether the purified *S. violacea* NtrC (SvNtrC)

216

recognizes DNA element containing NtrC consensus sequence. In the EMSA, as shown in lane 2 of Fig. 3B, DNA-protein complex was formed when the ^{32}P-labeled DNA probe containing the NtrC consensus sequence of the upstream region of *glnA* (from position –221 to +60, in Fig. 1A) was incubated with the purified SvNtrC protein. The complex was eliminated in the presence of non-radioactive excess amount (x 30) of target DNA (self competitor), as shown in lane 3 of Fig. 3B. Thus, the analysis demonstrated that the purified SvNtrC protein specifically recognizes the element containing the NtrC consensus sequence on the *S. violacea glnA* operon.

5. *In vitro* RECONSTITUTION AND CHARACTERIZATION OF THE NTR-B AND C

In order to reconstitute the two components regulatory system and characterize autophosphorylation of the SvNtrB and *trans*-phosphorylation of the SvNtrB~P to the SvNtrC *in vitro*, we constructed the recombinant-expression plasmids harboring *S. violacea ntrB* and *ntrC* genes. Overexpression experiments of the hexahistidine-tagged derivatives of the SvNtrB and SvNtrC were conducted and these fusion proteins were purified using the Ni^{2+}-NTA column. To confirm if the SvNtrB protein is a protein kinase capable of autophosphorylation, the SvNtrB was incubated with [γ-^{32}P] ATP from at 0 °C to 37 °C for 15 min. As shown in lanes 1 to 3 of Fig. 4A, the SvNtrB became marked, indicating that it was auto-phosphorylated. The autophosphorylation of the SvNtrB was observed only at low temperatures (maximum: 10 °C), while the activity was not detected at 37 °C. In *E. coli*, autophosphorylation activity of the NtrB at low temperatures (10 °C) was not detected. Furthermore, we have detected transcriptional activity at low temperatures in *S. violacea* (data not shown). Therefore, this bacterium adapts to the psychrosphere (low temperature environment) and may have evolved a low-temperature-adapted system in the deep-sea environment. This is also the first observation to detect autophosphorylation at low temperatures (20).

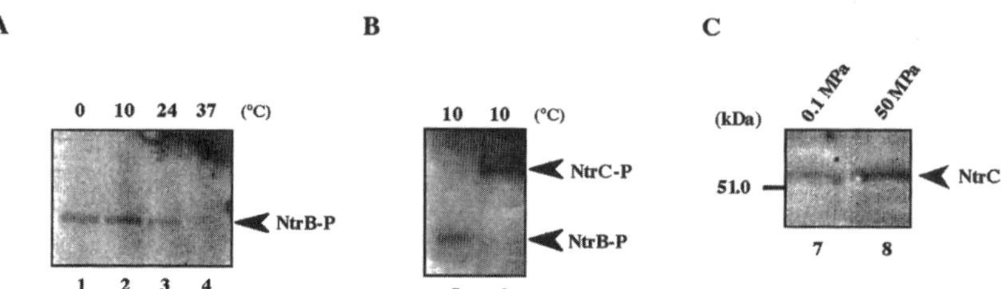

Fig. 4. Autophosphorylation of the SvNtrB protein, *trans*-phosphorylation of the SvNtrB~P to the SvNtrC protein *in vitro* and Western blot analysis of expression of the SvNtrC under different pressure conditions. (A) Autophosphorylation of the SvNtrB incubated in the presence of [γ-^{32}P] ATP at several temperature conditions. Lane 1, 0 °C; lane 2, 10 °C; lane 3, 24 °C; lane 4, 37 °C. (B) *Trans*-phosphorylation to the SvNtrC incubated with the phosphorylated SvNtrB~P at 10 °C for 1 min. Lane 5, the phosphorylated SvNtrB; lane 6, the phosphorylated SvNtrC. (C) Cell lysates, prepared from cells cultured at 0.1 or 50 MPa, were fractionated by 10 % SDS-PAGE and then blotted onto a PVDF membrane. The membrane was treated with antiserum against the SvNtrC.

To determine if the SvNtrB~P could phosphorylate the SvNtrC (*trans*-phosphorylation), the SvNtrB~P was incubated with the SvNtrC at 10 °C. As shown in lane 6 of Fig. 4B, the SvNtrC was labeled in the presence of the SvNtrB~P, indicating that it

SvNtrC was labeled in the presence of the SvNtrB~P, indicating that it is a substrate for SvNtrB. To develop an *in vitro* transcription system for the σ^{54}–dependent promoter, we must be cautious with the temperature dependency of the reaction, especially at low-temperature conditions. The phosphorylated relay of the two components system should also be performed under high-pressure conditions. Thus, we are now endeavoring to the assay systems at high-pressure condition.

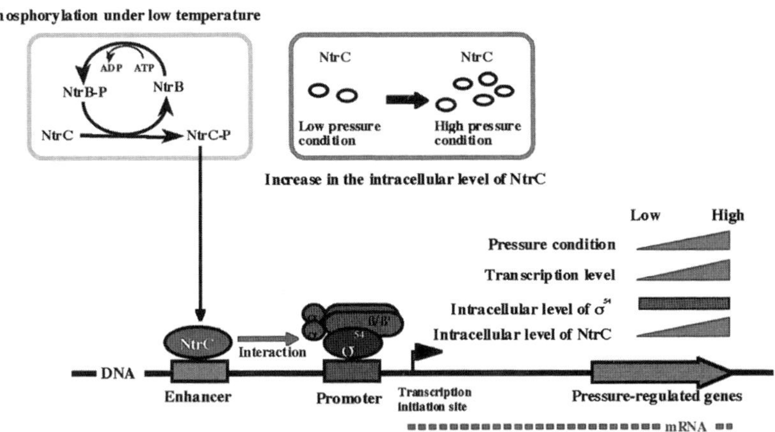

Fig. 5. Diagrammatic representation of the pressure-regulated transcription mechanisms in piezophilic *Shewanella violacea*.

6. POSSIBLE MODEL OF MOLECULAR MECHANISMS OF PRESSURE-REGULATED TRANSCRIPTION BY THE σ^{54} FACTOR

In transcription of the σ^{54}–dependent promoter such as *glnA* operon, σ^{54} containing RNA polymerase holoenzyme activates transcription at the promoter assisted by the activity of NtrC that is controlled by NtrB (21, 22). In previous studies, we suggested that these *trans*-acting factors (σ^{54}, NtrC or NtrB) might play an important role in pressure-regulated transcription at the σ^{54}–dependent promoter in this piezophilic bacterium (6, 7, 9). We discussed here that σ^{54} in *S. violacea* is expressed at a relatively consistent level under both atmospheric and high pressure conditions as determined by both primer extension and Western blotting analyses, suggesting that the level of functional σ^{54} molecules is possibly regulated by the availability of NtrC. In this regard, we tried to examine the expression of the NtrC protein in *S. violacea* by Western blot analysis (20). The cells were cultured at a pressure of 0.1 or 50 MPa, and cell lysates were prepared and fractionated by SDS-PAGE. The gel was then incubated with antiserum against the SvNtrC. As shown in Fig. 4C, the amount of this factor expressed under high-pressure condition (lane2 of Fig. 4C; 50 MPa) is greater than that at atmospheric pressure condition (lane1 of Fig. 4C; 0.1 MPa). Consideration of the results presented here together with previous work leads us to suggest a possible model for the mechanism of regulated expression of the *glnA* operon in the deep-sea piezophilic bacterium *S. violacea*, as shown in Fig. 5. *S. violacea* σ^{54}, expressed at

218

a consistent level at both atmospheric and high pressure, suggests the intracellular level of σ^{54} containing RNA polymerase holoenzyme under both conditions is consistent. This observation also strongly suggests that the transcriptional activity at this σ^{54}-dependent promoter is proportional to the amount of SvNtrC factors and it is regulating the gene expression of the *glnA* operon under high-pressure conditions. This model to explain the gene expression under high-pressure, should be tested by molecular-genetical approaches analyzing several pressure and low-temperature sensitive mutants in these transcription systems.

ACKNOWLEDGEMENTS

We thank Dr. James Hunt for assistance in editing the manuscript. This study was partially supported by a Grant-in-Aid for Scientific Research, Scientific Research on Priority Areas: Single-Cell Molecular Technology (Area number; 736).

REFERENCE

1. C. Kato, T. Sato and K. Horikoshi, Biodiv. Conserv., 4 (1995) 1.
2. Y. Nogi, C. Kato and K. Horikoshi, Arch. Microbiol., 170 (1998) 331.
3. K.Nakasone, A. Ikegami, C. Kato, R. Usami and K. Horikoshi, Extremophiles, 2 (1998) 149.
4. C. Kato, A. Ikegami, M. Smorawinska, R. Usami and K. Horikoshi, J. Mar. Biotechnol., 5 (1997) 210.
5. K. Nakasone, A. Ikegami, C. Kato, R. Usami and K. Horikoshi, FEMS Microbiol. Lett., 176 (1999) 351.
6. A. Ikegami, K. Nakasone, C. Kato, Y. Nakamura, I. Yoshikawa, R. Usami and K. Horikoshi, FEMS Microbiol. Lett., 192 (2000) 91.
7. A. Ikegami, K. Nakasone, M. Fujita, S. Fujii, C. Kato, R. Usami and K. Horikoshi, Biochim. Biophys. Acta, 1491 (2000) 315.
8. M.J. Merrick and R.A. Edwards, Microbiol. Rev., 59 (1995) 604.
9. A. Ikegami, K. Nakasone, C. Kato, R. Usami and K. Horikoshi, Biosci. Biotech. Biochem., 64 (2000) 915.
10. D.W. Cowing, J.C.A. Bardwell, E.A. Craig, C. Woolford, R.W. Hendrix and C.A. Gross, Proc. Natl. Acad. Sci. USA, 82 (1985) 2679.
11. R. Dixon, Nucleic Acids Res., 12 (1984) 7811.
12. L.J. Reitzer and B. Magasanik, Cell, 45 (1986) 785.
13. T.P. Hunt and B. Magasanik, Proc. Natl. Acad. Sci. USA, 82 (1985) 8453.
14. A.J. Ninfa, L.J. Reitzer and B. Magasanik, Cell, 50 (1987) 1039.
15. S. Lutz, R. Bohm, A. Beier and A. Bock, Mol. Microbiol., 4 (1990) 13.
16. T. Kohler, J. Harayama, J.L. Ramos and K.N. Timmis, J. Bacteriol., 171 (1989) 4326.
17. M. Jishage, A. Iwata, S. Ueda and A. Ishihama, J. Bacteriol., 178 (1996) 5447.
18. S. Sasse-Dwight and J.D. Gralla, Cell, 62 (1990) 945.
19. A. Alvarez-Morales, R. Dixon and M. Merrick, EMBO J., 3 (1984) 501.
20. A. Ikegami, K. Nakasone, H. Kawano, C. Kato, R. Usami and K. Horikoshi, FEMS Microbiol. Lett., (2001) in press.
21. J. Keener and S. Kustu, Proc. Natl. Acad .Sci. USA, 85 (1988) 4976.
22. M. Buck, M.T. Gallegos, D.J. Studholme, Y. Guo and J.D. Gralla, J. Bacteriol., 182 (2000) 4129.

Trends in High Pressure Bioscience and Biotechnology
R. Hayashi (editor)

The biological significance of tryptophan availability on high-pressure growth in yeast

Fumiyoshi Abe and Koki Horikoshi

The DEEPSTAR Group, Japan Marine Science and Technology Center (JAMSTEC), 2-15 Natsushima-cho, Yokosuka 237-0061, Japan

Abstract

Application of hydrostatic pressure of 15 to 25 MPa was found to cause arrest of the cell cycle in G_1 phase in the yeast *Saccharomyces cerevisiae*, whereas a pressure of 50 MPa did not. We found that a high concentration of L-tryptophan or a plasmid carrying the *TAT2* gene which encodes a high-affinity tryptophan permease enabled the cells to grow at pressures of 15 to 25 MPa. Hydrostatic pressure significantly inhibited tryptophan uptake into the cells, and the ability of the uptake was impaired during incubation of the cells at 25 MPa. The activation volume associated with overall tryptophan uptake was found to be a large positive value, 46.2±3.85 ml/mol, indicating that there was a net volume increase in a rate-limiting step in tryptophan import. The result suggest that the most pressure sensitive process in terms of cell growth is tryptophan uptake, and increasing tryptophan availability enables the cell to grow under high-pressure conditions.

1. INTRODUCTION

There has been a renewal of interest in the survival strategies employed by deep-sea, high-pressure adapted microorganisms as well as the effects of high-pressure on 0.1 MPa-adapted microorganisms (1, 2). Increasing hydrostatic pressure usually has a significant influence in biological activities of organisms inhabiting atmospheric pressure, causing growth inhibition, slow down of metabolism or disorganization of cytoskeleton. Deep-sea organisms would have developed the ability to survive under such high-pressure conditions. In terms of maintenance of appropriate membrane fluidity, such organisms increase the ratio of unsaturated fatty acids within their cell membranes (3). However, at present, it is still unclear whether there is one critical mechanism for high-pressure adaptation in deep-sea organisms or all of their cellular components i. e. proteins, membrane or nucleic acids, have been specialized for high-pressure adaptation.

We have studied the physiological effects of hydrostatic pressure at non-lethal levels (<100 MPa) in the yeast *Saccharomyces cerevisiae*, focusing on intracellular pH homeostasis (1, 4), and a potential application in flow cytometry (5). Recently we obtained a striking result suggesting that the availability of tryptophan may be of primary importance for high-pressure growth in yeast (6). Tryptophan is known to be transported into the cell via a high-affinity tryptophan permease encoded by *TAT2* (7). *TAT2* was originally identified as a gene that conferred resistance to an immunosuppressive drug, FK506 (7). Another immunosuppressant, rapamycin, is known to arrest the growth of yeast cells in early G_1 phase, causing them to express several physiological properties typical of starved (G_0) cells, and also causing significant reduction in protein synthesis. In this paper, we describe the biological significance of tryptophan availability on cell growth in *S. cerevisiae* under high-pressure conditions.

2. MATERIALS AND METHODS

2.1. Strains and media

The wild type haploid strain YPH499 was used in this study. The composition of YPD medium, synthetic minimal medium (SD) and synthetic complete medium (SC) supplemented with appropriate nutrients are described in ref. 8. SCv medium, which is SC medium lacking valine, was mainly used throughout this study because growth of strain YPH499 was found to be poor in SC medium containing 150 mg/l valine.

2.2. Culture conditions

Cells were grown in YPD or SCv medium at 24°C with shaking. For analysis of cell growth under various pressure conditions, experiments were performed using liquid media rather than solid agar media. An exponentially growing culture ($3 \times 10^6 - 5 \times 10^6$ cells/ml) was diluted to 1×10^6 cells/ml with fresh YPD or SCv medium. Aliquots of the resulting cell suspension were put into a series of sterilized polypropylene tubes. Each tube was placed in a stainless-steel hydrostatic chamber, and the appropriate hydrostatic pressure was applied using a hand pump. Viable cell numbers were determined on the basis of colony counts after decompression. To examine cell growth in the presence of a high concentration of individual L-amino acids, adenine sulfate or uracil, YPD medium supplemented with each of these test substrates at 1 g/l was prepared.

2.3. Tryptophan uptake assay

Radiolabeled L-tryptophan (L-[5-^{3}H]tryptophan, code no. TRK460, Amersham Pharmacia Biotech Inc.) was used in the assay. Cells from an exponentially growing culture in SCv medium were collected by centrifugation and resuspended in fresh SCv medium in a polypropylene tube. Then radiolabeled tryptophan was added to the cell suspension, immediately mixed, and the mixture was split into seven equal aliquots in sterilized microtubes. Each tube was put into a pressure chamber, then subjected to a hydrostatic pressure of 0.1 MPa, 25 MPa or 50 MPa, for 30 min or 60 min. The cells in the tube were collected by filtration on a glass filter, washed with distilled water, and the radioactivity on the filter was counted using a liquid scintillation counter (Wallac 1411). All experiments were performed three times and the mean values are shown as pmol tryptophan incorporated per 10^7 cells.

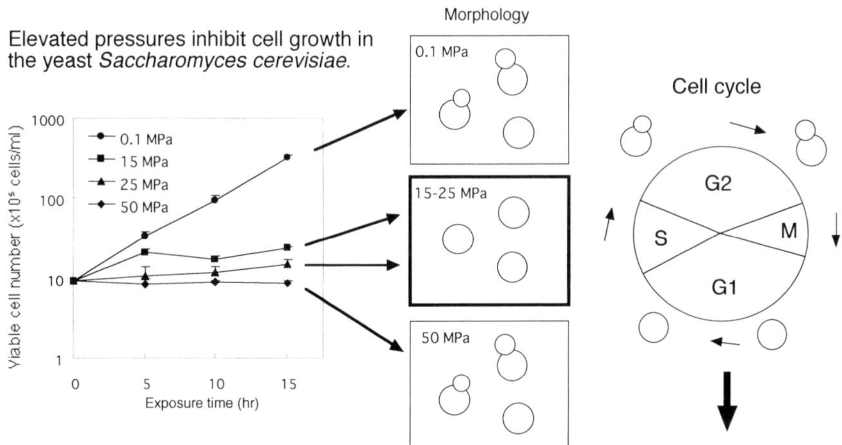

Fig. 1. Hydrostatic pressure induces cell cycle arrest in G_1 phase.

3. RESULTS

3. 1. Hydrostatic pressure causes arrest of the cell cycle in G_1 phase.

Fig. 1 shows the viable cell number derived from counting colony-forming units. Hydrostatic pressures at more than 15 MPa considerably reduced the growth rate and a pressure of 50 MPa completely prevented cell proliferation. Ten hours after pressurization, the cell number had increased 11.0-, 1.7-, 1.4- and 0.95-fold at pressures of 0.1, 15, 25 and 50 MPa, respectively. Interestingly, we observed that cells from culture at 15 to 25 MPa became unbudded with round shape under a microscope. However, the morphology of the cells from culture at 50 MPa was almost identical with that observed at 0.1 MPa. The result suggests that a pressure of 15 to 25 MPa induces cell cycle arrest at G1 phase. This hypothesis was confirmed by the analysis of nuclear DNA contents of individual cells using flow cytometry system (data not shown).

3.2. A high concentration of tryptophan enables high pressure growth.

YPH499 cells were incubated in YPD medium containing a high concentration of individual amino acids or other test substrates (final concentration, 1 g/l) for 24 hours at 25 MPa. Surprisingly, only L-tryptophan was effective to enable cell growth at 25 MPa (Fig. 2). The increase in OD_{600} was 22-fold. Whereas in the rest of the cultures, the increase in OD_{600} was less than 2-fold. When the concentration of each test substrate was reduced to 0.2 g/l, still, only L-tryptophan was effective to enable cell growth at 25 MPa and the OD_{600} reached 0.49. These results suggest that L-tryptophan is of primary importance for cell growth under high pressure conditions.

3.3. *TAT2* enables growth under high pressure conditions.

We examined whether expression of a high-affinity tryptophan permease gene *TAT2* conferred high-pressure growth in strain YPH499. The data presented are the mean values of the ratio of cell number after 24 hours of culture (N_{24}) to the initial cell number (N_0). Both the low-copy number plasmid carrying *TAT2* (YCplac33-*TAT2*) and the high-copy number plasmid carrying *TAT2* (YEplac195-*TAT2*) conferred the ability to grow at 15 MPa (Fig. 3B), whereas the vector alone (YCplac33or YEplac195) had no effect. At a pressure of 25 MPa, YCplac33-*TAT2* was not effective, but YEplac195-*TAT2* still conferred the ability to grow at this elevated pressure (Fig. 3C). Therefore, it seems that the upper limit of hydrostatic pressure for growth was dependent on the copy number of the *TAT2* gene, probably corresponding to the amount of Tat2 protein expressed and the extent of tryptophan uptake. YEplac112[*TRP1*], carrying the phosphoribosyl-anthranilate isomerase gene (*TRP1*), also allowed the cells to grow at 15 or 25 MPa (Fig. 3B, C). Our findings suggest that tryptophan uptake

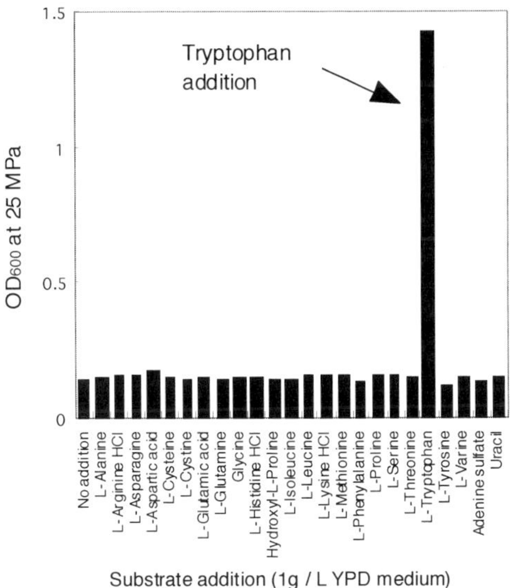

Fig. 2. Excess tryptophan enables cells to grow at high pressure.

via Tat2 is the cellular process most sensitive to elevated hydrostatic pressure, and reduced tryptophan availability would result in a decrease in protein synthesis, hence progression of the cell cycle is impaired in the G_1 phase. None of the strains examined was able to proliferate at 50 MPa; there was no change in cell numbers (Fig. 3D).

3.4. Hydrostatic pressure decreases and down-regulates tryptophan uptake activity.

We next analyzed tryptophan uptake under various pressure conditions. Fig. 4A shows the levels of tryptophan incorporation into cells under several pressure conditions. As expected, tryptophan uptake decreased to an extent depending on the magnitude of the pressure applied. The rate constants for tryptophan uptake at 0.1, 25 and 50 MPa were 4.20±1.15, 2.61±0.778 and 1.65±0.444 pmol 10^7 cells^{-1}min^{-1}, respectively. The activation volume associated with of overall process of tryptophan uptake (i.e.,ΔV^*) was a large positive value, 46.2±3.85 ml/mol, indicating that there was a net volume increase in a rate-limiting step in tryptophan import.

224

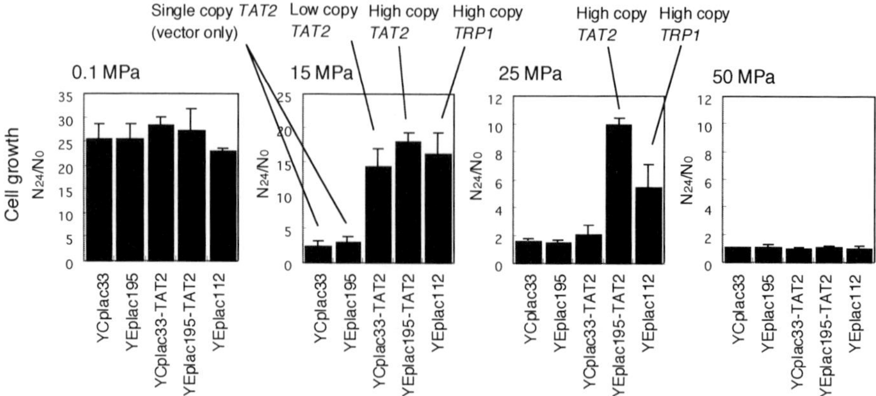

Fig. 3. *TAT2* and *TRP1* confers high-pressure growth in yeast.

To examine how exposure to elevated pressure for a long period affected tryptophan uptake, the cells were incubated in SCv medium at 25 MPa for 2 or 5 hours. After decompression, the assay of radiolabeled tryptophan uptake was performed. It was found that exposure to a pressure of 25 MPa for 2 hours markedly down-regulated tryptophan uptake (Fig. 4B). The rate constants for tryptophan uptake at 0.1, 25 and 50 MPa were 2.73±0.614, 1.80±0.872 and 0.937±0.526 pmol 10^7 cells^{-1} min^{-1}, respectively, as determined 2 hours after the exposure, and the activation volume was increased to 57.8±19.3 ml/mol. After 5 hours incubation at 25 MPa, the rate constants decreased, and, moreover, the activation volume increased to 83.0±28.4 ml/mol (Fig. 4C). These results indicate that tryptophan uptake became impaired during incubation of the cells at 25 MPa. Therefore, we conclude that (i) hydrostatic pressure decreases the rate constant for tryptophan uptake activity, and (ii) hydrostatic pressure impaired the ability of tryptophan uptake. The results are summarized in Fig. 5.

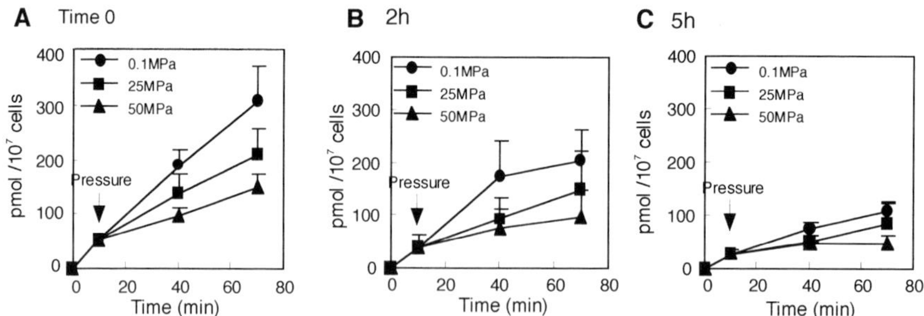

Fig. 4. Hydrostatic pressure inhibits and down-regulates tryptophan uptake.

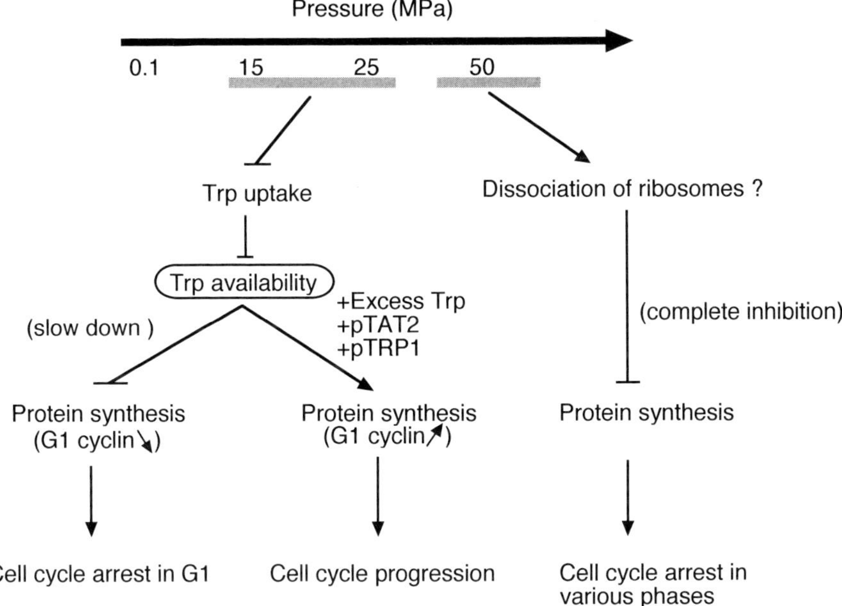

4. DISCUSSION

Studies with 0.1 MPa-adapted microorganisms have demonstrated that elevated hydrostatic pressure inhibits cell division, nutrient uptake, biosynthesis of DNA, RNA and proteins, and association of multimeric proteins (1, 2, 9, 10). The mechanisms of adaptation to high pressure are thought to be very complex because these organisms are likely to encounter all of the inhibitory effects listed above. Perhaps the presumed complexity is the reason why no attempt has been made to isolate mutants with high growth rates or to identify gene(s) allowing high-pressure growth in mesophiles. However, our results indicate that only one gene is sufficient to confer high pressure growth in yeast. We also demonstrated that the availability of tryptophan was key factor in the yeast high-pressure growth. Although we have no evidence for generalizing our model to deep-sea organisms, these findings provide a novel aspect in investigations of biological adaptation to high pressure environments as well as studies on cellular responses to environmental stresses. In our studies on piezophysiology, we try to adopt this thermodynamic parameter 'hydrostatic pressure' as a tool to understand complex physiological events via the 'volume change' associated with reactions or equilibria (1, 4).

REFERENCES

1. F. Abe, C. Kato and K. Horikoshi. Trends Microbiol. 7 (1999) 447.
2. D. H. Bartlett. Sci Prog. 76 (1992) 479.
3. E. E. Allen, D. Facciotti and D. H. Bartlett. Appl. Environ. Microbiol. 65 (1999) 1710.
4. F. Abe and K. Horikoshi. Extremophiles 2 (1998) 223.
5. F. Abe. Appl. Environ. Microbiol. 64 (1998) 1139.
6. F. Abe and K. Horikoshi. Mol. Cell. Biol. 20 (2000) 8093.
7. A. M. Schmidt, M. N. Hall and A. Koller. Mol. Cell. Biol. 14 (1994) 6597.
8. C. Guthrie and G. R. Fink. Guide to yeast genetics and molecular biology. Academic Press, Inc. (1991)
9. M. Gross, K. Lehle, R. Jaenicke and K. H. Nierhaus. Eur. J. Biochem. 218 (1993) 463.
10. J. L. Markley, D. B. Northrop and C. A. Royer. High-pressure effects in molecular biophysics and enzymology. Oxford Univ. Press, Inc. (1996)

Trends in High Pressure Bioscience and Biotechnology
R. Hayashi (editor)

227

Restoration of *Escherichia coli* from high hydrostatic pressure - A study of the FtsZ-ring formation using confocal laser microscopy-

T. Miwa, T. Sato, C. Kato, M. Aizawa and K. Horikoshi

The DEEPSTAR Group, Frontier Research Program for Deep-sea Extremophiles, Japan Marine Science and Technology Center, 2-15 Natsushima-cho, Yokosuka 237-0061, Japan.

Escherichia coli has been observed to abnormally elongate under highly pressure culture. Abnormally elongated *Escherichia coli* was recovered to normal form after decompression to the atmospheric pressure condition. We examined the recovered mechanism of elongated *Escherichia coli* using observation of formation of FtsZ ring.

1. INTRODUCTION

Escherichia coli is a representative gram-negative bacteria of about 2μm length. Maximum pressure for cell growth of *Escherichia coli* is about 50 MPa, however the multiplication is not generated above this pressure. Below this pressure *Escherichia coli* has been observed to abnormally elongate compared to the normal pressure culture, while the bacteria cell keeps the form of the rod[1-4]. Also, it is known that the elongated *Escherichia coli* returns to the original length when the pressure is released to atmospheric condition. Recently it has been confirmed that this elongation is due to the growth with weight and volume increase since the cell number remains the same[5]. It is believed that this is a kind of filament formation which is generated in response to the external stimulation of various bacteria. The detailed mechanism of this abnormal elongation of the cell by the pressurisation has not yet been proven, which may entail the inhibition of the cell division enzymes under pressure. Also, there is no data on how the cell recovers its original length and volume as the pressure is released back to atmospheric pressure.

In this paper we report a recently observed *Escherichia coli*, abnormally elongated by piezosensitivity, the behavior of which was recorded whilst recovering under atmospheric pressure. In addition, we examined the recovered mechanism of *Escherichia coli* from the pressurized environment using observation of formation of FtsZ ring. The process of FtsZ ring formation was given information of cell division[6].

2. EXPERIMENTS

Escherichia coli (JM109) was cultivated in a culture bottle with oxygen gas saturated fluorinert solvent. After incubation, the pressurization was carried out in the pressure-resistant vessel by using a hand pump at 50 MPa for 24 hours at 37 °C (Figure 1). *Escherichia coli* solution was then immediately observed through a dark field microscope (OLYMPUS, x40 objective lens), after the pressure was released.   For   the   observation, *Escherichia   coli*   solution   was sandwiched by coverslip and incubated at 37 °C.

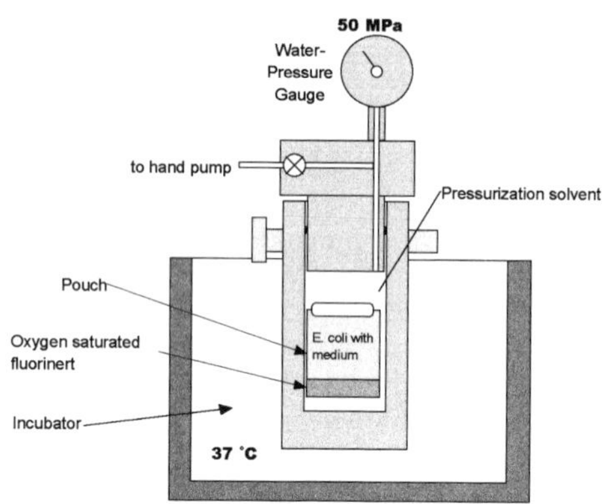

Figure 1. Schematic image of the pressure-resistant vessel.

In order to observe FtsZ-ring, we fixed *Escherichia coli* by 80% methanol after released pressure.   The fixed cells were collected by centrifuge.   Collected cells suspensions were spread on the lysine coated coverglasses for immunofluorescent label. Immobilized cells were treated lysozyme solution and PBST-BSA solution. After immunofluorescent label of FtsZ-ring, we observed *Escherichia coli* cells using conforcal laser microscope (OLYMPUS, FLUOVIEW FV300, x100 objective lens).

3. RESULTS AND DISCUSSION

3.1. *Escherichia coli* elongation

We observed a growth of around 100μm length, while *Escherichia coli* kept the tube shape in the pressurized culture. No   movement   of   this   extended *Escherichia coli* was detected.   Figure 2 shows relation of length and culture time at 50 MPa.   According to culture time, it was observed that *Escherichia coli* grow

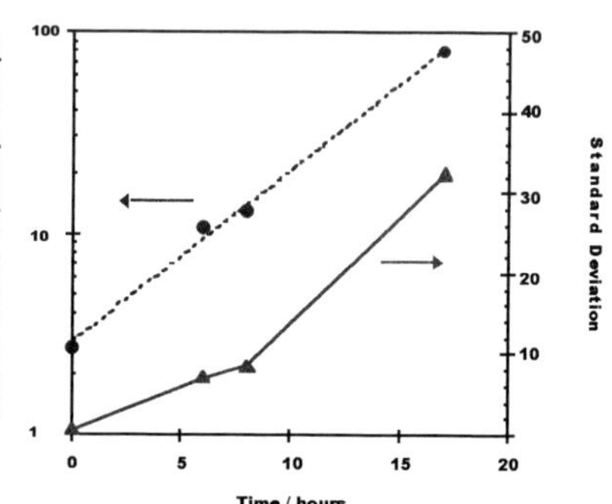

Figure 2. Elongation of E. coli at high pressure cultivation.

exponentially.   In addition, standard deviation of length was increased with cultivation time.   This result suggests that cell growth expected at high-pressure environment.

3.2.   After decompression process

When we study the cause that *Escherichia coli* elongates, it is important to observe behavior after reduced pressure of *Escherichia coli* that elongated at the high pressure. We examined behaviour at atmospheric pressure condition using *Escherichia coli* which we cultured with 50 MPa for 24 hours.   Using elongated *Escherichia coli* over 100 µm, we observed by dark-field microscope with the atmospheric pressure at 37 °C.   We also noticed that this elongated *Escherichia coli* after incubation forms several septums in the final stage of cell division.   After the pressure is released the elongated *Escherichia coli* divides itself into many normal size *Escherichia coli*s with rapid movement (Figure 3).   In Figure 3, septums of elongated cell were observed each 3 µm distance after 90 min from the decompression.

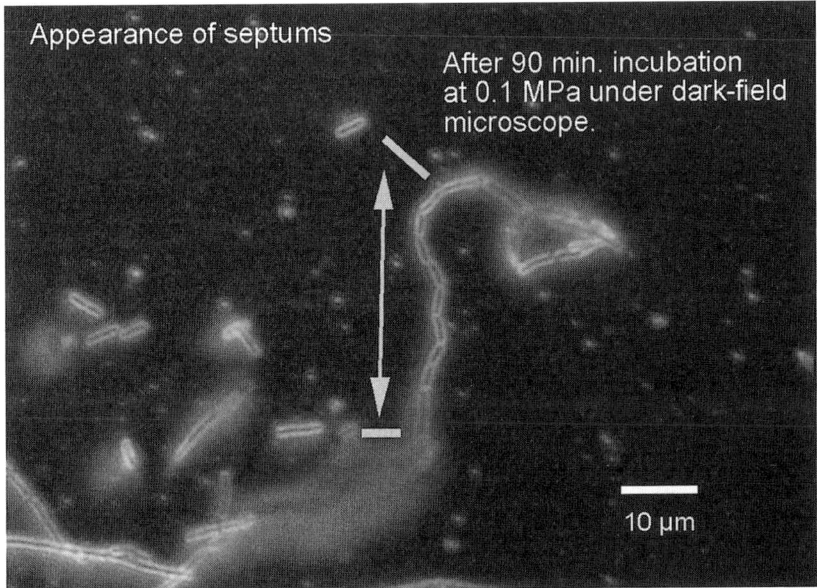

Figure 3. Dark-field microscope image of elongated *Escherichia coli*s (JM109) after 90 min. incubation from the decompression.

3.3 Observation by Conforcal laser microscopy

Recently, the sequencial process of cell division using assembled Fts-Z ring was reported[6].   The Fts-Z ring formation in the *Escherichia coli*s grown after high hydrostatic

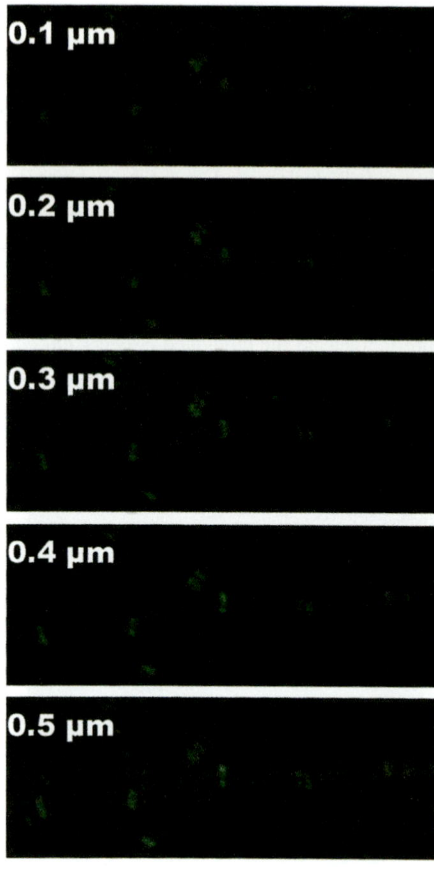

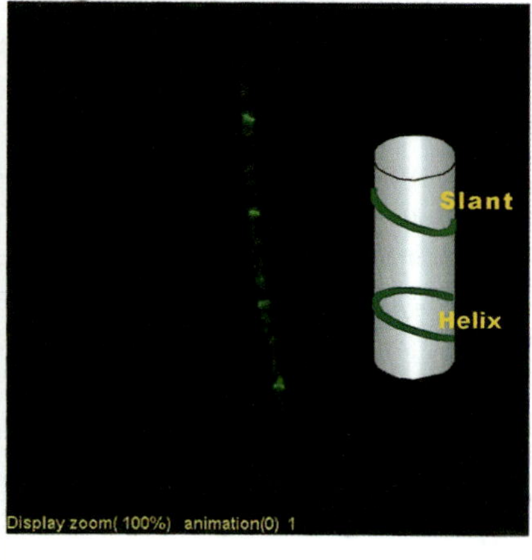

Figure 5. Conforcal laser scanning microscope image of fluorescence stained Fts-Z ring of JM109. Fixation time was 4 min. after decompression to atmospheric pressure. Ring formation was not completed on this elongated JM109. (x100 objection lens)

Figure 4. Conforcal laser scanning microscope images of fluorescence stained Fts-Z ring of JM109. Fixation time was 4 min. after decompression to atmospheric pressure. Images indicate sections of Z-axis at 0.1 μm.

pressure was studied by the conforcal laser scanning microscope. In the case of cell fixation at about 4 min after pressure release on ice bath, many FtsZ rings were observed in the elongated fixed-cells. Figure 4 show a high-resolution sectioning image of the elongated *Escherichia coli* JM109. The center of lot can observe a Fts-Z ring in a peripheral department of f *Escherichia colis*. On the other hand, a bright spot was divided into two points when I put focus together to a core of *Escherichia colis*. As a result it is suggested that Fts-Z is a ring. Figure 5 shows ring formation of elongated *Escherichia coli* after 4 min. from the decompression. Top of Fts-Z ring was slant form, lower ring was formed helical structure. Right after the decompression, though it irregularly existed FtsZ ring in the cell, it existed after 10 minutes comparatively regularly-interval.

The imperfect ring was mainly observed right after the decompression. Furthermore, it was able to be confirmed that the FtsZ ring was extinguished in the center of septum, when the septum formation is confirmed (show in figure 6.). We concluded that this elongated form is in fact an aggregation of *Escherichia coli*s.

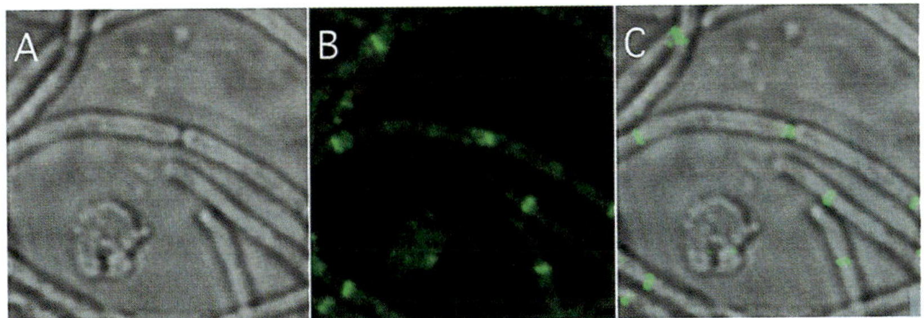

Figure 6. Transmission image (A) and fluorescence image (B). Septum was formed in image(A), FtsZ-rings were in image(B), and its superposition (C).

4. ACKNOWLEDGMENT

We thank Mr. Makoto Kato, Olympus Japan Co., Ltd. and Takumi Kuribayashi, Takeda Rika Inc., for providing conforcal laser microscopy.

REFERENCES

1. C.E. ZoBell, High pressure effects on cellular processes, (ed.) A.M.Zimmerman, Academic press, 1970.
2. Tamura, K., Shimizu, T., and Kourai, H. FEMS Microbiol Lett 78, (1992) 321.
3. Studer, D., Michel, M., and Muller, M. *Scanning Microsc Suppl 3*, (1989)253.
4. Gross, M., Kosmowsky, I., Lorenz, R., Molitoris, H., and Jaenicke, R. *Electrophoresis 15*, (1994) 1559.
5. Erijman, L., and Clegg, R. *Biophys J 75*, (1998) 453.
6. Addinall, S., Bi, E., and J, L. *J Bacteriol 178*, (1996) 3877.

Trends in High Pressure Bioscience and Biotechnology
R. Hayashi (editor)
© 2002 Elsevier Science B.V. All rights reserved.

The dynamism of *Escherichia coli* under high hydrostatic pressure – repression of the FtsZ-ring formation and chromosomal DNA condensation

T. Sato[a], T. Miwa[a],   A. Ishii[a], C. Kato[a], M. Wachi[b], K. Nagai[b], M. Aizawa[a] and K. Horikoshi[a]

[a]The DEEPSTAR Group, Japan Marine Science and Technology Center, 2-15 Natsushima-cho, Yokosuka 237-0061, Japan.
[b]Department of Bioengineering, Tokyo Institute of Technology, 4259 Nagatsuta, Midori-ku, Yokohama 226-8501, Japan.

The elongation of *Escherichia coli* cells, in response to high hydrostatic pressure growth, is a unique property of this bacterium. Our high pressure observation microscope system (HPOMS) has provided direct observation of living *E. coli* elongated cells under high pressure. To study elongation mechanisms of *E. coli* under high pressure, the formation of FtsZ ring which was an early event of cell division was examined by immuno-fluorescence microscopy (IFM), and the chromosomal DNA was also detected by DAPI staining. The FtsZ-ring formation and the condensed nucleoid structure was observed in the filamentous cells when the cells were fixed after releasing the pressure. Surprisingly, no FtsZ ring was observed in whole filamentous cells, when the cells were fixed without decompression. Also the chromosomal DNA was segregated but not condensed at the same condition. These results suggested that the cell division of *E. coli* was inhibited by high hydrostatic pressure at early stage in the cell cycle before the FtsZ-ring formation during cultivation, however, cells could grow continuously without septum formation, chromosomal DNA condensation.

1. INTRODUCTION

The morphological change of living organisms from bacteria to eukaryotic cells under high hydrostatic pressure has been studied in these decades (1, 2). In the case of eukaryote, after cells were exposed to high hydrostatic pressure, most of the cell shapes were changed to be round (3). The cytoskeletal structures of round-shaped cells, such as microtubles or actin, myosin filaments, were observed by staining (4). These structures seemed to be damaged as a result of high pressure response. *In vitro* study of self assembled polymerization with myosin or tubulin extracted from rabbit has also revealed the piezosensitivity of these filament formation under high pressure (5, 6). On the other hand, the shape of bacterial cells are generally piezotolerant for short time exposure to high pressure, however, the elongated filament cell with no septum formation was observed after high pressure cultivation in some rod shaped bacteria, but not in spherical shaped cocci (2). Terrestrial bacterium, *Escherichia coli*, is one of the moderately piezotolerant bacteria and able to grow even at 50 MPa with elongation, whereas deep-sea piezophilic bacterium which we isolated from a deep-sea sample shows normal cell growth at the same pressure (Fig. 1). We have reported that *E.*

coli was closely related to those deep-sea piezophilic and piezotolerant bacteria, according to the phylogenetic analyses of 16S rDNA sequences (Fig. 2) (7, 8). Thus, it is important to study the piezosensitive mechanism of cell division under high pressure in *E. coli* to understand pressure adaptation in comparison with deep-sea bacteria. We have also reported that *E. coli* has provided a unique response to high hydrostatic pressure, the induction of *lac* or *tac* promoter (8, 9) and inhibition of *malB* operon promoters (10) for gene expression under high pressure. Furthermore, it is also important to observe "real time" elongation of living *E. coli* cells under high pressure to know how *E. coli* grow under different pressure. Recently, the sequencial process of cell division was reported using immunofluorescence microscopy (IFM) (11), and the first step of this process is the formation of FtsZ ring, consisting of self assembled tubulin like protein, FtsZ (12). Therefore, we have studied whether FtsZ ring was formed or not in the filamentous cells grown under high pressure by IFM, to know the piezosensitive state in cell division. Moreover, since the cell division is well coordinated with chromosome replication and segregation, the form of chromosomal DNA was also studied by DAPI staining.

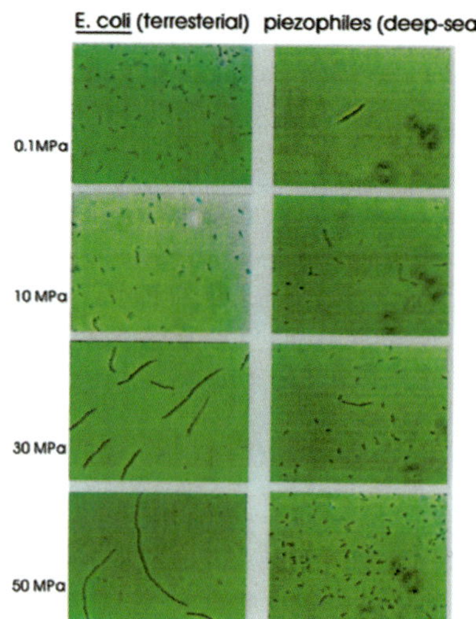

Fig. 1. Comparison of cell morphology between terresterial bacterium (*E.coli*) and deep-sea piezophilic bacterium (*Shewanella benthica* DB6705) grown under elevated pressure.

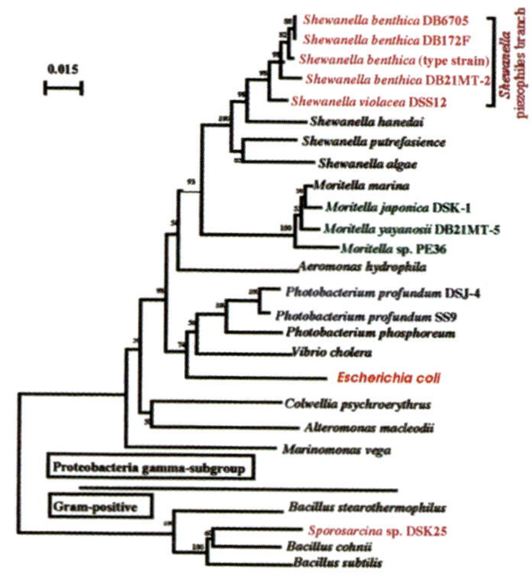

Fig. 2. Phylogenetic tree showing the relationships of *E.coli*, piezophilic, and piezotolerant bacteria within the Proteobacteria gamma subgroup, as determined by a 16S ribosomal DNA sequence comparison.

2. THE "REAL-TIME" ELONGATION OF ESCHERICHIA COLI DURING HIGH PRESSURE CULTIVATION

We have developed a high pressure observation microscope system (HPOMS) (Fig. 3), which provides direct observation of living cells under high pressure. In order to study the dynamism of any organisms under high pressure, it may be important to observe the cells without any pressure release, otherwise some phenomena may be missed as they are recovered at 0.1 MPa state quickly after decompression. Since some treatment of the cells to maintain any changes coming from high pressure effects may also cause artificial ones, it seems to be the best method to observe directly. The HPOMS consists of high pressure observation cell with heater for constant temperature incubation on an objective stage of microscope, pump system to pressurize the cell, and flow system to supply fresh medium in the cell without any decompression. Those systems are controlled by computer so that we can obtain exact reproducibility of any experiment and long time manipulations or observations of the cells under high pressure automatically. We observed elongation of *E. coli* cells during 20h cultivation at 48 MPa. *E. coli* was inoculated into the high pressure observation cell with LB medium at 37°C. The cells were quickly pressurized at 48 MPa supplying fresh LB medium slowly by the flow system. The form of *E. coli* cells was recorded every 30 min during high pressure cultivation. As a result, most of cells grew filamentous rod shapes with aging, without further morphological change (Fig. 4). Since some bacterial mutants, such as *Bacillus subtilis* bearing a morphogene, *mbl* alleles, showed bloated or twisted cellular morphology (13), it is possible that the cells grown under high pressure appear similar in morphological change due to damage to some cytoskeletal structure, such as the pressure response of eukaryotic cells. This is a first observation of living microorganisms in a high pressure environment.

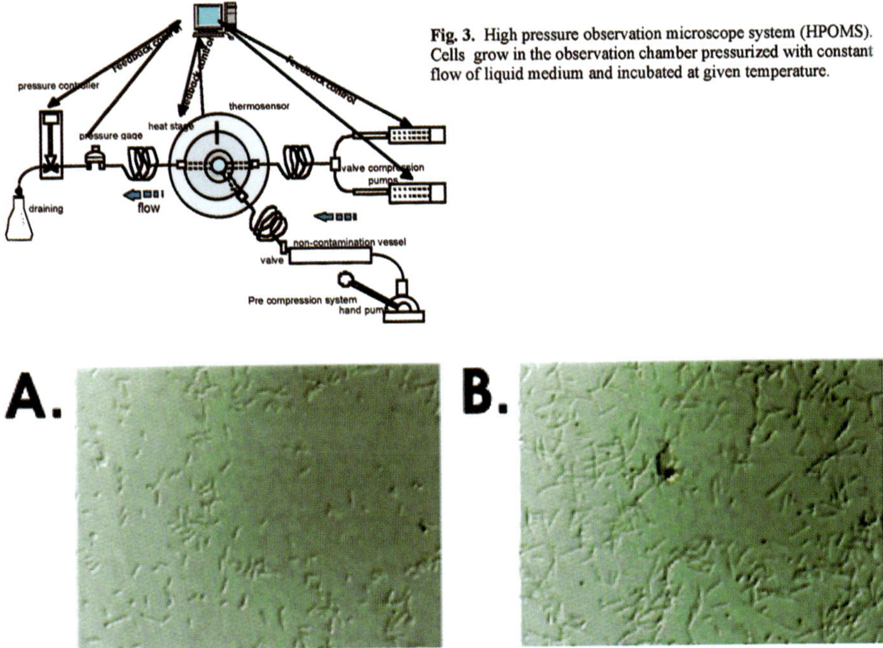

Fig. 3. High pressure observation microscope system (HPOMS). Cells grow in the observation chamber pressurized with constant flow of liquid medium and incubated at given temperature.

Fig. 4. Effect of pressure on the growth and cell morphology of *E. coli* observed by the high pressure observation microscope. *E. coli* was grown in the microscope cell with LB medium under 48 MPa at 37°C . Cell shapes were inspected at 48 MPa under phase contrast microscope after 0.5h (A) and 20h (B) incubation.

3. HIGH HYDROSTATIC PRESSURE REPRESSES THE FTSZ-RING FORMATION AND CHROMOSOMAL DNA CONDENSATION IN ESCHERICHIA COLI

First, we examined the formation of filamentous cells and FtsZ ring in the *E. coli* wild type strain W3110 and *recA⁻* strain JM109 grown under high hydrostatic pressure by IFM. After cell growth at 48 MPa for 7 and 17h, filamentous form was observed in the *recA⁻* strain as well as the wild type strain.   This result suggested that cell elongation was not induced by SOS response as RecA protein is essential for this response (14).   After high pressure cultivation, cells were fixed with methanol, harvested by centrifugation and mounted on slide glass, then treated with FtsZ antibody and second antibody with fluorescent dye.   In the case of cell fixation at about 5 min after pressure release on ice bath, many FtsZ rings were observed in the elongated cells with both strains.   To examine exact state of *E. coli* cells under high pressure, new cell fixation system has been developed (Fig. 5).   The cells grown under high pressure were fixed inside the vessel without pressure release using this system. Surprisingly, the result of IFM was completely different from the one of cells fixed after pressure release.   There were almost no FtsZ rings in the elongated cells (Fig. 6).   In order to assess the expression of *ftsZ* gene during high pressure cultivation at the transcriptional and translational level, the quantity of *ftsZ* mRNA and FtsZ protein from cells grown under elevated pressure were examined by primer extension analysis (data not shown) and Western blotting (Fig. 7).   There were no changes found due to pressure using either way of analysis. These results indicate that high hydrostatic pressure may affect before the FtsZ-ring formation, then the rings were formed immediately after depressurization, however pressure does not affect gene expression of *ftsZ*.

We also observed chromosomal DNA segregation by DAPI staining at the same time. The radio-labelled substrates uptake studies indicate that DNA synthesis does not completely become inhibited by pressure until around 50MPa (2).   The cells were grown and fixed inside and outside of pressure vessel as mentioned above.   The condensed nucleoid structure in the filamentous cells has been reported already (15), and we could confirm this result when the cells were fixed after releasing the pressure.   In contrast, the chromosomal DNA was segregated, but not condensed in whole filamentous cells, when the cells were fixed before pressure release with the new cell fixation system (Fig. 6).

In conclusion, *E. coli* may grow continuously without septum formation and be "frozen" at early stage in the cell cycle before the FtsZ-ring formation and DNA condensation, as in the case of the mitotic apparatus in eukaryote (2) under high hydrostatic pressure.   These inhibitions seem to be transient, as the FtsZ-ring formation and DNA condensation were restored within few minutes (Fig. 6, right column), and cell division has been occurred at once in many parts of the elongated cells, and almost completed within 20 min after pressure release (data not shown).   The finding is probably one of the general phenomena caused by pressure in atmospheric pressure cells and may be important in the study of piezo-adaptive mechanisms of organisms to high hydrostatic pressure.

ACKNOWLEDGEMENTS

We thank Dr. Mitra Sohirad for assistance in editing the manuscript.   This study was partially supported by a Grant-in-Aid for Scientific Research, Scientific Research on Priority Areas: Single-Cell Molecular Technology (Area number; 736).

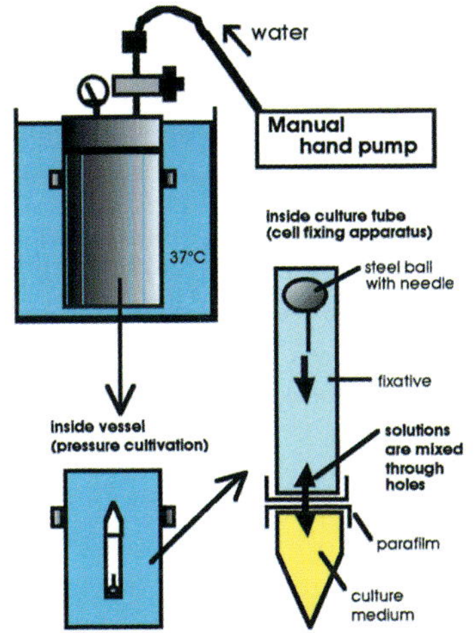

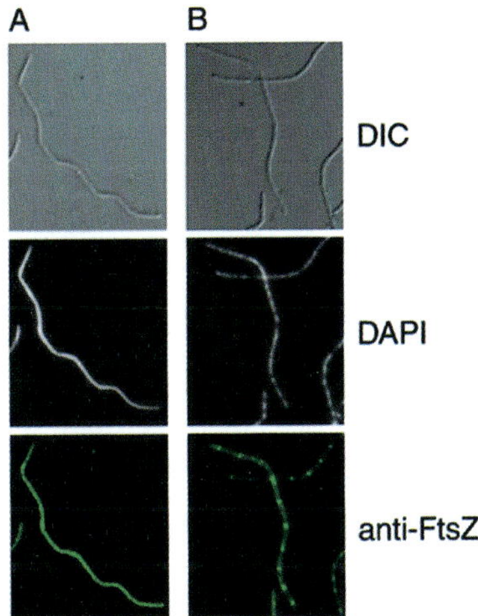

Fig. 5. Pressure apparatus for bacterial cultivation and fixation under high pressure. Cell culture chamber in sterilized tube was sealed by parafilm and connected with cell fixative chamber including steel ball with needle and incubated at 37°C inside of pressure vessel. After cell growth, pressure vessel was shaken upside down to break parafilm between two chambers with needle and fix cells under high pressure.

Fig. 6. Localization of chromosomal DNA and FtsZ by DAPI staining and IFM using FtsZ antibody in *E. coli* cells grown at 48 MPa fixed before (A) and after (B) pressure release.

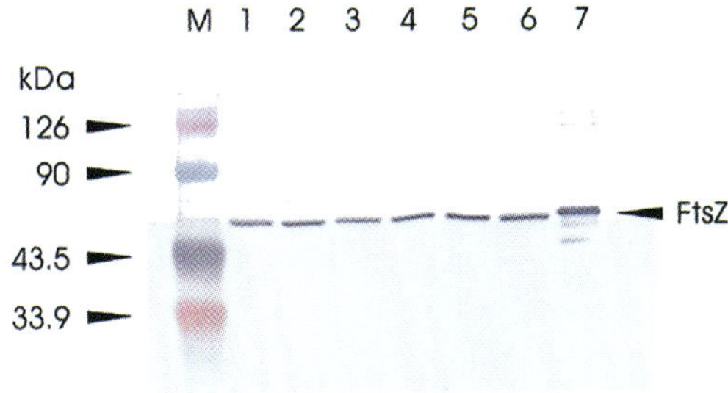

Fig. 7. Western blotting analysis of FtsZ expression in *E.coli* in response to elevated pressure. Lane 1: crude extract from the *E.coli* wild type strain grown at 0.1 MPa; lane 2: at 30 MPa; lane 3: at 48 MPa; lane 4: from *E.coli* JM109 grown at 0.1 MPa; lane 5: at 30 MPa; lane 6: at 48 MPa. Lane7: purified FtsZ protein. M: molecular weight marker.

REFERENCE

1. C. E. Zobell and A. B. Cobet, J. Bacteriol., 87 (1964) 710.
2. R. E. Marquis, Adv. Mic. Phys., 19 (1976), 159.
3. B. Bourns, S. Franklin, L. Cassimeris and E. D. Salmon, Cell Motil. Cytoskeleton, 10, (1988), 380.
4. H. C. Crenshaw, J. A. Allen, V. Skeen, A. Harris and E. D. Salmon, Experiment. Cell Res., 227, (1996), 285.
5. S. J. Tumminia, J. F. Koretz and J. V. Landau, Biochimi. Biophysi. Acta, 999, (1989), 300.
6. E. D. Salmon, Science, 189, (1975), 884.
7. C. Kato, T. Sato and K. Horikoshi, Biodiv. Conserv., 4 (1995) 1.

8. T. Sato, C. Kato and K. Horikoshi, J. Mar. Biotechnol., 3 (1995) 89.

9. C. Kato, T. Sato, M. Smorawinska and K. Horikoshi, FEMS microbiol. Lett., 122, (1994), 91.

10. T. Sato, Y. Nakamura, K. K. Nakashima, C. Kato and K. Horikoshi, FEMS microbiol. Lett., 135, (1996), 111.

11. R. Losick, and L.Shapiro, J. Bacteriol., 181 (1999) 4143.

12. E. Bi and J. Lutkenhaus, Nature, 354 (1991) 161.

13. Y. Abhayawardhane and G. C. Stewart, J. Bacteriol., 177 (1995) 765.

14. F. C. Neidhardt, R. Curtiss III, J. L. Ingraham, E. C. C. Lin, K. B. Low, B. Magasanik, W. S. Reznikoff, M. Riley, M. Schaechter and H. E. Umbarger (eds.), *Escherichia coli* and *Salmonella*-cellular and molecular biology, ASM Press, Washington, D. C., (1996) 1400.

15. D. H. Bartlett, Sci. Progress Oxford, 76 (1992) 479.

Trends in High Pressure Bioscience and Biotechnology
R. Hayashi (editor)

Barophysiology of *Saccharomyces cerevisiae* from the aspect of 6,000 gene-expression levels.

H. Iwahashi,[ab] H. Shimizu,[b] M. Odani,[b] Y. Komatsu[b]

[a]Human Stress Signal Research Center
National Institute of Advanced Industrial Science and Technology
Higashi 1-1, Tukuba, Ibaraki 305-8566, Japan

[b] International Patent Organism Depositary
National Institute of Advanced Industrial Science and Technology
Higashi 1-1, Tukuba, Ibaraki 305-8566, Japan

Hydrostatic pressure is one of the physical factors that affect cellular physiology. With hydrostatic pressure, a few hundred MPa decreases the viability of yeast cells and a few tens MPa decreases the growth rate. However, there is little information about how hydrostatic pressure affects viability and growth. To understand the effects of hydrostatic pressure, we have used yeast DNA chips and analyzed 6,000 gene-expression levels after four different conditions. A 180MPa treatment, 4℃ for zero minutes, and 0.1MPa, 25℃ for 60min, which was used for recovery, induced the genes of *BTN2, HSP26, SSC1*, and so on. One of 40MPa, 4℃ for 16h, and 0.1MPa, 25℃ for 60min, induced the genes of *SSA4, HSP12, TIP1*, and so on. The many genes induced by these treatments belong to the group of molecular chaperon and protein degradation. The genes of *YRO2, PST1*, and so on were induced by a 40MPa treatment, 25℃ for 2h, but the number of induced genes was not very high. The growing yeast cells under 10MPa, 25℃ induced the genes of *HFA1, PMD1*, and so on. The characteristic genes induced under these conditions were cyclins for the G1/S phase and have a chromosome duplication function. The genes induced by pressure treatment suggest that lethal pressure causes damage to the protein structure, and mild pressure inhibits chromosome duplication.

1. INTRODUCTION

Hydrostatic pressure is one of the physical factors that affects cellular physiology. With hydrostatic pressure, a few hundred MPa decreases the viability of yeast cells and a few tens MPa decreases the growth rate (Fig. 1). New technology for food sterilization is developing based on these facts, and some food products may be sterilized by using hydrostatic pressure (1).

As the technology develops, many efforts are being focused on the growth and survival of the organisms under hydrostatic pressure. In the range of pressure that causes growth inhibition, Abe and Horikoshi (2) found that hydrostatic pressure of 40 to 60MPa promoted acidification of the vacuoles in yeast cells, and Tamura *et al.* showed the induction of yeast heat-shock protein in the same range of pressure (3). In *Escherichia coli*, filament formation was observed up to 40MPa (4). From the pressure of 100MPa, organisms can be sterilized. Sonoike *et al.* showed that hydrostatic pressure sterilization is more effective at cold temperatures (around 0°C) than at room temperatures (5) (Fig. 1). For example, *Lactobacillus casei* is killed by hydrostatic pressure (300MPa) with a death rate of 0.32 at 0°C, 0.1 at 20°C, and 0.32 at 60°C (5). Iwahashi *et al.* showed that a pressure treatment of more than 100MPa decreased the CFU, but a mild heat-shock treatment of 43°C for 30min increased barotolerance (resistance to hydrostatic pressure) (6). This result was shown to be due to the accumulation of trehalose and Hsp104 (7). They also found that trehalose contributed independently of the temperature to the barotolerance and that Hsp104 was dependent on the temperature. Therefore, they concluded that effective hydrostatic pressure sterilization at a cold temperature occurred because of the low activity of Hsps (8). Hamada *et al.* (9) observed the induction of tetraploids or homozygous diploids in the industrial yeast *Saccharomyces cerevisiae* by hydrostatic pressure (above 100MPa). In the same range of pressure, Kobori *et al.* (10) also demonstrated with immunoelectron microscopy using frozen thin sections that there had been damage to the nuclei of *S. cerevisiae*.

On the other hand, recent biotechnology enables us to monitor cellular response through almost any level of gene expression. DNA chip (micro array) technology is used to monitor almost all levels of gene expression (11). On the yeast DNA chip, we have 6,000 DNA probes that correspond to almost all of the yeast open-reading frames, and we can estimate each expression level of the 6,000 open-reading frames (11). To understand the effect of hydrostatic pressure, we have used yeast DNA chips and analyzed 6,000 gene-expression levels using a variety of hydrostatic pressures and recovery treatments.

2. MATERIALS AND METHODS

2.1. Hydrostatic pressure treatment

Yeast cells were treated with A) 180MPa, 4°C for zero minutes, and 0.1MPa, 25°C for 60min, as recovery; B) 40MPa, 4°C for 16h, and 0.1MPa, 25°C for 60min, as recovery; C) 40MPa, 25°C for 2h, and D) 10MPa, 25°C for 12h. For A and B, the yeast cells were centrifuged and resuspended in the medium where they were pressurized as described previously (6). For C and D, the yeast cells were poured into 50ml of the medium, and the syringes were transferred to a stainless steel vessel for pressurization. The control cells were treated identically under all conditions with the exception of pressure.

2.2. RNA preparation, hybridization of the DNA chip, and characterization of induced genes

Total RNA was isolated using a hot-phenol method (12). Poly (A)+RNA was purified from total RNA with an Oligotex-dT30 mRNA purification kit (Takara, Kyoto, Japan). Fluorescently labeled cDNA was synthesized by oligo dT-primed polymerization using PowerScriptTM reverse transcriptase (Clontech, CA, USA). cDNA made from poly(A)+ RNA of control was fluorescently labeled with Cy3 and that of the pressure-treated sample was labeled with Cy5. Two-fourμg of poly(A)+ RNA was used for each labeling, and the same amount of each poly(A)+RNA was used in one slide. The two labeled cDNA pools were mixed and hybridized for 24-36 hr at 65°C. After hybridization, the labeled micro-arrays were washed and dried. Subsequently labeled micro-arrays were scanned using a conforcal laser ScanArray 4000 (GSI Lumonics, MA, USA) system. Resulting image data were quantified using QuantArray (GSI Lumonics, MA, USA). The fluorescence intensity of each spot on the images was subtracted from each background, and the ratios of intensity Cy5/Cy3 were calculated and normalized using *ACT1* as positive control using GeneSpring (Silicon Genetics, CA, USA).

3. RESULTS AND DISCUSSION

3.1. Conditions of the pressure treatment

Hydrostatic pressure affects the cellular physiology of yeast. This effect is observed as cellular death or growth inhibition. Generally, cellular death is observed under more than 150MPa at 4°C in a few minutes or more than 30MPa at 4°C in a few days (Fig. 1). For studying the damage caused in lethal conditions, the conditions were 180MPa for zero minutes and 40MPa for 16h. Under these conditions, the yeast cells showed 60% of CFU compared to the untreated control. Under the pressure noted above, ~~yeast cells~~ can not synthesize macromolecular such as DNA, protein, and cell wall. To induce the genes for repairing the damages, we made yeast cells grow at 25°C for 1h after the pressure treatment.

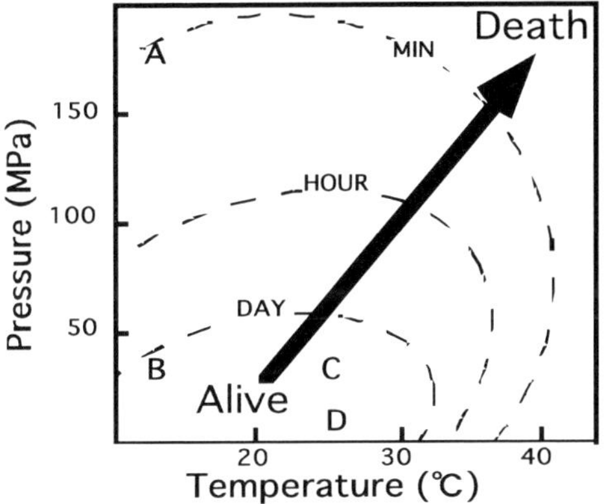

Figure1 Phase diagram between pressure and temperature
A,180MPa, 4 °C for zero minute and 0.1MPa, 25 °C for 60min
B, 40MPa, 4 °C for 16h and 0.1MPa, 25 °C for 60min
C, 40MPa, 25 °C for 2h
D, 10MPa, 25 °C for 12h

242

Hydrostatic pressure inhibits cellular growth with a few doses of MPa (Fig. 1). Thus, we treated yeast cells with 40MPa, 25°C for 2h, and with 10MPa, 25°C for 12h. With 40MPa, the yeast cells did not change CFU, but the absorbency increased two times to 660nm. At 10MPa, the yeast cells grew at one half the speed of 0.1MPa.

3.2. 6,000 gene responses with 180MPa, 4°C for zero minutes, and 0.1MPa, 25°C for 60min, as recovery

After treatment with 180MPa, 4°C for zero minutes (just increasing pressure to 180MPa and decreasing to 0.1MPa for four minutes), the yeast cells decreased their viability to around 60%. The functionary characterized genes induce by this treatment were summarized in Table 1.   More detailed results are described by Shimizu *et al.* (13).   There are many heat-shock proteins in Shimizu's list. The heat-shock proteins are molecular chaperons and contribute to protein metabolism. This suggests that treatment of 180MPa at 4°C damages proteins.

Table 1. The functionary characterized genes induced in the cells grown under 10MPa, 25°C for 12h

Fold	Gene	Character
12	BTN2	Gene/protein elevated in a btn1 minus/Btn1p lacking yeast strain
9.8	HSP26	Heat shock protein 26
8.4	SSC1	Mitochondrial matrix protein involved in protein import
8.0	CPR6	A cyclophilin related to the mammalian CyP-40
7.7	HSP104	Heat shock protein 104
7.5	ERO1	Required for protein disulfide bond formation in the ER
6.6	GIN3	C-terminal part starting with aa
6.5	KAR2	Homologue of mammalian BiP (GPR78) protein
6.0	UGP1	UTP--glucose-1-phosphate uridylyltransferase;
6.0	SSA4	Member of 70 kDa heat shock protein family;
5.9	HXK1	Hexokinase I (PI) (also called Hexokinase A);
5.8	AUT7	Essential for autophagy
5.6	RPN7	Subunit of the regulatory particle of the proteasome
4.6	RPN4	Involved in ubiquitin degradation pathway;
4.4	SHE10	mRNA that causes growth arrest when overexpressed
4.4	CDS1	CDP-diacylglycerol synthase
4.2	CYC7	iso-2-cytochrome c;
3.9	GPD1	Glycerol-3-phosphate dehydrogenase; up regulated in high salt"
3.9	RPT2	Probable 26S protease
3.8	MRP8	Mitochondrial ribosomal protein
3.6	UBP6	Ubiquitin-specific protease
3.6	YET1	Yeast endoplasmic reticulum 25 kDa transmembrane protein
3.1	RPN11	26S proteasome regulatory subunit

3.3. 6,000 gene responses with 40MPa, 4°C for 16h, and 0.1MPa, 25°C for 60min, as recovery

After the 40MPa treatment, 4°C for 16h, and 0.1MPa, 25°C for 60min, as recovery, the yeast cells decreased their viability to around 60%. The functionary characterized genes induce by this treatment were summarized in Table 2. The details are also in Table 2, as provided by Shimizu *et al* (13). As observed under 180MPa at 4°C, many heat-shock proteins were induced. Thus, the treatment with 40MPa at 4°C also activated protein metabolism.

The treatment with 40MPa, 4°C for 16h and 180MPa, 4°C for zero minute made yeast cells start to induced the genes concern to protein metabolism. This suggests that some proteins were denatured and were have to be degraded. Generally, enzymes without a complex structure are inactivated under conditions that are more severe than those we can duplicate. Thus, proteins that have a complex structure such as ribosome, ER, and so on were damaged by the pressure applied in this study.

Iwahashi *et al.* reported that heat-shock treatment increases barotolerance (6) and Hsp104 contributes to barotolerance (8). In this DNA chip analysis, the induction of many heat-shock proteins was detected including the *HSP104* gene. Thus, these results support previous findings. The induction of heat-shock proteins by heat treatment must be ready for protein metabolism.

Table 2. The functionary characterized genes induced in the cells grown under 10MPa, 25°C for 12h

Fold	Gene	Character
7.4	SSA4	Member of 70 kDa heat shock protein family
4.8	HSP12	12 kDa heat shock protein
4.4	TIP1	Cell wall mannoprotein
4.1	HSP26	Heat shock protein 26
4.0	CPR6 a	Cyclophilin related to the mammalian CyP-40
4.0	BIO2	Biotin synthase
3.8	MET17	O-Acetylhomoserine-O-Acetylserine Sulfhydralase;
3.5	MAG1	3-methyladenine DNA glycosylase
3.1	BTN2	Gene/protein whose expression is elevated in a btn1 minus/Btn1p lacking yeast strain
3.0	DDI1	DNA Damage Inducible
2.9	PRE3	Subunit of 20S proteasome
2.9	UBC5	Ubiquitin-conjugating enzyme
2.8	RPT2	26S proteasome regulatory subunit
2.7	PRE1	22.6 kDa proteasome subunit
2.5	AUT7	Essential for autophagy
2.4	NCA3	Involved in regulation of synthesis of Atp6p and Atp8p
2.4	NCE103	Involved in secretion of proteins that lack classical secretory signal sequences
2.3	RPN7	Subunit of the regulatory particle of the proteasome;
2.2	UMP1	20S proteasome maturation factor
2.1	CDC48	Microsomal protein of CDC48/PAS1/SEC18 family of ATPases
2.1	UFD1	Ubiquitin fusion degradation protein
2.1	HSP104	Heat shock protein 104

244

Table 3. The functionary characterized genes induced in the cells grown under 10MPa, 25°C for 12h

Fold	Gene	Character
8.0	YRO2	Homolog to HSP30 heat shock protein Yro1p
4.4	PST1	Strong similarity to SPS2 proteinstrong similarity to SPS2 protein
2.9	ADE1	Phosphoribosyl amino imidazolesuccinocarbozamide synthetase
2.9	GRX1	Glutaredoxin
2.5	HRT2	High level expression reduced Ty3 transposition
2.5	TY1A	TY1A protein; inactive Ty1 transposon according to L22015
2.4	GLN1	Glutamine synthetase
2.2	ADE12	Adenylosuccinate synthetase
2.2	SHM2	Serine hydroxymethyltransferase
2.1	MTD1	NAD-dependent 5,10-methylenetetrahydrafolate dehydrogenase
2.0	PCL5	PHO85 cyclin

3.4. 6,000 gene responses with 40MPa, 25°C for 2h

The functionary characterized genes induce by the treatment with 40MPa, 25°C for 2h were summarized in Table 3. Under these conditions, no significant gene induction was detected, but the induction of YRO2 that is a homolog to the *HSP30* heat-shock protein Yro1p was detected. As *Hsp30* were shown to contributed to ATPase (14) activity that may concern to the acidfication of cell. Abe and Horikoshi showed that a pressure of 40 to 60MPa reduced the vacuolar pH (2); therefore, the induction of YRO2 may reflect this acidification. It should be noted that the low levels of induction of mRNA may merely reflect the inhibition of macromolecular synthesis, since it has been shown that macromolecular biosynthesis is inhibited around the range of 30MPa (15). The growth inhibition under these conditions may reflect this.

3.5. 6,000 gene responses with 10MPa, 25°C for 12h

The functionary characterized genes induced (up to 5 fold) in the cells grown under 10MPa, 25°C for 12h were summarized in Table 4. More detailed results were described by Odani *et al* (16). The most interesting gene in the list is the *CLN2* gene. This gene is shown to regulate the cell cycle of the G1/S phase (17), and thus the induction of the *CLN2* gene implies that yeast cells were arrested in the G1/S phase or the population of the G1/S phase was dominant. Under the 10MPa treatment, yeast cells certainly grew, and the dominance of the G1/S phase is reasonable. This suggests that hydrostatic pressure inhibited the events in the G1/S phase. In the G1/S phase, the yeast cells synthesize the chromosome and the components for cell division. Indeed, the induction of genes that contributes to chromosomal duplication could be seen. HFA1, PMD1, STB1, HAT2, KAR3, YKT9, ZIP1, DOC1, RPP1, and STB3 are genes concerned with the events of chromosome duplication. In addition, the induction of PCL5 that is a PHO85 cyclin produced in the G1/S phase under the 40MPa treatment with 25°C for 2h. This result also supports the idea that hydrostatic pressure inhibits the event in the cellular phase of G1/S, probably duplication of chromosome under growing conditions.

Table 4. The functionary characterized genes induced in the cells grown under 10MPa, 25°C for 12h

Fold	Gene	Character
7.8	HFA1	Similar to acetyl-coenzyme A carboxylase
7.0	PMD1	Negative regulator of early meiotic expression
6.6	STB1	Binds Sin3p in two-hybrid assay
6.3	PGD1	Probable transcription factor, polyglutamine domain protein
6.3	HAT2	Subunit of a cytoplasmic histone acetyltransferase
5.9	KIN82	Putative serine/threonine protein kinase
5.7	SWH1	Probable NH-terminus of OSH1/SWH1
5.7	YAT1	Outer carnitine acetyltransferase, mitochondrial
5.6	MSY1	Tyrosyl-tRNA synthetase
5.6	KAR3	Kinesin-like nuclear fusion protein
5.5	MRK1	MDS1 related protein kinase
5.5	YKT9	PTK1-YKT9 overlapping sequences may arise from a gene duplication event
5.4	ATP7	ATP synthase d subunit
5.4	CTT1	Cytoplasmic catalase T
5.4	CMK2	Calmodulin-dependent protein kinase
5.4	ZIP1	Synaptonemal complex protein
5.4	TSL1	123 kD regulatory subunit of trehalose-6-phosphate synthase/phosphatase complex
5.4	DOC1	Doc1p and Cdc26p are associated with the anaphase-promoting
5.3	CLN2	G(sub)1 cyclin
5.2	RAD28	Protein involved in the same pathway as Rad26p, has beta-transducin (WD-40) repeats
5.2	YPT53	Rab5-like GTPase involved in vacuolar protein sorting and endocytosis
5.1	ECM23	ExtraCellular Mutant¥; similar to SRD1
5.1	RPP1	Protein subunit of nuclear ribonuclease P (RNase P)
5.0	STB3	Binds Sin3p in two-hybrid assay
5.0	EFT1	Translation elongation factor 2 (EF-2)
4.9	DAK2	Dihydroxyacetone kinase
4.8	TUP1	Glucose repression regulatory protein, exhibits similarity to beta subunits of G proteins
4.8	VPS27	Hydrophilic protein¥; has cysteine rich putative zinc finger esential for function
4.7	RNH35	RNase H(35), a 35 kDa ribonuclease H
4.7	PRP5	RNA helicase homolog
4.6	NIP29	Nuclear import protein
4.6	IXR1	Intrastrand crosslink recognition protein
4.6	PDS5	Precocious Dissociation of Sister chromatids
4.6	GCN3	Translation initiation factor eIF2B, 34 KD, alpha
4.6	HVG1	Homologous to VRG4
4.5	RIF1	RAP1-interacting factor, involved in establishment of repressed chromatin
4.5	CWP2	Cell wall mannoprotein
4.5	HCM1	Dosage-dependent suppressor of cmd1-1
4.5	FMC1	Formation of Mitochondrial Cytochromes 1
4.4	SEN54	54kDa subunit of the tetrameric tRNA splicing endonuclease
4.4	ROD1	O-dinitrobenzene,calcium and zinc resistance protein
4.4	RFC2	Subunit 2 of Replication Factor C¥; homologous to human RFC 37 kDa subunit
4.4	LPP1	Lipid phosphate phosphatase

REFFERENCES

1. R.Hayashi, In Use of High Pressure in Food, R.Hayashi, (ed.), Sanei, Kyoto, 1989.

2. F. Abe, K. Horikoshi, FEMS Microbiol Lett 130 (1995) 307.

3. K. Tamura, M. Miyashita, and H. Iwahashi, Biotech. Lett., 20 (1998) 1167.

4. K. Tamura In High Pressure Bioscience , R.Hayashi, (ed.), Sanei, Kyoto, 1994.

5. K. Sonoike, T. Setoyama, Y. Kuma, T. Shinno, K. Fukumoto, and M. Ishihara, In High Pressure Bioscience and Food Science, Hayashi R. (ed.) Sanei, Kyoto, 1993.

6. H. Iwahashi, K. Obuchi, S. C. Kaul, and Y. Komatsu, FEMS Microb. Lett., 325 (1991) 80.

7. H. Iwahashi, K. Obuchi, S. Fujii, and Y. Komatsu, .Let. Appl. Microbiol. 25 (1997) 43.

8. H. Iwahashi, K. Obuchi, S. Fujii, and Y.Komatsu, FEBS Lett. 416 (1997) 1.

9. K. Hamada, Y. Nakatomi, S. Shimada, Curr. Genet. 22(1992)371

10. H. Kobori, M. Sato, A. Tameike, K. Hamada, S. Shimada, M. Osumi, FEMS Microbiol. Lett. 132 (1995) 253.

11. H. Iwahashi, Bioscience and Bioindustry. 58, (2000) 27.

12. C. Guthrie and G. R. Funk (eds.) In Methods in enzymology Academic Press, New York, 1990.

13. H. Shimizu, M.Odani, Y. Komatsu, H. Iwahashi (this book)

14. P. W. Piper, FEMS Microbiol. Rev. 11 (1993) 339.

15. Y. Okami, In High Pressure processed Food, R.Hayashi, (ed.), Sanei, Kyoto, 1990.

16. M. Odani, H. Shimizu, Y. Komatsu, H. Iwahashi (this book)

17. F. R. Cross andA. H. Tinkelenberg Cell , 65 (1991) 875.

Trends in High Pressure Bioscience and Biotechnology
R. Hayashi (editor)

RESPONSES OF GROWING YEAST CELLS IN THE HYDROSTATIC PRESSURE STATUS MONITORING BY DNA MICROARRAY

M. Odani, [a] H. Shimizu, [a] Y. Komatsu, [a] H, Iwahashi [ab]

[a] International Patent Organism Depository
National Institute of Advanced Industrial Science and Technology
Higashi 1-1, Tukuba, Ibaraki 305-8566, Japan

[b] Human Stress Signal Research Center
National Institute of Advanced Industrial Science and Technology
Higashi 1-1, Tukuba, Ibaraki 305-8566, Japan

Hydrostatic pressure affects almost all physiological activities of living cells. Numerous studies on the effects of hydrostatic pressure on proteins, membranes, viability, and growth rates have been reported and reviewed. However, the mechanism that causes the damage when cellular systems are under hydrostatic pressure has not yet been clarified. We used DNA microarray technology to evaluate the differential expression of genes in cells with variances in hydrostatic pressure. Using a DNA microarray, we monitored the response of yeast cells as they were treated under the conditions of 10 MPa and 25 C for 12 h. The DNA microarray experiments were conducted comparing mRNAs isolated from yeast cells treated with 10 MPa and 0.1 MPa. For the genes of HFA1, YAT1, PMD1, YKT9, STB1, SWH1, HAT2, KAR3, ARF1, CLN1, CLN2, and so on, the induction of the gene expression for the treated yeast cells with 10 MPa was observed. When the induced genes were classified into a category with a database of MIPS (Munich Information Center For Protein Sequences), these genes were considered to be largely involved in cell growth, cell division, and DNA synthesis and transcription. Furthermore, the genes of CLN1 and CLN2 are cyclin specific to the G1/S phase and are induced after passing a DNA damage checkpoint. This suggests that yeast cells were affected inhibition during budding period.

248

Introduction

Hydrostatic pressure affects almost all physiological activities of living cells. Numerous studies on the effect of hydrostatic pressure on proteins, membranes, viability, and growth rates have been reported and reviewed. For example, in the case of yeast, Abe *et al.* have reported that influence appears in pH of vacuoles and in glucose metabolism etc. under high pressure of about 50 Mpa (1-3). Tamura *et al.* showed the induction of a yeast heat-shock protein in the same range of pressure (4). Sonoike *et al.* have analyzed the sterilization effect of temperature and pressure using *Lactobacillus casei* and *Escherichia coli* (5). Iwahashi *et al.* showed that a mild heat-shock treatment of 43 C for 30 min increased barotolerance (6). However, the mechanism that causes damage to cellular systems under hydrostatic pressure has not yet been clarified.

We used DNA microarray technology to evaluate the differential expression of genes in cells with a variance in hydrostatic pressure. DNA microarray technology provides us with the ability to measure the expression levels of thousands of genes in a single experiment. Furthermore, it enables us to monitor biological responses through almost all gene expression levels (7). Currently, most approaches for the analysis of data from microarray gene expressions attempt to study the functionality of a significant classification of genes. We have employed microarrays containing approximately 6,000 open reading frames from *S. cerevisiae* printed on glass slides and investigated the effects of hydrostatic pressure on growing yeast cells.

Materials and Methods

Growth conditions and hydrostatic pressure treatments

A treatment of cells using hydrostatic pressure was performed under the conditions of 10 MPa and 25 C for 12 h. Yeast cells in a growth medium (YPD medium : 2 % polypeptone, 1 % yeast extract, 2 % glucose) were poured into a 50 ml syringe and then transferred to a stainless steel vessel for pressurization. The control cells were treated under the same conditions except for the pressure.

RNA isolation and DNA microarray analysis

Total RNA was isolated by a hot acid phenol method. Quantitation of RNA was carried out by UV spectroscopy. mRNA was isolated from total RNA using an Oligotex^{TM}-dT30 mRNA purification kit from Takara.

Cy3-dUTP or Cy5-dUTP (Amersham Pharmacia Biotech) was incorporated during a reverse transcription of 3.0 µg of polyadenylated RNA, primed by a dT (12-18) oligomer. This mixture was heated to 70 C for 5 min. A premixed solution, consisting of 200 U Superscript II (Gibco), a buffer, 10 mM DTT, 5 U Rnasin, deoxyribonucleoside triphosphates, and fluorescent nucleotides, was added to the RNA. Nucleotides were used at these final concentrations: 500

µM for dATP, dCTP, and dGTP and 200 µM for dTTP. Cy3-dUTP and Cy5-dUTP were used at a final concentration of 100 µM. The reaction was incubated at 42 C for 40 minutes. After 40 minutes, this reaction mixture was added again 200 U Superscript II was incubated at 42C for 40 minutes. Furthermore, this reaction mixture was added 180 mM NaOH and 45 mM EDTA was incubated at 65 C for 1 hour. Unincorporated fluorescent nucleotides were removed by first diluting the reaction mixture with 750 µl of 10 mM tris-HCl (pH 8.0) / 1mM EDTA and then subsequently concentrating the mix to 20µl, using Microcon-YM30 (Amicon). Purified, labeled cDNA was resuspended in 10 µl of 20 x SSC and 2 µl of 10 % SDS. A suitable quantity of distilled water was added to the mixture, which was finally set at 40 µl. Before hybridzation, the solution was boiled at 95 C for 3 minutes and allowed to cool to room temperature. The solution was applied to a microarray under a covere glass, and the slide was placed in a hybridization chamber, which was subsequently incubated for 36 hours in a water bath at 65 C. Before scanning, the slides were washed twice with 2 x SSC, 0.1 % SDS for 20 minutes, and 0.2 x SSC, 0.1 % SDS for 20 minutes. They were then rinsed for 0.2 x SSC and 0.05 x SSC. The slides were dried before scanning centrifugation at 600 rpm.

Scanning was carried out using a conforcal laser ScanArray 4000 (GSI Lumonics, MA, USA) system. Scanned image data were analyzed using a QuantArray (GSI Lumonics, MA, USA). The ratio of fluorescence intensity of each spot on the image was calculated using GeneSpring (Silicon Genetics, CA, USA).

Results and Discussion

We monitored the response of yeast cells by DNA microarray as the cells were treated under 10 MPa at 25 C for 12 hours. Under these conditions, the yeast cells were grown at almost half the speed of those at 0.1 MPa (Table 1).

Table 1. growth rate of yeast under hydrostatic pressure conditions.

	0.1 MPa	10 MPa
doubling time	2.5 h	4.3 h

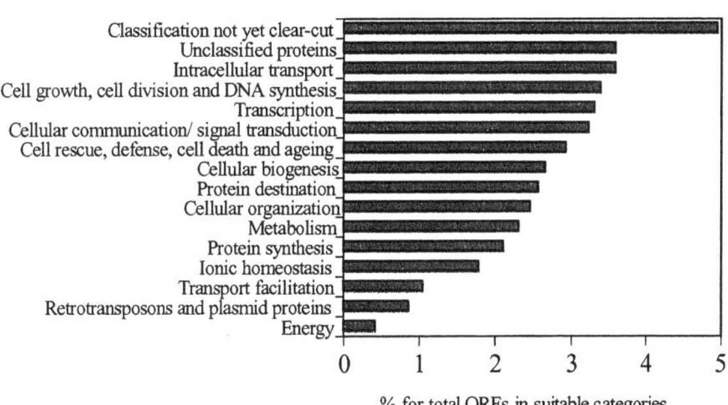

Fig. 2 Classification with functional categories of 4 or more times induced genes by 10 MPa treatment.

Table 2a. The classified 4 or more times induced genes by the functional categories

fold	gene	character
INTRACELLULAR TRANSPORT		
9.04	ERP3	p24 protein involved in membrane trafficking
5.75	YAT1	Outer carnitine acetyltransferase, mitochondrial
5.61	KAR3	kinesin-like nuclear fusion protein
5.51	YKT9	Protein of unknown function
5.46	ATP7	ATP synthase d subunit
5.21	YPT53	rab5-like GTPase involved in vacuolar protein sorting and endocytosis
4.82	VPS27	hydrophilic protein\; has cysteine rich putative zinc finger esential for function
4.70	NIP29	Nuclear import protein
4.62	GOS1	SNARE protein of Golgi compartment
4.61	HVG1	Homologous to VRG4
CELL GROWTH, CELL DIVISION AND DNA SYNTHESIS		
7.02	PMD1	negative regulator of early meiotic expression
6.31	PGD1	Probable transcription factor, polyglutamine domain protein
5.61	KAR3	kinesin-like nuclear fusion protein
5.45	GIC1	Cdc42 GTPase-binding protein
5.43	ZIP1	synaptonemal complex protein
5.32	CLN2	G(sub)1 cyclin
4.68	PDS5	Precocious Dissociation of Sister chromatids
4.44	RFC2	Subunit 2 of Replication Factor C\; homologous to human RFC 37 kDa subunit
4.38	SCC2	Sister chromatid cohesion protein
4.38	TPM1	tropomyosin I
4.33	RHO2	GTP-binding protein of the rho subfamily of ras-like proteins
4.30	SIS2	sit4 suppressor
4.29	BNS1	Bypasses Need for Spo12p
4.27	CNB1	Type 2B protein phosphatase\; regulatory B subunit of calcineurin
4.25	CDC15	protein kinase domain

DNA microarray experiments were conducted comparing mRNAs isolated from yeast cells treated with 10 MPa and 0.1 MPa. cDNA was labeled Cy3 and Cy5, and hybridized to microarrays. The ratio of fluorescence intensity was quantified from data from a scanned microarray image and calculated by dividing the intensity of Cy5 by the intensity of Cy3. Normalization was performed as positive control using ACT1, which was a housekeeping gene. Consequently, the gene expressed four or more times by pressure treatment was 190 ORFs

Table 2b. The classified 4 or more times induced genes by the functional categories

fold	gene	character
TRANSCRIPTION		
6.31	PGD1	Probable transcription factor, polyglutamine domain protein
6.31	HAT2	subunit of a cytoplasmic histone acetyltransferase
6.05	NUT2	negative transcription regulator from artifical reporters
5.15	RPP1	Protein subunit of nuclear ribonuclease P (RNase P)
5.09	STB3	Binds Sin3p in two-hybrid assay
4.70	IXR1	intrastrand crosslink recognition protein
4.59	RIF1	RAP1-interacting factor, involved in establishment of repressed chromatin
4.49	SEN54	54kDa subunit of the tetrameric tRNA splicing endonuclease
4.47	NRD1	involved in regulation of nuclear pre-mRNA abundance
4.40	MOT3	2 Cys2-His2 zinc fingers at c-terminus, glutamine and asparagine rich
4.36	MTR2	mRNA transport regulator
CELLULAR COMMUNICATION/SIGNAL TRANSDUCTION		
5.45	GIC1	Cdc42 GTPase-binding protein
5.44	CMK2	Calmodulin-dependent protein kinase
4.33	RHO2	GTP-binding protein of the rho subfamily of ras-like proteins
4.12	STT4	encodes a phosphatidylinositol-4-kinase, homologous to VPC34

among about 6,000 ORFs. Based on a yeast genome project database of MIPS of Germany (http://www.mips.biochem.mpg.de/), the gene induced four or more times were classified by functional categories (the ORFs applicable to the category of four higher ranks was shown in Table 2a, b). The classified category was made into a graph of the rate to the total ORF of the suiting category (Fig. 1). According to the classification by category, many ORFs dealing with cellular matters, such as intracellular transport, cell growth, cell division, and DNA synthesis, were induced. Moreover, the functional unknown genes occupied 49.5 % of the induced genes. Furthermore, these categories can be classified into a more detailed sub category. According to the classification by sub category, the majority of ORFs applicable to the cell cycle control and DNA repair were in the category of cell growth, cell division, and DNA synthesis. CLN2 is an especially interesting gene among the applicable ORFs. This gene is cyclin specific to the G1/S phase and is induced after passing the DNA damage checkpoint (8). The genes such as CLN1, CLB5, and CLB6 (data not shown) that act on G1/S specifically were similarly induced. Moreover, CLB5 and 6 genes are promoting some steps of DNA synthesis in the S phase, which is the duplicate phase of DNA, and are shortening the time of the S phase (9-10). Furthermore, in the induced genes, the genes relevant to the event of chromosome duplication, such as HFA1,

PMD1, STB1, HAT2, KAR3, YKT9, ZIP1, and DOC1, were also induced. The following things are suggested from these results. DNA synthesis is probably considered to have received inhibition with pressure in the stage of a DNA duplicate in the S phase. Therefore, the S phase is extended. In order to compensate it, CLB5 and 6 genes, which promote DNA synthesis of the S phase, are induced. Moreover, there were many genes of functional unknown. Probably, it is thought that these genes also belong to categoris of cell growth, cell division, and DNA synthesis. We want to clarify also about these genes from now on.

References

1. F. Abe and K. Horikoshi, FEMS Microbiol. Lett., 130 (1995) 307.
2. F. Abe and K. Horikoshi, Extreamphile, 2 (1998) 223.
3. F. Abe, Appl. Environ. Microbiol., 64 (1998) 1139.
4. K. Tamura, M. Miyashita, and H. Iwahashi, Biotechnol. Lett., 20 (1998) 1167.
5. K. Sonoike, T. Setoyama, Y. Kuma, T. Shinno, K. Fukumoto, and M. Ishihara, In High Pressure Bioscience and Food Science, Hayashi R. (ed.) Sanei, Kyoto, 1993.
6. H. Iwahashi, K. Obuchi, S. C. Kaul, and Y. Komatsu, FEMS Microbiol. Lett., 325 (1991) 80.
7. H. Iwahashi, Bioscience and Bioindustry, 58 (2000) 27.
8. J. M. Sidorova and L. L. Breeden, Genes Dev., 11 (22) 3032.
9. C. B. Epstein and F. R. Cross, Genes Dev., 9 (1992) 1695.
10. E. Schwob and K. Nasmyth, Genes Dev., 7 (1993) 1160.

Trends in High Pressure Bioscience and Biotechnology
R. Hayashi (editor)
© 2002 Elsevier Science B.V. All rights reserved.

Response of actin cytoskeleton on *Schizosaccharomyces pombe* to high pressure-stress

M. Sato[a], R. Kobayashi[b], S. Shimada[c], and M. Osumi[b]

[a]Laboratory of Electron Microscopy, [b]Department of Chemical and Biological Sciences, Faculty of Science, Japan Women's University, Tokyo 112-8681, Japan

[c]Oriental Yeast Co., Ltd., Tokyo 174-8505, Japan

We investigated response to pressure stress on the actin cytoskeleton and actin cytoskeleton-related protein Cdc8 tropomyosin of the cells of a cold-sensitive mutant *nda3*-KM311 *Schizosaccharomyces pombe* by rhodamine-conjugated phalloidin and the specific antibodies of the actin cytoskeleton-related protein Cdc8 tropomyosin. At below 100 MPa actin cytoskeleton and tropomyosin were equally distributed. Tropomyosin was localized at the actin ring and cables, and at 100 MPa it was localized in the long and fine actin cable. Above 150 MPa, however, tropomyosin was dispersed throughout the cytoplasm. We were unable to elucidate the relationship between the actin cytoskeleton and tropomyosin on the degradation pattern of the former. Using a *cdc8* temperature-sensitive mutant (*cdc8*-110) grown at a permissive temperature we studied response of ultrastructure of the cell to pressure stress. At 200 MPa septa of *cdc8* cells were drastically changed.

1. INTRODUCTION

Pressure stress causes ultrastructural changes in yeast cells, including changes in the actin cytoskeleton and microtubules. We have been investigating the response in yeast cells to pressure stress using fluorescence microscopy, conventional electron microscopy and immunoelectron microscopy [1,2]. We also found that the fission yeast *Schizosaccharomyces pombe* is more sensitive to pressure stress than budding yeast *Saccharomyces cerevisiae* [3].

To learn the influence of pressure stress on the actin cytoskeleton of *S. pombe* we used the cells of a cold-sensitive mutant, *nda3* KM-311 (*h⁻, leu1 nda3*-KM311) [4]. The *nda3* cells were arrested highly synchronously (about 80%) at a step similar to mitotic pro-metaphase under restrictive temperature of 20°C for 4h [5,6]. We identified the degradation pattern of actin cytoskeleton in the *nda3* cells grown at 20°C for 4h by pressure stress: the order of sensitivity of actin cytoskeleton to pressure stress was patches > cables > ring in the *nda3* cells [6,7]. Then, using the specific antibody of the actin cytoskeleton-related protein Cdc8 tropomyosin [8] we investigated the relationship between the actin and tropomyosin to learn the process of degradation on actin cytoskeleton.

2. MATERIALS AND METHODS

2.1. Yeast strain

S. pombe strains used were a cold-sensitive mutant KM311 (*h⁻ leu1 nda3*-KM311) [4] and a *cdc8* temperature-sensitive mutant *cdc8*-110 [9] .

2.2. Cultivation

The *nda3*-KM311 cells, aerobically grown at 30°C in YPD liquid medium to mid-exponential phase (5×10^6 cells/ml), were transferred to a restrictive temperature of 20°C for 4 h (5×10^7 cells /ml) [6].   The *cdc8*-110 cells were grown at a permissive temperature of 25°C in YPD liquid medium to mid-exponential phase.

2.3. Pressure treatments

Each cell suspension (approx. 5×10^7 cells/ml) was treated with pressure stress (50~200 MPa) for 10 min at room temperature by a high-pressure apparatus, TP-500 (Teramecs Co., Kyoto, Japan) and NKK-ABB (NKK Co., Tokyo, Japan).   The decompressed samples were immediately fixed to prepare specimens for fluorescence microscopy and conventional electron microscopy.

2.4. Fluorescence microscopy

Cell suspensions were fixed with 0.5% glutaraldehyde and 3% paraformaldehyde for 30 minutes or with 3.7% formaldehyde for 2 h at room temperature.   Actin cytoskeleton, septum and Cdc8 tropomyosin were visualized by fluorescence microscopy.

Actin cytoskeleton: F-actin was stained with rhodamine-conjugated phalloidin [3].

Septum: Septum was stained with Calcofluor white [10].

Cdc8 tropomyosin: Anti-Cdc8 tropomyosin monoclonal antibody and rhodamine-conjugated anti-rabbit or anti-guinea IgG were used as primary and secondary antibodies, respectively [8]

2.5. Preparation for conventional electron microscopy

Cell suspensions were fixed with 2.5% (w/v) glutaraldehyde in 0.1 M phosphate buffer, pH 7.2, for 2 h at 4°C.   Cells were washed three times with 0.1 M phosphate buffer, and postfixed with 1.5% potassium permanganate ($KMnO_4$) in distilled water for 17 h at 4°C. Specimens were embedded in 2% agarose, stained with 0.5% (w/v) uranyl acetate in distilled water for 1 h at 4°C. Dehydration was done by alcohol series and absolute acetone; cells were embedded in a Quetol 653 mixture.   Ultrathin sections were stained with 4% uranyl acetate and 0.4% lead citrate and examined with a JEM 1200 EXS transmission electron microscope at 120 kV [3].

3. RESULTS AND DISCUSSION

3.1. Response of pressure stress on actin cytoskeleton and tropomyosin in *nda3* cells

In *nda3* cells arrested at the restrictive temperature of 20°C for 4 h, actin patches were concentrated in the central region (←), and the actin ring (◄) and cables were also seen (Fig. 1a).   The specific actin distribution pattern in the cell cycle was lost even at 100 MPa.

Long and fine actin cables were seen all over the cells (Fig. 1b, ←), and at 150 MPa they became thick and short cables (Fig. 1c, ←).   At 200 MPa, the faint actin ring remained in the center of the cell (Fig. 1d, ◄), and at more than 200 MPa, the actin cytoskeleton decomposed (data not shown).   We identified the degradation pattern of actin cytoskeleton of *nda3* cells by pressure stress: the order of sensitivity to pressure stress was patches > cables > ring in these cells [6,7].   We investigated the relationship between the actin and tropomyosin to learn the process of degradation on actin cytoskeleton using the specific antibody of the actin cytoskeleton-related protein Cdc8 tropomyosin [8]. Figure 2 shows the effect of pressure stress on actin cytoskeleton (←) and localization of tropomyosin (◄) in *nda3* cells grown at 20°C for 4 h by fluorescence microscopy.   Tropomyosin was localized in the actin ring and cables.   After pressure stress below 100 MPa, it was localized in the long and fine actin cable (Fig. 2b), and at 150 MPa, it was dispersed throughout the cytoplasm (Fig. 2c).   The order of sensitivity to pressure stress was tropomyosin > actin.   Use of another specific antibody of the actin cytoskeleton-related protein may make it possible to elucidate the process of degradation on actin cytoskeleton.

3.2. Response of ultrastructure of *cdc8* cells to pressure stress

We are currently analyzing the relationship between the actin and tropomyosin to response of pressure stress using *cdc8* temperature-sensitive mutant cells. $cdc8^{ts}$ mutant cells grown at a permissive temperature(25°C) showed normal F-actin cables, but these cables disappeared 1 h after shift to 37°C [8].   To date we have studied the effect of pressure stress only on $cdc8^{ts}$ mutant cells grown at a permissive temperature (25°C).   Several changes in cell ultrastructure were caused by pressure stress.   Figure 3 shows the ultrastructure of *cdc8* cells fixed with $KMnO_4$ and septa were stained with Calcofluor white.   The cells without pressure stress showed the typical normal ultrastructure of organelles: nucleus, mitochondria and vacuoles (Fig. 3a$_1$).   After treatment at 50 MPa, the cells were little changed (data not shown).   At 100 MPa the nuclear membrane was damaged (Fig. b$_2$, ◄), vacuoles began to fuse into large pieces (Fig. b$_{1,3}$, ⇌) and cell membrane was damaged (Fig. b$_1$, ←).   At 150 MPa, the damage of cellular organelles had progressed, and nuclear membrane was disrupted (Fig. 3c$_{1,2}$, ◄).   At 200 MPa, the cellular organelles could hardly be detected, damage to the cell membrane had advanced further (Fig. 3d$_1$, ←) and septa were dramatically bent (Fig. 3d$_1$, ⇌).   From these facts, several ultrastructural changes in *cdc8* cells were similar to those of the wild type of *S. pombe* [3], except abnormal septum formation.   From fluorescence microscopic observation of septa in *cdc8* cells, it was also clear that the septa had changed: at less than 150 MPa, they were not changed, but at 200 MPa they were curved (Fig. 3d$_2$).   In several mutant cells of *S. pombe* there are abnormal septum formations [11,12], but this change is characteristic of pressure stress.   In wild-type cells, the septa were changed at 250 MPa.   The septa of *cdc8* cells are more sensitive to pressure stress than those of wild-type cells.   Change of actin cytoskeleton in *cdc8* cells by pressure stress was the same as that in wild-type cells and nda3 cells grown at 30°C.   We will now investigate using, either *cdc8* cells grown at 37°C or other mutant cells.

Acknowledgements

We would like to thank M. Yanagida of Kyoto University for the cold-sensitive mutant *nda3*-KM311 and I. Mabuchi of Tokyo University for the $cdc8^{ts}$ mutant *cdc8*-110.

256

REFERENCES

1. M. Sato *et al.*, FEMS Microbiol. Lett. **131**, 11 (1994)
2. H. Kobori *et al.*, FEMS Microbiol. Lett. **132**, 253(1995)
3. M. Sato *et al.*, Cell Struct. Funct. **21**, 167 (1996)
4. Y. Hirooka *et al.*, Cell, **39**, 349 (1984)
5.M. Osumi *et al.*, Biological Systems under Extreme Conditions,(eds., Taniguchi *et al.*) Springer, Heidelberg (2001)
6. M. Sato *et al.*, FEMS Microbiol. Lett. **176**, 31 (1999)
7. M. Sato *et al.*,Science and Technology of High Pressure (eds., Manghnani *et al.*) Vol.1, Universities Press (India) Ltd. (2000)
8. R. Arai *et al.*, Eur. J. Cell Biol. **76**, 288 (1998)
9. M. K. Balasubramanian *et al.*, Nature **360**, 84 (1992)
10. H. Kobori *et al.*, J. Cell Sci. 107, 1131 (1994)
11. K. L Gould *et al.*, GENES & DEVELOPMENT 11, 2939 (1997)
12. A. Grallert *et al.*, Yeast 15, 669 (1999)

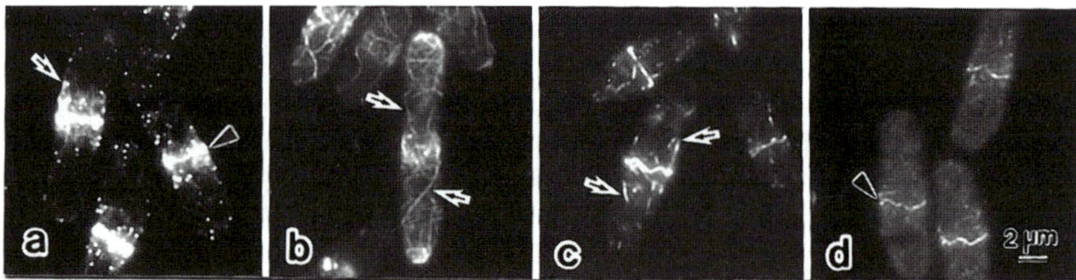

Fig. 1   Changes in actin cytoskeletons by pressure stress in *nda3* cells grown at 20°C. Images of fluorescence microscopy without (a) and with pressure of 100 (b), 150 (c), and 200 (d) MPa.

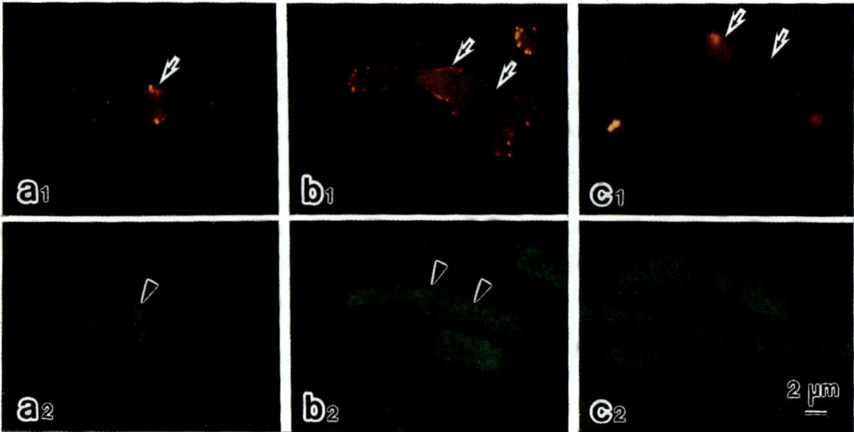

Fig. 2   Localization of actin cytoskeleton and tropomyosin in *nda3* cells grown at 20°C without (a) and with pressure of 100 (b), and 150 (c) MPa. a_1, b_1 and c_1, Rhodamin-conjugated phalloidin staining; a_2, b_2 and c_2, tropomyosin immunostaining.

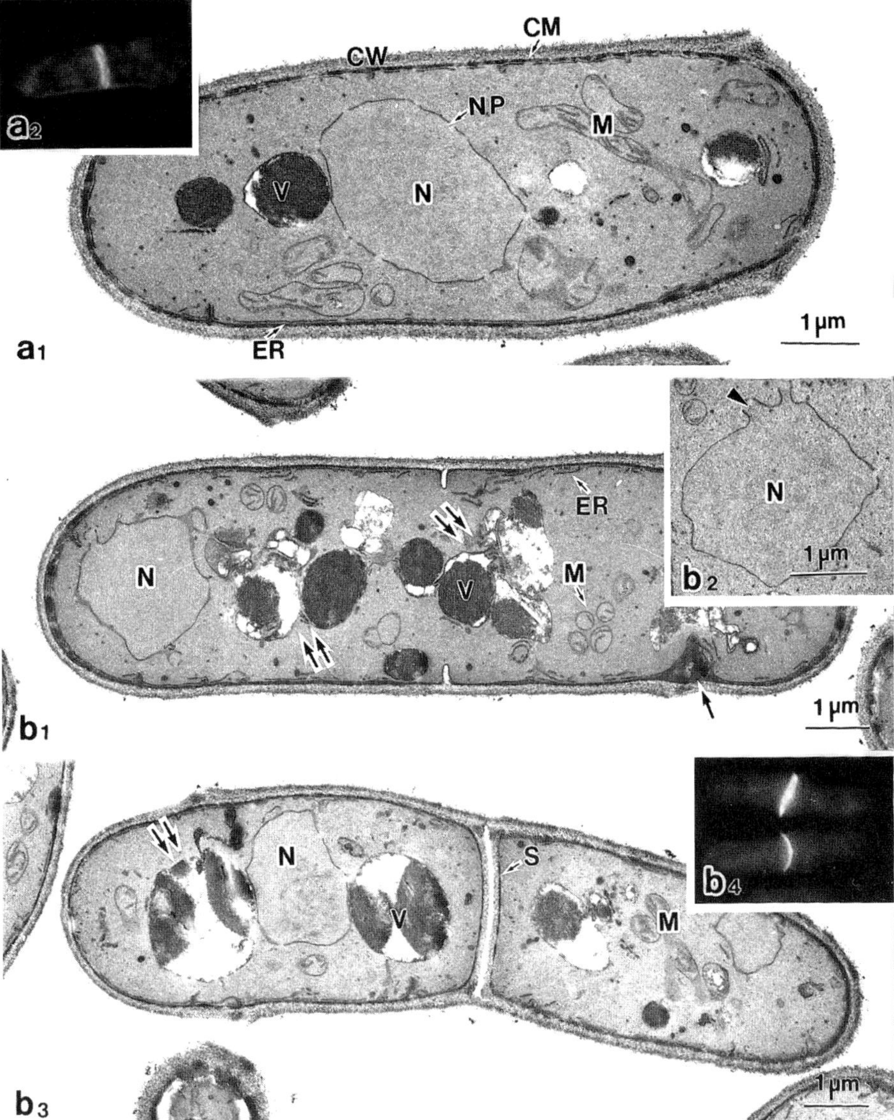

Fig. 3 Changes by pressure stress in ultrastructure of *cdc8* cells grown at 25 °C. Transmission electron microscopic images without (a₁) and with pressure of 100 (b₁₋₃) MPa. Fluorescence microscopic images of septa stained with Calcofluor white (a₂, b₄). CW, cell wall; CM, cell membrane; ER, endoplasmic reticulum; M, mitochondrion; N, nucleus; NP, nuclear pore; V, vacuole; S, septum.

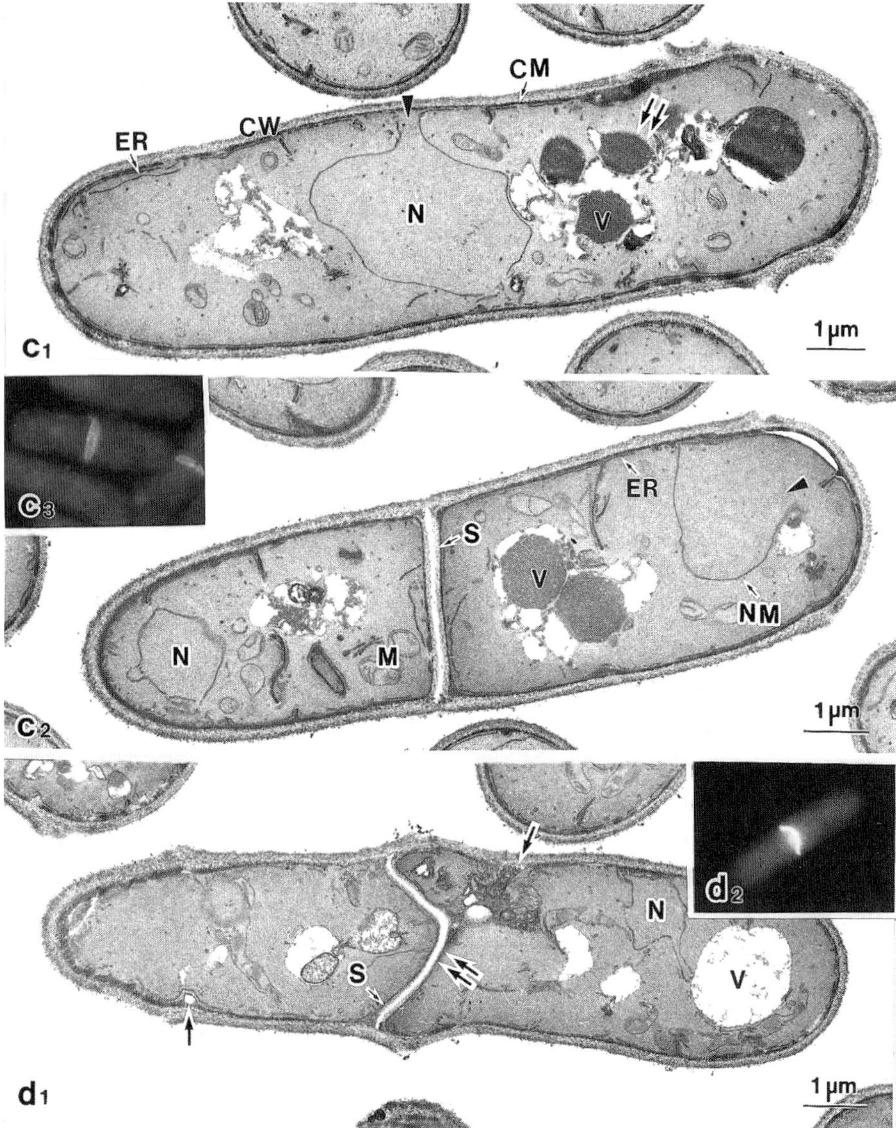

Fig. 3 Continued from the preceding page. Transmission electron microscopic images with pressure of 150 ($c_{1,2}$) and 200 (d_1) MPa. Fluorescence microscopic images of septa stained with Calcofluor white (c_3, d_2). CW, cell wall; CM, cell membrane; ER, endoplasmic reticulum; M, mitochondrion; N, nucleus; NM, nuclear membrane; V, vacuole; S, septum.

Trends in High Pressure Bioscience and Biotechnology
R. Hayashi (editor)
© 2002 Elsevier Science B.V. All rights reserved.

Cytoskeletal adaptation of living mammalian cells surviving under extremely high hydrostatic pressure

Mitra Sohirad[a], Tetsuya Miwa[a], Fumiyoshi Abe[a] and Masuo Aizawa[a,b]

(a)Frontier Research Program for Deep-sea Extremophiles, Japan Marine Science and Technology Centre,
2-15 Natsushima-cho, Yokosuka 237-0061, Japan.
(b)Department of Bioengineering, Tokyo Institute of Technology, Nagatsuta, Midori-ku,.
Yokohama 226-8501, Japan.

Abstract

We have visualised in real time live mammalian cells under extremely high hydrostatic pressure by using a newly developed computer controlled microscope pressure chamber system. The cytoskeletal adaptation and the morphological changes of several cell lines including NHDF, HMVEC, 3T3L1 and PTK1 have been observed. The result shows that cell lines demonstrate different rounding thresholds to extremely high hydrostatic pressure. And that the cells within the same cell line behave differently. Although they do mainly recover after decompression to various extent surviving even pressures as high as 100Mpa. Furthermore, the green fluorescent protein (GFP) tagging of Arp2/3 complex, CP and ß-actin of 3T3L1 and PTK1 cell lines was carried out in order to determine the effect of pressure on the cytoskeleton, and to understand how these proteins regulate the dynamics of actin filament assembly under high hydrostatic pressure.

1. Introduction

 Hydrostatic pressure in the range of hundreds of atmosphere is known to alter cytoskeleton assembly and reversibly induce significant changes in cell morphology ie rounding up of mammalian tissue cells [1, 2]. The complex molecular mechanisms responsible for these morphological and cytoskeletal changes are poorly understood.

Actin is a highly conserved and abundant cytoskeletal protein in eukaryotic cells. It is involved in a number of cellular activities, including reorganisation of cell shape and cell motility [3].
It has been shown that Arp2/3 complex and capping protein (CP), which regulate actin assembly in vitro, are concentrated in motile regions at the cell periphery and at small dynamic spots within the lamella of living cells [4].

 In order to determine the effect of pressure on the cytoskeleton, we need to understand how these proteins regulate the dynamics of actin filament assembly under high hydrostatic pressure. Hence we visualised the distribution of these proteins in living cells under high hydrostatic pressure using green fluorescent protein (GFP) tagging, whilst observing them in real time by a newly developed computer controlled microscope pressure chamber system.

2. Experimental Procedure

The expression plasmid for GFP-CP-beta2 subunit and GFP-Arp3 were donated courtesy of Dr D.A. Schafer., and the EGFP ß-actin construct by Prof. Beat Imhof. Various mammalian cell lines were used, epithelial and fibroblast. The cells were transfected using lipofectamine (GIBCO BRL, Gaithersburg, MD) and selected using 1mg/ml G418. Cell lines were maintained in media containing 0.5 mg/ml G418. Cells were plated on 2-mm thick collagen covered coverslips in Dulbecco's modified Eagle medium (DMEM) containing 10% fetal bovine serum (FBS), 50 units/ml penicillin, and 50 µg/ml streptomycin. The cells were grown for 1-2 days at 37°C in a humidified atmosphere of 5% CO_2. The coverslip was placed in the computer controlled high hydrostatic pressure chamber (Biott Corp) that was filled with Ham's F12 medium containing 10% FBS, 50 units/ml penicillin, and 50 µg/ml streptomycin. The cells were then exposed to hydrostatic pressure ranging from 40-70 MPa for 30 minutes. The compression and decompression speeds were varied from 20MPa/min to 70MPa/hr. Meanwhile live cell imaging was performed in real time using an inverted fluorescence video microscope (Olympus Corp), equipped with x20, x40/0.40 and CAP-G1.2 objective lenses (LC Plan, LC Plan Fl.) Single or time lapse pictures were acquired with a digital CCD camera (Olympus Colour chilled camera M-3204C) and data processed by Adobe Premier PhotoShop software. We used Ham's F12 medium in order to obtain low background fluorescence from the culture medium, since it contains low concentrations of fluorescent media components, such as Phenol red and riboflavin.

3. Results

Preliminary videos were obtained showing the dynamics of GFP-CP and GFP-Arp2/3 and EGFP ß-actin whilst the rounding effect occurs in live mammalian cells under extremely high hydrostatic pressure. Their re- distribution whilst pressure is released is also illustrated. The figures below show some samples of these results.

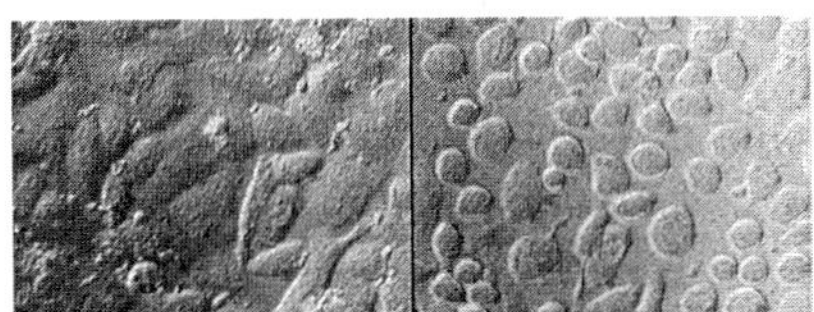

Figure 1   HeLa cell line normal microscopy demonstrating morphological changes due to the applied hydrostatic pressure, left frame 0.1M Pa , right frame
70M Pa after 30 minutes

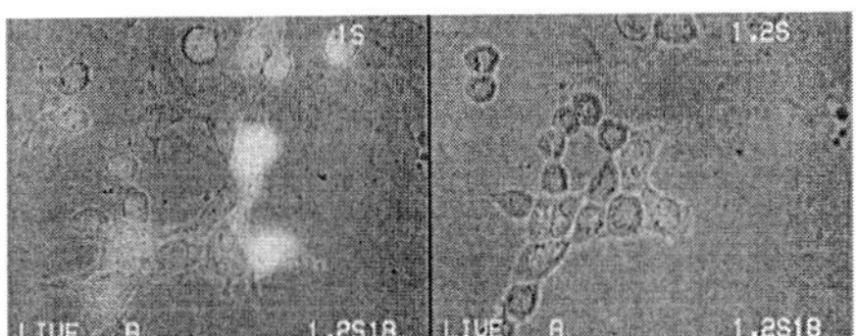

Figure 2   PtK1 cells expressing GFP-CP, fluorescence imaging left frame at 0.1 M Pa, right frame at 40 M Pa after 30 minutes

262

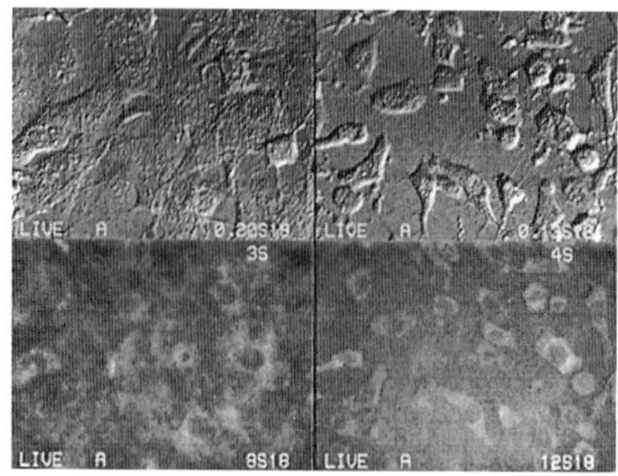

Figure 3   3T3 L1 cell line expressing GFP-Arp3 Left frames at 0.1M Pa , right frames at 40 M Pa after 30 minutes. Bottom frames show the fluorescent images

4. Discussion

We have observed the morphology changes of many live mammalian cell lines which have demonstrated different rounding thresholds to extremely high hydrostatic pressure, with 3T3- L1 and PtK1 being the most and least sensitive cell lines examined respectively. The human cell lines we have observed, NHDF, EC and HeLa, all fall in this range with EC being less sturdy than fibroblast and He La. Although these cells surprisingly survive under hydrostatic pressures of up to 100 M Pa for 30 minutes, they do not however return to their original shape and remain elongated after pressurisation at such high levels.
The rate of compression and decompression has no significant effect on these results.

The one common significant factor affecting the response of the cell lines to extremely high pressure was found to be their confluence ie the higher the confluence, the greater the resistance to rounding.   Hence, the cell -cell adhesion proteins are to be investigated next.

The GFP-tagged proteins are valuable tools for observing where actin assembles and investigating how actin assembly is controlled in living cells. We have observed the movement of these proteins and the change in their concentration under pressurisation. The quality of the measurements however has been limited by the long optical working distance caused by the necessity of a thick enough cover glass to withstand the chamber pressure.

References

[1] B. Bourns, S. Franklin, L. Cassimeris, E.D. Salmon. Cell Mot. Cytoskel. 15, 380 (1988)
[2] H.C. Crenshaw, J.A. Allen, V. Skeen, A. Harris, E.D. Salmon. Exp. Cell Res., 227, 28 (1996)
[3] T. P. Stossel, Science, 260, 1086 (1993)
[4] D. A. Schafer, M.D. Welch, etal. J. Cell Biol., 143, 7 (1998)

Trends in High Pressure Bioscience and Biotechnology
R. Hayashi (editor)

THE STRESS RESPONSE AGAINST HIGH HYDROSTATIC PRESSURE IN *SACCHAROMYCES CEREVISIAE*

H. Shimizu,[a] H. Iwahashi,[ab] Y. Komatsu[a]

[a]International Patent Organism Depositary

[b]National Institute of Bioscience and Human Technology

Higashi 1-1, Tsukuba, Ibaraki 305-8566, Japan

In cells of *Saccaromyces cerevisiae*, exposure to physical stress such as hydrostatic pressure may result in the enhancement of some particular categories of gene expression. To estimate the response to high hydrostatic pressure, we used DNA chip (microarray) technology. In this study, to investigate detailed mechanisms of the response of hydrostatic pressure, we monitored 6,000 gene expressions under three different conditions. In the conditions of 180MPa at 4℃ for zero min and 0.1MPa at 25℃ for 60min, genes of heat-shock proteins and a protein-degradation system were mainly induced. In the conditions of 40MPa at 4℃ for 16h and 0.1MPa at 25℃ for 60min, the tendency of gene expressions was similar to that under the conditions of 180MPa at 4℃ for zero min, and genes of the heat-shock protein family and protein-degradation system were mainly detected. In the conditions of 40MPa at 25℃ for 2h, we did not observe significant gene expression as we did under other conditions; however, some heat-shock protein was induced. Microarray technology should give us genome-wide information, and we will be able to have further discussion about the effects of hydrostatic pressure.

1. Introduction

In general, physical stress such as hydrostatic pressure affects physiological activities of living cells. There have been numerous studies on the effects of hydrostatic pressure (1). Abe and Horikoshi reported that hydrostatic pressure of 40 to 60MPa promoted acidification in yeast cells (2). Iwahashi *et al.*(3) reported that the treatment of more than 100MPa decreased the CFU of yeast cells; however, after a mild heat-shock treatment of 43℃ for 30min, the barotolerance of the cells increased. This barotolerance was shown to be due to the accumulation of trehalose and Hsp104. Iwahashi *et al.* reported that trehalose contributed to barotolerance in a temperature-independent manner and Hsp104 contributed temperature-dependently because of the thermal effect of enzyme activity (4). Hamada *et al.* (5) observed that under hydrostatic pressure above 100MPa, industrial yeast cells were induced to form tetraploids or homozygous diploids. These previous studies about physiological changes in cells allow us to estimate the mechanism for barotolerance; however, the genome-wide characterization of cells under hydrostatic pressure has little information. To estimate the genetic response against high hydrostatic pressure, we used DNA microarray technology. Yeast has over 6,000 open reading frames, and almost all gene-expression levels can be monitored with this technology.

In this study to investigate detail mechanisms of the response of hydrostatic pressure, we monitored three different conditions: A) 180MPa at 4℃ for zero min and 0.1MPa at 25℃ for 60min as recovery; B) 40MPa at 4℃ for 16h and 0.1MPa at 25℃ for 60min as recovery; and C) 40MPa at 25℃ for 2h.

2. Materials and Methods

2.1 Hydrostatic Pressure Treatment

Cells of *S. cerevisiae* S288C were grown in YPD medium (2% peptone, 1% yeast extract, and 2% D-glucose) to a mid-log phase at 30℃ and treated under A) 180Mpa at 4℃ for 0 sec and 0.1Mpa at 25℃ for 60min as recovery; B) 40MPa at 4℃ for 16h and 0.1MPa at 25℃ for 60min as recovery; and C) 40MPa at 25℃ for 2h. Under A and B, the yeast cells that were centrifuged and resuspended in the medium were pressurized as described previously (6). Under C, the cells were poured into a 50ml syringe and transferred to a stainless-steel vessel for pressurization. In all circumstances, the control cells were treated the same except for the pressure.

2.2 RNA Isolation and DNA Chip Analysis

Harvested cells were resuspended in a lysis buffer (50mM Sodium acetate, 10mM EDTA, 1%SDS, pH5.0), and total RNA was isolated using a hot-acid phenol method. Isolation of mRNA was carried out by Oligotex-dT30 (Super) (Takara Shuzo Co., Ltd.) and quantified by UV spectroscopy. 5-10 μg of mRNA was used to incorporate cDNA with fluorescence label Cy3-dUTP (control) or Cy5-dUTP (treated) by Super Script II reverse Transcriptase (GibcoBRL). Labeled cDNA was concentrated and hybridized to a Kuhara yeast DNA microarray. Microarrays were scanned using ScanArray5000 (GSI Lumonics, MA, USA), and image data were quantified with QuantArray (GSI Lumonics, MA, USA). The fluorescence intensity of each spot was subtracted from each background, and the ratio of fluorescent intensity Cy5/Cy3 was calculated using ACT1 as a positive control for normalization. The calculation was performed with GeneSpring (Silicon Genetics, CA, USA).

3. Results and Discussion

To investigate the genome-wide response to hydrostatic pressure, we used DNA microarray experiments under three different conditions. Under condition A, 180MPa at 4°C for zero min and 0.1MPa at 25°C for 60min as recovery, yeast-cell viability decreased to around 60%, and we detected that 309 genes had enhanced their intensity of fluorescence over 2.5-fold more than the controls. Under 180MPa, a long period of incubation causes a decrease in the number of living cells; therefore, a single pulse of this pressure could have a severe effect on a cell. The induced genes were summarized by classifying functional categories of the database of MIPS (Figure 1). Under these conditions, we detected 10 genes related to a heat-shock response. The heat-shock proteins contribute to protein metabolism as molecular chaperones. We also detected genes related to protein degradation such as the proteasome system, autophagy, and so on. This suggests that the treatment of 180MPa at 4°C damages protein and increases a protein-degradation system.

On the other hand, we detected a series of genes related to meiosis or sporulation. It may reflect a disturbance in the cell cycle.

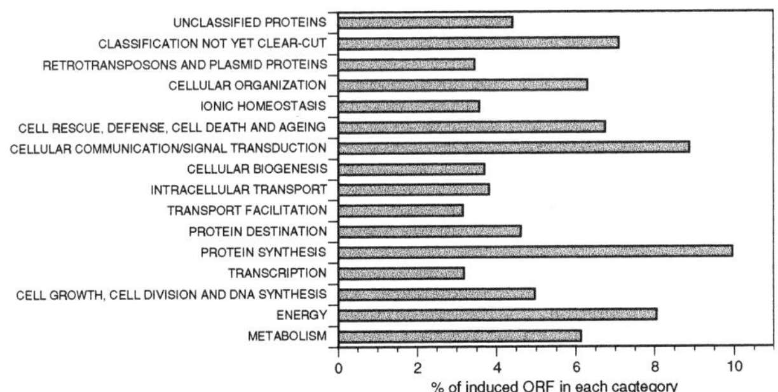

Figure 1. The percentage of induced genes in each category of MIPS under the treatment with 180MPa at 4°C for zero min

Table1. The functional category of induced genes under 180MPa at 4°C for zero min

	Fold	Gene		Fold	Gene
Proteasome/ubiquitin	7.8	UBI4	Heat shock proteins	13.4	SSA4
	5.9	UBI4		7.8	UBI4
	5.2	RPN1		7.4	HSP104
	3.5	PRE8		5.9	UBI4
	2.8	UBP5		5.6	SSA1
	2.6	UBC1		5.2	HSP26
	2.6	PEX4		3.9	SSA2
	2.5	DOA4		3.1	HSF1
	2.5	RPL40B		2.6	SBA1
				2.5	MRH1

Under condition B, 40MPa at 4℃ for 16h and 0.1Mpa at 25℃ for 60min as recovery, the viability of the cells was around 60%, and 70 genes were monitored with an intensity more than two-fold higher than the control (Figure 2). The tendency of the induced genes was similar to that under the A conditions. The categories of PROTEIN DESTINATION, CELL RESCUE, DEFENSE, CELL DEATH, AGING, and METABOLIZM were mainly detected. Fourteen genes were related to protein metabolism such as proteasome/ubiquitin, autophagy, cyclophilin, and so on. This suggests that, under these conditions, cells were severely affected in their protein-protein interaction or protein conformation because of the low activity of the heat-shock proteins. Six heat-shock protein genes or homologues were induced. Iwahashi *et al.* reported that the heat-shock treatment increases (6) barotolerance of yeast cells and Hsp104 plays an important role in barotolerance (4).

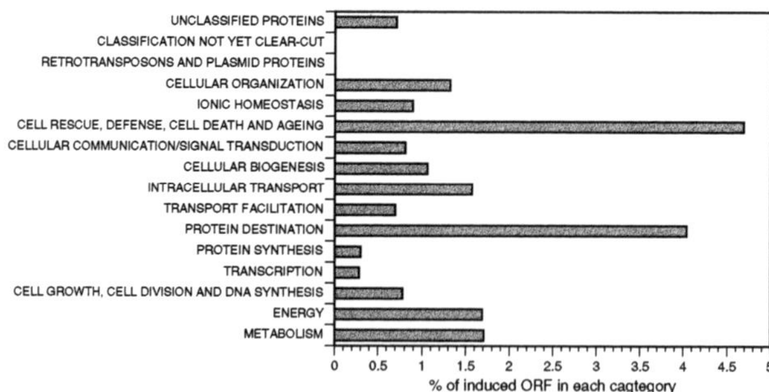

Figure 2. The percentage of induced genes in each category of MIPS under the treatment with 40MPa at 4°C for 16h

Table 2. The functional category of induced genes under 40MPa at 4°C for 16h

	Fold	Gene		Fold	Gene
Proteasome/ubipuitin	2.9	PRE3	Heat shock proteins	7.4	SSA4
	2.9	UBC5		4.8	HSP12
	2.8	RPT2		4.1	HSP26
	2.7	PRE1		2.3	YOR285W
	2.3	RPN7		2.1	SSA2
	2.2	UMP1		2.1	HSP104
	2.2	RPN8	Other protein destination	4.0	CPR6
	2.1	CDC48		2.4	AUT7
	2.1	RPT4		2.3	NCE103
	2.1	UFD1		2.1	CPH1
	2.1	PRE4		2.0	PRD1
	2.1	RPN13			
	2.0	RPN5			
	2.0	RPT5			

Under condition C, the treatment with 40MPa at 25°C for 2h enhanced gene-expression levels of 31 open-reading frames more than two-fold. Significant induction was not observed except for RETROTRANSPOSONS AND PLASMID PROTEINS. Six Ty Transposon genes were induced, and we detected an eight-fold gene expression of *YRO2*, a homologue to the Hsp30 heat-shock protein Yro1. It was reported that Hsp30 contributed to ATPase (7) activity and homeostasis of protons, and it may play a part in the acidification of the cell. Abe and Horikoshi showed that a treatment with 40 to 60MPa reduced

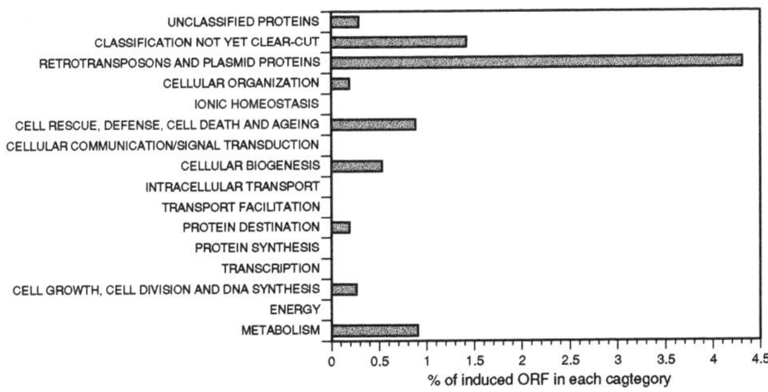

Figure 3. The percentage of induced genes in each category of MIPS under the treatment with 40MPa at 25°C for 2h

270

Table 3. The functional category of induced genes under 40MPa at 25°C for 2h

	Fold	Gene
Heat shock proteins	8.0	YRO2
	2.4	HSP10
Glutathione peroxidases	2.9	GPX2
	2.9	GRX1
Ty transposon	3.1	TY2B
	2.5	TY1A
	2.5	TY1A
	2.5	TY1B
	2.3	TY1B
	2.1	YBL005W-B

vacuolar pH (2). This may suggest that the induction of *YRO1* reflects acidification. On the other hand, we detected *GPX1* and *GPX2* genes that express under oxidative stress. Iwahashi *et al.* reported that high hydrostatic pressure (180MPa at 25℃) may have an effect that is essentially the same as oxidative stress (8), and a similar effect may be detected with a lower pressure of 40MPa at 25℃.

The reason for a weak response under these conditions may be due to a permissive temperature. The activity of many enzymes that are involved in heat-shock proteins may not decrease, and the response to the treatment may be delayed compared to conditions with low temperatures.

References:

1. Kato C, Inoue A, Horikoshi K, Trends Biotechnol., 14 (1996) 6.

2. Abe F, Horikoshi K, FEMS Microbiol Lett., 130 (1995) 307.

3. Iwahashi H, Kaul SC, Obuchi K, Komatsu Y, FEMS Microbiol Lett., 64 (1991) 325.

4. Iwahashi H, Obuchi K, Fujii S, Komatsu Y, FEBS Lett., 416 (1997) 1.

5. Hamada K, Nakatomi Y, Shimada S, Curr Genet., 5 (1992) 371.

6. Iwahashi H, Kaul SC, Obuchi K, Komatsu Y, FEMS Microbiol Lett., 64 (1991) 325.

7. Braley R, Piper PW, FEBS Lett., 418 (1997) 123.

8. Iwahashi H, Fujii S, Obuchi K, Kaul SC, Sato A, Komatsu Y, FEMS Microbiol Lett , 108 (1993) 53.

Trends in High Pressure Bioscience and Biotechnology
R. Hayashi (editor)

The immunoelectron microscopic analysis of Hsp104 under the hydrostatic pressure conditions

R. Matsumoto[a], H. Iwahashi[ab], K. Obuchi[c], and Y. Komatsu[a]

[a] International Patent Organism Depository
National Institute of Advanced Industrial Science and Technology
1-1, Higashi, Tsukuba, Ibaraki 305-8566, Japan

[b] Human Stress Signal Research center
National Institute of Advanced Industrial Science and Technology
1-1, Higashi, Tsukuba, Ibaraki 305-8566, Japan

[c] Molecular and Cellular Biology Laboratory
National Institute of Advanced Industrial Science and Technology
1-1, Higashi, Tsukuba, Ibaraki 305-8566, Japan

ABSTRACT

Hsp104 is one of the heat-shock proteins in the yeast *Saccharomyces cerevisiae*, and it works as a molecular chaperone. The chaperone function of Hsp depended on the temperature under hydrostatic pressure conditions. When yeast cells were exposed to mild heat of 43°C, the intracellular aggregates were appeared throughout the cells and induced Hsp104 to surround the aggregates. With immunoelectron microscopy, we observed that these cells moved under hydrostatic pressure. Under 100MPa hydrostatic pressure for 80min, Hsp104 dissappeared from intracellular aggregates at 35°C. However, on exposure to 100MPa at 4°C, the intracellular localization of Hsp104 did not change when compared to only mildly heat-shockd cells. From these results, it suggests that Hsp104 works as a molecular chaperone under hydrostatic pressure at 35°C, but not 4°C.

1. INTRODUCTION

When organisms are exposed to stress, stress proteins are induced that acquired stress tolerance. This phenomenon is called a "stress response"(1). A heat-induced stress protein is called a heat-shock protein (Hsp) (1). Hsp104 is one of the 100kDa heat-shock protein families in yeast *Saccharomyces cerevisiae* (2). A small amount of Hsp104 exists at a normal temperature. However, when yeast cells are exposed to mild heat, Hsp104 is strongly induced. The cells then acquire thermotolerance (2,3). Hsp104 also disaggregates the denatured protein in the manner of a molecular chaperone (4,5). Hsp104 works as a chaperone with co-chaperones Ssa1 (yeast Hsp70) and Ydj1 (yeast Hsp40) (6). Hsp104 is not highly induced under the stress of hydrostatic pressure; however, the cells that are induced Hsp104 by mild heat acquire barotolerance (7). Thus, the mechanism of barotolerance is similar to thermotolerance. In the Hsp104 null mutant strain, the barotolerane decreases compared to the wild-type at 35°C; however, the deletion of Hsp104 does not affect barotolerance at 4°C(8). From 2-D electrophoresis data of an insoluble fraction in pressurized cells that are mildly heat-shocked, the protein patterns between 35°C and 4°C are different (9). Moreover, in pressurized cells at 35°C, the amount of Hsp104 was lower than mild heat-shocked cells. However, in the pressurized cells at 4°C, Hsp104 level does not change compare with mild heat-shocked cells (9). These phenomena indicate that the chaperone function of Hsp104 also depends on temperature while under hydrostatic pressure. To be certain of the chaperone function of Hsp104 under hydrostatic pressure, we analyzed the chaperone function of Hsp104 by observing the intracellular localization of Hsp104 in the cells under hydrostatic pressure using immunoelectron microscopy.

2. MATERIALS AND METHODS

2.1 Strain and growth conditions

Saccharomyces cerevisiae IFO-0224 cells were grown to logarithmic phase in a YPD liquid medium (1% yeast extract, 2% peptone, and 2% glucose) at 30°C. To induce Hsp104, cells were treated with mild heat at 43°C for 90min.

2.2 Hydrostatic pressure treatment

After a mild heat-shock treatment, cells were treated with hydrostatic pressure under 100MPa at 35°C and 4°C for 80min. The system of hydrostatic pressure system is described in

Iwahashi *et al.*(1997)(9).

2.3 Immunoelectron microscopic observation

The procedure is described in Kawai *et al.* (1999)(10). For immuno-labeling, ultrathin sections were incubated anti- Hsp104 (1: 2000 dilution) antiserum for 90min and then incubated with goat anti-rabbit IgG conjugated to colloidal gold (British Biocell International Technical Services Department, 1: 40 dilution) for 60min at room temperature.

2.4 Quantitative analysis of immunoelectron micrographs

The area of intracellular aggregates and the colloidal gold particles that localized around aggregates were quantified using an NIH image 1.62 from each of the 40 immunoelectron micrographs. The ratio of colloidal gold particles against the intracellular aggregate area was calculated.

3. RESULTS

3.1 The intracellular localization of Hsp104 under hydrostatic pressure

We performed immunoelectron microscopic analysis of Hsp104. At first, we observed the cells under normal and mild heat conditions after immuno-labeling of Hsp104. At a normal temperature (30°C), a few gold particles indicated that Hsp104 existed in the nucleus and cytoplasm (Fig. 1). On exposure to mild heat of 43°C for 90 min, the electron-dense materials appeared throughout the cells (Fig.2). Speculation is that these materials are intracellular aggregated proteins that are denatured by mild heat-shock. As a result, Hsp104 is increased in nucleus and cytoplasm, and surround the intracellular aggregates (Fig.2, arrow).

Furthermore, to reveal the function of Hsp104 under hydrostatic pressure, we used immunoelectron microscopy to analyze the pressurized cells, which showed that Hsp104 was highly induced by mild heat-shock treatment. On exposure to 100MPa at 35°C for 80 min, Hsp104 dispersed from the intracellular aggregates and decreased (Fig. 3, arrow). In contrast, on exposure to 100MPa at 4°C for 80 min, the intracellular localization of Hsp104 did not change compared to mildly heat-shocked cells (Fig. 4, arrow).

3.2 The fixed quantity of colloidal gold particles and the intracellular aggregate area

To confirm the results of immunoelectron micrographs under hydrostatic pressure, we performed a fixed quantity of colloidal gold particles around the intracellular aggregates area

274

by using software for image analysis, NIH Image 1.62. The amount of colloidal gold particles under hydrostatic pressure at 35°C is less half of that at 4°C (Fig. 5). These results agree with those shown on immunoelecton micrographs.

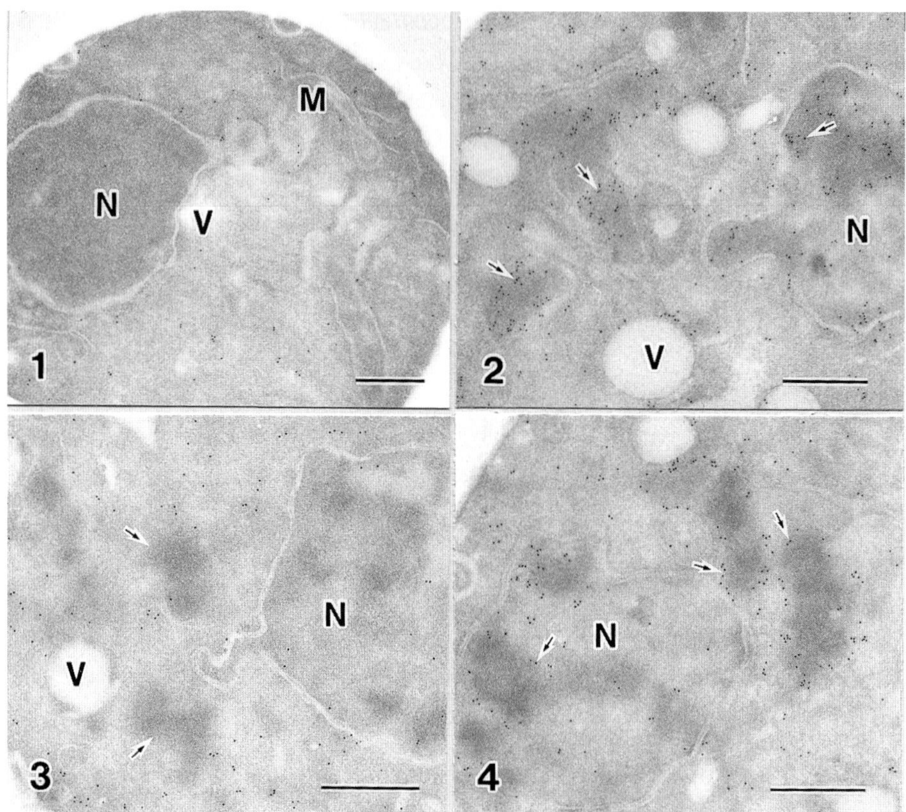

Fig. 1-4   Immunoelectron micrograph of *S. cerevisiae* IFO-0224 cells

Immunogold labeling of Hsp104. Cells were grown to logarithmic phase at 30°C (Fig. 1). Mild heat-shock treatment was applied at 43°C for 90 min (Fig. 2), and then hydrostatic pressure treatment (100MPa) was applied at 35°C (Fig. 3) and 4°C (Fig. 4) for 80 min. Cells were fixed with 3% paraformaldehyde and 0.5% glutaraldehyde were then treated with sodium metaperiodate and 50mM ammonium chloride, respectively. Ultrathin sections were incubated with the anti-Hsp104 antiserum and labeled with colloidal gold conjugating anti-rabbit IgG. N, nucleus; M. mitochondria; V, vacuole. Bars, 500 nm.

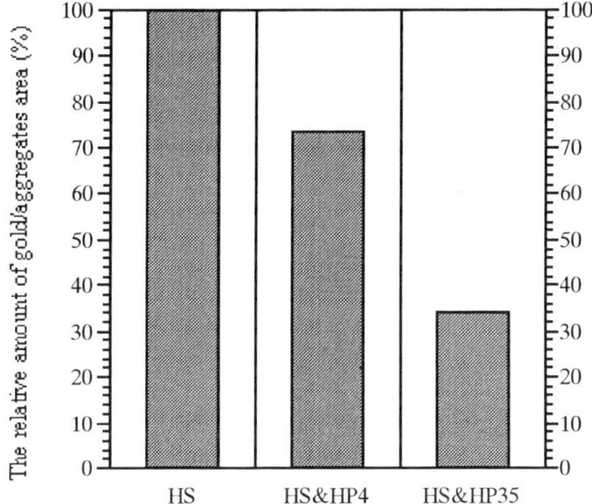

Fig. 5   Relative amount of colloidal gold particles around intracellular aggregates

Mild heat-shock treatment was applied at 43°C for 90 min (HS), and then hydrostatic pressure treatment (100MPa) was applied at 4°C (HS&HP4) and 35°C (HS&HP35) for 80 min.

4. DISCUSSION

In previous work, we reported that Hsp104 surrounds intracellular aggregates after a mild heat-shock treatment (10). It suggests that Hsp104 associates closely with protein aggregates under mild heat condition. In this paper, we showed the results under hydrostatic pressure as follows. On exposure to 100MPa at 35°C for 80 min, Hsp104 dispersed from intracellular aggregates and decreased (Fig. 3). In contrast, on exposure to 100MPa at 4°C for 80 min, the intracellular localization of Hsp104 did not appear any differently from mildly heat-shocked cells (Fig. 4).

The fixed quantity of colloidal gold particles suggests that the amount of Hsp104 under hydrostatic pressure at 35°C is less than half of that at 4°C (Fig. 5). This data agreed with that from the immunoelecton micrographs.

These results suggests that Hsp104 works as a chaperone at 35°C under hydrostatic pressure; however, cold temperature (4°C) is not an optimum temperature for a chaperone function of Hsp104. Therefore, it is possible that Hsp104 did not perform a chaperone function.   These different intracellular localizations of Hsp104 between 35°C and 4°C revealed that the chaperone function of Hsp104 depends on the temperature while under hydrostatic pressure conditions.

REFERENCES

1. L. Nover, "Heat Shock Response", L. Nover, CRC Press, Florida, 5 (1991).

2. Y. Sanchez and S. Lindquist, *Science* **248**, 1112 (1990).

3. Y. Sanchez, J. Taulien, K. A. Borkovich, and S. Lindquist, *EMBO J.*, **11**, 2357 (1992).

4. D. A. Parsell, A. S. Kowal, M. A. Singer, and S. Lindquist, *Nature*, **372**, 475 (1994).

5. Y. O. Charnoff, S. Lindquist, B. Ono, S. G. Inge-Vechtomov, and S. W. Liebman, *Science*, **268**, 880 (1995).

6. J. R. Glover and S. Lindquist, *Cell*, **94**, 73 (1998).

7. H. Iwahashi, K. Obuchi, S. Fujii, and Y. Komatsu, *Lett. Appl. Microbiol.*, 25, 43 (1997).

8. H. Iwahashi, S. Nwaka, K. Obuchi, and Y. Komatsu, in "High Pressure Biotechnology", S. Kunugi and R. Hayashi (eds.), Sanei Shuppan, Kyoto, 83 (1998).

9. H. Iwahashi, K. Obuchi, S. Fujii, and Y. Komatsu, *FEBS Lett.*, **416**, 1 (1997).

10. R. Kawai, K. Fujita, H. Iwahashi, and Y. Komatsu, *Cell Stress & Chaperones*, **4**, 46 (1999).

Trends in High Pressure Bioscience and Biotechnology
R. Hayashi (editor)
                                          277

Is there an influence of heat shock proteins on the pressure stable fraction of *Penicillium digitatum*? [*]

E. M. Sternberger and H. Ludwig

Institut für Pharmazeutische Technologie und Biopharmazie, Gruppe Physikalische Chemie
Universität Heidelberg, Im Neuenheimer Feld 366, 69120 Heidelberg, Germany

For all the different fungal spores investigated so far a typical pressure resistance was observed under certain conditions of pressure and temperature, specific to each strain. Penicillium digitatum, for example, can not completely be inactivated at 40 °C and 250 MPa even with very long pressure holding times. The pressure resistance is caused by a stable fraction of conidiospores.

The aim of this investigation was to find out if any relation could be found between the pressure resistance of the stable fraction of conidiospores and the expression of heat shock proteins in these spores. Therefore the spore suspension was exposed to different temperatures, and the pressure was established with different speeds to induce the expression of heat shock proteins. This different pretreatments were followed by the inactivating pressure of 250 MPa at 40 °C.

Independent of the pretreatment the stable fraction remained nearly constant. Pretreatments expected to induce heat shock proteins gave in some cases even smaller amounts of the stable fraction. It seems likely that heat shock proteins are not a factor for the stable fraction of conidiospores.

1. INTRODUCTION

All the various fungal species we investigated so far have in common a special feature that we call "pressure stable fraction". This stable fraction appears if we try to inactivate fungal conidiospores by pressure combined with temperatures higher than the optimal growth temperature of the fungus. Usually the fraction is about 10^{-5} of all the spores.

Penicillium digitatum is an ubiquitous fungus spoiling citrus fruits. The financial damage it causes in citrus plantations is enormous.

P. digitatum produces conidiospores. Sowed on agar plates they develop the mycelium after 1 day and the first blue-green spores after 3 - 4 days. The spores survive some unfavorable conditions, they prefer more acid media. The optimal growth temperature is 20-25 °C, the minimal -3 and the maximal 32-35 °C.

---

[*] This work was supported by the EU (FAIR-CT96-1175)

2. MATERIALS & METHODS

Culture conditions: *P. digitatum* (strain DSMZ 62840) was obtained from the Deutsche Sammlung von Mikroorganismen und Zellkulturen GmbH (DSMZ), Braunschweig, Germany. It was cultivated on Potato Dextrose Agar (PDA, DSM 129) at 24 °C. The spores were harvested after 14-21 days. For pressure treatments, the spores were freshly prepared and suspended in isotonic NaCl-solution. The solutions contained 0.1 % Polysorbate 80 to prevent spore clumping. The number of surviving cells was determined by counting colony forming units (cfu) on Malt Extract Agar (MEA, DSM 90) after 3 days of incubation at 24 °C.

High pressure treatment: The high pressure device consisted of ten pressure vessels. They were filled with 1 to 2.5 ml samples, enclosed in polyethylene tubes. All vessels could be simultaneously thermostated and pressurised. The maximum pressure used was 250 MPa, the pressure medium was a water/ethyleneglycol-mixture with a ratio of 9/1 (v/v). The single vessels could be opened at different times in order to measure the kinetics of inactivation.

The spores were pressurised in tubes made of polyethylene from Kronlab, Sinsheim (Germany). The tubes had an inner diameter of 8 mm, and were sealed at both ends using silicon plugs from Migge, Heidelberg (Germany).

3. RESULTS & DISCUSSION

Fig. 1 shows typical inactivation kinetics of *P. digitatum* conidiospores at 250 MPa and 10 or 40 °C. At 40 °C, a fast inactivation (D = 1 min) kills the majority of all the spores within 5 minutes. The remaining stable fraction is two spores out of one million, or %lg cfu = -5.68. At 10 °C, the inactivation runs in two phases (D_1 = 9.3 min and D_2 = 19.3 min) to a sterile solution.

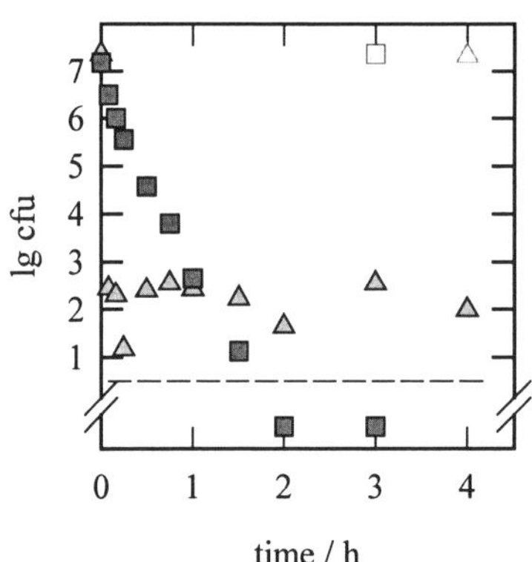

Figure 1. Inactivation of *P. digitatum* spores at 250 MPa and 10 or 40 °C. ■ 10 °C, ▲ 40 °C, open symbols are controls; − − − − detection limit.

Living cells, and most probably spores, too, contain heat shock proteins. It is difficult to imagine how ,in case of stress, heat shock proteins could be built by the slow metabolism in spores, but the proteins may already be present in an inactive form waiting for an activating signal like heat, or moderate pressure, or the short-time heat shock caused by adiabatic compression.

To examine the hypothesis given above, the time to pressurize the spores to 250 MPa was varied between < 30 seconds and 10 minutes. With < 30 s, the temperature rises immediatly from 40 to 48 °C and then goes back to 40 °C within some minutes. For slower pressurization times the temperature rise is much less, it vanishes in case of 10 min.

The expectation was to get more heat shock proteins by faster pressurisation and consequently a larger stable fraction.

The result is shown in fig. 2 and table 1: In contrast to the expectation, there is no positive correlation between the speed of pressurisation and the stable fraction.

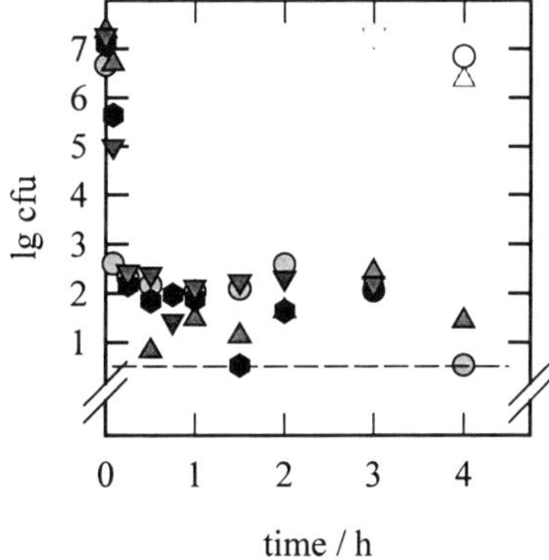

Figure 2. Inactivation of *P. digitatum* spores at 250 MPa and 40 °C after different pressure building times of ● 2 min, ▲ 5 min, ● 7 min, ▼ 10 min; open symbols are controls; at t = 0 h 250 MPa has been reached.

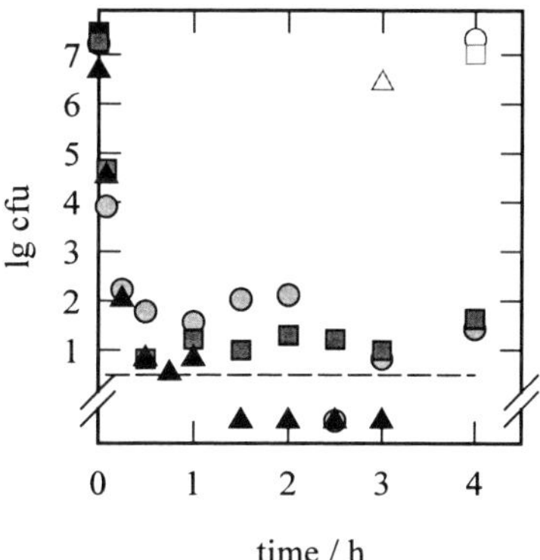

Figure 3. Inactivation of *P. digitatum* spores at 250 MPa and 40 °C after pre-incubation at different temperatures of ● 24 °C, ■ 32 °C, ▲ 40 °C. Open symbols are controls. The incubation time was ca. 14 h.

Table 1
The stable fraction in dependence on the pressure building time

Pressure building time / min	stable fraction / %lg cfu (calculated from data point 0 h)	stable fraction / %lg cfu (calculated from initial germ number)
< 0.5		-5.68
2	-5.53	-5.83
5	-5.78	-5.77
7	-4.76	-5.46
10	-4.87	-5.46

In a second experiment, the spores were stored at different temperatures of 24, 32 and 40 °C for ca. 14 hours before the high pressure inactivation started. Higher incubation temperature should produce more heat shock proteins.

The results, given in figure 3 and table 2, again do not show the expected sequence.

Table 2
The stable fraction in dependence on incubation at different temperatures

Temperature of incubation / °C	stable fraction / %lg cfu (calculated from data point 0 h)	stable fraction / %lg cfu (calculated from initial germ number before incubation)
no incubation		-5.68
24	-5.80	-5.88
32	-6.12	-6.12
40	-6.04	-6.61

4. CONCLUSIONS

The presented data contradict the hypothesis that heat shock proteins cause the pressure stable fractions of conidiospores. Other arguments could be obtained by direct analysis of the protein patterns of spores before and after pressure treatments.

REFERENCES

1. L.A. Weber, Cell Prolif. 25 (1992) 101-113.
2. V. Zimarino, C. Tsai and C. Wu, Mol. Cell. Biol. (1990) 752-759.
3. M.L. Deegenaars, K. Watson, Extremophiles 2 (1998) 241-249.
4. Tsiomenko, G.P. Ttuymetova , Biochemistry 60 (1995) 625-628.
5. Y. Auffray, E. Lecesne, A. Hartke, P. Boutibonnes, Curr. Microbiol. 30 (1995) 87-91.
6. J.L. Smilaneck et al, Plant Dis. 81 (1997) 379-382.
7. K.C. Shellie, Plant Dis. 82 (1998) 380-382.

Trends in High Pressure Bioscience and Biotechnology
R. Hayashi (editor)

DNA replication is suppressed by pressure in *Xenopus* egg cell-free system

H. Takahashi,[1] T.Yamaguchi,[1] M.Koga,[2] H.Kageura,[2] and S.Terada [1]

Department of [1]Chemistry and [2]Biology, Faculty of Science, Fukuoka University, Jonan-ku, Fukuoka, 814-0180 Japan

It is reported that when mouse cells are exposed to a pressure of 80 MPa, the cells in S-phase are sensitive to the pressure. To analyze the sensitivity of S-phase cells to pressure, we used a *Xenopus* egg cell-free system. The DNA replication was monitored by the incorporation of biotin-16-dUTP into sperm nuclei. The progress of DNA replication was completely inhibited by exposure to a pressure (80MPa) of a mixture of sperm nuclei and egg extracts or egg extracts only. On the other hand, 80 MPa-treated sperm nuclei incorporated biotin-16-dUTP, and entered M-phase. These results suggest that sperm nuclei are insensitive to a pressure of 80 MPa and some components in egg extracts are sensitive to the pressure.

1. INTRODUCTION

A pressure is an important variable as well as a temperature in natural science. As compared with temperature, however, pressure is less used due to technical problems concerning the generation and maintenance of high pressure. All biochemical reactions are characterized in more detail by examining their volume changes. Such volume changes are obtained from changes in pressures. Thus, it is expected that pressure provides useful information about biochemical reactions.

We have applied pressure to the investigation of the cell cycle. When murine erythroleukemia (MEL) cells are exposed to a pressure of 80 MPa and then cultured at atmospheric pressure, the progression of the cell cycle is delayed in S-Phase and then arrested in G2/M phase. On the other hand, the cells in G1 or G2-phase progress the normal cell cycle [1]. Similar results have been obtained by using synchronized HeLa cells. To investigate the sensitivity to the pressure of S-phase cells, we have used a *Xenopus* cell-free system. In this system, the process of DNA replication is carried out when sperm nuclei are incubated with S-phase extracts. In this paper, we describe that sperm nuclei are insensitive to a pressure of 80 MPa, whereas egg extracts are sensitive to the pressure.

2. EXPERIMENTAL

2.1. Preparation of *Xenopus* egg extracts

Adult females of *Xenopus laevis* were injected with 600 IU of human chorionic gonadotropin to induce ovulation. After 16h, unfertilized eggs were squeezed into 0.1 M NaCl. The S-phase extracts were prepared by the methods previously described [2,3]. The eggs were dejellied with a solution containing 5 mM dithiothreitol, 110 mM NaCl, and 20 mM Tris-HCl (pH 8.5). Dejellied eggs were washed with 1/4 MMR (MMR: 0.1 M NaCl, 2 mM KCl, 1 mM $MgSO_4$, 2 mM $CaCl_2$, 0.1 mM EDTA, 25 mM Hepes-KOH pH 7.5), and activated with 0.2 µg/ml calcium ionophore A23187 in MMR for 5 min. After activation, these eggs were washed with ice-cold S buffer (0.25 M sucrose, 50 mM KCl, 2.5 mM $MgCl_2$, 2 mM ß-mercaptoethanol, 50 mM Hepes-KOH pH 7.5). The washed eggs were packed into tubes by centrifugation at 6000 g for 20 sec, and an excess buffer was removed. To obtain the crude extracts, these eggs were centrifuged at 15000 g for 10 min. The resultant supernatant between lipid cap and yolk pellet was collected, mixed with 10 µg/ml cytochalasin B, and recentrifuged. The supernatant was stored on ice and used within 1h (cycling extracts) or mixed with 200 mM sucrose and frozen in liquid nitrogen (frozen extracts). Frozen extracts and cycling ones were suspended with 60 mM creatine phosphate, 150 µg/ml creatine phosphokinase, and 1 mM ATP.

2.2. Preparation of sperm nuclei

Adult males of *Xenopus laevis* were injected with 600 IU of human chorionic gonadotropin. The dissected testes were mingled, homogenized in SuNaSp (250 mM sucrose, 75 mM NaCl, 0.5 mM spermidine, 0.15 mM spermine), and centrifuged at 1000 g for 5 min at room temperature. The Pellet was suspended in SuNaSp, and lysolecithin (0.5 mg/ml) was added to the suspension at room temperature. The process of demembranation was monitored by staining the sperm with 10 µg/ml Hoechst 33258. When most of sperms were stained, the reaction of the demembranation was stopped by adding of ice-cold SuNaSp containing 3 % BSA. The sperm suspension was centrifuged at 1000 g for 5 min at room temperature. The pellet was washed with SuNaSp containing 0.3 % BSA and resuspended in SuNaSp containing 30 % glycerol. The concentration of sperm nuclei was determined by hemocytometer. The sperm nuclei were frozen in liquid nitrogen, and stored at –80 °C [4].

2.3. Pressure treatment

The samples were laid on EB in a pressure cell and subjected to a pressure of 80 MPa for 30 min at 23 °C. After decompression, samples were placed on ice.

2.4. Measurement of DNA replication

DNA replication was examined by measuring the incorporation of biotin-16-dUTP into DNA. The frozen extracts (50 µl) containing biotin-16-dUTP (8 µM) were incubated with sperm nuclei (final concentration 1 x 10^6 nuclei/ml) at 23 °C. At 60 min after incubation, aliquots (15 µl) were removed and mounted onto S buffer containing 25 % glycerol [2]. The samples were centrifuged at 1200 g for 15 min at 4°C, and sperm nuclei were collected onto poly-L-lysine-coated cover glass. The sperm nuclei were fixed with fixing buffer (3% formaldehyde, 80 mM KCl, 15 mM NaCl, 50 % glycerol, 15 mM Pipes pH 7.2), washed three times with S buffer, and incubated with avidin-FITC for 30 min at room temperature. After incubation, the sperm nuclei were washed three times with EB (100 mM KCl, 2.5 mM MgCl$_2$, 50 mM Hepes-KOH at pH 7.5), stained with 100 µg/ml propidium iodide (PI) for 5 min, and washed with EB. The cover glass was mounted on glass slides. To observe the nuclear morphology, the cycling extracts (5 µl) containing nuclei (1 x 10^6 nuclei/ml) were mixed with 5 µl of fixing buffer containing Hoechst 33258 (5 µg/ml). Sperm nuclei were observed by a fluorescence microscope.

3. RESULTS

3.1. DNA replication is suppressed by exposure to 80 MPa of a mixture of sperm nuclei and frozen extracts

We analyzed the effect of pressure on DNA replication. In *Xenopus* cell-free system, DNA replication was examined by measuring the incorporation of biotin-dUTP into sperm nuclei. The mixture of frozen extracts and sperm nuclei was subjected to a pressure of 0.1 (atmospheric pressure) or 80 MPa for 30 min. Then, the mixture was incubated for 30 min at atmospheric pressure. The incorporation of biotin-dUTP was seen in the pressure-untreated mixture, but not in the pressure-treated one (Fig.1).

3.2. DNA replication is suppressed in 80 MPa-treated frozen extracts.

The progress of DNA replication was suppressed in 80 MPa-treated mixture, as described above. To examine which of extracts and sperm nuclei is affected by pressure, only the frozen extracts were exposed to a pressure of 80 MPa. The pressure-treated frozen extracts and pressure-untreated sperm nuclei were mixed and incubated for 60 min. In this case, no incorporation of biotin-dUTP was observed (Fig.2).

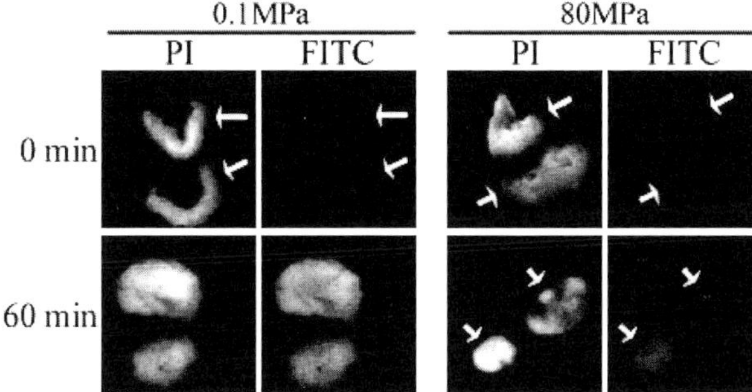

Figure 1. Suppression of DNA replication by exposure to 80 MPa of a mixture of sperm nuclei and frozen extracts. The mixture of frozen extracts and sperm nuclei was subjected to a pressure of 0.1 or 80 MPa for 30 min and then incubated for 30 min at atmospheric pressure. Sperm nuclei and biotin-dUTP were detected by PI and avidin-FITC, respectively. Arrows indicate sperm nuclei.

284

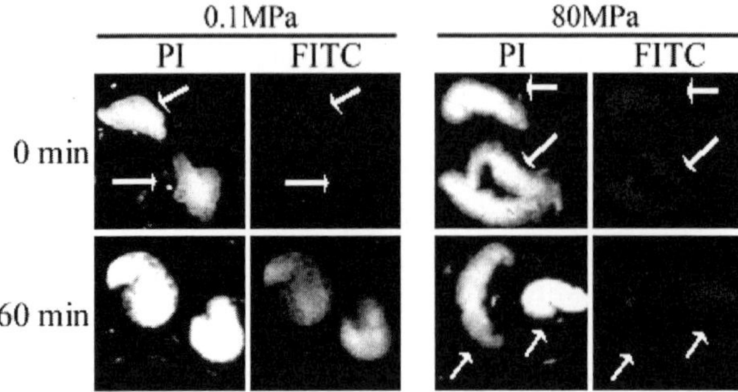

Figure 2. Suppression of DNA replication by exposure to 80 MPa of frozen extracts. Pressure-treated or –untreated frozen extracts were mixed with pressure-untreated sperm nuclei and incubated for 60 min.

3.3. Cell cycle progression is inhibited by pressure treatment of cycling extracts

In cycling extracts, the cell cycle progress a few times. After addition of pressure-untreated cycling extracts, most of the sperm nuclei entered M-phase at 45-60 min and S-phase at 75 min. In the case of the pressure-treated cycling extracts, however, all sperm nuclei remained spherical and did not enter into M-phase during the incubation time (Fig.3).

3.4. Sperm nuclei are insensitive to a pressure of 80 MPa

Sperm nuclei were subjected to a pressure of 80 MPa and then mixed with pressure-untreated frozen extracts containing biotin-dUTP. After incubation for 60 min, the incorporation of biotin-dUTP was observed in pressure-treated sperm nuclei, as seen in pressure-untreated ones (Fig.4).

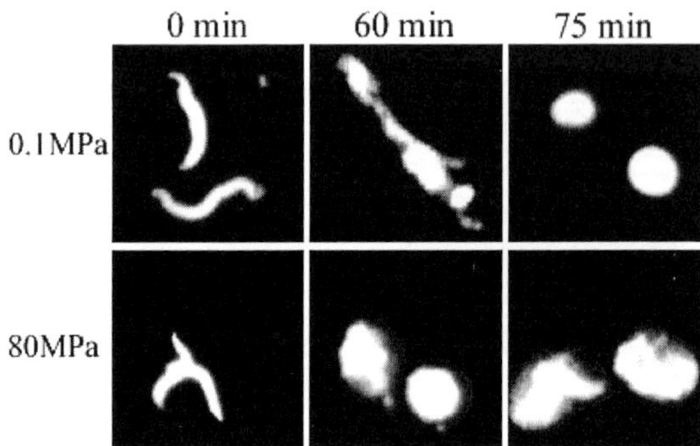

Figure 3. Effects of pressure on the cell cycle progression in cycling extracts. Pressure (80 MPa)-treated or -untreated cycling extracts were mixed with pressure-untreated sperm nuclei. After 0, 60, and 75 min, aliquots were removed and sperm nuclei were stained with Hoechst 33258.

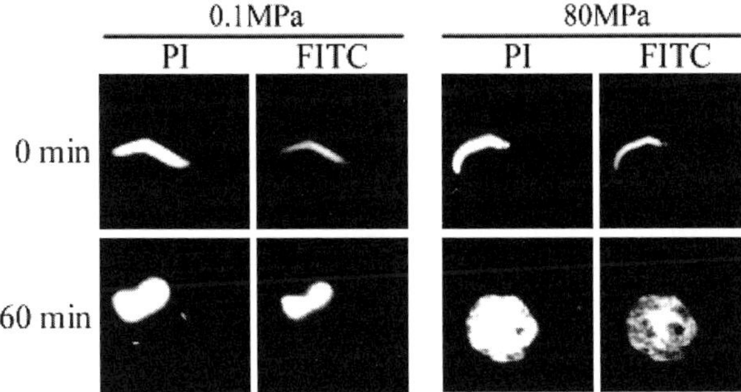

Figure 4. Pressure treatment of sperm nuclei in DNA replication. The 80 MPa-treated or –untreated sperm nuclei were incubated with pressure-untreated frozen extracts containing biotin-dUTP for 60 min.

3.5. Pressure-treated sperm nuclei enter M-phase

The influence of pressure to sperm nuclei was investigated by using cycling extracts. Sperm nuclei were subjected to a pressure of 80 MPa, and mixed with cycling extracts. As with pressure-untreated sperm nuclei, 80 MPa-treated sperm nuclei entered M-phase at 60 min, and then entered S-phase at 75 min (Fig.5). Thus, we conclude that sperm nuclei are insensitive to a pressure of 80 MPa.

4. DISCUSSION

In this paper, we have shown that when a mixture of extracts and sperm nuclei is exposed to a pressure of 80 MPa, the process of DNA replication and progression of the cell cycle are

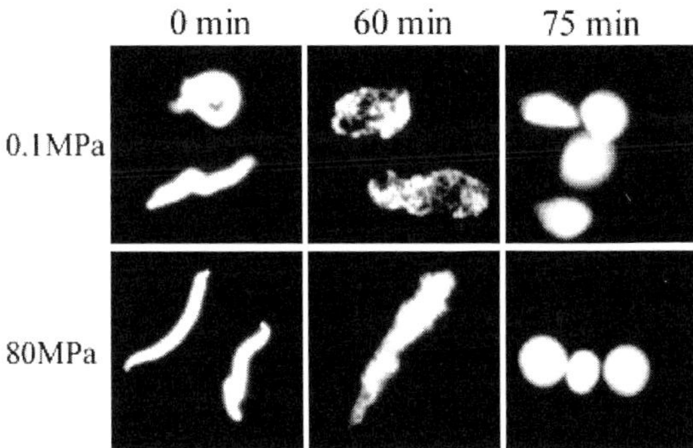

Figure 5. Entrance of pressure-treated sperm nuclei into M-phase. Pressure-treated or –untreated sperm nuclei were incubated with pressure-untreated cycling extracts for 0, 60, and 75 min. Sperm nuclei were stained with Hoechst 33258.

inhibited. Similar results are obtained by aphidicolin, an inhibitor of DNA polymerase α. Namely, Upon addition of aphidicolin to mixture of extracts and sperm nuclei, DNA replication is inhibited and the cell cycle arrests in S-phase [5]. In our system, what is affected by pressure? When sperm nuclei are exposed to a pressure of 80 MPa, both DNA replication and cell cycle progression are normal. This suggests that sperm nuclei are insensitive to a pressure of 80 MPa. On the other hand, the egg extracts are sensitive to pressure. Many multimeric components such as MCM (minichromosome maintenance) proteins and ORC (origin recognition complex) proteins in extracts play an important role in the replication of DNA [6]. In general, oligomeric proteins dissociate under pressure [7]. Thus, it seems likely that some components in extracts are affected by pressure. Further studies are required to determine what kinds of proteins in extracts are affected by pressure.

REFERENCES

1. M. Matsumoto and T. Yamaguchi and Y. Fukumaki and R. Yasunaga and S. Terada, J Biochem (Tokyo), 123 (1998) 87.
2. Y. Kubota and H. Takisawa, J Cell Biol, 123 (1993) 1321.
3. A. Shimada and K. Ohsumi and T. Kishimoto, Biol Cell, 90 (1998) 519.
4. A. W. Murray, Methods Cell Biol, 36 (1991) 581.
5. A. Kumagai and W. G. Dunphy, Mol Cell, 6 (2000) 839.
6. S. Mimura and T. Masuda and T. Matsui and H. Takisawa, Genes Cells, 5 (2000) 439.
7. L. King and G. Weber, Biochemistry, 25 (1986) 3637.

Trends in High Pressure Bioscience and Biotechnology
R. Hayashi (editor)

Metabolism of *Lactobacillus sanfranciscensis* under high pressure: investigations using stable carbon isotopes

Maher Korakli[a], Michael G. Gänzle[a], Ruth Knorr[b], Manuela Frank[b], Andreas Rossmann[b], and Rudi F. Vogel[a]

[a]TU München, Lehrstuhl für Technische Mikrobiologie, D-85350 Freising, Germany
[b]TU München, Lehrstuhl für Biologische Chemie, D-85350 Freising, Germany

1. Abstract

The effect was studied of high pressure on the maltose metabolism of *Lactobacillus sanfranciscensis* in the presence or absence of fructose at pressures ranging from 0.1 to 300 MPa. Substrate consumption and product formation was determined by HPLC. To elucidate mechanisms of pressure induced effects on carbohydrate metabolism, the ratio of ^{12}C / ^{13}C isotopes in maltose and ethanol was measured by IRMS and GC-c-IRMS, respectively.

L. sanfranciscensis tolerated pressures up to 150 MPa for 3 h without appreciable loss of viability and incubation at 200 MPa decreased viable cell counts by 2 log. Application of 50 MPa did not affect maltose uptake. Incubation at 100, and 150 MPa resulted in a decrease of maltose consumption by 67% and 82%, respectively. The molar ratio of lactate produced to maltose consumed was unchanged by pressures of 100 – 150 MPa. Incubation at 100 – 150 MPa resulted in formation of lactate and acetate only, ethanol formation was not observed. These results conform with the preferential production of acetate observed also at comparable maltose turnover levels at ambient pressure. *L. sanfranciscensis* treated at 100 MPa exhibited normal metabolic activity after pressurization but treatment with 150 MPa resulted in an inhibition of maltose metabolism up to 3 h post-treatment.

The isotope ratio analysis of carbon in ethanol produced from maltose by *L. sanfranciscensis* revealed that incomplete maltose consumption resulted in an enrichment of ^{12}C in the ethanol due to a kinetic isotope effect. This kinetic isotope effect was enhanced upon metabolism under high pressure conditions with the same maltose turnover levels.

1. Introduction

Lactic acid bacteria are widely used in food biotechnology for the production of fermented foods and food additives. Studies on the physiology and genetics of these organisms are the basis for their successful use in biotechnological applications. A key element to the application and stable preparation of lactic acid bacteria is their cellular stress response to environmental stressors, e.g. temperature or high salt conditions. Hydrostatic pressure affects the viability, metabolic activity, and gene regulation of biological systems. It was shown that sublethal high pressure results in an acidification of the vacuolar pH in *Saccharomyces cerevisiae* (Abe and Horikoshi, 1998), which was attributed to the ionization of phosphates and carbonates as well as the inactivation of membrane ion pumps. The stress response of *Escherichia coli* to hydrostatic pressure was shown to include the synthesis of pressure inducible proteins (Welch et al., 1993). More recently, differential gene expression as response to elevated hydrostatic pressure was demonstrated for *Escherichia coli* and

288

Saccharomyces cerevisiae (Abe and Kato, 1999, Iwahashi et al., 2000). High pressure processes have found commercial application in the food industry in the past years, however, few data are available on high pressure effects on metabolism and genetic regulation of lactic acid bacteria. It was therefore the aim of this work to investigate the effects of high pressure on the metabolism of *L. sanfranciscensis*.

L. sanfranciscensis is a heterofermentative lactic acid bacterium with industrial use in sourdough fermentations and has a potential for biotechnological production of food additives (Hammes et al., 1996). *L. sanfranciscensis* degrades hexoses via the pentose-phosphate-shunt; an overview of the metabolic pathways is shown in **Figure 1** (Stolz et al., 1995b, Vogel et al., 1999). Maltose is cleaved by maltose phosphorylase to glucose-1-phosphate and glucose. Glucose is either phosphorylated at the expense of ATP to yield glucose-6-phosphate, or excreted into the medium. Glucose-6-P is converted to xylulose-5-P with concomitant production of CO_2 and reduction of $NADP^+$. Acetyl phosphate represents a major branching point of this metabolism where the carbon flux is directed towards the alternative end products acetate or ethanol. If additional substrates for cofactor regeneration are unavailable, acetyl phosphate is used to regenerate the NADH formed upstream and quantitative conversion of acetyl-phosphate to ethanol is observed. In the presence of electron acceptors, acetate formation is favored over ethanol formation since one additional ATP is gained in the acetate branch of metabolism. Substrates that are used by *L. sanfranciscensis* to regenerate NADH include oxygen, citrate, and fructose (Stolz et al., 1995a).

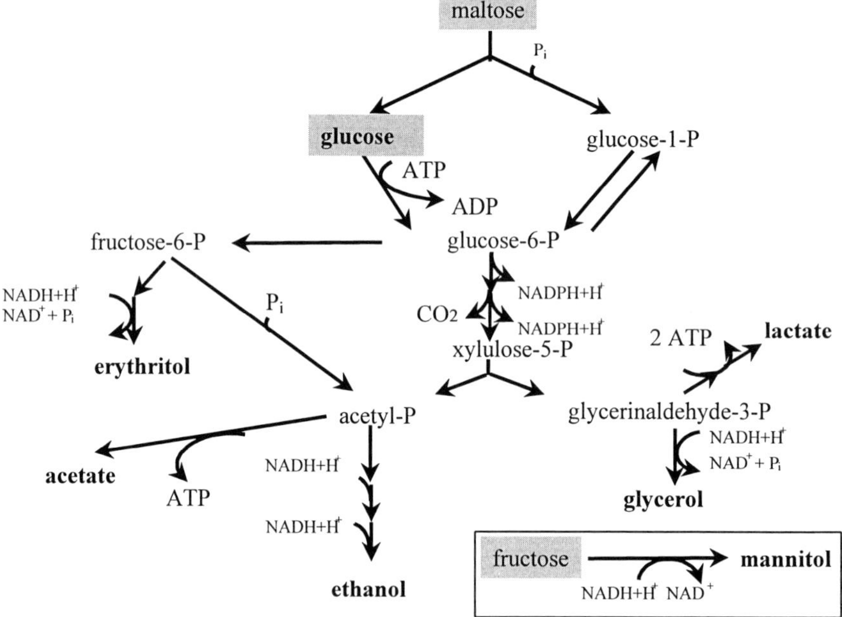

Figure 1. Overview over the metabolic pathways of maltose utilization in *L. sanfranciscensis*. (Stolz et al., 1995, Vogel et al., 1999, modified). Substrates are marked in gray, products are printed in bold letters.

2. Materials and Methods

Organism and culture conditions: *L. sanfranciscensis* LTH2581 and LTH1729 were cultured at 30°C in mMRS4 containing the following components per liter: peptone from casein, 10g; yeast extract, 5g; meat extract, 5g; $K_2HPO_4.3H_2O$, 2.6g; KH_2PO_4, 4g; cystein-HCl, 0.5g; NH_4Cl, 3g; maltose, 10g; fructose, 10g; Tween 80, 1 ml, $MgSO_4$ x H_2O 0.1g, $MnSO_4$ x $4H_2O$, 0.05 g. The pH was adjusted to 6.2, and the medium was sterilized at 121°C for 20 min. Sugars were autoclaved separately. A vitamin mix containing biotin, folic acid, nicotinic acid, pyridoxal phosphate, thiamine, riboflavin, cobalamin and panthothenic acid (0.2 g/l each) was sterilized by filtration and 1 ml vitamin mix was added per 1 mMRS4. For the isotope experiments C_3-MRS4 was used. This medium had the same composition as mMRS4 but did not contain fructose. The maltose used for C_3-MRS4 was derived from a C_3-plant (Sigma, USA) and had a δ^{13}C-value of -25.20‰. Plates were incubated at 30°C under controlled atmosphere (76% N_2, 20% CO_2, 4% O_2).

High pressure treatment. An overnight culture of *L. sanfranciscensis* was sub-cultured with 1% inoculum in mMRS4 or C_3-MRS4. Late stationary cells were harvested by centrifugation and resuspended in an equal volume of mMRS4 or 0.5 volumes of C_3MRS4, respectively. This cell suspension was transferred to 2 ml Eppendorf reaction tubes (ERT), sealed with silicon stoppers avoiding enclosure of air, and pressurized. The high pressure metabolism and inactivation kinetics of *L. sanfranciscensis* were investigated in HP-autoclaves at 30°C for 3 hours. Compression and decompression rates were 200 MPa min^{-1}. Samples were taken after decompression for determination of viable cell counts on mMRS4 agar and the determination of metabolites by HPLC.

Determination of metabolites: Cells from culture samples were separated by centrifugation and the supernatants were analyzed using HPLC. The concentrations of maltose, lactic acid, acetic acid and ethanol in the supernatant were determined using polyspher® OA KC column (Merck, Germany); mobile phase was H_2SO_4 5 mmol/l and temperature of the column 70°C. For detection a refractive index detector (Gynkotek, Germany) was used.

Determination of the ^{12}C / ^{13}C carbon isotope ratio: Ethanol was extracted from medium using liquid-liquid extraction with ethylacetate, adding a mixture of sodiumsulphate / sodiumcarbonate (99:1) for the removal of water. The isotope ratio measurements were performed using a Finnigan MAT δS isotope mass spectrometer (Finnigan MAT, Bremen, Germany) on-line coupled to a Varian 3400 GC via a combustion interface. The GC was equipped with a Poraplot U fused silica capillary column (25m x 0.32 mm; 10 μm film thickness). Helium was used as carrier gas; the sample was applied by 0.3 – 1.0 μl split injection at an injector temperature of 250 °C, the column was held at 140 °C for 3 min followed by a 8°C/min increase to 190 °C. The isotope ratios are expressed as δ^{13}C-values [‰] versus the PDB Standard. The working reference gas was calibrated versus NBS-22 (IAEA, Vienna) using dual inlet IRMS as described previously (Koziet et al. 1993).

3. Results

Inhibition of metabolism by high pressure. The inhibitory effect of high pressure in the range of 0.1 to 250 MPa towards *L. sanfranciscensis* LTH2581 and LTH1729 was evaluated. After 3 h of incubation, cell counts and the maltose consumption were determined (**Fig. 2**). Cell counts and maltose consumption of *L. sanfranciscensis* LTH2581 remained unaffected by 50 MPa. Incubation at 100 and 150 MPa did not result in appreciable inactivation of the strain, however, the maltose consumption was reduced to 32 and 18 % of the control,

290

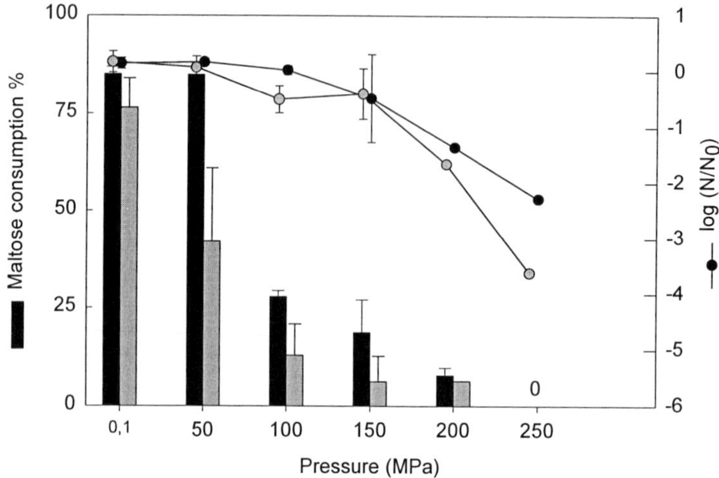

Figure 2. Effect of high pressure on viability and metabolic activity of two strains of *L. sanfranciscensis* in mMRS4 after a 3h incubation period. Shown is the maltose consumption (bar chart) and the viable cell counts (symbols). Black bars and symbols: *L. sanfranciscensis* LTH2581; gray bars and symbols: *L. sanfranciscensis* LTH1729. Results represent means ± standard deviation of two independent experiments

respectively. Pressures of 200 MPa or greater resulted in almost complete inhibition of metabolism, however, this inhibition can largely be attributed to a decrease of cell counts by 90% or greater. The strain *L. sanfranciscensis* LTH1729 was more sensitive to pressure than LTH2581; with the former strain, the application of 50 MPa sufficed to reduce the metabolic activity by more than 50% and a significant reduction of viable cell counts was observed already at 100 MPa.

Composition of metabolites formed during high pressure metabolism. The consumption of substrates and the formation of products during HP metabolism are shown in Table 1. Metabolites other than those shown in Table 1 were not formed during HP metabolism and carbon and electron balances generally accounted for more than 90% of the substrates consumed. At any pressure, maltose was converted to lactate and acetate or ethanol; fructose was not used as carbon source but reduced to mannitol. In accordance with the metabolic pathway shown in Fig. 1, one mole of lactate was formed for each glucose moiety from maltose and reduction of two moles fructose to mannitol resulted in formation of one mole acetate instead of ethanol. High pressure resulted in a decrease of maltose consumption with a concomitant decrease of lactate and ethanol formation. In contrast, the formation of acetate was unaffected by application of 80 MPa. The molar ratio of lactate to acetate was 1.8 after metabolism at ambient pressure but these metabolites were produced in equal amounts during high pressure metabolism. In accordance with the effects of HP on acetate formation, fructose conversion to mannitol was unaffected by 80 MPa and was reduced to 73 and 10 % at 100 and 150 MPa, respectively. The small amounts of glucose present in the medium were utilized during incubation at 0.1, 80 and 100 MPa. However, during incubation at 150 MPa, glucose accumulation equivalent to maltose consumption was observed.

Table 1. Composition of metabolites upon 3 h incubation of *L. sanfranciscensis* LTH2581 in mMRS4 at pressures from 0.1 to 150 MPa.

pressure	[metabolites] (mmol/l)						
[MPa]	maltose	glucose	fructose	mannitol	lactate	acetate	ethanol
0.1	-16	-1.3	-45	+42	+33	+18	+14
80	-8.8	-1.3	-44	+39	+18	+17	+2.9
100	-5.0	+0.1	-33	+29	+13	+12	+2.5
150	-3.9	+4.1	-6	+5	+2.3	+2.6	+0.6

Values are representative of two independent experiments. Negative values indicate consumption, positive values indicate production of metabolites relative to the medium.

Reversible versus irreversible inhibition of metabolic activity. In order to determine whether HP inactivation of metabolic activity is a reversible or irreversible process, cultures of *L. sanfranciscensis* LTH2581 were incubated for 3 h at 100 and 150 MPa. After this initial incubation time under HP conditions, cells were harvested by centrifugation, resuspended in fresh medium and incubated for further 3 h at ambient pressure. The maltose consumption at the various conditions is shown in Table 2. The maltose consumption at HP conditions were in accordance with the data shown in Fig. 2. During the second incubation step at ambient pressure, the metabolic activity of 100 MPa treated cells was almost fully restored to the activity of untreated cells, indicating that the HP mediated inhibition of metabolic activity is almost fully reversible. However, incubation at 150 MPa resulted in an irreversible inhibition of metabolism since less than 20% maltose consumption was observed during the second incubation at ambient pressure.

High pressure effects on the isotope distribution of metabolic products. The kinetics of maltose consumption was observed and the distribution of the ^{13}C isotope in the metabolic product ethanol was determined. Maltose was the only substrate and lactate and ethanol were the sole products of metabolism. The kinetics of maltose consumption are shown in **Fig. 3**. At ambient pressure, maltose consumption ceased after 4 h. At 100 and 150 MPa, an almost

Table 2. Metabolic activity of *L. sanfranciscensis* during and after a 3 h incubation at high pressure

	Maltose consumption %		
Pressure [MPa]	First incubation at high pressure		Second incubation at ambient pressure
0.1	89 ± 4	harvesting of cells and incubation in fresh media	90 ± 7
100	32 ± 1		70 ± 13
150	22 ± 9		18 ± 11

Results represent means ± standard deviation of two independent experiments

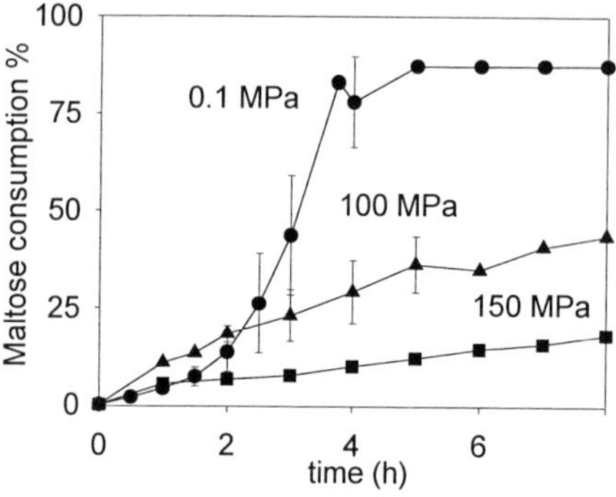

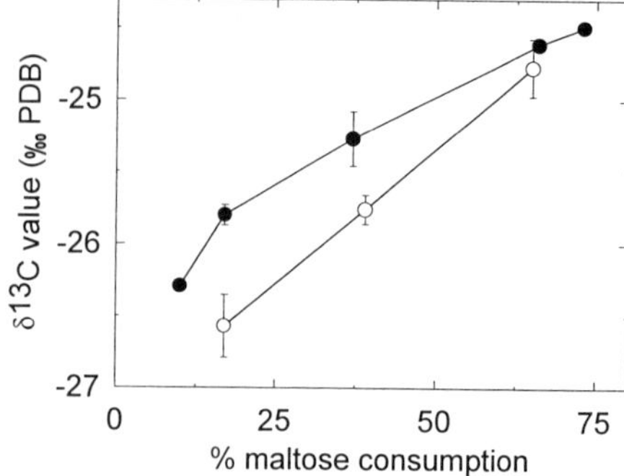

Figure 3. High pressure effects on maltose consumption by *L. sanfranciscensis*. Shown is the maltose consumption of cultures in C_3-MRS during incubation at 0.1, 100, and 150 MPa. Results represent means ± standard deviation of two independent experiments

Figure 4. High pressure effects on isotope distribution of ethanol. Shown are the $\delta^{13}C$ values of ethanol from cultures grown in C_3 MRS at 0.1 MPa (●) and 100 MPa (○).Results represent means ± standard deviation of two independent experiments

linear maltose consumption was observed and after 8 h 40 and 20% of the maltose were utilized, respectively. The $\delta^{13}C$ value of ethanol formed during metabolism at 0.1 and 100 MPa was determined and the values were compared based on equal maltose consumption (Fig 4.). The maltose used in the experiments had a $\delta^{13}C$ value of − 25.2 ‰ and 73% maltose consumption at ambient pressure resulted in an $\delta^{13}C$ value of − 24.8 ‰ in the metabolite ethanol. Because only carbon atoms 2 and 3 from the glucose moiety are recovered in the metabolic product ethanol this discrepancy is attributed to the non-statistical carbon isotope distribution in glucose (Rossmann et al., 1991). At low maltose consumption levels a preference for ^{12}C substrates was apparent. This decreased $\delta^{13}C$ values in the ethanol indicate a kinetic isotope effect. Comparison of the $\delta^{13}C$ values of ethanol formed during metabolism at 0.1 and 100 MPa on the basis of comparable maltose consumption levels demonstrates that this kinetic isotope effect is more pronounced during HP metabolism. For example, at 17% maltose consumption at 0.1 MPa a $\delta^{13}C$ value of − 25.8 ‰ was determined whereas ethanol formed at 100 MPa had a $\delta^{13}C$ value of − 26.6 ‰.

4. Discussion

The response of micro-organisms to increasing hydrostatic pressure includes growth arrest, synthesis of stress proteins, inhibition of metabolism, and cell death. In this communication, we provide a preliminary assessment of the metabolic response of *L. sanfranciscensis* to sublethal high pressure. Whereas up to 50 MPa were tolerated without adverse affects on metabolism, 100 MPa resulted in a reversible inhibition of metabolic activity without adverse effects on cell viability. The inhibitory effect of 150 MPa on metabolism was irreversible up to 3 h post treatment without adverse effects on cell viability. These observations indicate that several targets are involved in inhibition of metabolism. Reversible processes that may account for the decreased metabolic activity include the inhibition of membrane bound transport enzymes due to phase transitions of the cytoplasmic membrane (Ulmer et al., 2000, Wouters et al., 1998), and dissociation of enzymes consisting of several subunits (Deville-Bonne and Else, 1991). High pressure arrest of growth of *Saccharomyces cerevisiae* was reported to relate to the inhibition of tryptophan transport (Abe and Horikoshi, 2000). Previous work of Wouters et al. (1998) with *L. plantarum* have shown that the glycolytic activity remained unaffected by HP treatment whereas the ability of the organism to transport protons across the membrane was impaired. The irreversible inhibition of metabolic activity in *Lactococcus lactis* and *Lactobacillus plantarum* as determined with a formazan dye occurred concomitant with the loss of viability (Ulmer et al., 2000; Molina-Guttierrez et al., 2000).

High pressure incubation of *L. sanfranciscensis* resulted in a major shift in products formed from maltose and fructose. Whereas the alternative metabolites ethanol and acetate were formed in roughly equal amounts at ambient pressure, only acetate was formed after high pressure metabolism. In the presence of electron acceptors such as fructose, acetate production, associated with the gain of one additional ATP, is favored over ethanol formation. Cofermentation of maltose and fructose results in production of lactate and acetate until the fructose is quantitatively reduced to mannitol, thereafter, lactate and ethanol are produced (Stolz et al., 1993). Therefore, the composition of metabolites observed at high pressure conditions is not markedly different from that observed at comparable maltose consumption levels at ambient pressure. These results imply that the formation of acetate from acetyl-phosphate is not a rate limiting step during high pressure metabolism.

Isotope fractionation during enzymatic reactions reflects kinetic carbon isotope effects because of differences in rate constants k_{12} and k_{13} for ^{12}C and ^{13}C substrates, respectively (O'Leary, 1982). Carbon isotope fractionation effects during photosynthetic CO_2 fixation and decarboxylation reactions are well documented (O'Leary, 1993; Swanson et al., 1998) During ethanol formation from maltose a pronounced kinetic isotope effect was observed resulting in an enrichment of the ^{12}C isotope at incomplete maltose consumption levels. Because the $\delta^{13}C$ values of lactate could not be determined by the analytical setup used in this study, it remains unclear whether this isotope effect is the result of cumulative fractionation over the entire metabolic pathway or may be attributed to a single enzymatic activity. Most remarkably, high pressure metabolism did enhance the kinetic isotope effect. Pressure effects on enzymatic isotope fractionation have so far only been reported for deuterium / hydrogen hydride transfer catalyzed by yeast dehydrogenases (Quirk and Northrop, 2001; Northrop and Cho, 2000). Pressure mediated isotope effects may contain highly specific information about chemical and kinetic mechanisms of enzymatic transformations and may furthermore be relevant for the interpretation of stable isotope data used to trace the carbon flux in deep sea environments (Valentine and Reeburgh, 2000).

5. Literature

Abe, F., and K. Horikoshi. 2000. Tryptophan permease gene TAT2 confers high-pressure growth in *Saccharomyces cerevisiae*. Mol. Cell. Biol. 20:8093-8102.

Abe, F., C. Kato. Barophysiology (Piezophysiology). in: K. Horikoshi, K. Tsujii (eds), Extremophiles in deep-sea environments, pp. 227-248. Springer, 1999.

Deville-Bonne, D., A.J. Else. 1991. Reversible high hydrostatic pressure inactivation of phosphofructokinase from *Escherichia coli*. Eur. J. Biochem. 200:747-750.

Hammes, W.P., P. Stolz, M.G. Gänzle. 1996. Metabolism of lactobacilli in traditional sourdoughs. Adv. Food Sci.. 18:176-184.

Iwahashi, H. H. Shimizu, M. Odani, Y. Komatsu. 2000. Barophysiology of *Saccharomyces cerevisiae* from the aspect of 6000 gene expression levels. HPBB-2000, Kyoto, Japan.

Koziet, J., A. Rossmann, G.J. Martin, P.R. Ashurst. 1993. Determination of carbon-13 content of sugars of fruit and vegetable juices. Anal. Chim. Acta 271:31-38.

Molina-Gutierrez, A., B. Rademacher, M.G. Gänzle, R.F. Vogel. Effect of sucrose and sodium chloride on the survival and metabolic activity of *Lactococcus lactis* under high pressure conditions. Proceedings of the first conference on High Pressure Bioscience and Biotechnology, Kyoto, 26. - 30. November 2000, in press.

Northrop, C.B., Y.-K. Cho. 2000. Effect of pressure on deuterium isotope effects of yeast alcohol dehydrogenase: evidence for mechanical models of catalysis. Biochem. 39:2406-2412

O'Leary, M.H. Biochemical basis of carbon isotope fractionation. in: J.R. Ehleringer, A.E. Hall, J.D. Farquhar (eds), Stable isotopes and plant carbon-water relations, pp. 19-28, 1993 Academic Press.

O'Leary, M.H. Heavy isotope effects on enzyme catalyzed reactions. In: Schmidt, H.L., H. Förstel, and K. Heinzinger (eds), Stable isotopes. pp. 67-75. Analytical Chemistry Symposia Series Vol 11, Elsevier Science 1982.

Quirk, D.J., and D.B. Northrop. 2001. Effect of pressure on deuterium isotope effects of formate dehydrogenase. Biochem. 40:847-851.

Rossmann, A., M. Buthenlechner, and H.-L. Schmidt. 1991. Evidence for a nonstatistical carbon isotope distribution in natural glucose. Plant. Physiol. 96:609-614.

Stolz, P., G. Böcker, W.P. Hammes, R.F. Vogel. 1995a. Utilization of electron acceptors by lactobacilli isolated from sourdough. I. *Lactobacillus sanfrancisco.* Z. Lebensm. Unters. Forsch. 201:91-96.

Stolz, P., G. Böcker, W.P. Hammes, R.F. Vogel. 1995b. Utilization of electron acceptors by lactobacilli isolated from sourdough. II. *Lactobacillus pontis, L. reuteri, L. amylovorus,* and *L. fermentum.* Z. Lebensm. Unters. Forsch. 201:402-410.

Stolz, P., G. Bpöcker, R.F. Vogel, W.P. Hammes. Utilisation of maltose and glucose by lactobacilli isolated from sourdough. FEMS Microbiol. Lett. 109:237-243.

Swanson, T., H.B. Brooks, A.L. Osterman, M.H. O'Leary, M.A. Phillips. 1998. Carbon-13 isotope effect studies of *Trypanosoma brucei* ornithine decarboxylase. Biochem. 37:14943-14947.

Ulmer, H.M., M.G. Gänzle, R.F. Vogel. 2000. Effects of high pressure on survival and metabolic activity of *Lactobacillus plantarum.* Appl. Environ. Microbiol. 66:3966-3973.

Valentine, D.L., W.S. Reeburgh. 2000. New perspectives on anaerobic methane oxidation. Environ. Microbiol. 2:477-484.

Vogel, R.F., R. Knorr, M.R.A. Müller, U.Steudel, M.G. Gänzle, M.A. Ehrmann. 1999. Non-dairy lactic fermentations: The cereal world. Antonie van Leeuwenhoek 76:403-411.

Welch, T.J., A. Farewell, F.C. Neidhardt, D.H. Bartlett. 1993. Stress response of *Escherichia coli* to elevated hydrostatic pressure. J. Bacteriol. 175:7170-7177.

Wouters, P.C., E. Glaasker, and J.P.P.M. Smelt. 1998. Effects of high pressure in inactivation kinetics and events related to proton efflux in *Lactobacillus plantarum.* Appl. Environ. Microbiol. 64:509-514.

Trends in High Pressure Bioscience and Biotechnology
R. Hayashi (editor)
© 2002 Elsevier Science B.V. All rights reserved.

Effect of sucrose and sodium chloride on the survival and metabolic activity of *Lactococcus lactis* under high-pressure conditions

Adriana Molina-Gutierrez[b], Britta Rademacher[a], Michael G. Gänzle[b] and Rudi F.Vogel[b]

[a]Institute for Food Process Engineering, TU München, D-85350 Freising
[b]Lehrstuhl für Technische Mikrobiologie, TU München, D-85350 Freising

Abstract

High-pressure processing affects the metabolism and the viability of *Lactococcus lactis*. Treatments at 200 MPa did not affect the viability of the cells, but reduced their metabolic activity. The analysis of the cells treated at 300 MPa showed that high-pressure initially affects metabolic activity and subsequently damages membrane integrity of *Lactococcus lactis*. These treatments reduced the number of viable cells by 4 log. Sucrose and NaCl protected the viability of the cells from high-pressure inactivation at 300, 400 and 600 MPa. Sucrose preserved the metabolic activity and membrane integrity of the cells during the high-pressure treatments, salt preserved the membrane integrity, but not the metabolic activity. These results showed that ionic and non-ionic have a different mechanism of protection the micro-organisms against the high-pressure inactivation.

1. Introduction

Lactococcus lactis plays a major role in the production of sour milk, sour cream and cheese. Currently, special attention is paid to improve the flavor, ropiness and probiotic properties of dairy products by selection of appropiate strains (23). In ripened cheeses, lactococci are initially involved in lactic acid production, which lowers the pH. They are further involved in proteolysis (11) and, by their intra- and extracellular enzyme activities, in the production of aroma compounds during ripening.

Treatment with high pressure could be an alternative to thermal processes in order to control the activity of *Lactococcus lactis* in the milk industry. Choosing the appropiate parameters of treatment, the high pressure process could i) inactivate the micro-organisms, ii) liberate their endoenzymes without inactivating the cells, iii) change the structure of the product without inactivating the starter culture. Casal and Gómez (1) suggested an addition of cells of *Lactococcus lactis* treated at 300 MPa during the cheese making to enhance the amount of enzymes with potential debittering properties. Since salts and sugars are normally used during the manufacture of dairy products, it is important to examine the effect of salts and sugars during the pressure-inactivation of *Lactococcus lactis.*

The composition of the medium where the micro-organisms are dispersed during pressurization affects the efficiency of inactivation. A protective effect of sucrose against pressure inactivation of *Rhodoturula rubra* was reported by Oxen and Knorr (15). Rademacher (18) showed a marked protective effect of sodium chloride and sugars against the pressure inactivation of *Escherichia coli* in milk. Baroprotective effects of NaCl or glucose were also observed with suspensions of *Zygosccharomyces rouxii* and *Saccharomyces cerevisiae* (5). These effects were suggested to relate to the decrease in water activity. However, the mechanism of protection remains unknown.

The aim of this investigation was to determine the possibility of applying high pressure for controlling the metabolic activity of *Lactococcus lactis* and the effect of addition of salts and sugars on the physiological reactions of the cells during pressurization.

2. Material and Methods

Bacterial strain and culture conditions. Lactococcus lactis ssp. *cremoris* MG1363 was grown at 30°C in M17 broth (Merck, Darmstadt, Germany) supplemented with 1% glucose.

Pressurization of cell suspensions. The cells of an overnight culture were harvested by centrifugation (15 min at 5500 rcf·g), washed and resuspended in milk buffer to about 10^9 cfu·ml^{-1}. The milk buffer was chosen to contain the same amounts of minerals and lactose as whey from rennet casein; the buffer contained the following compounds (g·l^{-1}): KCl, 1.1; $MgSO_4 \cdot 7H_2O$, 0.7110; $Na_2HPO_4 \cdot 2\,H_2O$, 1.874; $CaSO_4 \cdot 2\,H_2O$, 1; $CaCl_2 \cdot 2\,H_2O$, 0.99; citric acid, 2; lactose, 52. The pH was adjusted to 6.5 with KOH (2 M). Milk filtrate was prepared by ultrafiltration. The cells were suspended in 2 ml portions in sterile plastic micro test tubes, sealed with silicon stoppers and stored on ice until they were pressurized. The pressure chamber was heated/cooled to a desired level prior to pressurization with a thermostat jacket connected to a water bath. The pressure level, time and temperature of pressurization were controlled by a computer program. The compression/decompression rate was 200 MPa min^{-1}, the temperature was 20°C and the temperature rise due to compression was 12 °C or less. Samples were placed in the pressure chamber 5 min prior to treatment to equilibrate the sample temperature. Cells were exposed to a pressure of 200, 300, 400 or 600 MPa for various time intervals (0-120 min). Following the release of pressure the samples were stored on ice for determination of viable counts and membrane integrity. For each HP inactivation kinetics, untreated cultures and cultures sterilized by treatment with 800 MPa for 10 min or 15 min at 80°C were used for preparation of "calibration samples" containing 100, 50, 25, 12.5, 6.25, 3.125 and 0% viable cells.

Enumeration of viable cells. The cell suspensions from each vial were serially diluted with saline immediately after the pressurization treatment and were surface plated on M17 agar (Merck, Darmstadt, Germany). The plates were incubated for 24 h at 30°. Data presented are mean plus/minus standard deviation obtained from two to three independent experiments.

Membrane integrity assay. 1 ml pressure-treated cell suspensions were harvested by centrifugation at 6000 rcf·g for 10 min. The supernatant was removed and the pellet resuspended in 1 ml of phosphate buffer. A stock solution of the LIVE/DEAD®BacLight™ was prepared, the final concentration of each dye was 33.4 µM SYTO® 9 and 200 µM propidium iodide (PI). 100 µl of each of the bacterial cell suspensions were mixed in 100 µl of the stock solution, mixed thoroughly and incubated at 30°C in the dark for 5 min. The fluorescence intensities of SYTO® 9 and PI were measured with excitation and emission wavelengths of 485 and 520 nm, and 485 and 635 nm, respectively, using a spectraflour microtiter plate reader (TECAN, Grödig, Austria). The ratio of SYTO® 9 to PI flurecence intensity was used as measure for membrane integrity. A calibration curve was established for each inactivation kinetics using the "calibration samples" described above and the results are reported as % intact membranes.

Metabolic activity assay. The method developed by Ulmer et al. (22) was used to determine the metabolic activity. 1 ml pressure-treated cell suspension were harvested by centrifugation. The supernatant was removed and the pellet resuspended in 1 ml of phosphate buffer. A stock solution of tetrazolium was prepared mixing 4-iodonitrotetrazolium violet (INT; 2-(4-iodophenyl)-3-(-4-nitrophenyl)-5-phenyltetrazolium chlorid) and glucose in phosphate buffer. The final concentration of each was 4 mM and 20 mM respectively. 100 µl of each of the bacterial cell suspensions were mixed with 100 µl of the stock solution of tetrazolium salt. The absorbance was measured at 590 nm during 30 min with a spectraflour microtiter plate reader (TECAN, Grödig, Austria). A calibration curve was established for each inactivation kinetics using the "calibration samples" described above and the results are reported as % metabolic activity.

Determination of water activity (a_w). The water activity was determined by measuring the freezing point depression with a Crioscope (Type A0284, Knauer GmbH, Berlin).

3. Results

Inactivation of Lactococcus lactis in milk buffer. To verify the suitability of the milk buffer as food system, cells suspensions of the *Lactococcus lactis* ssp. *cremoris* MG1363 were exposed to a treatment at 300 MPa for different times intervals in milk buffer and in milk filtrate, the results are shown in Fig. 1. The inactivation curves of the bacteria in both medium showed sigmoid asymmetric shapes when plotted in logarithmic scale. Within the inactivation curves two parts can be distinguished: an initial shoulder and an inactivation phase followed by tailing (6) (Fig. 1). The curves in milk buffer and in milk filtrate did not differ significantly.

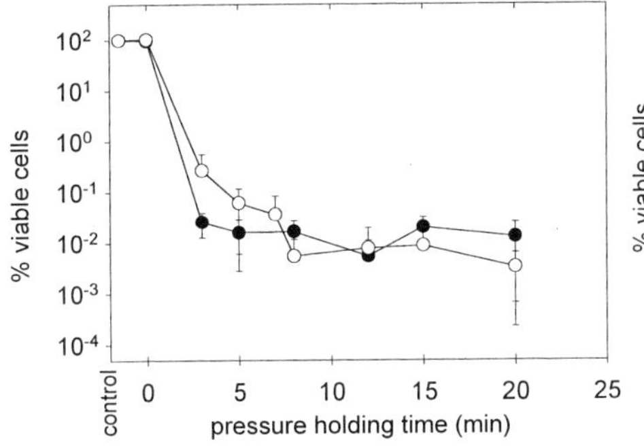

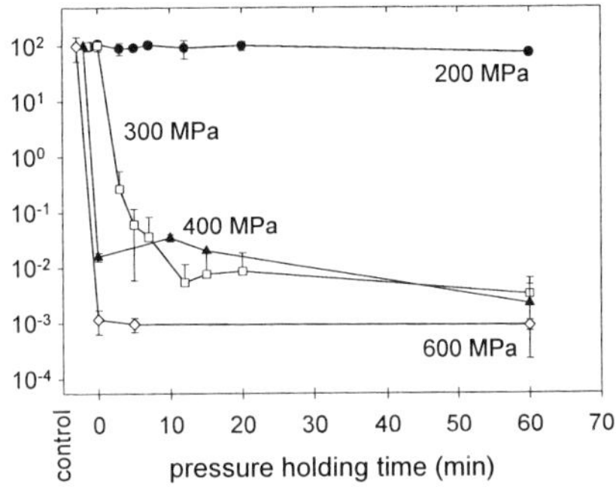

Fig. 1 Kinetics of inactivation of *Lactococcus lactis* ssp. *cremoris* MG1363 cells in milk buffer (O) and milk filtrate (●) after pressure treatments at 300 MPa and 20°C

Fig. 2 Effect of pressure on inactivation of *Lactococcus lactis* ssp. *cremoris* MG1363 in milk buffer at 20 °C

298

Cells suspensions of *Lactococcus lactis* ssp. *cremoris* MG1363 were exposed to 200, 300, 400 and 600 MPa at 20°C for different time intervals (Fig. 2). Pressure treatment at 200 MPa upon 60 min did not affect the viability of *Lactococcus lactis*. The inactivation curves of the bacteria at 300, 400 and 600 MPa showed the typical sigmoid asymmetric shape. The shoulder of the inactivation curves was reduced with the increase of the pressure but the level of survivors after 60 min was the same for all pressure treatments.

Metabolic activity and membrane integrity after high-pressure treatments. The cells from the treatments at 200 and 300 MPa were analyzed to determine their metabolic activity and membrane integrity (Fig. 3 A, B). After a treatment for 5 min at 300 MPa the metabolic activity is 10-12% of the activity of untreated microorganisms, and after 12 min of treatment the cells did not show any metabolic activity. Treatments at 200 MPa up to 60 min did not inactivate the micro-organisms, but reduced their metabolic activity to 50%. This reduction occurred in the first 20 min of treatment (Fig. 3A). During treatment at 300 MPa, cell death was closely followed by the loss of metabolic activity, however cultures retained about 25% of the metabolic activity, even after a 3 log reduction in cell counts (Fig. 3B).

The cytoplasmic membrane of *Lactococcus lactis* remained impermeable to PI even after cell death. Treatment at 200 MPa did not result in a reduction of membrane integrity after 20 min and cultures retained 30% of membrane integrity after 20 min at 300 MPa, corresponding to 4 log reduction of viable cell counts. These results show that the treatment of *Lactococcus lactis* ssp. *cremoris* MG1363 with high-pressure initially affects metabolic activity and subsequently damages membrane integrity.

Effect of low molecular solutes during high-pressure treatments. The protective effect of a reduction of the water activity (a_w) against pressure inactivation was determined. Since *Lactococcus lactis* confronted with decreased a_w responds differently if the reduction was achieved with an ionic or nonionic solute, the effect of NaCl and sucrose were compared. 1 and 1.5 mol·l^{-1} sucrose provided complete protection against cell death at either 400 or 600 MPa (Fig. 4B, Fig.5).

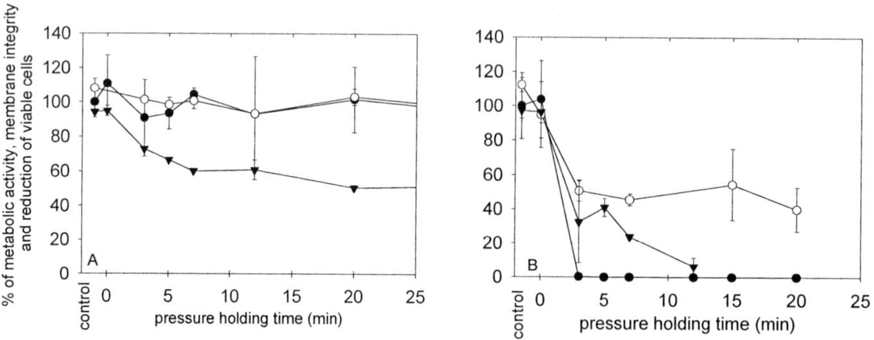

Fig. 3 Effect of pressure treatment and 20 °C on the metabolic activity (▼), membrane integrity (○) and viability (●) of *Lactococcus lactis* ssp. *cremoris* MG1363. A: treatments at 200 MPa, B: treatments at 300 MPa.

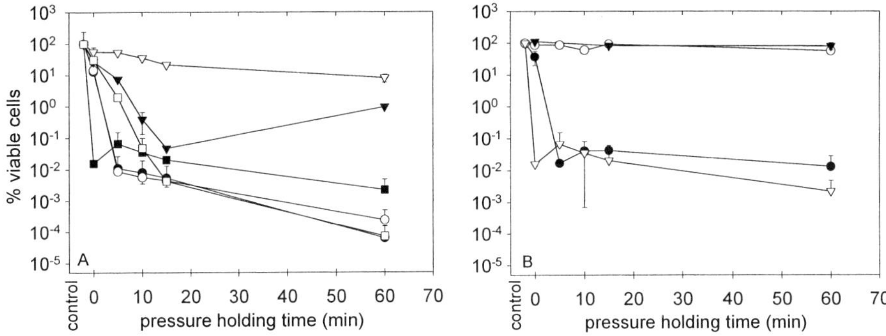

Fig. 4 A. Effects of NaCl in milk buffer at concentration of 1 M (●), 2 M (○), 2 M with 2.5 mM glycine betaine (□), 3 M (▼), 4 M (▽) and control (■), B. Effects of sucrose in milk buffer at concentration of 0.5 M (●), 1 M (○), 1.5 M (▼) and control (▽) on the pressure resistance of *Lactococcus lactis* ssp. *cremoris* MG1363 at 400 MPa (20°C).

No difference was observed between the baroprotective effect of 0.5 and 1 mol·l^{-1} and 0.5 and 1 mol·l^{-1} lactose at 300 MPa during 0, 15 and 30 min pressure holding time, respectively (data not shown). Due to the low solubility of lactose further experiments were carried out with sucrose.

A protective effect of 2, 3 and 4 mol·l^{-1} NaCl was observed at 300 MPa (data not shown) as well as 400 and 600 MPa (Fig. 4 A, Fig. 5). A comparison of the baroprotective effect of 2 and 3 mol·l^{-1} NaCl and 2 and 3 mol·l^{-1} KCl at 300 MPa during 0, 15 and 30 min pressure holding time showed no difference between these salts. The cytoplasmic accumulation of exogenous glycine betaine upon an osmotic upshift restores the cellular volume and increases the hydration of the cytoplasm of *Lactococcus lactis* (14). If the compatible solute glycine betaine was added at a concentration of 2.5 mM, the baroprotective effect of 2 mol·l^{-1} NaCl was increased in treatments at 400 MPa (Fig. 4A).

The baroprotective effect of solutes was proportional to the increase of their concentration in the buffer, but it is not proportional to the decrease of the a$_w$. The buffers with NaCl showed lower a$_w$ than the buffers with sucrose, however the protective effect from sucrose at the same molar level was much higher (Tab. 1). In contrast to sucrose, even the highest NaCl concentration failed to achieve complete protection against cell death at 600 MPa. .

Effect of low molecular solutes on metabolic activity and membrane integrity. The cells from the treatments at 300 MPa with 4 mol·l^{-1} NaCl and 1.5 mol·l^{-1} sucrose were analyzed to determine the effect of the solutes on the metabolic activity and membrane integrity. The results with sucrose showed a protective effect on the metabolic activity and membrane integrity which decreased to 90 and 50%, respectively within 60 min of pressure holding time (Fig. 6A). NaCl protected the membrane integrity to the same level, but the metabolic activity decreased rapidly (Fig. 6B).

Table 1. Comparision of water activity an baroprotective effect of solutes in milk buffer

Solute	CFU ml^{-1} after a treatment at 300MPa, 15 min, 20°C	a_w -Value
Milk buffer	2.0×10^4	0.9960
NaCl		
1 mol	5.6×10^4	0.9659
2 mol	1.0×10^7	0.9408
3 mol	1.7×10^8	0.9167
4 mol	2.5×10^8	0.8954
Sucrose		
0.5 mol	1.6×10^8	0.9849
1 mol	4.8×10^8	0.9760
1.5 mol	2.7×10^8	0.9663

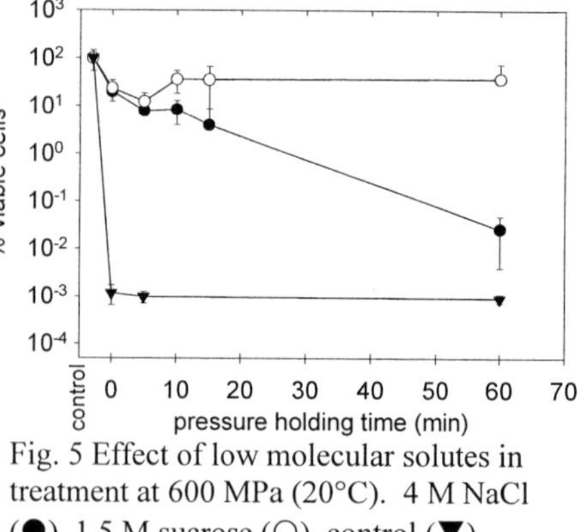

Fig. 5 Effect of low molecular solutes in treatment at 600 MPa (20°C). 4 M NaCl (●), 1.5 M sucrose (○), control (▼)

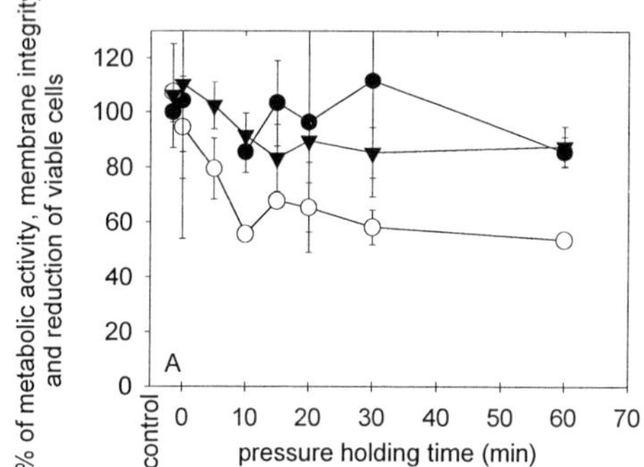

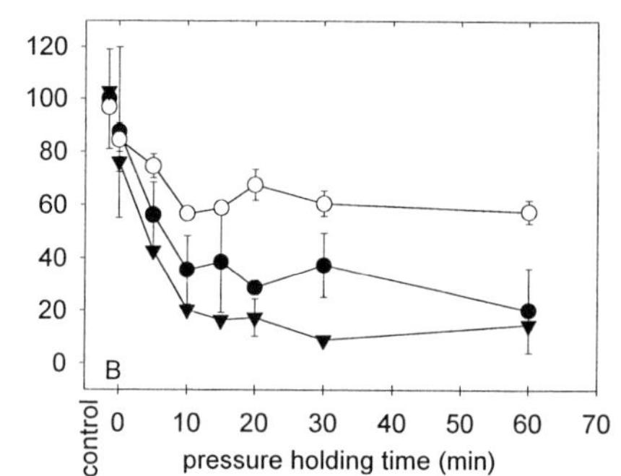

Fig. 6 Effect of solutes on the metabolic activity (▼), membrane integrity (○) and viability (●) of *Lactococcus lactis* ssp. *cremoris* MG1363 after treatment with 300 MPa and 20 °C. A: 1.5 M Sucrose, B: 4 M NaCl.

4. Discussion

Microbial inactivation by high pressure is a complex phenomenon, which involves more than one physiological reactions and is still not completely understood. Micro-organisms are inactivated owing mainly to membrane damage, but the changes observed after a pressure treatment, further involve ribosome denaturation, coagulation of cytoplasmic protein, enzyme inactivation, inhibiton of transcription and protein synthesis (2, 7, 16).

In this study the potential of applying high pressure to control the survival and metabolic activity of *Lactococcus lactis* was evaluated. Furthermore, the baroprotective effect of ionic and nonionic solutes on the viability and metabolic activity was compared.

The composition of the medium where the micro-organisms are dispersed during pressurization affects significantly the efficiency of inactivation. The loss of viability is lower in food systems than in phosphate buffer (8). Our results indicate that the milk buffer used in our work is a suitable, chemically defined milk model system.

Studies with bacteria have revealed that the loss of cell viability increased with increase in pressure and time and up to 200 MPa, even after 30 min, cell death was ≤1 log cycle (7, 8, 15). This is corroborated by our finding that treatments at 200 MPa up to 60 min did not reduce the number of micro-organisms A higher pressure reduced the length of the shoulder and increased the rate of inactivation, but it did not affect the tailing.

Methods involving dyes have been successfully used to determine the loss of membrane integrity after treatment with high-pressure. Pagan and Mackey (16) used propidium iodide to determine the loss of membrane integrity in pressure-treated *Escherichia coli*, in our laboratory Ulmer et al. (22) used propidium iodide to analyze *Lactobacillus plantarum* after high-pressure treatment. Tetrazolium salt offers a rapid, culture–independent means to determine viability and relative rate of reducing equivalent production in micro-organisms (20, 24). This method has been used to determine the metabolic activity of pressure-treated *Lactobacillus plantarum* (22).

Treatment of *Lactococcus lactis* at 200 MPa did not affect viability but reduced metabolic activity to 50%. Cell death observed during treatment at 300 MPa was followed closely by loss of metabolic activity whereas dead cells conserved membrane integrity. These results suggested that the high-pressure treatment of *Lactococcus lactis* affects metabolic activity and subsequently damages membrane integrity.

The addition of NaCl and sugars protects micro-organisms during high-pressure treatments. This increased barotolerance was attributed to cell shrinkage which probably causes a tickening in the cell membrane reducing membrane permeability and fluidity (5, 10, 15, 19). However, the effect of NaCl or sugars on the osmotic pressure of foods or suspension media is not sufficient to explain the marked baroprotective effects of these solutes on microbial inactivation (2). Our work supports this statement because 1.5 M sucrose with a higher water activity had a stronger baroprotective effect than 4 M NaCl.

In general, sugars had a stronger protective effect than salts on the inactivaton of *Lactococcus lactis* and this effect did not only rely on the lowering of a_w. High concentrations of sugars (lactose or sucrose) impose only a transient osmotic stress in *Lactococcus lactis,* because the cells are able to equilibrate the extra- and intracellular concentrations of sucrose and lactose (4). Disaccharides stabilize proteins in their native state and preserve the integrity of membranes during stresses (12). Adding a disaccharide lowers the transition temperature Tm of the membranes by replacing the water between the lipid headgroups, preventing the phase transition, and inhibiting the fusion between the liposomes due to glass formation (vitrification) (3). Furthermore, the addition of sucrose raises the internal pressure of the solvent, forcing the protein to overcome a greater energy barrier in unfolding (21) and stabilizes the proteins due to a preferential exclusion of water. We suggest that the intracellular sucrose protected the metabolic enzyme and the lipid of the membrane during the high-pressure treatment.

High salt concentration is detrimental for micro-organisms, however the addition of NaCl was shown to protect the cells during the high-pressure treatments. Micro-organisms accumulate compatible solutes upon challenge by decreased water activity. The intracellular accumulation of compatible solutes maintains the osmotic balance with the extracellular environment, the integrity of biological membranes and enhances the stability of enzymes (9). *Lactococcus lactis* accumulates glycine betaine when subjected to salt stress. However, since *Lactococcus lactis* is unable to synthesize this compound, it is replaced by unknown osmolytes, unless it is added to the medium. The increase of the baroprotective effect of 2 NaCl in presence of glycine betaine indicated that the intracellular accumulation of compatible solutes played a major role in the baroprotective effect of NaCl. In the treatments with NaCl without glycine betaine the baroprotective effect could be explained by an accumulation of unknown osmolytes from the milk buffer.

5. Conclusions

High-pressure treatments were effective for inactivating *Lactococcus lactis*. Prior to cell death ocurred a loss of metabolic activity and membrane damage. Sucrose and NaCl exhibited a baroprotective effect against the loss of viability of *Lactococcus lactis*. The baroprotective effect of sucrose is stronger compared on a molar basis. While sucrose protected the metabolic activity and membrane integrity, salt did not show any protection on metabolic activity. This work suggested two mechanisms to understand the baroprotective effect of low molecular solutes.

References

1. V. Casal and R. Gómez, J. Dairy Sci. 82 (1999)1092.
2. J.C. Cheftel, Food Sci. Technol. Int., 1 (1995) 75.
3. J.H. Crowe, A. E. Oliver, F.A. Hoekstra and L.M. Crowe, Criobiology, 35 (1997) 20.
4. E. Glaasker, F.S.B. Tjan, P.F. Ter Steeg, W.N. Konings and B. Poolman, J. Bacteriol., 180 (1998) 4718.
5. K. Hayakawa, Y. Ueno, S. Nakanishi, Y. Miyano, S. Kikushima and S. Shou, High Pressure Bioscience (R. Hayashi, S. Kunigi and A. Suzuki, eds.), Kyoto, Japan, 1994.
6. V. Heinz and D. Knorr, Food Biotechnol., 10 (1996) 149.
7. N. Kalchayanand, A. Sikes, C.P. Dunne and B. Ray , Food Microbiol., 15 (1998) 207.
8. N. Kalchayanand, A. Sikes, C.P. Dunne and B. Ray, J. Food Protect., 61 (1998) 425.
9. E.P.W. Kets, M.N. Groot, E.A. Galinski and J.A.M. De Bont, Appl. Microbiol. Biotechnol., 48 (1997) 94.
10. D. Knorr, New Methods of Food Preservation, London, England, 1994.
11. J. Law and A. Haandrikman, Int. Dairy J., 7 (1997) 1.
12. S. B. Leslie, E. Israeli, B. Lighthart, J.H. Crowe and L.M. Crowe, Appl. Env. Microbiol., 61 (1995) 3592.
13. Molecular Probes, Product information. Eugene, USA (1999).
14. D. Obis, A. Guillot, J.C. Gripon, P. Renault, A. Bolotin and M.Y. Mistou, J. Bacteriol., 181 (1999) 6238.
15. P. Oxen and D. Knorr, Lebensm. Wiss. Technol., 26 (1993) 220.
16. R. Pagán and B. Mackey, Appl. Env. Microbiol., 66 (1993) 2829.
17. E. Palou, A. López-Malo, G.V. Barbosa-Cánovas, J. Welti-chanes and B.G. Swanson, Lett. Appl. Microbiol., 24 (1997) 417
18. B. Rademacher, Dissertation, TU-Munich, Germany, 1999.
19. K. Takahashi, H. Ishii and H. Ishikawa, High Pressure Bioscience and Food Science (Hayashi R, ed.), Kyoto, Japan, 1993.
20. S.M. Thom, R.W. Horobin, E. Seidler and M.R. Barer, J. Appl. Bacteriol., 74 (1993) 433.
21. S.N. Timasheff, J.C. Lee, E.P. Pittz and N. Tweedy, J. Colloid Inter. Sci., 55 (1976) 658.
22. H. Ulmer, M. Gänzle and R Vogel, Appl. Env. Microbiol., 66 (2000) 3966.
23. E.W.J. Van Niel and B. Hahn-Hägerdal, Appl. Microbiol. Biotechnol., 52 (1999) 617.
24. S. Walsh, H.M. Lappin-Scott, H. Stockdale and B.N. Herbert, J. Microbiol. Methods, 24 (1995) 1.

Trends in High Pressure Bioscience and Biotechnology
R. Hayashi (editor)

High pressure induced alterations in morphology and cell characteristics of the bacterium Bacillus thuringiensis

H. Ludwig, K.G. Werner, E. Schattmann and M. Schauer

Institut für Pharmazeutische Technologie und Biopharmazie, Universität Heidelberg, INF 366, 69120 Heidelberg, Germany

Adaptation of microorganisms to high hydrostatic pressure is well illustrated by the existence of piezophilic deep-sea bacteria. This adaptation is assumed to be the result of a long-lasting evolutionary process. There are only a few examples, however, of breeding a pressure resistant strain from terrestrial bacteria in the laboratory.

In this paper we present evidence for a pressure induced transmutation of the bacterium *Bacillus thuringiensis subsp. israelensis*. For that effect, only a single high pressure treatment is needed. The attributes of the newly emerging bacteria are compared with those of the original strain.

1. INTRODUCTION

Bacillus thuringiensis is an ubiquitous, spore forming bacterium living in soils and flat water. Associated with the sporulation process, a glycoprotein is synthesized appearing as a crystal near the endospore. Spores and crystals are eaten by the larvae of insects. The crystalline protein then dissolves in the alkaline midgut of the insect and is converted to a powerful toxin killing the insect which can now be invaded by the bacillus. This procedure is very specific and restricted to certain insect larvae. Different strains of *B. thuringiensis* are effective against different hosts. Spore- and crystal preparations are therefore used, worldwide, as biological pesticides [1]. In Germany, however, it is forbidden to distribute such amounts of living organisms into the environment and therefore the spores in these preparations have to be inactivated. This is usually done by γ-radiation with the effect that half of the protein crystals are destroyed as well. That was the reason why we tried to kill the spores by high pressure having in mind the fact that crystalline proteins should be rather stable to pressure. Some years ago we had already succeeded in case of *B. thuringiensis subsp. tenebrionis* which is used against the Colorado potato beetle [2]. Now we tried it again with commercially available spore preparations of *B. thuringiensis subsp. israelensis* used to combat the larvae of mosquitos. And in this case we failed due to the following reason: During the pressure treatment the spores are converted to more pressure resistant specimens.

2. MATERIALS AND METHODS

The commercial spore suspension „VectoBac 12AS" from Abbott Laboratories, Chemical and Agricultural Division, North Chicago, IL, 60064, USA was used in the first experiments. Then, a clone was isolated from that suspension to start the following experiments from a homogeneous population. In the third stage, a specified strain of *B. thuringiensis subsp. israelensis* (DSMZ-Nr.5724) was provided by the Deutsche Sammlung für Mikroorganismen und Zellkulturen, Braunschweig, Germany.

With the exception of the initial experiments with „VectoBac 12AS", all spore suspensions used in our experiments were prepared starting from one single colony of microoganisms. The spores were prepared as described elsewhere [3] and harvested after 14 days. They were washed and suspended in 0.9 % aqueous NaCl solution, containing 0.1 % Polysorbate 80 to prevent spore clumping, and thus used in the high pressure experiments. The number of surviving organisms was determined by counting colony forming units (cfu) on suitable agar plates.

The experiments were done in a multivessel pressure device consisting of 10 vessels with an inner diameter of 14 mm. The samples were enclosed in polyethylene tubes, the pressure transmitting medium was water with 10 volume percent ethyleneglycol. The temperature of the pressure vessels was controlled by a thermostate.

3. RESULTS AND DISKUSSION

The Figures 1 to 4 give some examples of the unsuccessful attempts to inactivate the spores and to preserve the protein crystals simultaneously. The results shown in the figures were obtained with spores from one clone of the DSMZ-strain, but the same was found using the other sources (VectoBac 12AS). The conditions, high pressure combined with high temperature of 70 °C, are too hard for the crystalline protein. But even under these conditions sterility could not be reached in reasonable time. Pressure cycles (Fig. 3 and 4) are more effective than constant pressure is (Fig. 1 and 2). In all cases, however, a strange phenomenon appears: The surviving organisms split up into two groups, the members of one group being more pressure resistant than the others.

Figure 5 shows the usual colonies growing from spores of *B. thuringiensis subsp. israelensis* on agar plates. The colonies from spores surviving a pressure treatment are shown in Fig. 6. Here, the two groups of colonies can clearly be distinguished. From the detail given in Fig. 7 it can be realized that the newly emerging organisms have surface properties very different from those of the original bacteria. The significant changes of growth behaviour and habitus are surely based on altered interactions between the single cells. This conclusion is confirmed by a comparative electron microscopic study [4].

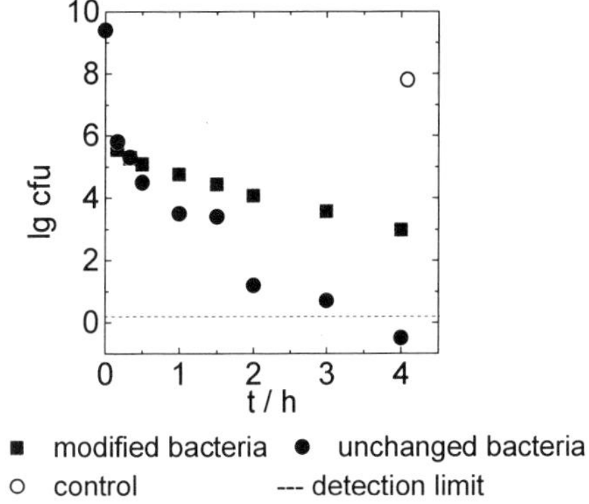

Figure 1. Inactivation of spores of *B. thuringiensis* at 70 °C and 400 MPa

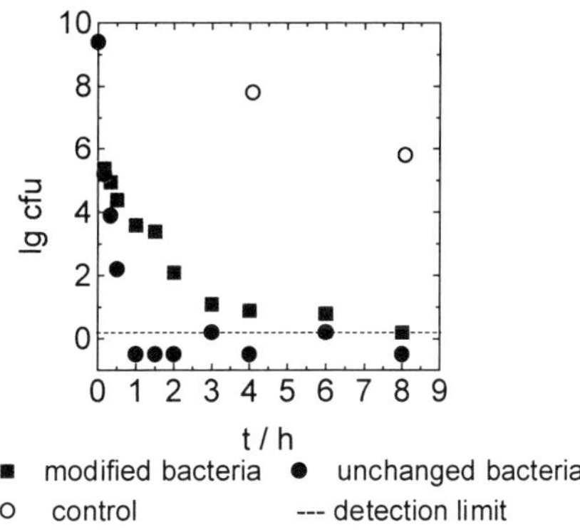

Figure 2. Inactivation of spores of *B. thuringiensis* at 70 °C and 500 MPa

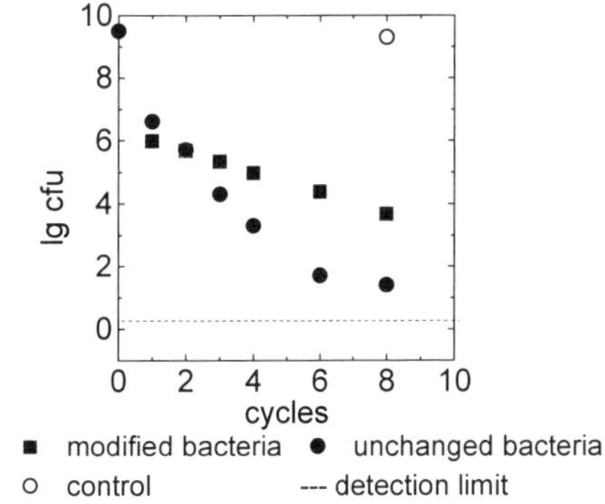

Figure 3. Inactivation of spores
by pressure cycles at 60 °C
One cycle: 5 min 0.1 MPa /
10 min 400 MPa

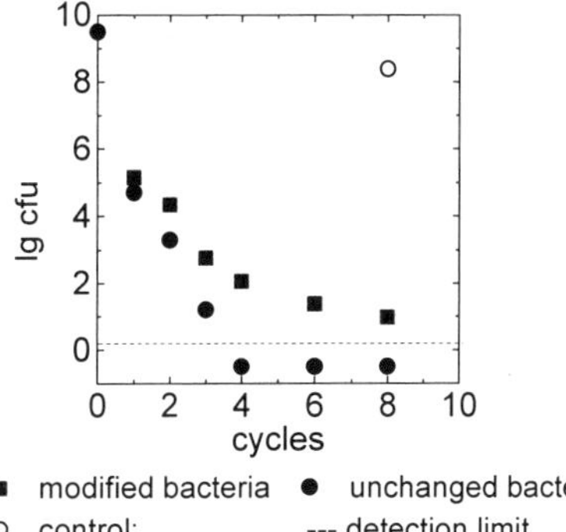

Figure 4. Inactivation of spores by
pressure cycles at 70 °C
One cycle: 5 min 0.1 MPa /
10 min 500 MPa

Figure 5. Colonies of *B. thuringiensis subsp. israelensis*

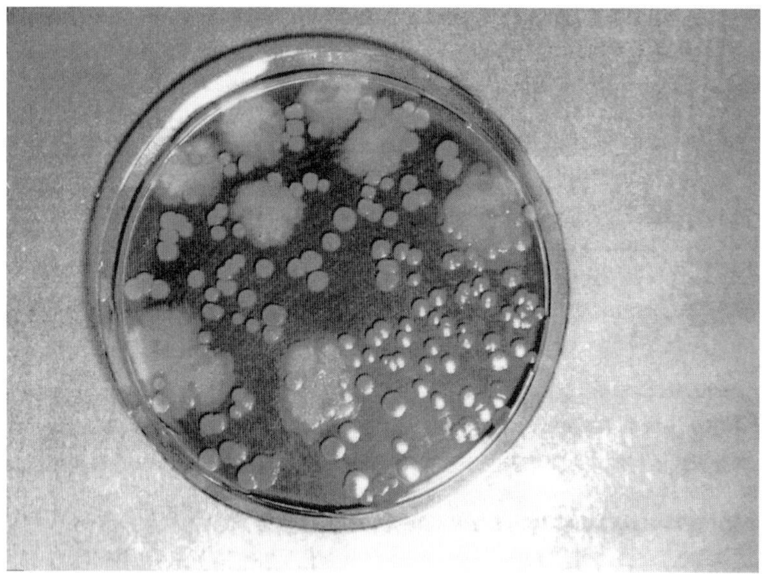

Figure 6. Colonies of *B. thuringiensis subsp. israelensis* and of modified bacteria

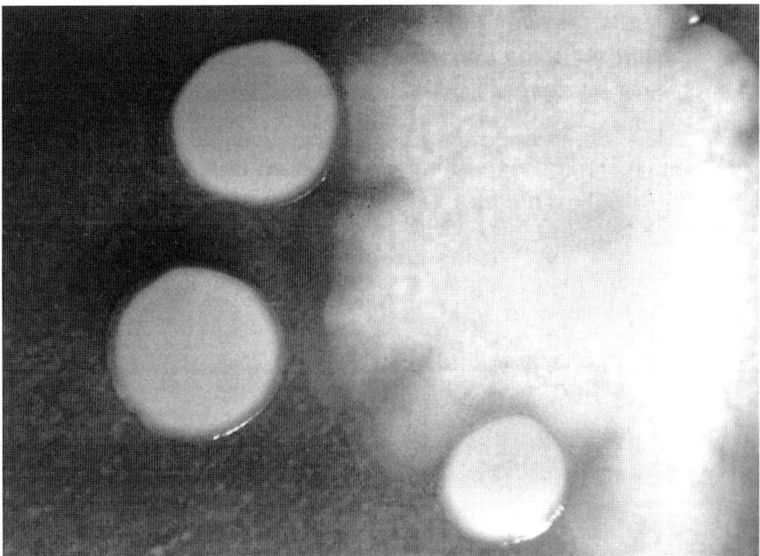

Figure 7. Colonies of *B. thuringiensis subsp. israelensis* and of modified bacteria

Regarding the Figures 1 to 4 one could think that a small fraction of resistant germs were already present in our spore populations from the beginning. Such a small fraction remained hidden until the majority of the sensitive spores were killed by pressure. Only then, the more resistant germs would become visible, just as it is seen in our pictures. This interpretation seems obvious, but is not correct. It is rejected by the following arguments and observations. 1) The spores used in the experiments were grown from a single bacterial clone and are therefore suggested to be a homogenous population. 2) The procedures of spore preparation and high pressure treatment have carefully been repeated several times, always with the same results. 3) Pressurizing the spores at room temperature yields a much larger "fraction" of the modified specimens (Figures 8 and 9).

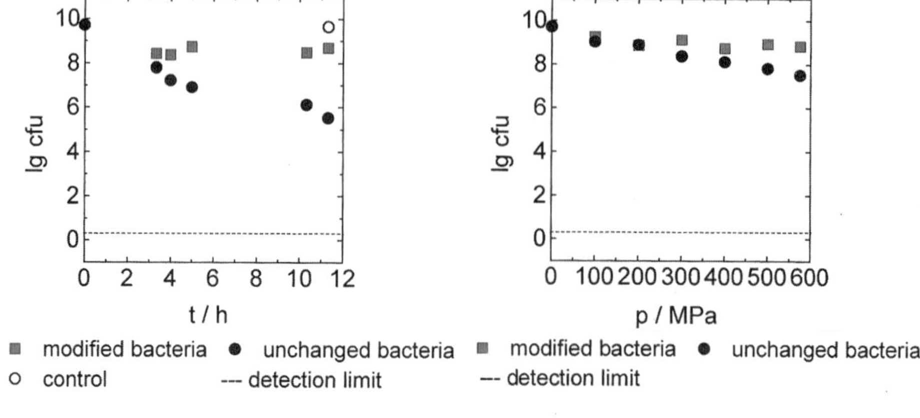

Figure 8. Inactivation of spores of
B. thuringiensis at 20 °C, 300 MPa

Figure 9. Pressure-tolerance of
B. thuringiensis spores, 45 min, 25 °C

At room temperature, almost no inactivation occurs even at pressures up to 600 MPa, but a transmutation to modified spores. The number of original spores is reduced and the new spores emerge in nearly the same quantity as the original ones had been there before.

The bacteria grown from the high pressure modified spores are stable. They can be cultivated over many generations, including sporulation steps, without any further visible changes. The high pressure induced modification seems irreversible. In Figure 10 the growth rates of original and high pressure changed bacteria are compared. Figure 11 compares the inactivation kinetics of spores produced from the original and the high pressure modified strains. Colonies from the pure high pressure strain are shown in Figure 12.

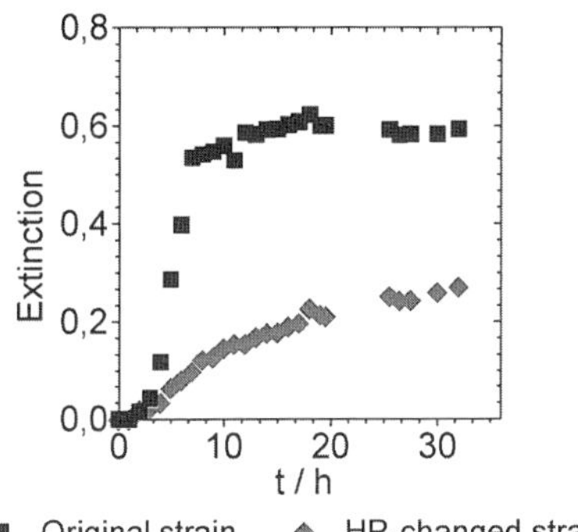

Figure 10. Growth curves of *Bacillus thuringiensis*

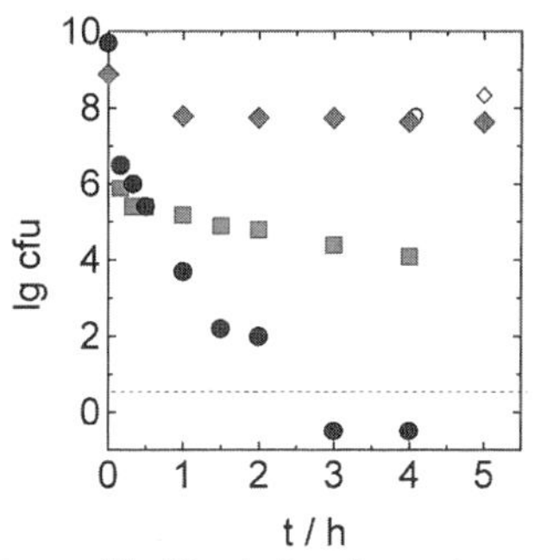

Figure 11. Comparison of the inactivation kinetics of different Spores of *Bacillus thuringiensis* at 70 °C and 300 MPa

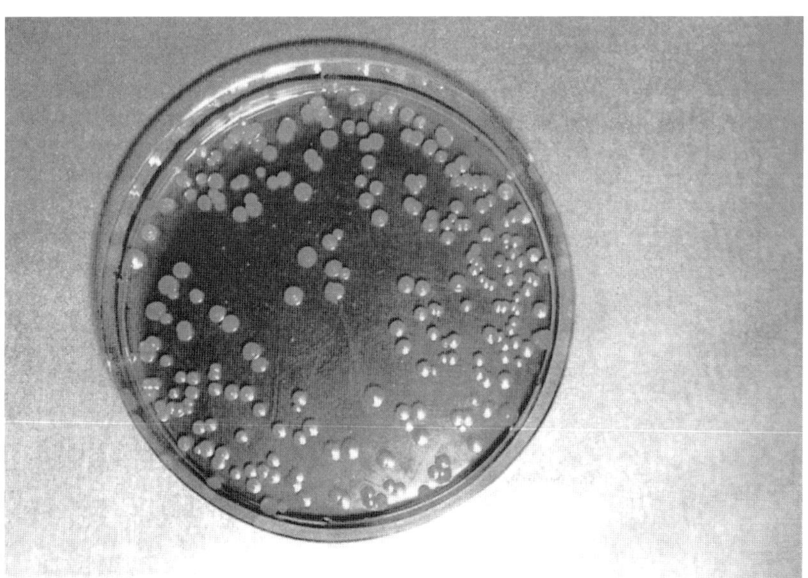

Figure 12. Colonies of HP-changed *Bacillus thuringiensis*

The transmutation can not be induced by temperature and it is not possible to do it with vegetative bacteria, only pressurisation of the spores leads to success.

The high pressure changed bacteria are spore-forming gram-positive rods and build protein crystals during sporulation. They are *Bacillus thuringiensis*. Some morphological differences to the original bacteria can be seen in the electron microscope [4]. More detailed characterisation will follow.

The question arises why spores are the suitable targets for the pressure induced transmutation but not the vegetative bacteria. The reason could be that repair mechanisms which are effective in vegetative cells do not work in spores. To reveal the mechanisms of how pressure is causing the conversion more experiments have to be done.

4. CONCLUSION

A single pressure treatment of spores of the bacterium *B. thuringiensis subsp. israelensis* leads to a heritable transmutation of these spores. The resulting specimens are more pressure resistant. This is, to our knowledge, the first example of such a spontanous transmutation induced by pressure.

REFERENCES

1. E. Schnepf, N. Crickmore, J. VanRie, D. Lereclus, J. Baum, J. Feitelson, D.R. Zeigler and D.H. Dean, Microbiology and molecular reviews, 62 (1998) 775.
2. P. Butz, E. Fritsch, J. Huber, B. Keller, H. Ludwig, B. Tauscher, Biocontrol Science and Technology 6 (1995) 243.
3. B. Sojka and H. Ludwig, Pharmazeutische Industrie 56 (1994) 660.
4. K.G. Werner and H. Ludwig, in Trends in High Pressure Bioscience and Biotechnology, edited by R. Hayashi.

Trends in High Pressure Bioscience and Biotechnology
R. Hayashi (editor)
© 2002 Elsevier Science B.V. All rights reserved.

High pressure experiments with the porins from the barophile Photobacterium profundum SS9

A.G.Macdonald. Department of Biomedical Sciences, University of Aberdeen, Zoology Building, Aberdeen AB24 2TZ Scotland, UK.

B.Martinac. Department of Pharmacology, Queen Elizabeth II Medical Centre, University of Western Australia, Nedlands, W.A 6907, Australia.

D.H.Bartlett.Centre for Marine Biomedicine and Biotechnology, Scripps Institution of Oceanography, University of California, San Diego, La Jolla 92093-0202, USA.

1. INTRODUCTION

Porins are ion channels in the outer membrane of Gram negative bacteria, and their incidence of opening/closing is thought to regulate the composition of the periplasm (1). In the case of *Photobacterium profundum* strain SS9, regulation of the expression of two porin genes, *ompH* and *ompL*, is determined by hydrostatic pressure. When SS9 is grown at 28MPa the porin OmpH is produced in abundance whilst, conversely, at normal atmospheric pressure OmpL predominates (2, 3). SS9 grows over the temperature range 2-20°C and the pressure range 1-80 MPa, with 28 MPa being optimal.

Although OmpH production does not confer tolerance to high pressure (mutants lacking OmpH are not particularly sensitive to pressure (2)), it is reasonable to assume that OmpH functions adequately at high pressure. Equally, there is no obvious reason why OmpL should be pressure-tolerant. Clearly a comparison of these two porins is of considerable interest.

Bacterial porins have a distinct structure (4) with which we assume OmpH and OmpL comply. The more familiar porins, such as OmpF and OmpC extracted from *E.coli*, can be reconstituted in lipid bilayers and hence studied by single channel recording techniques. Reconstitution in liposome bilayers enables patch clamp recordings to be carried out, revealing the porins' opening/closing kinetics, and other properties, in well-defined conditions. Patch clamp recording can be carried out at high pressure (5) and OmpC from E.coli has been studied at pressures up to 90 MPa (6). It was found that the main effect of pressure was to induce the open state without affecting the porins' conductance.

The susceptibility of OmpC to pressure was one factor which stimulated us to undertake the present experiments. More generally high hydrostatic pressures, in the range found in the oceans (up to 100 MPa), exert profound effects on many proteins, both structural and enzymic, and increase the order of lipid bilayers. Experiments with deep-sea animal material have produced some evidence for the adaptation of proteins and bilayers to their normal ambient high pressure (7) and deep-sea bacteria have also revealed their lipid bilayers to be adapted to offset the ordering effect of high pressure (8). Rather less is known about the effects of high pressure on ion channels in general (9) and thus far no channels (animal or bacterial) which normally function at high pressure have been studied. The availability of OmpH and OmpL has tempted us to undertake high pressure experiments using the same reconstitution procedure which proved successful with Omp C. Here we report limited success and show that conversion of the reconstitution procedure for marine species is required for future progress.

2. MATERIALS AND METHODS

2.1. Porins

Growth of *Photobacterium profundum* strain SS9 and the extraction of outer membrane proteins were as previously described (2), except that 1 or 2 liter culture volumes were used and the methodology was scaled up accordingly. The strains employed were DB110 (OmpH$^+$, OmpL$^+$), transposon Tn5 mutant DB100 (OmpH$^-$, OmpL$^+$) and transposon mini-MudI1681 mutant TW10 (OmpH$^+$, OmpL$^-$,

2.2. Reconstitution in liposome bilayers

In the case of OmpC from *E.coli* it has been shown that the porins uniformly adopt an orientation in the lipid bilayer such that the normally outer-facing end of the porin is on the outer face of the liposome (11). When a Giga-ohm seal is formed with a patch pipette and the patch excised from the liposome, it creates a patch analogous to an "inside-out" patch from a plasmamembrane. The outer end of the porin, supported in the "patched" bilayer, is bathed by the solution in the patch pipette, and the inner end of the porin is bathed by the recording bath solution. It is assumed that OmpH and L behave in the same obliging way as OmpC. The solution used for patch clamp recording OmpH and L here is one of several used with OmpC, namely, in mM, 150KCl, 0.1 KEDTA, .01 CaCl$_2$ and 5 Bis-Tris (Sigma) adjusted to pH 7.2 at 23°C (some experiments used 5mM Hepes, pH 7.2). In vivo, marine porins would be bathed with sea water on their outer face and periplasm (of uncertain composition) on their inner face. It should be emphasized that the arbitrary solution used for examining the properties of OmpH and L avoids the problematical conversion of the reconstitution procedure to sea water, see Discussion.

The method of extracting and reconstituting OmpH and L in liposomes followed that used for *E.coli* (11, 12). In brief, liposomes were made of phosphatidylcholine, extracted from soybeans, and cholesterol (10% by weight), both from Sigma. The phosphatidyl choline contained various carbon chains, C16:0, C18:0, C18:1, C18:2, C18:3 of which the last named but one contributed 60%. The main phospholipid head groups were ethanolamine, inosityl and choline. A dehydration, hydration and osmotic swelling procedure produced liposomes which form Giga-ohm seals readily with patch pipettes, and excised patches usually lasted for an hour or more at room temperature and high pressure.

2.3. Experimental design

Patches were usually subjected to a steady state applied voltage (V$_H$) of between 60 and 100 mV, pipette side first + then - , each for 20 secs, separated by 5 secs of OmV and followed by a 5 secs train of ± 5mV pulses each of 10msec duration. During this time the current flowing across the patch was recorded. These readings were repeated at 15-20 min intervals, and were intended to reveal the patch conductance at low (5mV) voltage and at higher voltage, (60-100mV), the latter liable to activate any voltage-dependent gating (Fig 1.). Standard patch clamp recording methods were used, namely an Axon 200A amplifier in conjunction with a Vetter PCM recorder, the signal being low-pass filtered at 1KHz. Off-line analysis used Dempster software (13). All the experiments were at room temperature, 23 ± 2°C. Pressure was applied stepwise to 30 MPa and then to 83 MPa, corresponding to abyssal (3000m), and trench, (8300m), depths, respectively (Fig 1).

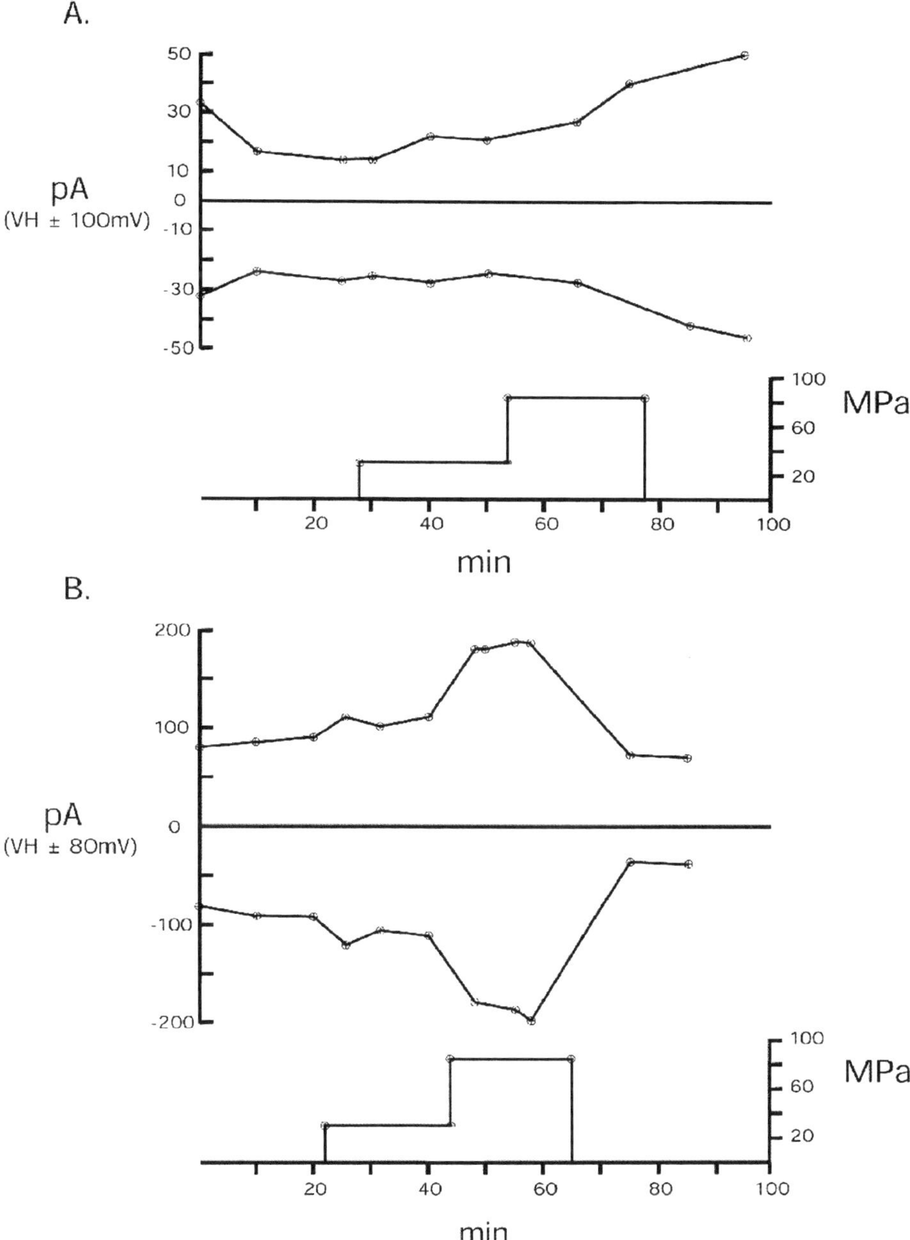

Fig.1   Effect of high hydrostatic pressure on porins from PSS9.

A.The data show the steady state current (vertical scale on left) flowing across a porin-free bilayer patch over 20 secs. at +/- 100 mV at time intervals over a total of 95 mins. The recordings were initially at normal atmospheric pressure,followed by 30 then 83 MPa,and finally at atmospheric pressure (vertical scale on right). The conductance of the patch is approximately 25 pS rising to 50 pS.

B.The steady state current flowing across a bilayer patch containing porins OmpH and OmpL from DB110. The current was recorded over 20 secs.at +/- 80 mV,initially at normal atmospheric pressure and then at high pressure, as in A above. The conductance of the patch is initially 1000 pS and peaks at 2250pS.

Three types of experiment were carried out, using excised patches. Those from liposomes (a) lacking porins and (b) supporting porins, were pressurised, as above. Patches supporting porins were also studied at room pressure, (c), ie not subjected to high pressure. As explained above, 3 sets of porins were extracted from mutants of PSS9; OmpL (from DB100), OmpH (from TW10) and a mixture of OmpL and H (from DB110, the wild type cell).

3. RESULTS

3.1. Experiments with porin-free patches

A total of 9 experiments of 30-130 mins duration, and based on 4 different batches of liposomes showed that 30 and 83 Mpa had no effect on the steady state conductance of the patch (Fig1A). Other experiments at normal atmospheric pressure also showed that the patch conductance was relatively constant. Rarely, and not obviously related to pressure, there were small steprise conductance changes, typically seen as 10pA changes in steady state current, which were also apparent in patches containing porins.

3.2. Experiments with porins at normal atmospheric pressure

Six experiments of 50-70 mins duration were carried out with Omp L. Four experiments of 65-100 mins duration were carried out with OmpH; with OmpH mixed with Omp L, from DB110, six experiments (60-164 mins duration) were carried out. The results were uniformly similar. All experiments showed a fairly constant patch conductance and none showed step changes which might mimic the effect of high pressure. A few experiments in all three groups showed occasional, brief (5-100 ms) stepwise reductions in the steady state current, typically of 5-20pA. These micro events may arise from the bilayer itself, but could also reflect changes in the conductance of the porins. Larger changes in conductance, more typical of porin gating, were only rarely recorded. An example of OmpH and L from DB110 took the form of transient decreases in conductance of 700 pS lasting 1-6 msec, occuring in bursts over several seconds at V_H-100mV. Generally the steady state conductance of the patches containing porins was consistent with permanently open porins being present.

3.3. Experiments with porins at high pressure

Five experiments with OmpL were carried out, and as shown in Table 1, only one revealed any effect of pressure on the steady state conductance of a patch. In contrast, of seven experiments with OmpH, five showed an increased patch conductance with pressure (Table 1). Of seven experiments with OmpH mixed with OmpL, four showed an increased patch conductance at high pressure.

Table 1 **Effect of high hydrostatic pressure on the steady state conductance of excised patches containing porins from Photobacterium Species strain SS9. V_H 60-100mV. (0.1 MPa is normal atmospheric pressure. Each line refers to an individual experiment).**

Conductance pS (Pressure MPa)

OmpL from DB100	600 (·1)	1600 (30)
	170 (·1)	150 (83)
	1640 (·1)	1780 (83)
	230 (·1)	214 (30)
	187 (30)	175 (83)
OmpH from TW10	162 (·1)	450 (83)
	100 (·1)	440 (83)
	383 (·1)	350 (83)
	220 (·1)	200 (30)
	5750 (30)	8750 (83)
	260 (30)	450 (83)
	200 (30)	300 (83)
OmpH and OmpL from DB110	1000 (·1)	2250 (83)
	750 (·1)	1750 (83)
	200 (·1)	300 (83)
	1330 (·1)	1283 (83)
	1366 (·1)	1333 (83)
	94 (30)	110 (83)
	82 (.1)	82 (30)

DISCUSSION

As already mentioned the porins studied here would normally function with their outer surface in contact with sea water (pH 8.0) and their inner surface bathed by periplasm. The composition of the latter in marine bacteria is not particularly well characterized but it is probably similar to sea water with a somewhat reduced pH. The ionic composition of the solution used in the present patch clamp recording experiments differed markedly from that of sea water, and its pH was 7.2, either of which might account for the lack of gating in the recordings.

Although voltage gating may not be physiological,it reveals interesting properties and may well act via the charged state of the constriction loop,L3 (4). The high conductance of liposome patches containing porins strongly suggests that a proportion of the porins are in a permanently open state. The increase in conductance on compression to 30 or 83 MPa would therefore suggest that high pressure increased that proportion. It could be argued that pressure is merely increasing the conductance of a protein-lipid or glass-lipid interface but there is no

316

theoretical or practical support for this. The data in Table 1 suggest that the porin expected to function normally at high pressure, OmpH from TW10, tends to open at high pressure whereas OmpL from DB100 does not. Future work may show this to be an adaptive property of OmpH. DB110, the wild type cell, expresses both OmpH and OmpL and it is significant that these mixed porins also confer an increased conductance of the patches at high pressure, presumably due to the presence of OmpH.

The main conclusion from this work is that OmpH and OmpL can be reconstituted in liposomes and appear to function. Differences between OmpH and OmpL are apparent, with the former tending to adopt an open state at high pressure. However if the porin reconstitution procedure can be adapted to marine bacteria and patch clamping carried out with a solution more closely resembling periplasm, much more interesting data will be obtained.

References

1. Delcour, A.H. (1997) Function and modulation of bacterial porins: insight from electrophysiology. FEMS Microbiol Letters 151, 115-123.

2. Chi, E., and Bartlett, D.H. (1993) Use of a reporter gene to follow high-pressure signal transduction in the deep-sea bacterium Photobacterium sp Strain SS9. J.Bacteriol. 175, 7533-7540.

3. Welch, T.J., and Bartlett, D.H. (1996) Isolation and characterization of the structural gene for OmpL, a pressure regulated porin-like protein from the deep sea bacterium Photobacterium species strain SS9. J.Bacteriol. 178, 5027-5031.

4. Jap, B.K., and Walian, P.J. (1996) Structure and functional mechanisms of porins. Physiol.Rev. 76, 1073-1088.

5. Macdonald, A.G. (1997) Effects of high hydrostatic pressure on the BK channel in bovine chromaffin cells. Biophys. J. 73, 1866-1873.

6. Macdonald, A.G., and Martinac, B. (1999) Effect of high hydrostatic pressure on the porin OmpC from Escherichia coli. FEMS Microbiology Letters. 173, 327-334.

7. Behan, M.K., Macdonald, A.G., Jones, G.R., and Cossins, A.R. (1992) Homeoviscous adaptation under pressure: the pressure dependence of membrane order in brain myelin membranes of deep sea fish Biochim.Biophys. Acta 1103, 317-323.

8. Yano, Y., Nakayama, A., Isbrihara, K., and Saito, H. (1998) Adaptive changes in membrane lipids of barophilic bacteria in response to changes in growth pressure. Appl.Environ.Microbiol. 64, 479-485.

9. Kendig, J.J., Grossman, Y. and Heinemann, S.H. (1993) Ion channels and nerve cell function in Effects of High Pressure on Biological Systems (Macdonald, A.G. Ed..) pp 88-124. Springer-Verlag, Heidelberg.

10. Bartlett, D.H., and Chi, E. (1994) Genetic characterization of ompH mutants in the deep-sea bacterium Photobacterium species strain SS9. Archives Microbiol 162, 323-328.

11. Delcour, A.H., Martinac, B., Adler, J. and Jung, C. (1989) Modified reconstitution method used in patch-clamp studies of Escherichia coli ion channels. Biophys.J. 56, 631-636.

12. LeDain, A.C., Häse, C.L., Tommassen, J., and Martinac, B. (1996) Porins of Escherichia coli: unidirectional gating by pressure. EMBO J. 15, 3524-3528.

13. Dempster, J. (1993) Computer analysis of electrophysiological signals. Academic Press. London.

Trends in High Pressure Bioscience and Biotechnology
R. Hayashi (editor)

317

Effect of compressed gases on the high pressure inactivation of *Lactobacillus plantarum* TMW 1.460

H. M. Ulmer[a], D. Burger[b], M. G. Gänzle[a], H. Engelhardt[b] and R. F. Vogel[a]

[a]Lehrstuhl für Technische Mikrobiologie, TU München, D-85350 Freising, Germany

[b]Instrumentelle Analytik / Umweltanalytik, Universität des Saarlandes, D-66123 Saarbrücken, Germany

The efficiency of high pressure (HP)- treatment as preservation method in foods depends on environmental conditions. The presence of gases may affect the behaviour of micro organisms during HP-treatment. These additional inactivation effects were investigated in model beer (MB) at 200 MPa and 15°C. Oxygen (O_2), nitrogen (N_2) and carbon dioxide (CO_2) were dissolved by two different methods. The effect of CO_2 was additional tested at 12 MPa under conditions where a liquid CO_2 phase was present. Gases in dissolved state had no additional inactivation effect on *L. plantarum*. In contrast, the application of liquid CO_2 induced a fast inactivation depending on the ratio of the CO_2 - aqueous phase interface to the volume of the aqueous phase. A subcritical extraction of *L. plantarum* cells with CO_2 followed by GC-analysis of the extracts demonstrated that fatty acids in form of triglycerides or phospholipids were extracted off the cellular cytoplasmic membrane. The bactericidal effect of liquid and critical CO_2 thus appears to involve the extraction of membrane compounds.

1. Introduction

The application of high pressure (HP) in the range of 200 – 800 MPa as preservation method of foods requires insight into mechanisms of HP-mediated cell injury and death. Gases present during HP-treatment may affect microbial survival based on their general effects on Eh- and pH-values or specific effects. Oxygen dissolved during HP treatment may accelerate oxidation of membrane components and thus result in sublethal injury or cell death [1]. Supercritical and subcritical CO_2 in the range of 5 – 20.5 MPa is bactericidal [2, 3, 4]. Several mechanisms were suggested to account for this bactericidal effect of CO_2 [3, 4, 5]. (i) CO_2 may diffuse into the cell and dissipate the transmembrane pH-gradient. This pH-lowering effect should act synergistic with HP-treatment. (ii) Liquid CO_2 may act as a non-polar solvent and dissolve cell compounds, especially membrane phospholipids, and thus results in membrane perturbation and cell death. It was the aim of this study to determine whether dissolved gases affect HP-mediated inactivation of *L. plantarum*, a beer spoiling bacterium, at 200 MPa in the presence of different concentrations of N_2, O_2 and CO_2. To examine whether the bactericidal effect of CO_2 requires a liquid CO_2 phase, CO_2 was applied additionally at 12 MPa and 15°C.

318

The effects on microbial inactivation as well as the ability to extract membrane compounds from *L. plantarum* were investigated.

2. Material and Methods

Effect of dissolved gases in the HP inactivation of L. plantarum. The HP-inactivation kinetics of *L. plantarum* were investigated in MB, a yeast fermented malt extract prepared according to previously published procedures [6]. The pH was adjusted to 4.0. Cells were grown at 30°C for 24h, harvested and resuspended in new MB. The HP-treatment was done at 15°C and 200 MPa. Compression and decompression rates were 200 MPa min^{-1}. The concentrations of N_2, O_2 and CO_2 dissolved in MB were adjusted by two methods. (i) Equilibration method: MB was equilibrated with a head space pressure of 0.4 MPa air, N_2 or CO_2 at 5°C for 3 d before pressure treatment. (ii) Cylinder-piston method: a gas filled head space of defined volume O_2, N_2 or CO_2 was pressed into MB during pressure treatment as shown in Fig. 1. The maximum concentration of dissolved gases with these methods was (i) Equilibrium method, 0.048 g l^{-1} O_2, 0.082 g l^{-1} N_2, and 13.5 g l^{-1}CO_2, and (ii) cylinder-piston-method 49 g l^{-1} O_2, 49 g l^{-1} N_2, 67 g l^{-1} CO_2.

Bactericidal effect of liquid CO2 on L. plantarum. The bactericidal effect of liquid CO_2 at 12 MPa and 15°C was determined with a system shown in Fig. 2. At these conditions, CO_2 remains fluid and a phase interphase between liquid CO_2 and MB is present. A portion of CO_2 is dissolved in MB and the CO_2 concentration is 69 g l^{-1} [7]. By opening the outlet valve of the gas cylinder CO_2 rises up in the pressure vessel until pressure is equilibrated to 12 MPa. Increasing the volume of MB in the reaction tube containing the sample leads to decreasing surface-volume ratios. The surface-volume ratio was varied in the range of 121 to 7770 m^{-1}.

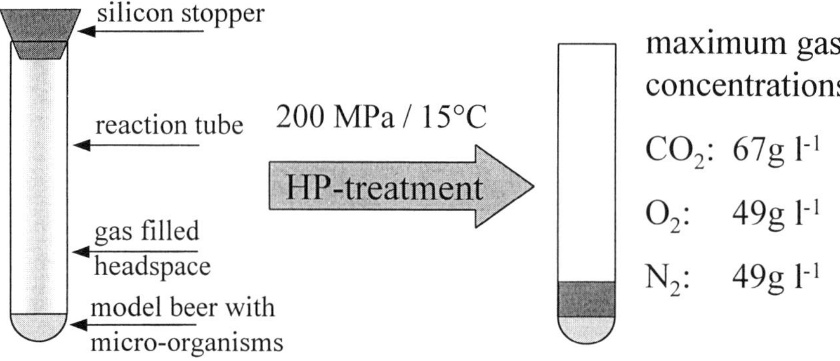

Figure 1. High pressure treatment with the cylinder-piston method

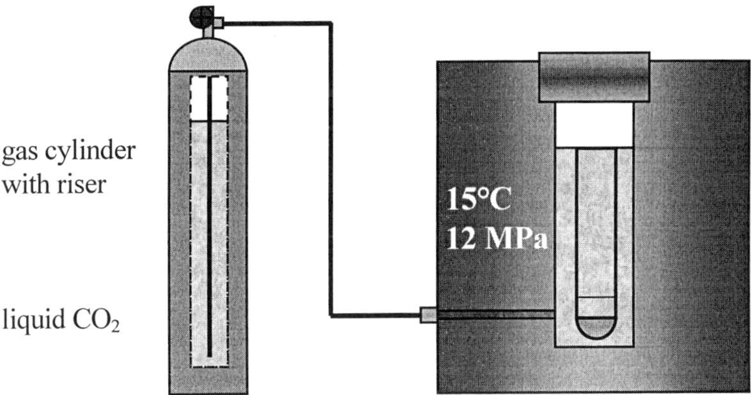

Figure 2. Apparatus used for HP-treatment of *L. plantarum* with liquid CO_2

Viable cell counts were determined by plating pressure treated samples on MRS agar, the degree of sublethal injury was estimated by plating on MRS + 4% NaCl. Membrane permeability to propidium iodide (PI) was determined by staining with the DeadLive® Kit (Molecular Probes, Eugene, USA) and reported as % membrane integrity as described previously [7].

Extraction of membrane fatty acids by liquid CO_2. The subcritical CO_2 extraction and GC-analysis of membrane compounds was executed by filling 633 mg of *L. plantarum* cells and 1 g of water in a thimble (extraction chamber: 79 mm x 11 I.D. mm). The thimble was kept at a temperature of 4°C for 12 hours. After that the extraction chamber was treated in an analytical SFE system (HP SFE Module 7680 T, HP Waldbronn) for 5 hours at 25 °C and 11.5 MPa with liquid CO_2. The extracted analytes were collected on an RP18 solid phase trap and eluted with 1 ml heptane. For the GC analysis of the extract on membrane compounds, triglycerides or phospholipids were transesterified to give the corresponding fatty acid ethyl esters (ee). This was done by refluxing 1 ml of the heptane fraction for 1 hour in 20 ml (H_2SO_4/Ethanol, 1/9, v/v) followed by three extractions with 20 ml heptane. Heptane was removed by distillation to result in a solution of 100 µl containing fatty acid ethyl esters. Analysis was performed on an DANI GC 6500 system, column: SUPELCOWAX TM 10 (30 m x 0,25 mm I.D., 0,15 µm), FID: 250 °C, injector: 250 °C, temperature program: 140 °C (1 min), 8 °C/min to 230 °C (15 min). Tricosanic acid ethyl ester (23:0) was used as internal standard. The membrane fatty acid composition of *L. plantarum* TMW 1.460 was determined by the DSMZ, Braunschweig, Germany.

3. Results

Effect of dissolved gases in the HP inactivation of L. plantarum. The inactivation of *L. plantarum* in the presence of 0.048 g l^{-1}O$_2$, 0.082 g l^{-1} N$_2$, and 13.5 g l^{-1} CO$_2$ was determined by the equilibration method.

320

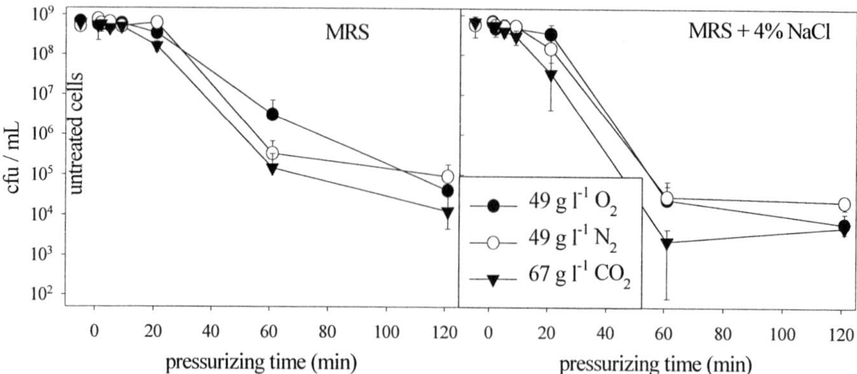

Figure 3. HP-inactivation kinetic of *L. plantarum* in presence of different gases at 200 MPa and 15°C by cylinder-piston-method.

These gas concentrations had no effect on the inactivation kinetics at 200 MPa (data not shown). The inactivation of *L. plantarum* TMW 1.460 at 200 MPa in the presence of N_2, O_2, and CO_2 in concentrations of 49, 49, and 67 g l^{-1}, respectively, is shown in Fig. 3 (cylinder-piston method). Plate counts on MRS and MRS + 4% NaCl showed the typical sigmoidal shape of microbial inactivation by HP-treatment. Curves exhibited pronounced shoulders and tailing. During the exponential inactivation phase, the cell counts on MRS + NaCl were lower by about 2 log cycles than the viable cell counts, indicating sublethal injury. Neither N_2, O_2 and CO_2 affected HP-mediated cell death or sublethal injury of *L. plantarum*. The inactivation curves presented in Fig. 3 do not differ from previously published death-time data using air saturated MB but otherwise identical conditions [6].

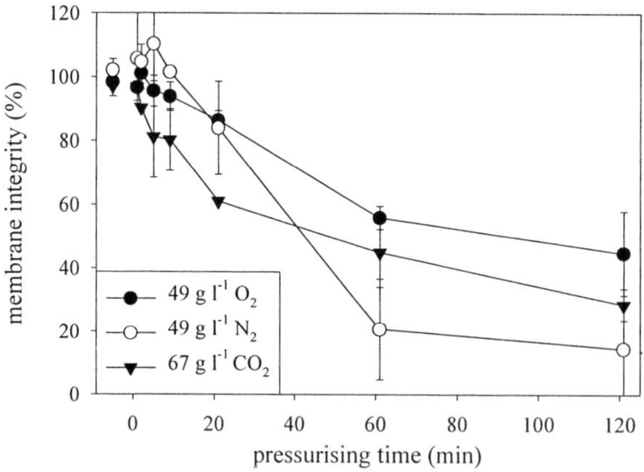

Figure 4. HP-destruction kinetic of *L. plantarum* membranes in presence of different gases at 200 MPa and 15°C by the cylinder-piston-method.

The membrane integrity was determined with PI assay (Fig. 4). The cytoplasmic membranes were disrupted by HP-treatment, however, the loss of membrane integrity was observed only after cell death. The presence of dissolved N_2, O_2, or CO_2 had no additional detrimental effect on the membrane.

Bactericidal effect of liquid CO_2 on L. plantarum. The experimental setup in these investigations ensured the presence of a liquid CO_2-phase during pressure treatment. Pressure and temperature – 12 MPa, 15°C – were chosen to obtain levels of dissolved CO_2 that are comparable to those obtained with the cylinder-piston method. In order to assess the relevance of the CO_2 – MB phase interface, experiments were performed at different surface-volume ratios of MB (Fig. 5). In contrast to the experiments with dissolved CO_2, pressurization of *L. plantarum* with liquid CO_2 had a strong bactericidal effect. The death time curves were linear and did not exhibit a shoulder or tailing because of barotolerant cells, which argues in favour for different inactivation mechanism compared to HP-inactivation at 200 MPa in the presence of dissolved CO_2 only. The bactericidal effect strongly depended on the surface to volume ratio. To visualize this relationship of bactericidal effect vs. the surface-volume ratio, the bactericidal effect was estimated by a first order kinetics (log-linear regression) and the rate constants are plotted against the surface-volume ratio in Fig. 6. The variation of the surface-volume ratio revealed that the inactivation rate of *L. plantarum* is proportional to this ratio ($r^2 = 0.996$) and thus corresponds to the amount of cells exposed to liquid CO_2 at the phase interface.

Extraction of membrane fatty acids by liquid CO_2. To determine whether cellular compounds are extracted by treatment with liquid CO_2, cultures of *L. plantarum* were subjected to a continuous flow of liquid CO_2 at 25°C and 11.5 MPa. Fatty acids present in the extract as free fatty acids, phospholipids or triglycerides were quantified with GC (Fig. 7).

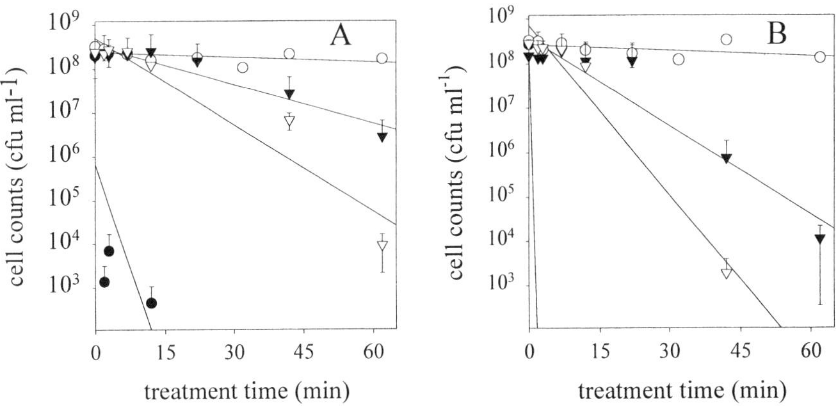

Fig. 5. Inactivation of *L. plantarum* in MB at 12 MPa, 15°C in the presence of liquid CO_2 at different surface-volume-ratios of MB. A: cell counts on MRS. B: Cell counts on MRS + NaCl. The surface to volume rations were set to 121 m^{-1} (O), 304 m^{-1} (▼), 1216 m^{-1} (▽) and 7800 m^{-1} (●). Lines represent results of log-linear regression.

322

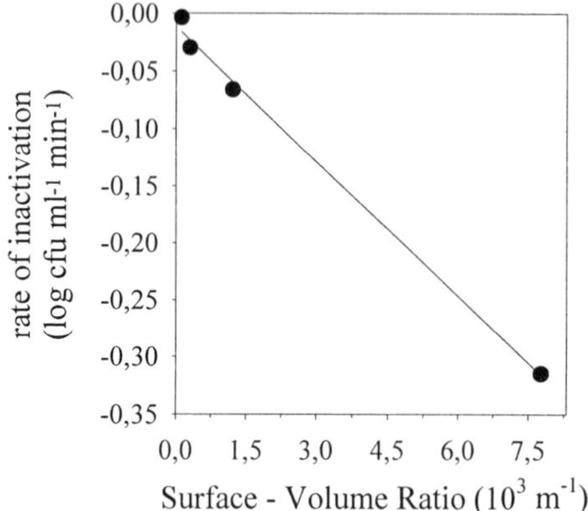

Fig. 6. Plot of the inactivation rate, estimated by log-linear regression of the death-time data on MRS against the surface – volume ratio.

This chromatogram was compared with membrane fatty acids obtained by a conventional extraction procedure (Fig. 8). The analytical chromatogram (Fig. 7) of the cell extract shows clearly that 14:0, 16:0, 18:0 and 18:1 fatty acids have been extracted. The fatty acid pattern obtained by subcritical CO_2 extraction of *L. plantarum* corresponds to the fatty acid pattern of the cytoplasmic membrane (Fig. 8). High amounts of 16:0 fatty acid were found in both cases, however, the peaks for the oleic acid (18/1 cis9) and lactobacillic acid (19/cy 9,10) ethyl esters were not resolved on our GC set up.

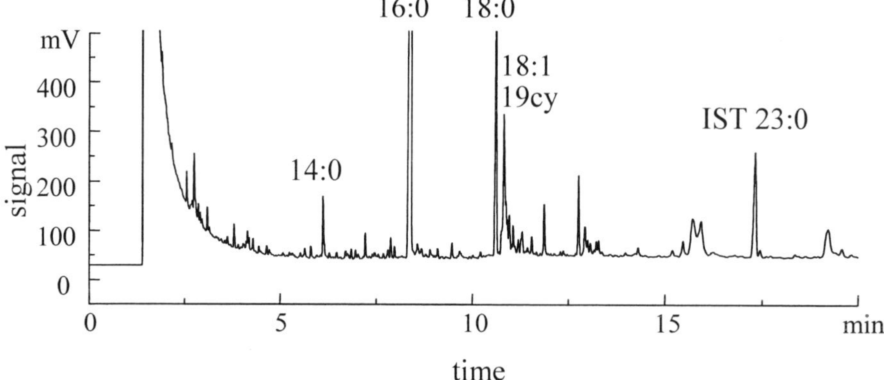

Figure 7. Chromatogram of cell extract (CO_2 extract obtained at 11.5 MPa) with 23:0 ethyl ester as internal standard.

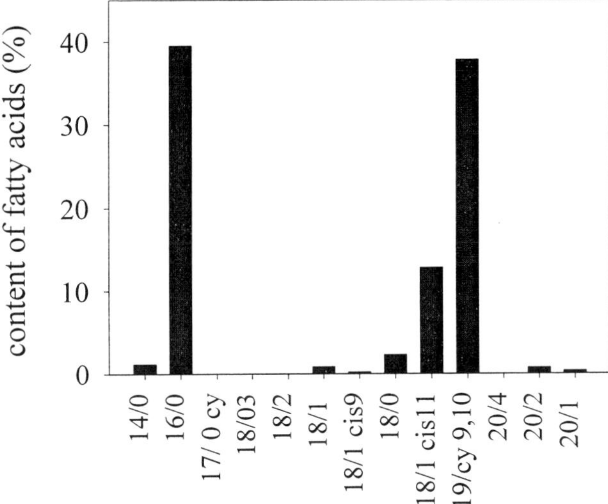

Figure 8. Percental portions of membrane fatty acid contents after conventional extraction method of *L. plantarum*.

4. Discussion

Oxygen, nitrogen and carbon dioxide dissolved in MB did not accelerate the pressure mediated inactivation and inactivation of *L. plantarum*. Even at the high concentrations reached with the cylinder-piston-method, these gases showed no additional inactivation or membrane disrupting effect.

The oxidation of membrane phospholipids was reported to contribute to cell death during drying of lactic starter cultures [1]. However, O_2 present during HP treatment did not oxidise membrane phospholipids in a way accelerating cell death.

In the presence of excess dissolved gases, gas bubbles formed during explosive decompression may mechanically disrupt cells. However, compression and decompressing rates of 200 MPa min^{-1} as used in our work maintained the system in thermo dynamical equilibrium, allowing dissolved gases to diffuse out off the cells before gas bubbles burst the cells.

Dissolved CO_2 may act bactericidal by its pH-lowering effect. At pH-values around the first pK_a of CO_2 (6.35) inside the cell, it lowers the intracellular pH by about 0.3 pH units [8]. It was furthermore suggested that carbonic acid penetrates membranes and dissipates transmembrane pH gradients [4, 9]. In our work we did not observe a synergistic effect of dissolved CO_2 with HP inactivation of *L. plantarum*. The inactivation kinetics with HP-treatment and gases did not differ from previous published data without CO_2 [6]. It must be noticed that the pH of MB used in our work (pH 4.0) is well removed from the pK_a of carbonic acid, thus excluding an effect on the medium pH.

The results on the effect of liquid CO_2 conform with previously published data on bacterial inactivation by subcritical or supercritical CO_2 [9, 4, 5]. These authors used an experimental setup allowing for a phase interface of aqueous suspension of micro organisms and liquid or supercritical CO_2 as it was used in this work. A proportional relationship of the inactivation rate of *L. plantarum* to the surface-volume ratio was found. CO_2 thus appears to inactivate *L. plantarum* mainly at the interface of aqueous phase to liquid CO_2 phase. This suggests a cell inactivation mechanism involving membrane disturbance by liquid CO_2 rather than a bactericidal effect of dissolved CO_2. The subcritical CO_2 extraction of *L. plantarum* demonstrated that fatty acid esters can be extracted from the culture. Fatty acids extracted with CO_2 included mainly those fatty acids predominating in the membrane phospholipids of *L plantarum*. It can thus be concluded that membrane phospholipids are extracted off the membrane causing cell death. Based on the amount of free fatty acids in the extracts, it can be estimated that about 10% of the membrane fatty acids have been extracted by the liquid CO_2. This organism is readily inactivated by the application of liquid CO_2, however, the applicability of this preservation method in food technology is limited because it may result in co-extraction of components relevant for food quality.

5. Acknowledgement

Thanks are due to Mr. Luciano Horn and Mr. Carsten Hinrichs for their excellent collaboration in microbial laboratory, Dr. Müller and Mr. Piest, Linde Gase, for providing data on phase behaviour and solubility of CO_2, and Mr. Matthias Jochum for his support concerning the GC measurements.

References

[1] H. P. Castro, P. M. Teixeira and R. Kirby, J. Appl. Microbiol. 82 (1997) 87-94.

[2] E. Debs-Louka, N. Louka, G. Abraham, V. Chabot and K. Allaf, Appl. Environ. Microbiol. 65 (1999) 626-631.

[3] S.-I. Hong and Y.-R. Pyun, J. Food Sci. 64 (1999) 728-733.

[4] A. K. Dillow, F. Dehghani, J. S. Hrkach, N. R. Foster and R. Langer, Proc. Natl. Acad. Sci. 96 (1999) 10344-10348.

[5] S.-I. Hong and Y.-R. Pyun, Int. J. Food Microbiol. 63 (2001) 19-28.

[6] H. M. Ulmer, M. G. Gänzle and R. F. Vogel, Appl. Environ. Microbiol. 66 (2000) 3966-3973.

[7] Linde Gase, Germany, Technical Manual

[8] F. Abe and K. Horikoshi, Extremophiles 2 (1998) 223-228.

[9] S.-I. Hong, W.-S. Park and Y.-R. Pyun, J. Food Sci. Technol. 34 (1999) 125-130.

Trends in High Pressure Bioscience and Biotechnology
R. Hayashi (editor)
© 2002 Elsevier Science B.V. All rights reserved.

Thermotolerance and barotolerance of alcohol-shocked yeast

Kazuhiro Hisada, Yoshihisa Suzuki, and Katsuhiro Tamura

Department of Chemical Science and Technology, Faculty of Engineering, The University of Tokushima,
Minamijosanjima-cho, Tokushima 770-8506, Japan

The acquisition of stress tolerance in response to a preconditioning alcohol shock treatment at various concentrations was studied using eight kinds of alcohols in the yeast *Saccharomyces cerevisiae*. Alcohol shock treatment induced tolerance against stresses such as high temperature and high pressure in yeast. The optimum alcohol shock concentration that induces maximal tolerance against these stresses was dependent on the nature of alcohols. The optimum concentration was inversely proportional to the number of carbon atom in the alcohol molecules. Moreover, the position of the hydroxyl group in the alcohol molecules affected the degree of stress tolerance. Alcohol-shocked yeast also acquired barotolerance. These results show that the acquisition of stress tolerance is closely related with alcohol preconditioning and the nature of alcohols.

Microcalorimetric measurement was done to estimate the strength of alcohols as stress substances or inhibitors for the growth of the yeast. From the data on the growth thermograms of yeast, the minimum inhibitory concentration (MIC) of alcohols was determined and correlated with the optimum alcohol concentration for alcohol shock.

1. INTRODUCTION

The cells from microbes synthesize heat shock proteins (hsps) in response to various stresses such as heat [1-3], pressure [4] and alcohols [5, 6], and acquire resistance to various subsequent stresses that would normally be lethal [5]. For example, yeast cells exposed to ethanol up to 1.75 mol/l^{-1} immediately response by induction of hsps, with increase in tolerance against lethal stress. It is known that in the yeast *Saccharomyces cerevisiae* cells exposed to a temperature of about 40°C, the induction of hsp104 often contributes to the acquisition of thermotolerance of them. We have studied the stress tolerance (thermotolerance and barotolerance) of alcohol-shocked yeast using a series of alcohols (methanol, ethanol, 1-propanol, 2-propanol, 1-butanol, 2-butanol, *iso*-butanol, *tert*-butanol) to get more information about chemical shock effects on yeast cells. The results obtained on alcohol-shocked yeast were compared with those on the grow thermograms of yeast in the medium containing alcohols.

2. MATERIALS AND METHODS

2.1. Growth conditions and induction of stress tolerance by alcohol shock

The yeast *Saccharomyces cerevisiae* IFO 10149 was incubated at 30℃ for 2 days in YPD medium. The yeast cells (1×10^8 cells/ml) were diluted to 20 times with fresh YPD medium and then incubated for 4 hours at 30℃. These cultures in the logarithmic phase of growth were used, because thermotolerance depends on the stage of growth or cell cycle. Logarithmic phase yeast cells were resuspended in fresh YPD medium containing various concentrations of alcohols. The yeast cells were exposed to the alcohols for 10 min at 30℃, and then subjected to high temperature (51℃) for 10 min and high hydrostatic pressure (150 MPa) for 60 min.

2.2. Effects of alcohol on the growth thermograms

Alcohols of various concentration were added to YPD medium including yeast cells (culture density 2.0×10^4 cells/ml), and the 5 ml of yeast suspensions in 50 ml vials was set in a multiplex batch isothermal calorimeter. The calorimeter was employed to monitor the growth activity of yeast. Yeast suspensions in 25 vials were incubated at 30℃ until the observed thermograms returned to the base line. It took 2 - 6 days.

3. RESULTS AND DISCUSSION

3.1. Stress tolerance of alcohol-shocked yeast cells

Plesset *et al* have reported that ethanol is an inducer of heat shock proteins and thermotolerance, and 1.55 mol/l^{-1} is the most effective [5]. Generally alcohols with long hydrocarbon chains have stronger sterilizing power, this means that these alcohols act as stressors against yeast cells. So, we investigated the effect of alcohol as an inducer of stress tolerance on the viability of yeast using alcohol homologs.

Figures 1 and 2 show thermotolerance and barotolerance of alcohol-shocked yeast at 30°C. Yeast cells acquired resistance against both high temperature and high pressure after the shock treatment. This indicates that the addition of alcohol to the cultures contributes to the acquisition of tolerance. The degree of stress tolerance was different by the nature of alcohols used.

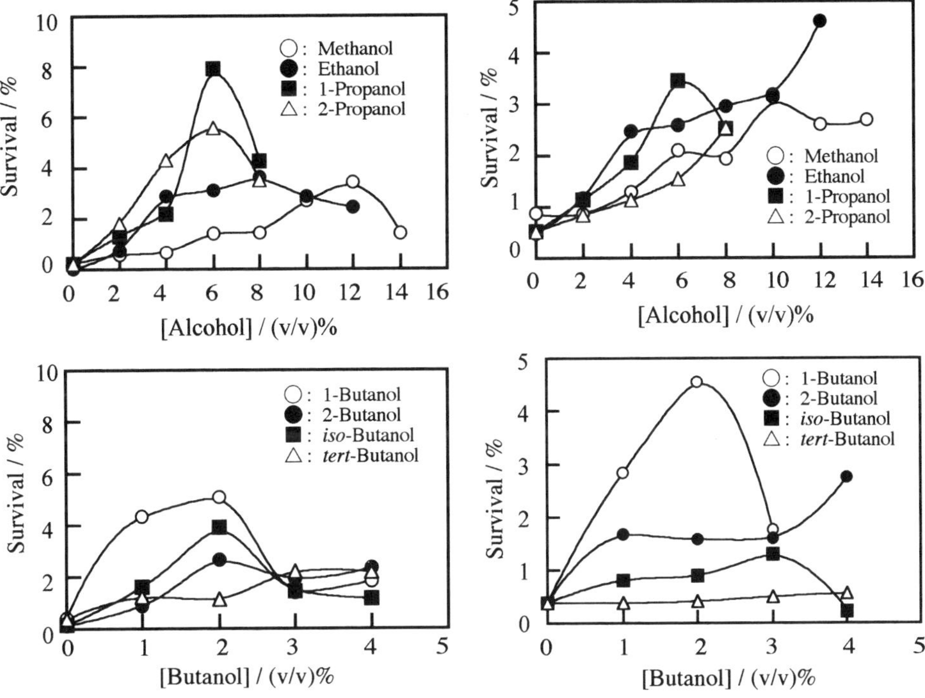

Fig.1 Thermotolerance (51°C, 10 min) of alcohol-shocked yeast.

Fig.2 Barotolerance (150 MPa, 1 h) of alcohol-shocked yeast.

Table 1 Concentration of alcohols inducing maximum thermotolerance (51°C, 10 min) and barotolerance (150 MPa, 1 h)

Alcohols	Number of carbon	Concentration of alcohols (v/v %)	
		Thermotolerance	Barotolerance
Metanol	1	12	10
Etanol	2	8	12
1-Propanol	3	6	6
2-Propanol	3	6	8
1-Butanol	4	2	2
2-Butanol	4	2	4
iso-Butanol	4	2	3
tert-Butanol	4	3	4

The optimum concentrations of alcohols that induce maximal tolerance against these stresses were inversely proportional to the number of carbon atoms in the *n*-alcohol molecules (Table 1).

3.2. Effects of alcohol on yeast metabolic process

Figure 3 shows some examples of the growth thermograms obtained during incubation of yeast cultures in multiplex batch isothermal calorimeter (Biothermo Analyzer). The pattern of the thermograms depends on the alcohol concentration. As a general characteristic, the growth thermograms significantly broadened with increasing the amount of alcohols in the medium and the peak shifted to longer incubation times.

The method to determine the minimum inhibitory concentration (MIC) of alcohol has already described elsewhere [7]. All of the values of MIC determined by two methods are summarized in Table 2. The action mechanism of the antimicrobial activity of alcohols is assumed to be based on the chemical structures of alcohols used. To characterize them more quantitatively the MIC values were plotted against the number of carbon atoms of *n*-alcohol molecules in Fig.4. It is obvious that the larger the number of carbon

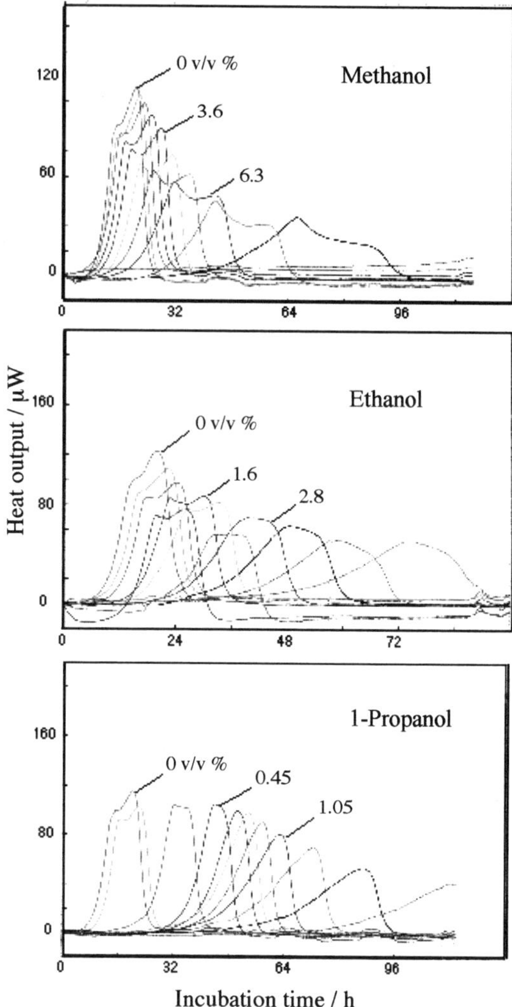

Fig.3 Effect of alcohol on the growth thermograms of yeast at atmospheric pressure and 30°C.

Table 2 Inhibitory parameters determined by a multiplex batch isothermal calorimeter for action of alcohols at 30℃

Alcohols	Number of carbon	determined from μ_i/μ_m		determined from $t_a(0)/t_a(i)$	
		K_μ (mol/l)	MIC_μ (mol/l)	K_θ (mol/l)	MIC_θ (mol/l)
Methanol	1	1.71	2.83	1.29	2.46
Ethanol	2	0.404	0.638	0.239	0.778
1-Propanol	3	0.0712	0.301	0.0363	0.190
2-Propanol	3	0.775	1.53	0.279	0.352
1-Butanol	4	0.0478	0.105	0.0598	0.118
2-Butanol	4	0.0714	0.112	0.0484	0.121
iso-Butanol	4	0.0562	0.110	0.0714	0.164
tert-Butanol	4	0.0857	1.12	0.0828	0.824
1-Pentanol	5	0.0301	0.0884	0.0252	0.0385
1-Hexanol	6	0.00449	0.0121	0.00699	0.0982

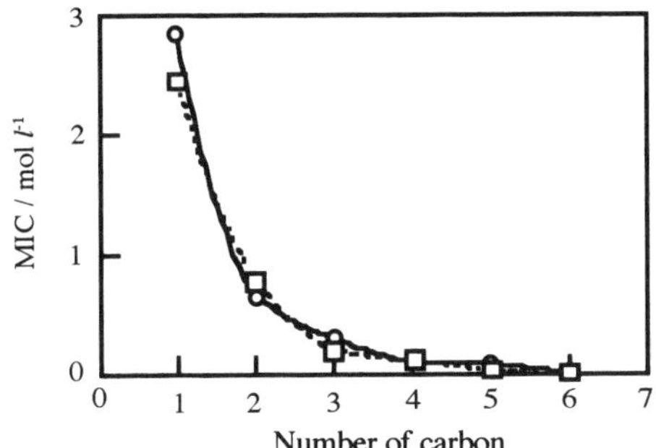

Fig.4 Values of MIC_μ (○ , solid line) and MIC_θ (□ , broken line) for *Saccharomyces cerevisiae* plotted against the number of carbon atoms of *n*-alcohols.

atoms of the alcohols, the stronger their antimicrobial effects. Comparison of the MIC of isomers of propanol (1-propanol and 2-propanol) and butanol (1-butanol and 2-butanol, *iso*-butanol, *tert*-butanol) given in Table 2 suggest that their antimicrobial effects are dependent on not only the length of alkyl chains but also the position of hydroxyl and alkyl groups.

REFERENCES

1. K. Watson and R. Cavicchioli, Biotech. Lett., 5 (1983) 683.
2. Y. Komatsu, S. C. Kaul, H. Iwahashi, and K. Obuchi, FEMS Microbiol. Lett., 72 (1990) 159.
3. H. Iwahashi, S. C. Kaul, K. Obuchi, and Y. Komatsu, FEMS Microbiol. Lett., 80 (1991) 325.
4. K. Tamura, M. Miyashita, and H. Iwahashi, Biotech. Lett., 20 (1998) 1167.
5. J. Plesset, C. Palm, and C. S. McLaughlin, Biochem. Biophys. Res. Commun., 108 (1982) 1340.
6. K. Tanji, T. Mizushima, S. Natori, and K. Sekimizu, Biochim. Biophys. Acta, 1129 (1992) 172.
7. T. Arao and K. Tamura, Prog. Aneth. Mechanism, 6 (2000) 496.

Effects of saccharide in medium on stress tolerance of yeast

Toshiaki Arao, Yoshihisa Suzuki, and Katsuhiro Tamura

Department of Chemical Science and Technology, Faculty of Engineering, The University of Tokushima,
Minamijosanjima-cho, Tokushima 770-8506, Japan

Effects of saccharides (glucose, fructose, galactose, mannose, lactose, maltose, sucrose, trehalose) in medium on stress tolerance of yeast (*Saccharomyces cerevisiae*) were studied by colony counting method. The protection effect of various saccharides for yeast cells against high pressure and high temperature was estimated with survivals. Saccharides at the concentration of 0.5 mol/l or 1.0 mol/l added to YPD culture medium apparently increased the survivals of the yeast during subjected high pressure and high temperature. Barotolerance (150 MPa, 1 hour) of the yeast cell increased with increasing concentration of saccharides. Trehalose induced maximum barotolerance to the yeast cells and disaccharides endowed the yeast cells with higher tolerance compared with monosaccharides. Thermotolerance (51℃, 10 min) also showed a similar pattern of survival vs. saccharide concentration to barotolerance. The protection effects of saccharides against high pressure and high temperature were explained by the mean number of equatorial OH groups in saccharide molecule. Stress tolerance of pressure-shocked yeast in the medium containing saccharides was also estimated.

1. INTRODUCTION

The organisms cells from microorganisms to mammals subjected to several stresses have common mechanisms to protect their life from lethal environment, such as high temperature [1-3], high hydrostatic pressure [4], ultraviolet, and chemicals [5, 6]. Yeast cells exposed to high pressure of 20 - 80 MPa, immediately response and induce a small set of proteins called stress proteins. Furthermore, yeast cells exposed to slightly higher temperatures than an optimum growth temperature, induce heat shock proteins and trehalose.

Trehalose was generally thought to be accumulated as energy source of cells, since trehalose was not found in logarithmic phase cells and was found in stationary phase cells. However, trehalose is recently regarded as a protection substance for living cells, because the amount of trehalose was closely related with stress tolerance.

It is also known that several saccharides are significantly effective in providing protection against hydrostatic pressure damage in cells [7]. Similarly, saccharides confer protection against dry stress [8]. The phenomena against dry stress were explained by the OH groups in saccharide molecule which serve as substitutes of water molecules [9, 10]. Namely the saccharide prevents yeast cells from dry stress disrupting components of cells such as protein [11] and biomembrane [12, 13]. But the mechanisms for the acquisition of stress tolerance against high pressure stress or high temperature stress was not well known. Recently, it is proposed that protection effects of saccharides are closely related to the mean number of equatorial OH groups in saccharide molecule, which significantly improve the structurization of water [14-17]. Furthermore, at 4°C the mean number of equatorial OH groups is proportional to the ability that saccharides prevent protein from denaturation induced by pressure stress or thermal stress [7]. However, we can not conclude whether yeast cells acquired high-pressure and high-temperature stress tolerance under these conditions, since 4°C is very low from optimum growth temperatures for yeast.

In this study, eight kinds of saccharides (glucose, fructose, galactose, mannose, lactose, maltose, sucrose, trehalose) were used to estimate the protection effect of the saccharides for yeast in further detail. We have already found that pressure shock treatment induces shock proteins only and not trehalose [18], so, studied whether the added saccharides in the medium affect the thermotolerance and barotolerance of pressure-shocked yeast.

2. MATERIALS AND METHODS

2.1. Yeast strains and growth conditions

The yeast *Saccharomyces cerevisiae* IFO10149 was incubated at 30°C for 2 days in YPD medium containing (g/l) glucose, 20; polypeptone, 20; yeast extract, 10. The yeast cells (1 $\times$ 10^8 cells/ml) were diluted to 20 times with fresh YPD medium and then incubated for 4 hours at 30°C. These cultures in the logarithmic phase of growth were used, since stress tolerance varies according to the stage of growth or cell cycle.

2.2. Effects of saccharide in medium on stress tolerance of yeast

Saccharides (glucose, fructose, galactose, mannose, lactose. maltose, sucrose, trehalose) at the concentration of 0.5 mol/l and 1.0 mol/l were added to the suspension of logarithmic phase yeast cells and a part of the suspension was used as control. Then, the cells suspended in fresh (YPD + saccharide) medium were exposed to lethal environments (high temperature: 51°C, 10 min and high hydrostatic pressure: 150 MPa, 1h). For the experiment of pressure-shocked yeast, logarithmic phase cells suspended in fresh (YPD + saccharide) medium were incubated for 1 hour at 30°C under high pressure (pressure-shock treatment) and then exposed to lethal environments. Survival was determined by standard dilution plate counts on YM nutrient agar medium containing (g/l) glucose, 20; agar, 20; polypeptone, 5; yeast extract, 3; malt extract, 3.

Colony forming units were counted after incubation for 3 days at 30°C. All experiments were carried out at least three times and mean value was calculated.

3. RESULTS AND DISCUSSION

3.1. Toxic effect of saccharide on yeast cells

At first, we estimated the toxic effects of eight saccharides on the yeast cells in the (YPD+saccharide) suspension. As shown in Table 1, it is obvious that saccharides have toxic effects on yeast cells. The effects at the concentration of 1.0 mol/*l* was greater than those of 0.5 mol/*l* and disaccharides were more effective than monosaccharides. Concerning exposed time, the effects of 2.5 hours were slightly greater than those of 1.5 hours.

3.2 Protection effects of saccharides on pressure and heat stresses

In order to confirm the contribution of the saccharide to gain barotolerance and thermotolerance of yeast, we estimated the change of these tolerances due to the addition of saccharides. As an index of each tolerance, the survival of yeast under stressed conditions was used. Figure 1 shows

Table 1. Toxic effects of saccharide on the yeast cells in the (YPD+saccharide) medium

	Saccharides	Concentration	Survivals [a]	
			1.5 hours	2.5 hours
	Without saccharides (control)		100%	100%
Monosaccharide	Glucose	0.5 mol/*l*	77%	74%
		1.0 mol/*l*	49%	37%
	Fructose	0.5 mol/*l*	84%	80%
		1.0 mol/*l*	43%	46%
	Galactose	0.5 mol/*l*	73%	74%
		1.0 mol/*l*	52%	39%
	Mannose	0.5 mol/*l*	87%	80%
		1.0 mol/*l*	48%	39%
Disaccharide	Lactose	0.5 mol/*l*	73%	67%
	Maltose	0.5 mol/*l*	75%	62%
		1.0 mol/*l*	39%	25%
	Sucrose	0.5 mol/*l*	74%	77%
		1.0 mol/*l*	29%	21%
	Trehalose	0.5 mol/*l*	88%	73%
		1.0 mol/*l*	34%	31%

a) The yeast cells were exposed to monosaccharides and disaccharides in the medium for 1.5 or 2.5 hours at 30°C .

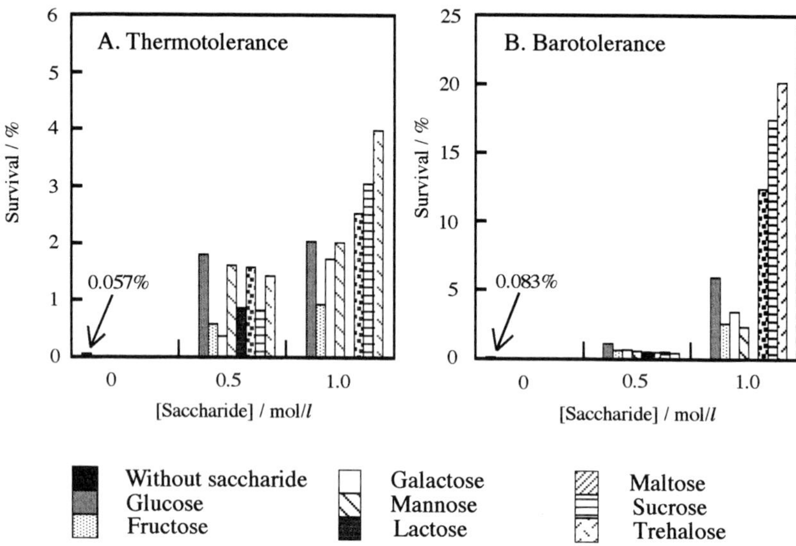

Fig.1 Thermotolerance (51°C, 10 min) (A) and Barotolerance (150 MPa, 1 h) (B) of yeast in medium containing monosaccharides and disaccharides.

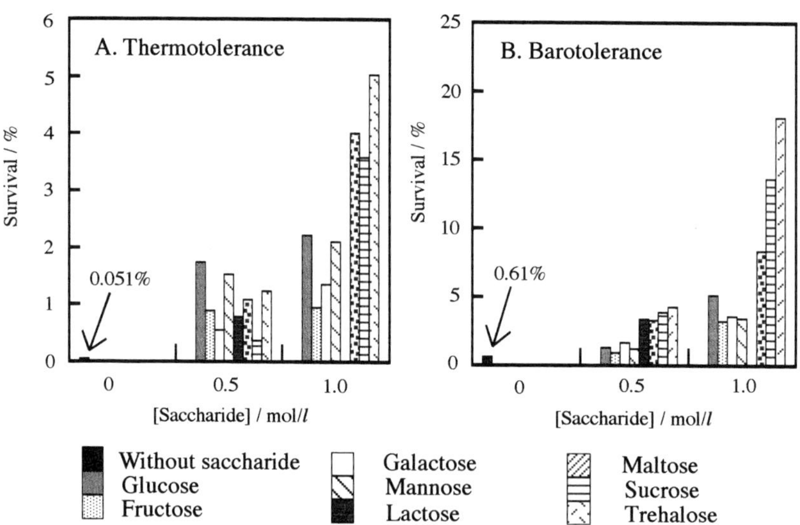

Fig.2 Thermotolerance (51°C, 10 min) (A) and Barotolerance (150 MPa, 1 h) (B) of pressure-shocked (50 MPa, 1h) yeast in medium containing monosaccharides and disaccharides.

that thermotolerance (51°C, 10 min) and barotolerance (150 MPa, 1 h) of yeast in medium containing four kinds of monosaccharides and four kinds of disaccharides. For barotolerance, the survival after the pressure stress of 150 MPa for 1 hour without saccharide (YPD only) was very low (0.08%), and the tolerance of the yeast increased with increasing concentration of saccharides. Added saccharides at the concentration of 0.5 and 1.0 mol/l were apparently effective in the survival and the protection against high pressure stress. These results also indicate that trehalose induced maximum barotolerance to the yeast cells and disaccharides confered larger tolerance compared with monosaccharides on yeast. Thermotolerance (51°C, 10 min) also showed a similar pattern of survival vs. saccharide concentration to barotolerance.

In Fig. 3, the survival of yeast was plotted against mean number of equatorial OH groups in saccharide molecules. The mean number of equatorial OH groups of each saccharides was 4.6 (glucose), 3.0 (fructose), 3.3 (galactose), 3.6 (mannose), 6.3 (sucrose), 7.2 (maltose), 8.0 (trehalose) [15,16,19]. The results show that both barotolerance (R=0.926) and thermotolerance (R=0.943) were proportional to the mean number of equatorial OH groups. The equatorial OH groups in saccharides play an important role in the acquisition of both thermotolerance and barotolerance. It has already been reported that the equatorial OH groups of saccharides significantly increase the ability of improving the structurization of water [14-16]. Therefore, the structurization of water seems to be related to the protection of biomembrane and proteins in yeast against heat and high-pressure stresses.

Subsequently, similar experiments were done using pressure-shocked yeast cells (50 MP, 1 h) for thrmotolerance and barotolerance. In Fig. 2, very similart results to untreated yeast were obtained with pressure-shocked yeast. Concerning YPD medium (without added saccharide),

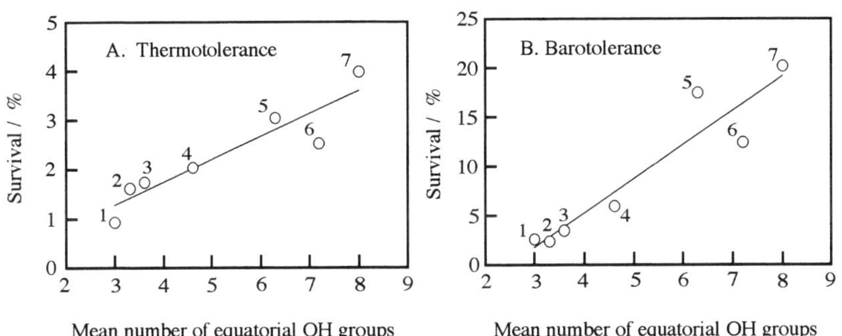

Fig.3 Effects of mean number of equatorial OH groups in saccharide molecules in the medium on the survival of yeast (A. Thermotolerance, B. Barotolerance). The concentration of added saccharides is 1.0 mol/l.

Monosaccharide : 1. Fructose, 2. Galactose, 3. Mannose, 4. Glucose

Disaccharide : 5. Sucrose, 6. Maltose, 7. Trehalose.

the thermotolerance was 0.057% for untreated yeast, and 0.051% for pressure-shocked yeast, while the barotolerance was 0.083% for untreated yeast, and 0.61% for pressure-shocked yeast. The barotolerance was about 8 times higher than that of untreated yeast. On the other hand, the thermotolerance did not change by pressure shock treatment. It is known that pressure shock treatment for yeast cell induces shock proteins, but not trehalose [18]. Accordingly, these results suggest that pressure shock proteins were effective for the barotolerance of yeast.

REFERENCES

1. K. Watoson and R. Cavicchioli, Biotech. Lett., 5 (1983) 683.
2. Y. Komatsu, S. C Kaul, H. Iwahashi, and K. Obuchi, FEMS Microbiol. Lett., 72 (1990) 159.
3. H. Iwahashi, S. C. Kaul, K. Obuchi, and Y. Komatsu, FEMS Microbiol. Lett., 80 (1991) 325.
4. K. Tamura, M. Miyashita, and H. Iwahashi, Biotech. Lett., 20 (1998) 1167.
5. J. Plesset, C. Palm, and C. S. McLaughiln, Biochem. Biophys. Res. Commun., 108 (1982) 1340.
6. K. Tanji, T. Mizushima, S. Natori, and K. Shimizu, Biochem. Biophys. Acta, 1129 (1992) 172.
7. S. Fujii, K. Obuchi, H. Iwahashi, T. Fujii, and Y. Komatsu, Biosci. Biotech. Biochem., 60 (1996) 476.
8. J. H. Crowe, J. F. Carpenter, L. M. Crowe, and T. J. Anchordoguy, Cryobiology, 27 (1990) 219.
9. J. H. Crowe, L. M. Crowe, J. F. Carpenter, and C. A. Wistrom, Biochem. J., 242 (1987) 1.
10. J. H. Crowe, L. M. Crowe, J. F. Carpenter, A. S. Rudolph, C. A. Wistorm, B. J. Spargo, and T. J. Anchordoguy, Biochem. Biophys. Acta, 947 (1988) 367.
11. T. Hottiger, M. S. Webb, G. Bryant, and D. V. Lynch, Biochem. Biophys. Acta, 1193 (1994) 143.
12. J. H. Crowe, L. M. Crowe, and D. Chapmann, Science, 223 (1984) 701.
13. S. B. Lesiel, S. A. Teter, L. M. Crowe, and J. H. Crowe, Biochem. Biophys. Acta, 1192 (1994) 7.
14. H. Uedaira and H. Uedaira, Bill. Chem. Soc. Jpn., 53 (1980) 2451.
15. H. Uedaira and H. Uedaira, J. Sol. Chem., 14 (1985) 27.
16. H. Uedaira, M. Ishimura, S. Tsuda, and H. Uedaira, Bull. Chem Soc. Jpn., 63 (1990) 3376.
17. H. Uedaira, M. Ikura, and H. Uedaira, Bull. Chem. Soc. Jpn., 62 (1989)1.
18. M. Miyashita, K. Tamura, and H. Iwahashi, Progress in Biotechnology Vol.13, "High Pressure Bioscience and Biotechnology", R. Hayashi and C. Balny Eds, Elsevier, (1996) p.101.
19. H. Kawai, M. Sakurai, Y. Inoue, R. Chujo, and S. Kobayashi, Cryobiology, 27 (1992) 219.

Trends in High Pressure Bioscience and Biotechnology
R. Hayashi (editor)

A comparative electron microscopic study of cell growth and ultrastructure from a regular and a HP-changed type of *Bacillus thuringiensis ssp. israelensis*

K. G. Werner and H. Ludwig

Institut für Pharmazeutische Technologie und Biopharmazie, Universität Heidelberg, INF 366, 69120 Heidelberg, Germany

Spores of *Bacillus thuringiensis ssp. israelensis* in isotonic Sodiumchloride-solution were treated with high pressure. As an effect of this treatment a changed type of microorganisms appeared.
The bacteria grown from the original and the high pressure changed spores were compared in the electron microscope.
The results show the taxonomic identity in terms of both species as well as the changes within the cells caused by high pressure.

1. INTRODUCTION

During the last years *Bacillus thuringiensis* gained importance as a biological insecticide. The primary insecticidal activity of *Bacillus thuringiensis* resides in a glycoprotein crystal that is synthesized and crystallised within the parent cell during sporulation [1]. Therefore it is not necessary and also not desirable to spread the living microorganisms into the environment. High pressure treatment was investigated concerning its capability in killing the living cells by simultaneously preserving the pesticidal crystal protein [2].

Spores of *Bacillus thuringiensis ssp. israelensis* were treated with high pressure. As an effect of this treatment a changed type of microorganisms appeared. The aim of this investigation was to find out if the new type of bacteria were also *Bacillus thuringiensis ssp. israelensis* and if so, whether the changes in growth and habitus were the result of the high pressure treatment.

For that an overnight culture of both strains, changed and unchanged, were investigated by electron microscopy, first to prove the identity of both microorganisms and second to investigate the changes in ultrastructure caused by high pressure treatment.

The authors gratefully acknowledge the kind support of Mrs. Gorgas, Institut für Pathologie, Universität Heidelberg and Ms. Schattmann.

338

2. MATERIALS AND METHODS

2.1. Organisms

The starting strain for the study was isolated from the commercial product "VectoBac 12 AS", an insecticidal spore suspension from Abbott Laboratories, Chemical and Agricultural Division, North Chicago, IL, 60064, USA.

For proofing the results a specified strain of *Bacillus thuringiensis ssp. israelensis* was provided by the Deutsche Sammlung für Mikroorganismen und Zellkulturen (DSMZ-Nr.5724).

2.2. Medium and cultural conditions

Starting from one single colony of microorganisms an overnight culture was grown in fluid standard nutrient medium under permanent stiring at 30 °C. Subsequently 500 µl of the overnight culture were put on each of 80 standard nutrient agar plates. After 14 days of incubation at 30 °C all plates were scraped off and a spore suspension was prepared as described in [3].

2.3. Preparation of the changed species

To obtain the changed species of *Bacillus thuringiensis*, 2 ml of the described spore suspension were pressurized in polyethylene tubes at 70 °C and 300 MPa for 180 min. Afterwards 100 µl were plated on the standard nutrient agar and incubated for 2 days at 30 °C. Finally starting from one single colony of HP-changed bacteria a spore suspension was prepared as described above.

2.4. Electron microscopy

To carry out the electron microscopy an overnight culture of both strains were made. Subsequently 100 µl were plated on standard nutrient agar and incubated at 30 °C for 3 days. With an inoculating loop one colony was taken into 1 ml 0,9 % NaCl solution. After centrifugation at 2000 g the bacteria were fixed with 2 % Diglutaraldehyde solution and washed 4 times with phosphate buffer. The contrast was generated with 1 % Osmiumtetroxide in phosphate buffer. Afterwards the samples were drained with aceton in increasing concentrations. The embedding in synthetic resin and the preparation of the ultrathin cuttings were carried out as described before [4, 5].

3. RESULTS AND DISCUSSION

When we started our pressure experiments with *Bacillus thuringiensis* we realized that depending on the experimental conditions two different types of colonies appeared. As you can see in Figure 2 this new type was significant different in growth and habitus from the known strain, shown in Figure 1.

So we decided to take a look at the ultrastructure of both strains to find out if the macroscopic changes depended on visible alterations in the bacteria.

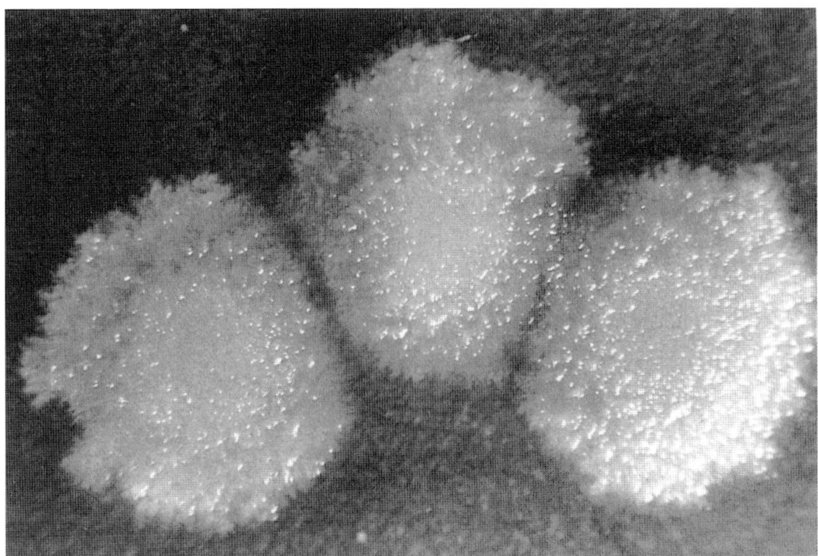

Figure 1

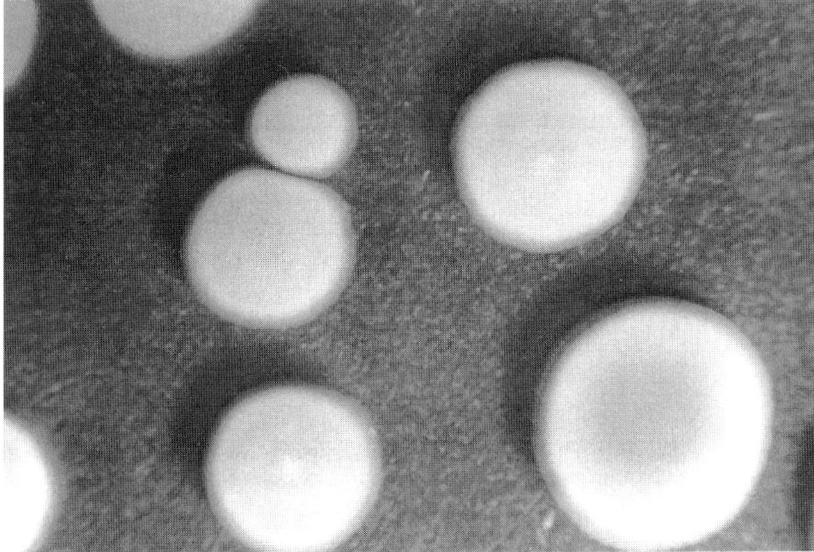

Figure 2

The first recognizable difference between both strains is the size. Both strains are gram-positive, spore-forming, rod-shaped bacteria but as you can see in Figure 4 the modified species is on average shorter and thicker than the unmodified *Bacillus thuringiensis* shown in Figure 3.

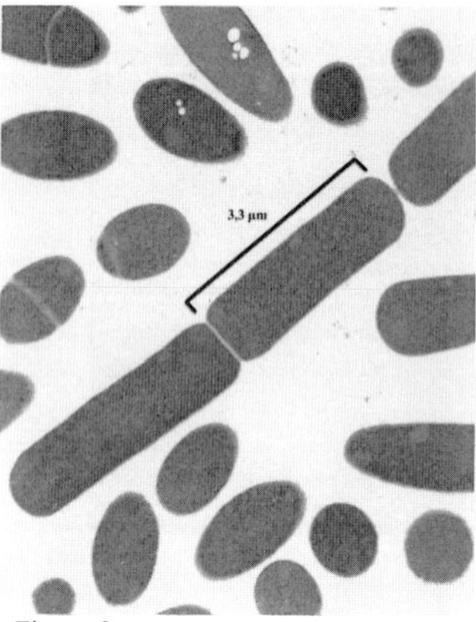

Figure 3

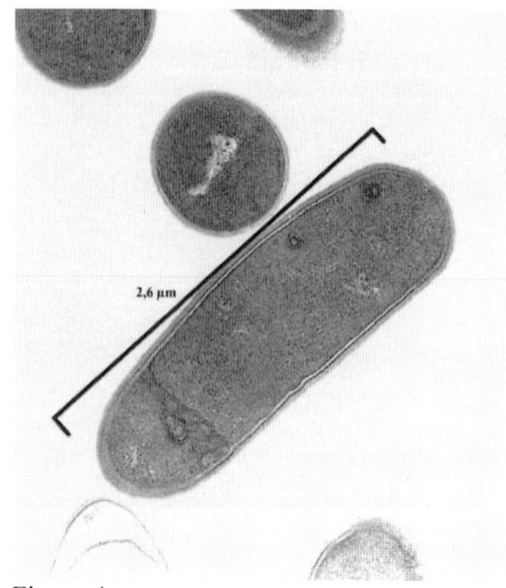

Figure 4

The cross-sections in Figure 5 and Figure 6 show that there is a difference in structure and surface of the cell wall. As you can see the cell wall of the modified species (Fig.6) looks like separated from the cell plasma and the glycocalix is much thinner than on the unchanged bacteria.

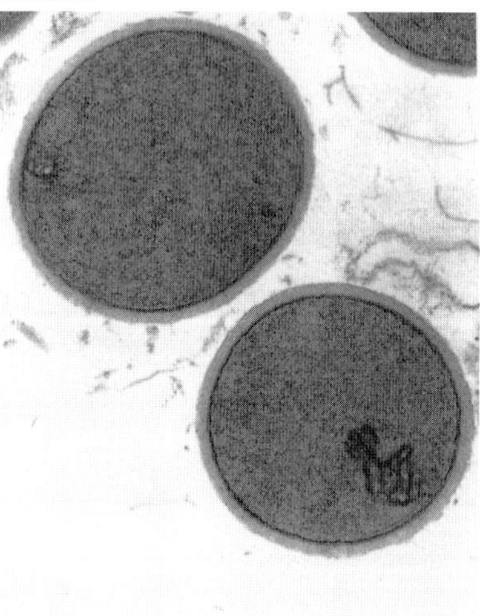

Figure 5

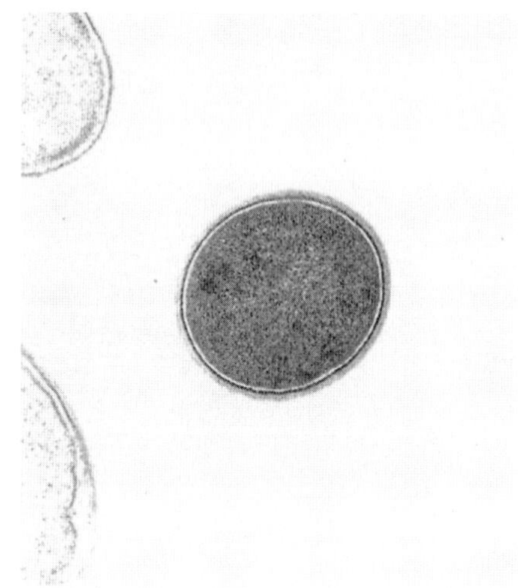

Figure 6

In Figure 7 and Figure 8 you can see an important step in the life cycle of *Bacillus thuringiensis*: vegetative cell division. Cell division is characterized by the formation of division septa which are initiated midway along the plasma membrane. Normally the gap between the two new cell walls is approximately 12 nm but when you take a look at the modified species in Figure 8 you can see that the distance is much shorter there.

Figure 7   Figure 8

The end of cell division is shown in Figure 9 and Figure 10. Here, too, the gap between the new born cells is much closer in case of the changed strain than it is in the unmodified *Bacillus thuringiensis.*

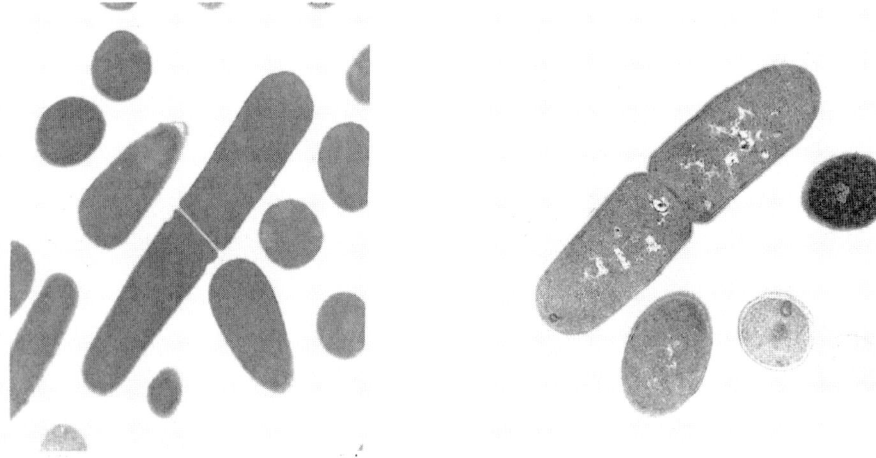

Figure 9   Figure 10

The close connection of the changed bacteria during cell division is not found in case of mature cells. On the contrary, the changed cells appear single (Figure 4), whereas the original bacteria often chain themselves to their neighbours (Figure 3). The colonies of the changed bacteria (Figure 2) show a rounded appearance like a substance of high surface tension.

3. CONCLUSION

The results give the evidence that it is possible to change the growth and habitus of a viable bacterium by a single pressure treatment. This makes *Bacillus thuringiensis* a new model for studying the effects of high pressure on microorganisms.

REFERENCES

1. E. Schnepf, N.Crickmore, J. VanRie, D. Lereclus, J. Baum, J. Feitelson, D.R. Zeigler and D.H. Dean, Microbiology and molecular biology reviews, 62 (1998) 775.
2. P. Butz, E. Fritsch, J. Huber, B. Keller, H. Ludwig, B. Tauscher, Biocontrol Science and Technology 6 (1995) 243.
3. B. Sojka and H. Ludwig, Pharmazeutische Industrie 56 (1994) 660.
4. A.R. Spurr, Journal of ultrastructure research, 26 (1969) 31.
5. E.S. Reynolds, Journal of cell biology, 17 (1963) 208.

Trends in High Pressure Bioscience and Biotechnology
R. Hayashi (editor)
                        343

Effect of High Pressures on the Antibacterial Properties of Lactic Bacteria

I.Warmińska-Radyko*, Ł.Łaniewska-Moroz*, A. Reps**, A. Krzyżewska**
* Chair of Industrial and Food Microbiology, University of Warmia and Mazury, Pl. Cieszyński 1, 10-957, Olsztyn, Poland
** Institute of Food Biotechnology, University of Warmia and Mazury, Heweliusza 1 str., 10-724 Olsztyn, Poland

Introduction

The use of ultra high pressures (UHP) in food preservation instead of thermal methods allows to avoid negative changes caused by high temperatures. The advantage of high pressures is that food preserved in this way maintains its full nutritive value and desirable sensory properties. The pressure used should inactivate microorganism cells which cause food contamination. The sensitivity of microorganisms to high pressures is different and depends on numerous factors. Vegetative forms of bacteria in the phase of logarithmic growth are among the most sensitive ones. Gram-positive bacteria, protected by a glycopeptide multi-layer network, are more resistant to high pressures than Gram-negative bacteria characterized by thin cell walls. Bacterial spores are resistant to high pressures, but the process of pressurization increases considerably their sensitivity to the effect of physicochemical environmental factors, e.g. temperature. Inactivation of large yeast and mildew cells takes place at pressures that are much lower than those needed to inactivate bacteria [1, 2]. This indicates that the degree of sensitivity to high pressures depends also on the size and shape of cells. The effect of antibacterial activity of high pressures on cells increases in acid environments and in the presence of some substances, e.g. enzymes, chitosane or bacteriocins [3, 4].

  High pressures at a proper level, after braking down technological barriers, may be applied for partial reduction in the number of bacterial populations, e.g. in fermented drinks. This method may replace the process of thermization commonly used in the dairy industry [5]. It enables the elimination or minimization of enzymatic processes 'carried out' by technological microflora, taking place during food storage, as well as longer preservation of typical sensory properties of a given product.

  One of the properties of lactic acid bacteria (LAB) used in the production of fermented foodstuffs is their ability to inhibit or limit the growth of harmful microorganisms, including pathogenic bacteria [6, 7]. The antibacterial activity of lactic fermentation bacteria is caused by the secretion of such metabolites as lactic acid, acetic aldehyde, hydrogen peroxide, and bacteriocis synthesized by some strains. It was proved experimentally that the application of high pressures in the presence of bacteriocins allowed to reduce the number of populations of *Listeria monocytogenes* and *Escherichia coli O157:H7* strains to a much greater degree that the use of high pressures only [8, 9]. The results of research work indicate the possibility of employing the UHP method for preserving certain kinds of food or extending its keeping quality, after introducing appropriate production technologies.

Aim of Research

The aim of the research work was to determine the antibacterial activity of *Lactobacillus delbrueckii ssp. bulgaricus* and *Streptococcus thermophilus* strains in relation to selected pathogenic bacteria found in food, and the effect of high pressures on this activity.

Material and Methodology

The following strains of lactic acid bacteria (LAB) used for yogurt production were analyzed: 9 *Lactobacillus delbrueckii ssp. bulgaricus* strains: 60, 162, 168, 144, B12, B13, V4, S11, 256 and 5 *Streptococcus thermophilus* ones: 7, 72, 79, 149, 150. All strains were cultured in milk for 24 hours at a temperature of 37°C; streptococci were additionally cultured on a liquid medium M17 (Merck) under the same conditions.

The level of pH was measured in cultures, population size was determined using solid media provided by the Merck company: MRS (*Lactobacillus*) and M17 (*Streptococcus*); their antibacterial activity was analyzed towards 18 test strains: *E. coli* 16, 22, 94, 402; *Y. enterocolitica* 2, 8; *P. vulgaris* 1, 16; *S. gallinarum, S. typhi; S. aureus* 1, 1a, 14, 16; *L. monocytogenes* 1, 2, 38; *B. cereus* J4.

The antibacterial activity of LAB was determined employing the diffusion method, using wells 9 mm in diameter, made in a solid nutritive medium inoculated with individual test strains in the amount of $10^4 - 10^5$ cfu/ml. 0.125 ml of cultures of individual LAB strains characterized by acid reaction and – in the case of streptococci – also neutralized to pH 7.0, were put into wells. The process of incubation was carried out for 24 hours at temperatures optimum for individual test strains. The antibacterial activity was evaluated on the basis of measuring the diameters of growth inhibition zones of test strains around wells.

All cultures were subjected to a pressure of 800 MPa for 15 minutes in a high-pressure chamber Liquid Vessel LU/30/16 (Laboratory Hydraulic Press C1 01). After pressurization the count of live cells was determined again in cultures and the antibacterial activity was analyzed according to the methodology described above.

Results

Lactobacillus delbrueckii ssp. bulgaricus strains were characterized by different acidic capacities and growth rate in milk. After 24 hous of incubation at a temperature of 37°C, most strains formed populations whose size usually amounted to 10^8 cfu/ml, in three cases to 10^9 cfu/ml. One strain showed a lower growth rate. Those populations produced lactic acid with different efficiency, acidifying cultures to pH 3.9 – 4.62 and 5.46 – 5.64 in the case of two strains characterized by weaker acidifying properties. (Tab.1)

A pressure of 800 MPa applied for 15 minutes resulted in inactivation of a large number of cells. Depending on the strain, population reduction by 5 - 9 log was observed. The most significant reduction was noted in populations of strains showing the strongest acidifying capacities. More cells survived pressurization in cultures whose pH was above 4.3.

All strains of LAB rods were characterized by antibacterial activity towards the majority of test strains. Individual strains in milk cultures formed metabolites which inhibited the growth of 11 - 18 (61 - 100%) strains of pathogenic bacteria. Some *E. coli, P. vulgaris* and *S. typhi* strains were resistant to metabolites of five *Lactobacillus* strains.

Pressurization caused inactivation of metabolites inhibiting the growth of test strains. All cultures of LAB rods lost their antibacterial activity in relation to most test bacteria, inhibiting the growth of 1 – 4 strains only.

Table 1. Effect of 800 MPa pressure on the survivability and antibacterial activity of 24-hour milk cultures of *L. delbrueckii ssp. bulgricus* strains

	L. delbrueckii ssp. bulgricus strains								
	60	162	168	144	B12	B13	V4	S11	259
Culture size before pressurization	$2.0*10^9$	$3.0*10^8$	$1.0*10^9$	$1.9*10^8$	$5.2*10^8$	$1.3*10^9$	$1.2*10^8$	$3.0*10^7$	$6.9*10^8$
Number of test strains whose growth was inhibited	18	18	17	16	18	18	16	17	14
Culture size after pressurization	np. in 0.1	np. in 0.1	np. in 0.1	$1.1*10^2$	$2.8*10^2$	$1.5*10^2$	$7.5*10^2$	$4.0*10^1$	$1.8*10^3$
Number of test strains whose growth was inhibited	3	11	12	4	2	4	1	1	1
pH before and after pressurization	4.12	3.90	4.00	5.64	4.48	4.42	4.62	4.37	5.46

*np. – not present

After pressurization, none of *Lactobacillus* strains inhibited the growth of 3 staphylococcus strains and 2 *E. coli* ones. As the acidity of cultures did not change after pressurization, the main antibacterial factor were probably bacteriocin-like substances inactivated under conditions of the pressure applied.

Streptococcus thermophilus strains cultured in milk were characterized by a high growth rate and the size of their populations was by ten times larger than that of populations cultured on M17. An exception was here strain no. 7, which proliferated better on a synthetic medium. All streptococci acidified milk to pH 3.9 – 4.9, depending on the strain. During growth on a synthetic medium streptococci produced less acid, which was confirmed by the reaction of those cultures: pH 4.55 - 5.72. (Tab. 2)

Strains of streptococci were characterized by similar sensitivity to pressurization conditions, regardless of the culture medium. Only cells of *Streptococcus thermophilus* 150 cultured in milk and *Streptococcus thermophilus* 7 – on a synthetic medium - survived the process. No live cells were found in 0.1 ml of the other cultures.

The antibacterial activity of *Streptococcus* strains was much weaker compared with *Lactobacillus* rods, and its level differed depending on the culture medium. Individual streptococci cultured in milk inhibited the growth of 5–11 test strains.5 test strains: 2 *E. coli*, 2 *S. aureus*, and *S. typhi* turned out to be resistant to metabolites of all streptococci used in the experiment. Cultures on a synthetic medium inhibited a smaller number of test strains and to a lower degree. None of them inhibited the growth of staphylococci. On this medium all streptococci formed metabolites which were especially active towards *L. monocytogenes* and *B. cereus* strains, whose growth was inhibited by both acid and neutralized cultures.

Table. 2. Effect of 800 MPa pressure on the survivability and antibacterial activity of milk and M17 cultures of *Streptococcus termophilus* strains

	Streptococcus termophilus strains				
	7	72	79	149	150
	MILK CULTURES				
Culture size before pressurization	$3.1*10^8$	$2.7*10^8$	$1.1*10^8$	$1.7*10^9$	$1.81*10^8$
Number of test strains whose growth was inhibited	9 (5)	10 (8)	11 (7)	5 (1)	10 (4)
Culture size after pressurization	np. in 0.1	np. in 0.1	np. in 0.1	np. in 0.1	np. in 0.1
pH before and after pressurization	4.26	4.02	4.41	3.94	4.98
	M17 CULTURES				
Culture size before pressurization	$2.3*10^9$	$8.5*10^7$	$1.7*10^8$	$4.3*10^8$	$6.0*10^7$
Number of test strains whose growth was inhibited	5 (4)	8 (4)	10 (6)	6 (5)	8 (4)
Culture size after pressurization	$1.0*10^2$	np. in 0.1	np. in 0.1	np. in 0.1	np. in 0.1
pH before and after pressurization	4.89	5.23	4.92	4.55	5.72

*digits in brackets concern the number of test strains whose growth was inhibited by neutralized cultures
*np. – not present

The other test strains were in most cases resistant to the effect of neutralized cultures, and the growth of some of them was even stimulated. The process of pressurization caused total loss of antibacterial activity in relation to all test strains of all *Streptococcus* strains in all cultures on both media.

S. aureus strains used in the research were sensitive to the antibacterial activity of metabolites produced in cultures of all rods, and resistant to metabolites of most streptococci. (Tab. 3) Only two *S. aureus* strains showed sensitivity to metabolites of two or three streptococci formed in milk cultures. *E. coli* strains differed from one another in their sensitivity to metabolites of LAB strains, but – compared with the other test strains – turned out to be the most resistant, also as concerns metabolites of some *Lactobacillus* strains. The most sensitive were *L. monocytogenes* and *B. cereus,* whose growth was inhibited by almost all LAB strains used in the experiment.

Table. 3. Number of LAB strains inhibiting the growth of test strains.

	Number of LAB strains inhibiting the growth of test strains							
	Lactobacillus		*Streptococcus*					
	on milk		on milk			on M17 medium		
	BP	AP	BP	N	AP	BP	N	AP
E. coli 16	6	4	2	0	0	1	0	0
E. coli 22	7	2	0	0	0	0	0	0
E. coli 94	6	0	3	1	0	1	0	0
E. coli 402	7	0	0	0	0	0	0	0
Y. enterocolitica 2	9	4	4	2	0	3	3	0
Y. enterocolitica 8	9	2	3	1	0	3	1	0
Proteus vulgaris 1	8	2	4	2	0	3	0	0
Proteus vulgaris 16	9	2	4	1	0	3	1	0
S. gallinarum	9	4	4	2	0	3	3	0
S. typhi	7	1	0	0	0	0	0	0
S. aureus 1	9	0	2	0	0	0	0	0
S. aureus 1 a	9	0	0	0	0	0	0	0
S. aureus 1 4	9	4	0	0	0	0	0	0
S. aureus 1 6	9	0	3	2	0	0	0	0
L. monocytogenes 1	9	7	2	1	0	5	3	0
L. monocytogenes 2	9	3	3	3	0	5	2	0
L. monocytogenes 38	9	3	5	4	0	5	5	0
B. cereus J4	9	2	5	5	0	5	5	0

BP – cultures before pressurization, AP – cultures after pressurization, N – neutralized cultures

Conclusions
1. *Lactobacillus* strains used in the studies synthesized metabolites inhibiting the growth of test pathogenic bacteria.
2. Treatment of *Lactobacillus* cultures with a pressure of 800 MPa for 15 minutes caused, in most cases, inactivation of metabolites which had an antibacterial effect on test strains.
3. Each of pressurized *Lactobacillus* cultures maintained its antibacterial activity in relation to one, several or more test strains.

4. *Streptococcus* strains in cultures characterized by acid reaction showed antibacterial activity only towards some test strains. Most of them lost this activity after culture neutralization.
5. In all *Streptococcus* strains cultured in milk and a synthetic medium, the application of a pressure of 800 MPa for 15 minutes resulted in total inactivation of metabolites showing antibacterial activity towards test strains.

References

1. T. Tanaka, K. Hatanaka, J. Jpn. Soc. Food Sci. Technol. 39 (1992) 173.
2. P. F. Steeg, J. C. Hellemons, A. E. Kok, Appl. Envirom. Mikrobiol. 65 (1999) 4148.
3. P. C. Wonters, E. Glaasker, J. Smelt, Appl. Envirom. Mikrobiol. 64 (1998) 509.
4. D. G. Hoover, Food Technlol. 47 (1993) 150.
5. M. Mitsuiki, Y. T. Kido, T. Awao, S. Toba, 6th ICEF, Congress in Chiba, Japan (1993).
6. M. Bielecka, A Majkowska, E. Biedrzycka, Pol. J. Food Nutr. Sci. 4 (1994) 34.
7. B. V. Balasubramanyam and M. C. Varadaraj, J. Appl. Mikrobiol. 84 (1998) 97.
8. L. Laniewska-Moroz, I. Warmińska-Radyko and A. Reps, Med. Fac. Landbouww. Univ. Gent 65/3b (2000) 551.
9. M. F. Styles, D. G. Hoover D.F. Farkas, J. Food Sci. 54 (1991) 1401.

Trends in High Pressure Bioscience and Biotechnology
R. Hayashi (editor)

The influence of high hydrostatic pressure on the adduct formation of Patulin with Cysteine

N. Merkulow and H. Ludwig

Institut für Pharmazeutische Technologie und Biopharmazie, Gruppe physikalische Chemie, Universität Heidelberg, Im Neuenheimer Feld 346, D-69120 Heidelberg
horst.ludwig@urz.uni-heidelberg.de

The influence of high hydrostatic pressure on the adduct formation of patulin with cysteine was investigated using different pressure/temperature combinations. The reaction was found to be promoted by pressure. The activation energies were calculated for 0.1, 400 and 500 MPa and the activation volumes at 4, 25 and 40 °C were determined.

1. INTRODUCTION

Patulin (figure 1) is a fungal secondary metabolite produced by various *Aspergillus* and *Penicillium* strains and by *Byssochlamys nivea* [1]. It is toxic to animals [2, 3] and presumably carcinogenic [4]. Patulin was isolated from various fruits [5], but is mainly found in apples and apple products. Patulin is stable in apple and grape juice but not in orange juice [6]. Probably, in orange juice patulin reacts with sulfur containing groups.

Patulin builds adducts with sulfhydryl compounds like cysteine or glutathione [7, 8]. Adduct formation with cysteine diminishes strongly the toxicity of patulin [9]. Ciegler et. al. reported the loss of toxicity after the adduct formation, though teratogenicity to chicken embryos of the adducts was retained [10]. Lindroth and Wright supposed it to be unlikely that the adducts can be metabolized to free patulin in mammals [11]. The toxicity of patulin is attributed to the reaction with sulfhydryl groups in enzyme systems [7]. Krivobok et. al. found an antitumoral activity of patulin that is reduced after adduct formation with cysteine [12].

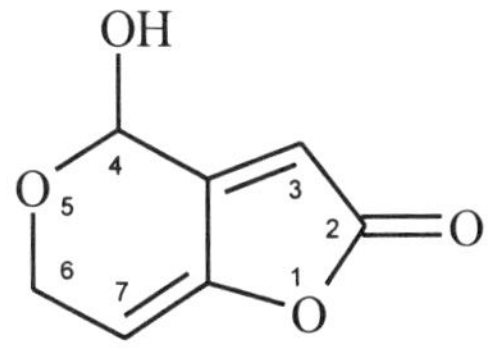

The chemical components responsible for the toxic activity of patulin are the lactone group and the C-atom in position 6 which has to be in plane with all the other C-atoms, otherwise the toxicity is lost [13]. Recently, Fliege and Metzler reported the ability of patulin to induce crosslinks of proteins involving cysteine, lysine, and histidine in vitro [14]. The adduct formation was found to be dependent on the environmental conditions [10] but kinetic data characterising this reaction are lacking.

Figure 1. Patulin

The aim of this study was to investigate the influence of high hydrostatic pressure on the adduct formation of patulin with cysteine. The activation energy and the activation volume were calculated for this reaction.

2. MATERIALS AND METHODS

Reaction conditions: 5 mg patulin were dissolved in 50 ml ethanol and stored at -22 °C. For the experiments, 10 ml of the solution were evaporated to dryness at 40 °C under reduced pressure. The residue was dissolved in 20 ml of distilled water, adjusted to pH 4 with phosphoric acid, to a final concentration of about 50 mg patulin/l.

100 mg cysteine were dissolved in 100 ml of distilled water. Both solutions were filtered through a 0.22 μm membrane filter and mixed in a ratio of 1:1 for the experiments. The resulting molecular ratio of patulin to cysteine was 1:25. 20 μl of the reaction solution were injected into the HPLC. The amount of the patulin-cysteine-complex was calculated using the reduction of patulin.

High pressure treatment: The high pressure device consisted of ten pressure vessels. They were filled with 1-2 ml samples enclosed in polyethylene tubes and sealed with silicon plugs. All vessels could be simultaneously thermostated and pressurised. The maximum pressure used was 500 MPa, the pressure medium was a water/ethyleneglycol mixture with a ratio of 9/1 (v/v). The single vessels could be opened at different times in order to measure the kinetics of the reaction.

HPLC conditions: A LC-10AS HPLC system from Shimadzu, Griesheim, Germany with a Rheodyne 7125 injector and a SPD-2A diodearray detector from LATEK, Heidelberg, Germany was used. The analytical column (150 x 4.6 mm ID) was packed with 125 Å, 5 μm, reversed phase C18 (polar endcapped aqua column) with a C18 guard column (4 x 3 mm ID). The colums were purchased from Phenomenex, Aschaffenburg, Germany. The eluent was a mixture of phosphate buffer (0.2 M, pH 2.1) : methanol in a ratio of 7 : 3, filtered through a 0.22 μl membrane filter. The final pH of the liquid phase was 2.4. The measurements were carried out at 276 nm with a flow rate of 0.5 ml/min. After the experiments, the column was floated with a acetonitrile : water mixture (65 : 35) as proposed by the producer.

3. RESULTS AND DISCUSSION

The reaction between patulin and cysteine was found to be pressure dependent. Figure 2 shows the reduction of patulin during the adduct formation at 4, 25, and 40 °C and different pressures. The reaction is promoted through increasing pressure and temperature. For the determination of the activation energy and the activation volume, kinetic measurements were carried out at 4, 25, and 40 °C and pressures of 0.1-500 MPa.

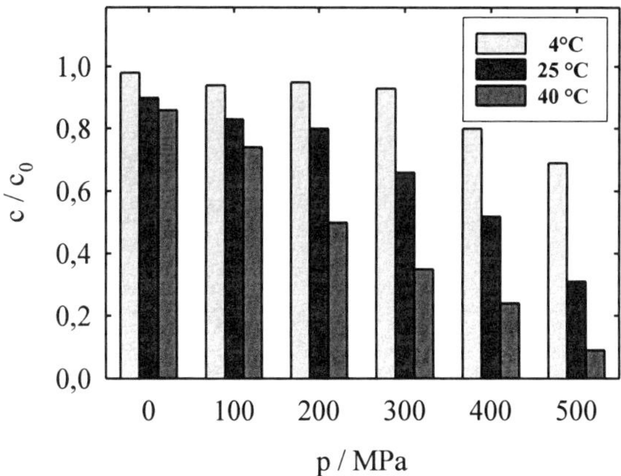

Figure 2. Reduction of patulin during the adduct formation with cysteine at 4, 25 and 40°C in 30 min at different pressures.

In figure 3, the kinetics of the adduct formation is shown at atmospheric pressure and 4, 25, and 40 °C. The linear course in the semilogarithmic plot indicates a pseudo first order reaction (the concentration of cysteine is nearly constant).

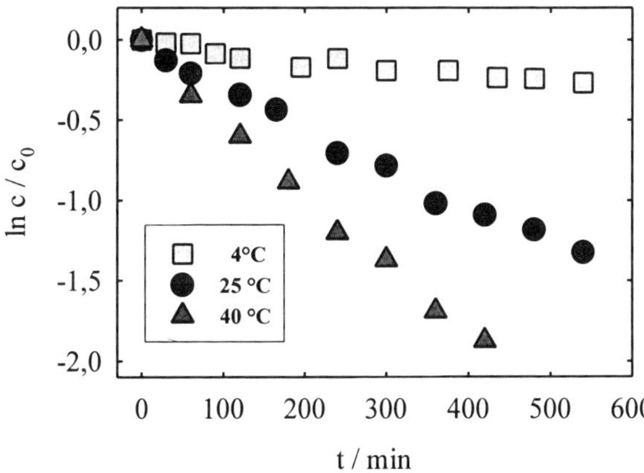

Figure 3. Reduction of patulin during the adduct formation with cysteine at 4, 25 and 40 °C at atmospheric pressure.

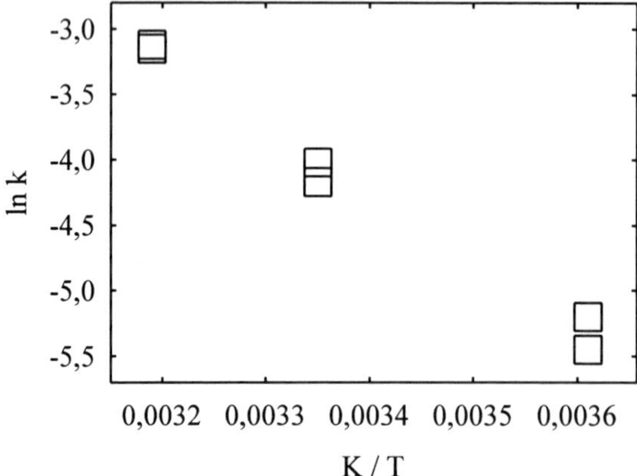

Figure 4. Determination of the activation energy at 400 MPa.

The activation energies were determined for 0.1, 400 and 500 MPa. As an example, the determination of the activation energy at 400 MPa is shown in figure 4.

The calculated values for the activation energies at different pressures are given in table 1 including the standard deviations obtained from the Arrheniusplot. The activation energy for this reaction diminishes with increasing pressure.

Table 1
Activation energies at different pressures

	0.1 MPa	400 MPa	500 MPa
Activation energy [kJ/mol]	49.5 ± 4.4	43.1 ± 2.5	40.3 ± 2.2

The activation volume of the adduct formation was determined for 4, 25, and 40 °C. Figure 5 shows a semilogarithmic plot of the rate constants at 40 °C and different pressures. The activation volume is negative, the absolute value being smaller in the higher pressure region than at normal pressure. Similar results are found at 4 and 25 °C. The values of the activation volumes are given in table 2.

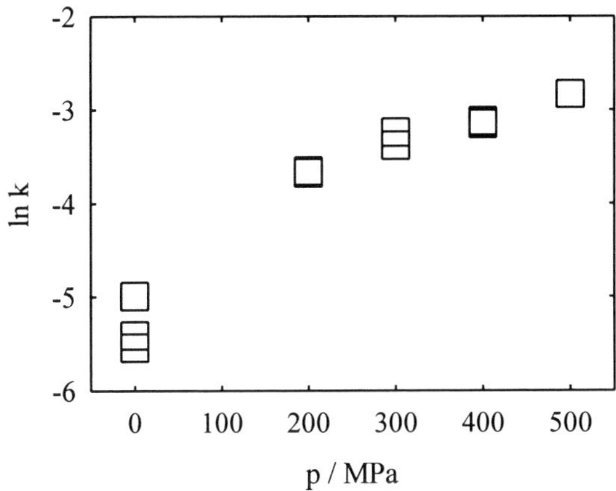

Figure 5. Determination of the activation volume at 40 °C.

Table 2
Activation volumes for 4, 25 and 40 °C

	4 °C	25 °C	40 °C
Activation volume [cm^3/mol]	-10.3 ± 5.0 (400-500 MPa)	-5.6 ± 1.8 (300-500 MPa)	-6.9 ± 0.6 (200-500 MPa)
	< -14.3 ± 1.0 (< 400 MPa)	< -15.0 ± 1.3 (< 300 MPa)	< -21.6 ± 2.9 (< 200 MPa)

The adduct formation of patulin with cysteine is promoted by high hydrostatic pressure at all temperatures and pressures. The reaction results in different products. Lindroth and Wright found seven patulin-cysteine-adducts and supposed that the number of adducts built during the reaction is dependent on the environmental conditions [11]. Ciegler et. al. suggested that one molecule patulin reacts with more than one molecule cysteine [10]. The various reaction products could be favoured at different pressures. This would explain the differing activation volumes.

354

REFERENCES

1. IARC monographs, (1986) 83.
2. W.A. Broom, E. Bülbring, C.J. Chapman, J.W.F. Hampton, A.M. Thomson, J. Ungar, R. Wien and G. Woolfe, Br. J. Exp. Pathol., 25 (1944) 195.
3. L. Escoula, M. Thomsen, D. Bourdiol, B. Pipy, S. Peuriere and F. Roubinet, Int. J. Immunopathol., 10/8, (1988) 983.
4. F. Dickens and H.E.H. Jones, Br. J. Cancer, 15 (1961) 85.
5. H.K. Frank, R. Orth and A. Figge, Z. Lebensm.-Untersuch. -Forsch., 163 (1977) 111.
6. P.M. Scott and E. Somers, J. Agr. Food Chem., 16/3 (1968) 483.
7. W.B. Geiger and J.E. Conn, J. Am. Chem. Soc., 67 (1945) 112.
8. K. Hofmann, H.J. Mintzlaff, I. Alperden und L. Leistner, Die Fleischwirtschaft, 51 (1971) 1534.
9. S. Lindroth and A. v. Wright, J. Environ. Pathol. Toxicol., 10/4-5 (1990) 254.
10. A Ciegler, A.C. Beckwith and L.K. Jackson, Appl. Environ. Microbiol., 31/5 (1976) 664.
11. S. Lindroth and A. v. Wright, Appl. Environ. Microbiol., 35/6 (1978) 1003.
12. S. Krivobok, F. Seigle-Murandi, R. Steiman, J.-L. Benoit-Guyod and M.-H. Bartoli, Die Pharmazie, 49 (1994) 277.
13. F. Seigle-Murandi, R. Steiman, S. Krivobok, H. Beriel and J.-L. Benoit-Guyod, Die Pharmazie, 47 (1992) 288.
14. R. Fliege and M. Metzler, Chem.-Biol. Interactions, 123 (1999) 85.

Trends in High Pressure Bioscience and Biotechnology
R. Hayashi (editor)
© 2002 Elsevier Science B.V. All rights reserved.

Inactivation of viruses in plasma by cycled pulses of high pressure*

S. Dusing, C. Li, J. Behnke, M. Manak and R. Schumacher

BBI BioSeq, 217 Perry Parkway, Gaithersburg, MD 20877

Preliminary studies have demonstrated that alternating pulses of high and atmospheric pressure can be used to inactivate both enveloped and non-enveloped viruses under conditions that are compatible with retention of therapeutic plasma protein activity.

1. INTRODUCTION

Establishing simple, effective, inexpensive, and widely applicable methods for inactivation of both enveloped and non-enveloped viruses in plasma is an important approach for ensuring a consistent supply of safe plasma and plasma-derived therapeutics worldwide. Pressure cycling technology (PCT) is an unique and innovative physical methodology that meets this need without reliance on possibly denaturing heat treatment or use of potentially toxic or mutagenic chemical additives.

PCT achieves virus inactivation through the application of repeated pulses of hydrostatic pressure to plasma or other protein solutions. These pressure pulses, which are uniformly transmitted at the speed of sound through any sample volume, can be applied to material in closed containers such as plasma collection bags thus precluding the need for unnecessary handling and eliminating the likelihood for introduction of extraneous contaminants.

The specific pressure ($\leq$570 MPa) and temperature (-60°C to +60°C) requirements as well as the duration and number of cycles of applied pressure necessary for virus inactivation are determined empirically to optimize both virus inactivation and retention of functional activity of the therapeutic protein(s) of interest. To meet Regulatory Agency requirements for virus inactivation in plasma, objectives for PCT technology include inactivation of a minimum of 4 logs of enveloped viruses (e.g., HIV, hepatitis B, hepatitis C), inactivation of a minimum of 4 – 6 logs of non-enveloped viruses (e.g., human parvovirus) and preservation of the integrity and biologic activity of plasma components such as clotting factors and immune globulins.

2. MATERIALS AND METHODS

2.1. Pressure Cycling Technology

PCT treatment was performed using a BaroCycler™, a high pressure computer-controlled research instrument developed jointly by BBI BioSeq and BBI Source Scientific (Garden Grove, CA) that is capable of generating 540 MPa working pressure in a pressure chamber of sufficient size to hold up to 3 ml samples. Ramp times were of approximately 10 seconds duration. Temperature control was maintained by means of an external chiller/heater circulating fluid through the pressure chamber jacket. Pooled plasma, serum or protein solution samples were aliquoted into sample tubes or sealed in sample bags appropriate for PCT treatment.

2.2. Virus Inactivation

High titer virus stock was added to sample material prior to sealing the sample bags. In initial studies, MS2 phage (ATCC 15597-B1), a 24 – 26 nm diameter, non-enveloped, single-stranded RNA virus, was selected as the initial model for viruses that are difficult to inactivate or remove from therapeutic products such as the human B19 parvovirus. This single-stranded DNA viral contaminant of human plasma is extremely difficult to remove or inactivate by current methodologies due to its small size (18 – 26 nm diameter) and non-enveloped icosohedral structure. Because the B19 virus cannot be propagated *in vitro*, the closely related porcine parvovirus (PPV) was used as a surrogate in these studies. PCT virus inactivation studies were also performed using the Human Immunodeficiency Virus (HIV), a non-enveloped RNA virus of public health concern that may contaminate plasma.

To measure MS2 bacteriophage titer after processing, serial 10-fold dilutions of the samples were prepared in phosphate buffered saline and mixed with a broth culture of *Escherichia coli* (ATCC 15597) and plated in soft agar. The plaques were enumerated and the titer expressed as plaque forming units (PFU) per ml.

PPV (ATCC VR-742) was titered by *in vitro* assay on PK13 swine kidney cells (ATCC CRL-6489) in a 96-well plate format. Ten-fold serial dilutions of PCT-treated samples and controls were inoculated into 6 replicate wells and maintained in culture for a 7 day period. Cells, fixed and stained with fluorescein-conjugated anti-PPV serum (VMRD, Inc.) were scored microscopically for the presence of intracellular fluorescence and these results were used to determine the $TCID_{50}$ titer by the method of Reed and Muench (1).

Inactivation of the $HIV-1_{IIIB}$ laboratory strain was assessed in the human lymphocytic H9 cell line in a 96-well plate format as described for PPV. Culture medium harvested at 14 and 28 days post-infection was tested for the presence of HIV p24 protein by antigen capture immunoassay (Coulter Corp.) and the results were used to determine HIV $TCID_{50}$ titer.

Results were reported as logs of virus inactivation relative to a virus-spiked control plasma sample maintained at atmospheric pressure for the period of time required for PCT treatment.

2.3. Plasma/Serum component biological activity

Factor VIII (fVIII) is one of the most labile of the therapeutic proteins present in human plasma and, for this reason, efforts were concentrated upon establishing PCT treatment conditions that would result in the loss of not more than 20% of plasma fVIII activity.

Factor VIII activity was measured by microplate ELISA assay (COAMATIC® Factor VIII, Chromogenix-Instrumentation Lab.) or by activated partial thromboplastin time (APTT) assay using an Amelung KC4 A™ micro coagulation analyzer (Sigma) in which fVIII activity is measured as a function of clotting time. Results were reported as percentage of fVIII activity following PCT treatment relative to either a plasma control maintained at atmospheric pressure for the period of time required for PCT treatment or an unthawed plasma control.

3. RESULTS

3.1. MS2 phage inactivation

MS2 phage was rapidly inactivated when exposed to three repeated cycles of pressure in excess of 400 MPa at -20°C with approximately 4 logs of MS2 inactivation at 500 – 540 MPa (Fig. 1). Cycle durations were of 60 seconds at high pressure and 300 seconds at ambient pressure. Using these pressure conditions, fVIII activity ranged from 53% of control at 400 MPa to 32% of control at 540 MPa. The optimum conditions for inactivation of MS2 with retention of fVIII activity lies at about 520 MPa.

To evaluate the effect of pressure pulsation, the time that the plasma sample was held at -20°C and 540 MPa was varied between 15 and 240 seconds while the time at ambient pressure was maintained at 300 seconds for each of the three cycles. Short (<60 second) pulses of pressure favored fVIII retention whereas MS2 inactivation was not appreciably affected by the duration of high pressure application in each cycle (data not shown). Attempts to increase retention of fVIII activity by further reducing the cycle time at 500 MPa were effective in retaining 82% of fVIII activity with three 1 second pulses (Fig. 2). However, cycle times of less than 10 seconds at high pressure resulted in a significant decrease in MS2 inactivation.

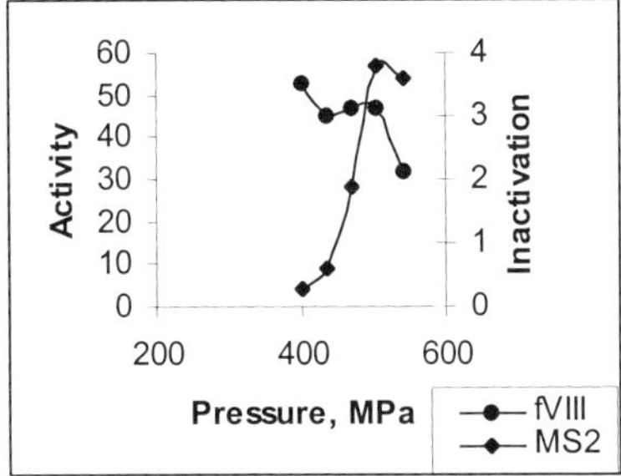

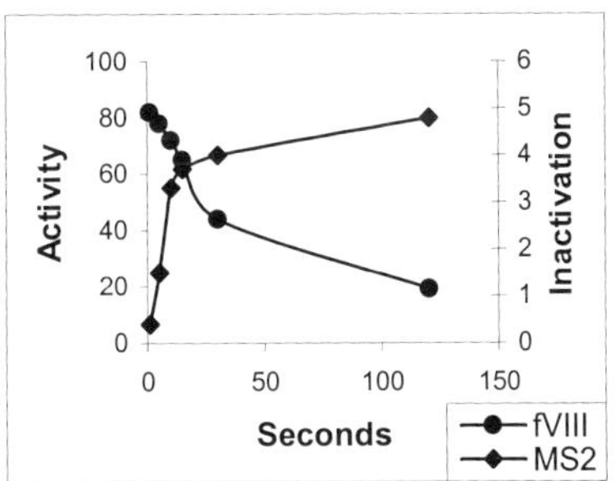

Figure 1. Inactivation of MS2 and stability of fVIII as a function of pressure. fVIII activity is denoted as the % of a plasma control maintained at atmospheric pressure. Inactivation of MS2 is reflected by the logs of reduction in PFU/ml titer.

Figure 2. MS2 inactivation and fVIII activity relative to high pressure pulse length. FVIII activity is denoted as the % of a plasma control maintained at atmospheric pressure. Inactivation of MS2 is reflected by the logs of reduction in PFU/ml titer.

3.2. Retention of Factor VIII activity

Based upon initial experiments demonstrating the critical importance of cycle time at high pressure on fVIII activity, experiments were performed using three alternating cycles of various lengths of either 540 or 335 MPa pressure and 60 seconds at atmospheric pressure at a temperature of -20°C. In this series of experiments, <40% of the fVIII activity was retained when plasma was exposed to three cycles of 540 MPa pressure (Fig. 3). However, nearly 80% of fVIII activity was present in plasma treated under the same conditions of cycle time and temperature at 335 MPa.

Retention of fVIII activity also depended upon the temperature at which PCT was performed. Plasma treated with 20 cycles consisting of 30 seconds at 335 MPa plus 60 seconds at atmospheric pressure contained a significantly higher level of fVIII activity at PCT temperatures of ≤ -20°C than that treated at 0 or -10°C (Fig. 4). In this experiment, control plasma samples were maintained in the pressure chamber at the selected temperature for a period of time equivalent to that required for the PCT treatment.

These results demonstrate that lower pressure and lower temperature as well as short cycle time at high pressure are most suitable for retention of fVIII activity.

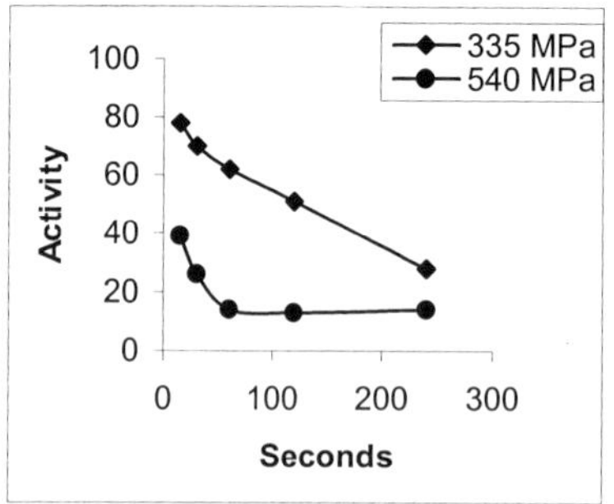

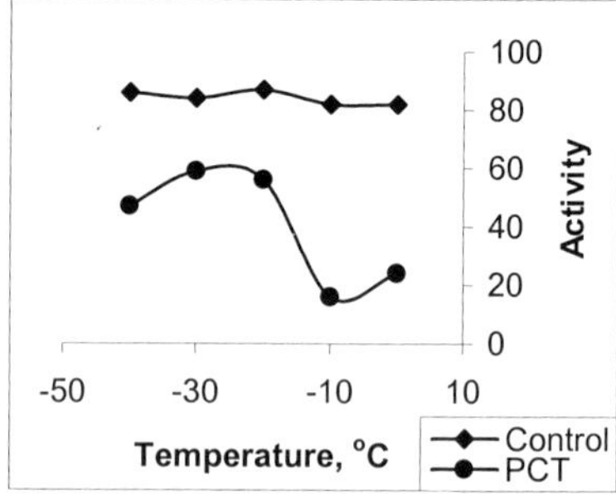

Figure 3.   fVIII activity relative to high pressure pulse length at two pressures. fVIII activity is reported as the % of a plasma control maintained at atmospheric pressure.

Figure 4. Effect of temperature on retention of fVIII activity. fVIII activity is denoted as the % of a previously unthawed and unmanipulated plasma control.

3.3. PPV and HIV inactivation

Viral inactivation studies in which PPV was added to a therapeutic protein solution that was subsequently exposed to 540 MPa of either static or cycled pressure at various temperatures demonstrated that PCT-mediated inactivation of this virus was most effective at high (60°C) and low (-30°C, -40°C) temperatures (Fig. 5). The duration of the 60 pressure cycles used for this study was 30 seconds at high pressure and 60 seconds at ambient pressure. Virus-spiked plasma samples were maintained at static pressure for a period of 30 minutes.

To evaluate the effect of PCT on a laboratory strain of human immunodeficiency virus, HIV_{IIIB} culture supernatants were exposed to five cycles consisting of 1 minute at elevated pressure and 1 minute at atmospheric pressure at -10°C. Under these PCT treatment conditions, 6 logs of HIV inactivation were obtained at 335 MPa (Fig. 6). Efforts are currently underway to maximize PPV and HIV inactivation in plasma and serum while maintaining fVIII functional integrity.

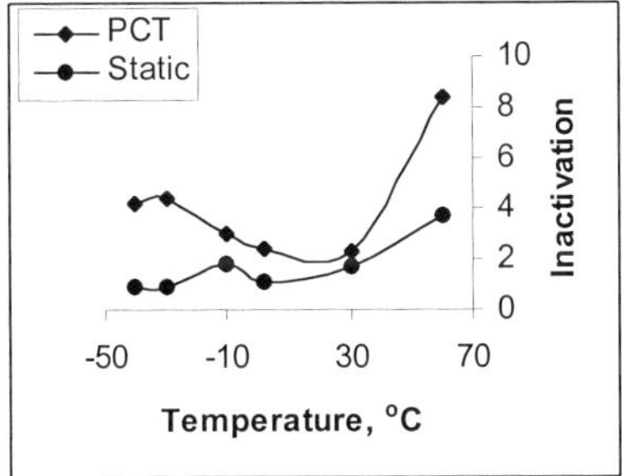

Figure 5. Effect of temperature on inactivation of PPV by 540 MPa of cycled or static pressure. Inactivation is reported as the logs of reduction in $TCID_{50}$ titer relative to an untreated control.

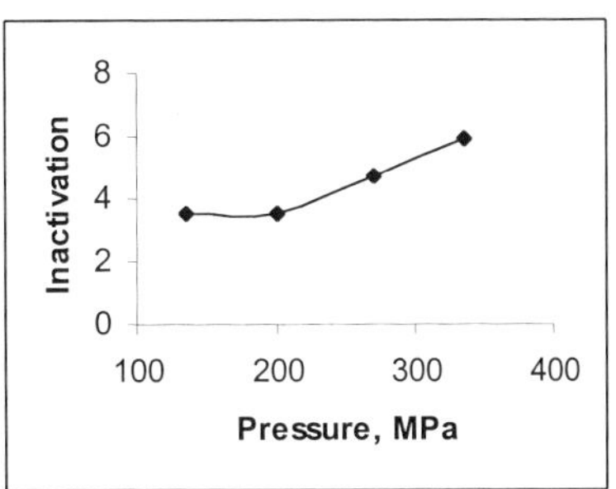

Figure 6. PCT-mediated inactivation of HIV as a function of pressure. The reduction in $TCID_{50}$ virus titer relative to an untreated control was used to calculate logs of virus inactivation.

4. SUMMARY

Initial studies have demonstrated that PCT inactivates both enveloped and non-enveloped viruses under conditions that may be compatible with the retention of therapeutic protein activity. Short pulses of high pressure were shown to be optimal for maintenance of protein function; pulse length did not appear to have a significant effect upon MS2 inactivation. The role of high pressure pulse length and cycle number in PCT-mediated inactivation of other viruses remains to be established.

To maximize the application of this new technology, studies are ongoing to establish PCT conditions that both preserve therapeutic protein function and integrity and provide maximum inactivation of a wide range of relevant and model viruses.

REFERENCE

1. Dulbecco, R. Endpoint method – measurement of the infectious titer of a viral sample *in* Virology, 2nd edition, Lippincott, Philadelphia, PA, 1998.

*This work was funded in part by The Consortium for Plasma Science and the US NIH National Heart, Lung and Blood Institute.

Inactivation of HIV-1 by the freeze pressure generation method (FPGM)

T. Otake*[a], T. Kawahata*, H. Mori*, Y. Kojima*, I. Oishi*, K. Hayakawa**

*Osaka Prefectural Institute of Public Health, 1-3-69, Nakamichi, Higashinari-ku, 537-0025 Osaka, Japan, **Kyoto Prefectural Comprehensive Center for Small and Medium Enterprises, 17 Chudoji Minami-machi, Shimogyo-ku, 600-8813 Kyoto, Japan
[a]Tel:+81.6.6972.1321   Fax:+81.6.6972.2393   E-mail:otake@iph.pref.osaka.jp

It has been reported that the infectivity of HIV-1 was lost by high hydrostatic pressure treatment (HHPT) at 400-650 MPa, and decreased the reverse transcriptase activity (RT) as a mechanism. It was shown recently that the high pressure generated when water is frozen inactivates microorganisms. In this study, we investigated the effect of this novel high-pressure method, that is, the freeze pressure generation method (FPGM), on HIV-1 infectivity. After FPGM at 250 MPa, some T-cell tropic and macrophage tropic HIV-1 strains completely lost their infectivity. A possible inactivation mechanism for this is that FPGM causes functional changes in the viral envelope protein.

1. INTRODUCTION

To ensure safety of blood products, blood materials are tested for anti-HIV antibody. To increase the safety of blood, an additional process of eliminating pathogens by heat or filtration is usually performed. However, most of these methods may inactivate active substances in blood such as coagulating factors, and contamination with pathogens is possible although it is very rare. We previously reported that HIV-1 lost the infectivity after high-pressure treatment at 400 MPa or higher[1]. This finding suggested that a high-pressure method may be useful for detoxication of blood products.

It was reported recently that when water was frozen in a container, a high pressure was produced due to an increase in the water volume, and this pressure inactivated bacteria[2]. Here, we report the effect of a freeze pressure generation method (FPGM) on HIV infectivity.

2. MATERIAL AND METHODS

HIV-1 strains used in this study were four T-cell tropic strains (LAI, RF,KK-1, KK-2) and a macrophage tropic strain (HTLV-IIIBa-L). A 3-ml plastic tube filled with viral suspension was placed in a high-pressure vessel. The vessel was filled with distilled water, and tightly sealed. The high pressure chamber was kept in a freezer at -30°C for 24 hours, then immersed in tap water for 20 minutes to thaw the ice in the chamber. The infectivity of T-cell tropic strains was determined based on the measurement of the 50% tissue culture-infectious doses (TCID50) in MT-4 cells. The infectivity of macrophage tropic strain was measured by counting blue cells in MAGIC-5a cells (kindly provided by Dr. M.Tatsumi). The reverse transcriptase activity in the samples was measured using RT assay kit (Asahi Chemical Industry, Co.,Ltd.,Japan). The amount of HIV-1 core protein (p24) was measured using EIA (RetroTec, Zeptro Metrics, U.S.A) for p24 antigen of HIV-1 detection. To examine the effect of FPGM on the function of HIV-1 envelope protein, viral samples were concentrated by ultracentrifugation and then reacted with MT-4 cells. These cells were reacted with anti-envelope(gp120) monoclonal antibody (0.5β, kindly provided by Dr. S.Matsushita), and the viral amount bound to the cells was measured by flowcytometry. To investigate the effect of FPGM on the function of immunoglobulin, anti-HIV-1 antibody-positive serum was treated with a high pressure, and the anti-HIV-1 antibody titer was measured by the particle agglutination (PA) method (Fuji Rebio).

Table 1

The effects of freeze pressure generation method on HIV-1

HIV-1Strain	Infectivity[1]		RT acivity[2]		p24[3]	
	N[4]	T[5]	N	T	N	T
LAI	5.6	0	688.6	81.7	258	246
RF	4.5	0	231.2	13.4	258	246
KK-1	2.75	0	324.8	46.3	737	833
KK-2	4.25	0	95.7	12.4	218	245
HTLV-IIIBa-L	5.11	0	282.2	31.4	209	188

1;LAI,RF,KK-1 and KK-2:log 50% tissue culture-infectious doses (TCID50/mL), HTLV-IIIBa-L:log blue focus unit(BFU/mL) 2; mU/mL 3;ng/mL 4; PTHPM not treated 5; PTHPM treated

3. RESULTS AND DISCUSSION

Hayakawa et al. reported that when water placed in a high pressure vessel was kept at -30°C, the pressure in the vessel reached 250 MPa[3]. After FPGM treatment, the infectivity titers of the four T-cell tropic HIV-1 strains and 1 macrophage tropic HIV-1 strain were below the detection limit. FPGM treatment did not affect the p24 protein level, and the antigenicity of core protein was retained. The RT activity was decreased to 1/10 by the high-pressure treatment (Table 1). The fluorescence intensities of cells reacted with the viruses treated with the high pressure were almost the same as that of the non-reacted control, showing that the viruses lost the cell surface-binding activity (data not shown).

The sensitivity to the high hydrostatic pressure treatment (HHPT) markedly varied among the HIV-1 strains. We previously reported that the pressure required for virus inactivation varied from 400 to 650 MPa. and a decrease in the RT activity was shown to be the major factor of the inactivation mechanism[4]. In the new FPGM using 250 MPa pressure induced in freezing, a wide range of HIV-1 strains lost the infectivity. The decrease in the reverse transcriptase activity caused by HHPT was associated with the decrease in the viral infectivity. In FPGM of this study, although the viral infectivity was completely lost, RT activity corresponding to 10% of that of non-treated virus was retained. The T-cell-binding activity of the viral samples was examined by flowcytometry. FPGM made the viruses unable to bind to T-cells. This treatment did not affect the anti-HIV-1 antibody titer, and the immunoglobulin function was not changed (data not shown).

The above findings showed that FPGM completely inactivates HIV-1 at a pressure of 250 MPa, at which the effect was not exhibited in HHPT at room temperature. Regarding the inactivation mechanism, changes in the viral envelop protein function rather than a decrease in the reverse transcriptase activity may be involved. To establish FPGM as a detoxication method of blood products, it is necessary to investigate the effects of FPGM on active substances such as coagulating factors in blood plasma.

REFFERENCES
1.T.Nakagami, H.Ohno, T.Shigehisa, T.Otake, H.Mori, T.Kawahata, M.Morimoto, N.Ueba, *Transfusion*, **36** (1996) 475-476
2. K.Hayakawa, Y.Ueno, S.Kawamura, T.Kato, R.Hayashi, *Appl Microbiol Biotechnol*, **50** (1998) 415-418
3.K.Hayakawa, N.Miyajima, K.Saho, T. Komai, M.Matsumoto and R. Hayashi, R.Hayashi (eds.),Trend in High Pressure Bioscience and Biotechnology, Elsevier Science, Amsterdam, 2001
4. T.Otake, H.Mori, T.Kawahata, Y.Kojima, H.Hishimura, I.Oishi, T.Shigehisa, *Biocontrol Science*, **5** (2000) 127-129

Trends in High Pressure Bioscience and Biotechnology
R. Hayashi (editor)

High pressure-processed foods in Japan and the world

Atsushi Suzuki

Department of Applied Biological Chemistry, Faculty of Agriculture, Niigata University, Ikarashi, Niigata 950-2181, Japan

ABSTRACT

It has passed over 10 years since the application of high hydrostatic pressure on food processing was successfully demonstrated as non-heat food processing.

In a decade, various kinds of food materials had been exposed to high pressure and several kinds of high pressure-processed foods were tested to launch on the market. However, the pressure-processed foods still remaining in the market are small in number.

The present situation and future view of the high pressure-processed foods in Japan and the world are described.

1. INTRODUCTION

It has passed over 10 years since the application of high hydrostatic pressure on food processing was successfully demonstrated as non-heat food processing. Because the high pressure processing provides a new methods of avoiding off-flavor and deterioration of food components and nutrients, producing unique texture on food, and saving total amounts of energy required for food processing, the high pressure processing attracted the attention of food industry in Japan. This attention quickly spread through the world.

In a decade, various kinds of food materials had been exposed to high pressure and several kinds of high pressure-processed foods were tested to launch on the market. In Japan, the pressure-processed jams first appeared on the food market at 1990, followed by high pressure-processed fruit jellies and sauces. After a little while, high pressure-processed jams and juices were also commercialized in the Europe. However, the pressure-processed foods still remaining in the market are small in number. September 2000, a mass-productive line of pressure-processed rice products has been developed in Japan, and the trials to make pressure-processed foods are still underway through the world.

This paper describes the present situation and future view of the high pressure-processed foods in Japan and the world, especially focusing on the products in Japan.

2. COMMERCIAL PRESSURIZED FOODS IN JAPAN

High pressure-processed foods on the market in Japan are shown in Table 1. This table was originally prepared by Professor Cheftel [1] and now revised on the basis of recent information.

2.1. Fruits based products (Jams, Jellies, Purees and Yogurt sauces) [2]

Probably you may know, the high pressure-processed jam is the world-first merchandized pressure-processed food. When a mixture of raw materials (fruits, sugar and pectin) filled in plastic container is subjected to high-pressure treatment of 400~600 MPa for 10~30 min, penetration of sugar into fruits as well as sterilization can be conducted simultaneously. The fresh color and flavor are retained in the finished product, but the shelf life of the product is 2 months at 4 °C due to the residual enzyme. Now the container is changed from plastic to special made glass with a rid from aluminum.

The quality of the pressure-processed jam has been highly evaluated at the market after introduction in April 1990. At present, the production of jellies, purees and yogurt sauces are stopped.

2. 2. Grapefruit juice [3]

Grapefruit juice treated with high pressure appeared on the market at 1991. The process is that high pressure of 200 MPa is applied to freshly squeezed grapefruit juice for 10 min at 5 °C and then conducted with convent ional manufacturing process. The purpose of the pressurization is to inactivate the enzyme, not to sterilize microbes. The high pressure prevents precursor, limonin A-ring lactone, converting into limonin, which causes uncomfortable bitterness of grapefruit juice. The shelf life of this product is 3 months at room temperature.

2.3. Pressure-processed pork meat [4]

Pressure-processed pork meat has been developed and tested on the market. High pressure of 250 MPa at 20~40 °C for 3 hr was applied to cured pork after smoking at 65 °C for 90 min. High pressure treatment improved the organoleptic score, the microbial quality during storage at low temperature and water holding capacity. The texture of this product is like "nama" ham. "Nama" in Japanese means "raw".

The company which developed this product have obtained the official approval April, 2000 from the Ministry of Health & Welfare in Japan, but is now wondering to expand the production because of huge sum of money upon equipment and facilities.

2.4. Mochi (rice cake) [5]

Introducing high pressure to manufacturing process has succeeded improvement of the quality of rice cakes. The high pressure treatment provides rice cakes with novel properties, such as increased elasticity, extensibility, quick toasting and cooking properties, and reduced hot water dispersion. This improvement is derived from the decrease of the air bubbles in the rice cakes through the application of high pressure.

When the sub materials necessary for rice cakes containing mugwort leaves, green soybeans or dried green laver are pressurized, can be effectively sterilized without impairing the color and flavor as compared with the manufacturing process with heat sterilization. The condition of the pressurization is dependent on the raw materials. The shelf life of the product is 3 months at 4 ℃.

2.5. Hypoallergenic precooked rice [6]

The numbers of food-allergic patients in Japan have been increasing in recent years.

The numbers of rice-allergic patients are estimated to be about 400,000 among 2 million of food-allergic patients. It has been reported that cereal grains contain allergenic protein. Rice also contains allergenic proteins which are classified into albumin and globulin based on their solubility [7,8]. Some research groups have exploited new technology to eliminate these allergenic proteins. For example, Watanabe *et al.* [9] have succeeded in eliminating 16 kDa allergenic component in rice by enzymatic treatment and are now supplying hypoallergenic rice for rice-allergic patients.

Instead of the enzymatic treatment, the elimination of allergenic protein by high pressure was developed. As shown in the paper by Yamazaki and Sasagawa [6], the extractable proteins with D-isoascorbic acid increased with the increase of pressures applied up to 400 MPa. The salt-extractable allergenic proteins were eliminated due to the enhancement of rice porosity induced pressurization. The elimination of the allergenic proteins (about 95 %) was observed on the SDS-PAGE profile [6]. From the result of RAST (radioallergosorbent test)-inhibition test [6], it is cleared that the allergen in the cooked rice was eliminated by high pressure treatment.

This allergen free rice is steam-cooked, aseptically packed and now distributed to the patients through mail order on the advices of medical doctor. And the bread made from this

Table 1   Commercial Pressurized Foods in Japan

Firm	Product	Process	Role of HP	Sales
A	Fruits-based product (pH<4.5); Jams (apple, kiwi strawberry, blueberry)	400 MPa 20 ℃ 30 min	Pasteurization Improved gelation Faster sugar penetration	2.5×usual price Since 1990, with HP advertising Shelf-life: 2 months at 4 ℃ (1 week, if open)
B	Grapefruit juice	200 MPa 5 ℃ 10-15 min	Reduced bitterness	2×usual price Started August 1991, with HP advertising Shelf-life: 3 months at room temp
C	Mandarin juice Only=20% of HP juice in final juice mix	300-400 MPa 20 ℃ 2-3 min	Reduced odor of dimethyl sulfide Reduced thermal degradation of methyl methionine sulfoxide	Since October 1992, with HP advertising Shelf-life: 3 months at 4 ℃
D	Sugar-impregnated tropical fruits (kept at -18℃ without freezing) For sherbets and ice creams	50-200 MPa pieces of pineapple, papaya and mangoes immersed in a fructose syrup	Faster sugar penetration and water removal	Since October 1992, without HP advertising No recent information
E	Raw pork ham (1/4 the usual NaCl content), Pink and tender slices	250 MPa 20 ℃ 3 hr	Faster maturation (reduced from 2 weeks to 3 hr); Faster tenderization by internal proteases Effect on Microorganisms and parasites Improved water retention & shelf-life	Test market only Selling permit given at April, 2000

F	"Raw" sake (rice wine),turbid, with specific flavor	400 MPa Pressurization of insoluble particles, then reintroduced into the microfiltrated juice	Yeast inactivation; fermentation stopped (without heating)	Since 1994, with HP advertising Small market
G	Mochi, rice cake containing 40% water Yomogi fresh aromatic herbs	400 MPa 45 ℃ or 70 ℃ 10 min (depending on the raw material)	Microbial reduction of the cake and/or of some ingredients (seaweeds, moist aromatic herbs) Fresh flavor & taste	Since 1994, Shelf-life: 3 months at 4 ℃ 100 million yen/y
	Hypoallergenic precooked rice	400 MPa 20 ℃ 10 min	Enhance rice porosity & salt-extraction of allergenic proteins	100 million yen/y
	Easy-to-cook brown Rice	400 MPa	Enhance GABA content	200 million yen/y
	Precooked rice	400 MPa	Prevent retrogradation of gelatinized starch	1 billion yen/y
H	Ice-nucleating bacteria (used for the juice & milk concentration)	300 MPa 5 ℃ 5 min	Inactivation of *xanthomonas* bacteria without loss of ice-nucleating properties	Approved as a food-additive
I	Kipper products	185 MPa -15 ℃ 30 min	Microbial reduction Improved texture	Shelf-life: One year at - 20 ℃
J	Salmon products	400 MPa 15-25 ℃ 10 min	Microbial reduction Improved texture	Shelf-life: 3 weeks at 10 ℃

allergen-free rice powder is also on the market. The amount of the sales is about 100 million yens / year (one million US dollars / year).

2.6. Pressure-processed easy-to-cook brown (unpolished) rice [10,11]

In Japan there are many people loving brown rice, because of its high nutritious components. Therefore a lot of paper describing the functional properties of the unpolished rice has been published. But home-cooking by the conventional method is impossible due to the disturbance of water absorption by the seed coat of the grain.

This problem was now resolved by high pressure treatment. Adjust the water content of unpolished rice to 57.5 %, then expose to high pressure of 400 MPa. The pressurized rice is stored for a while and then dried. After 400 MPa pressure treatment, the water content in the brown rice increased in comparison with the untreated rice (control). During this process, the unpolished rice turned to easy-to-cook rice product by conventional method, and the accumulation of GABA (γ-amino butyric acid) known to be blood pressure-lowering compound was observed.

The amount of the sales of this is about 200 million yens / year (one million US dollars / year).

2.7. Precooked rice

A mass-productive line of pressure-processed precooked rice has been developed. This product has appeared on the market September 2000. The quality of this product is now highly evaluated, and the amount of the sales is estimated to be one billion yens / first one year (one thousand million US dollars / year). The details of this product are described in the following paper (Yamazaki and Sasagawa).

2.8. Ice-nucleating bacteria [12,13]

Ice-nucleating bacteria (*Xanthomonas campestris* INXC-1) is known to be able to freeze water at a subzero temperature higher than - 5 °C. The experiment was carried out in order to save the energy cost required for concentration or freeze-drying of egg white by introducing this ice nucleating bacteria.

As the ice nucleating activity was lost at a temperature higher at 27 °C, it is impossible to add this bacteria to foods after the heat sterilization. However it is desirable to add these bacteria to foods after sterilization. It was proved that *Xanthomonas campestris* INXC-1 was killed at 300 MPa at 5 °C for 5 min without losing the ice nucleation activity. It was observed that the pressure-sterilized cells retained the ice nucleating activity even if it was highly diluted, and that caused no super cooling as distilled water did. The ice nucleating activity is derived from the ice nucleating protein in the cell.

The high pressure-processed *Xanthomonas campestris* INXC-1 was approved to be one of the food additives at June 1993 [14], and used to concentrate egg white, juice, soy sauce and wine. But the market of this product is small.

2.9. Pressure-processed kipper

Pressure-processed kipper products have been commercialized early November 2000. The sliced kippers which have been mixed with NaCl and trehalose or marinated with wine vinegar, vegetable oil and various kinds of seasonings such as lemon extract, sucrose, wine, amino acids, meat extract etc. are vacuum packed, and then exposed to high pressure of 185 MPa at - 15 °C. These products are distributed in a frozen state, and shelf life is about one year in the refrigerator. High pressure reduces the microorganisms up to 1/10~1/100 of the initial number, and pathogenic bacteria such as *coliforms*, *staphylococcus*, *salmonella* and *vibrio* were not detected in the products.

The quality of pressure-processed kipper products are highly evaluated due to their new texture profiles.

2.10. Pressure-processed salmon

Pressure-processed salmon products have also been introduced to the market early November 2000. The sliced red-salmon and salmon-trout which have been smoked for 2 hr at 10 °C in fibrous casing after the seasoning for 60 min are vacuum packed, and then exposed to high pressure of 400 MPa for 10 min at 15~20 °C. High pressure treatment reduces the microbial counts and improves the texture and flavor. Shelf life of these products are about 3 weeks at 10 °C

Besides the above-mentioned pressure-processed foods, many kinds of foods were pressurized and appeared on the test market. However, most of all could not succeed in remaining in the market due to their high running cost.

3. COMMERCIAL PRESSURIZED FOODS OUTSIDE JAPAN

The commercial pressurized foods outside Japan are shown in Table 2.

3.1. Pressure-Processed Avocado sauce (pressurized guacamole)

Sliced avocado is vacuum packed in plastic bag and pressurized at 580 to 600 MPa for 5~10 min at 25 °C. Before pressurization, citric acid or some acid may be added to the product to reduce the pH. High pressure is applied mainly to prevent the acidic microbes and to inactivate polyphenol oxidase [15].

This product has a minimum of 3 weeks shelf life at 25 °C. Upon opening, it is to be refrigerated. This product is widely distributed around west part of USA and Mexico.

3.2. High Pressure processed sliced ham

Pressure processed sliced cooked ham has been commercialized in Spain at 1999. A cooked ham, which has been cooked in the conventional manner, is sliced and vacuum-packaged with

Table 2　Commercial Pressurized Foods Outside Japan

Firm	Product	Process	Role of HP	Sales
A	Avocado sauce	580-600 MPa 25 ℃ 5-10 min	Prevent acidic bacteria	Shelf-life: 3 weeks at 25 ℃
B	Sliced ham	400 MPa 17 ℃ 20 min	Microbial reduction Improve the quality	Shelf-life: 8 weeks at 4 ℃
C	Oyster	400 MPa 7 ℃ 10 min	Microbial reduction	Shelf-life: 5 weeks at 2 ℃

a microfilm between slices. After packaging, it is de-contaminated by means of high pressure of 400 MPa for 20 min at 17°C. The high pressure-processed ham is said to have superior organoleptic qualities, and has a shelf life of 8 weeks at 4 ℃.

3.3. High-pressure processed oyster [16]

Oysters are high-value products and in Europe are generally eaten raw from the shell. Hence, it is limited to the time they survive out of water, for which purpose they are kept around 5 ℃ in a saturated moisture ambient. It would, therefore, be highly advantageous economically to prolong their shelf life as much as possible without adversely affecting the sensory properties.

All shelled oysters were washed in cold water (5 ℃) with 3.5 % marine salt for 30 min. After draining excess drip solution, oysters were wrapped individually in plastic bags, vacuum-packed in oxygen permeable bags, and then pressurized at 400 MPa for 10 min.

High-pressure reduced the number of all target microorganism (total viable count, H_2S-producing microorganisms, lactic acid bacteria, *Brochothrix thermosphacta*, and *coliforms*), in some case 5-log unit.

According to the information of Dr. Raghubeer, (Flow International Co., Ltd.), the pressure-processed oyster has been already commercialized at New Orleans, USA.

Besides these products, the sales of the pressurized orange juice are continuing in France according to the information from the Professor Cheftel.

3.4. Pressure-processed foods in the USA

Finally, I would like to mention the present situation of the pressure processed foods in the USA. This information is given by professor B. Ray, University of Wyoming.

1) Under USA Army funding, Oregon State University has developed several acid products: Spanish rice, Spaghetti with meat sauce, Yogurt with fruit, Chicken-vegetable rice, Lemon rice with vegetables.

All the products have pH 4.5 and pressurized at 580 MPa for 15 min at 25 $^{\circ}$C.

2) University of Wyoming has worked on pressure-processed meat products: Roast beef, Summer sausage, Hot dogs and Other processed meats.

3) Some commercial companies are working on fruit juice.

4) One company has marketed guacamole (avocado dip).

I am not sure whether these products except guacamole, have been already commercialized or not. However, I am sure these products would be commercialized in near future. Because the agreements to develop the use of high pressure technology to produce high quality, shelf-stable food products have been signed among the Food Processing and High-pressure Equipment Processing Companies, the U.S. Army Soldier Biological Chemical Command, Natick Soldier Center and the National Center for Food Safety.

4. CONCLUSION

In a decade, commercialized pressure processed-foods have been increasing little by little. However, the pressure-processed foods easily available from the market are small in number. It seems that the companies hesitate to expand the production of pressure-processed foods because of the sanitary regulation and / or huge sum of money upon equipment and facilities, and high running cost. Concerning the running cost, the story of the development of a mass-productive line of pressure-processed packed rice, which will be described in the following paper, is an attractive.

I would like to end my report in expectation of the further increase of pressure-processed foods in the market.

REFERENCES

1. C. Cheftel, In High Pressure Food Science, Bioscience and Chemistry, (ed. N. S. Isaccs), The Royal Soc. Chem., Cambridge, pp.506, 1997.
2. Y. Horie, K. Kimura, M. Ida, Y. Yoshida, and K. Ogame, In High Pressure Science for Foods (in Japanese), (ed. R. Hayashi), San-Ei Publishing Co., Kyoto, pp.33, 1991.
3. N. Yuge, H. Mieda, Y. Mushika, and H Tamaki, In High Pressure Food Science and Bioscience (in Japanese), (ed. R. Hayashi), San-Ei Publishing Co., Kyoto, pp.350, 1993.

4. M. Nose, A. Iwaki, and M. Hattori, In High Pressure Bioscience (in Japanese), (eds. S. Kunugi, S. Shimada A. Suzuki, and R. Hayashi), San-Ei Publishing Co., Kyoto, pp.272, 1994.

5. A. Yamazaki, A. Sasagawa, M. Kinebuchi, and A. Yamada, In High Pressure Bioscience (in Japanese), (eds. S. Kunugi, S. Shimada, A. Suzuki, and R. Hayashi), San-Ei Publishing Co., Kyoto, pp.328, 1994.

6. A.Yamazaki and A. Sasagawa, The Review of High Pressure Sci. and Technol. (in Japanese), 6(1997), 236.

7. T. Matsuda, R. Nokura, M. Sugiyama, and R. Nakamura, Agric. Biol. Chem., 55 (1991), 509.

8. T. Adachi, A. M. Alvarez, N. Aoki, R. Nakamura, V. V. Garcia, and T. Matsuda, Biosci. Biochem. Biotech., 59 (1995), 1377.

9. M. Watanabe, J. Miyakawa, Z. Ikezawa, Y. Suzuki, T. Hirao, and S. Arai, J. Food Sci., 55 (1990), 781.

10. A. Yamazaki, The Review of High Pressure Sci. and Technol. (in Japanese), 6 (1997),182.

11. M. Kinebuchi, Chemical Industry (in Japanese), 61 (1997), 327.

12. M. Watanabe, In High Pressure Science for Foods (in Japanese), (ed.R.Hayashi), San-Ei Publishing Co., Kyoto, pp.111, 1991.

13. K. Honma and A. Takeuchi, In High Pressure and Biotechnology, (eds. C.Balny, R. Hayashi, K. Heremans and P. Masson), John Libbey Eurotext, Montrouge. pp.291, 1992.

14. K. Honma, M. Ogura, and M. Higaki, In High Pressure Bioscience (in Japanese), (eds. S. Kunugi, S. Shimada A. Suzuki, and R. Hayashi), San-Ei Publishing Co., Kyoto, pp.211, 1994.

15. C. A. Weemaes, L. R. Ldukhuyze, I. Van den Broeck, and M. E. Hendrickx, Biotechnology and Bioengineering, 60 (1998), 292.

16. M. E. López-Caballero, M. Pérez-Mateos, P. Montero, and A. J. Borderéias, J. Food Protection, 63 (1999), 196.

Trends in High Pressure Bioscience and Biotechnology
R. Hayashi (editor)
                                                            375

Development and Industrialization of Pressure-Processed Foods

A. Sasagawa and A. Yamazaki

Research Institute, Echigo Seika Co., Ltd.
1003-1 Takanashi-machi, Ojiya, Niigata 947-0102

In response to current social needs for security, safety, and health, we would like to take this opportunity to present the utilization of high pressure in food processing. In view of our consumption-oriented society, which shows a preference for convenience, a new awareness is developing in the administration of food-related industries. And, as a perpetual theme of mankind, we continue to seek new forms of food processing which take the environment into consideration.

Here, on the assumption that product development results from the innovation cycle, four virtual platforms are assumed; namely, the innovation platform, the concept platform, the engineering development platform, and the industrialization platform. The operations carried out on these platforms are introduced in the order of the processing. In addition, by giving an example of the development of our company's cooked high-pressure processed rice, the pressure energy recycling system and cost accounting are mentioned from the viewpoint of economics. As for the total cost relating to the high-pressure process, in U.S. Dollars, it costs approximately 1.25 cents for one package (300 ml) of product as the sum of depreciation expenses for machinery and building, electric power costs, maintenance expenses, and operation management expenses. Furthermore, there is the possibility that, in the future, a cost of less than 0.6 cents for 300 ml (0.2 cents per 100 ml) can be obtained by installing large-scale equipment. The same benefits are expected to become a reality in the case of applying high-pressure processing to other food products.

In only the last year of the 20th century, it is significant that "pressure," as one of the state-transforming factors, be put into practical use in the field of dietary staples, as opposed to "heat" with a history of use by mankind for 3,000,000 years. In the field of dietary staples where high prices are not acceptable, our product of cooked high-pressure processed rice has been successively marketed. This demonstrates that the consumer has economically supported high pressure processing. It is certain that utilization of high-pressure processing will be extended to a wide range of food processing. For example, the fields of noodles, bread, daily household dishes, liquor, vegetables, fruits, and frozen foods will be brought to market in the future as high-pressure processed foods.

1. Introduction

It has been eleven years since we started the development and industrialization of high-pressure foods. As our society requires security, safety and healthiness in food, we can now say that we have an extremely good chance to utilize high pressure for food processing. The technical base for this is the fact that high-pressure treatment makes sterilization possible without using heat. It also lessens decomposition of nutrients and pigments, and, high-pressure denatures starch and protein. In addition, the recent decrease in cost of high-pressure treatment may realize an almost free-of-charge exploitation of the treatment by taking advantage of scale merit through mass processing.

As a private enterprise, we tend to make proposals for new products supported and based on a dream in the early stages of development, and only thereafter, at the stage of industrialization, do we need to pursue the profitability of the business. Next, we should take risk into consideration. Development always entails a considerable risk. There is always some risk as an uncertain element before a great success, but nothing is generated from fear of risk. We hope more engineers come to realize the feeling of achievement by avoiding such risk through using their knowledge and making an effort.

In the field of dietary staples where high prices are not acceptable, our cooked high-pressure processed rice product has been successively marketed. This shows that the consumer has economically supported high-pressure processing. We hope the present paper will make a contribution as to the development of various kinds of high-pressure processed foods, and we expect that various industries dealing with high pressure will be established.

2. Process for industrialization

Fig. 1 shows the correlation between the cycles of innovation relating to product development and social factors. Innovation of product development is achieved by tracing these four platforms in the order of the arrows, as follows:
(1) the first platform of innovation that is supported by the requirements of
society or by the needs of the consumer,
(2) the second platform to create the concept by using information on technologies
from universities or technical institutes,
(3) the third platform of technical development supported by the current
capability for technical development or by production technologies, and
(4) the last platform of industrialization of the product supported by the power of
management and marketing.

The figure shows that it is possible to see a permanent cycle or evolution of product development as long as we can obtain new needs or seeds from the platform of innovation. Generally speaking, the "maturity of the product" forms a link in a cycle. As long as we sustain the desire for development, we may find a new requirement. In this way, going through numerous cycles, "those products required by society" will be perpetually generated.

We would like to verify the above-mentioned statements with reference to some examples by tracing the platforms as indicated above.

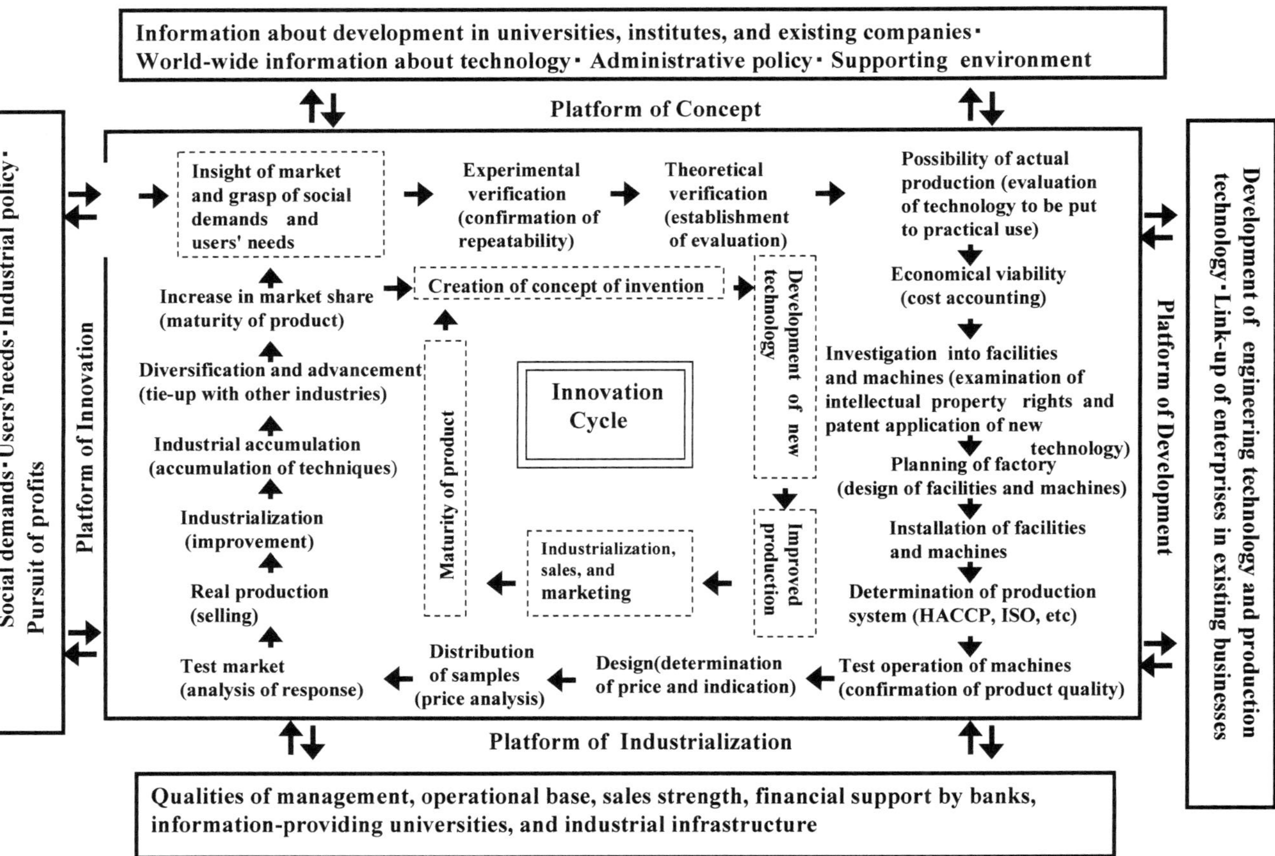

Fig.1 Correlation between the cycles of innovation concerning product development and social factors.

378

3. Insight of market and grasp of requirements of society and consumers

The process of "insight of the market and a grasp of the requirements of society and consumers' needs" is the origin and a starting point for the development of each product and, at the same time, a benchmark for the final evaluation. The current social requirements and consumers' demands will be considered by referring to an example of our company's development on the assumption that the current social requirements and consumers' demands are security, safety, and healthy nutrition. The properties and advantages obtained by pressure application treatment, although well known by readers who conduct research on high pressure, are summarized as follows[1]:

i). Pressure application has a sterilizing effect on parasites, germs, mold, yeast, and viruses, and the resulting sterilizing effect is by far more uniform than the effect obtained by application of heat or medications.

ii). The pressure application treatment can minimize deterioration of taste and decomposition of nutrients.   There is little possibility that toxicity will be generated after high-pressure treatment.   This proves that the high-pressure treatment is much safer than other treatments.

iii). High-pressure treatment can improve the congelation and texture of a material by utilizing denaturation of protein without use of an additive.

iv). New food products with surprising texture can be created from conventional cereal products when the pastiness and other properties of starch that cause retrogradation can be controlled by high- pressure treatment.

v). The application of high pressure to cereal products can allow water to smoothly penetrate to the inside of the products through the pericarp, seed coat, cell wall, or cell membrane.   In another way, the content of the food products can be efficiently extracted by the application of high pressure.

vi). Simply closing a valve connected to a sealed vessel regardless of retention time can retain high-pressure application. This contributes to saving energy.

Here, items i), ii), and iii) mentioned above relate to the social requirements and consumers' demands; that is, security, safety, and healthy nutrition. Items ii) and iii), which are also referred to in the above-mentioned platform, and items iv) and v) in combination can form characteristics of a newly developed product. Item vi)   is involved with the problem of environmental and economical viability. The condition as described in item vi) is advantageous for enterprises dealing with high- pressure treatment. In consideration of the advantages as mentioned above, we are sure that the use of high pressure can create new food products applicable to the social requirements and consumers' demands.

We have purchased high-pressure testing equipment with a maximum capacity of 1.5GPa[2], and effected high-pressure treatment on various materials.   As a result of the basic research, we were able to confirm phenomena such as sterilization and denaturation of starch with our own eyes at an early stage.

Most surprisingly, as shown in fig. 2, we discovered that starch containing excessive water became so cohesive that we could play catch with it as a result of applying high-pressure treatment[3].

This fact should show that excessive water was instantly integrated into the starch by the application of high pressure. We can utilize this technology for the dough or base for bread or pasta, which contains abundant water. We can use this dough to produce easily digestible bread and pasta with a new taste. As for rice, we are able to utilize its merit for cooked rice, rice cakes, and rice crackers.

I believe it is desirable that, for industrializing a product, we select a typical indispensable food of high importance. We should therefore study staple foods. Then, and only then, other foods and luxury foods, in that order of priority, should be considered.

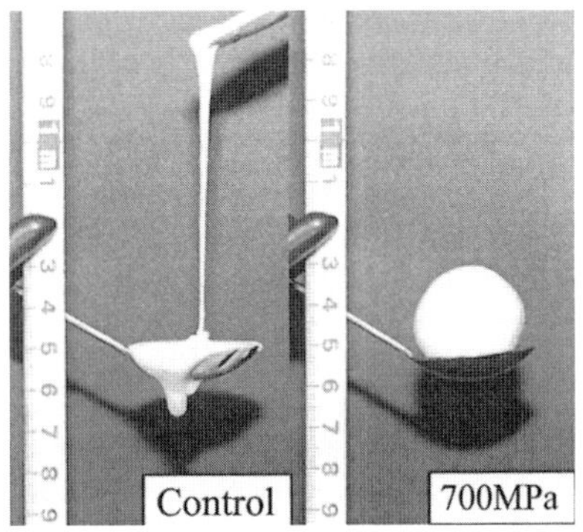

Fig.2 Photograph of dough ball containing 52 % water. Dough ball was made from rice flour pressurized at 0 or 700 MPa.

In this demonstration, I would like to take up "cooked rice" which is a popular staple food in Asia, as an example. The same as shown in fig. 1 can also be applied to "bread" or "pasta," which is popular worldwide.

The market in Japan for its staple food, rice, is a 100-billion-dollar business. Most of the rice is used in private homes. The production of the food processing industry accounts for only about two percent of the annual consumption. When compared to the other types of prepared rice dishes, such as frozen, canned, retorted, and dried rice, sterilized packaged rice items are superior when it comes to taste and ease of preparation and distribution. The production of sterilized packaged rice has increased by one hundred times over the last ten years. Therefore, if a greater variety of tasty and reasonably priced items were to be produced, such items would find a ready market quite easily. Our company's research has found that the application of high pressure to rice can produce the following properties[4~10]:

I). When high pressure is applied to polished rice immersed in water, water-enhancement necessary for cooking of rice can be instantaneously completed.[4,9,10]

II). The cooked high-pressure processed rice that is prepared by cooking rice permeated with water by osmosis after high pressure treatment shows a high degree of pastiness and gelatinization, and good digestibility[5,7].

III). The cooked high-pressure processed rice has a rich, flexible texture, and keeps its shape after cooking. Namely, the rice retains a shape similar to the original rice grains and exhibits high gloss[1,6].

IV). The restoration properties of the cooked high-pressure processed rice after microwaving are excellent. The cooked high-pressure processed rice can revert by microwaving it, and it can have an even higher degree of gelatinization than normal cooked rice right after cooking. Even after cooling, the cooked high-

pressure processed rice is still tasty, and free of a retrograded texture[5,7].
V). By strictly selecting the raw material of the rice, many kinds of bacteria excepting spore cells can be sterilized by high-pressure treatment. So, the manufacture of a perfectly sterilized packaged rice item, long regarded as a dream in the food industry, can be realized without subjecting rice to retorting or chemical treatment[8,9].

We have developed cooked high-pressure processed rice in light of the above-mentioned properties obtained by the application of high pressure to rice, and set a goal of realization of a sterilized packaged cooked rice product.

4. Experimental verification (confirmation of evaluation)

When verifying a newly developed item, it is very important to numerically show the degree of change due to the new processing method. The first step in development should be not only to make a sensory evaluation, but also to identify the best conditions to make the evaluation. Statements like "tastes good somewhat" or "if you eat carefully, you will find it tasty" are not enough to achieve added value in the marketplace.

We tried to demonstrate the tastiness of the high-pressure processed rice in terms of pastiness and degree of gelatinization[10], and texture.   Fig. 3 shows the change of the degree of pastiness and gelatinization after keeping the rice for a certain number of days after cooking, and its restoration properties[7] after microwaving.   This proves that the rice has a higher degree of gelatinization, not only right after cooking, but also anytime, when compared with normal cooked rice.   Even after many days of storage, the high-pressure processed rice does not show any properties of retrogradation.   On the contrary, after reheating it in a microwave oven, the high-pressure processed rice reverts and has an even higher degree of gelatinization than normal cooked rice.

As far as the texture of high-pressure processed rice is concerned, we were able to express its properties numerically by showing the stickiness and balance ratio. Fig. 4 shows the results when checking the stickiness and balance ratio one hour after cooking[13].   In Japan, rice with a high degree of stickiness and balance ratio is considered tasty.   It is found from the figure that the high-pressure method shows high numbers for both stickiness and balance ratio.

Fig. 5 is a picture showing the appearance of cooked rice. With the usual method of cooking rice, the rice becomes vertically long.   In the case of the high-pressure process method, it is significant that the rice retains a form similar to the original rice grains.   It is shown by using a clinical MRI that when cooking the rice by the traditional method, the rice contains bubbles filled with air. However, it is reported that the sticky, good-tasting rice tends to have few air bubbles. Cooked high-pressure processed rice contains only little air. The surface shows only a little unevenness but a high degree of flexibility, and thus tastes good.

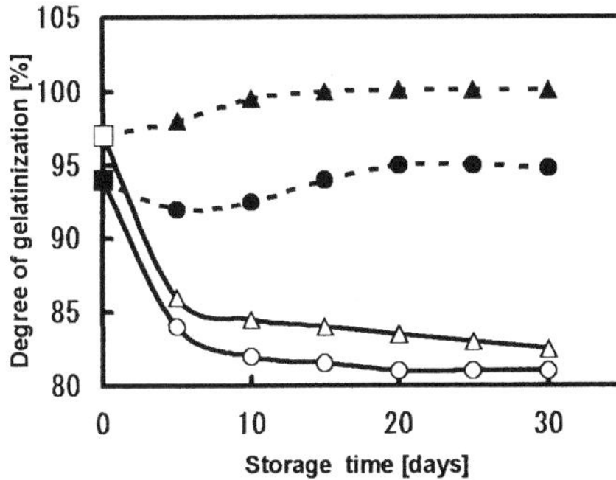
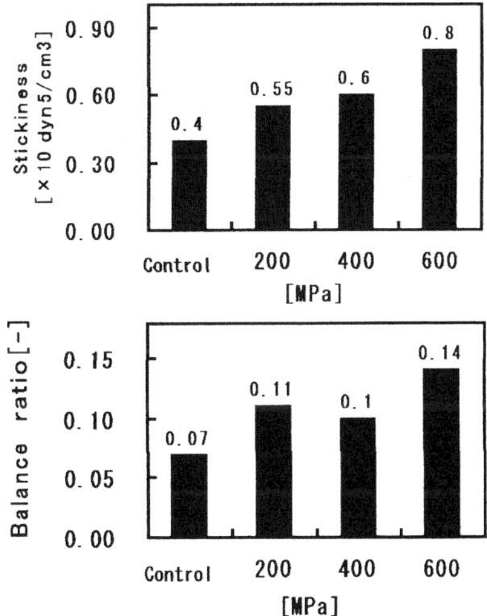

Fig.3   Changes in degree of gelatinization of cooked high-pressure processed rice with storage time, and the restoration properties by heating with a microwave oven.

□, cooked high-pressure processed rice (1 hour after cooking); ■, cooked untreated rice (1 hour after cooking); △, retrogradation of cooked high-pressure processed rice with time; ▲, restoration properties of cooked high-pressure processed rice by heating with a microwave oven; ○, retrogradation of cooked untreated rice with time; ●, restoration properties of cooked untreated rice by heating with a microwave oven.   High-pressure processing conditions: A pressure of 400 MPa was applied to rice for 10 min at 30℃ after soaking: Strong conditions: stored at 5℃; Restoration method: 100 g of cooked rice was heated in a microwave oven of 550 W for 2 minutes; Gelatinization measuring method: After heating with a microwave oven, the cooked rice was rapidly frozen at 80 ℃, and freeze-dried to have a water content of 3 %.The freeze-dried cooked rice was pulverized using a sample mill. The degree of gelatinization was measured by the BAP method.

Fig.4   Physical properties of cooking high-pressure processed rice.

Measurement: Three-grain method, by use of a tensipresser "TTP-50BX", made by taketomo Electric Co.,Japan(clearance: 0.4min).The average value of five measurement is shown. Rice grains: 90% polished rice grains(Oryza sative L.Japonica, cultivar Koshihikari harvested in Niigata prefecture in 1994) prepared using a small rice-polishing machine " KG-1000" made by Matsushita Electric Industrial Co., Ltd. High-pressure processing conditions: Pressure increasing or decreasing time = 2 min. Processing temp. = 30 ℃.

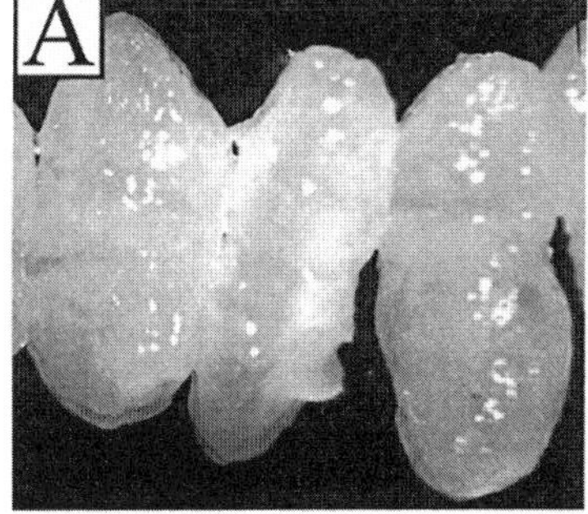

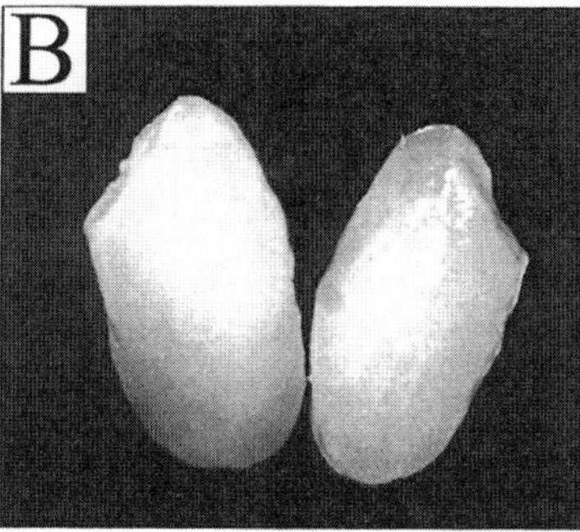

Fig.5   Appearance of the cooking rice

A: Cooked control rice; The grains of the cooking control rice are elongated.

B: Cooked high-pressure-treated rice; Grains of cooked high-pressure-treated rice are similar in to raw rice grain.

5. Economic viability (cost accounting)

The high-pressure process becomes cheaper with the advance of technology.
At present, the cost is about half that of ten years ago. Because adding value in the case of foods is difficult, we think that only two to four percent of the ex-factory cost could be allowed for the high-pressure process. Table 1 shows the cost of the high-pressure process for one 300 ml package in the case of a 100% utilization rate of the equipment in our factory.

More than ninety percent of the energy necessary for the high-pressure system is used in the intensifier. Because we use two pressure vessels alternately, as shown in fig. 6, we do not allow the built-up-pressure to go to waste. By utilizing the pressure differential retrieving method, we use the pressure again for the next build-up of pressure in the second pressure vessel and are thus able to save energy. In this way, one-third of the energy can be retrieved. Since, as is the case in Japan, there is a basic charge on top of the actual charge for electricity used, our factory can save 60% of the total cost of electricity compared to not using this pressure differential retrieving method.

At a 100%-utilization rate of the equipment, the cost of the high-pressure comes to 1.25 cents per 300 ml. However, even if we operate at a 50%-utilization rate of our capacity, it still only costs 2.5 cents per 300 ml. This represents approximately two to four percent of our ex-factory price, the allowable ex-factory cost for the high-pressure process as mentioned before.

Table 1   Economic Viability

1. Depreciation Cost of High Pressure Equipment		0.15[cent／pack]
2. Depreciation Cost of The Peripheral Machinery and Building		0.08[cent／pack]
3. Electrical Energy (Basic Charge Include		0.57[cent／pack]
4. Cost of Write-Off Equipment and		0.15[cent／pack]
5. Cost of Operation and Admini		0.30[cent／pack]
TOTAL	At 100 % of Utilization	1.25[cent／pack]
	At 50 % of Utilization	2.50[cent／pack]

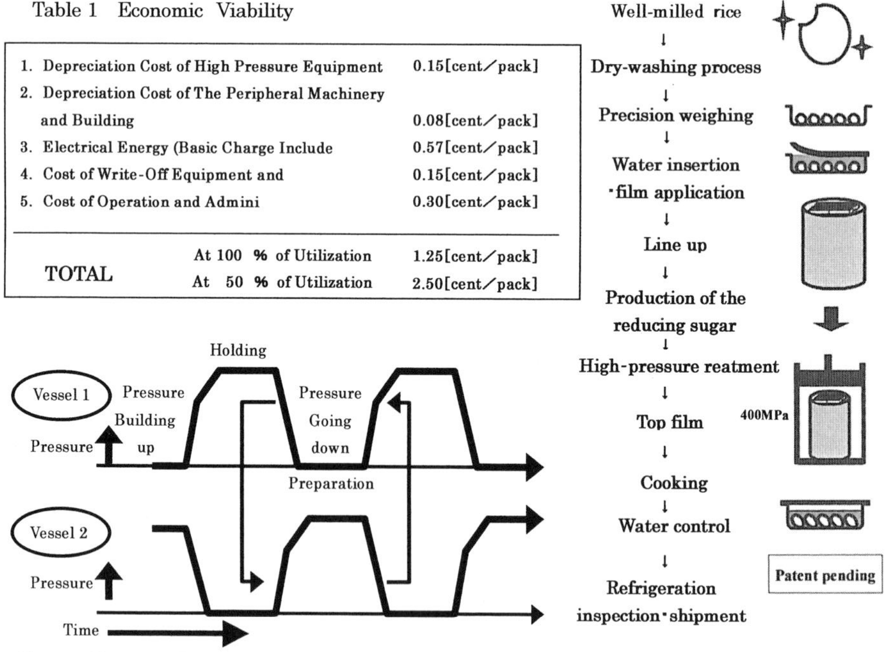

Fig.6   Alternate Operation with re-use of Pressure.

FIg.7   The manufacturing process of cooked rice by high pressure.

The cost of the high-pressure equipment is approximately one million U.S. Dollars in the case of 400 MPa with a 0.2 cubic meter capacity, three million dollars for a one cubic meter capacity, and ten million dollars for a ten cubic meter capacity. However, the cost per capacity decreases with the larger size of the equipment. Even if one adds the depreciation cost and the running cost, if large quantities are processed and a high degree of utilization is maintained, the cost of processing 100 ml should decrease to approximately 0.2 cents and add little to the total cost.

6. Facilities and factory design

In bringing the development of a new machine or technology to completion, the final problem is what materials are employed on the processing line for manufacturing the machine or product in a factory. To solve such problems on a wide scale, it is necessary to gather not only internal technologies, but also numerous pieces of information from the outside. Making use of patent gazettes, technical reports, technological journals, newspapers, and advertisements, a technical tie-up with other companies should be taken into consideration. Furthermore, it becomes important to research whether the developed technology or method would interfere with any granted patents, or whether the newly developed technology could be entitled to be classified as intellectual property. We have filed twenty or more patent applications concerning the high-pressure treatment over the last three years. In designing a factory, it is of great significance to recognize a factory as "a piece of large-scale mechanical equipment." Apart from a small-scale plant, automated facilities for mass production are required to prioritize the flow of products and the flow of raw materials. In our case, continuous and completely automated operation without any supervision has been accomplished by uniting the independent high-pressure application equipment with the peripheral technologies.

Fig. 7 shows the processing line for manufacturing the cooked high-pressure processed rice, which has been intensively discussed with respect to the above-mentioned points. The weighed rice is put into a tray. Then, the tray is filled with water to displace the air, and the water is used to transmit pressure to the rice. Next, the tray is covered with film. Since the pressure vessel is cylindrical, we use a special bucket in order to utilize the full space. Therefore, even numerous multi-edged trays are easily arranged in the bucket and moved. As a pre-process for more water-enhancement, the bucket is put into hot water for 30 minutes, and, because of the enzymes contained in the rice itself, the reducing sugar, which acts as a sweetener, is formed. As the next step, by applying high pressure, the sweetener is pushed into the rice together with the water, and the rice is completely permeated with water by osmosis. After the high-pressure process, the excess water is removed and the rice is cooked. Since all the steps of handling the rice are perfectly automated, from putting the rice into the tray until the cooking is completed, the system can be used without supervision. Since the cooked high-pressure processed rice is sterilized, it can be kept at room temperature and can be eaten after being put into the microwave oven for about two minutes.

7. Conclusion

In closing, when you look again at the innovation cycle, you will find that our high-pressure rice processing line clearly demonstrates one phase of our company's industrialization. Again, in only the last year of the 20th century, it is significant that "pressure," as one of the state-transforming factors, be put into practical use in the field of dietary staples, as opposed to heat with a history of use by mankind for 3,000,000 years. We were ultimately able to substantiate the efforts of the researchers under the guidance of Professor Rikimaru Hayashi, the members of the Japanese Research Group of High-pressure Bioscience & Food Science, and the various makers of high-pressure equipment. In the field of dietary staples where high prices are not acceptable, our product of cooked high-pressure processed rice has been successively marketed. This exemplifies that the consumer has economically supported high-pressure processing. And, it is certain that utilization of high-pressure processing will be extended to a wide range of food processing. The fields of noodles, bread, daily household dishes, liquor, vegetables, fruits, and frozen foods will most assuredly be brought to market in the future as high-pressure processed foods.

References
1) R.Hayashi (eds.), Use of High Pressure in Food, Kyoto, San-Ei Shuppan (1989).
2) N.Yamagishi, T.Ogata, H.Kasukabe, N.Matsuda, A.Sasagawa and A.yamazaki, High Pressure Bioscience and Technology, A.Suzuki, R Hayashi (eds.), Kyoto, San-Ei Shuppan, 195-202 (1997).
3) A.Sasagawa, M.Kinefuchi, A.Yamazaki and A.Yamada, High Pressure Bioscience, S.Kunugi, S.Shimada, A.Suzuki, and R.Hayashi, (eds.), Kyoto, San-Ei Shuppan, 335-343 (1994).
4) A.yamazaki and A.Sasagawa, *Nippon Shokuhin Kagaku Kougaku Gakkaishi* 45, 526 (1998).
5) M.kinefuchi, K.Watanabe, S.Komiya, A.Yamazaki and K.Yamamoto, *Oyo Toushitsu Kagaku* 46, 31 (1999).
6) A.Yamazaki, M.Kinefuchi, K.Yamamoto and A.Yamada, *Reprinted from The Review of High Pressure Science and Technology,* 5,168 (1996).
7) A.Yamazaki and A.Sasagawa, *Nippon Nogeikagaku Kaishi* 74, 619 (2000).
8) A.Yamazaki, A.Sasagawa, M.Kinefuchi and A.Yamada, High Pressure Bioscience, S.Kunugi, S.Shimada, A.Suzuki, and R.Hayashi, (eds.), Kyoto, San-Ei Shuppan, 328-335 (1994).
9) A.Yamazaki, *The Review of High Pressure Science and Technology,*6,182 (1997).
10) A.Yamazaki, A,Sasagawa and A.Yamada, *Netsu Sokutei,* 23, 149 (1996)
11) K.Kainuma, A.Matsunaga, T.Itagawa and S.Kobayashi, *Denpun Kagaku,* 28, 235 (1981).
12) H.Isono, K.Ohtsubo, T.Iwasaki and A.Yamazaki, *Nippon Shokuhin Kogyo Gakkaishi,* 41, 485 (1994).
13) A.Yamazaki and A.Sasagawa, Shin Shokkan Jiten, K.Nishinari, F.Nakazawa, K.Katsuta and J.Toda (eds.) Tokyo, Science Forum,371-383 (1999).

Trends in High Pressure Bioscience and Biotechnology
R. Hayashi (editor)

Commercial use of High Hydrostatic Pressure in Sliced Cooked Ham in Spain

Narcís Grèbol

R&D Department, Esteban Espuña, S.A., Mestre Turina 39-41, 17800 Olot, Spain

Sliced Cooked Ham, produced by Esteban Espuna, pressurized at 400 MPa during 10 minutes, in a 320 litters industrial unit made by ACB, is commercialized in Spain since October 1998. Shelf life is extended to 60 days, with fresher flavor. Risks associated to pathogens are reduced significantly. Higher pressures (600 MPa) would be necessary to efficiently process fresh meats or meat products with low water activity.

1. PRODUCT DESCRIPTION

Product name: Jamón Cocido Reserva Especial
Comercial category: Extra Sin Fosfatos Añadidos (no polyphosphates added)
Weight: 200 g – 5 slices per pack
Ingredients: Ham, water, salt, dextrose, carrageen, sodium citrate, sodium ascorbate, spices and sodium nitrite.
Humidity: 71.89 %
Protein: 18.50 %
Salt: 17.20 g/Kg
Water activity: 0.980
Best before: 60 days
Vacuum Darfresh skin packaging, with polypropylene interleaver.
Labeling: Product pasteurized by High Pressure. Keep between 0°C. and 5°C.

2. HIGH HYDROSTATIC PRESSURE SYSTEM DESCRIPTION

Made in France by ACB.
Horizontal pressure vessel.
Capacity: 320 litters.
Vessel diameter: 280 mm (int.); 800 mm (ext.).
Total length: 18.5 m (including automatic loading/unloading system).
Total weight: 44 000 Kg.
Maximum pressure: 400 MPa.

3. PRODUCTION CAPACITY

6 baskets per cycle.
20.8 Kg. sliced packed cooked ham per basket.
125 Kg. per cycle
500 Kg. per hour
HHP Sliced Cooked Ham processed by January 2001: 1 200 000 Kg.

4. HHP PROCESS DESCRIPTION

Target pressure: 400 MPa.
Cycle time: 15 – 16 minutes (Figure 1).
    30 sec. loading/unloading.
    4 min. pressurizing
    10 min. at 400 MPa.
    1 min. depressurizing.
Initial water temperature: + 8°C.
Water temperature at 400 MPa: +13°C.
Final water temperature: + 9°C.

5. JUSTIFICATION OF THE USE OF HHP IN SLICED COOKED HAM

5.1. Lower risk of pathogens:
After inoculation at 10E5 level and HHP 10 min. 400 MPa, we obtained:

Salmonella: absence in 25 g.
Listeria monocytogenes: 3 log reduction.
Staphylococcus aureus: less than 2 log reduction.

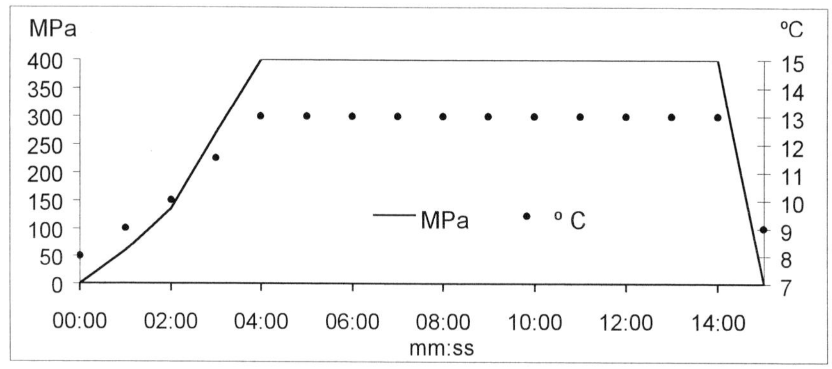

Figure 1. HHP standard process for sliced Cooked Ham: pressure and temperature.

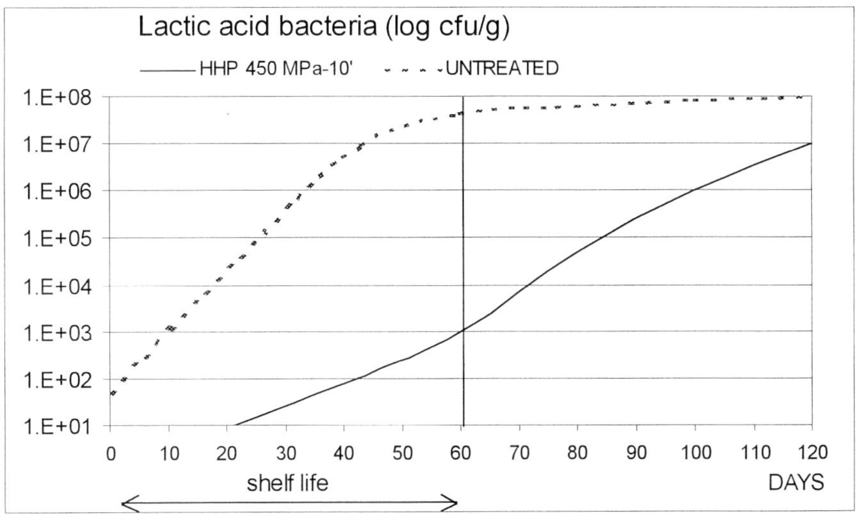

Figure 2. Compared growth of LAB in HHP and untreated samples.

5.2. Fresher flavor.

10 min at 400 MPa in sliced cooked ham allows a significant reduction of population, activity and growth capacity of microorganisms producing unpleasant flavors in sliced cooked ham (lactic acid bacteria, Enterobacteriaceae, yeasts, etc.).

That means higher consumer satisfaction and longer shelf life. Figure 2.

Consumer tests show no differences on flavor, color, texture and taste between treated and untreated samples.

6. HHP COSTS.

The main costs of using this technology are financial costs and maintenance costs. Using higher pressure, meaning shorter processing time and higher production capacity, financial costs will be reduced. Maintenance costs depend on the reliability of every component of the system, and they will tend to be lower in the future. Labor costs depend on the degree of automation, and energy costs are extremely low.

Because of financial costs, present standard for HHP in most food products could be 600 MPa with short processing time, to combine effectiveness, processing capacity and reasonable cost.

388

7. TESTS WITH OTHER MEAT PRODUCTS AT HIGHER PRESSURES.

7.1. Dry cured ham.
Dry cured ham is a low water activity meat product with a microbial natural flora containing big numbers of *Microccoccus* and *Staphylococcus*, highly resistant to High Pressure Processing.

But, using 600 MPa, preliminary tests suggest that HHP could be the right technology for effective cold pasteurization of dry cured meat products against pathogens.

Consumer tests show no differences on flavor, color, texture and taste between treated and untreated samples if the dry cured ham has been previously aged more than six months.

7.2. Fresh or marinated meats.
Fresh meat free from pathogens has been produced also in experimental tests using HHP at 600 MPa.

Raw meat changes in color and texture when exposed to high pressure, because of protein denaturation.

In the case of raw beef meat, natural unprotected meat pigment changes to grey-brown color when pressurized, but if nitrite is present in the marinade, the pigment remains pink instead of the red color of raw beef meat.

8. CONCLUSIONS

HHP 400 MPa, as an additional final processing step, is a suitable method to produce safer and fresher sliced meat products.

A real field test of more than two years, with sliced cooked ham produced by Esteban Espuna in Spain, demonstrates that this technology is safe, efficient, and accepted by consumers.

Higher-pressure systems are needed to optimize HHP in meat products with lower activity, to obtain higher pathogen reduction levels, and to minimize treatment costs.

Trends in High Pressure Bioscience and Biotechnology
R. Hayashi (editor)

Effects of high-pressure processing on the quality of green beans

B. Krebbers [a], M. Koets, F. van den Wall, A.M. Matser, R. Moezelaar,
S.W. Hoogerwerf

Agrotechnological Research Institute (ATO), P.O. Box 17, 6700 AA, Wageningen,
The Netherlands

[a] Corresponding author. Tel.: +31-317-477566; fax.: +31-317-475347. E-mail address:
b.krebbers@ato.wag-ur.nl

ABSTRACT

High-pressure processing is a mild preservation method, currently applied for pasteurization of foods. In this work a pulsed high-pressure method for achieving sterilization of whole intact products was under investigation. The effects of several high-pressure processes on the product quality of intact green beans were compared to conventional preservation techniques. The effects of these preservation technologies on texture, color and ascorbic acid content were evaluated. In addition reduction of natural flora was determined, also after storage of green beans. High-pressure conditions, 500 MPa at 20°C, showed inactivation of vegetative cells, only spores remained. Two-pulse pressure treatment, 1000 MPa at 75°C, showed a reduction of all natural flora below the confidence threshold. After one month storage there was no significant outgrowth of microorganisms. Both 500 MPa high-pressure and two-pulse high-pressure treatment showed a large retention of firmness for green beans compared to conventional preservation methods. High-pressure and two-pulse pressure treatments resulted in changes in greenness (a*-value). 500 MPa high-pressure treatment at ambient temperature showed no effect on ascorbic acid content, whereas two-pulse high-pressure treatment, 1000 MPa, 75°C showed a small decrease.

1. INTRODUCTION

High-pressure as a pasteurization process has been applied for mild food preservation (Cheftel, 1992). Meyer (2000) has patented a high-pressure process consisting of at least two pressure-pulses at a minimum product temperature of 70°C aimed to sterilize foods, without adversely affecting flavors and minimal changing texture and color.
Due to the extensive thermal treatment caused by slow heat penetration and subsequent slow cooling, conventional processing induces off-flavors, softens the

texture and destroys color and nutrients. This will affect food acceptability determined by these major quality attributes.

High-pressure techniques create new opportunities for replacing conventional processes, such as heat sterilization and freezing, to meet the consumer demand of fresh-like minimal processed foods.

Effects of high-pressure preservation techniques are generally marked as retention of color, flavor and fresh appearance as a result of the mild processing character. However, pressure may cause several adverse biochemical changes, such as activation of enzymes (Hendrickx et al., 1998), enhanced enzyme-substrate reactions and undesired browning reactions (Matser, 2000). As a result of complex biochemical changes, pressure conditions applied to obtain a microbial stable product may be in conflict with conditions to maintain preservation of color, texture or vitamins. An integrated approach is therefore necessary to study the effect of high pressure on the total quality of intact whole products, involving aspects such as microorganisms, nutrients, endogenous enzymes, color, texture and storage stability.

The aim of the present study was to investigate the effects of high-pressure on the product quality and storage stability of green beans compared to conventional preservation techniques. We studied the effect of high-pressure processing at pasteurization conditions (HPP, 500 MPa, ambient temperature) and sterilization conditions as described by Meyer (pulsed high-pressure processing pHPP; two-pulse treatment at a minimum temperature of 70°C) (Meyer, 2000). The effects of these preservation technologies on texture, color and ascorbic acid content of green beans were evaluated directly after processing. In addition, reduction of natural flora, vegetative cells as well as spores, were determined after processing and one month storage.

2. MATERIALS AND METHODS

2.1 Processing conditions

High-pressure treatments

Green beans (variety Amstel, stored at 6°C before processing) were washed in tap water and vacuum packed in polyethylene bags. High-pressure treatments were performed in duplicate in a Resato high-pressure apparatus (Resato, Roden, The Netherlands) at several pressure-temperature combinations. Pressure buildup rate was 10-15 MPa/s. High-pressure treatment (HPP) was done at ambient initial temperature for 60 sec. at 500 MPa. Pulsed high-pressure treatment (pHPP) was done according to the Meyer patent (Meyer, 2000). The samples were preheated at 75°C for 2 min., transferred to the high-pressure apparatus and processed at 75°C, holding time 80 sec. at 1000 MPa, second pressure pulse of 1000 MPa after 30 sec. at 0.1 MPa. Treatments were performed in duplicate. Due to adiabatic compression, the maximum temperature at the first pulse in the vessel for HPP and pHPP was respectively 45°C and 105°C.

Conventional treatments
For the canning process (CS), portions (360 g) of blanched (4 min. at 90°C) beans were packed into glass jars (720 mL) and a 1% NaCl and glucose solution was added. Closed jars were sterilized at 118°C for 30 min in duplicate. For the freezing process (BF), beans were frozen at −20°C after blanching under forced air in 10 min. For firmness and color measurements, beans were defrosted by microwave (4 min. 180 W, 2450 MHz).
After different processing treatments, the samples were either used for firmness and color measurements or frozen in liquid nitrogen and kept at −50°C until biochemical analyses. Processed, packaged samples were also stored for 1 month at 20°C for CS and pHPP treated samples or at 6°C for raw (R) or HPP treated samples.

2.2 Analysis
Firmness measurement
The firmness of raw and treated beans was measured in 15-fold fold using a Texture Analyser TA-XT2i (Stable Micro Systems, Godalming, UK) equipped with a Warner-Bratzler Blade. A green bean was placed on two parallel bars with a gap of 10 mm between them. The bean was fractured by a downward motion (10 mm/min.) of a steel blade with a thickness of 3 mm. The maximum force (top value in N) applied to break the beans was used to quantify the firmness.

Color measurement
The color of the beans was calculated from fifteen measurements with a Minolta Chroma meter, using the L*,a*,b* color space.

Microbial determinations
Reductions of the natural flora, sporal and total plate count, were investigated in whole green beans, processed under different conditions.

Ascorbic acid analysis
Extraction of ground green beans was performed in line with Veltman et al., (1999). Total ascorbic acid (AA) and dehydroascorbic acid content were measured using a Waters (Milford, USA) HPLC system model 510 with a Waters 486 UV-VIS detector (215 nm) described by (Keijbets and Ebbenhorst Seller, 1990).

3. RESULTS AND DISCUSSION

3.1 Effect of high-pressure and conventional treatments on microbial reduction

In Table 1 results are shown for the counts of naturally present flora, vegetative cells and spores, of the raw, sterilized and pressure-treated green beans.

Table 1
Number of natural flora (log cfu/ g product) compared to raw green beans

	Viable count	Spore count
Raw	6.1	3.5
HPP (500 MPa, 20°C)	1.7	1.7
Conventional sterilization	<CT[a]	<CT
pHPP (2* 1000 MPa, 75°C)	<CT[b]	<CT[c]

[a] CT= confidence threshold
[b] 1 cfu on 1 plate out of 6 plates
[c] 6 cfu on 1 plate out of 6 plates

Conventional sterilization indeed showed complete inactivation of viable cell number. HPP-treatment of beans (500 MPa, 20°C) resulted in complete inactivation of vegetative cells, ie. a reduction of 4 log cycles, resembling a pasteurization treatment. Pulsed high-pressure treatment (2 pulses of 1000 MPa, 75°C) showed a reduction of spores and vegetative cells below the detectable levels, ie. a reduction of viable cell number of 6 log cycles of the natural flora. These results confirmed the Meyer patent (Meyer, 2000), claiming a commercial sterilization of foods under applied conditions.

Throughout storage at respectively 6°C and 20°C for HPP and pHPP samples no significant outgrowth of total mesophilic (an)aerobic count or spores could be detected. The raw untreated beans were spoiled (> 10^6 cfu) after 1 week storage at 6°C.

3.2 Effect on color

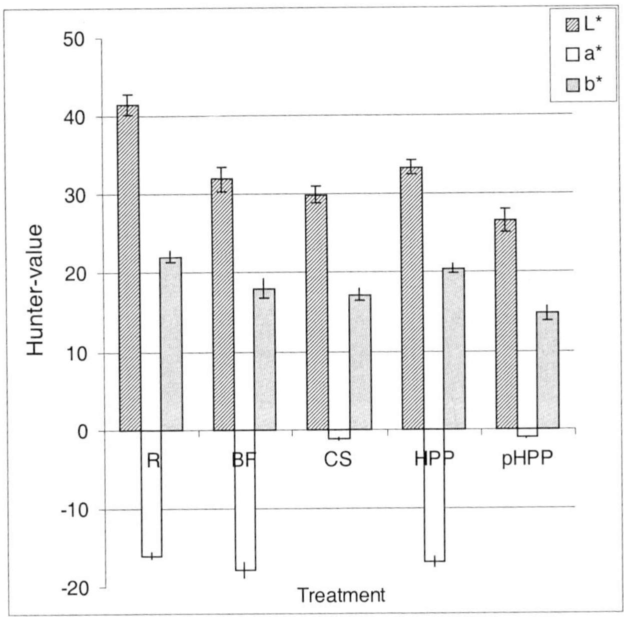

Figure 1. Effects of high-pressure and conventional treatments on L*, a*, b* values of green beans. R= raw; BF= freezing after blanching; CS = conventional sterilization; HPP= high-pressure processing at 500 MPa, 20°C; pHPP= pulsed high-pressure treatment at 75°C, 1000 MPa.

Color differences between conventionally and pressure-processed beans were determined directly after processing (L*, a* and b*-value in Figure 1). L* and a*-values of blanched and frozen beans were significantly lower than the raw green beans, representing a more intense green color. This decrease was also measured for the HPP-treated beans, resembling the appearance of mildly heat treated beans. An increase in permeability and leakage of chlorophyll of the cell wall is a possible explanation for the observed bright green color.

During conventional sterilization the beans showed a clear increase in a*-value to − 1.2; a visible color change from green to olive-green. Similar changes were measured for pHPP treatment of green beans. These results indicate that high-pressure in combination with higher temperatures, even for a short time, can already be detrimental for greenness. Temperatures higher than 50°C will cause conversion of chlorophylls to yellow-olive colored pheophytin.

394

3.3 Effect of high-pressure and conventional treatments on firmness

In Figure 2., the firmness of green beans as a result of different processing conditions is shown. The firmness of green beans decreased dramatically after conventional sterilization; the firmness was only 3 % of the original firmness of fresh beans.
The most important softening process during sterilization is believed to be temperature-dependent β-elimination of pectin (Saijaanantakul, 1989).

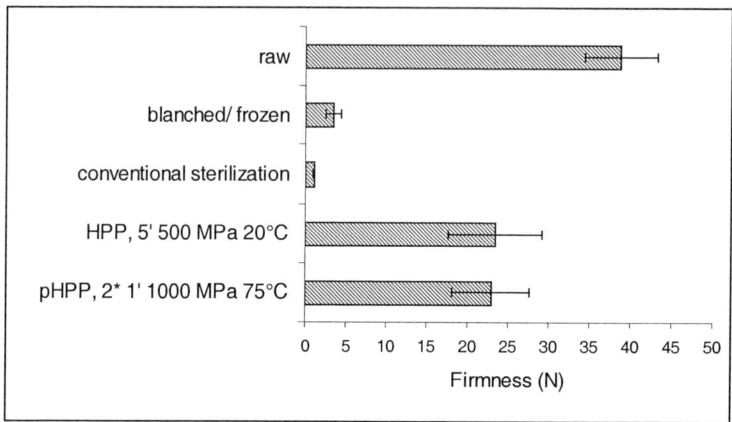

Figure 2. Effects of high-pressure and conventional treatments on firmness of green beans.

Pulsed high-pressure treatment (pHPP, 2 pulses of 1000 MPa, 75°C) and high-pressure treatment (HPP, 500 MPa, 20°C) resulted in a significant higher firmness compared to the conventional sterilization process (3% of initial firmness), up to 60% of the original firmness of the raw beans. This relatively large retention of firmness after pressurization may partly be due to the lower temperature and shorter time to which the beans have been exposed, resulting in less β-elimination of pectin. Besides the milder conditions applied, the larger retention of the firmness may also be attributed to activation or retention of pectin methylesterase (Stute et al., 1996). Blanching, freezing and subsequent defrosting by microwave also appeared detrimental for firmness, as less than 10% of the initial firmness remained.

3.4 Effect on ascorbic acid content

The influence of pressurization, freezing and sterilization on the ascorbic acid content is represented in Figure 3.

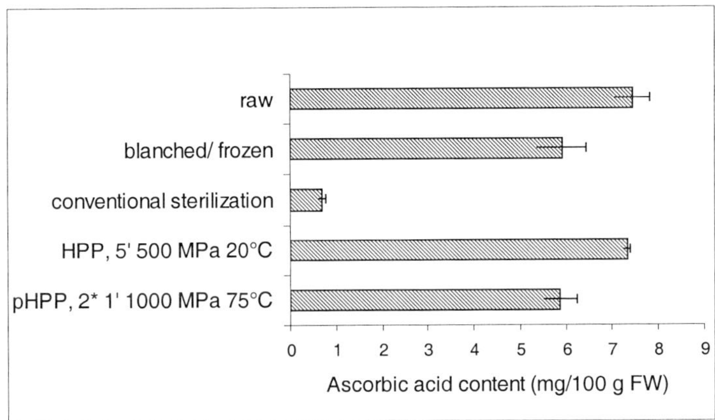

Figure 3. Effects of high-pressure and conventional treatments on the changes in the average ascorbic acid content for green beans during storage.

Losses in ascorbic acid were largest after conventional sterilization (up to 90%). HPP processing did not significantly affect ascorbic acid content compared to the raw beans. This implicated that pressure (500 MPa) at ambient temperature had no effect on this nutrient. Ascorbic acid retention of green beans after pHPP treatment was 75%; this decrease was probably due to (short) exposure to heat and pressure. Blanching followed by freezing resulted in an equal loss of ascorbic acid as in pHPP-treated beans. Trends for vitamin C appeared to be similar to those measured for ascorbic acid.

4. CONCLUSIONS

The effects of high-pressure processing on color, texture, microbial load and ascorbic acid content compared to conventional processing, have been examined.
Results showed that by use of HPP (500 MPa, 20°C) color, firmness and ascorbic acid were retained after HPP treatment compared to raw beans.
Pulsed high-pressure processing at elevated temperature (double pulse of 1000 MPa, 75°C) showed similar microbial reductions as in sterilized green beans. Firmness and ascorbic acid were largely retained (respectively 60% and 75%). Green color loss was comparable to sterilized green beans.
HPP and pHPP (double pulse of 1000 MPa, 75°C) can be used for production of green beans with extended microbial shelf-life (at least 30 days) resulting either in pasteurization (stored at 6°C) or sterilization (stored at 20°C), while retaining firmness, color and ascorbic acid.

Despite the fact that retention of texture, ascorbic acid and microbial stability was advantageous in the pasteurization and sterilization process of green beans by high-pressure, changes in color appeared to be a limiting factor in enhancing shelf-life.

Both high-pressure and pulsed high-pressure treatments have potential to be an alternative for conventional sterilization or freezing processes, with improved texture and nutritional attributes. Consumer studies and cost-yield studies should prove the merits of double pulse pressurization combined with higher temperatures aimed to sterilize, such as quality, safety, energy and water savings are acceptable for replacing conventional freezing or sterilization techniques in future.

ACKNOWLEDGEMENTS

This work was a joint action of Unilever and Stork, funded by the Dutch ministry of Economical Affairs (EET program). With special thanks for their practical and relevant contribution: L.M. van Mourik, R.W. van den Berg, P.V. Bartels and R. Meyer.

REFERENCES

1. Cheftel, J. C. (1992). Effects of high hydrostatic pressure on food constituents: an overview. High-pressure and Biotechnology. C. Balny, R. Hayashi, K. Heremans and P. Masson: 195-209.
2. Hendrickx, M., Ludikhuyze, L., Van den Broeck, I. and Weemaes, C. (1998). "Effects of high-pressure on enzymes related to food quality." *Trends in Food Science and Technology. 9*: 197-203.
3. Keijbets, M. J. H. and Ebbenhorst Seller, G. (1990). "Loss of vitamin C (L-ascorbic acid) during long-term cold storage of Dutch table potatoes." *Potato Research 33*: 125-130.
4. Matser, A. M., Knott, E.R., Teunissen, P.G.M., Bartels, P.V. (2000). "Effects of high isostatic pressure on mushrooms." *Journal of Food Engineering 45*: 11-16.
5. Meyer, R. S. (2000). US 6017572. Ultra high-pressure, high temperature food preservation process. Us, Meyer, Tacoma, WA 98443, USA.
6. Saijaanantakul, T., Van Buren, J.P., Downing, D.L. (1989). "Effects of methyl ester content on heat degradation of chelator-soluble carrot pectin." *J. Food Science 5*: 1273-1277.
7. Stute, R., Eshtiaghi, M., Boguslawski, S. and Knorr, D. (1996). High-pressure treatments of vegetables. High-pressure Chemical Engineering. R. Von Rohr, Trepp, C. New York, Elsevier Science.
8. Veltman, R. H., Sanders, M.G., Persijn, S.T., Peppelenbos, H.W., Oosterhaven, J. (1999). "Decreased ascrobic acid levels and brown core development in pear (Pyrus comminis L. vc. Conference)." *Physologia Plantarum 107*: p. 39-45.

Trends in High Pressure Bioscience and Biotechnology
R. Hayashi (editor)

397

High pressure advantages for brewery processes

S. Fischer, W. Russ, R. Meyer-Pittroff

Lehrstuhl für Energie- und Umwelttechnik der Lebensmittelindustrie,
Technische Universität München – Weihenstephan
Weihenstephaner Steig 22, 85350 Freising – Weihenstephan, Germany

The intention of these determinations was to improve the stability of beer without thermal influence. After the awareness, that it is possible, we started screening the whole brewery processes, beginning with high pressure treatment of mash, wort and green beer. The most interesting points of these trials were the filterability of green beer and the isomerisation of hop ingredients. First determinations in filterability showed, that an improvement is possible. Especially after a treatment with 500 MPa beer has a good filterability. Following we analyzed the influence of the high pressure treatment on poly phenols, proteins, the particle size and β-Glucan-gel. These measurements figured out, that only the content of β-Glucan-gel is changing, so it seems to be possible that β-Glucan-gel can be transformed in β-Glucan-sol, a state with more integrated water. Isomerisation of α-acids of hop using high pressure is also very interesting, because there is a loss of volatile compounds during the wort boiling. First measurements showed, that isomerisation of α-acids is only possible at 700 MPa and a minimum of 30 minutes. The problem is the solution of the hop ingredients and the diffusion into the wort. In further trials we determined, that different isomerisation products arise by using high pressure or temperature. The identification of these products is not finished yet. The high pressure treatment of wort, green beer and beer is very promising. The shown effects can be first steps for solving problems in different brewery processes, either in technological or in analytical way.

1. INTRODUCTION

The stabilization of beverages with low pH-value is a well known application of high pressure. Also some determinations were done in stabilization of beer [1]. The stabilization of beer is a very interesting field for breweries. The larger distribution areas and the greater competition for market shares make it necessary to produce beer with better stability and taste during storage in liqueur stores. Thermal processes for improving stability lead to poor sensoric quality. So it is necessary to search for new, innovative techniques for solving this assignment. Several determinations pointed, that high pressure is very effective in stabilizing food and beverages with low pH-value. High pressure improves the microbiological, colloidal and flavor stability [2]. These requirements seem to be perfect for the stabilization of beer. But this is not the only possibility of using high pressure in breweries. During our screening of the whole brewing process, we detected that the influence of high pressure on the filterability of beer and the isomerisation of hop ingredients will be very promising. Disposal of kieselguhr, which is used for the filtration, becomes even more expensive. Thus it is very interesting for breweries to improve filtration duration, for getting less of the problematic kieselguhr waste. Isomerisation of hop ingredients is important for the characterization of different beer types. The conventional technology is boiling. But there is an enormous loss of volatile ingredients which are flavoring agents for a fine bitterness. Thus it is very interesting to find a possibility to isomerise hop ingredients without thermal influence and without losing valuable substances.

2. BASICS

Figure 1 gives a short overview of the brewing process. First malt must be crushed, the so called milling, then it has to be mixed with water. During mashing the enzymatic processes take place, e.g. degradation of proteins and carbohydrates.

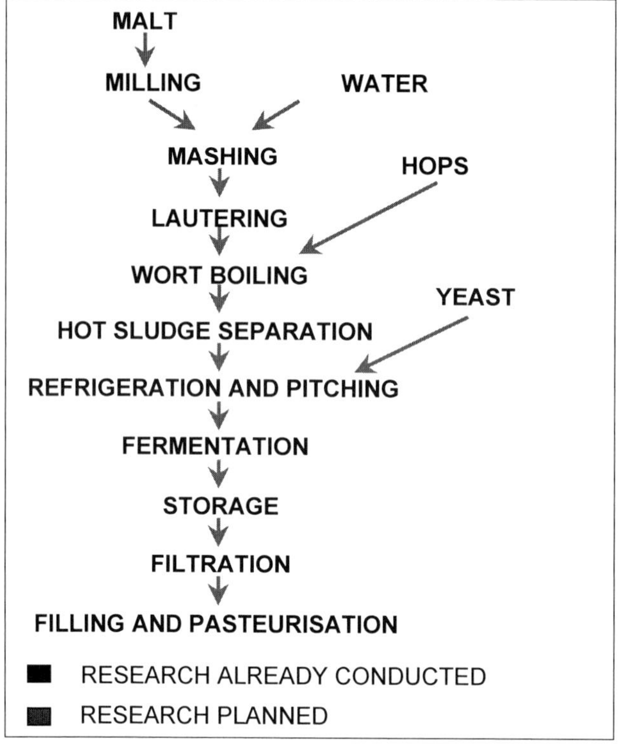

Figure 1. The brewing process

After mashing the solid particles have to be separated from the liquid, which happens during the lautering. The hop is added while the wort boiling. Next step in the process is the separation of hot sludge, then cold sludge. At ca. 8 °C the yeast will be added and fermentation starts. The yeast convents low molecular carbohydrates (starch) to alcohol. The storage is at 0 °C, it is necessary for the carbon dioxide content, the limpidity and maturation of beer. Filtration is not absolute necessary, it is possible to consume beer with a natural turbidity, but for good stability the filtration is inevitable. The last steps are filling in bottles or barrels and pasteurisation.

3. MATERIALS & METHODS

For determinations in the different brewery processes different requisitions were needed. For the analyses of isomerisation "pfannevollwürze" of the "Weihenstephaner Staatsbrauerei" was mixed with hop products. The best results were achieved with hop extract, so it was used for the determinations. First the mash was mixed with a defined amount of hop extract. Second the mixture was high pressure treated and third the analyses according to the "Mitteleuropäische Brautechnische Analysenkomission (MEBAK)" were performed. For getting closer information of the arising substances an identification using HPLC was employed.

Measurements of the filterability were realized with two different beers. First we took a beer with no filtration problems, just to ascertain, if there is any influence. Second we used a beer with enormous filtration problems. These beers were high pressure treated and filterability was measured in different systems. The analyses were done on a laboratory filter with a cellulose shift, and on a filtration system with kieselguhr. Additional the influence of high pressure on protein-, polyphenol-, anthocyanogen-, β-Glucan-gel-content and particle size was detected.

The influence of high pressure on the stability of beer was determined with unstabilized beer samples which were high pressure treated and pasteurized. Analyses were realized with an alcohol-chill-test and the detection of the turbidity.

All samples were treated in PE-bags or PET-bottles. The compression of the fluid was compensated by the package.

4. RESULTS

The results will be presented in diagrams, if there is some high pressure induced influence remarkable. Otherwise the results will only be mentioned in the discussion.

4.1. Isomerisation of hop ingredients

Isomerisation of α-acids is very important for the development of the typical bitterness of the different beer types. Isomerisation is accomplished by boiling the wort. This technology has a negative effect, the volatile substances which are very important for the taste steam out during the boiling.Thus it is necessary to lace hops with better quality at the end of the boiling.

400

Using high pressure for accomplishing the isomerisation the loss of volatile components will decrease, because there is no steaming out. First measurements showed, as in figure 2 demonstrated, that it is possible to isomerise α-acids in wort.

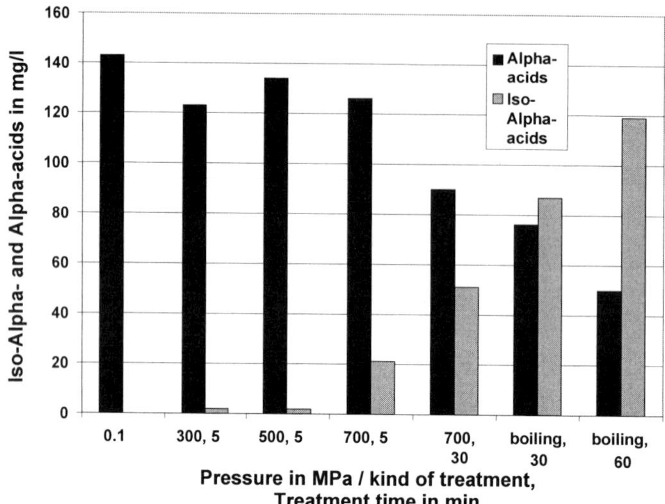

Figure 2. Iso-α-acids and α-acids in mg/l in dependence of pressure and treatment time resp. boiling time

Isomerisation depends on a significant time relation. Only at 700 MPa a clear increase of iso-α-acids is detected. Longer treatment time of 30 minutes leads to even better results. The decreasing content of α-acids depends on the isomerisation and verifies the result of the analyses. In comparison to boiled samples the high pressure induced isomerisation is less. This depends on worse solution of hop ingredients because of the missing motion during the high pressure treatment. Thus the bitter units of high pressure treated samples are less than with the boiled samples. Additionally treatment time is shorter, because there is no heating and cooling time by using high pressure. In consideration of these facts, the isomerisation at 700 MPa is very good, it is near to the 30 minutes boiled sample.

The next step was the identification of the iso-α-acids, and these determinations lead to very astonishing results. The only suggestive pressure range for this determinations was at 600 MPa or higher. The following treatments were carried out at 600 MPa for 30 minutes. As shown in figure 3, a significant high pressure induced isomerisation occurs, but only one peak is in the area of the thermal isomerisation products. The other high pressure induced isomerisation products fulfill the characteristics of iso-α-acids, but they are not identical with the thermal ones. Next step of this determination has to be the identification of the single iso-α-acids.

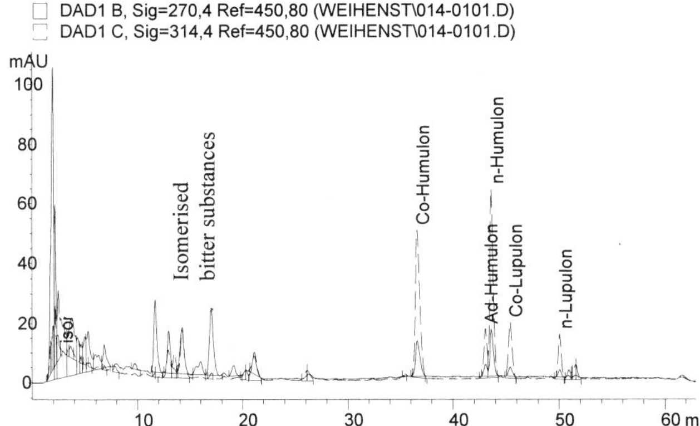

Figure 3. Isomerised bitter substances at 600 MPa, 30 minutes treatment time

The isomerisation of α-acids by using high pressure is possible. It is necessary to understand the difference between thermal and high pressure isomerisation. But the possibilities of a high pressure induced isomerisation are very promising, because there is no loss of volatile substances and perhaps it is possible to produce specific iso-α-acids with specific pressure-temperature ranges.

4.2. Filterability of beer

The filtration of beer is very important for stability, foam and limpidity. Conventional filter techniques are shift filters and kieselguhr filters. The problem with these filtration techniques is the waste disposal. It is very expensive to dispose kieselguhr, because in the newest waste regulations it is characterized as carcinogenic. Thus it is very interesting for breweries to increase the filtration duration of kieselguhr, because the longer the filtration duration, the less is the amount of waste.

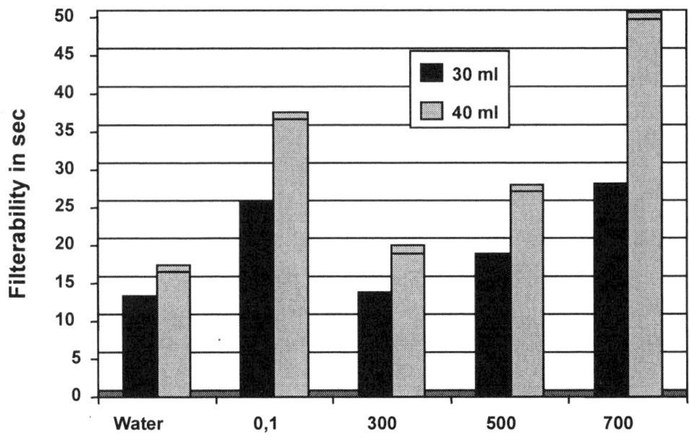

Figure 4. Filterability of beer in sec. in dependence of the pressure in MPa (shift filter)

402

The influence of high pressure on the filterability of beer was first measured with beer that has not shown any filtration problems. This measurement was done with a shift filter. The result is demonstrated in figure 4. The filterability becomes better, the filtration time decreases to 50 %. The sample treated with 300 MPa is even comparable with water. Next step was the high pressure treatment of beer with filtration problems. The result was equal, filtration time decreases to ca. 50 % by using 300 MPa. Of course, the level of filtration time was a bit higher than with the other beer. For the practical application it is necessary to verify these determinations on a kieselguhr filter. Figure 5 shows the influence of high pressure on the filterability of beer on a kieselguhr filter.

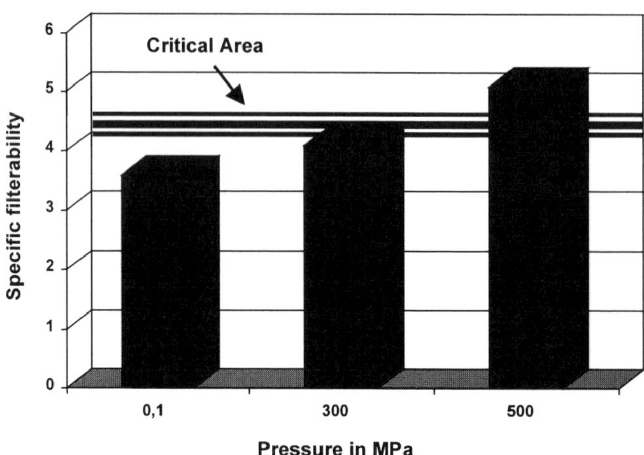

Figure 5. Specific filterability in dependence of the pressure in MPa (kieselguhr filter)

The result is significant, the high pressure treatment effects an improvement of the specific filterability. The only difference is that the best effect is achieved at 500 MPa. The filterability of beer can be improved by high pressure treatment, with shift filters and also with kieselguhr filters. The inevitable question was now, what is the reason for this phenomena? To figure out this problem, we determined behavior of several substances which are responsible for the filterability. The determination of poly phenols, anthocyanogens, protein fractions and particle size showed no change. Only the content of β-Glucan-gel is decreasing. As demonstrated in figure 6, the β-Glucan-gel-content after high pressure treatment with 300 and 500 MPa is unverifiable. There are two possible reasons, first β-Glucan-gel is completely transformed into sol, a state with more integrated water. This state does not allow detection and it has no negative influence on the filterability. Second, only a part of the β-Glucan-gel is transformed, the detection limit of the analytical system is 10 mg/l. So it is not possible, to commit if the β-Glucan-gel is transformed completely or not. Fact is that its content is decreasing and the filterability becomes better.

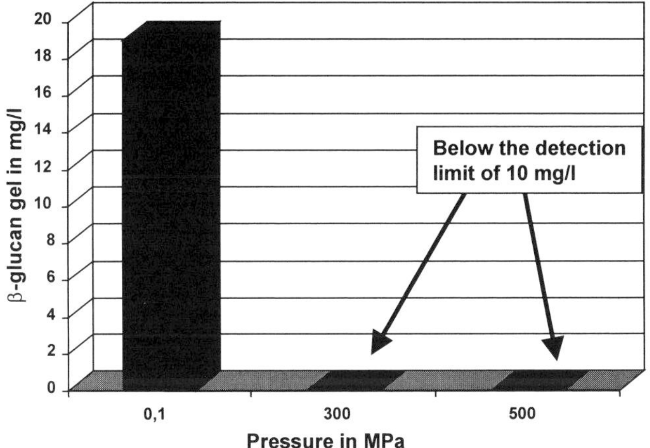

Figure 6. β-Glucan-gel in mg/l in dependence of the pressure in MPa

4.3. Colloidal stability of beer

The stability of beer is especially very important for breweries with a wide trading area. The problem is stabilization without any negative influence on the flavor. Conventional thermal processes cause a negative effect on the flavor. As demonstrated in figure 7 it is possible to stabilize beer of different types (Lager and Pilsener) by using high pressure.

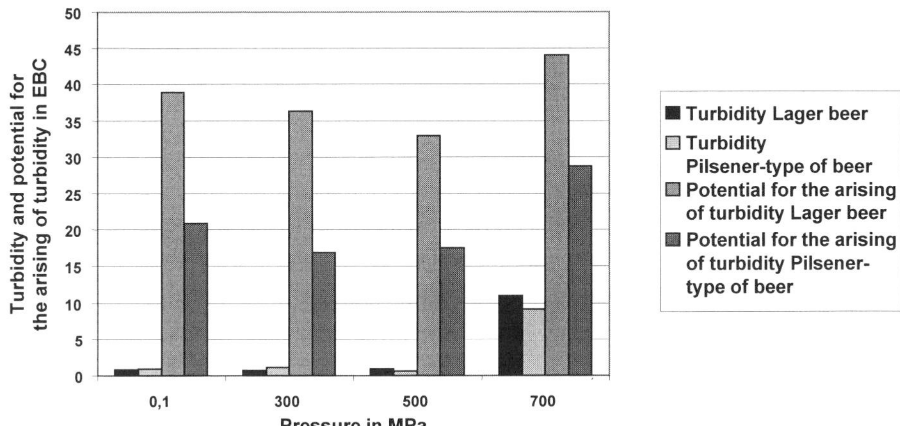

Figure 7. Turbidity and potential for the arising of turbidity in EBC in dependence of the pressure in MPa

The potential for the arising of turbidity is decreasing, this means that the stability becomes better. Pilsener type beer is more stable because of the higher content of hop ingredients. It is astonishing that turbidity is not influenced by high pressure. The particles responsible for turbidity are proteins and carbohydrates. They show no reactions under high pressure, but their tendency to arise turbidity during the storage becomes lower. Proteins and carbohydrates are influenced in that way that their reactivity becomes lower.

5. CONCLUSIONS

The high pressure technology shows very interesting possibilities in the brewing area, especially the isomerisation of hop ingredients without the loss of valuable, volatile substances seems to have a very good perspective. This application is also interesting for other industries like the pharmaceutical or food industry. The improvement of the filterability will be very interesting when it is possible to realize a continuos high pressure process. But also the discontinuous variation shows interesting possibilities for the analytical interpretation of filtration problems. The stabilization of beer by using high pressure is possible, but at the moment beer is to cheap or high pressure to expensive for using it in this way. The possibilities of high pressure will become better, when a continuos process is realized.

High pressure treatment of beer shows some interesting rudiments. At the moment the technology is not far enough developed for producing beer. But for analytical methods and in industries which are able to sell their product more expensive high pressure will have very good perspectives, because products are mostly of better quality and good stability.

REFERENCES

1. Fischer, S., Schöberl, H., Russ, W., Meyer-Pittroff, R., The Effects of Hydrostatic High Pressure on the Brewing Process and on Beer, Advances in High Pressure Bioscience and Biotechnology, Ed. Ludwig, H.. Springer Verlag, p. 419 - 422
2. Cheftel, J. Cl., Effects of High Pressure on Food Constituents: an Overview, High Pressure and Biotechnology, Ed. Balny, C.; Hayashi, R.; Heremans, K.; Masson, P.. John Libbey Eurotext Ltd., Vol. 224, 1992, p. 195 – 209.

Trends in High Pressure Bioscience and Biotechnology
R. Hayashi (editor)

STARCH-AMPHIPHILE COMPLEX FORMED BY HIGH PRESSURE

K. Yamamoto,[a] S. Handschin,[b] B. Conde-Petit[b] and F. Escher[b]

[a]National Food Research Institute, Kannondai 2-1-2, Tsukuba, Ibaraki 305-8642, Japan

[b]Institute of Food Science, Swiss Federal Institute of Technology Zurich, CH-8092 Zürich, Switzerland

1. ABSTRACT

Treated by high pressure, starch can gelatinize in the presence of water. On the other hand, starch can form starch-amphiphile complex when gelatinized by heating. In this study, we have treated potato starch in the presence of SDS (sodium dodecyl sulfate), an amphiphilic compound, by high pressure. X-ray diffraction and thermal analysis indicated the formation of starch-SDS complex under high pressure.

2. INTRODUCTION

Starch, comprised of linear chain amylose and branched chain amylopectin, is one of major food components, and it plays an important role in food texture like viscosity, roughness, and crispiness. Heated under the atmospheric pressure in the presence of water, starch gelatinizes and forms gel due to absorption of water and swelling of starch granules. This is termed "heat-gelatinization". After starch is gelatinized by heating, starch can form complexes with hydrophobic or amphiphilic compounds such as flavors and emulsifiers. The most known example is starch-iodine complex, of which purple is used for detecting starch. In this event the existence of amylose is indispensable, as an amylose molecule can form a helical complex with iodine molecules. The complex can crystallize and show in X-ray diffraction measurement different diffraction patterns from those of native or gelatinized starches such as A-type, B-type, and amorphous starches.

By high-pressure treatment, starch shows similar behavior to that in heat-gelatinization. This is termed "pressure-gelatinization". By incorporating amphiphile molecules into heat-gelatinized starch, amylose-amphiphile complex is formed. Based on this, it would be probable that such a complex is also prepared by high-pressure treatment. Therefore, we investigated on a ternary system comprised of starch, water, and an amphiphile (SDS: sodium dodecyl sulfate) in order to prove the feasibility. After adjusting the amounts of starch and SDS solution, the mixture was applied to the high-pressure equipment (Johannes-type piston cylinder) that can reach 2.0GPa. X-ray diffractometry, differential scanning calorimetry (DSC) and microscopy analyzed physico-chemical properties of the products.

3. MATERIALS AND METHODS

3.1. High-pressure treatment

Commercial potato starch was purchased from Blattmann (Switzerland). Sodium dodecyl sulfate (hereafter, SDS; Merck) was used as a model amphiphile. Starch was put into a copper capsule of approximately 0.1ml capacity. Water or SDS aqueous solution was added to the starch in the capsule to reach the water content of 50%w/w. The capsule was designed to seal wet sample well with a help of Teflon seal tape. The capsule was inserted into a salt assembly, which works both as solid medium and as a graphite furnace. The salt assembly was set to a Johannes-type piston cylinder that can theoretically pressurize solid medium up to 4.0GPa (practical max. 2.0GPa for safety). Temperature was monitored with a thermocouple that is inserted into the assembly and reached the capsule. After that, temperature was set at 40±1°C and automatically controlled during compression. Compression up to a constant pressure 1.5GPa was manually achieved in about 3-5min while monitoring the temperature so that adiabatic heating was minimized. After holding the high pressure for 1 hour, the pressure was released down to atmospheric pressure in about 2 min.

3.2. DSC analysis

Samples were weighed into DSC pressure pans (Perkin Elmer Ltd., USA) and water was added to the ratio of 3/7 (dry matter / water). Measurements were carried out using a computer-controlled Thermal Analyst 2000 (DSC 2910, TA Instruments, UK). The samples were heated at 10°C/min from 4 to 160°C under a nitrogen gas atmosphere

$(30cm^3 / min)$. Enthalpy change (ΔH) of endothermic peak was calculated by a software installed on the instrument, and it was evaluated by starch dry base (J/g dry starch). DSC data was not affected by storage time corresponding to the X-ray measurement time (about 4 hour) in this study.

3.3. X-ray analysis

After pressed into pellets or cut into pieces, the sample was mounted on a sample holder. X-ray diffractograms were measured under ambient condition by using an X-ray powder diffractometer (Siemens Kristalloflex D500, Germany) that generated monochromatic Cu $K\alpha$ beam (wavelength = 1.542Å) at 40kV and 35mA. The scanning regions of the diffraction angle were 5 to $30°$ 2θ, covering the significant diffraction peaks of starch crystallites. Measurement was done on the following condition: transmission mode; step interval, 30s; scan rate, $0.1°$ 2θ/min; divergence slit, 2mmϕ; receiving slit, $1°$; anti-scattering slit, $0.15°$. Native starch was moistened over saturated $BaCl_2$ aqueous solution to reach the water content of 24%w/w, and it was used for X-ray measurement.

4. RESULTS AND DISCUSSION

High-pressure treated starch (HP-starch) prepared without SDS appeared sticky translucent gel-like material, which was microscopically composed of tightly packed distorted starch granules. It showed a gelatinization enthalpy change of approximately 1J/g (Fig. 1) and basically unaltered X-ray diffractogram compared with that of native starch (Fig. 2).

On the other hand, HP-starch prepared in the presence of SDS was white non-sticky powder. Observed by light microscope, the resulting powder was composed of separate, slightly distorted starch granules, which developed brown by iodine staining, indicating that amylose molecules, potent iodine-complex (purple), were consumed by SDS-complex formation. Over the SDS concentration of 30mmol / mol AGU(anhydro-glucose unit: monomer unit of starch chain), the enthalpy change for SDS-amylose melting reached the maximum and the enthalpy change for gelatinization became zero (Fig. 1). The resulting specimen did not show typical native feature in X-ray diffraction but that of SDS-amylose complex, V-amylose (Fig. 2). These results suggest that amylose-SDS complex is formed by high-pressure treatment without preliminary heat-gelatinization.

In this study, SDS was tested as one of amphiphiles in order to investigate the feasibility of amphiphile complex formation with potato starch. As grain starch gelatinizes at lower pressure (0.2 – 0.4GPa)[1] than potato starch (1.0GPa)[2], it is probable that amylose-SDS complex would be formed in granular state at lower pressure by using grain starch such as wheat and corn starches. Further study employing the other amphiphiles would enlarge the application of this technique for preparing granular starch complex with amphiphiles of industrial interest such as flavors and emulsifiers.

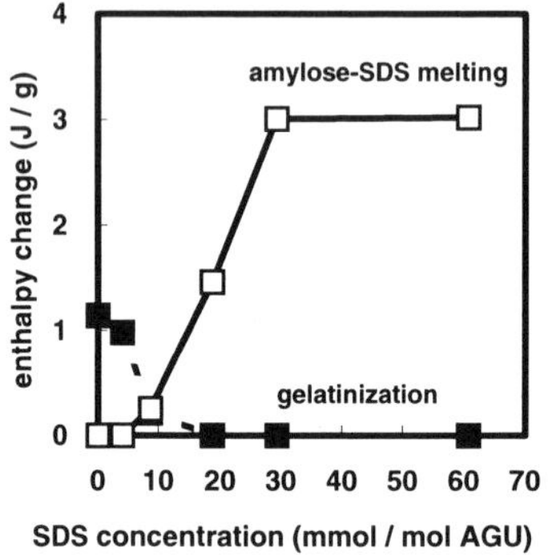

Fig. 1 Effect of SDS concentration on the enthalpy changes for amylose-SDS melting and gelatinization.

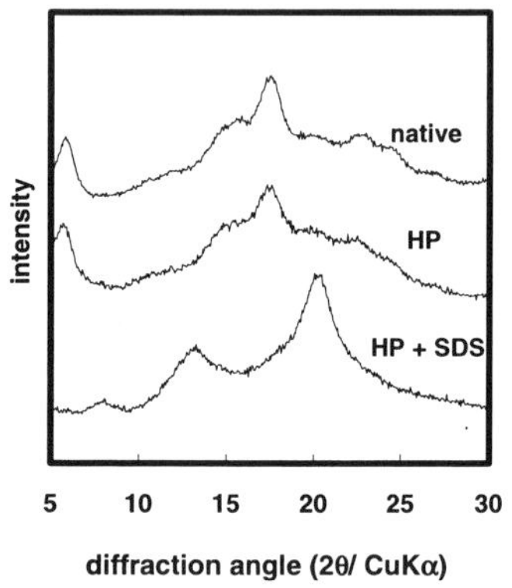

Fig. 2 X-ray diffractograms of high-pressure (HP) treated starches.

5. CONCLUSIONS

-   Gelatinization was completed and starch-SDS complex was maximally formed over the SDS concentration of 30mmol/mol AGU at 1.5GPa.
-   The complex was powdery and microscopically granular.
-   Our results suggest that starch-flavor/emulsifier complexes in granular state could be formed by high-pressure treatment.

REFERENCES

1. S. Ezaki and R. Hayashi, High Pressure and Biotechnology (Eds C. Balny, R. Hayashi, K. Heremans, and P. Masson). Colloque INSERM/ John Libbey Eurotext

1992, Vol. 224, pp.163-165.

2. K. Yamamoto, P. Ulmer, S. Handschin, B. Conde-Petit, and F. Escher, 1999 American Association of Cereal Chemists (AACC) Annual Meeting Program Book, 1999, pp256-257.

ACKNOWLEDGMENTS

Authors wish to thank the JST Overseas Fellowship by Japan Science and Technology Corporation (JST) for supporting K.Y. and this work. Authors are grateful to Dr. P. Ulmer (Institute of Mineralogy and Petrography, Swiss Federal Institute of Technology Zurich) for allowing the use of the high-pressure instrument.

Trends in High Pressure Bioscience and Biotechnology
R. Hayashi (editor)
                                                    411

Effect of High-Pressure Storage on the Processing Quality of Tilapia Meat

W.C. Ko and K.C. Hsu

Department of Food Science, National Chung-Hsing University,
250 Kuokuang Road, Taichung, Taiwan, ROC.

Abstract

The influence of high hydrostatic pressure-normal temperature storage on the processing quality of tilapia fillets was examined. Pressure ranged from 500 to 3,000 atm was applied to tilapia fillets at 25°C for up to 12 hr. The gel strength of the meat paste prepared from the pressurized tilapia showed slight increase for storage at 500 atm but showed obvious decrease for storage beyond 2,000 atm. The extraction and Ca-ATPase activity of actomyosin from tilapia fillets stored over 2,000 atm showed apparently falling trend and the protein denatured evidently. Low-pressure was useful for fish storage. Though relatively high pressure resulted in denaturation of tilapia proteins but the liability to denaturation indicates that pressure treatment was applicable to tilapia processing.

1. INTRODUCTION

Fish are the most perishable foodstuffs due to the action of intrinsic enzymes and the growth of extrinsic microorganisms (1). Both autolytic and microbial spoilages are dependent upon the temperature at which the fish are stored. Most deteriorative factors are usually retarded by temperature control. Though freezing and frozen storage remain primary means for cold preservation, several new methods such as controlled freezing point storage and partial freezing are developed for storing marine products to prevent texture from destruction.

Hydrostatic pressure exerts the same action as heating to inactivate microorganisms, influence biopolymers and affect activities of enzymes (2). Changes in the tertiary and quaternary structures are associated with volume changes and may therefore be affected by pressure (3, 4). Protein denaturation is caused by rearrangement and/or destruction of non-covalent bonds such as hydrogen bonding, hydrophobic interaction, and ionic bonding of tertiary structure of protein, while covalent bonds are not affected. Protein tends to denature, dissociate, or precipitate reversibly at hydrostatic pressures from 1,000 to 3,000 atm (5-7). Coagulation and gelation of protein under pressure are due to protein denaturation with SH groups still reactive after the pressurization. However, hydrophobic interactions are associated with a volume decrease and tend to be stabilized at pressures over 1,000 atm (8). The objective of this study was to store tilapia fillets at normal temperature and try to find out other means besides cold storage for maintaining the processing quality of fish meat.

2. MATERIALS AND METHODS

2.1. Materials.   Live round tilapia weighing about 600 g was purchased from a local retail market.   The fish was kept in ice before sample preparation.   The fish were filleted immediately and were packaged in polyethylene pouches and then vacuum-sealed.   All chemicals were of reagent grade.

2.2. Pressure treatment.   A high-pressure apparatus (CIP UNIT, Mitsubishi Heavy Industries Ltd., Japan) was employed.   Throughout the pressure treatment, the vessel temperature was controlled at 25℃.   The tilapia fillets were pressurized at 500-3,000 atm for 0-12 hr.

2.3. Gel preparation and gel strength measurement.   For the preparation of gels and measurement of gel strength, the method of Ko et al. (9) was used.

2.4. SDS-PAGE analysis.   The extracts of water- and salt-soluble proteins were prepared by the method of Dyer et al. (10).   10% acrylamide gels were used for the electrophoresis.

2.5. Assay of actomyosin Ca-ATPase activity.   Actomyosin preparation was done by using the method of Taguchi et al. (11).   For determining the activity of actomyosin Ca-ATPase, the method of Ko and others (12) was used.

2.6. Scanning electron microscopic (SEM) observation.   Pressurized fillets were fixed by freezing at -65℃ for 24 hr.   Frozen fillets were dehydrated by a sublimator (Va Co I, Zirbus, Germany) for 48 hr.   The microstructure of the fillets was observed by SEM (Bausch and Lomb, Nanolab 2100, USA).

3. RESULTS AND DISCUSSION

3.1. Gel strength measurement.
It is believed that pressure can induce protein denaturation and gelation as a processing method.   Gel strength of the meat gel prepared from the tilapia fillets via pressure storage at 500, 1,000, 2,000 and 2,500 atm for 1 hr were 2,780, 2,540, 1,470 and 650 g×mm, respectively, while that of fresh tilapia meat was about 2,200 g×mm (Table 1).   There existed no obvious relationship between storing time and gel strength when the storing pressure was over 2,500 atm.   The gel strength of the meat gel was promoted by storing the meat below 1,000 atm within 3 hr.   Because the rate of protein denaturation is reduced due to pressure enhancement of hydrogen bonds which are responsible for maintaining the helical structure of the peptides (13).   Conversely, the treatment at and over 2,000 atm for more than 3 hr would result in a decrease in the gel strength due to the denaturation of actomyosin (14).   The processing functions of the meat were kept under the pressure below 1,000 atm.

Table 1　Change in gel-forming property (breaking force×breaking strain; g×mm) of tilapia muscle during storage under high pressure at 25°C.

Pressure (atm) \ Time (hr)	1	2	3	6	12
1	2,316[g]	2,489[h]	2,493[h]	1,917[f]	1,986[f]
500	2,781[i]	2,628[h]	2,534[h]	2,329[g]	2,416[h]
1,000	2,537[h]	2,300[g]	2,197[g]	2,054[g]	1,863[f]
2,000	1,466[e]	1,610[e]	1,551[e]	1,067[d]	868[c]
2,500	645[b]	708[b]	633[b]	588[a]	560[a]
3,000	579[a]	632[b]	563[a]	560[a]	550[a]

Fresh meat: 2,202[g] g•mm
[a-i]: Means with the same letter are not significantly different (P<0.05).

3.2. SDS-PAGE analysis.

Extraction of water-soluble proteins was affected insignificantly by pressure but significantly by storage time, however, after pressurization at 2,000 and 3,000 atm, proteins with molecular weight above 97 kDa disappeared (Fig. 1). The results were attributable to the small molecular weight of the proteins and the low neutral lipid (e.g. triglyceride) content (15-17). The extraction rates of salt-soluble proteins of tilapia meat stored behind 2,000 atm were barely affected. But the extraction rates decreased rapidly due to the denaturation of proteins with larger molecular weight. Extraction of salt-soluble proteins was affected by pressure higher than 2,000 atm because actomyosin coagulated at the situation. The phenomenon demonstrates that the extraction of actomyosin from tilapia meat decreased rapidly when the meat was stored at 2,000 atm.

3.3. Actomyosin Ca-ATPase activity measurement.

The activity of actomyosin ATPase is an index for processing quality of some kinds of fish (18, 19). Table 2 shows the effect of pressure on the activity of actomyosin Ca-ATPase extracted from tilapia meat. The activity decreased from 0.46 to 0.28μmole Pi/min×mg protein when the meat was stored at 1 atm for 12 hr. When the meat was pressurized, the activity of actomyosin which could be extracted was maintained over 90%. Although the extraction of actomyosin from tilapia meat stored at 2,000 atm was below 10%, the activity of Ca-ATPase was kept above 0.40 μmole Pi/min×mg protein. Milkfish actomyosin was also not affected by pressure below 1,000 atm but its Ca-ATPase reduced below 20% of the fresh meat by 2,000-atm treatment due to denaturation (12). The result was different from this study. It was thought that the extractable actomyosin belonged to the undenatured protein. Compared to the fresh meat, the actomyosin Ca-ATPase activity almost does not change when the meat is stored at 2,000 atm. The structure of the actomyosin was not disrupted at the pressure lower than 1,000 atm, so its Ca-ATPase activity remained. The results indicate that tilapia meat stored under lower pressures still kept its processing quality.

414

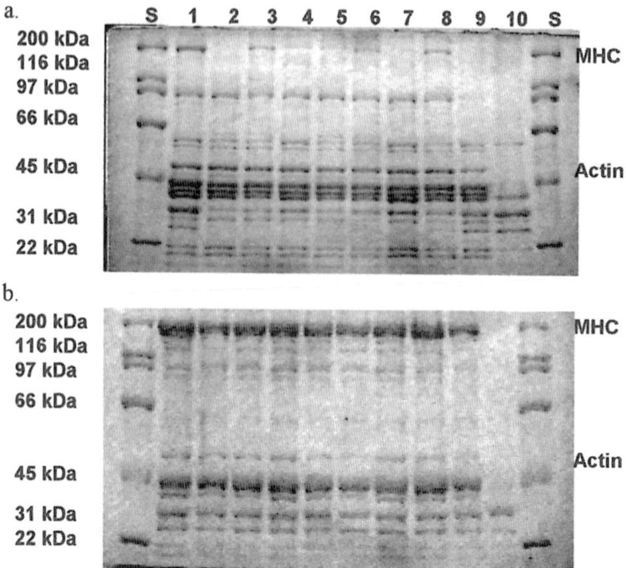

Fig. 1   SDS-PAGE patterns of dissolved protein of tilapia stored under high pressure at 25℃.

a. water-soluble proteins, b. salt-soluble proteins

S : standards; 1 : fresh tilapia meat; 2, 3 : stored at 1 atm for 6 and 12 hr; 4, 5, 6 : stored at 500 atm for 3, 6, and 12 hr; 7, 8 : stored at 1,000 atm for 3 and 12 hr; 9 : stored at 2,000 atm for 3 hr; 10 : stored at 3,000 atm for 3 hr.

3.4. SEM observation.

The muscle fibers of fresh tilapia meat presented an orderly and complete structure. After storing at 1 atm for 3 hr, the meat fibers decayed but still were orderly, however, after pressurizing at 500 and 1,000 atm for 3 hr, meat fibers did not change apparently (Fig. 2).   Furthermore, meat fibers collapsed completely after pressurization at 2,000 and 3,000 atm for 3 hr.   The result showed that pressure above 2,000 atm destroyed the fiber structures and therefore the processing quality of tilapia meat.

REFERENCES

1. Tomiyama, T., Kobayashi, K., Kitahara, K., Shiraushi, E., and Nobuyoshi, O. 1966. A Study on the change in nucleotides and freshness of carp muscle during the chill-storage. Bull. Japan. Soc. Sci. Fish. 32:262-266.
2. Hoover, D.G., Metrick, C., Papineau, A.M., Farkas, D.F., and Knorr, D.. 1989. Biological effects of high hydrostatic pressure on food microorganisms. Food Technol. 43:99-107.

Table 2　Change in actomyosin Ca-ATPase activity (μmole Pi/min×mg protein) of tilapia muscle during storage under high pressure at 25℃.

Time (hr) Pressure (atm)	1	2	3	6	12
1	0.446^e	0.437^d	0.424^c	0.364^b	0.281^a
500	0.461^g	0.459^g	0.449^{ef}	0.450^f	0.446^{ef}
1,000	0.464^g	0.463^g	0.461^g	0.450^f	0.444^{ef}
2,000	0.426^c	0.426^c	NE*	NE	NE

Fresh meat: 0.462^g μmole Pi/min×mg protein
*NE: Not extractable
$^{a\text{-}g}$: Means with the same letter are not significantly different (P<0.05).

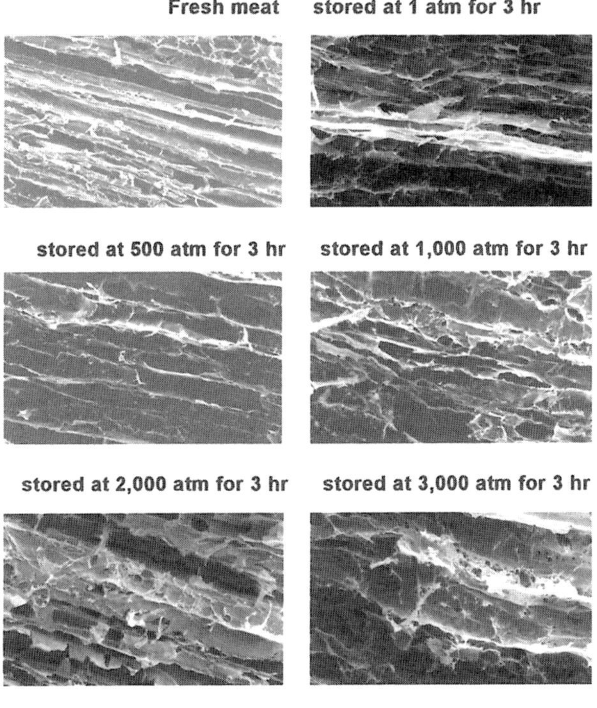

Fig. 2　Scanning electron microscopy image (× 100) of muscle filament of tilapia stored under high pressures at 25℃.

416

3. Heremans, K. 1992. From living systems to biomolecules. In High Pressure and Biotechnology; Balny C, Hayashi R, Heremans K, Masson PJ, Eds. London: Libbey Eurotext. p 37-44.

4. Cheftel, J.C. 1992. Effect of high hydrostatic pressure on food constituents: an overview. In: High Pressure and Biotechnology, Balny C, Hayashi R, Heremans K, Masson PJ, Eds. London: Libbey Eurotext. p 195-209.

5. Morild, E. 1981. The theory of pressure effects on enzymes. Adv. Protein Chem. 34:93-107.

6. Heremans, K. 1982. High pressure effects on proteins and other biomolecules. Ann. Rev. Biophys. Bioeng. 11:1-17.

7. Weber, G. and Drickamer, H.G. 1983. The effect of high pressure upon proteins and other biomolecules. Q. Rev. Biophys. 16:89-108.

8. Suzuki, K. and Taniguchi, Y. 1972. Effects of pressure on biopolymers and model systems. In The Effects of Pressure on Living Organisms; Sleigh MA, Macdonald AG. Eds. New York: Academic Press. p 103-131.

9. Ko, W.C., Tanaka, M., Nagashima, Y., Taguchi, T., and Amano, K. 1990. Effect of high pressure treatment on the thermal gelation of sardine and Alaska pollack meat and myosin pastes. Nippon Shokuhin Kogyo Gakkaishi 37:637-642.

10. Dyer, W.J., French, H.V., and Snow, J.M. 1950. Protein in fish muscle. 1. Extraction of protein fraction in flesh fish. J. Fish Res. Bd. Can. 7:585-593.

11. Taguchi, T., Kikuchi, K., Oguni, M., Tanaka, M., and Suzuki, K. 1978. Heat change of myosin B Mg-ATPase and "setting" of fish meat paste. Bull. Japan. Soc. Sci. Fish. 44:1363-1368.

12. Ko, W.C., Tanaka, M., Nagashima, Y., Taguchi, T., and Amano, K. 1991. Effect of pressure treatment on actomyosin ATPase from flying fish and sardine muscles. J. Food Sci. 56:338-340.

13. Johnson, F.H. and Campbell, D.H. 1945. The retardation of protein denaturation by hydrostatic pressure. J. Cell. Comp. Physiol. 26:43-49.

14. Shigehisa, T., Ohmori, T., Saito, A., Taji, S., and Hayashi, R. 1991. Effect of high pressure on characteristics of pork slurries and inactivation of microorganisms associated with meat and meat products. Int. J. Food Microbiol. 12:207-216.

15. De Koning, A.J., Milkovitch, S., and Mol, T. 1985. Quantitative quality tests for frozen fish: part I. Changes in the proteins and lipids of frozen hake fillets and mince and stored at -5 °C In: Thirty-ninth Annual Report of the Director. Rondebosch: Fishing Industry Research Institute. p 49-53.

16. Srikar, L.N. and Reddy, G.V.S. 1991. Protein solubility and emulsifying capacity in frozen stored fish mince. J. Sci. Food Agric. 55:447-453.

17. Pastoriza, L. and Samperdro, G. 1994. Influence of ice storage on ray win muscle (Paja clavata) . J. Sci. Food Agric. 64:9-18.

18. Kawashima, T., Arai, K., and Saito, T. 1973. Studies on muscular proteins of fish-X. The amount of actomyosin in frozen "surimi" from Alaska-pollack. Bull. Jap. Soc. Sci. Fish. 39:525-532.

19. Matsuda, Y. 1979. Influence of sucrose on the protein denaturation of lyophilized carp myofibrils during storage. Bull. Jap. Soc. Sci. Fish. 45:573-579.

Trends in High Pressure Bioscience and Biotechnology
R. Hayashi (editor)

Influence of high pressure treatment on sensorial and nutritional quality of fruit and vegetables

P. Butz, A. Fernández García and B. Tauscher

Institute for Chemistry and Biology, Federal Research Centre for Nutrition,
Haid-und-Neustr. 9, 76131 Karlsruhe, Germany

The influence of high pressure treatment on the sensory and nutritional quality of some fruit and vegetable products was assessed. The effect of packaging and high pressure treatment on sensory attributes of orange juice was studied using the triangular ‚Forced Choice' technique. No change caused by the packaging material was found. With increasing storage time odour, taste and harmony of high pressure treated juice changed much less than in non-treated. After 21 days at 4 °C, no remarkable changes in the content of fructose, glucose, sucrose, total sugar and titrable acids were observed, L-ascorbic acid content of orange juices decreased about 10 % in both, the pressurised and the untreated samples. Orange juice, apples, carrots and tomatoes were assessed for pressure effects on antioxidant activity which proved to be pressure insensitive in most cases.

1. INTRODUCTION

Increasing consumer demands for high-quality food with 'fresh-like' characteristics resulted in the introduction of several non-thermal minimal processing methods. Hydrostatic ultrahigh-pressure (UHP) above 100 MPa is used to inactivate micro-organisms and to produce changes in food texture without thermal degradation [1]. The UHP process is said to preserve not only flavour and taste, but also the nutritional properties of fruit and vegetable products [2] but information on packaging effects, on pressure effects on antioxidant activity and on changes during storage is still very rare [3].

2. MATERIALS AND METHODS

Orange juice was squeezed from ripe Salustiana (Spain) and Valensina, late (Morocco) oranges using a household electric squeezer (Braun) and then immediately chilled to 4 °C.
Juices were packaged in glass, polypropylene (PP) Teflon (TF) and Barex 210 (modified acetonitrile methyl acrylate copolymers) flasks of 0.5 litre in volume or in polyethylene (PE) pouches coated by aluminium. The packaged juices were chilled and stored until treatment at 4 °C. 800 MPa for 5 min (and 500 MPa for 5 min) were reached in a QFP6 Flow International autoclave system as described by [4]. 600 MPa (5 min) were achieved in a high pressure device described by [5].

418

Panels for sensory analysis comprised 6 to 11 assessors. The juices, at room temperature, were tested within 1 hour. The effect of packaging and high pressure treatment on odour and taste was analysed by means of the ,forced choice' technique [6] in a triangle test.

Sensory quality of the juices as a function of high pressure treatment and storage was analysed by means of descriptive assessment according to a 9-point scale by at least six assessors on the basis of colour, odour, taste, harmony and overall quality [7].

Acid content was determined by titration and calculated as citric acid following [8]. Vitamin C content (L-ascorbic and L-dehydroascorbic acid) were analysed fluorometrically by HPLC using on-line post column oxidation of L-ascorbic acid to L- dehydroascorbic acid and derivatisation to fluorescent quinoxaline described by [9]. Fructose, glucose and sucrose were determined by HPLC using an amino-column and refraction index detector according to [10]. The antioxidative capacity of the water soluble fraction was determined applying the $ABTS^+$ (2,2'-azino-bis-(ethylbenzothiazoline-6-sulfonic acid) diammonium salt [Sigma]) radical cation assay according to [11].

3. RESULTS AND DISCUSSION

Directly after the pressure treatments (500 MPa and 800 MPa) of freshly squeezed juices of Valensina and Salustiana oranges they showed significant differences in the sensory triangle tests in their odour and aroma compared to the untreated controls (Fig. 1). The differences

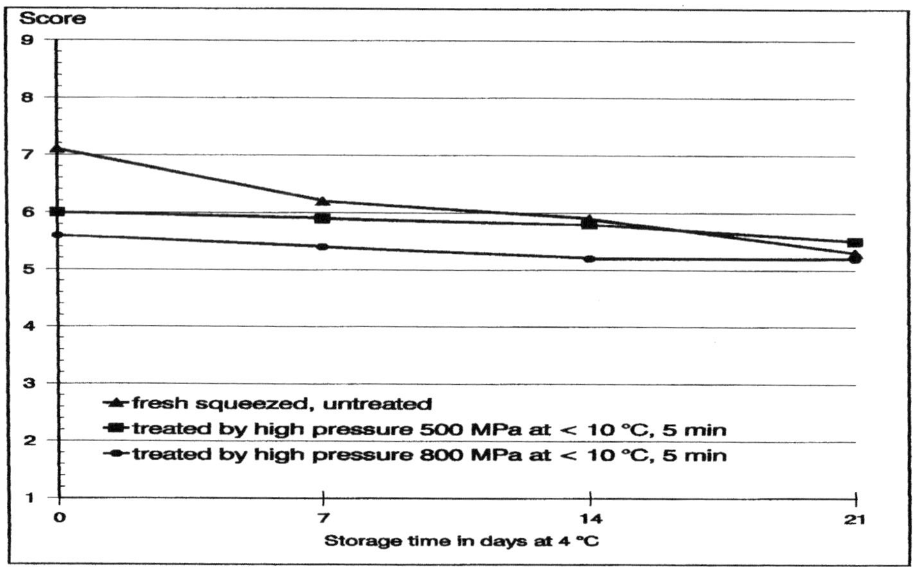

Fig. 1: High pressure treatment and storage of orange juice: Overall sensory quality. Scores correspond to following descriptions: 9.0 to 7.0 perfect, optimal to typical with slight deviations; 6.9 to 5.5 noticeable deviations to noticeable detractions and slight defects; 5.4 to 3.0 distinct to strong defects.

were the same for all packaging materials used (glass, Teflon, polypropylene, Barex, polyethylene). The pressure treated orange juices were described as non-typical, artificial, metallic and somewhat bitter and were rated only as moderate to satisfactory. The defects in odour and aroma were somewhat stronger in the juices pressurised by 800 MPa than in those treated by 500 MPa

On the other hand odour and taste of commercial pasteurised orange juice in which enzyme activity is nearly completely lost did not change during high pressure treatment. Thus it is suggested that the untypical odour and taste changes due to high pressure treatment of fresh orange juice are induced enzymatically during and maybe also shortly after the treatment. Flavour changes in fresh orange juice after treatment at 500 MPa (90 sec) and 700 MPa (60 sec) were also reported by [12].

The sugar content of orange juices (fructose, glucose, sucrose, total sugar) was not or only slightly affected by high pressure treatment at 500 and 800 MPa. After storage for 21 days at 4 °C the sugar content was fully retained in both controls and pressurised orange juices. Total acid contents and pH were also independent of treatment and storage. This confirms results of [13] who examined the respective subjects during storage of Valencia late orange juice after pressure treatment and also found them to remain unaltered.

Vitamin C content (both, L-ascorbic acid and dehydroascorbic acid) of the orange juices processed at 500 and 800 MPa was not or only insignificantly reduced confirming other reported results on the stability of L-ascorbic acid in orange juice when pressurised at mild temperatures [14]. During subsequent three weeks storage, L-ascorbic acid content of orange juices decreased about 10 % in both, the pressurised and the untreated samples.

This loss of ascorbic acid in stored juices suggests oxidative degradation. As a consequence a decrease of the antioxidant capacity may be expected. The pressure stability of antioxidants is

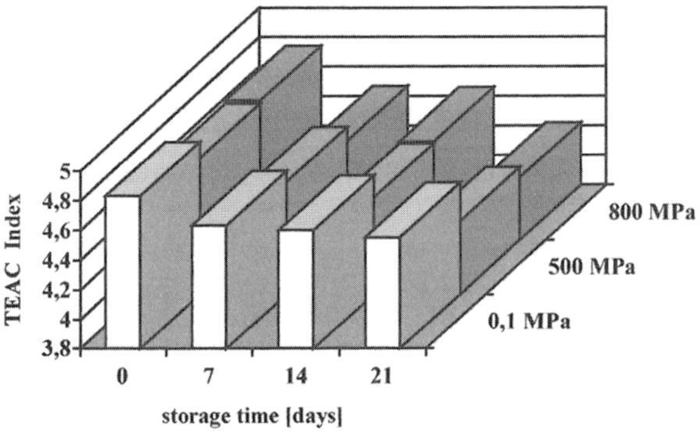

Fig. 2:   Water soluble antioxidative capacity of high pressure treated Salustiana orange juice during storage at 4 °C measured by ABTS$^+$ radical cation assay. Treatments applied at 20 °C for 5 min. Results are means of three independent replications.

420

of interest since they play an important role in reducing the risk of free radical-related oxidative damage associated with a number of diseases. Fig. 2 shows that the water soluble antioxidative capacity of high pressure treated orange juice during three weeks storage at 4°C is indeed also reduced and also to the same extent (about 10 % loss for the pressurised samples). In the untreated sample the loss is only about 6 % but that might depend only on the cultivar since there were no differences detected between pressurised and untreated Valencia late juices during storage (data not shown).

The antioxidative potential of apple homogenate (Jona Gold) was measured in the presence and absence of oxygen; the samples were pressurized in aluminium coated polyethylene bags. This food system was significantly influenced by the presence of oxygen during sample preparation for treatment (Fig. 3). The initial antioxidative capacity of oxygen-free processed

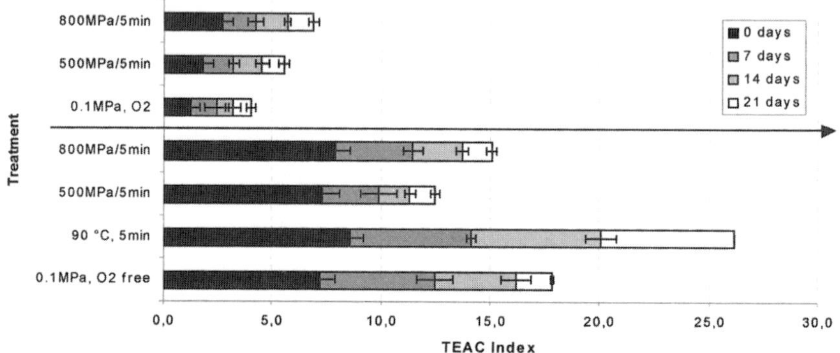

Fig. 3. Water soluble antioxidative capacity of chilled stored apple homogenate measured by the ABTS+ radical cation assay (upper part: in presence of oxygen; below: oxygen exclusion).

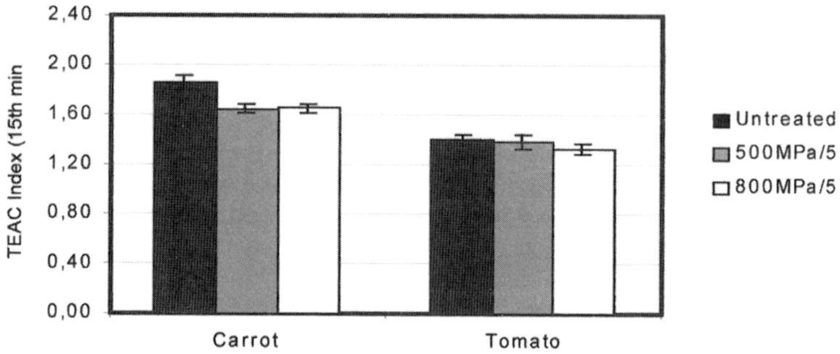

Fig. 4: Antioxidative capacity of pressure treated (5 min, 500 and 800 MPa at 25 °C) carrot homogenate and tomato pulp against peroxyl radicals measured using the ABTS$^+$ radical cation assay (water soluble fraction). Data are means of three independent replications.

homogenate was indeed up to four times higher as in the samples treated in presence of oxygen, but after 21 days chilled storage the antioxidant capacity had the same low level. Only heating was able to preserve a high TEAC index over 3 weeks chilled storage.

Measurements of the antioxidant capacity of the water soluble fraction of tomato pulp and carrot homogenate against $ABTS^+$ radicals had the following results (Fig. 4) : Immediately after pressure processing no great differences were observed. Pressurisation had shown only a small effect on the water soluble fraction of carrot homogenate, but also tomato lost only a minimal amount of antioxidant capacity, although only evident after the 800 MPa treatment.

4. CONCLUSIONS

High pressure treatment did not result in mayor changes of nutrient content. Also the antioxidative capacity remained fairly unaltered. Sensory changes due to processing may occur depending on the used fruits and the process conditions but not on the kind of packaging.

5. ACKNOWLEDGEMENTS

FAIR projects CT96-1113, CT96-1175 and CT98-5031 supported this investigation.

REFERENCES

1. B. Tauscher. Z. Lebensm. Unters. Forsch. 200 (1995) 3-13.
2. P. Butz and B. Tauscher. High Pressure Research 19 (2000) 11-18.
3. A. Fernández García, P. Butz, B. Tauscher. High Pressure Research 19 (2000) 153-160.
4. A. Maggi, P. Rovere, S. Gola, G. Dall'Aglio. Industria Conserve 68 (1993) 232-235.
5. P. Butz, W.-D. Koller, B. Tauscher, S. Wolf, Lebensm-Wiss u Technol. 27 (1994) 463.
6. DIN ISO 4120, 1995. Sensory analysis-methodology. Beuth Verlag, Berlin.
7. DIN 10952, 1983. Sensory testing methods (scoring). Beuth Verlag, Berlin.
8. L-31.00/3, 1997. Food Analysis: Determination of titrable acid in foods. Collection of official methods under article 35, German Food Act. Beuth Verlag, Berlin.
9. A. Bognar and H.G. Daood. J. of Chromatography Sci. 8 (2000) 162-168.
10. T. Van Den, C.J. Biermann, J.A. Marlett. J. Agric. Food Chem.34 (1986) 421-425.
11. N.J. Miller, C. Rice-Evans, M.J. Davies, MJ., V. Gopinathan, and A. Milner. Clinical Science 84 (1993) 407-412.
12. M.E. Parish. Lebensmittel-Wissenschaft und Technologie; 31 (5) (1998) 439-442.
13. M.E. Parish. In: K. Heremans (ed) High pressure research in the Biosciences and Biotechnology pp. 443-446. Leuven University Press, Leuven, Belgium1997.
14. G. Donsi, G. Ferrari, M. Di Matteo. Ital. J. Food Sci. 8 (2) (1996): 99-106.

Trends in High Pressure Bioscience and Biotechnology
R. Hayashi (editor)
© 2002 Elsevier Science B.V. All rights reserved.

An investigation into the transient movement of browning front through high pressure treated potatoes

A. Sopanangkul[*] , K. Niranjan and D.A. Ledward

School of Food Biosciences, University of Reading,
P.O. Box 226, Whiteknights, Reading RG6 6AP, United Kingdom

1. INTRODUCTION

The application of high pressure processing in foods is of great interest to the food industry due to its advantages over thermal methods (Cheftel, 1992; Thakur & Nelson, 1998). However, in an oxygen containing environment, the risk of increasing the extent browning in the treated material is a major problem. Browning in fruits and vegetables requires of two reactants; polyphenol oxidase (PPO) and oxygen. Under high pressure (up to 600 MPa) PPO can be inactivated (Anese et al., 1995; Gomes & Ledward, 1996) and also cells are damaged, which, combined with the high solubility of oxygen at these pressures, may lead to enhanced oxygen penetration; thus browning is more pronounced. In this study, the penetration of browning was investigated in high pressure treated potato cylinders, as a function of time after treatment.

2. MATERIALS AND METHODS

Potato cylinders (20 mm diameter, 70 mm in length) were subjected to pressures of 200, 400 and 600 MPa for 5, 10, 20 and 30 min (Stansted Fluid Power Ltd., Stansted, UK). After treatment, the samples were held at room temperature and a transverse cut was made at the time 0 (immediately after treatment), and after 3, 6, 9, 12, 15, 18, 21 and 24 hours. Each experiment was done in triplicate and an untreated sample was used as a control.

---

[*] The author would like to thank The Royal Government of Thailand for the financial support.

The depth of penetration of the browning front *(a)* moving from the surface to the centre of the potato cylinder was measured. The rate of the browning front movement was characterised as a quantity that was proportional to the diffusion coefficient *(D*)*. *D** was calculated from the slope of the graph of a^2 versus *time*.

3. RESULTS AND DISCUSSION

The movement of the browning front was observed in the pressure treated samples during storage, whilst, in the untreated sample, only a little browning was seen adjacent to the skin (Fig. 1). Increasing pressure and treatment time increased the depth of penetration of the browning front during subsequent storage (Table 1).

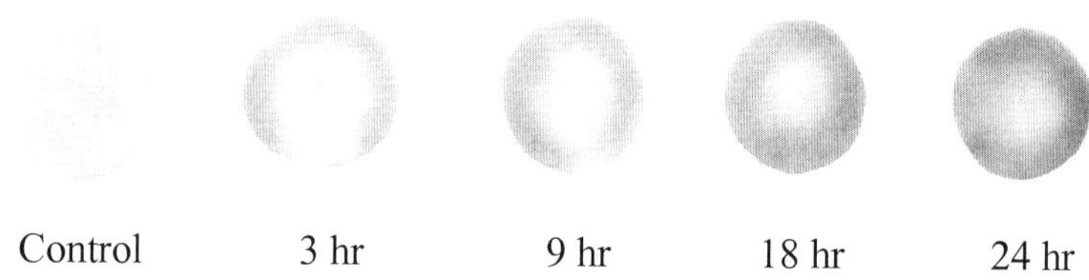

Figure 1. Transversely cut samples of high pressure treated potatoes (400 MPa, 20 min) at different storage times when stored at room temperature compared with an untreated (control) sample.

Table 1. Penetration of browning as a function of storage time after different pressure treatments

Storage		Penetration depth of browning front; a (mm)											
Time	Treatment Time :		5 min			10 min			20 min			30 min		
(hr)	MPa :	0.1	200	400	600	200	400	600	200	400	600	200	400	600
0		0	0	0	0	0	0	0	0	0	0	0	0	0
3		0	0	2.50	1.90	0	2.20	2.40	1.30	1.84	2.44	0.60	2.48	3.00
				(0.12)	(0.09)		(0.11)	(0.14)	(0.05)	(0.24)	(0.08)	(0.06)	(0.03)	(0.18)
6		0	0	3.74	3.98	0	2.00	3.12	1.30	1.52	2.72	0.62	2.90	3.24
				(0.13)	(0.07)		(0.08)	(0.14)	(0.13)	(0.15)	(0.04)	(0.07)	(0.09)	(0.15)
9		0	0	3.28	5.08	0	2.36	4.32	1.64	1.78	3.78	1.02	3.00	3.56
				(0.22)	(0.19)		(0.19)	(0.05)	(0.10)	(0.06)	(0.05)	(0.20)	(0.09)	(0.22)
12		0	1.52	3.98	6.00	0.60	3.94	5.42	2.14	2.78	4.64	2.64	3.58	5.36
			(0.04)	(0.13)	(0.18)	(0.02)	(0.08)	(0.18)	(0.06)	(0.08)	(0.12)	(0.14)	(0.15)	(0.12)
15		0	1.58	3.70	6.66	0.60	4.00	5.04	2.72	3.18	5.42	3.64	4.14	6.88
			(0.02)	(0.05)	(0.14)	(0.01)	(0.18)	(0.15)	(0.08)	(0.06)	(0.13)	(0.05)	(0.12)	(0.12)
18		0	1.60	4.26	6.80	1.00	4.30	4.70	2.40	3.84	5.56	3.94	4.32	7.02
			(0.09)	(0.04)	(0.13)	(0.13)	(0.18)	(0.13)	(0.28)	(0.10)	(0.06)	(0.09)	(0.12)	(0.08)
21		0	1.84	4.88	7.50	1.52	5.16	6.38	2.68	3.92	5.22	4.30	4.36	7.14
			(0.09)	(0.08)	(0.28)	(0.02)	(0.08)	(0.03)	(0.03)	(0.08)	(0.11)	(0.05)	(0.03)	(0.12)
24		0	2.20	4.86	8.74	1.52	5.32	7.18	2.84	4.24	5.50	4.52	4.82	8.70
			(0.11)	(0.12)	(0.19)	(0.05)	(0.18)	(0.10)	(0.03)	(0.06)	(0.39)	(0.03)	(0.08)	(0.13)

Values are the means of 3 determinations (±SD).

426

Figure 2 shows the effect of high pressure treatment on the diffusion coefficient ($D*$) in the potato cylinder. It can be seen that for all pressures, $D*$ decreased initially with treatment time but subsequently increased. It appears that for short treatment times, the rate of oxygen penetration slows but this is presumably the result increased loss of enzyme activity with time. At longer treatment times, sufficient cells are broken down to enable oxygen penetration to increase. The enhancement in oxygen penetration rate is much higher than the rate of loss of enzyme activity. Hence browning rates and $D*$ increase.

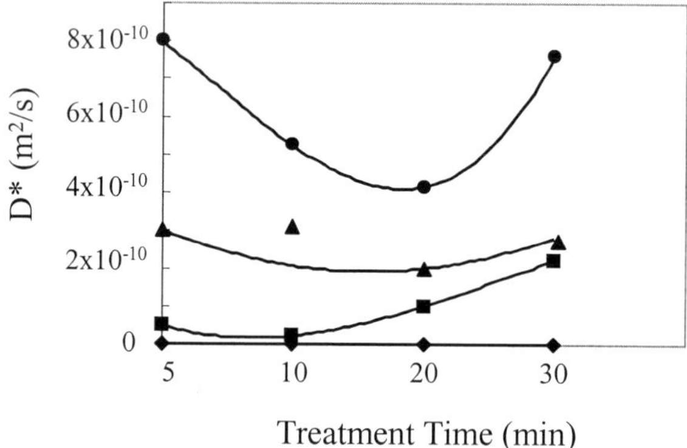

Figure 2. Effect of high pressure treatment on characteristic diffusion coefficient $(D*)$ as a function of treatment time and pressure (0.1 MPa, --◆--; 200 MPa, --■--; 400 MPa, --▲-- and 600 MPa, --●--).

It is also interesting to note that the initial drop in $D*$ is sharper at the higher treatment pressures. This suggests that, for short treatment times, enzyme inactivation rates increase more sharply with pressure than the oxygen diffusion rate.

4. CONCLUSION

The net rate of movement of browning front in any high pressure treated vegetable is a balance between the extent of enzyme inactivation (in this case PPO) caused by the applied pressure, and the enhancement in oxygen diffusion rates caused by the opening up of tissue structure and higher solubility.

REFERENCES

1. Anese, M., Nicoli, M.C., Dall'aglio, G. and Lerici, C. (1995) *J. Food Biochem.* **18**, 285-293.
2. Cheftel, J.C. (1992) In *High Pressure and Biotechnology*, (Ed) Balny, C., Hayashi, R., Heremans, K. and Masson, P. Editions John Libbey Euro-text, Montrouge, p. 195-209.
3. Gomes, M.R.A and Ledward, D.A. (1996) *Food Chem.,* **56**, 1-5.
4. Thakur, B.R. and Nelson, P.E. (1998) *Food Rev. Int.,* **14(4)** , 427-477.

Trends in High Pressure Bioscience and Biotechnology
R. Hayashi (editor)
© 2002 Elsevier Science B.V. All rights reserved.

Experimental Investigation on Thermofluiddynamical Processes in Pressurized Substances

M. Pehl, F. Werner and A. Delgado

Technical University of Munich, Chair of Fluid Mechanics and Process Automation, Weihenstephaner Steig 23, D-85350 Freising, Germany

In previous investigations it has been shown that thermofluiddynamical processes occur in food related substances at high hydrostatic pressure (HHP) treatment. In order to gain insight into these processes an in-situ measuring technique for visualizing temperature and velocity fields has been developed [1]. Based on extensive test series the thermal and fluiddynamical processes in water have been analyzed. After pressurizing water onto 230 MPa within 12 seconds temperature gradients of 0.1 K/mm are generated in the visible field of the optical cell.

The present study comprises investigations on liquids of higher viscosity. Sucrose solution of 50% (weight percent) has been chosen as model fluid because the viscosity data are available up to 600 MPa [2]. The results show that the temperature gradients are six times higher than those in water under the same test conditions. This in turn may cause heterogeneities of biochemical processes in foods. This correlation has been confirmed by numerical simulations of an enzyme inactivation process [3].

In the present contribution visualized inhomogeneous temperature fields arising in sucrose solution are presented and compared to the thermofluiddynamical processes in water.

1. INTRODUCTION

In a large body of publications the influence of pressure p and temperature t on food systems has been studied. However, it has been shown [2] that the processes occuring during the HHP treatment depend not only on p and t but also on the induced temperature and velocity fields. Thermofluiddynamical processes depend on the one hand on the process parameters such as initial temperature, time, pressurizing ramp, maximum pressure level, and the geometry and material of the pressure vessel. On the other hand the properties of the substance (viscosity, density, heat conductivity, compressibility, and heat capacity) have to be taken into consideration. The number of parameters and the lack of knowledge concerning the properties under the influence of HHP make clear that thermofluiddynamical processes can not be predicted excluding uncertainty especially in case of complex food related substances.

With regard to the pressure processing of foods there are two items of principal interest: the magnitude of the temperature gradients and the time for the temperature compensation. Only by knowing these magnitudes, micro-biological and chemical processes can be interpreted accurately. Additionally, HHP processing in a technical scale necessitates knowledge about the thermofluid-dynamical interrelations in order to achieve products of high quality and safety and in order to realize high-level process uniformity and economic processing.

Experimental investigations with water have shown the thermofluiddynamical processes occuring during a HHP treatment; forced convection is induced by pressurizing due to the mass flux into the cell. In this stage temperature gradients occur because of heat exchange with the walls of the vessel. While the forced convection dominates in the first stage, density differences due to temperature gradients drive the convection in the stage of pressure holding. In the last stage of depressurizing, forced convection dominates again.

This means that temperature fields and convection fields are coupled. A high convection velocity indicates high temperature gradients in the stage of pressure holding. Also, the velocity determines the thermal dwell of particles in the regions of different temperature (e. g. thermal stress "history" of a micro-organism). For this reason, temperature and velocity fields have to be considered.

2. METHODS

The topic of this contribution is the investigation of temperature gradients occuring in food components with a viscosity higher than that of water. As model substance 50% sucrose solution has been chosen. According to the results of [3], the viscosity of this solution is approximately one order of magnitude higher than that of water in the pressure and temperature range applied in the present study. The sucrose solution is pressurized at 59 °C up to 280 MPa within 15 seconds (see figure 2 in the chapter of results and discussion).

The thermofluiddynamical processes in water have been analyzed in a previous study [4]. At a pressurizing ramp of 230 MPa within 12 seconds temperature differences of about 0.6 K have been found in the vsible area of the optical cell ($\varnothing$ = 6 mm) at the beginning of pressure holding. Because of the higher viscosity of the sucrose solution obviously higher temperature differences can be expected in the present study. The reason for this is the lower intensity of convective heat transfer.

Thermochromic liquid crystals reflect selectively light as a function of temperature. This effect is utilized in fluid testings by illuminating the liquid crystals with white light and analyzing the reflected light (color or wavelength) afterwards. These temperature sensors can also be employed in high pressure investigations as it has been shown in previous investigations [1]. While temperature causes a decrease of the wavelength, pressure induces the opposite effect. Performing experiments in a wide range of pressure and temperature with different types of liquid crystals have shown that they can be used in a wide temperature range and at pressures at least up to 700 MPa. The practical use for HHP experiments requires the measurement of calibration curves, i. e. the reflected color or wavelength is estimated as a function of temperature and pressure.

The liquid crystal particles used in this study are encapsulated in a polymeric material for protection. Furthermore, the particles are disposed for tracing the fluid motion in the substance. Visualizing temperature and velocity fields in liquids necessitates only a low particle concentration of about 10^{-2} to 10^{-3} volume percent.

Figure 1 gives an illustration of the experimental setup. The optical cell has a volume of 2 ml and is equipped with sapphire windows of 6 mm optical width. The test substance is illuminated by means of a xenon light source and the depth of the visualized field is 0.3 mm. For generating reproducible pressurizing ramps the pressure pump is driven by a controlled stepmotor.

For preparation of the experiments presented in this contribution, the liquid crystal particles are dispersed into the sucrose solution and the liquid is pumped into the pressure system afterwards.

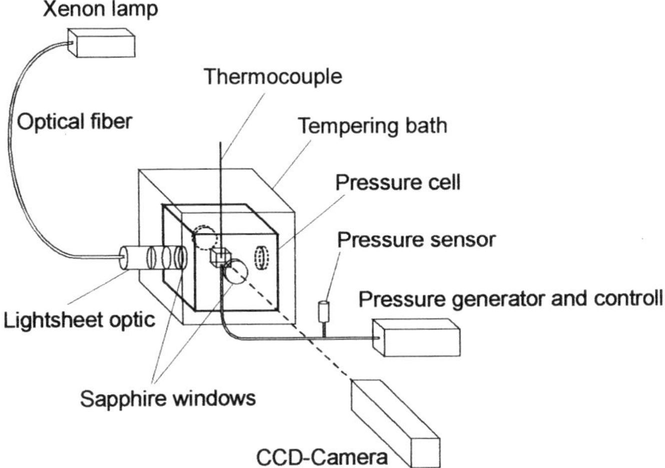

Figure 1. Arrangement of the experimental setup.

The pressure chamber is heated up to the initial temperature of 59 °C. The pressurization starts from thermal equilibrium. By means of a CCD-Camera and a video recorder images of the distributions are recorded. In order to evaluate the temperature fields at the respective pressure, the reflected colors are compared to the calibration curves of the liquid crystal type. The evaluation can be done by human eye or by digital thermography [3]. In this study, results have been analyzed by eye. This seems justifiable as the color fields are clearly determined and all visible colors appear.

3. RESULTS AND DISCUSSION

The sucrose solution has been pressurized to a maximum pressure of 280 MPa within 15 seconds. As it can be seen from figure 2, the pressure increase is approximately linear. By means of a thermocouple placed near the wall in the cell the local temperature is measured. In this HHP process the maximum temperature increase occurs straight after pressurizing and it is about 7 K. This increase fits very good to the estimate for adiabatic compression of water (2.5 K per 100 MPa). From this, it can be derived that the properties of the sucrose solution (density, heat capacity, and coefficient of expansion) are nearby to those of water. In the stage of pressure holding the temperature decreases exponentially as it is typical for thermal compensation processes. After 90 seconds of pressure holding the locally measured temperature seems to remain constant. The fact that the initial temperature is not reached even after 340 s might be due to a failure of the thermocouple due to the effect of HPP on the sensing metal device. Considering the equation

$$\tau_{th} = \frac{l^2}{a} \tag{1}$$

the thermal decay time τ_{th} can be estimated by a characteristic length l (diameter of the optical cell) and the temperature conductivity a.

432

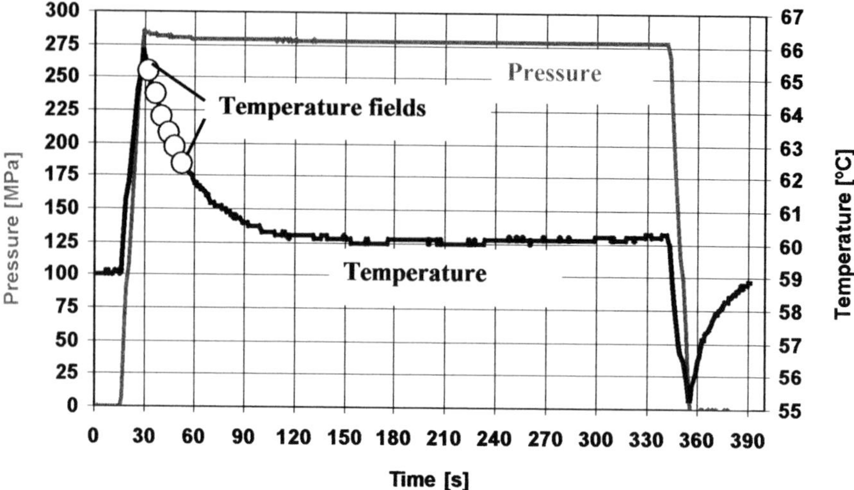

Figure 2. Pressure and temperature profile vs. time.

Taken from figure 2, the thermal decay time is about 90 seconds. Therefore, the thermal conductivity of the sucrose solution amounts approximately $2 \cdot 10^{-6}$ m²/s, so being one order of magnitude higher than the thermal conductivity of water.

At six consecutive sequences in the stage of pressure holding, see figure 2, the temperature fields in the optical cell have been recorded. These chronology is shown in figure 3 where red color indicates colder regions and blue color the warmer ones. The distance between the red and blue particle region ($\approx$ 2 mm) is nearly constant and corresponds to 1.2 K. The visible temperature zone moves from the bottom to the top of the cell with time. In this stage of pressure holding, convective flow is low whereby heat conduction dominates. The temperature difference over the whole field ($\varnothing$ = 6 mm) amounts about 3.5 K presuming a linear gradient. The gradient in the overall cell is supposed to be higher because only 20 % of the cell is in sight and the lowest temperature is at the wall. This has been confirmed by numerical simulations [4] regarding the geometry of the used optical cell. In this study, maximum temperature differences of about 6 K have been calculated in the stage directly after pressurizing the water. While these differences decrease rapidly with time in water, figure 3 indicates longer times for the sucrose solution.

Referring to the problem of temperature homogeneity only very few numerical investigations have been performed [3, 5]. Though numerical calculations can be validated by experiments, they are restricted to known physical properties of food related substances under high pressure. This demands the developement of adequate measuring devices.

The comparison of the thermofluiddynamical processes in sucrose solution with those in water shows some fundamental differences: in sucrose solution the forced and thermal convection as well as the fluiddynamical decay time are lower due to the higher viscosity. In water heat transfer processes are dominated by convection that is increasing with the pressurizing slope, whereas heat conductivity dominates in the used sucrose solution.

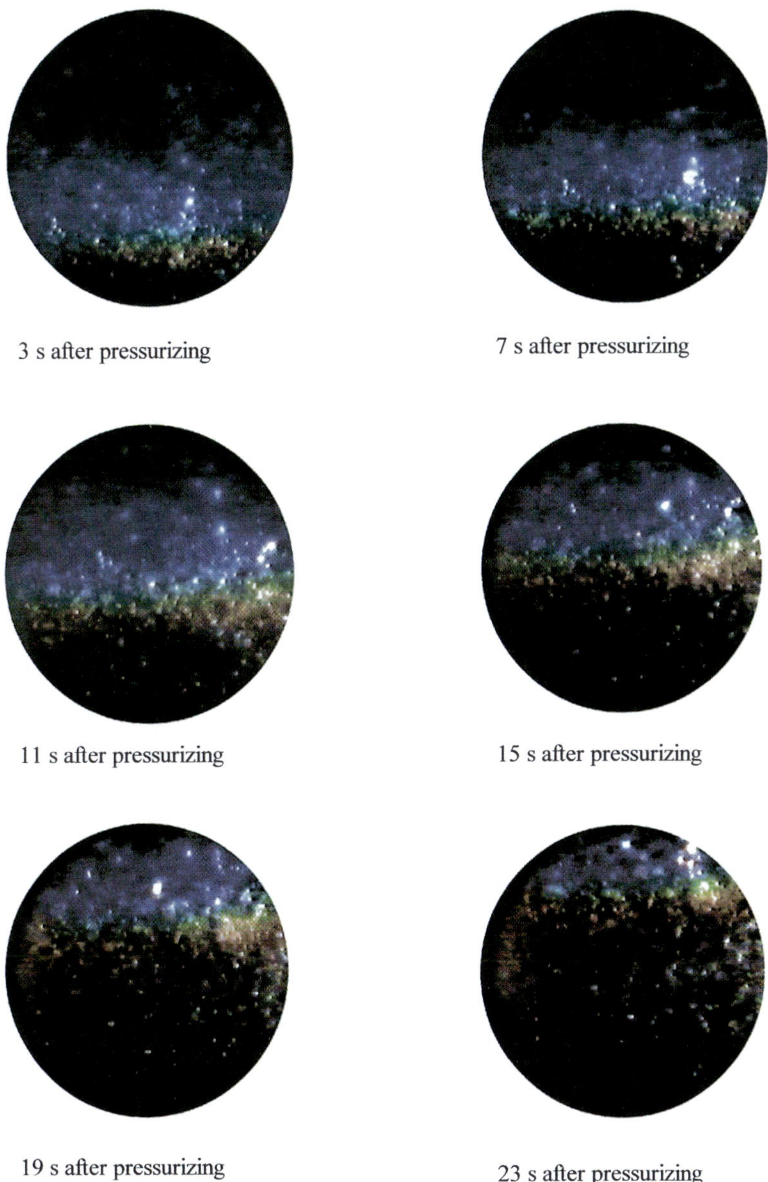

Figure 3. Time dependency of the temperature fields in 50%-sucrose solution.

The thermal decay time is in the same order of magnitude for both cases. While the heat transfer in water is enforced by convection, the higher thermal conductivity increases the heat transfer in sucrose solution The most important difference has been found in the magnitude of the temperature gradients that are six times greater in sucrose solution than in water. The reason for this is that the gradients can not be removed by convective flow in the sucrose solution. Concerning HHP processing of foods, this means that in liquids of low viscosity temperature gradients are damped quickly due to convective flow. As temperature homogeneities are of importance in substances of middle or high viscosity thermofluiddynamical processes have to be taken into consideration. While the present studies have been performed in an optical cell of 2 ml volume this becomes significantly more relevant in middle and high capacity vessels. Numerical simulations [4] confirm these considerations. Even 20 minutes after pressurizing the activity retention of *Bacillus subtilis a-amylase* (BSA) varies between 28 % and 48 % in a vessel of 0.8 l capacity. These data have been calculated assuming that the viscosity of the matrix fluid is hundred times higher than that of water.

4. CONCLUSION AND PERSPECTIVES

In the present contribution the temporal evolution of inhomogeneous temperature fields occuring in sucrose solution after pressurizing has been studied by analyzing experiments using thermochromic liquid crystals. An aqueous sucrose solution of 50 % is pressurized up to 280 MPa within 15 seconds. In the stage of pressurizing, forced convection is generated. Due to the high viscosity thermal convecion is damped out nearly completely in the stage of pressure holding. In the visible field of the optical cell the temperature difference is about 0.6 K/mm, measured directly after pressurizing. The thermal decay time amounts approximately 90 seconds what can be deduced from the locally measured temperature profile.

With regard to HHP processing of food related substances, heterogeneities of bio-chemical processes have to be considered in substances of middle or high viscosity as it has been shown by [4]. Therefore, the viscosity under HHP has to be a known magnitude and thermofluiddynamical processes have to be investigated. Due to the lack of knowledge of the relevant properties of complex substances, numerical simulations have to be supplemented by experimental investigations. A possible industrial application of the results found in this study is the HHP treatment of jam (Japan) that contains 50% to 60% sucrose.

In future works the experiments will be extended on more complex substances. Furthermore, studies on dimensional analysis are planed in order to estimate thermofluiddynamical processes in vessels of higher volumetric capacity. Finally, biological processes which exhibit a pronounced temperature sensitivity at HHP are going to be analyzed considering the thermofluiddynamical processes occuring at the same time.

REFERENCES

[1] M. Pehl, F. Werner, A. Delgado: First Visualization of Temperature Fields in Liquids at High Pressure Using Thermochromic Liquid Cristals, Experiments in Fluids, 29/3, (2000), 302 – 304.

[2] P. Först, F. Werner, A. Delgado: A model for the pressure-viscosity behaviour of watery solutions of food ingredients, Trends in High Pressure Bioscience and Biotechnology, R. Hayashi (ed.), accepted for publication (2001).

[3] Chr. Hartmann, A. Delgado: Numerical Simulation of Thermofluiddynamics and Enzyme Inactivation in a Fluid Food System under High Hydrostatic Pressure, Trends in High Pressure Bioscience and Biotechnology, R. Hayashi (ed.), accepted for publication (2001).

[4] M. Pehl, F. Werner, A. Delgado: Digital Particle Image Thermography & Velocimetry in kompressiblen Flüssigkeiten, Proc. of Lasermethoden in der Strömungsmesstechnik, 12.-14. 09. (2000), Technische Universität München, Germany.

[5] S. Denys, A.M. Van Loey, M.E. Hendrickx: A modeling approach for evaluating process uniformity during batch high hydrostatic pressure processing: combination of a numerical heat transfer model and enzyme inactivation kinetics, Innovative Food Science & Emerging Technologies, H. Ludwig (ed.), Springer Verlag, Heidelberg (2000), 5-19.

Trends in High Pressure Bioscience and Biotechnology
R. Hayashi (editor)

A model for the pressure-viscosity behaviour of aqueous solutions of food ingredients

P. Först, F. Werner and A. Delgado

Lehrstuhl für Fluidmechanik und Prozessautomation, Weihenstephaner Steig 23, D-85350 Freising, Germany

In the present paper new data on the pressure dependence of the viscosity of aqueous solutions of sugars and whey protein isolate up to a pressure of 700 MPa are presented and discussed in terms of concentration and temperature.

An existing model valid for the concentration dependence of the viscosity at ambient pressure is applied to aqueous sugar solutions. An dimensionless equation describing the dependence of the viscosity of the solution on temperature, concentration and pressure is presented and applied to aqueous solutions of sucrose and glucose. It is shown that for sugar solutions the model is able to predict the pressure dependence of the viscosity of the solution if the behaviour of the solvent is known. The model is also applied to whey protein solutions and pressure induced structural changes are detected. The behaviour of whey protein solutions is compared to sugar solutions under pressure.

1   INTRODUCTION

The knowledge of the thermophysical properties of basic and more complex food ingredients is important in order to understand the effects of pressure on molecular and cellular systems. An important thermophysical property is the viscosity because it provides or supplements information on pressure induced structural changes, on transport processes like diffusion processes under pressure or fluiddynamic processes. In literature there is a lack of data concerning the viscosity of food ingredients under pressure; e.g. the viscosity of aqueous solutions [1]. In the current work new data for the pressure dependence of the viscosity of sugar and whey protein solutions are presented and discussed. A model is presented to describe the viscosity in a wide parameter range. The model is also applied to whey protein solutions and discussed in terms of pressure-induced changes.

1.1   Materials and Methods

The sugars used in the experiments were glucose and sucrose. The sucrose solutions were prepared from commercial sugar. The glucose (D(+)-Glucose-Monohydrate) was supplied by J. T Baker, Holland. The purity of the glucose used in the experiments was higher than 99%. The whey protein isolate (MILEI Isolac) was from Milei GmbH, Stuttgart, and had a whey protein content of 90%. The solutions were prepared using demineralized water as solvent. The solid content refers to a mass fraction w (g saccharine[1]/g solution and g whey protein

---

[1] Referred to dry substance

isolate/g solution). The solutions were prepared by heating them up to about 50 °C and stirring it with a magnetic stirrer for about 30 min. The whey protein solutions were stored for at least 12 h in the refrigerator. The viscosity and density data at ambient pressure for the different saccharide solutions for the mass fraction range of 0.005 to 0.8 and temperature range of 5 °C to 80 °C are taken from literature [2]. The viscosity and density values for the whey protein solutions at ambient pressure were measured with a stress controlled rotational rheometer (SR 5000, Rheometric Scientific) using a concentric cylinder geometry and an oscillating u-tube method, respectively.

The viscometer developed for the use at high pressures up to 700 MPa and a temperature range between -20 °C and 80 °C is of the rolling ball type as described in [3] for ambient pressure, but with variable inclination angle between 0 and 90° to allow for different viscosity ranges. The proportional control factor between sphere velocity, which is measured by an inductive method and viscosity is found by calibration at ambient pressure. At high pressure the elastic expansion of the pressurised viscometer tube and the compression of the sphere are calculated. The pressure dependence of the proportional control factor can therefore be determined. The system was kept at constant temperature by a circulating water jacket connected to a thermostat. The method as well as an optical method to measure the pressure dependence of the viscosity of liquids are described in more detail in [4] and [5].

2   RESULTS

2.1   The pressure dependence of the viscosity of sugar and whey protein solutions

The viscosity of sucrose solutions was determined in the parameter range between 0.1 and 700 MPa, 5 and 60 °C and mass fractions between 0.01 and 0.6. The viscosity of glucose solutions was determined between 5 and 40 °C and from 0.1 to 600 MPa for w = 0.2. The pressure dependence of the viscosity of sugar solutions can in most cases - like for pure liquids [6] - be described by an exponential function

$$\eta_p = \eta_{p_0} e^{\alpha p} \quad , \tag{1}$$

where η_{p0} is the viscosity at ambient pressure, and α an exponential factor, that is dependent on temperature and concentration. α increases with increasing concentration and decreasing temperature. Therefore the pressure dependence of the viscosity is strongest for low temperature and high sugar concentration. For low temperatures and low concentration the anomalous behaviour of the solvent water [5] dominates the system and the pressure dependence of the viscosity is not longer exponential. The pressure dependence of the viscosity of sucrose solution at constant mass fraction w = 0.5 and for different temperatures is shown in Fig. 1.

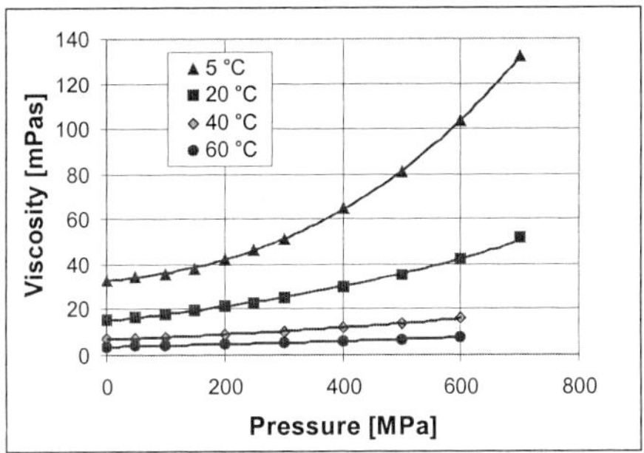

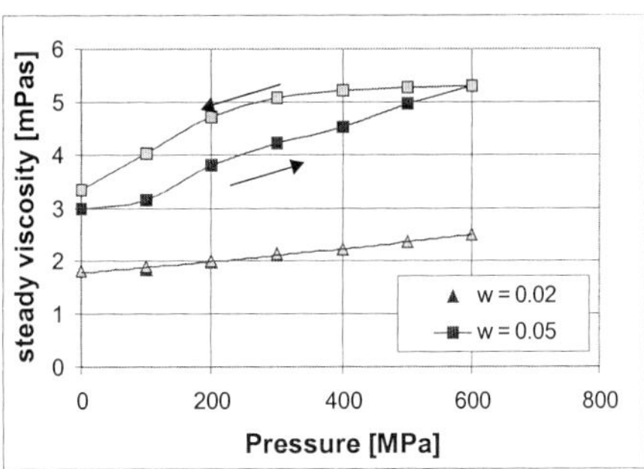

Fig. 1: The pressure dependence of the viscosity of sucrose solution at constant mass fraction (w=0.5) and for different temperatures. The measured values are presentd by the symbols, the lines are an exponential fit

Fig. 2: The pressure dependence of the viscosity of solutions of whey proteine isolate for T = 20 °C and two different mass fractions. For w = 0.02 an exponential function (line) is fitted to the measured values (symbols)

For the sugar solutions no irreversible effects of pressure on the viscosity could be found. The pressure dependence of the viscosity of whey protein solutions was studied in the pressure range between 0.1 and 600 MPa for 20 °C and for w = 0.02 and w = 0.05. The viscosity was determined during pressurisation and depressurisation. After the pressure change the time dependent viscosity was recorded until steady state was reached. During pressurisation the viscosity is at first increasing at each pressure step for pressures higher than 100 MPa until steady state is reached. During depressurisation no time dependence was found within the measurement time. The maximum time at each pressure step was 40 min. The results for the steady state viscosity are plotted in Fig. 2. The viscosity during depressurising was higher than during pressurising; the effect being much stronger for the higher concentration. For w = 0.05 irreversible pressure effects could be detected. The viscosity was slightly higher after the pressure treatment. This irreversible effect of pressure is assumed to be due to the denaturation of the β-Lactoglobulin fraction of the protein [7]. For w = 0.02 the pressure dependence of the viscosity can be described by eq (1). For w = 0.05 the pressure dependence is no longer exponential.

2.2 A model for the description of the viscosity of sugar solution

At ambient pressure the viscosity of sugar solutions for constant temperature can be described according to a hydrodynamic model [8] with the following power series:

$$\frac{x}{\ln(\eta_r)} = q_0 + q_1 x + q_2 x^2 + \dots \quad with \quad x = \rho w \quad and \quad \eta_r = \frac{\eta}{\eta_s} \quad . \tag{2}$$

The relative viscosity is defined as ratio between the viscosity of the solution to that of the solvent. x is the mass concentration of the solute that can be calculated from the mass fraction by multiplication with the density ρ of the solution. The coefficient q_0 provides information

440

on hydration and shape of the molecules, q_1 and q_2 on intermolecular interactions and on the change of the flow field due to the presence of the particles. By application of the model it is also possible to estimate the temperature and concentration dependent hydration of the sugar molecules [8].

If ambient pressure data for the viscosity and the density of sucrose solution with different concentrations and temperatures are plotted in terms of $x/ln(\eta_r)$ vs. x the points yield parallel, straight lines for different temperatures. The temperature dependence of q_0 can be described by an Arrhenius function:

$$q_0 = q_0^* \exp\left(\frac{E_a}{RT}\right) \quad . \tag{3}$$

q_1 does not significantly change with temperature and is therefore assumed to be constant. As the curves are linear, q_2 can be neglected. For sucrose solution the numerical value of q_0^* is 1.1234 g/cm^3, of $E_a = 2609$ J/mol and q_1 is -0.248. The activation energy E_a is related to the hydration of the molecules. Neglecting the second power term and introducing eq. (3) into eq. (2) the model equation yields

$$\frac{x}{\ln\eta_r} = q_0^* \exp\left(\frac{E_a}{RT}\right) + q_1 x \quad . \tag{4}$$

By coordinate transformation eq. (3) can be brought into an dimensionless form:

$$\frac{x^*}{\ln(\eta_r)} = 1 + q_1 x^* \quad with \quad x^* = \frac{x}{q_0^* \exp\left(\frac{E_a}{RT}\right)} \quad . \tag{5}$$

Separation of variables yields:

$$\ln(\eta_r) = \frac{x^*}{1 + q_1 x^*} \quad . \tag{5a}$$

The high pressure relative viscosity is defined as the quotient of viscosity between solution and solvent at the same pressure and temperature. x is calculated with the pressure dependent density [9]. If the relative viscosity is determined from the high pressure data the values at ambient and elevated pressures coincide and can both be described by the dimensionless model equation (5). The molecular interpretation is that there are no major pressure induced structural changes in the parameter range considered here and that the pressure increase is equivalent to an increase in mass fraction at ambient pressure. The above equation is able to describe the dependence of the relative viscosity on temperature, concentration and pressure in a wide parameter range. This is shown in Fig. 3 where the measured values are compared to eq. (5a).

It can be seen that the model equation is capable to predict the viscosities of sucrose solution at high pressures with good approximation if the pressure dependence of the solvent viscosity is known. The maximum relative deviation is 10 % in the temperature range from 20 °C to 40 °C and from 0.1 to 700 MPa. For T = 60 °C the relative deviation slightly exceeds 10%. This is assumed to be due to uncertainties in the density.

For glucose solutions the same procedure as for sucrose solution has been carried out and it was found out that the same characteristics are valid for glucose solution. Due to the fact that glucose is a monosaccharide, the model constants are slightly different:
$E_a = 2761$ J/mol , $q_0^* = 1.2305$ and $q_1 = -0.2380$.

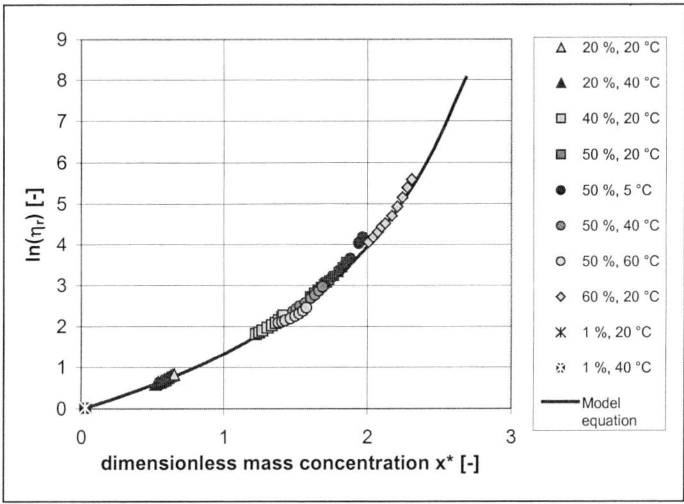

Fig. 3: High pressure relative viscosities for different temperatures and mass fractions of sucrose compared to the model

2.3   Application of the model to protein solutions

In the next step a first attempt is made to apply the model to protein solutions. For proteins it is known from literature that they undergo modifications with pressure that can be reversible or irreversible [10, 11]. The model is tested for whey protein solutions at one temperature (T = 20 °C) and for mass fractions from w = 0.02 and w = 0.13 at ambient pressure and for w = 0.02 and w = 0.05 in the pressure range of 0.1 to 600 MPa. For the calculation of the high pressure relative viscosities the steady state values of the pressurising curve have been used (see 2.1). If $x/\ln(\eta_r)$ is plotted vs. x for ambient pressure in the same way as for sugar solutions the points yield a non-linear curve with positive slope (see Fig. 4). It can be seen that for whey protein solution the second power term of eq. (2) can not be neglected. If $\ln(\eta_r)$ is plotted vs. x for both the concentration dependent ambient pressure relative viscosities and the high pressure relative viscosities (see Fig. 5) it can be seen that up to 100 MPa the ambient and high pressure values coincide and that for higher pressures the high pressure relative viscosities are higher than the ambient pressure values for the same concentration. This behaviour is much more pronounced for the higher concentration (w = 0.05) suggesting pressure induced structural changes like protein unfolding.

442

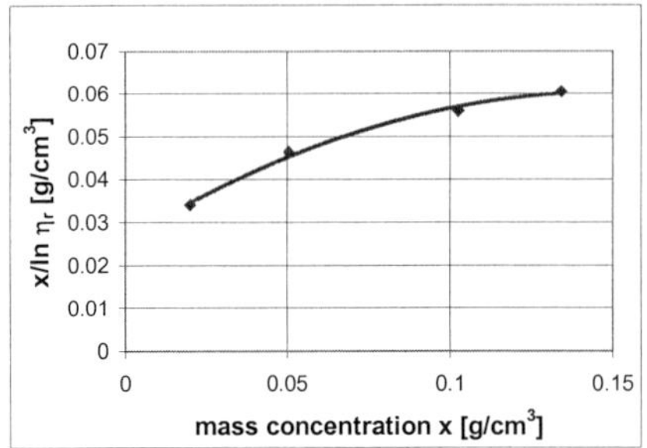

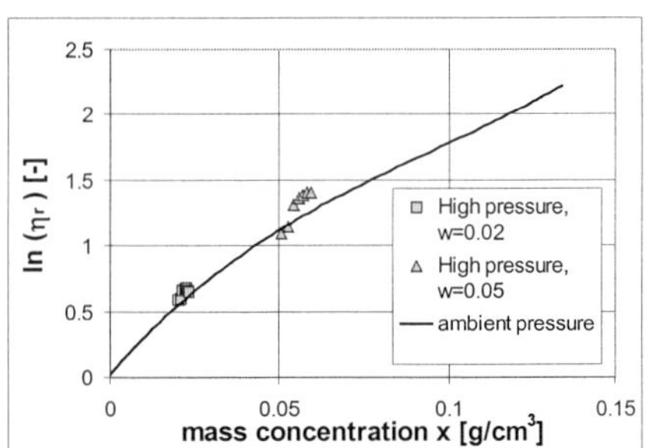

Fig. 4: x/ln(η_r) vs. x for whey protein solutions with different concentration at ambient pressure

Fig. 5: ln(η_r) vs. x for whey protein solutions. Concentration dependent values at ambient pressure: lines; Pressure dependent values at constant mass fraction: symbols.

3   DISCUSSION AND FUTURE ASPECTS

The application of the model shows a very different behaviour for sugar and whey protein solutions. For aqueous solutions of glucose and sucrose the pressure dependent relative viscosities coincide with the ambient pressure relative viscosities. This suggests that a pressure increase is equivalent to an increase in mass concentration and no pressure induced structural changes do occur in the pressure range regarded here. The model can therefore be used to predict unknown high pressure viscosities if the behaviour of the solvent is known.

For whey protein solution up to 100 MPa the high pressure relative viscosities coincide with the ambient pressure data with good approximation. For pressures of more than 100 MPa the pressure dependent relative viscosities are systematically higher than the ambient pressure values.

Whey proteins are mainly composed of ß-Lactoglobulin, α-Lactalbumin and minor contents of immunoglobulins and serum albumens. They partly unfold for pressures higher than 150 MPa. After unfolding they can aggregate or form gels [12]. It is also known for proteins that with unfolding water penetrates into the hydrophobic core for higher pressures [13] causing the volume fraction of the hydrated protein to increase. The macroscopic consequence of these structural changes can be detected by the viscosity measurements and the application of the model. The results could be used to supplement information on the behaviour of proteins under pressure which is obtained by other *in-situ* techniques like spectroscopic techniques. To check this potentiality further measurements on defined protein solutions which have been already studied by spectroscopic techniques have to be carried out.

Acknowledgements. The work was supported by the DFG (German Research Foundation) as part of the joint project "Effect of Pressure on Molecular and Cellular Structures in Foods"

References

[1] Matsuo, S.: Thermophysical properties of water and alcohol mixtures. In: Taniguchi, Y.; Senoo, M.; Hara, K. (ed.). High Pressure Liquids and Solutions, Elsevier Science, 1994.

[2] Mageean, M.P.; Kristott, J.U.; Jones, S.A. Scientific & Technical surveys. Physical Properties of sugars and their solutions, 1991, Leatherhead

[3] Höppler, F. Der exzentrische Fall von Kugeln in Hohlzylindern mit Flüssigkeiten oder Gasen. Z. Tech. Physik, 1933, 14, 165-169

[4] Först, P.; Delgado, A. In-situ viscosity measurement during the ultra high pressure treatment of fluid food systems. In: Ludwig, H (ed.). Advances in High Pressure Bioscience and Biotechnology , 1999, Springer Verlag, Heidelberg, 511-515

[5] Först, P.; Werner, F.; Delgado, A.: The viscosity of water at high pressures - especially at subzero degrees centigrade. Rheol. Acta 39 (2000), 566-573

[6] Dow, R.B.: In: Eirich, F.R. (ed.): Rheology: Theory and Applications. Vol. 1. Academic Press, 1959.

[7] Hinrichs, J.: Ultrahochdruckbehandlung von Lebensmitteln mit Schwerpunkt Milch und Milchprodukte – Phänomene, Kinetik und Methodik. Habilitation, TU München, 2000

[8] Vand, V.: Viscosity of solutions and suspensions. III. J. Phys. Colloid Chem. 52 (1948) 314:321

[9] Först, P.; Werner, F.; Delgado, A.: Rheological behaviour of mono-and disaccharide solutions at ambient and elevated pressures. Proceedings 2nd International Symposium on Food Rheology and Structure, Zürich, März 2000

[10] Heremans, K.: High Pressure Effects on Proteins and other Biomolecules. Ann. Rev. Biophys. Bioeng. 11 (1982), 1-21

[11] Balny, C.; Masson, P.; Travers, F.: Some recent aspects of the use of high-pressure for protein investigations in solutions. High Pressure Research 2 (1989), 1-28

[12] Buchheim, W.: Ultra high pressure technology (Review). In: Werner, H. (ed.). Proc. of the 25th Int. Dairy Congress. The Danish National Comittee of the Int. Dairy Federation. Aarhus (DK), 1999

[13] Weber, G.; Drickamer, M.G.: The effect of high pressure upon proteins and other biomoloecules. Quart. Rev. Biophys. 16 (1983), 89.

Trends in High Pressure Bioscience and Biotechnology
R. Hayashi (editor)

PRESSURE INDUCED CHANGES IN THE GELATION OF MILK PROTEIN CONCENTRATES

B.J. Briscoe, P.F. Luckham and K.U. Staeritz

Department of Chemical Engineering and Chemical Technology
Imperial College of Science, Technology and Medicine
Prince Consort Rd, London SW7 2BY, United Kingdom, e-mail: K.Staeritz@ic.ac.uk

Gelation of aqueous milk protein concentrates made up of spray dried milk was monitored both at atmospheric pressure and at moderate pressures (up to 1000 bar) using rheological methods. A Paar Physica UDS 200 rheometer was used to measure the gelation time at atmospheric pressure as a function of the temperature for a protein concentration of 16% [w/w]. The temperature dependence of the gelation time is well approximated by the Ross-Murphy model which describes the concentration dependence of the gelation time and has been used in the past to describe the concentration dependence of the gelation time for some biological polymers. Using a purpose built Haake High Pressure High Temperature Rheometer, the effects of moderate pressures upon the gelation of milk protein concentrates (13% [w/w] and 16% [w/w] protein) were investigated. It was found that the gelation time reduces significantly (by up to almost an order of magnitude), when pressures of up to 1000 bar were applied. These findings, combined with the work of others indicate, that these pressures bring about two effects: 1) casein micelle dissociation and as a result facilitate reassociation of those smaller micelle fragments, and 2) denaturation of beta-lactoglobulin. These two mechanisms contribute to different degrees to the accelerated gelation process brought about by elevated pressure in the range investigated. It is not clear if association of whey proteins with caseins occur at moderate pressures around $50°C$. We have modeled the dependence of the rate of gelation on pressure by a Eyring reaction rate process type equation. This description is not inconsistent with the experimental results obtained.

1. INTRODUCTION

It was at the end of the nineteenth century that the effects of the application of hydrostatic pressure on milk was originally investigated [1]. Hite' s work was instigated from the shortcomings that were being experienced with sterilisation and Pasteurisation Processes. The problem experienced at that time was the rapid temperature jumps associated with those processes. He was able to show that the application of 13.9 kbar, 1390 MPa applied for one hour was able to postpone the souring of milk for around 4 days. Despite this success, however, technical difficulties associated with the process meant that commercially this was a non-viable process. In the developed world, in the last fifteen to twenty years, there has been a reluctance by the consumer to purchase heavily processed foods or foods containing additives. Thus food scientists have revisited the potentially valuable effects of pressure on foods. In the hundred years since Hite's study there have been considerable developments in process technology and the application of high pressures in the engineering of plastics, ceramics, metals, etc., has become routine. Now in Japan, it is possible to purchase high pressure processed foods such as fruit juices and sauces. It has been noted that compared to temperature processed products these foods retain a remarkable degree of perceived "freshness". The reason for this is simple. In a heating process, inevitably some part of the food, that in contact with the heat source is hotter for longer than the food in the middle and so becomes overheated and, to some extent "spoils". The application of a hydrostatic pressure to a fluid is uniform and hence localised "spoiling" is considerably reduced.

Let us consider the role of pressure here. From Le Chatelier's principle we recall that when a system is subject to a constraint then it will adapt to remove the influence of the constraint. When applying pressure, that constraint will be volume, thus processes with a negative change in volume ($-\Delta V$) will be enhanced and those with a positive change in volume ($+\Delta V$), will be suppressed. The effect of pressure on milk is complex because of the complexity of milk itself. It must be remembered that milk is largely water, which is a hydrogen bonded liquid, whose structure breaks down at high pressures. The inter and intra molecular interactions between the proteins and water are also subject to the influence of pressure. For example, hydrogen bonding, which stabilises the α helix and β sheet structures, is influenced by pressure but this effect is probably less than ionic and hydrophobic interactions. For example, the ionisation of charged groups on a protein would involve a volume change, because the water molecules close to the charge reorient. This ordered alignment of water would give a dense layer of water around the charge and hence there would be a net volume decrease. As a result, application of pressure should increase the ionisation of charged groups on biological molecules.

There have been many studies probing the effects of high pressures on milk proteins, particularly on the casein micelles. These all show that pressure breaks-down these micelles into smaller aggregates and that this changes the properties of the milk (see for example: [9-15, 19]). Studies have also been carried out on β-lactoglobulin (Møller (1998) [17], Tanaka (1996) [18]) and again some structural changes in the protein are seen to occur. There have also been studies to determine the effect that pressure has upon the physical properties of milk. For example Johnson et al [16] have shown that a more rigid product is produced in milk which has been treated at high pressures compared to untreated milk. However, this study itself was not formed at pressure; only the milk was held at a high pressure and the test performed at ambient pressure.

Recently Zhu et al have developed a rheometer that probes the rheological properties of a material whilst it is still at high pressure. In his work, on aqueous polymer solutions, he showed that increasing pressure was rather analogous to increasing temperature, or adding electrolyte, in that it changed the solvency of the water for the polymer from a "good" solvent to a "poor" one. In this study, we report similar experiments on reconstituted skimmed milk, and in particular have investigated the gelation of reconstituted skimmed milk. Preliminary results of our findings have been published elsewhere [8, 5].

2. EXPERIMENTAL METHODS

Milk protein concentrates were made up from skimmed milk powder (Premier Baverages, UK), which was produced by spray drying and had undergone a special agglomeration process. In the experiments reported here, the milk was reconstituted at two different protein contents (13%w/w and 16%w/w). In the first step the milk powder was stirred with the help of a magnetic stirrer and in each case (in order to reach the final protein concentration), a Silverson (UK) high shear stirrer was used while no heat was supplied. It was assumed, that the milk proteins in the concentrates are either dissolved or in the case of casein micelles, are well dispersed.

In order to reduce drying-out of the sample surface, the sample was covered with a thin layer of liquid paraffin (the tetradecane) during the experiments using the Paar rheometer in ambient air. The Haake High Pressure High Temperature (HPHT) Rheometer (Germany/UK), a purpose built viscometer based upon a Haake RV 100 viscometer adopts the principle of a standard rotational viscometer and uses a special magnetic coupling to transmit the driving torque from the sample which is housed in a pressure chamber [2], [3], [4], [5]. Pressures of up to 1000 bar can be applied to the material under investigation and the instrument measures the material properties while pressure is applied. The Haake HPHT Rheometer is unable to precisely measure the viscoelastic material properties but is able to measure the shear stress in a continuous rotational mode. The maximum measureable shear stress is about 60 Pa and is limited by the coupling strength of the magnets imbedded in the rotor and of the turning magnets on the opposite side of the (stainless steel) diaphragm which separates the driving parts of the rheometer from the parts which are inside the autoclave. The sample feeding and pressure generating unit has previously been modified by Huxley-Bertam Engineering (Cambridge, UK). The pressure vessel or autoclave, which was initially developed for flow operation and process control, was replaced by a special construction, which facilitates the operation in the discontinuous modes. The total volume of the autoclave was about 300 ml. A special pressure tube connected the pressurisation cylinder with the autoclave and valves allowed the regulation of the flow of the sample from the pressurisation cylinder into the autoclave. Inside the autoclave the flow was altered by some deviations of the instrument from the ideal Searle Type cylindrical rotational viscometer. The stator contained some slits in the outer cylinder, thus allowing the sample to enter the geometry laterally during the actual measurements. The temperature control was achieved by heating electrically the whole metal block of the autoclave, into which the cover of the instrument holding the geometry was placed. For the temperature control a Haake temperature controller was used. The commercially available Paar Physica rheometer was operated both in steady state and in oscillatory mode using the Z2 Searle-Type geometry with a sample volume of 100 ml.

3. EXPERIMENTAL RESULTS AND DISCUSSION

The changes of the dynamic mechanical properties of milk protein concentrates reconstituted at 16% protein content were measured in order to monitor the temperature dependence of the gelation time. In Figure 1 typical courses of the measured values of G' and G" , at a frequency of 1 Hz during gelation at 50°C are represented, monitored at a displacement of 0.04 units. These measurements were carried out at ambient pressure on the Paar Physica

448

UDS 200 rheometer. Initially the storage modulus is close to zero, whilst the loss modulus is larger than the storage modulus, although still small in value. For around 10,000 s little change is observed, after this point both the storage and loss modulus increase, and the rate of increase in the storage modulus is faster than that of the loss modulus, such that at about 19,300 s the cross over point of the moduli occurs. G' reaches its steepest slope at about 28,000 s and reaches a plateau above 40,000 s.

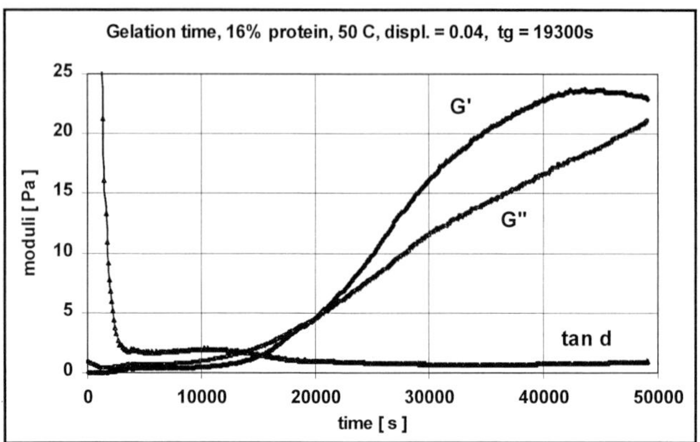

Figure 1   Dynamic mechanical properties of 16% protein concentrates as a function of the elapsed time

In Figure 2 the dependence of the gelation time on the temperature, determined by the cross-over-point of the loss and storage moduli is represented. The chosen frequency of oscillation was 1 Hz and the displacement was 0.04 units. The data were obtained from measurements similar to those presented in figure 1.

Based upon a model developed by Ross-Murphy [6] a non linear regression was carried out according to the following equation

$$\log t_g = k - n \log\left(\frac{T}{T_c} - 1\right)$$

The dotted lines in figure 2 correspond to this model. The full lines correspond to the Ross-Murphy model without logarithmic conversion, i.e.

$$t_g = \frac{k}{\left(\dfrac{T}{T_c} - 1\right)^n}$$

The regression is based on the Levenberg-Marquardt method, a commonly used algorithm for non-linear regression. The Ross-Murphy model [6] represents the behaviour of the gelation process as a function of the temperature for the current substance rather well. In this equation the analogy with the critical behaviour near a thermal phase transition was pointed out by Ross-Murphy [7] .

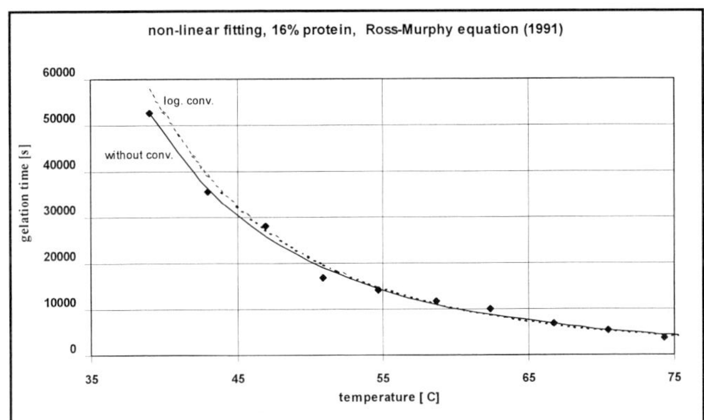

Figure 2   Dependence of the gelation time on temperature for 16% protein content at ambient pressure

The gelation time as measured in the steady state shear measurements on the Paar Physica rheometer and on the Haake instrument, was defined as the steepest part of the shear stress-time curve. [8, 5]. In a previous paper [8] the two different measures of the gelation time were compared and found to be similar: In order to directly compare the gelation time measured with the Haake instrument with the gelation time measured with the Paar Physica rheometer in the oscillatory mode, a strain of $\gamma = 0.1$ was chosen as this induces an amount of shear disruption and hence gel time delay, comparable to that produced by the Haake system [8, 5].

Figures 3 and 4 show the shear stress plotted as a function of time for various pressures at two milk protein concentrations, 13% and 16% respectively. It can be seen from both graphs that increasing the pressure decreases the gelation time. (The gelation time is now taken to be the steepest slope of the shear stress time curves). This effect is rather large, a factor of over 7 for the more dilute case. The question is, why does this occur?

As was mentioned earlier, pressure will encourage any process which results in the reduction in volume of the system. Gelation is such a process as any particles, casein micelles, in the milk are attracted to each other in a gelation process. It has been determined by many workers [9-15, 19] that increased pressure encourages the casein micelle to fragment into smaller units, which would gel then more easily [5, 8]. Studies carried out on β-lactoglobulin at 500 bar [17, 18] prove, that at moderate pressures these globular proteins partly denature. We conclude, that acceleration of gelation is mainly due to micelle fragmentation and to a lesser extend to a limited denaturation of β-lactoglobulin.

450

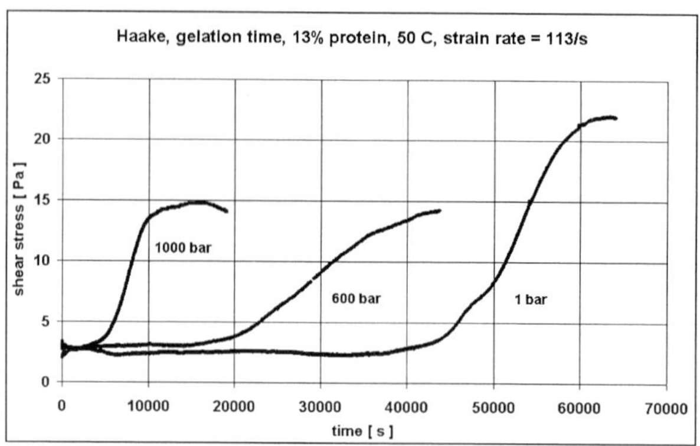

Figure 3   Dependence of the gelation time on pressure: Stress versus time for aggl. rec.milk (13% protein) at 50°C for three different pressures. Haake HPHT rheometer ($\dot{\gamma}$ = 113 /s). The gelation time is defined as the maximum slope and is at   1 bar : t_g = 53,300s (± 200s), at 600 bar: t_g = 29,000s (± 200s), at 1000bar: t_g = 7,500s (± 200s).

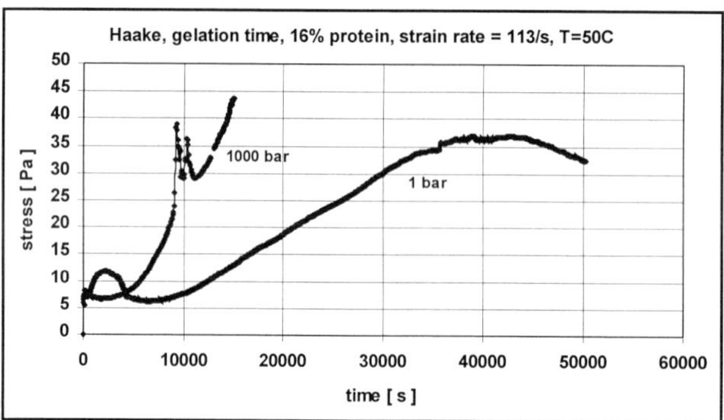

Figure 4   Dependence of the gelation time on pressure: Stress versus time for aggl. rec.milk (16% protein) at 50°C for two different pressures. Haake HPHT rheometer ($\dot{\gamma}$ = 113 /s). The gelation time is defined as the maximum slope and is at   1 bar : t_g = 23,500 s (± 200s), at 1000bar: t_g = 8,900 s (± 200s).

We shall now speculate upon the effects of pressure on the rate of gelation. As a simple first step the kinetic approach introduced simultaneously by Eyring [20] and Evans and Polanyi [21] will be adopted. According to Evans et al [21] the specific rate constant k_p can be written as:

$$k_p = k_o \exp(-\frac{P\Delta V^{\neq}}{RT})$$

where k_0 is the rate constant at zero pressure, k_p the rate constant under pressure P and $\Delta V^{\neq}$ is the pressure (as opposed to thermal) activation volume, which is the difference between the partial molar volume of the activated complex minus that of the reactants.
Taking the rate of gelation k_t (proportional to $1/t_g$) as the specific rate constant k_p, we find for our system, that the experimental results are reasonably consistant with the above equation.
Figure 5 shows the fitting of the experimental results for $c_p = 13$ % protein and a temperature of T = 50°C. Note that during the experiments the temperature fluctuated slightly ($\Delta T \approx \pm 2$°C), due to the large surface area of the pressurization vessel. It appears, that the aggregation of milk proteins under moderate pressures is a process, that can be treated to a first approximation by this first order kinetic approach. Further work will need to be performed to determine whether this approach is truly and generally applicable, or whether one only observes an effect of pressure on the gelation rate once the casein micelles have begun to fragment, which according to Schmidt [10], is around 100 bar.

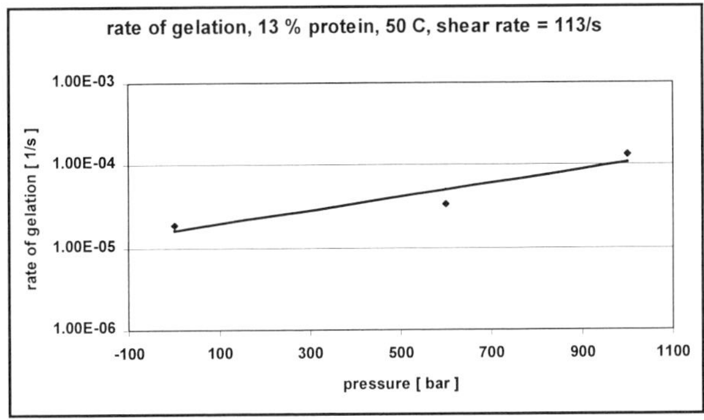

Figure 5   Rate of gelation vs pressure. Fitting according to the Eyring reaction rate process type equation

It is sensible, to also derive an equation for the combined influences of temperature and pressure on the gelation time. The temperature dependence on gelation time at atmospheric pressure is governed by unfolding of globular proteins, as the caseins are very heat resistant. At moderate pressures we concluded, that the aggregation is mainly due to casein micelle dissociation which, as a result, facilitated gelation. The nature of the dependence of the gelation time on temperature at moderate pressures may be different, as association of caseins with whey proteins may occur. Further experiments will need to show, whether gelation caused by the action of moderate pressures is a combined effect of heat and pressure (and involves a combined effect on caseins and whey proteins), or whether moderate pressures alone may bring about gelation at temperatures, where whey protein denaturation does not occur at atmospheric pressures.

REFERENCES

[1]   Hite, B.H., *Bull. West Virginia Univ. Agric. Exp. Statn.* 58(1899)15-35.

[2]   Briscoe, B.J.; Luckham, P.F.; Zhu, S., *Macromolecules* 29 (1996) 6208-6211.

[3]   Briscoe, B.J.; Luckham, P.F.; Zhu, S., *Proc.Roy.Soc.*, A455(1999)737-756.

[4]   Zhu, S.: *"Rheology of Polymer Solutions at High Temperature and High Pressure"*. In: PhD-Thesis (1997), Imperial College, London.

[5]   Staeritz, K.U.; Briscoe, B.J.; Luckham, P.F., *High Pressure Research*, Vol.19(2000)55-60.

[6]   Ross-Murphy, S.B., *Food Polymers, Gels and Colloids* - E.Dickinson (ed.), Cambridge: Royal Society of Chemistry (1991) 357-368.

[7]   Ross-Murphy, S.B.; Tobitani, A.: *"The Gel-Time - Theory and Practice"*. In: The Wiley Polymer Networks Group Review Series, Volume one, K. te Nijenhuis and W.J.Mijs (editors) (1998)39-49.

[8]   Briscoe, B.J.; Luckham, P.F.; Staeritz, K.U.: *"Dynamic and Steady State Measurements of Gelation Time of Milk Protein Concentrates at Moderate Pressures"*. In: Proceedings of the 2nd International Symposium on Food Rheology and Structure, March 12-16, 2000, Zuerich - P. Fischer, J.Marti, E.J. Windhab (eds.), 317-320.

[9]   Payens, T.A.J.; Heremans, K., *Biopolymers* 8(1969)335-345.

[10]   Schmidt, D.G.; Buchheim, W., *Milchwissenschaft* 25(1970)10, 596-600.

[11]   Buchheim, W.; Prokopek, D., *Dtsch. Milchwirtschaft* 43(1992) 1374-1376.

[12]   Shibauchi, Y.; Yamamoto, H.; Sagara, Y., Conformational Change of Casein Micelles by High Pressure Treatment, in: High Pressure and Biotechnology, C. Balny; R.Hayashi; K. Heremans; P.Mason (eds.) Colloque INSERM/John Libbey Eurotext Ltd. (1992) 239-242.

[13]   Lopez-Fandiño, R.; Ramos, M.; Olano, A., *J.Dairy Res.* 65(1998)69-78.

[14]   Desobry-Banon, S.; Richard, F.; Hardy, J., *J. Dairy Sci.* 77(1994)3267-3274.

[15]   Ohmiya, K.; Fukami, K.; Shimizu, S.; Gekko, K., *J. Food Sci.* 52(1987)1, 84-87.

[16]   Johnston, P.E.; Austin, B.A.;, Murphy, R.J., *Milchwissenschaft* 4(1993)48, 206-209.

[17]   Møller, R.E., Stapelfeldt, H., Skibsted, H.L., *J.Agric.Food Chem.* 46(1998)425-430.

[18]   Tanaka, N.; Kunugi, S.; *J.Biol.Macromol.* 18(1996)33-39.

[19]   Schmidt, D.G.; Payens, T.A.J., *J.Coll.Interf.Sci.* 39(1972)3, 655-662.

[20]   Eyring, H., *J.Chem.Phys.* 3(1935)107-115.

[21]   Evans, M.G.; Polanyi, M., *Trans. Faraday Soc.* 31,(1935)875-894.

Trends in High Pressure Bioscience and Biotechnology
R. Hayashi (editor)
© 2002 Elsevier Science B.V. All rights reserved.

Gel formation of individual milk whey proteins under hydrostatic pressure

Choemon Kanno and Tai-Hau Mu

Department of Applied Biochemistry, Utsunomiya University,
Utsunomiya 321-8505, Japan

The pressure-induced denaturation and gelation of β-lactoglobulin (β-Lg), α-lactalbumin (α-La) and bovine serum albumin (BSA) at various concentrations under 200-800 MPa at 30°C for 10 min were studied. Pressure-induced gels were formed from 10% β-Lg with 800 MPa and from 18% β-Lg with 400 MPa. The concentration of BSA required to form a gel was more than 16% at 800 MPa. However, no gel was induced by up to 24% α-La at 800 MPa. The α-helix content of β-Lg was decreased by increasing pressure and the random structure was increased, while the secondary structure of α-La and of BSA was not affected. The microstructure of the β-Lg gels showed a porous network, while the BSA gels had a plate-like structure. The addition of β-Lg or cysteine to α-La induced a stronger gel at 800 MPa with a similar microstructure to that of the gel from BSA. SDS-PAGE with or without 2-mercaptoethanol demonstrated that a disulfide-bonded high-molecular-weight aggregate was contained in the pressure-induced gels from β-Lg, BSA and α-La with added cysteine. These results suggest that an SH/S-S interchange reaction between β-Lg and α-La through the reduction of disulfide bonding with free SH groups greatly contributed to the gelation of the individual whey proteins.

1. INTRODUCTION

Whey of bovine milk is a by-product from cheese manufacturing and contains about 13% protein on a dry matter basis. One of the functional properties of milk whey proteins is their gelation and binding. They are, therefore, widely used as functional ingredients in many formulations of bakery, dairy, and sausage products (1, 2). The advent of improved and more cost-effective processing technology and production procedures such as ultrafiltration, reverse osmosis, and electrodialysis have resulted in a dramatic increase in the production of whey protein products. Industrial whey proteins are classified into whey protein concentrate (WPC) and whey protein isolate (WPI) which respectively contain less or more than 90% protein.

The gelation of milk whey proteins is induced by means of heating or hydrostatic pressure. High-pressure treatment has received considerable attention for food processing and preservation as a new technique for energy saving in recent years (3-6). There are a limited number of published reports on the pressure-induced gelation of whey protein and of individual whey proteins, although most of these studies are concerned with the pressure-induced gelation of β-Lg (5,7-10). However, the mechanisms responsible for the pressure-induced denaturation and gelation of whey protein are still not fully understood because of the complicated interactions existing among the different proteins of whey. Thus, we need to study the denaturation and gelation of the individual whey proteins under hydrostatic pressure. It may provide a better understanding of the mechanisms for the pressure-induced denaturation and gelation of whey protein.

The aim of this study is to report on 1) the gel formation of the individual whey protein components, β-Lg, α-La and BSA, 2) the effect of the protein concentration and applied pressure on the rheology and microstructure of gels induced from the individual whey protein components, 3) the interaction among the individual whey protein components in gel formation, and 4) the mechanism for the gel formation of whey proteins under hydrostatic pressure.

2. METHODS

2.1. Preparation of the whey protein solutions

β-Lg, BSA and α-La (from Sigma) were each dissolved in 0.1 M Tris-HCl at pH 7.2 or in distilled water adjusted to pH 7.0 to give the desired concentration. Cysteine was included in the protein solution before pressurization, if necessary.

2.2. Measurement of the viscosity

The viscosity of each fluid sample before and after pressurization was measured at 30°C and at a conical rotor speed of 50 rpm.

2.3. Pressurization of the samples

Each whey protein solution was put into a Teflon tube of 4 ml in volume and then pressurized in a range from 200 to 800 MPa with a hand-operated oil pressure generator (HR15-B2, Hikari Koatsu) at 30°C. The indicated pressure was achieved within 0.5-2 min, maintained for 10 min, and then released to atmospheric pressure within 0.5 min.

2.4. Measurement of the hardness and breaking stress

The hardness and breaking stress of each gel were individually measured at about 25°C with a Fudoh rheometer as indicated previously (10).

2.5. SDS-polyacrylamide gel electrophoresis (SDS-PAGE).

SDS-PAGE in the presence or absence of 2-mercaptoethanol was conducted according to the method of Laemmli (11), using a linearly graded separating gel from 4% to 20% in polyacrylamide.

Each gel sample was homogenized for 1 min at 19,000 rpm in a 0.086 M Tris-0.09 M glycine buffer (pH 8.0) containing 4 mM EDTA (buffer A) and then separated into the supernatant and precipitate by centrifuging for 15 min at 20,000 x g. The precipitate was then dissolved in 2 ml of buffer A containing 8 M urea and 0.5% SDS. The supernatant and precipitate fractions were diluted with 4 volumes of a 50 mM Tris-HCl buffer (pH 6.8) containing 15 mM EDTA, 5% SDS, and 60% sucrose, with or without 5% 2-mercaptoethanol, and applied to SDS-PAGE. The protein concentration was determined according to the method of Peterson (12) with 2.0% sodium deoxycholate.

2.6. Circular dichroic measurements

CD spectra were measured, after releasing the pressure, at 25°C with a Jasco J-700 spectropolarimeter, using a quartz cell with a path length of 1 mm. Each protein at 0.08% (w/v) in a phosphate buffer (pH 7.0) was pressurized for 10 min before being diluted to 0.016%. The molar ellipticity of the protein was used to estimate the α-helix, β-sheet, β-turn and random coil contents according to the method of Yang *et al.* (13).

2.7. Scanning electron microscopy

Small pieces of each gel induced from α-La containing cysteine were fixed at 4°C in 3.5% glutaraldehyde in a 0.05 M sodium phosphate buffer (pH 7.0) for 12 h, rinsed for 1 h in distilled water, dehydrated in a series of ethanol mixtures of 30%, 50%, 70%, 85%, 95% and 100%, and finally immersed in 97% isoamyl acetate. After dehydration, critical point drying was done in liquid CO_2. The dried sample was sputtered with a carbon-coating Hitachi E1020 10N unit. The pressure-induced gels from 16% (w/w) β-Lg and 20% (w/w) BSA were also cut into small pieces without any chemical fixation. All observations were conducted at an acceleration voltage of 15 or 25 kV.

3. RESULTS AND DISCUSSION

3.1. β-Lactoglobulin

The viscosity of β-Lg increased with increasing pressure at a concentration of more than 6%. The viscosity of the pressurized 10% β-Lg solution was 21 mPa·s at 600 MPa. However, the 10% β-Lg solution began to form a soft and watery gel with a white appearance at 800 MPa. The concentration of β-Lg required for inducing gelation was 14% at 400 MPa, 12% at 600 MPa, and 10% at 800 MPa. A sufficiently firm gel was formed at a 14% β-Lg concentration and 600 MPa and at a 12% β-Lg concentration and 800 MPa.

The hardness and breaking stress of gel samples with different β-Lg concentrations were both enhanced by increasing the protein concentration and increasing the hydrostatic pressure to 800 MPa. The gels formed from the 12% β-Lg solution at 600 MPa and from the 14% β-Lg solution at 400 MPa were too soft to measure their rheological properties with a rheometer. Pressure-induced syneresis was observed, the gel volume increasing with increasing concentration of β-Lg and applied pressure.

SDS-PAGE patterns for the supernatant of the proteins extracted with a Tris-glycine-EDTA buffer from pressure-induced gels of β-Lg showed that, without 2-mercaptoethanol, the dimer, trimer, tetramer and a high-molecular-weight aggregate were increased, while the β-Lg monomer was decreased with increasing pressure. Similar results were found for the precipitate which was further extracted with the same buffer containing 8 M urea and 0.5% SDS. On the other hand, with 2-mercaptoethanol, most of the β-Lg was detected as the monomer, suggesting that disulfide bridges contributed to the gel formation.

The typical microstructure of the pressure-induced gels from 16% β-Lg at 800 MPa by SEM resemble a honeycomb in the neutral pH region (Fig. 1-A and -B). A larger pore size was apparent in the gels induced at 400 MPa. The pore size in the microstructure of the β-Lg gel increased with increasing pressure and seemed to correspond to their hardness and breaking stress.

3.2. Bovine serum albumin

The concentration of BSA required to form a gel was more than 16% at 800 MPa, and a sufficiently firm gel was induced at a concentration of more than 20%. The obtained gels were pale yellows in color. No gel was induced from 14% BSA even at 800 MPa.

The hardness and strength of the gels from BSA were smaller than those of the gels from β-Lg under the same pressure and concentration.

SDS-PAGE patterns for the proteins extracted from the pressure-induced gels with a Tris-glycine-EDTA buffer showed no appreciable difference by increasing the pressure. Without 2-mercaptoethanol, the dimer, trimer, tetramer and a high-molecular-weight aggregate were detected, in addition to the BSA monomer, whereas with 2-mercaptoethanol, most of the BSA was detected as the monomer.

The microstructure of the gels induced from BSA showed a plate-like structure, and there was no porous network that was apparent with the β-Lg gels (Fig. 1-C). A change in the microstructure was observed for the BSA gels by increasing the pressure to 800 MPa

3.3. α-Lactalbumin

No relationship between the viscosity and concentration of α-La at 800 MPa was found. The viscosity of α-La increased with increasing concentration, but the difference in viscosity between the pressurized and unpressurized α-La solutions was very small. The viscosity of an untreated α-La solution was 1.68 mPa·s at 6% and 2.65 mPa·s at 18%, whereas the viscosity of the treated α-La solution increased to 1.74 mPa·s and 3.13 mPa·s, respectively. No gel was induced until a 24% α-La concentration was used at 800 MPa.

3.4. Secondary structure

The effect on the secondary structure of soluble protein extracted with the Tris-glycine-EDTA buffer from the gel produced by pressurization for 10 min at 800 MPa was studied. In β-Lg, the α-helix and β-turn structural content decreased at a pressure of more than 200 MPa, while the random and β-sheet content increased at a pressure of more than 200 MPa. Thus, the α-helix and β-turn structure of the β-Lg molecule was changed into a random and β-sheet conformation with increasing pressure. The β-turn content of BSA was slightly decreased, while the α-helix and random structure was slightly increased at a pressure of more than 200 MPa. The pressure did not affect the secondary structure of α-La.

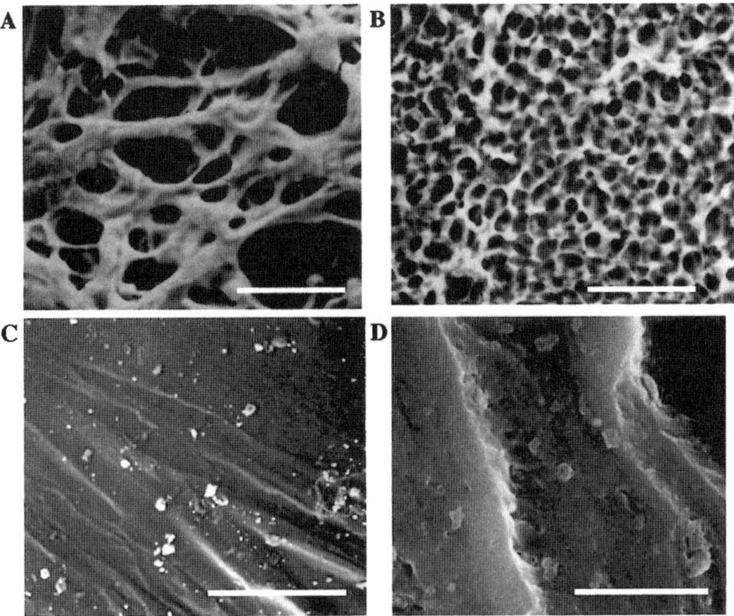

Figure 1. Scanning electron microscopic pictures of the pressure-induced gels from a 16% β-lactoglobulin (A and B), 20% bovine serum albumin (C) and 24% α-lactalbumin solution plus 3 mM cysteine (D). The solution was pressurized for 10 min at 400 MPa for A and at 800 MPa for B-D. The acceleration voltage of 15 kV for A, B and D, and 25 kV for C.

3.5. Effect of reducing agents

The addition of a reducing agent to WPI increased the hardness and breaking stress of the pressure-induced gels from WPI. Ascorbic acid, which has no thiol group, did not affect the texture of the gels; However, cysteine, 2-mercaptoethanol, reduced glutathione and dithiothreitol, which have one or more thiol groups, increased the hardness and breaking stress of the WPI gels with increasing concentration up to 6 mM, while dithiothreitol drastically decreased it at a concentration of more than 6 mM.

At pH 4.0, there was no effect of cysteine on the gel formation of 20% BSA. The effect of 18 mM of cysteine attained the maximum hardness and breaking stress at pH 7 and 9, but 30 mM cysteine decreased these levels. The effect of 18 mM cysteine on BSA was higher at pH 7 than at pH 9, but cysteine had no affect on the breaking stress at pH 4.

The addition of cysteine induced the gelation of α-La under hydrostatic pressure. A gel was induced by pressurizing α-La with 3 mM added cysteine for 10 min at 800 MPa. This gel was harder than the gel from 20% BSA containing 3 mM cysteine and from WPI containing 6 mM cysteine. Interestingly, the breaking stress of this gel was about 72% of the hardness. The microstructure observed by SEM was plate-like, as was the case of the gels from BSA (Fig. 1-D). SDS-PAGE without 2-mercaptoethanol showed no difference between the pressure-treated and untreated samples with 3 mM cysteine. Bands larger than 21 kDa were detected, in addition to the α-La monomer. With 2-mercaptoethanol, the monomer of α-La was most commonly detected.

Adding 5% β-Lg to 24% α-La that had not formed a gel at 800 MPa also induced gelation. The pressure-induced gel was transparent. The texture of this gel was harder and more fragile than that induced from 24% α-La with 3 mM cysteine. This gel had a smooth plate-like microstructure by SEM. Our results indicate that free SH groups were essential among proteins for the formation of a gelation matrix under hydrostatic pressure. The supernatant of the proteins extracted with a Tris-glycine-EDTA buffer from the pressure-induced gels of 5% β-Lg added to 24% α-La was analyzed by SDS-PAGE. Without 2-mercaptoethanol, the bands of the dimer, trimer, tetramer and a high-molecular-weight aggregate were observed, as well as that of pressured β-Lg alone. Similar results were found for the precipitate which had been further extracted with the same buffer containing 8 M urea and 0.5% SDS. With 2-mercaptoethanol, most of the β-Lg and α-La was commonly detected as the monomer, indicating that the sulfhydryl group of β-Lg and disulfide bridges of α-La contributed to the gel formation.

These results indicate that the degree of pressure-induced gelation and the firmness of their gels was in the order of β-Lg>BSA under the same conditions. This is opposite to the order from heat treatment of BSA>β-Lg>α-La at a low concentration (14,15). This reason can be explained by the different molecular structure between β-Lg and BSA, since β-Lg contains one free SH and two disulfide bonds, while BSA carries 17 disulfide bonds and one free SH for one molecule (16,17). To break these bonds, BSA needed a higher energy than β-Lg. However, the energy produced by pressure was smaller in magnitude than the thermal energy, such as pressure energy of 200 MPa equal to $\Delta 33°C$ of thermal energy (18).

The present results clearly show that when α-La alone was pressurized, the viscosity only slightly increase as compared with the unpressurized control at the same concentration, and did not form gel even at a higher concentration of 24% at 800 MPa. We found in this study, that a stronger gel could be induced by adding 3 mM cysteine to α-La. Bovine α-La is a globular whey protein of low molecular weight and has four disulfide bonds and no free thiol. Furthermore, adding 5% β-Lg to 24% α-La that did not form a gel at 800 MPa could also induce gelation. The addition of 3 mM cysteine to 20% BSA and 6 mM cysteine into WPI also accelerated the gelation. Therefore, the mechanism for the gel formation of α-La seem to involve the cysteine residue or free SH group participating in SH/S-S interchange reactions through the reduction of disulfide bonding to form intermolecular cross-links in the gel network. The existence of covalent interactions has been found by non-reducing SDS-PAGE, some oligomers and high-molecular-weight aggregates appearing to be involved in the formation of the 24% α-La gel. The formation of these covalent bonds was also confirmed in the pressure-induced gel from a lower concentration α-La (12%) plus a 3-mM cysteine solution, although the pressure could not induce a gel.

Gelation was not induced by heat treatment even in 20% α-La at neutral pH, indicating α-La had poor thermal gelation properties (19). However, it is known that the existence of other proteins allows the interaction to occur between α-La and the added protein (20,21). Furthermore, it was demonstrated that only the incorporation of β-Lg into α-La could induce the formation of gels via disulfide bonds (22). Therefore, our results certify that gels were induced from a α-La solution depending on the sulfhydryl groups have in this solution by both heat and pressure treatment.

The β-Lg molecule has five cysteine residues, of which four form two disulfide linkages, the remaining residue being an unpaired free SH group at the 121 position. This free SH group would become more exposed at higher pH and higher pressure. The free thiol groups exposed on β-Lg or on cysteine added as a reducing agent were oxidized and reduced the disulfide bonding, and then underwent intermolecular association. However, oxidation of the free thiol group only induced the formation of the dimer of β-Lg. Furthermore, for polymerization and final gelation, reduction of the intermolecular disulfide bridge was required by the free SH group. At pH 7 or higher, oxidation of the SH group, the reduction of S-S bonding, and exchange of SH-SS in the β-Lg and/or α-La molecules were each essential for the pressure-induced gelation of β-Lg and α-La, resulting in the porous network structure like a honeycomb and in higher water-holding capacity.

REFERENCES

1. W. J. Harper, Int. Dairy Fed., Brussels, Belgium, (1991) 77.
2. B. Thomsen and M. C. Pedersen, Eur. Food Drink Rec. (Winter), (1993) 61.
3. R. Hayashi, Report of Res. Institute for Food Science, Kyoto University, 51 (1988) 30.
4. V. B. Galazka, and D. A. Ledward, Food Technol. Int. Eur. (1995) 123.
5. J. C. Cheftel, Food Science and Technology International, 1 (1995) 75.
6. K. Heremans, J. Van Camp and A. Huyghebaert, Food proteins and their applications (S. Damodaran and A. Paraf, eds.), Marcel Dekker, New York, (1997) 473.
7. J. van Camp and A. Huyghebaert, Lebensm. Wiss. Technol., 28 (1995) 111.
8. J. van Camp and A. Huyghebaert, Food Chem., 54 (1995) 357.
9. D. V. Zasypkin, E. Dumay and J. C. Cheftel, Food Hydrocolloids, 10 (1996) 203.
10. C. Kanno, T. H. Mu, T. Hagiwara, M. Ametani and N. Azuma, J. Agric. Food Chem., 46 (1998) 417.
11. U. K. Laemmli, Nature, 227 (1970) 680.
12. G. L. Peterson, Anal. Biochem, 193 (1977) 265.
13. J. T. Yang, C. S. C. Wu and H. M. Martinez, Methods Enzymol, 130 (1989) 208.
14. M. Donovan, and D. M Mulvihill, Irish J. Food Sci. Technol., 11 (1987) 87.
15. N. Matsudomi, D. Rector, and J. E. Kinsella, Food Chem., 40 (1991) 55.
16. J. R. Brown, Federation Proc., 35 (1976) 2141.
17. M. Z. Papiz, L. Sawyer, E. E. Eliopoulos, A. C. T. North, J.B.C. Findlay, R. Sivaprosadarao, T. A. Jones, M. E. Newcomer, and P. J. Kraulis, Nature, 324 (1986) 383.
18. I. Hayakawa, M. Oda, and J. Kajihara, Res. Rep. Ito Foundation, 7 (1988) 391.
19. M. Paulsson, P-O. Hegg, and H. B. Castberg, J. Food Sci., 51 (1986) 87-90.
20. M. M. Calvo, J. Leaver, and J. M. Banks, Int. Dairy J., 3 (1993) 719.
21. D. G Dalgleish, V. Senaratne, and S. Francois, J. Agric. Food Chem., 45 (1997) 3459.
22. J. Gezimati, H. Singh, and L. K. Creamer, J. Agric. Food Chem., 44 (1996) 804.

Trends in High Pressure Bioscience and Biotechnology
R. Hayashi (editor)

461

Hydrostatic pressure-induced solubilization and gelation of chicken myofibrils

Katsuhiro Yamamoto, Takuji Yoshida, and Tomohito Iwasaki

Department of Food Science, Rakuno Gakuen University,
Ebetsu, Hokkaido 069-8501, Japan

Hydrostatic pressure effect on chicken myofibrils was studied. Some of myofibrillar proteins were solubilized by pressure application above 200 MPa. Most of them were derived from thin filament; namely, actin, tropomyosin, and troponin components. Myosin heavy chain was hardly solubilized in 0.1 M NaCl, while its solubilization occurred in 0.2 M NaCl at 200–300 MPa. Solubilization of myosin depended on magnitude of applied pressure and duration of treatment, and also salt concentration and pH. The maximal solubilization of myosin occurred at 200 MPa in 0.2 M NaCl at pH 7.

Myofibrils in 0.1–0.2 M NaCl with a protein concentration of 40 mg/ml formed a gel by pressure application. The gel strength at 0.1 M NaCl increased with pressure, while it remained almost at the same level in 0.2 M NaCl. The intrinsic structure of myofibril was retained after pressure treatment in 0.1 M NaCl; however, pressure-induced disruption of myofibrillar structure occurred in 0.2 M NaCl.

1. INTRODUCTION

Pressure treatment of meat is expected to improve meat quality such as meat tenderness, in addition pressure-induced gelation of myofibrillar proteins is useful for production of novel processed meat products [1, 2]. Since myofibril is a fundamental unit of muscle tissue, it is important to know the pressure-induced changes in myofibrils to understand pressure effect on meat.

Solubility of myofibrils is known to depend on ionic strength as well as pH. Myofibril retains its architecture at physiological salt concentration and pH, and no solubilization substantially occurs at low ionic strength. When salt is added to myofibrils to a final concentration of approximately 0.3 M, myosin is dissociated from thick filament. Most of myofibrillar proteins are solubilized in 0.5–0.6 M KCl at alkaline pH. Besides salt effect on solubilization of myofibrils, pressure treatment is known to induce dissociation of thin filament [3, 4] and thick filament [5, 6]. We had reported that myosin filaments formed a gel by pressure application [7] and monomeric myosin in high salt concentration failed to form a gel [8]. The present study concerns pressure effect on solubilization and gelation of myofibrils at low ionic strength, in which myofibrillar proteins are not soluble under the atmospheric pressure.

2. MATERIALS AND METHODS

Myofibrils were freshly prepared from chicken pectoral muscle. The muscle was ground and homogenized in 0.1 M NaCl and 20 mM phosphate (pH 7) with a Waring blender. The homogenate was centrifuged. The precipitate was suspended in 0.1 M NaCl and 20 mM phosphate (pH 7), and centrifuged again. Washing of myofibrils was repeated several times to remove sarcoplasmic proteins.

Myofibrils were suspended in 0.1 or 0.2 M NaCl with a protein concentration of 10 mg/ml, and pH was adjusted to 6.0 or 7.0 for a solubility measurement. Pressure application was done up to 500 MPa and 30 minutes at room temperature. After pressure release, the samples were centrifuged, then the supernatant and the precipitate were collected. Protein concentration in the supernatant was measured, and the protein composition was analyzed by SDS-PAGE. The precipitate was also analyzed by SDS-PAGE to elucidate unsolubilized protein components. The morphology of unsolubilized myofibrils was observed by phase contrast microscopy and also by transmission electron microscopy.

Pressure-induced gel was prepared from myofibrils suspended in 0.1–0.2 M NaCl at pH 6.0 or 7.0 with a protein concentration of 40 mg/ml. A myofibril suspension was poured into a plastic tube having diameter of 14 mm to give a height of 25 mm and plugged with silicon rubber without trapping air bubbles. Pressure was applied for 15 minutes. After pressure release, a plug was removed and a spherical plunger (diameter of 7 mm) was penetrated into a gel with a constant speed of 0.5 mm/sec. A force was recorded during penetration, and the maximum peak force was used as gel strength in this study. We also investigated morphology of gel structure by scanning electron microscopy.

3. RESULTS AND DISCUSSION

Figure 1 shows pressure-induced solubility change of myofibrils at 0.1 M and 0.2 M NaCl. When the applied pressure was 100 MPa, pressure-induced solubilization was very low regardless of NaCl concentration and pH. With increasing applied pressure, solubilization increased with extending time of pressure treatment. In case of 0.1 M NaCl and pH 6, solubility gradually increased at 200 MPa, while those at 300 to 500 MPa reached at maximal values at 2.5 minutes of pressure application and they remained at a constant level up to 30 minutes. The solubility at pH 6.0 was slightly higher than that at pH 7.0. The maximal protein concentration was about 0.8 mg/ml in pH 6 and 0.6 mg/ml in pH 7, indicating that 6–8% of myofibrillar protein was solubilized by pressure treatment in 0.1 M NaCl.

A solubility change in 0.2 M NaCl at pH 6 was comparable to that at 0.1 M NaCl except 200 MPa, in which case the solubility was similar to those above 300 MPa. The maximal protein concentration was about 1 mg/ml, which was slightly higher than that at 0.1 M NaCl. There was a notable solubility change in 200MPa

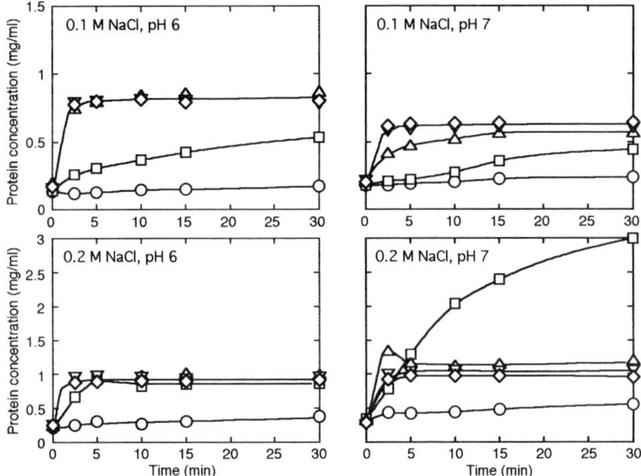

Figure 1. Pressure-induced changes in solubility of myofibrils in 0.1 M and 0.2 M NaCl. Myofibril suspension (10 mg/ml) was pressurized at a designated pressure and time. After pressure release, the suspension was centrifuged and the protein concentration of the supernatant was measured. O; 100 MPa, □; 200 MPa, △; 300 MPa, ▽; 400 MPa, and △; 500 MPa.

at pH 7. The solubility of myofibril treated at 200 MPa markedly increased with extending time of pressure application. The protein concentration reached at 3 mg/ml after 30 minutes of pressure application, which indicated approximately 30% of myofibrillar proteins were solubilized by pressure application of 200 MPa. Another interesting phenomenon was observed in the sample of 300 MPa. In this case, the protein concentration reached at maximal value at 2.5 minutes, while it decreased at 5 minutes of pressurization and remained at a constant level up to 30 minutes. No such temporal increase and following decrease in protein concentration was observed in 400 and 500 MPa. The protein concentrations in 300–500 MPa at 0.2 M NaCl were comparable to those at pH 6. Pressure-induced solubilization of chicken myofibrillar proteins is consistent with the result obtained from sheep [9] and rabbit [10] myofibrils.

As described above, it was clear that some of myofibrillar proteins were solubilized by pressure application. We performed SDS-PAGE to clarify what kinds of myofibrillar proteins were solubilized by pressure application. Figure 2a shows SDS-PAGE pattern of solubilized fraction at 0.1 M NaCl. Actin, tropomyosin and troponin, which are the components of thin filament, were solubilized by pressure treatment in both of pHs 6 and 7. The densities of these protein bands seemed almost the same from 2.5 to 30 minute of pressurization at 300 and 400 MPa. On the other hand, those in 200 MPa were less compared to 300–400 MPa, although the densities of bands of thin filament components increased with extending time of pressure treatment. Solubilization of thin filament components at 100 MPa was a little. Thin filaments in myofibril are known to be sensitive

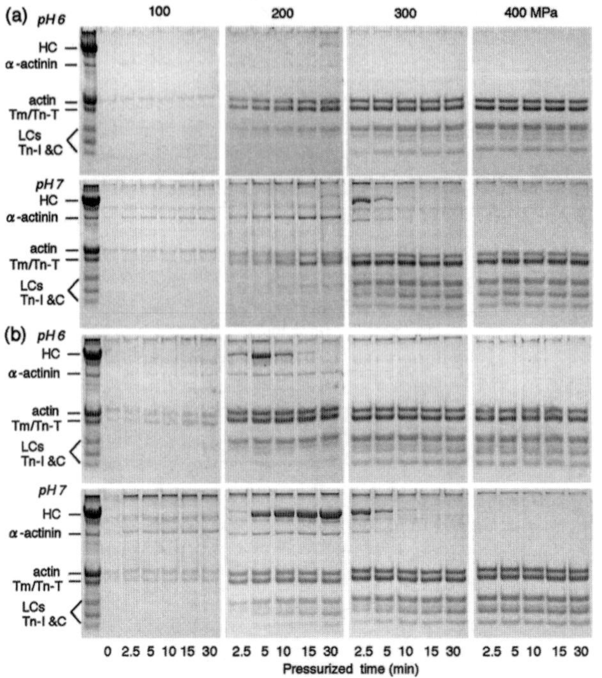

Figure 2. SDS-PAGE profiles of pressure-induced solubilized fraction of myofibril in 0.1 M (a) and 0.2 M (b) NaCl. HC, Tm, Tn, and LC represent myosin heavy chain, tropomyosin, troponin, and myosin light chain, respectively. The left lanes in 100 MPa represent myofibrils.

against pressure, and they depolymerize into G-actins by pressure application [10, 11]. Myosin heavy chain was observed in the supernatant pressurized at 300 MPa for 2.5 to 5 minutes at pH 7, but the band was not observed in the samples pressurized for 10 to 30 minutes. α-actinin, which is a component of Z-line of myofibril, was detected in the solubilized fraction of 0.1 M NaCl and pH 7 pressurized at 100 and 200 MPa. The density of α-actinin band in 200 MPa was higher than that in 100 MPa, and it increased with time of pressure application.

Figure 2b shows SDS-PAGE pattern of solubilized fraction at 0.2 M NaCl. Solubilization of thin filament components occurred as well as the case of 0.1 M NaCl. The bands of these proteins at 200 MPa were denser than those in 0.1 M NaCl regardless of pH, and they were comparable to those in 300–400 MPa. There was solubilization of myosin heavy chain in pH 6 at 200 MPa from 2.5 minutes to 10 minutes. There were no myosin heavy chain bands in 100, 300, and 400 MPa at pH 6. As shown in solubility change (Figure 1), solubility in 0.2 M NaCl notably increased at 200 MPa. There was marked increase of myosin heavy chain band with extending time of pressure application at 200 MPa and

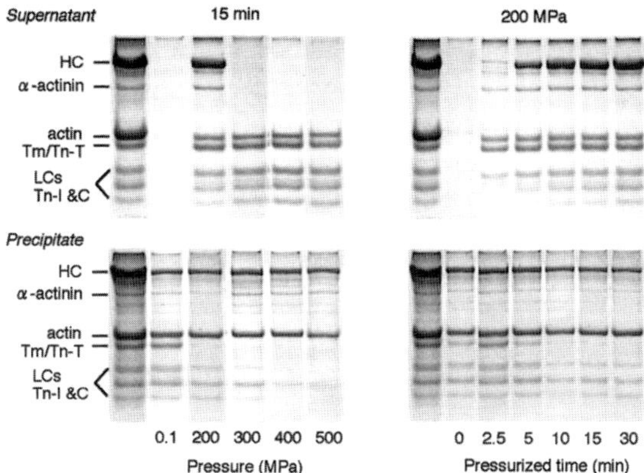

Figure 3. SDS-PAGE profiles of pressure-induced solubilized and unsolubilized fractions of myofibril in 0.2 M NaCl and pH 7. The most left lane in each electrophoretogram represents myofibril.

pH 7. In case of 300 MPa, the maximal solubilization of myosin heavy chain was observed at 2.5 minutes, and the density of the band decreased with extending time, and it was not detectable in the solubilized fraction treated for 15–30 minutes. This suggests that solubilized myosin became insoluble, in other words, myosin forms aggregates with extending duration of pressure application at 300 MPa. Myosin heavy chain was not observed in 400 MPa. Solubilization of α-actinin in 0.2 M NaCl was observed in 100 and 200 MPa from 2.5 to 30 minutes, and also in 300 MPa at 2.5–5 minutes as in the case of 0.1 M NaCl.

Figure 3 shows SDS-PAGE profiles of the solubilized and unsolubilized fractions in 0.2 M NaCl and at pH 7 with varying pressure and time of application. When pressure treatment was done with fixed time of 15 minutes with varying pressure, myosin heavy chain was only observed in 200 MPa. No heavy chain was observed in 300 to 500 MPa. Besides myosin heavy chain, thin filament components and myosin light chains were solubilized in the pressurized myofibrils regardless of applied pressure. In the precipitate, which is unsolubilized myofibrillar fraction, myosin heavy chain and actin bands were observable, indicating that not all of these proteins were solubilized by pressure. The bands of tropomyosin, troponin components, and light chains became faint with increasing pressure. When the applied pressure was fixed at 200 MPa, solubilization of myosin heavy chain increased with time of pressure application. A dense myosin heavy chain band at 200 MPa explains the increase in solubility shown in Figure 1. The density of troponin and light chain bands also increased with time in the supernatants, on the contrary, the bands of tropomyosin/troponin T and light chains, and troponin I and C in the precipitates decreased with time.

A morphological observation of pressure-treated and heat-treated myofibrils was done using phase contrast microscope (Figure 4). Applied pressure was 200 MPa. A typical striation pattern derived from A band and I band was observed

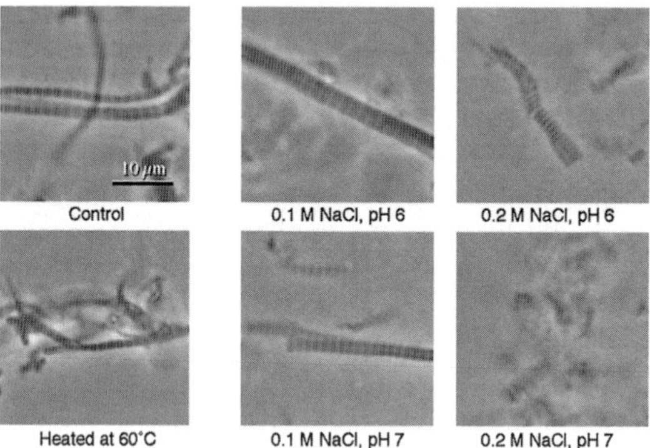

Figure 4. Phase contrast micrographs of myofibrils.

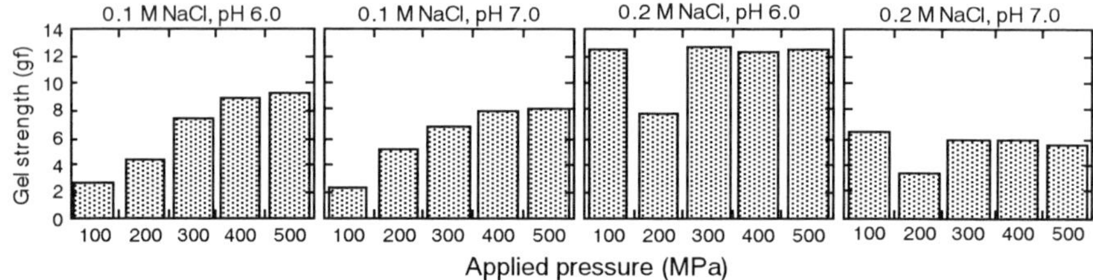

Figure 5. Gel strength of pressure-induced myofibrillar gel. Pressure was applied for 15 minutes at room temperature.

in the untreated myofibrils. When myofibrils were pressurized in 0.1 M NaCl, striation pattern was observable as that in control regardless of pH. On the other hand, periodic striation of dark and light bands in the myofibrils in 0.2 M NaCl at pH 6 was narrower than that in the untreated myofibrils, and the striation pattern became faint at pH 7. Obscure striation in the myofibrils at 0.2 M NaCl and pH 7 might be due to solubilization of myosin as well as thin filament components. These pressure-induced structural alteration of myofibrils is compatible with the electron microscopic observations [10–12].

The high pressure effect on binding of comminuted meat is reported [13], however, pressure-induced gelation of myofibrils is not well known. We made pressure-induced gels with the two protein concentrations, 30 and 40 mg/ml. Although a gel was formed at 30 mg/ml, it was very weak. No gel was formed below 30 mg/ml. When the protein concentration was 40 mg/ml, a firm gel was formed except 100 MPa in 0.1 M NaCl at pHs 6 and 7. Five minutes of pressure application was enough to form a gel, and there was no substantial difference in gel strength between 5 and 15 min of pressurization. The gel strength of the gel formed in 0.1 M NaCl almost lineally increased with applied pressure, whereas the gels formed in 0.2 M NaCl had almost the same gel strengths regardless of

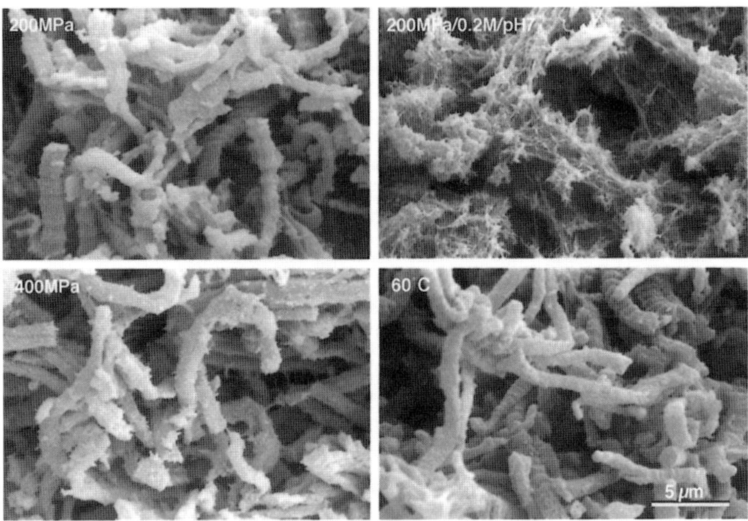

Figure 6. Scanning electron micrographs of pressure- and heat-induced myofibrillar gel. Pressure-induced gels were formed from myofibrils suspended in 0.1 M NaCl at pH 6 with applied pressure of 200 MPa and 400 MPa (left). The upper right is a pressure-induced gel formed in 0.2 M NaCl at pH 7 and 200 MPa. The lower right is a thermal gel heated in 0.1 M NaCl at pH 6.

applied pressure except 200 MPa, in which case the gel strength was rather low than the others. Pressure-induced gels formed at pH 6 were firmer than those at pH 7, and a such tendency was similar to that observed in myosin gel. The gel strengths of pressure-induced gels were comparable to that of the heat-induced gel formed in the same solvent condition. Those of thermal gel formed in 0.1–0.2 M NaCl at 60 °C were 8–10 gf at pH 6 and about 6 gf at pH 7.

Microstructure of pressure-induced myofibrillar gel was observed with scanning electron microscopy (Figure 6). Pressure- and also heat-induced myofibrillar gels were composed with a network of myofibrils. The structure of myofibril was retained in 0.1 M NaCl, while disruption of myofibril was observed in 0.2 M NaCl at pH 7 (upper right) and pH 6 as well (not shown). The disruption of myofibrils corresponds to solubilization of myofibrillar proteins. The surface of myofibrils in heat-induced gel in 0.1 M NaCl seemed to be smooth, while that in pressure-induced gel was spiny. Spiny structure was few in 200 MPa, however, it was obvious in the gels formed at higher pressure such as 400 and 500 MPa. It is possible that these spines are formed by solubilized myofibrillar proteins, and they play a role in the network formation of myofibrils.

It is concluded that high hydrostatic pressure induces solubilization of myofibrillar proteins and induces gelation of myofibrils at high protein concentration. Thin filament components were easily solubilized with pressure application; however, the solubilization of myosin depended on the magnitude and duration of pressure application, salt concentration and pH.

REFERENCES

1. J.J. Macfarlane, Developments in Meat Science, pp. 155-184 (ed.) R. Lawrie, Elsevier Applied Science Publishers, London 1985.
2. J.C. Cheftel and J. Culioli, Meat Sci., 46 (1997) 211.
3. T. Ikkai and T. Ooi, Biochemistry, 5 (1966) 1551.
4. R.R. Swezey and G.N. Somero, Biochemistry, 24 (1985) 852.
5. J.S. Davis, Biochem. J., 197 (1981) 301.
6. S.J. Tumminia, J.F. Koretz and J.V. Landau, Biochim. Biophys. Acta, 1040 (1990) 373.
7. K. Yamamoto, T. Miura and T. Yasui, Food Structure, 9 (1990) 269.
8. K. Yamamoto, S. Hayashi and T. Yasui, Biosci. Biotech. Biochem., 57 (1993) 383.
9. J.J. Macfarlane, J. Food Sci., 39 (1974) 542.
10. J.J. Macfarlane and D.J. Morton, Meat Sci., 2 (1978) 281.
11. A. Suzuki, M. Watanabe, K. Iwamura, Y. Ikeuchi and M. Saito, Agric. Biol. Chem., 54 (1991) 3085.
12. A. Suzuki, N. Suzuki, Y. Ikeuchi and M. Saito, Agric. Biol. Chem., 55 (1991) 2467.
13. J.J. Macfarlane, I.J. McKenzie and R.H. Turner, Meat Sci., 10 (1984) 307.

Trends in High Pressure Bioscience and Biotechnology
R. Hayashi (editor)

Effects of high pressure and salts on frozen egg custard gel

A. Teramoto and M. Fuchigami

Department of Nutritional Science, Faculty of Health & Welfare Science,
Okayama Prefectural University, 111 Kuboki, Soja, Okayama 719-1197, Japan

Our objective was to determine the effect of high pressure and/or the addition of salts on the improvement in texture of frozen egg custard gel. Stress and strain of gel frozen at 0.1, 100, 600 and 686 MPa increased, while those of gel frozen at 200 ~ 500 MPa changed only slightly. The pore size in gel frozen at 200 ~ 500 MPa was smaller than in gel frozen at 0.1, 100, 600 and 686 MPa. The 0% salt-gel, NaCl-gel and KCl-gel were supercooled at -20°C during pressurization at 200 ~ 400 MPa, 200 ~ 600 MPa and 200 ~ 500 MPa, respectively. When pressure was released, the supercooled gel froze quickly by pressure-shift-freezing and small ice crystals were dispersed throughout keeping the texture more suitable when it was thawed.

1. INTRODUCTION

Egg custard gel is formed by heat-induced aggregation of egg protein particles. Smooth taste is a very important quality of egg custard gel. However, egg custard gel frozen at atmospheric pressure (0.1 MPa) is unsuitable for consumption because it is spongy. When water is frozen at 0.1 MPa, its volume increases. Ice I at 0.1 MPa is the only ice less dense than liquid water. This seems to cause tissue damage during freezing at 0.1 MPa. Conversely, under high pressure, several kinds of ices (ice II ~ IX) with different structures and physical properties are formed. The densities of these high pressure ices are greater than water. Also, a non-freezing region (liquid phase) below 0°C exists under high pressure [1][2]. When water was pressurized at 200 MPa at ca. –20°C, it did not freeze. However, when pressure was released, water froze quickly by pressure-shift-freezing. If this happens to frozen gels, damage of gels may be reduced. In our previous studies, results suggested that the damage of texture and structure was reduced by freezing at 200 ~ 400 MPa [3][4][5]. Moreover, the addition of salts may depress the freezing point. Therefore, our objectives were to determine the effect of high pressure on the improvement in texture of frozen egg custard gel and to determine if the supercooling temperature decreased from the addition of salts under high pressure.

2. MATERIALS & METHODS

2.1. Sample preparation and method of high-pressure-freezing

The resulting egg mixture was 50% egg and 50% salt water. The mixture (20 g) was vacuum packed in heat-sealed polyethylene bags, then heated for 15 min at 85°C in a water bath. Three packs of egg custard gel with 0% (0% salt-gel), 1% sodium chloride (NaCl-gel) or 1.28% potassium chloride (KCl-gel) were put into a pressure vessel and immediately pressurized for 90 min at ca. -20°C at 100 MPa (ice I), 200 MPa (liquid phase), 340 MPa (ice III), 400, 500 or 600 MPa (ice V), 686 MPa (ice VI) using a high pressure food processor (Kobe Steel Co. Ltd.). After decompression, gel was stored for 1 day at -30°C, then thawed at 20°C. Changes in the temperature of the samples during freezing were measured. This gel was compared with untreated gel and gel frozen in freezers (-20°C, -30°C or -80°C) at 0.1 MPa.

2.2. Texture and structure measurement

The rupture stress and strain of samples were measured using a creepmeter (Rheoner RE-33005, Yamaden Co. Ltd.). Thickness of samples was measured using a sample high counter (HC-3305, Yamaden), then the samples were punctured by using a plunger (cylindrical shape: 5 mm diameter, 22 mm long) at 1 mm/sec, then stopping at 99% thickness using a loadcell of 200 g.

Structures of samples were observed with a cryo-scanning electron microscope (S-4500, Hitachi Ltd.). The magnifications used to observe ice crystals and gel networks were x100 and x20,000, respectively. The ice pore size was analyzed by a mac-scope (Mitani Co. Ltd.).

3. RESULTS & DISCUSSION

3.1. Temperature

Changes in temperature of gels during freezing are shown in Figure 1. The 0% salt-gel, NaCl-gel and KCl-gel were supercooled at -20°C during pressurization at 200 ~ 400 MPa, 200 ~ 600 MPa and 200 ~ 500 MPa, respectively. Thus, depression of freezing point was greatest to least, NaCl-gel > KCl-gel > 0% salt-gel. When pressure was released, the supercooled gel froze quickly by pressure-shift-freezing. On the other hand, at 0.1, 100 and 686 MPa, an exothermic peak was detected during pressurization, so the phase transition of water from liquid to ice in all gels occurred under high pressure.

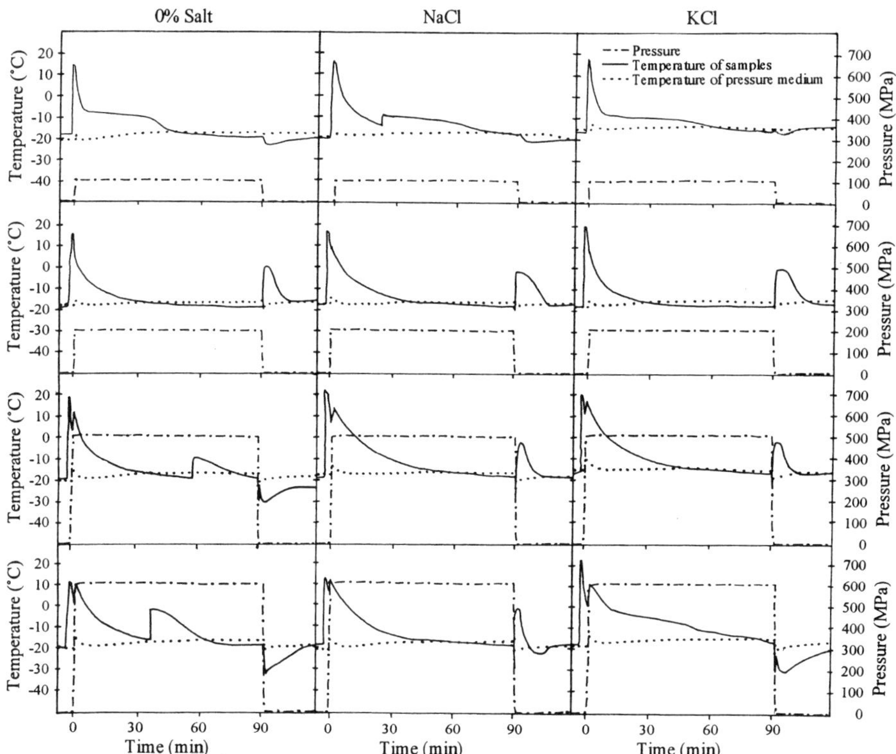

Figure 1. Changes in temperature of egg custard gel during freezing.

3.2. Texture

The averages of rupture stress and strain of pressurized gels are compared in Figure 2. Stress and strain of original gel increased by the addition of salts, but those of unfrozen gel pressurized at 20°C did not change. However, stress and strain of gel frozen at 0.1, 100, 600 and 686 MPa increased, while those of gel frozen at 200 ~ 500 MPa changed only slightly. Strain of 0% salt-gel frozen at 0.1 MPa in freezers was increased while stress of NaCl-gel and KCl-gel frozen at -30°C and -80°C changed only slightly.

3.3. Structure

The structure of frozen gels is shown in Figures 3 and 4. Moreover, ice-pore size of high-pressure-frozen gels is compared in Figure 5. Small ice crystals formed in the gels frozen at 200 ~ 500 MPa. However, ice crystals in gels frozen at 0.1, 100, 600 and 686 MPa were larger than those in gels frozen at 200 ~ 500 MPa. The size of ice crystals in gel frozen at 0.1 MPa in freezers at -20°C was the largest.

472

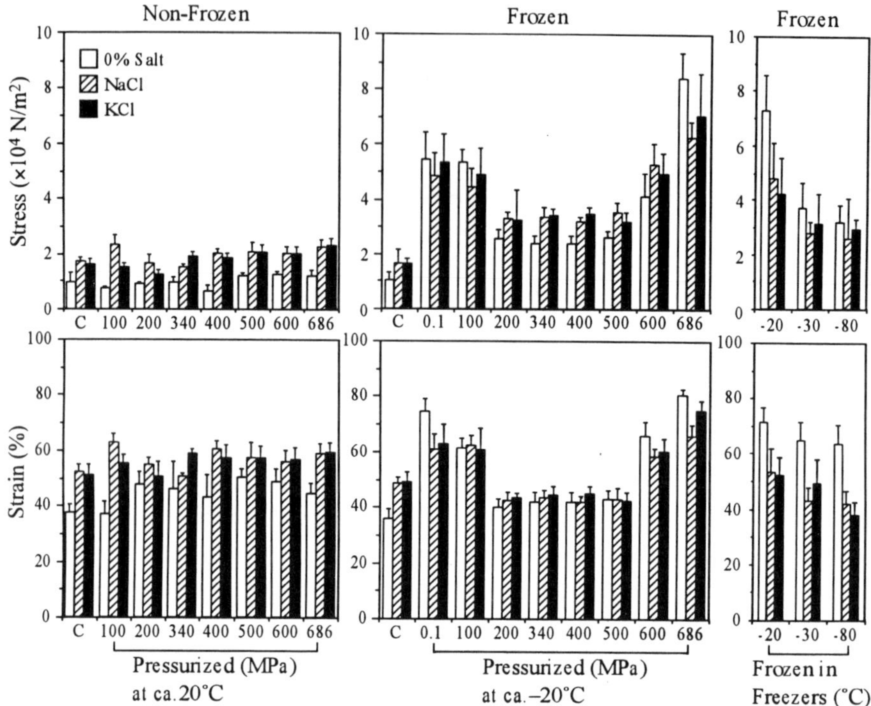

Figure 2. Effects of high pressure and salts on textural quality of egg custard gel.
C: control.

Finally, during pressurization at 200 ~ 400 MPa (0%-salt gel), 200 ~ 600 MPa (NaCl-gel) and 200 ~ 500 MPa (KCl-gel), gels were supercooled. However, when pressure was released, the supercooled gel froze quickly by pressure-shift-freezing and small ice crystals of a granular shape were dispersed throughout the gel keeping texture more suitable when thawed. Therefore, it was found that pressure-shift-freezing was effective in improving the quality of frozen egg custard gel.

REFERENCES

1. N. H. Fletcher, The Chemical Physics of Ice, Cambridge University Press, 1970.
2. P. V. Hobbs, Ice Physics: Oxford University Press, London, 1974.
3. M. Fuchigami and A. Teramoto, J. Food Sci., 62, (1997) 828.
4. M. Fuchigami, A. Teramoto, and N. Ogawa, J. Food Sci., 63, (1998) 1054.
5. M. Fuchigami and A. Teramoto, Rev. High Pressure Sci. Technol., Vol. 7, (1997) 826.

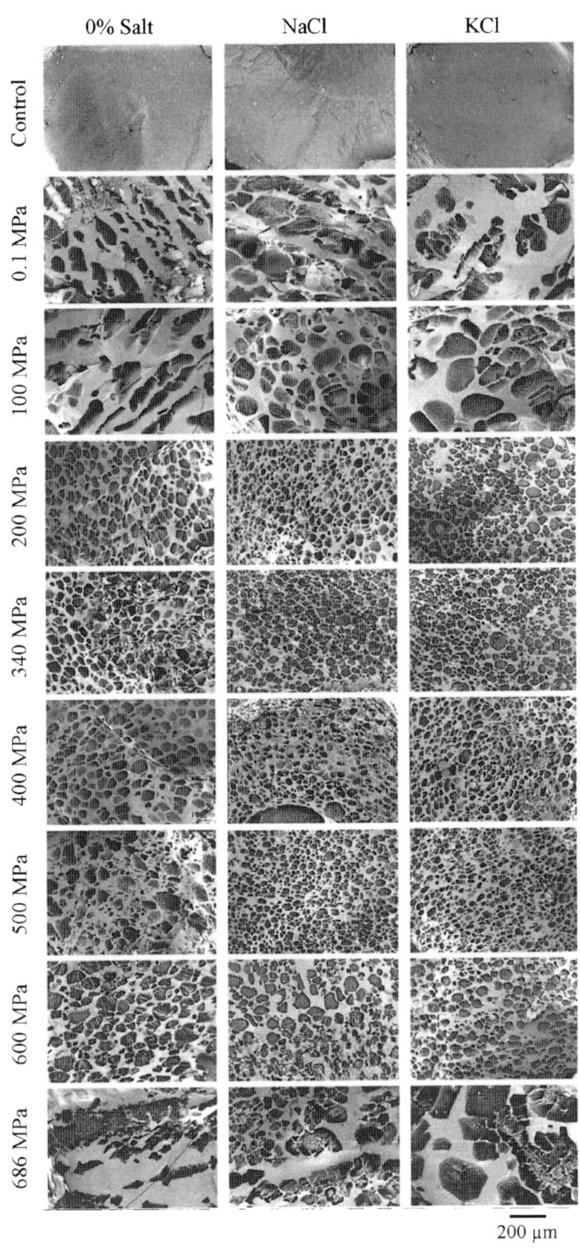

Figure 3. Cryo-scanning electron micrographs of egg custard gel frozen at high pressure.

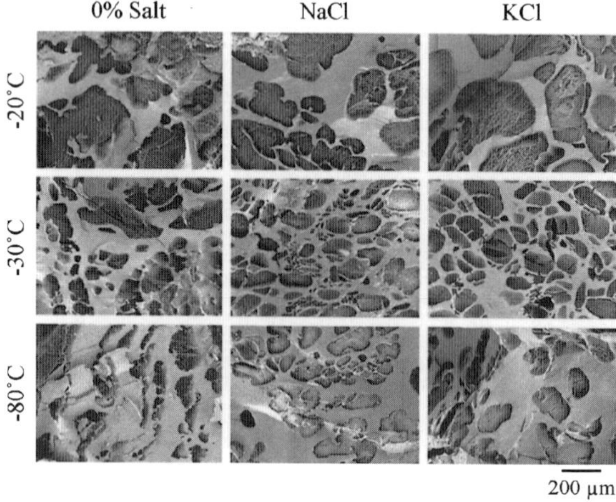

Figure 4. Cryo-scanning electron micrographs of egg custard gel frozen in freezers (–20°C, –30°C, –80°C).

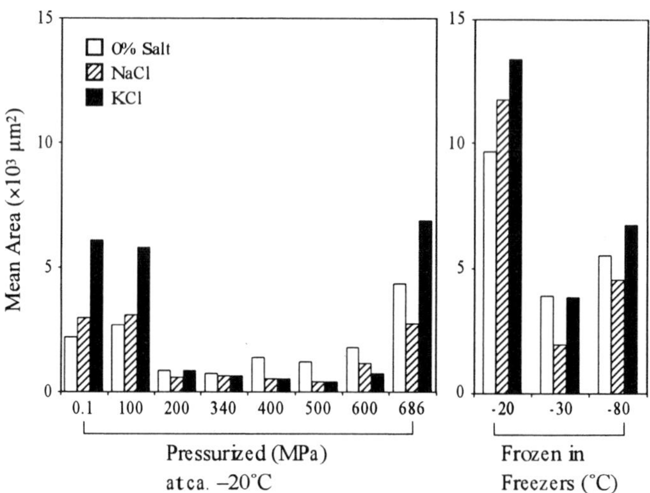

Figure 5. Pore size of frozen egg custard gel analyzed by mac-scope.

Trends in High Pressure Bioscience and Biotechnology
R. Hayashi (editor)
© 2002 Elsevier Science B.V. All rights reserved.

TEXTURAL PROPERTIES AND SENSORY EVALUATION OF SOFT SURIMI GEL TREATED BY HIGH PRESSURIZATION

K. Yoshioka and A. Yamada

Department of Food and Nutrition, Faculty of Home Economics, Nakamura Gakuen University, Befu, Jonan-ku, Fukuoka 814-0198, Japan

Heat and high pressure-treated fish meat surimi gels and soft surimi gels were prepared and their characteristics were examined on their textural properties, their fine structure by scanning electron microscopy and a sensory evaluation. A high pressurized gel containing 1.5% NaCl had an elastic and fine texture, compared to the heated one. In addition, the pressurized soft surimi gels with the ratio of the distilled water to fish meat surimi of 0.5 to 1 and 1 to 1 showed 1.6 and 1.4 times respectively, the value of the heated soft gels for the hardness of penetrating, and were formed into smooth, soft and elastic gels.

1. INTRODUCTION

The use of high pressure for food processing has been significantly developed and its utilization extends into the range of marine products. The high pressure treatment of fish gel produces a new type of excellent taste with luster, density and elasticity in the gels, which is different from those treated by heating. In a previous paper, we reported the effects of the amount of NaCl addition on fish meat surimi gel treated by heating and pressurization using fillets of horse mackerel and found possible gel formation of fish meat with a low percent of NaCl addition under high pressurization. In this study, we tried to prepare soft surimi gels with the addition of distilled water and investigated the effects of high pressure on each surimi gel. We made an attempt to apply the high pressurization and prepared these soft surimi gels as a food for persons who find it difficult to masticate.

2. MATERIALS AND METHOD

The fillets of horse mackerels stored at -30℃ for 30 days were thawed by running water at 15℃, then ground into surimi. The samples were prepared with no addition of NaCl (without NaCl) and 1.5% NaCl (with 1.5% NaCl). For the soft surimi gels, two

476

types of surimi gels were prepared in the ratio of the distilled water to the weight of surimi of 1 to 1 and 0.5 to 1. During the heat treatment, each sample was heated in boiling water for 7-8 mim to internal temperature 80℃. For the high pressure treatment, the samples were exposed to a 500MPa high pressure for 10 min in a food pressurized testing machine (Mitsubishi Heavy Industries, Ltd., NFP-7000).

The textural properties of the pressurized and heated surimi gels were measured using a creepmeter (Yamaden RE-3305). The hardness of these samples was determined by penetrating and compression tests. All the samples for the textural measurements were cut into 15 mm length and 6~8 test pieces were used for each experiment. For the creep test, the sample size of each group was $20 \times 20 \times 15$mm and the compression test was carried out using a 40mm ϕ plunger. All the samples examined by the creepmeter were analyzed using the creep compliance curves and the viscoelastic modulus was also calculated.

Both the heated and pressurized surimi gels were cut into small pieces and then fixed in half-strength Karnovsky fixative buffered with 0.1M phosphate at pH 7.2 for 1 h, followed by post-fixation with 1% osmium tetroxide in 0.1M phosphate buffer for 1 h. After dehydration in graded ethanols, they were freeze-dried and examined using a scanning electron microscope (Hitachi N-3000).

For the sensory evaluation test, the samples were evaluated by 20 panelists using the 5 score scaling method. The evaluation items were luster, well-heated, transparency, flavor, well-seasoned, salty, softness, elasticity, smoothness, density and juiciness.

3. RESULTS AND DISCUSSION

3.1. Textural properties

For the textural properties of the surimi gels and soft surimi gels, the hardness of penetrating showed significantly high values in the surimi gels both with and without NaCl by high pressure treatment, compared to those with heat treatment (Figure 1). On instantaneous deformation, the elastic modulus increased with the addition of NaCl but the viscosity showed little change between the high pressure and the heat treatments. (The data is not shown.) For the textural properties of the soft surimi gels of 0.5:1 and 1:1, the pressurized 0.5:1 surimi gel showed 4.6 and 1.6 times the hardness of penetrating in the surimi gels without NaCl and with 1.5% NaCl, respectively, compared to the heated one. For the 1:1 surimi gel, the pressurized one showed 1.7 and 1.4 times higher values in without NaCl and with 1.5% NaCl than the corresponding heated surimi gels. Furthermore, both types of pressurized soft gels showed 5 times the hardness of compression compared to the heated soft gels, and the pressurized gels were soft and elastic. Though the heat treated surimi without NaCl hardly had a gel consistency, the high pressure treated one formed a very good gel.

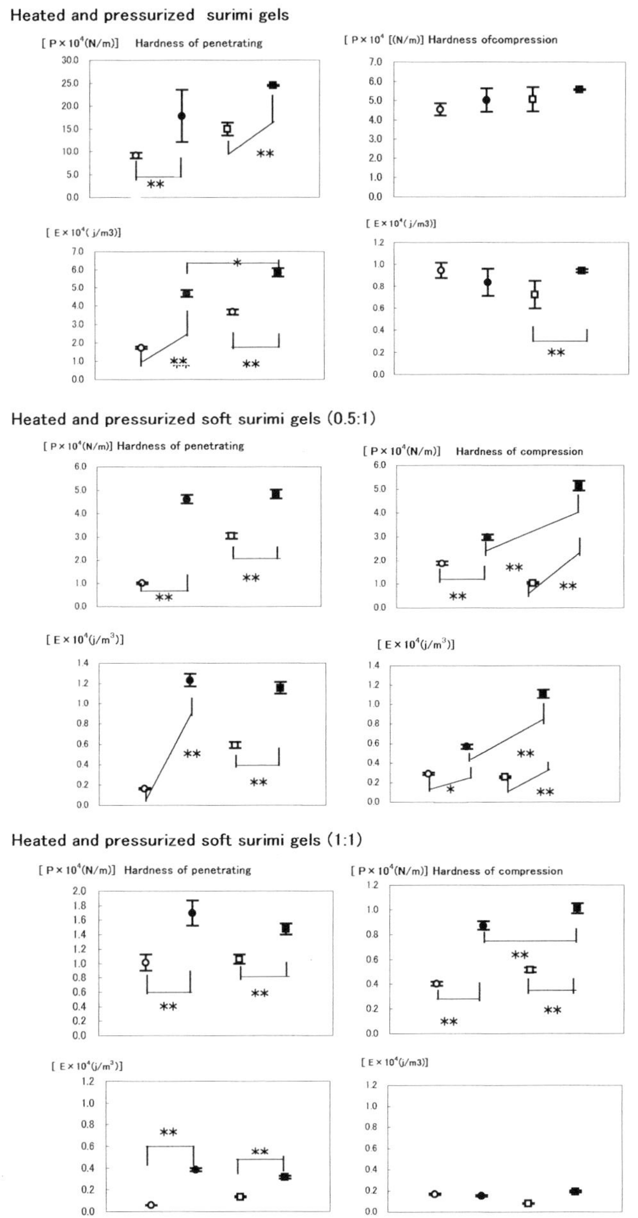

Figure 1. Texture properties of heated and pressurized surimi gels and soft surimi gels

Measuring conditions: penetrating test; plunger, 10mm ϕ, deformation rate, 85%; compression test; plunger, 40mm ϕ; test speed, 1mm/sec; P, breaking stress; E, breaking energy, ○; heated without NaCl, ●; heated with 1.5% NaCl, □; pressurized without NaCl, ■; pressurized with 1.5% NaCl, each values represents the mean $\pm$ SD of 6 determinations. * $p<0.05$, ** $p<0.01$.

478

3.2. SEM observation on surimi gel

Structural changes in the surimi gels and soft surimi gels treated by heating and high pressurization were observed by SEM and these images are shown in Figure 2.

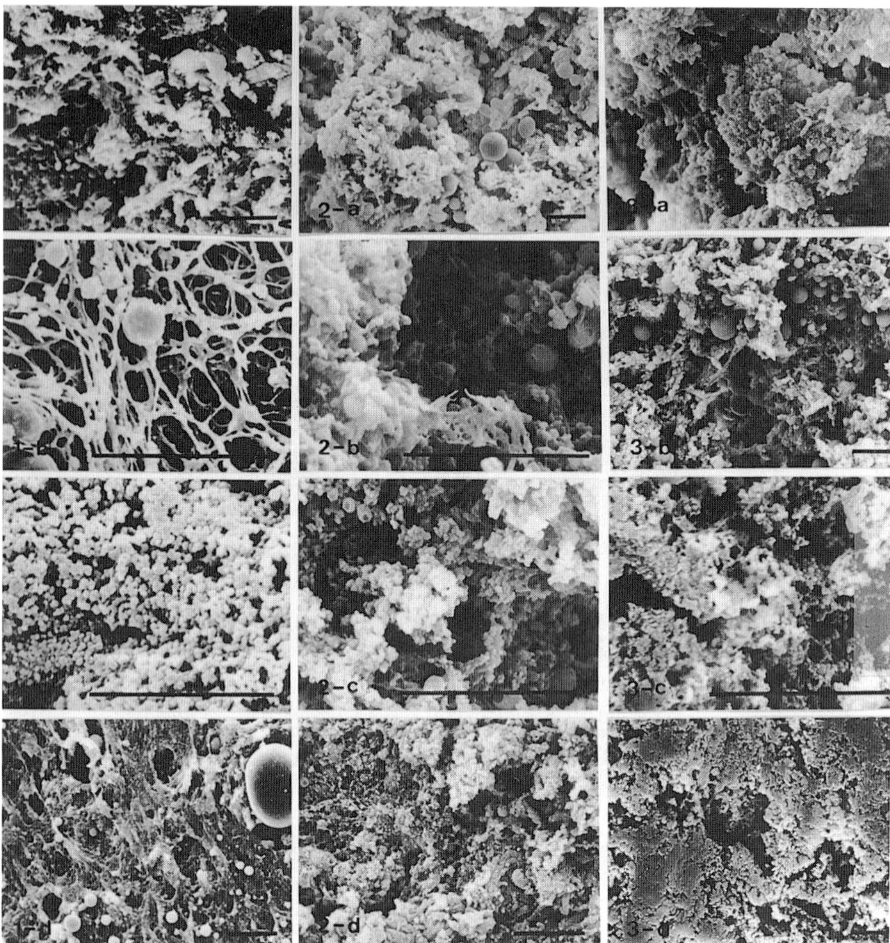

Figure 2. Scanning electron micrographs of heated and pressurized surimi gels and soft surimi gels

Sample gels: 1; fish meat surimi, 1-a; heated surimi gel without NaCl, 1-b; pressurized surimi gel without NaCl, 1-c; heated surimi gel with 1.5% NaCl, 1-d; pressurized surimi gel with 1.5% NaCl, 2; surimi:distilled water (1:0.5), 2-a; heated without NaCl, 2-b; pressurized without NaCl, 2-c; heated with 1.5%NaCl, 2-d; pressurized with 1.5%NaCl, 3; surimi:distilled water (1:1), 3-a; Heated without NaCl, 3-b; pressurized without NaCl, 3-c; heated with 1.5% NaCl, 3-d; pressurized with 1.5%NaCl. ——— expresses 10 μ m in each photograph.

The heat treated surimi gel showed an aggregation of granules and the absorbed air bubbles in the gel structure and the surimi gel with the addition of NaCl showed an elaborated structure (1-a, 1-c). The surimi was denatured into a gel by heating and furthermore, with the addition of NaCl, the network structure of the protein was quite minute. It is considered that the irreversible heat-induced gelation of myosin was formed by heating and a three-dimensional network structure was produced. In the surimi gel treated by high pressure, the fibrous latticed and granular structures of the protein were equally scattered and the porosities produced by the absorbed air bubbles and the fat globules are observed (1-b, 1-d). In the pressurized gel without NaCl, the protein was denatured by the pressurization, and the skeleton of the fibrous protein was observed, which is considered a sol transformed into a gel (1-b). On the other hand, with the addition of NaCl, the protein was uniformly dispersed and a porous and dense gel was produced in an elaborate and elastic form (1-d). For the heated 0.5:1 and 1:1 soft surimi gels, they showed a structure similar to the surimi gels, and with the addition of NaCl, the granules aggregated and a more dense gel was formed in each soft surimi gel (2-a, 2-c, 3-a and 3-c). In the case of the one treated by high pressure (2-b, 2-d, 3-b and 3-d), aggregation of the protein granules and fat globules were observed in the gels without NaCl, while in the NaCl- added gel, the protein granules were equally scattered and a very minute structure was formed.

3.3. Sensory Evaluation

The results of the sensory test in the heated and pressurized surimi gels with the addition of NaCl are shown in the radar chart of Figure 3. For each evaluation item, the pressurized surimi gel with 1.5% NaCl had a significant difference in luster, transparency, elasticity and density. Furthermore, in both the 0.5:1 and 1:1 soft surimi gels, the pressurized surimi gel with 1.5% NaCl had a significant difference in hardness, density and juiciness, compared to the heated one. These gels were favorably evaluated to be soft and smooth and suitable for persons who cannot masticate well.

4. CONCLUSION

Based on these results, by high pressure treatment, a fine and elastic gel was formed even with addition of 1.5% NaCl. It is suggested that it is possible to produce fish meat gel products with the addition of lower percent of NaCl compared with the fish surimi products currency on the market. Furthermore, in the soft surimi gel, high pressurization provides a soft, juicy and smooth taste to the gel. It is expected that these gels will be used as a food suitable for older persons who have difficulty with mastication.

480

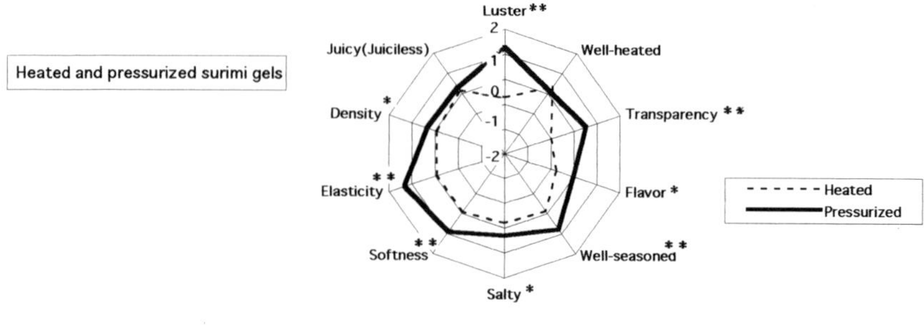

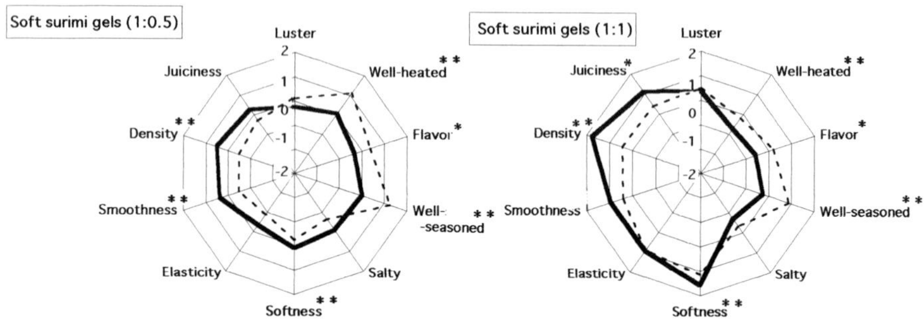

Figure 3. Rader charts of sensory evaluation in comparison of heated and pressurized surimi gels and soft surimi gels

Sensory test were examined by 20 panelists using the 5 score scaling method in regard to heated and pressurized surimi gels, soft surimi gels (1:0.5, and 1:1). * $p<0.05$, ** $p<0.01$.

REFERENCES

1. T. Shoji, H. Saeki, M. Nakamura, Y. Sasamoto, and M. Nonaka: Nippon Suisan Gakkaishi, **55**, 1841-1844 (1989).

2. K. W. Ching, M. Tanaka, Y. Nagashima, T. Taguchi, and K. Amano: J. Food Sci., **56**, 338-340 (1991).

3. K. Yoshioka and T. Yamamoto: Fisheries Science, **64**, 89-94 (1998).

4. T. Shoji, H. Saeki, A. Wakameda, M. Nakamura, and M. Nonaka: Nippon Suisan Gakkaishi, **56**, 2069-2076 (1990).

5. M. Okamoto, Y. Kawamura, and R. Hayashi: Agric. Biol. Chem., **54**, 183-189 (1990). (ed.

6. M. Ishikawa, K. Sasaki, Y. Yoshida, and K. Ohki: "High pressure Science for Food" by R. Hayashi), San-Ei Shuppan Co., Kyoto, 1991, pp. 309-316.

7. A. Yamada and K. Yoshioka: Bulletin of Nakamura Gakuen University and Nakamura Gakuen Junior College, **31**, 223-228 (1999).

8. S. Sato, T. Tsuchiya, and J. Matsumoto: Nippon Suisan Gakkaishi, **50**, 1869-1876 (1984).

9. E. Okazaki: "High pressure Science " (ed. by R. Hayashi, et al.), San-Ei Shuppan Co., Kyoto, 1994, pp. 296-303.

Trends in High Pressure Bioscience and Biotechnology
R. Hayashi (editor)

Influences of saccharides on the pressure-induced gels from a whey protein isolate

Choemon Kanno, Jin-Song He, Michiko Ametani, and Norihiro Azuma

Department of Applied Biochemistry, Utsunomiya University, Utsunomiya 321-8505, Japan

The influence of saccharides on the pressure-induced gel (800 MPa, 10 min, 30°C) from a 20% (w/v) whey protein isolate (WPI) solution was investigated for various concentrations of lactose (0-20%) or 10% glucose at various pH values. The addition of lactose increased the protein solubility and water-holding capacity, decreased the rheological properties, and changed the microstructure of the WPI gel. Glucose had a more significant effect on the rheological properties and water-holding capacity of the pressure-induced gel from the WPI at acidic and neutral than at alkaline pH values, whereas the protein solubility was higher at neutral and alkaline than at acidic pH values. The addition of saccharides weakened the protein-protein interaction under hydrostatic pressure, its effect being related to the pH value of the initial WPI solution.

1. INTRODUCTION

Whey from bovine milk is a by-product from cheese manufacturing and is widely used as functional ingredient. The aggregation and gelation of whey proteins are induced by heating and hydrostatic pressurization. The participation of intermolecular disulfide bond formation through an SH/SS interchange reaction and of hydrophobic interaction has been reported to be important for the gelation of whey proteins under hydrostatic pressure (1, 2).

The physical properties and microstructure of protein gels are generally susceptible to such environmental factors as the pH value, ionic strength, sugar, and fat. The heat treatment of a β-lactoglobulin (β-Lg) solution formed an opaque, white gel in the neutral pH range, and a transparent gel in the acidic and alkaline pH ranges (3-6). The microstructure of these gels displayed a particulate network in the pH 4-6 range, and a fine-stranded network below or above this pH range. An increase in pH value in the alkaline range of the isoelectric point (pI) resulted in an increase in the strength of the pressure-induced gel of whey protein (7). This effect of pH value on the gelation of whey protein has been ascribed to the increase in electrostatic repulsion force

482

between proteins with increasing pH and to the reactivity of SH groups in the alkaline range, where the formation of intermolecular S-S bonds is promoted through SH/SS exchange (6, 7).

The influence of saccharides has recently been studied on the aggregation and gelation of whey protein (1, 8-10). Sucrose in a 2.5% β-Lg solution reduced protein unfolding and the subsequent aggregation under hydrostatic pressure (8). The addition of sucrose decreased the pore size and strand thickness of the network and lessened the textural behavior of pressure-induced gels from 10-14% β-Lg solutions, this might having resulted from the protective effect of sucrose against protein unfolding and subsequent aggregation under high pressure (1). Although some work has demonstrated the effect of saccharides on the pressure-induced gelation of protein, it was not clear whether the effect was related to the kind of saccharide or to the pH value and ionic strength of the initial WPI solution.

The aim of the present study is to investigate the effect of saccharides on the protein solubility, water-holding capacity, rheological properties, and microstructure of pressure-induced gel from WPI at various pH values. We here deal with pressure-induced gels from a WPI solution in the presence of lactose at pH 6.8 or in the presence of glucose at various pH values.

2. METHODS

2.1. Preparation of pressure-induced gels

WPI consisted of about 6.1% moisture, 89.8% protein, 1.8% ash, 1.3% lactose, and 0.5% lipids. A WPI solution (20%, w/v) was prepared in a 50 mM sodium phosphate buffer (pH 6.8), and then lactose was added to make a final concentration of 0-20% (w/v). To make solutions at different pH values, WPI was dissolved in a 50 mM acetate buffer at pH 5.0, a sodium phosphate buffer at pH 6.8 and a Tris-HCl buffer at pH 8.0, and to each solution was added glucose to 10% (w/v).

Each WPI solution containing a saccharide was put into a Teflon tube (4 ml) and then pressurized for 10 min at 800 MPa and 30°C with a hand-operated oil pressure generator. The indicated pressure was achieved within 1.5 min, held for 10 min, and then released to atmospheric pressure within 0.5 min.

2.2. Protein solubility

The resulting WPI gel (500 mg) was dispersed in 10 ml of a 0.086 M Tris-glycine buffer pH 8.0 (buffer A), or in 10 ml of buffer A containing 8 M urea and 0.5% SDS (buffer B). The gel dispersion was homogenized at 19,000 rpm for 2 min, and then centrifuged at 13,000 rpm for 15 min. The supernatant was analyzed for its protein solubility, calculated as the percentage of the protein content of the supernatant to the total protein content.

2.3. Water holding capacity (WHC) of gels

In the case of the pH 5.0 condition, the non-incorporated liquid (NIL) and the remaining gel were carefully separated and weighed. At pH 6.8 and 8.0, so little NIL had been formed as to be ignored. A gel (0.1 g) sample was placed on a polyethylene mesh membrane fixed in the middle of a 1.5-ml centrifuge tube, and the centrifuge was run for 5 min at 120 x g. WHC was calculated from the following formula: WHC (%) = (Wa - W_{NIL}) / Wb x 100, where Wb is the weight of the gel sample before centrifugation, Wa is the weight of the gel sample after centrifugation, and W_{NIL} is the weight of NIL. W_{NIL} was effectively zero at pH 6.8 and 8.0.

2.4. Rheological properties

The hardness and breaking stress of each gel sample were individually measured at room temperature with a Fudoh rheometer. Data were analyzed with the Rheosoft program attached to the rheometer.

2.5. Microstructure

The microstructure of each gel sample was examined without any chemical fixation and photographed with a scanning electron microscope according to the method of Kanno *et al.* (2).

3. RESULTS AND DISCUSSION

3.1. Influence of lactose on the pressure-induced gels from WPI

The gels from 20% (w/v) WPI solutions containing 0-20% (w/v) lactose at pH 6.8 were induced by pressurizing for 10 min at 800 MPa and 30°C. After releasing the pressure, an opaque, white gel was obtained for each sample, and no significant NIL was apparent.

Protein solubility

The protein solubility in the buffer A of WPI gel increased with increasing lactose content, while no significant change was detected in the buffer B (Figure 1). This increase indicated that lactose decreased protein-protein interactions. High and constant protein solubility in the buffer B suggested that lactose increased hydrophobic interaction in the gelation and weakened the formation of intermolecular S-S bonding at neutral pH range

Water-holding capacity

A concentration of lactose in the range from 0% to 6% resulted in WPI gels with high WHC (Figure 1). Lactose of more than 8% decreased the WHC, and the resulting gel easily lost water when compressed by hand. The decrease in water retention with increasing lactose concentration seems to have influenced the gel microstructure.

484

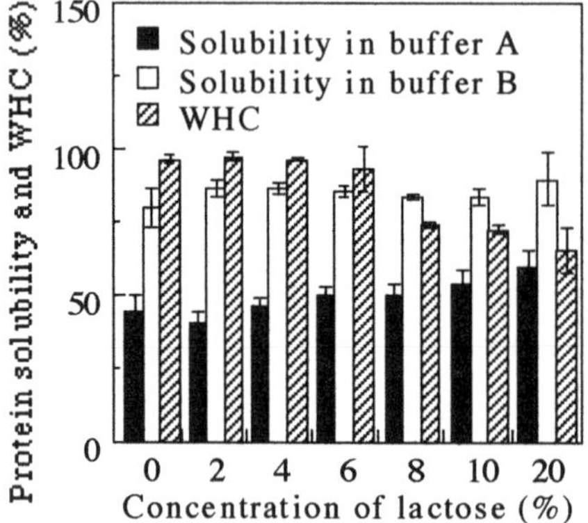

Figure 1. Effect of lactose on the protein solubility and WHC of WPI gel.

Rheological properties

The hardness and breaking stress of the pressure-induced WPI gels were decreased by increasing lactose content. This decrease was most marked in the WPI gel containing 20% lactose and may be attributable to weakened interaction between proteins by the addition of lactose.

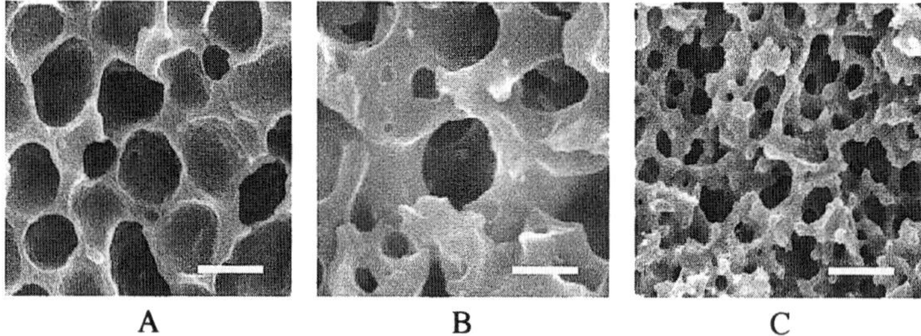

Figure 2. Microstructure of WPI gel samples without (A) or with 10% (B) and 20% (C) lactose. The bar shows 2 μm.

Microstructure

The gel samples from WPI containing less than 6% lactose displayed a honeycomb-like network structure which was similar to that of the WPI gel without lactose (Figure 2 A), while a stranded network structure was observed in the gel with a lactose content of 10% (Figure 2 B). Moreover, the pore size and strand thickness of the stranded network were both strikingly decreased by the addition of 20% lactose (Figure 2 C). The change in structure from a honeycomb-like to a stranded network was due to weakened protein-protein interaction. It was also confirmed that the extent of protein-protein interaction was directly related to the pore size and strand thickness of this stranded network (1).

In a stranded network, the ability to immobilize the liquid component of gel is weaker than that in a honeycomb-like network, in where the liquid is confined to small hermetic cells. However, the decrease in water retention did not lead to syneresis, except for the gel containing 20% lactose, because of the high capillary forces existing in the dense and fine stranded network.

3.2. Influence of glucose at various pH values on pressure-induce WPI gel

Gel was induced from 20% WPI with or without 10% glucose at pH 5.0, 6.8 or 8.0 by pressurizing for 10 min at 800 MPa. The obtained gel samples were white and opaque at pH 5.0 and 6.8, but transparent at pH 8.0. The gel formed at pH 5.0 with or without glucose exhibited a large degree of syneresis with a central gelled cylinder surrounded by NIL. No or negligible exudation was apparent at pH 6.8 and 8.0.

Protein solubility

Increasing pH value significantly decreased the protein solubility in buffers A and B of WPI gel samples with or without glucose. In the absence of glucose, the difference between the protein solubility in buffers A and B was marked at pH 6.8, but not significant at pH 5.0 and 8.0. A lower protein solubility in the buffer A at pH 8.0 and a small difference between that in buffer A and B indicates that a high level of S-S bonding contributed to the formation of gel, while the gel at pH 5.0 seemed to be mainly formed by hydrophobic interaction, since there was no significant difference in the protein solubility between buffers A and B. An increase in the pH value promoted the intermolecular formation of proteins through S-S bonding, this being confirmed by SDS-PAGE.

In the presence of glucose, the difference in protein solubility between the gels with and without glucose was marked at pH 6.8 and 8.0. The protein solubility in buffer A of the gel samples with added glucose at various pH values was similar to that of WPI alone in buffer B. Glucose was thus revealed to significantly weaken the intermolecular formation of proteins under hydrostatic pressure.

Water holding capacity

With or without glucose, WHC of the gel increased with increasing pH. At pH 8.0, the gel samples had the highest level of WHC (about 100%), the difference between the gel with or without glucose being negligible. At pH 6.8, glucose decreased the WHC level, while it increased the level at pH 5.0. The gel samples at pH 8.0 with or without glucose and at pH 6.8 without glucose had complete water retention and did not lose water when compressed by hand. Significant syneresis was observed at pH 5.0 both with and without glucose. Glucose had more effect on WHC of WPI gel in the neutral and acidic pH ranges than in the alkaline range.

Rheological properties

The addition of glucose increased the hardness and breaking stress of the gel with increasing pH. At pH 6.8, glucose markedly decreased the hardness and breaking stress, while the effect was not significant at pH 5.0 and 8.0. The relative difference (+/- of glucose x 100) in the hardness and breaking stress was less than 4% at pH 8.0, but 10.8% and 19.9%, respectively, at pH 5.0 and 8.0. The influence of glucose on the rheological properties of WPI gel was thus dependent on the pH value.

An increase in pH value above pI enhances not only the reactivity of SH groups to promote the formation of intermolecular S-S bonds, but also the electrostatic repulsive force, thus preventing interaction between proteins. The electrostatic repulsive force disadvantages hydrophobic interaction but not S-S bonding, since the energy of electrostatic interaction approximates to that of other noncovalent bonds, but is lower than that of S-S bonds. In the alkaline pH range, WPI gel would therefore be mainly formed by intermolecular S-S bonds, while, in the acidic pH range, gel would be mostly formed by noncovalent bonds.

Saccharides have been found to exert a protective action against protein unfolding, subsequent aggregation, and gelation of protein (1, 8). In the neutral pH range, saccharides stabilize the native protein structure due to their preferential exclusion from the protein domain, i.e., the balance of the affinity between protein and water is in preference to that between protein and saccharides (11, 12). In the alkaline and acidic pH ranges, however, the net charge of protein molecules will be changed and affect this balance by electrostatic interaction. In addition, at pH 8.0, the enhanced reactivity of SH groups could be weakened by the protective effect of saccharides.

REFERENCES

1. E. M. Dumay, M. T. Kalichevsky and J. C. Cheftel, Lebensm.-Wiss. u.-Technol., 31 (1998) 10.
2. C. Kanno, T-H. Mu, T. Hagiwara, M. Ametani and N. Azuma, J. Agric. Food Chem., 46 (1998) 417.
3. V. R. Harwalker and M. Kalab, Milchwissenschaft, 40 (1985) 65.
4. D. M. Mulvihill and J.E. Kinsella, J. Food Sci., 53 (1988) 231.
5. D. M. Mulvihill, D. Rector and J.E. Kinsella, Food Hydrocolloids, 4 (1990) 121.
6. M. Langton and A-M. Hermansson, Food Hydrocolloids, 5 (1992) 523.
7. J. V. Camp and A. Huyghebaert, Lebensm.-Wiss. u.-Technol., 28 (1995) 111.
8. E. M. Dumay, M. T. Kalichevsky and J. C. Cheftel, J. Agric. Food Chem., 42 (1994) 1861.
9. L. M. Rich and E. A. Foegeding, J. Agric. Food Chem., 48 (2000) 5046.
10. A. Kulmyrzaev, C. Bryant and D. J. McClements, J. Agric. Food Chem., 48 (2000) 1593.
11. J. C. Lee and S. N. Timasheff, J. Biol. Chem., 256 (1981) 7193.
12. S. N. Timasheff, Annu. Rev. Biophys. Biomol. Struct., 22 (1993) 67.

Trends in High Pressure Bioscience and Biotechnology
R. Hayashi (editor)

Physiological Aspects of Pressure Decontamination in Building Inactivation Models.

J.P.P.M. Smelt[a,b], J.C. Hellemons[b] and S. Brul[b,c]

[a]MCB EMSA University of Utrecht Padualaan 8 Utrecht, The Netherlands*,

[b]Unilever Research, Olivier van Noortlaan 120 3120 AC Vlaardingen.

[c]Swammerdam Institute for Life Sciences University of Amsterdam 1098 SM Amsterdam, The Netherlands

Predictive models describing growth or inactivation of microbial cells are mostly purely descriptive and mechanistic elements are hardly included. In most cases these so called 'black box' models are used in food processing. However, it should be borne in mind that even in descriptive models, knowledge of the physiology of the microorganism(s) to be eliminated is also necessary to ensure fail-safe predictions. In this overview various aspects related to microbiologically safe pressure processing will be addressed: Preparation of the inoculum, environment during pressure treatment, recovery conditions, effect of compression and decompression and model selection. Preparation of the inoculum seems one of the most important elements. Sufficient circumstantial evidence has been collected that compression and decompression has no effect other than the result of the integrated time-temperature-pressure history. Consequently, kinetic models describing inactivation of micro-organisms can be used in dynamic situations such as generation of adiabatic heat and heat dissipation and coming up times of pressure..

1. INTRODUCTION

Increasing consumer demand for minimally processed additive free, shelf stable products prompted the exploration of physical treatments other than traditional heat treatments as potential alternatives. Among these, pressure is one of the most promising techniques. Ideally pressure should result in microbiologically safe and stable foods over a wide range of products formulations. To predict safety and stability, reliable kinetic models are necessary that describe the fate of the micro-organism to be eliminated. As temperature and pressure are not constant during processing, the models should be applicable in dynamic situations. Although most inactivation models hardly include physiological characteristics, knowledge of the physiology of the micro-organism to be eliminated is necessary to build a fail-safe model. A model based on sound physiological principles will facilitate the identification of future optimised preservation strategies and will provide a basis for estimates of microbial behaviour outside the measured data pointsIn this overview various microbiological aspects related to model building will be addressed: Preparation of the inoculum, environment during pressure treatment, recovery conditions, effect of compression and decompression. In addition a few remarks will be made on the choice of the mathematical model itself.

2. ASPECTS TO BE TAKEN INTO ACCOUNT

2.1. Preparation of inoculum

2.1.1. Mixed cultures vs pure cultures

The choice of test strain(s) is of paramount importance and no single answer can be given to the question whether cocktails of strains or single strains should be used. Cocktails have the advantage that they always contain the strain that is most resistant in any particular situation. Consequently fail-safe situations are guarenteed by models based on cocktails of strains.The problem with cocktails is that survivor curves are difficult to interpret, especially when the inactivation cannot simply be described as a first order reaction. The situation becomes even more complicated by the fact that in practice most situations are dynamic due to adiabatic heat, compression and decompression time. As temperature effects always play a role in pressure treatments the test strain should be both resistant to heat and to pressure. A satisfactory solution to the problem is a careful selection of one test strain, followed by validation with other strains at a later stage.

2.1.2. Selection of strains

Pressure resistance can be screened by enumeration of colony forming units (cfu's), but that can be time consuming. Two examples of screening experiments conducted in this way are shown in Figs. 1 and 2.

a   b

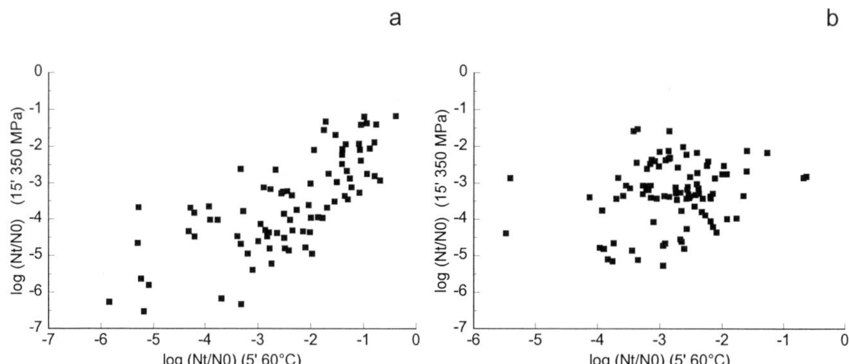

Figs 1a and 1 b. Heat resistance and pressure resistance of 83 strains of *Listeria monocytogenes* (a) and 93 strains of *Salmonella* ssp (b). (Smelt and Hellemons, 1999)

Whereas a weak positive correlation exists (0.44) between heat resistance and pressure resistance of *Listeria monocytogenes* no correlation (0.02) was found for *Salmonella* ssp.

This is consistent with earlier findings that *Salmonella senftenberg* was extremely resistant to heat but not to pressure (Metrick et al, 1989).

Selection strategy.
Although screening by enumeration of the survivors e.g by plate counts is a quite acceptable selection method, it is labour intensive and it requires some days before the results can be read. There are in principle several options to improve the efficiency of the selection procedure by additional screening methods. Besides, these methods have in common that they assess in more detail the physiological state of the cell. Three methods will be discussed here:

1. Measurement of ATP
2. Staining with life/dead stains
3. Measurement of plasmolysis

The concentration of intracellular ATP seems to be an indication of the physiological state of the cell. When the cell is dead (i.e. not able to form colonies) the content of internal ATP is low either due to consumption of the remaining ATP or due to leakage..

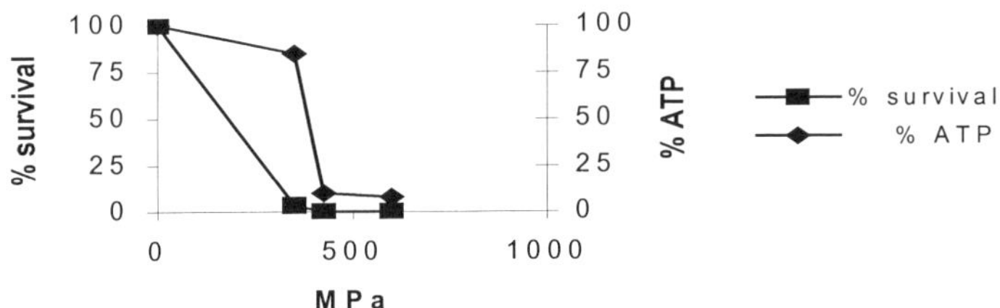

Fig. 3. Relation between survival of *Escherichia coli* and ATP content. (Tholozan et al, 2000)

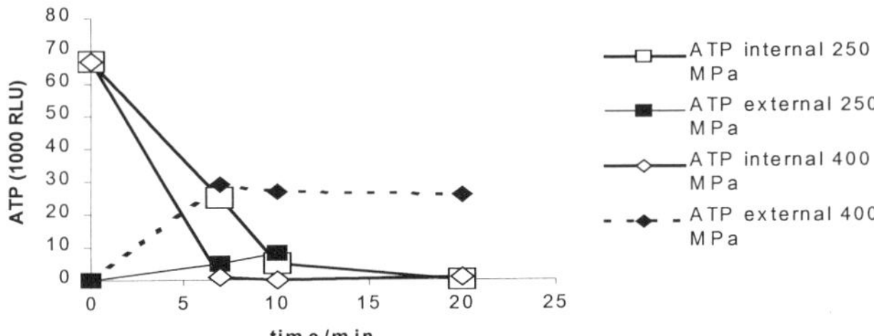

Fig. 4. Effect of pressure on intracellular and extracellular ATP in *Lactobacillus plantarum*. When no more than 10%internal ATP was left or when significant amounts of external ATP were observed, the number of colony forming units was reduced by more than a factor 10^5 (Smelt et al., 1994)

As a viable cell tries to save ATP, leakage of ATP is an indication of severe damage of the cell. Two examples are give in Figs 3 (Tholozan et al. 2000) and 4 (Smelt et al., 1994). Fig. 3 shows that the decrease of ATP coincides well with the decrease in cfu. Fig 4 shows that both internal ATP decreases and external ATP increases. ATP measurements are very fast and do not require high skills

Table 1. Staining of healthy cells and pressure damaged cells
of *Lactobacillus plantarum* with propidium iodide.
(flowcytometry)

Treatment ...MPa 15 min	Viability	% Cells not stained	% Cells stained	Total
0.1 MPa	Viable	93	0	93
	Dead	0	7	7
200 MPa	Viable	20	60	80
	Dead	2	18	20

Table 2. Staining of *Saccharomyces cerevisiae* VW k43 with Propidium iodide.
(flowcytometry, Brul et al., 2000)

Treatment ..MPa 15 min	Viability	% cell no stain	% low cell stain	% high stain	Total
0.1 MPa	Viable	93.2	0.0	0.0	93.2
	Dead	0.0	6.7	0.2	6.9
100 MPa	Viable	78.7	1.0	0.2	79.8
	Dead	19.9	0.2	0.0	20.1
200 MPa	Viable	0.0	0.0	9.1	9.1
	Dead	84.8	1.2	4.9	90.9

Propidium iodide and ethidium iodide can stain DNA. As long as the cell membrane is intact the stain cannot penetrate the cell. However as illustrated by Tables 1 and 2, the results with propidium iodide (PI) staining should be interpreted with care. At the one hand viable, sublethally damaged cells of *L. plantarum* are stained with PI (Smelt and Dutreux, unpublished results) and on the other hand dead cells of *Saccharomyces cerevisiae* (Brul et al., 2000) are not always stained by PI. In the former case it is likely that the cell membrane of *L. plantarum* is affected, allowing penetration of PI, whereas in the latter case DNA has probably leaked out the cells.

Pagan and Mackey (2000) measured the ability of the cell of *E. c*oli to plasmolyse as an indication of viability by measuring the change in optical density in high osmotic environment.It might be a universal indication of viability (Fig. 5). The procedure is easy to perform and fast. This method deserves certainly more investigation as a prescreening method.

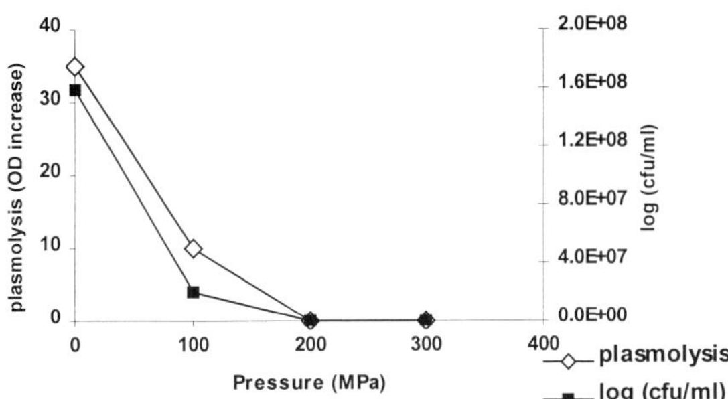

Fig. 5. Effect of pressure on inactivation and osmotic response of 3 strains of E. coli (Pagan and Mackey, 2000)

2.1.2. Culture conditions.

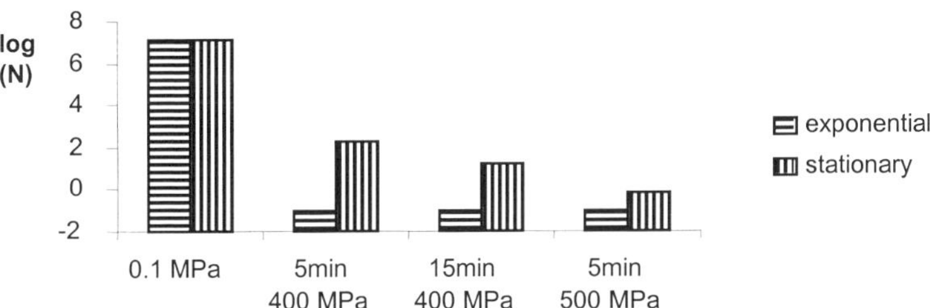

Fig. 6. Effect of growth phase on subsequent pressure resistance of *Lactobacillus plantarum* (Smelt et al., 1998)

As shown in Fig 6, exponentially growing cells are more susceptible to pressure than stationary phase cells. These observations are consistent with the observation that stationary phase cells are generally also more stress resistant e.g. (Pagan and Mackey, 2000;

492

McMahon et al., 2000). Contrary to our expectation prolonged incubation did result in increased resistance to pressure (Fig 7.). So far there is no conclusive explanation for this increased resistance. It has been frequently observed that the response of the living cell is similar (but not identical) to different stress conditions, such as starvation, sublethal heat and low pH, resulting in increased resistance to otherwise lethal environmental conditions such as severely elevated temperature.

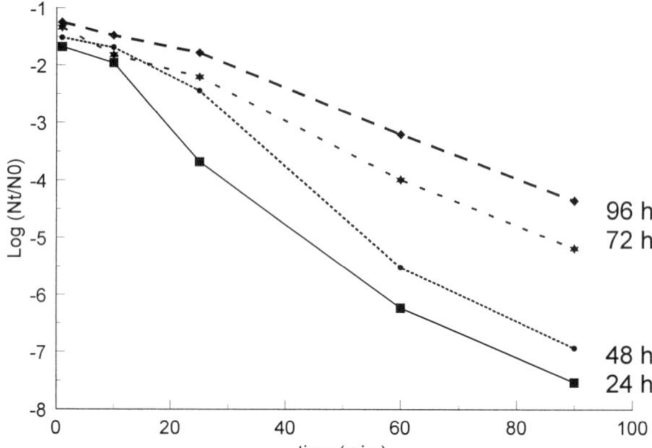

Fig. 7. Effect of duration of the stationary phase during cultivation on subsequent pressure resistance (400 MPa) of stationary cells of *Listeria monocytogenes* (Smelt et al., 1999)

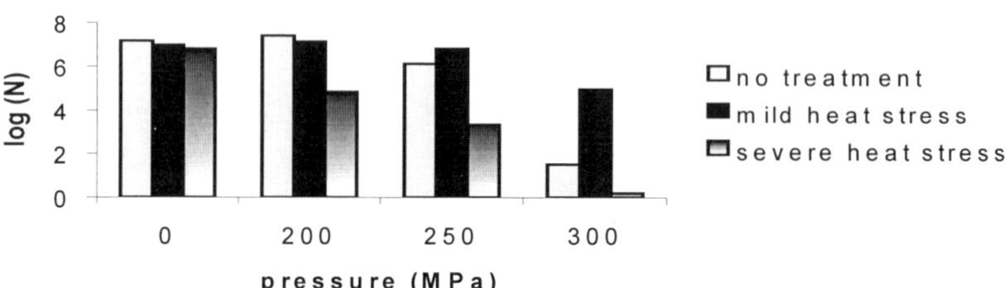

Fig. 8. Effect of previous heat stress (30 min 48 °C, 'mild' or 5 min 51 °C, 'severe') on subsequent pressure resistance (15 min at indicated pressure) of exponentially growing cells of *Lactobacillus plantarum* (Smelt et al., 1998)

As shown in Fig 8 heat stress conditions cause an enhanced resistance to pressure. It might be possible that starvation caused by prolonged incubation has induced increased resistance to pressure. The effect of incubation temperature is equivocal. We found that exponentially growing cells were more resistant to pressure when incubated at low temperature, stationary phase cells seemed to be more sensitive when incubated at low temperature (Fig 9). It might be possible that induction of starvation needs more incubation time at low temperature than at high temperature.

2.2. Environment during pressure treatment

It is well known that many factors play a role in pressure and heat resistance The main factors determining resistance are pH, water activity and antimicrobial agents such as lacticin, nisin, pediocin and chitosan. (Papineau et al., 1991; Morgan, 2000; Hauben, 1996; Alpas 2000; Garcia-Graells et al., 1999; ter Steeg, 1999). There are also several reports on the protective action of milk against pressure (Patterson, 1995; Smelt 1998).

2.3. Recovery conditions

As stated in the introduction microbial inactivation is often described by first order kinetics. It is difficult to reconcile a first order model with the fact that sublethal damage either by heat or by pressure is often observed. Sublethal damaged cells need more time to recover or are more susceptible to adverse environmental conditions. An example of sublethal damage is shown in Fig 10, where viable, pressure treated cells are more susceptible during recovery at low pH. This observation is not unexpected as the ability to maintain the internal pH is decreased after pressure treatment (Wouters et al., 1998)

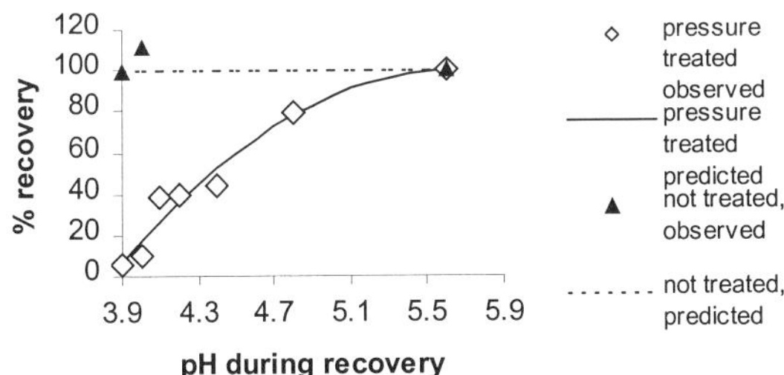

Fig. 10. Effect of recovery conditions on pressure treated cells compared to untreated cells.

2.4. Effect of compression and decompression

A specific lethal effect of compression and decompression on yeast cells has been reported (Palou et al 1998). The effect of decompression on the cell seems to be overestimated. It is known that piezophilic (barophilic) organisms are not able to multiply at atmospheric pressure but they can resume growth when they are again put under high pressure. (Chastain and Yayanos, 1991)) Apparently, decompression has not affected the cell. It seems that in many cases the effect of coming up time and the combined effect of adiabatic heat and heat dissipation during pressure treatment is responsible for the discrepancy between results of intermittent and continuous pressure treatment. To investigate whether *Salmonella* is particularly sensitive to compression the cells were subjected to different coming up times to the point where still 100 % or 10 % survivors were present.

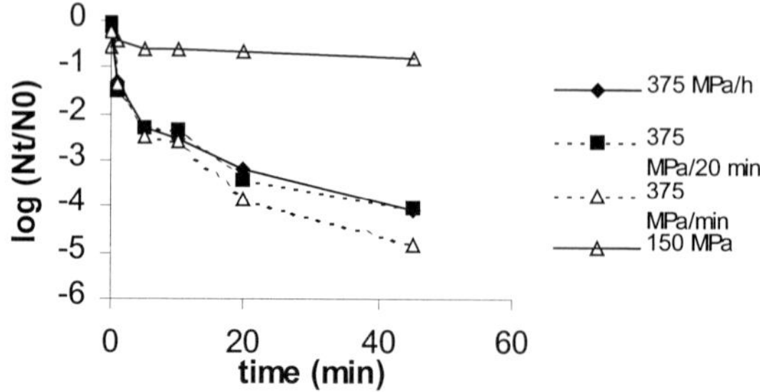

Fig. 11. Effect of previous pressure treatment as indicated on subsequent pressure treatment at (325 MPa) times varying from 0 to 50 min

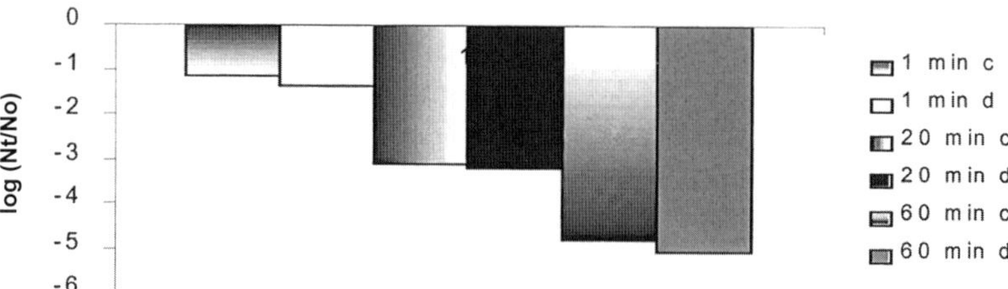

Fig. 12. Effect of ramp rate (time to reach 375 MPa) during compression (c) and on decompression (d) on inactivation of Salmonella enteriditis (Smelt and Hellemons, 1999)

After this treatment the cells were subjected to various pressure treatments, As shown in Fig. 11 hardly any difference was found between cells subjected to different coming up times. It seems that the effect of coming up time can be fully explained by the integrated effect of pressure, temperature and time. Fig 12 shows that the effect of decompression is almost identical to the effect of compression. From the latter effect we can conclude that decompression has no specific effect on *Salmonella* either. This means that the effect of compression and decompression does not play a role other than the integrated time/temperature/pressure history during compression and decompression.

2.5. Model selection

Although in this overview the emphasis lies on factors outside the mathematical description itself some remarks will be made. In processing, especially thermal processing first order kinetics is often assumed and it has proven its value for many situations in practice. However there are many deviations from first order kinetics. It seems that these deviations are even more pronounced in pressure inactivation. It is difficult to understand how first order models are consistent with the assumption that inactivation is a multiple hit event, as shown by sublethal damage. Distribution models are based on the assumption that the resistance in every non-synchronised culture is not identical. Several distribution models have been investigated. Augustin et al. (1998) showed that heat inactivation could be described by log normal distributions. Peleg and Cole (1998) investigated several pressure and heat inactivation curves and they were consistent with Weibull distributions. Whereas the Weibull distribution gives more freedom the correct physiological reasons for the choice of a particular model should be further investigated.

REFERENCES

H. Alpas and F. Bozoglu (2000) The combined effect of high hydrostatic pressure, heat and bacteriocins on inactivation of foodborne pathogens in milk and orange juice. World J. Microbiol. Biotechnol. 16: 387-392.

J.C. Augustin, V. Carlier and J. Rozier (1998) Mathematical modelling of the heat resistance of *Listeria monocytogenes.* J Appl. Microbiol. 84: 185-191

S. Brul, A.J.M. Rommens, and C.T. Verrips, (2000) Mechanistic studies on the inactivation of Saccharomyces cerevisiae by high pressure. Inn. Food Sci. Emerg. Technol. 1, 99-108.

R. Chastain, and A. Yayanos, (1991) Ultrastructural changes in an obligately barophilic marine bacterium after decompression. Appl. Environ. Microbiol. 57: 1489 - 1497.

C. Garcia-Graells, C. Valckx and C.W. Michiels (2000) Inactivation of Escherichia coli and Listeria innocua in milk by combined treatment with high hydrostatic pressure and the lactoperoxidase system. Appl.Environ. Microbiol. 66: (10) 4173-4179.

K.J.A. Hauben, E.Y. Wuytack, C.C.F. Soontjens and C.W. Michiels (1996) High-pressure transient sensitization of Escherichia coli to lysozyme and nisin by disruption of outer-membrane permeability. J. Food Prot. 59: 350-355.

C.M.M. McMahon, C.M. Byrne, J.J. Sheridan, D.A. McDowell, I.S. Blair andT. Hegarty (2000) The effect of culture growth phase on induction of the heat shock response in *Yersinia enterocolitica* and *Listeria monocytogenes.* J. Appl. Microbiol. 89: 198-206.

C. Metrick, D.G. Hoover, D.F. Farkas (1989) effects of high hydrostatic-pressure on heat-resistant and heat-sensitive of *Salmonella.* J. Food Sci. 54: 1547-

S.M. Morgan, R.P. Ross, T. Beresford, C. Hill (2000) Combination of hydrostatic pressure and lacticin 3147 causes increased killing of *Staphylococcus* and *Listeria* J.Appl. Microbiol. 88: 414-420.

R. Pagan and B. Mackey (2000) Relationship between membrane damage and cell death in pressure-treated Escherichia coli cells: Differences between exponential- and stationary-phase cells and variation among strains. Appl. Environ. Microbiol. 66: (7) 2829-2834.

496

E. Palou, A. Lopez-Malo, G.V. Barbosa-Canovas, J. Welti-Chanes and B.G. Swanson
(1998) Oscillatory high hydrostatic pressure inactivation of *Zygosaccharomyces bailii*
J. Food Prot. 61: 1213-

A.M. Papineau, D.G. Hoover and D. Knorr (1991) Antimicrobial effect of water-soluble
chitosans with high hydrostatic -pressure. Food Biotechnol 5: (1) 45-57 1991

M.F. Patterson, M. Quinn, R. Simpson, and A. Gilmour (1995) Sensitivity of vegetative
pathogens to high hydrostatic-pressure treatment in phosphate-buffered saline and foods
J. Food Prot. 58: 524-529.

M. Peleg and M.B. Cole (1998) Reinterpretation of microbial survival curves Crit. Rev.
Food Sci. Nutr. 38: 353-380.

J.P.P.M Smelt J.C. Hellemons and M.F. Patterson (2001) Effect of pressure on vegetative
microorganisms. Chapter in book to be published (M. Hendrikcx and D. Knorr eds).

J.P.P.M Smelt and J.C. Hellemons (1999) Final individual report for EU project FAIR CT
1175.

J.P.P.M Smelt, A.G.F.Rijke and A. Hayhurst (1994). Possible mechanisms of high pressure
High Pressure Research 12: 199 – 203.

J.P.P.M Smelt, P.C. Wouters and A.G.F. Rijke (1998) 18. Inactivation of microorganisms
by high pressure. *In*: The properties of water in foods ISOPOW 6 pp 398 – 413. D.S. Reid
ed. Blackie Academic & professional London UK.

P.F.ter Steeg, J.C.Hellemons and A.E. Kok. (1999) Synergistic actions of nisin, sublethal
ultrahigh pressure, and reduced temperature on bacteria and yeast. Appl. Environ.
Microbiol. 65: 4148-4154.

J.L. Tholozan, M. Ritz, F. Jugiau, M. Federighi and J.P. Tissier (2000) Physiological effects
of high hydrostatic pressure treatments on *Listeria monocytogenes* and *Salmonella
typhimurium*. J. Appl.. Microbiol. 88: 202-212 .

H.M. Ulmer, M.G. Ganzle and R.F. Vogel (2000) Effects of high pressure on survival
and metabolic activity of *Lactobacillus plantarum* TMW1.460. Appl. Environ. Microbiol.
66: 3966-3973.

P.C. Wouters, E. Glaasker and J.P.P.M. Smelt (1998) Effects of high pressure in inactivation
kinetics and events related to proton efflux in *Lactobacillus plantarum*. Appl Environ.
Microbiol. 64:509 –514.

Trends in High Pressure Bioscience and Biotechnology
R. Hayashi (editor)
                                497

Effect of high pressures on microflora of commercial kefir culture

A. Krzyżewska, A. Reps, A. Proszek, M. Krasowska, I. Warmińska-Radyko

Institut of Food Biotechnology, Warmia and Mazury University in Olsztyn
Olsztyn 10-957, Poland
e-mail: mada@uwm.edu.pl

Abstract: Kefir obtained from the commercial kefir culture, and its lyophilised components, after 24-hour cultivation in milk, were subjected to high pressures in the range of 200-1000 MPa, at an ambient temperature, for 15 minutes. Lactic streptococci were then added to skim milk and incubated at 25°C, whereas kefir and yeasts were added to milk with various additives, commonly used in dairy products, and stored at 4°C or incubated at 25°C, respectively. In all samples, before and after pressurization, the number of living microorganisms and acidity were periodically determined. The decrease of both viability and acidifying activity of kefir grain microflora together with the rise of the applied pressure was observed. It was also stated that additives used in the experiment had a protective effect on kefir microflora and yeasts exposed to the pressure of 400 MPa.

1. INTRODUCTION

Kefir has become more universally consumed dairy product due to its dietetic and therapeutic properties [1]. The starter for kefir production had a special form, called kefir grains. They contain a complex microflora, including different species of bacteria: Lactobacillus, Leuconostoc, Lactobacillus, yeasts and acetic acid bacteria [2,3]. As the precise microbiological composition of kefir grains and a mutual symbiosis of such a system has been not completely known, the industrial manufacture of kefir is conducted by the traditional method.

In Poland, Rhodia Food Biolacta company has developed a new production technology of kefir starter culture witch is directly added to milk. to develop the optimal conditions for kefir preservation, it is necessary to determine the effect of the pressure on the particular groups of microorganisms in the commercial kefir culture.

2. EXPERIMENTAL

2.1. The research material
The research material was received from Rhodia Food Biolacta (Olsztyn – Poland) and included the commercial kefir culture and its components in the lyophilised form:
- kefir grains
- mixture of acidifying and aromatizing streptococci
- yeasts

2.2. The run of the experiment

- Kefir grains after 24-hour cultivation in milk at a temperature of 25°C were subjected to pressurization of 200-1000 MPa for 15 min. and stored at a temperature of 4°C.
- Kefir obtained from the commercial kefir culture, before and after pressurization (400 MPa / 15 min) was added, in quantity of 10%, to milk with various additives (skim milk, 3% fat milk, condensed milk (20 % d.m.), milk with the addition of 2.5% of sugar and milk with the addition of 0.8% Meyhall stabiliser YO 750) and stored at a temperature of 4°C.

    During the storage, the following determinations were periodically performed:
    - the number of yeasts, streptococci and lactobacilli
    - acidity (pH and °SH).

- Lactic streptococci, after 24-hour cultivation in milk, were subjected to pressurization of 700 and 1000 MPa / 15 min. The cultivated bacteria, were added to the milk, before and after pressurization, in amount of 10% and incubated at a temperature of 25°C.

    During the incubation, the following determinations were periodically performed:
    - the number of acidifying bacteria
    - acidity (pH and °SH).

- After 24-hour incubation in milk, 1% of the yeasts was added to the milk, containing various additives, and incubated at a temperature of 25°C for 24h. Before and after pressurization (400 MPa / 15 min.), the culture was added to the milk in quantity of 10% and incubated at a temperature of 25°C.

    During the incubation, the following determinations were periodically performed:
    - the number of yeasts
    - acidity (pH and °SH).

2.3. Equipment for pressurization

The samples were subjected to pressurization in hydraulic pressure generator produced by UNIPRESS EQUIPMENT (Warsaw – Poland).

3. RESULTS

Figure 1 shows the effect of pressure on streptococci in kefir grains. In the kefir grains, entering the composition of the commercial kefir culture, any presence of yeasts was not found.

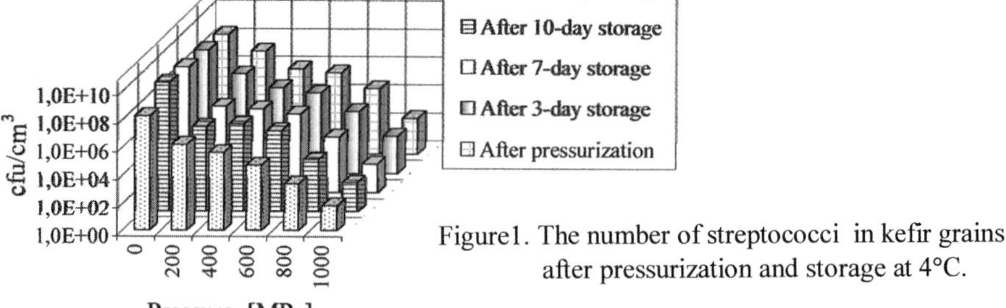

Figure1. The number of streptococci in kefir grains after pressurization and storage at 4°C.

Together with the rise of pressure, the decrease of survival of kefir grain microflora was observed. The number of lactobacilli decreased more quickly than the number of streptococci (Figure 2).

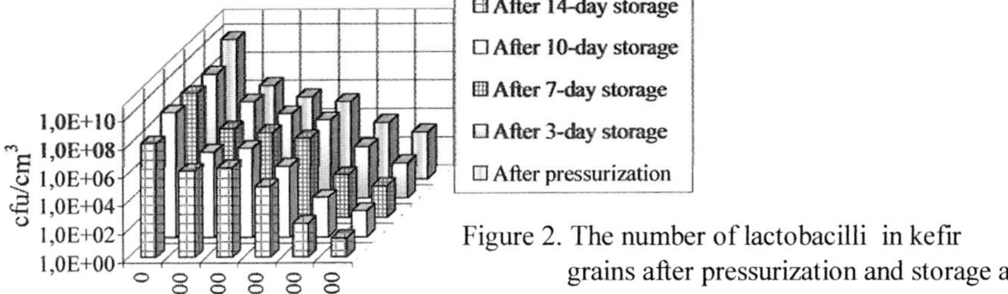

Figure 2. The number of lactobacilli in kefir grains after pressurization and storage at 4°C.

After pressurization under 400 MPa, the number of streptococci decreased by 1.2 size order and the number of lactobacilli –by 3.5 size orders. Small differences in the degree of microorganisms' survival under the pressure in the range of 400-600 MPa were found.

The pressure of 1000 MPa lowered the number of streptococci by 6.1 size orders and that of lactobacilli – by 6.5 size orders.

During the storage at a temperature of 4°C after pressurization, the decrease in the number of microorganisms by 0.5-2 size orders was observed.

The changes in acidity of kefir grains are demonstrated in Figure 3. The acidity of non-pressurized kefir grains during 14-day storage at 4°C increased from pH 4.30 to 3.71. After pressurization of kefir grains under 200 and 400 MPa, the gain of their acidity was considerably smaller and it reached pH 3.88 and 4.06, respectively. The acidity of the grains, subjected to pressurization at 600, 800 and 1000 MPa was unchanged during the storage.

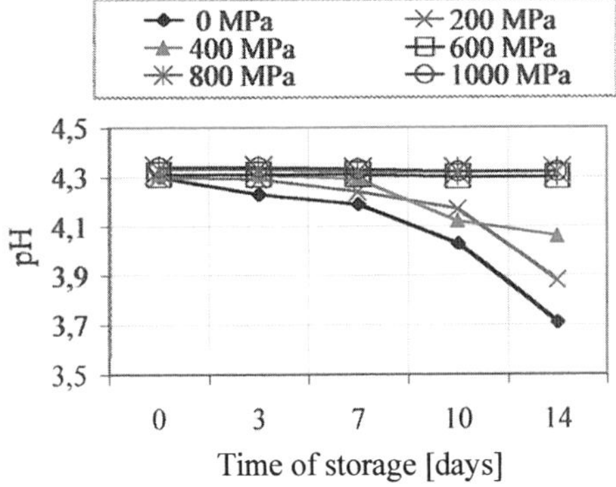

Figure.3 Acidity of kefir grains after pressurization and storage at 4°C.

One of the factors deciding on the effectiveness of the microorganisms' inactivation during pressurization, is the environment in which they are found.

In skim milk with the addition of kefir microflora, directly after pressurization (400 MPa /15min.), a similar degree of reduction of streptococci and lactobacilli was observed (by ca. 3.6 size orders) (Table 1).

Table1.
The survival of microorganisms in skim milk, inoculated with kefir, after pressurisation at 400 MPa (15 min) and storage at a temperature of 4°C.

	Time of storage [days]	Streptococci			Lactobacilli		
		cfu/cm^3	Survival [%]	log cfu/cm^3	cfu/cm^3	Survival [%]	log cfu/cm^3
Skim milk	Before pressurization	$9,7\times10^7$	100,0	7,98	$9,4\times10^7$	100,0	7,97
	After pressurization	$2,3\times10^4$	0,024	4,36	$2,1\times10^4$	0,022	4,32
	3	$5,0\times10^4$	0,052	4,70	$6,5\times10^4$	0,069	4,81
	14	$4,4\times10^5$	0,45	4,63	$3,0\times10^7$	31,9	7,48
Milk with 2.5% of sucrose	Before pressurization	$9,9\times10^7$	100,0	7,99	$7,4\times10^7$	100,0	7,87
	After pressurization	$1,9\times10^6$	1,9	6,28	$1,3\times10^6$	1,76	6,11
	3	$7,4\times10^7$	74,8	7,87	$6,0\times10^8$	810,8	8,78
	14	$3,4\times10^8$	343,4	8,53	$2,7\times10^8$	364,9	8,43
Concentrated skim milk (20% d.m.)	Before pressurization	$6,8\times10^7$	100,0	4,83	$5,2\times10^7$	100,0	7,72
	After pressurization	$5,3\times10^5$	0,8	5,72	$8,0\times10^4$	0,15	4,90
	3	$5,3\times10^5$	0,8	5,72	$2,1\times10^5$	0,40	5,32
	14	$6,4\times10^6$	9,4	6,81	$8,9\times10^6$	17,1	6,95

Any presence of yeasts in 0.1 cm^3 of milk after pressurization was not found. It should be mentioned that the milk with the addition of 10% of kefir contained before pressurization 7.3×10^4 cfu/cm^3 of yeasts.

The similar degree of kefir-acidifying microflora inactivation was observed after pressurization of 3% - fat milk and the milk with the addition of 0.8% of stabiliser.

Also, in the discussed milks and in the milk with the addition of 2.5% of sucrose, any presence of yeasts after pressurization was not found.

In the milk with the increased dry solids content after pressurization the number of streptococci was decreased by 2.1 size orders and that of lactobacilli by 2.8 size orders.

The greater survival of lactic acid bacteria was observed in the milk with the addition of sucrose. The number of streptococci and lactobacilli decreased by ca. 1.7 size orders

In all milk samples with the addiction of 10% of kefir, before and after pressurization, the increase of the number of streptococci and lactobacilli during the storage at 4°C was found.

During the storage of kefir in milk with various additives, except for skim milk, a considerable increase of yeasts' content was observed in spite of the fact that directly after pressurization, the yeasts were absent in the milk. This fact indicates that a small quantity of the yeasts could remain in the milk after pressurization and the additives to milk may play a protective role in relation to microflora.

The inhibitory effect of pressurization on the acidifying activity of kefir microflora in milk was found (Figure 4 and 5).

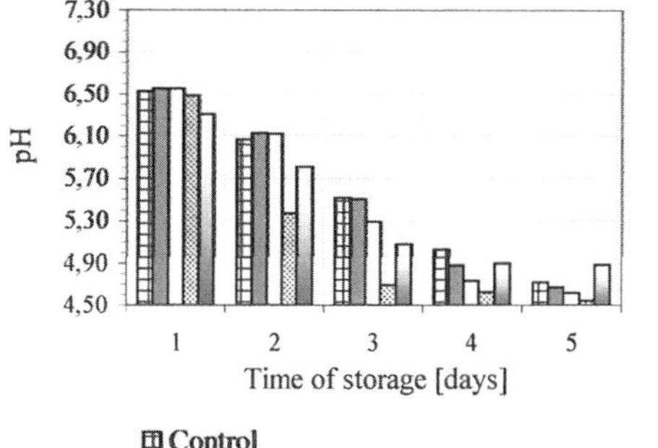

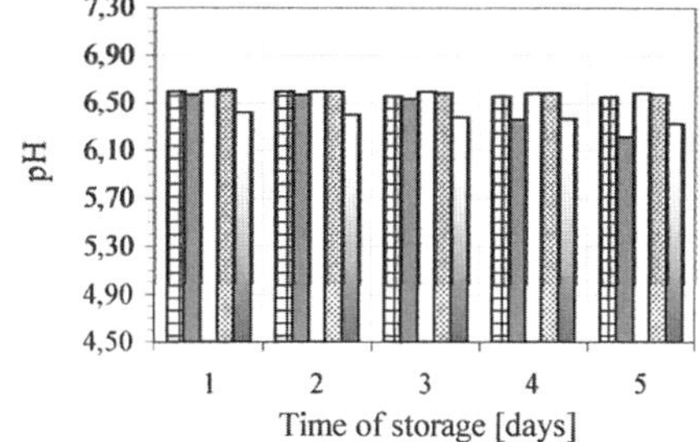

Control
Milk+2,5% of sucrose
Whole milk
Milk +0,8% of YO750
Concentrated whole milk (20% d.m.)

Figure 4. Acidity of milk, with addition of 10% of kefir, during storage at 4°C

Control
Milk+2,5% of sucrose
Whole milk
Milk +0,8% of YO750
Concentrated whole milk (20% d.m.)

Figure 5. Acidity of milk, with addition of 10% of kefir pressurized at 400MPa, during storage at 4°C

During 14-day storage of the non-pressurised milk with the addition of kefir, the increase of acidity, from 1.42 to 1.94 pH unit, was observed. On the other hand, in the milk with various additives, the acidity after pressurization increased from 0.01 to 0.09 pH unit, except for the milk with sucrose, where the acidity increased by 0.36 pH unit.

Also, many other authors indicate that many substances, being found in the environment may play a protective role in relation to microorganisms.

The changes in bacterial number and acidity of milk to which the mixture of streptococci after pressurization of 700 and 1000 MPa was added, have been presented in Table 2.

Table 2.
The number of bacteria in milk inoculated with the mixture of pressurized streptococci, during 44-h incubation at a temperature of 25°C.

Incubation time [h]	Control		After pressurization at 700 MPa		After pressurization at 1000 MPa	
	cfu/cm^3	pH	cfu/cm^3	pH	cfu/cm^3	pH
0	$1,2\times 10^2$	6,44	$1,1\times 10^3$	6,47	$1,2\times 10^2$	6,49
4	$3,3\times 10^3$	6,43	$9,4\times 10^3$	6,46	$2,2\times 10^3$	6,48
8	$2,5\times 10^5$	6,42	$7,0\times 10^4$	6,41	$4,0\times 10^4$	6,45
12	$6,5\times 10^6$	5,70	$9,0\times 10^5$	5,89	$5,4\times 10^5$	5,80
20	$1,7\times 10^9$	5,00	$8,9\times 10^8$	5,11	$1,0\times 10^8$	5,10
24	$1,6\times 10^9$	4,66	$9,9\times 10^8$	5,00	$7,1\times 10^8$	4,95
28	$1,3\times 10^9$	4,52	$1,1\times 10^9$	4,93	$9,0\times 10^8$	4,88
32	$2,3\times 10^9$	4,42	$1,6\times 10^9$	4,88	$1,2\times 10^9$	4,85
36	$3,2\times 10^9$	4,35	$3,4\times 10^9$	4,83	$7,4\times 10^8$	4,80
44	$5,0\times 10^9$	4,30	$6,0\times 10^9$	4,79	$4,0\times 10^9$	4,77

During incubation at a temperature of 25°C, the highest increase of streptococci number was observed in the control sample and then, in the milk inoculated with bacteria being pressurised at 700 MPa.

In the logarithmic growth phase, the time of streptococci generation in the control sample was 0.48h and after pressurization 1.02h. However, after the 32-hour incubation the number of bacteria in all samples was similar.

In all samples, between 8 and 20 hour of cultivation, the similar changes of acidity were observed.

Further prolongation of the incubation time of milk, containing bacteria after pressurization, caused smaller changes in acidity as compared to the milk incubated with the non-pressurised mixture of streptococci. After the 44-hour incubation, milk acidity in the control sample equalled to pH 4.30 and in the samples with bacteria after pressurization it amounted to pH 4.77-4.79.

The obtained results confirm that the pressurization caused lowering of a fermentation activity of lactic acid bacteria.

Table 3 shows the changes in the number of yeasts and milk acidity after inoculation with the yeasts pressurised at 400 MPa / 15 min after incubation in milk with various additives.

Table 3.
The number of living microorganisms and acidity of milk, inoculated with yeasts pressurized at 400 MPa, incubated for 48h at a temperature of 25°C

Incubation time [h]	Skim milk control		Skim milk		Concentrated milk (20% d.m.)		Whole milk (3%)	
	cfu/cm^3	pH	cfu/cm^3	pH	cfu/cm^3	pH	cfu/cm^3	pH
Before pressurization	$6,8\times10^4$	4,49	$6,8\times10^4$	4,49	$5,9\times10^5$	4,92	$1,2\times10^5$	4,51
0	$2,0\times10^0$	6,86	absent in 1cm^3	6,40	$1,1\times10^1$	6,40	$2,8\times10^1$	6,35
6	$3,5\times10^0$	6,85	absent in 1 cm^3	6,39	$1,6\times10^2$	6,39	$7,1\times10^3$	6,34
12	$3,5\times10^1$	6,78	absent in 1cm^3	6,37	$2,5\times10^3$	6,38	$3,9\times10^4$	6,28
24	$3,6\times10^4$	4,48	absent in 1 cm^3	4,50	$4,2\times10^4$	5,43	$2,9\times10^5$	4,52
30	$2,6\times10^5$	4,34	absent in 1 cm^3	4,35	$7,6\times10^4$	4,61	$4,4\times10^5$	4,41
36	$6,6\times10^3$	4,29	absent in 1 cm^3	4,33	$1,7\times10^5$	4,52	$2,3\times10^5$	4,37
48	$3,1\times10^5$	4,20	absent in 1 cm^3	4,22	$1,3\times10^5$	4,36	$1,5\times10^5$	4,27

A small number of yeasts cells was found in the milk after inoculation with the pressurised yeasts, incubated in milk with 20% d.m. and 3%-fat milk. Any presence of living yeast cells in 0.1 cm^3 of skim milk, inoculated with the pressurised mixture of yeasts, cultivated on skim milk, was not observed. Also, during the 48-hour incubation at a temperature of 25°C, any presence of the yeasts in the milk was not found.

During the incubation, a quick gain of yeast cells was also observed between 12 and 24h of the incubation.

The highest rise of milk acidity was stated up to 24 h of cultivation. The increase of milk acidity in the cultivation in which the presence of the yeasts was not found, may support the fact that under the pressure of 400 MPa / 15 min. inactivation if lactose-fermenting enzymes did not take place.

4. CONCLUSIONS

- The degree of reduction of streptococci and lactobacilli in kefir grains is dependent on the level of pressure.
- In lyophilised kefir grains, any presence of yeasts was not found.
- Acidity of kefir grains subjected to pressurization at 600, 800 and 1000 MPa at a temperature of 4°C for 14 days remained unchanged.
- Fat, sugar, addition of stabilisers and the increased level of milk dry solids lowers the degree of inactivation of kefir microflora and of the yeasts, subjected to pressurization.
- The pressure of 400 MPa / 15 min inactivates the yeasts; their enzymes are however capable of milk acidification at a temperature of 25°C.

REFERENCES

1. N.S. Koroleva.,. Bul. FIL., 227 (1988), 96-100
2. W. Kneifel, H.K.Mayer, Int. J. Food Sci. Tech., 26 (1991),423-428
3. V.M. Marschall, W.M. Cole, J.Dairy Res., 52 (1985),451-456

Trends in High Pressure Bioscience and Biotechnology
R. Hayashi (editor)

Effect of high pressure on mikroflora of kefir.

Reps A., Krzyżewska A., Łaniewska-Moroz Ł., Iwańczak M., Krasowska M.,

Instytut of Food Biotechnology
Warmia and Mazury Univercity
Heweliusza 1 str., 10-718 Olsztyn, Poland
e-mail:mada@uwm.edu.pl

1. Abstract

The aim of the studies was the effect of high pressure (200-800 MPa for 15 min.) on microflora of kefir. The number of streptococci, lactobacilli and yeasts was determined before and after pressure treatment and after 1, 2 and 3 weeks of storage at a temperature 4°C and 20°C. A considerable reduction of the number of streptococci, being 39%, lactobacilli – 45%, and completely inactivation of yeasts was found in kefir, subjected to the pressure of 400 MPa.

As affected by the pressure of 800 MPa, the complete inactivation of acidifying bacteria was not found. The pulsatory pressurization of kefir caused higher inactivation of microflora The decrease of the number of acidifying bacteria was observed during 1, 2 weeks storage of pressurized kefir , a small increase of bacterial count was found but not exceed, the number of lactic bacteria directly after pressurization. The acidity increased only slightly.

2. Introduction

Kefir is a cultured milk beverage of ancient lineage obtained by combined acidic and alcoholic fermentation . It is manufactured using kefir " grains" which contain a complex microflora including lactic acid bacteria, acetic bacteria, and yeasts. This gives it organoleptic and nutritional properties and beneficial effect on health. However, this characteristic also implies the same constrains with regard to storage and shelf life. [1, 2, 3]

High temperature processing or thermization , which reduces the population of lactic bacteria, makes it possible to prolong the period during which they can be consumed. The impact of high temperatures is not limited to the viability of the ferments and it can lead to a reduction or even the total loss of the same specific benefits.[4]

In the experiment, the effect of high pressure on microflora of kefir was studied.

506

3. Methods

Kefir was prepared, using kefir culture DIRECT, manufactured by Rhodia Biolacta Texel in Olsztyn. Pressurization was conducted at room temperature, employing the following pressures : 200, 400, 600, and 800 MPa for 15 min ; 400 MPa for 60, 120and 180 min., 400 MPa for 15, 30 and 45 min. and 400 MPa , by pulsatory method , for 3x15, 3x10 and 3x15 min. In non-pressurized kefir directly after pressurization and after 1, 2 and 3 weeks of storage at a temperature of 4 and 20°C, pH value was measured and the number of the selected groups of microorganisms was determined. In addition, to determine the effect of high pressure on capability of milk acidification by kefir microflora , the milk was inoculated with the microflora of kefir, being not subjected to pressurization and with the pressurized one; the number of acidifying microorganisms and the changes in pH directly after inoculation and after 4, 8, 12, 16 and 24 h of incubation at a temperature of 25°C, was determined.

4. Results

A considerable decrease of bacterial number occurred as late as under the pressure above 200 Mpa. In the range of the employed pressures of 200-800 MPa, the complete inactivation of bacteria was not obtained. After application of the pressure equal to 800 MPa, the number of streptococci decreased from $1,3x10^9$ CFU/cm^3 to $1,2x10^3$ CFU/cm^3 and that of bacilli from $5,6x10^8$ CFU/cm^3 to $5,2x10^2$ CFU/cm^3. During the 3-week storage of non-pressurized and pressurized kefir , at a temperature of 4 and 20°C, the number of streptococci and lactic acid bacteria in most of the cases was decreased by 0,5-2 size orders.
The greater decrease of the number of these bacteria occurred in kefir, stored at a temperature of 4 °C.(Fig. 1-2).

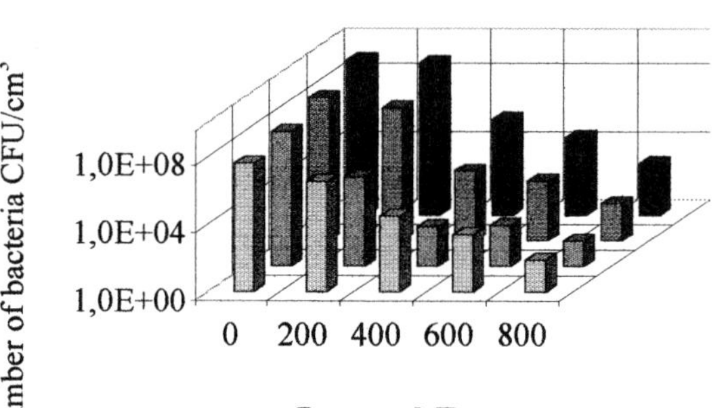

Fig. 1 The number of streptococci after pressurization and 1, 2, 3 weeks of storage at a temperature of 4°C.

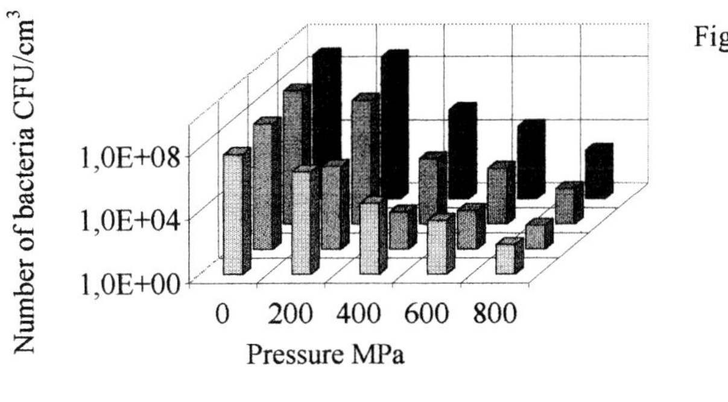

Fig. 2 The number of lactobacilli after pressurization and 1, 2, 3 weeks of storage at a temperature of 4°C.

■ after 3-week storage   ▒ after 2-week storage
▒ after 1-week storage   ▢ after pressurization

After pressurization , a small rise of active acidity of kefir was observed; pH of non-pressurized kefir was 4,51; after pressurization under 400 MPa-4,52; after pressurization under 600 and 800 MPa, it increased to 4,56 and 4,59, respectively. In kefir, pressurized under 400 MPa for 15 min., the reduction of streptococci and lactic acid bacteria by 4 size orders ie. to 4,4 $\times 10^5$ CFU/cm^3 and 6,5x10^4 CFU/cm^3, respectively, and the complete inactivation of yeasts, was found.

Tab.1. The number of living microorganisms in kefir , after pressurization at 400 Mpa for 60, 120, 180 minutes.

	Streptococci	Lactobacilli	Yeasts
		CFU/cm^3	
Control	$1,1\times10^9$	$1,6\times10^8$	$9,2\times10^4$
60	$2,3\times10^4$	$1,8\times10^2$	absent in 0,1 cm^3
120	$3,4\times10^2$	1×10^2	absent in 0,1 cm^3
180	$3,7\times10^2$	$1,3\times10^2$	absent in 0,1 cm^3

Prolongation time of pressure of 400 MPa to180 min. did not cause the complete reduction of the studied microorganisms (Tab.1.).

The following results were obtained : after 60 min. of pressurization, the number of streptococci was lowered from $1,1\times10^9$ CFU/cm^3 to $2,3\times10^4$CFU/cm^3, and that of lactobacilli from $1,4\times10^8$ CFU/cm^3 to $1,8\times10^2$ CFU/cm^3, after 120 min. – to $3,4\times10^2$ CFU/cm^3 and $1,0\times10^2$CFU/cm^3, respectively. Pressurization under 400 MPa for 15 min. caused reduction of streptococci and lactic acid bacteria by 4 size orders whereas pulsatory pressure of 400 MPa for 3x5 min. in 5 intervals caused reduction of these bacteria by 5 size orders (Tab.2).

Tab.2. The number of living microorganisms in kefir after pressurization at 400 Mpa (tree 5-minute cycles) and after 1, 2, 3 weeks storage.

Temperature [^{o}C]	Storage Time [week]	pH	Streptococci	Lactobacilli	Yeasts
				CFU/cm^3	
4	0	4,54	$7,5\times10^4$	$4,2\times10^3$	absent in 0,1 cm^3
	1	4,49	$2,2\times10^4$	$2,6\times10^3$	absent in 0,1 cm^3
	2	4,46	$5,5\times10^3$	2×10^2	absent in 0,1 cm^3
	3	4,45	$8,4\times10^3$	$9,3\times10^2$	absent in 0,1 cm^3
20	0	4,54	$7,5\times10^4$	$4,2\times10^3$	absent in 0,1 cm^3
	1	4,46	$2,1\times10^5$	$3,7\times10^3$	absent in 0,1 cm^3
	2	4,48	$2,6\times10^4$	$2,8\times10^3$	absent in 0,1 cm^3
	3	4,45	4×10^3	$3,1\times10^3$	absent in 0,1 cm^3

During the storage of kefir it was found that the higher was the pressure, the smaller was the decrease of acidity. In kefir pressurized under 400 MPa and stored for 3 weeks at a temperature of 4 and 20°C, pH was lowered from 4,54 to 4,43 and 4,39; respectively. After storage of kefir, pressurized under 600 and 800 MPa, any significant changes of pH were not found (Fig.5.)

The activity of milk acidification by microflora of kefir, subjected to pressure was decreasing together with the rise of the employed pressure. After 24 hours of incubation at a temperature of 25°C in milk, inoculated with kefir, pressurized under 400 and 800 MPa, lowering of pH value was observed and in case of continuous pressurization from 6,75 to 5,39 and from 6,78 to 6,31; respectively; in case of pulsatory pressurization – 3x10 min., the decrease of pH value was as follows: from 6,96 to 5,71 (400 MPa) and from 6,98 to 6,52 (800 MPa).(Fig.6)

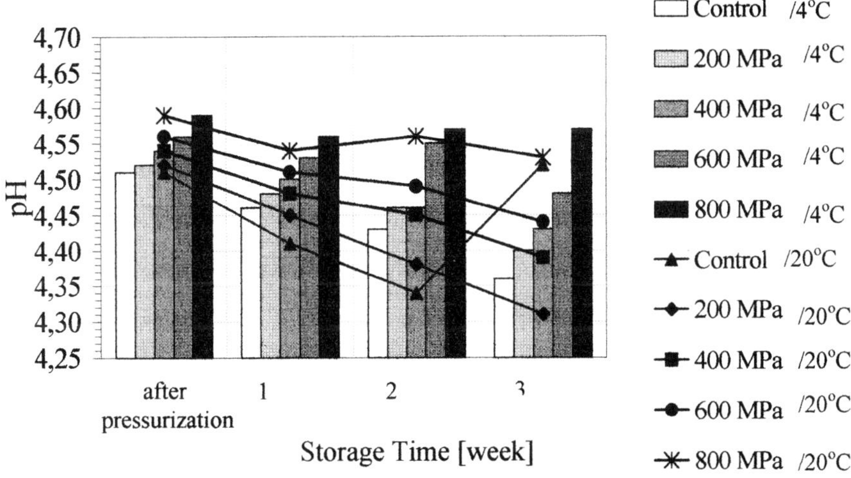

Fig.5. Acidify of kefir after pressurization and storage.

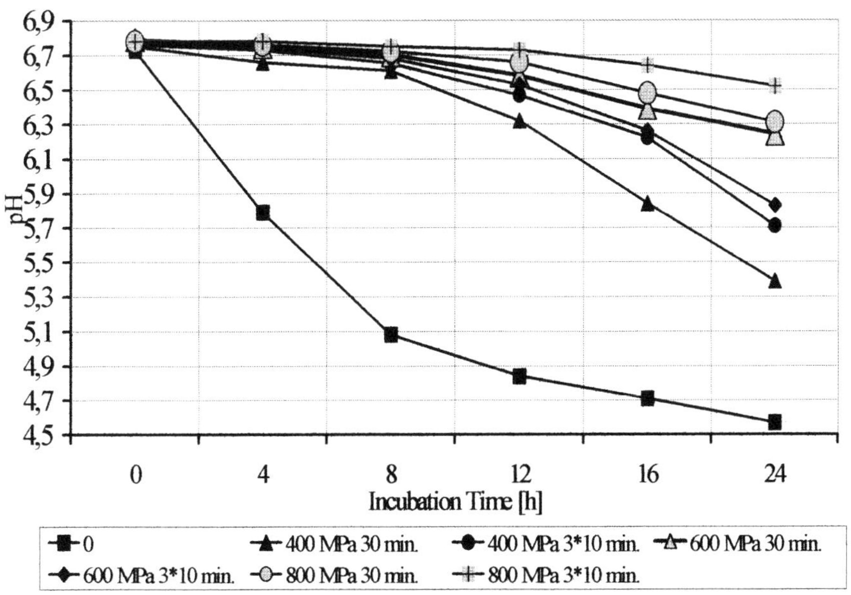

Fig. 6 Acidity of milk inoculated with kefir after constant and pulse pressurization at 400,600 and 800 Mpa , during incubation at 25 °C

5. Conclusions

The conducted studies and the obtained results allow to formulate the following conclusions :

- considerable reduction of microorganisms number was observed in the range of pressures from 400 to 800 MPa for 15 min. ;
- the yeasts were completely inactivated at 400 MPa, but the pressure of 400-800 MPa for 180 minutes did not cause the complete inactivation of acidifying bacteria ;
- prolongation of the pressurization time in the range of the pressures 400-800 MPa to 180 minutes did not cause the complete inactivation of acidifying bacteria ;
- the pulsatory pressurization of kefir caused higher inactivation of microflora ;
- activity of milk acidification, at a temperature of 25°C, by kefir microflora, subjected to pressurization was decreasing gradually with the employed pressure..

The obtained results allow to state that in the highly pressurized kefir not only the number of microorganisms was changed, what may be utilized in the industrial preservation of kefir.

6. References

1. C.J. Saloff , World Newsletter Danon , No. 11 (1996) 1-12.
2. W. Kneifel, H.K. Mayer, Int. J. Food Sci. Tech., 26 (1991) 423-428.
3. N. S. Koroleva , Bulletin of the IDF, 227 (1998) 96-100.
4. R. Negri , Food Qual. Nutr.,12 (1978) 511-515.

Trends in High Pressure Bioscience and Biotechnology
R. Hayashi (editor)

Effect of ultra-high pressure on fruit juices contaminant yeasts

A. Rosenthal[a], B. MacKey[b], A. Bird[b]

[a] Embrapa (Brazilian Company for Agricultural Research) - Food Technology Center, Av. das Américas 29501 - CEP 23.020-470, Rio de Janeiro-RJ, Brazil

[b] Food Science and Technology Department, University of Reading, Whiteknights, Reading RG6 3AZ, United Kingdom

The pressure resistance of a number of yeasts associated with spoilage of fruit products (*Zygosaccharomyces bailii, Saccharomyces cerevisiae, Pichia anomola, Candida magnoliae Rhodotorula glutinis*) was examined. No especially pressure resistant strains were apparent. Pressure treatment at 300 MPa for 10 min reduced viable numbers of a cocktail of *Zygosaccharomyces bailii* strains by 4 log units in orange, apple or pineapple juices. A smaller decrease occurred in tomato juice but, after storage for 24h at room temperature (*ca* 20 °C), the level of reduction had increased to equal or exceeded that seen in the other juices.

1. INTRODUCTION

The increasing demand for products that maintain the quality of fresh foods has led to a search for new technologies of food preservation. These technologies aim mainly at replacing those that involve heat treatment, which can result in a decrease in the nutritional value and a change in the sensory characteristics of the products (1,2). In this sense, high pressure has been investigated in order to create products of high quality and microbiological stability, individually or in combination with moderate heat treatment (3-5).

Unlike heat, high pressure affects only a few covalent bonds and therefore has little effect on ester molecules and other compounds responsible for food flavour and taste. Pressure also has negligible effects on vitamins. However pressure does cause denaturation or unfolding of proteins by disrupting hydrophobic interactions and causes phase changes in biological membranes which explain its efficiency in inactivating spoilage microorganisms and some enzymes (6,7).

Furthermore, ultra-high pressure (UHP) can also disrupt cell wall and inner structures of microorganisms. In this sense, many intracellular organelles cells such as nucleus, mitochondria, endoplasmic reticulum and vacuoles from *Saccharomyces cerevisae* showed deformation and disorganization under UHP treatment (8). UHP induces physiological

512

imbalances due to the internal and external structural damages, which are responsible for killing the microorganisms (4) .

The effect of UHP was also associated to a permanent shrinkage of cell volume, which was associated to an irreversible mass transfer that occurs between the cell and pressure medium, due to the disruption or increase in membrane permeability. Adiabatic expansion of water was also reported to be associated with microorganisms' inactivation (4).

In spite of being commercially used only recently, high-pressure technology has already originated a row of products in Japan, United States and Europe. Those are mainly low pH products, such as fruit juices, jams, jellies, fruit yogurts and salad sauces (9).

However, many aspects of the technology still have to be clarified, mainly relating to the variation in pressure resistance of foodborne pathogenic and spoilage organisms, according to the type and composition of the product, and the microbial species and strain.

This work aimed at evaluating the baroresistance of different spoilage yeasts, individually or in combination, inoculated in different fruit juices and in growth media, in order to evaluate the different factors that might influence process definition, and perspectives for future use of the technology.

2. MATERIALS AND METHODS

2.1.Preparation of fruit juices

The skin and core of fresh pineapple fruits were removed and the flesh pulped in a sterile food processor. Pulp was centrifuged at 8 000 rpm for 20 min and the supernatant used as the juice. Orange juice was obtained by squeezing fresh oranges and centrifuging as above. Tomato and apple juices were purchased at a local supermarket (organic tomato juice containing 0.5% sea salt and pressed apple juice containing L-ascorbic acid).

2.2.Yeast cultures

The following strains were obtained from the National Collection of Yeast Cultures (Institute of Food Research, Norwich UK): *Zygosaccharomyces bailii* (NCYC 1416, 1400, 1766, 2406, 1427, 1601), *Saccharomyces cerevisiae* (NCYC 505), *Candida magnoliae* (NCYC 765), *Rhodotorula glutinis* (NCYC 377), *Pichia anomola* (NCYC 2257). *Zygosaccharomyces bailii* is a common cause of spoilage of fruit products and *Rhodotorula, Candida* and *Pichia* species have been isolated from pressure-treated tropical fruits and derivatives.

2.3.Growth of organisms

Strains were maintained frozen on glass beads at -70°C and on slopes of yeast malt (YM) agar at 5°C. Cultures were prepared by inoculating 20 ml YM broth from a YM slope and incubating on a shaking platform for 6 days at 25°C. Cells were harvested by centrifugation at 6000 rpm for 20 min and resuspended in one tenth the volume of fresh YM broth to give a concentration of between 10.7 to 10.8 cells per ml. Concentrated suspensions were mixed together to produce a 'cocktail' of strains, as indicated, or were used singly to inoculate fruit juice or medium prior to pressure treatment.

2.4.Pressure treatment

Suspensions to be pressurized were divided into 1.5-ml portions and placed in polyethylene bags, which were heat-sealed. Samples were pressure treated in a 300 ml pressure vessel (Stanstead Mark II Enhanced mini-food lab, Stanstead Fluid Power, Stanstead, United Kingdom). The pressure transmitting fluid was ethanol: castor oil (80:20). The temperature increase in the pressurization fluid caused by adiabatic heating was monitored using a thermocouple and was about 25°C at 600 MPa and lasted about 30 sec. The samples were kept in ice before and after pressurization. All the pressure experiments were carried out in duplicate.

2.5.Counting of viable cells

Samples were diluted in maximum recovery diluent (Oxoid Ltd, Basingstoke UK) and appropriate dilutions were spread onto YM agar. Plates were incubated at 25°C and colonies counted after incubation for up to 4 days. All the counting procedure was done in duplicate.

3. RESULTS AND DISCUSSION

3.1.Inactivation of different spoilage yeasts

In preliminary experiments the pressure resistance of several yeast species associated with spoilage of fruit juice was compared in broth. Suspensions of yeast in YM broth were pressurised at ambient temperature for 10 min at 0.1 to 300 MPa (Figure 1). It is noteworthy that where there is no point corresponding to a certain combination of the parameters in any one of the figures, for any particularly yeast, it means that no viable cell was detected in the previous combination (point) considered.

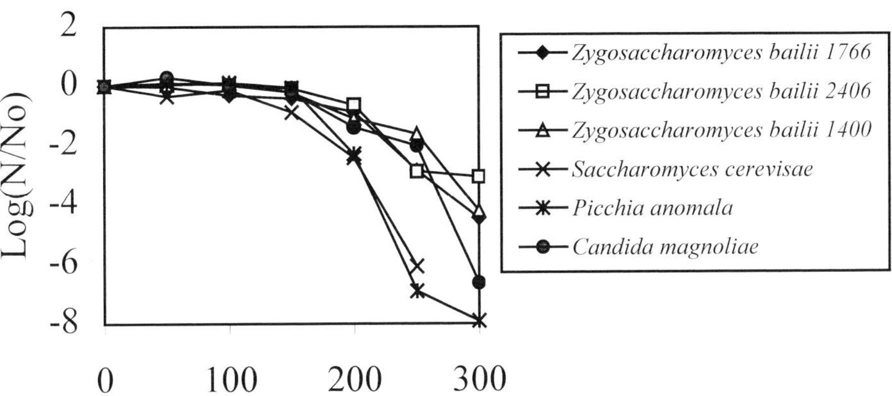

Figure 1. Effect of pressure on spoilage yeasts inactivation (time of pressurisation = 10 min)

514

Saccharomyces cerevisiae and *Pichia anomola* were more sensitive than the other spoilage yeasts (*Zygosaccharomyces bailii, Candida magnoliae*). Numbers of the more resistant species were reduced by 3-4 log units after 10 min at 300 MPa.

3.2.Time course of yeast inactivation

Differences in pressure resistance between strains of spoilage yeast were seen more clearly in time-course experiments at 250 MPa. The fast-growing pink yeast *Rhodotorula glutinis* was included in these experiments and its resistance matched that of the *Zygosaccharomyces bailii* strains (Figure 2).

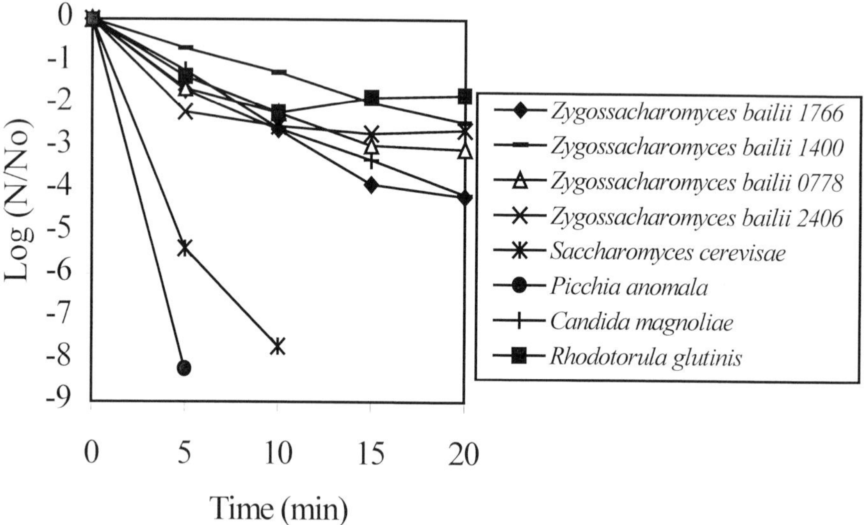

Figure 2. Inactivation of yeasts with pressurisation time at 250 MPa

3.3.Inactivation in fruit juice

The preliminary trials indicated that *Zygosaccharomyces bailii* was relatively resistant to pressure so a cocktail consisting of 6 strains of *Zygosaccharomyces bailii* was tested in different fruit juices (Figure 3). After 10 min at 300 MPa viable number decreased by 4 log units in orange, apple and pineapple juice, but by only 2.3 log units in tomato juice. However, after storing for 24h at ambient temperature, viable numbers in the tomato juice samples that had been treated at 250 or 300 MPa decreased by a further 3 log units, resulting in a net reduction of nearly 5 log units (Figure 4). No further reduction in count occurred in the other juices.

The study has shown that is possible to greatly reduce the number of viable cells of the yeasts usually responsible for product spoilage. In this way, it may be possible to use the technology as a non-conventional method for preserving fruit juices. Several factors such as pressure levels, product storage after the treatment and time of treatment affect the lethality of the process to yeasts, and the response to the treatment may greatly depend on the type of fruit juice, species and strain of yeast being considered. *Zygosaccharomyces bailii* was amongst the more resistant group of organisms but was not unusually resistant. A relatively short treatment at 300 MPa was sufficient to reduce numbers in fruit juices by 4 to 5 log units. Others have found that treatment at 300 MPa for 10 min was sufficient to reduce numbers of *Zygosaccharomyces bailii* in citrate buffer by 7 log units (10). The differences between that work and ours may be due to differences in the suspending media or strains used. A very slight elevation of temperature to 45 °C was found to increase inactivation substantially (10). In agreement with our results, ALEMÁN et al. (4) obtained decimal reductions of 0.7 and 5.1 in pineapple juice inoculated with *Saccharomyces cerevisae* and treated under 270 MPa for 40 and 400 sec, respectively.

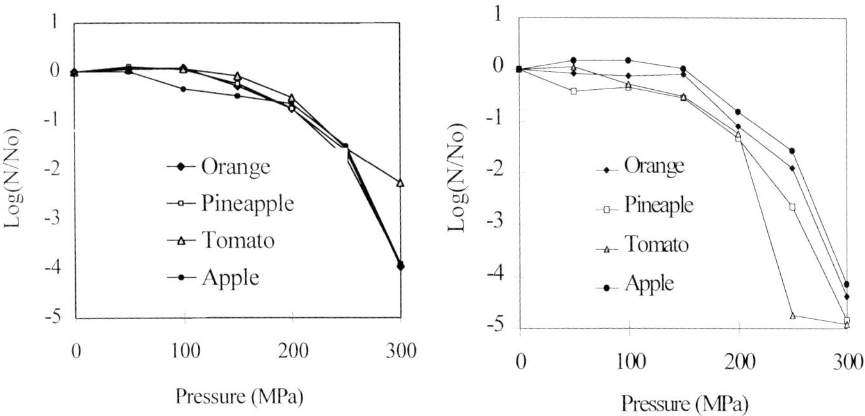

Figure 3. Inactivation of *Zygosaccharomyces bailii* in fruit juices (time = 10 min)

Figure 4. Inactivation of *Zygosaccharomyces bailii* after post-treatment storage(25°C/24 hr)

Of particular interest was the observation that the initial level of yeast inactivation in tomato was less than in the other juices but viable counts subsequently declined on storage. It is worth noting that the pH value of the tomato juice was 4.07, while the other juices presented values in the range between 3.34 to 3.67. Tomato and apple juices bought from commercial suppliers contained salt (0.5 %) and L-ascorbic acid, respectively. In the case of the tomato juice, the higher pH value and/or the presence of salt seemed to have contributed to a decrease in the number of cells after the storage period. Inactivation of spoilage yeasts may be influenced by the product composition and processing conditions. For example reduced water activity has a protective effect whereas relatively small increases in temperature, up to 40°C, increase inactivation (11-13).

516

Future studies related to processing and storing fruit juices may clarify such aspects and provide opportunities for enhancing the preservative effects of pressure.

4. CONCLUSION

The study has shown that is possible to greatly reduce the number of viable cells of yeasts usually responsible for the fruit juices spoilage. In this way, it is potentially viable to use the technology as a non-conventional method of fruit juices preservation.

Some factors such as pressure levels, product storage after treatment and time of treatment are relevant to the process, and the response to the treatment may greatly depend on the type of fruit juice, specie and strain of the yeasts being considered.

In particular the responses patterns greatly varied among the species and strains studied, what turns it difficult to generalise the results for all spoiling microorganisms.

REFERENCES

1.   R.G. Earnshaw; J. Appleyard and R.M. Hurst., Intern. J. Food Microb, 28 (1995) 197.
2.   J.P.P.M. Smelt., Trends Food Sci. Technol., 9 (1998) 152.
3.   A. Williams, Nut. & Food Sci., 1 (1994) 20.
4.   G.D. Alemán; E.Y. Ting; S.C. Mordre; A.C.O. Hawes, M. Walker; D.F. Farkas and J.A. Torres., J. Food Sci., 61 (1996) 388.
5.   J.C. Cheftel, Food Sci. Technol. Int., 1 (1995) 75.
6.   D. Knorr. Food Technol., 47 (1993) 156.
7.   T. Vardag; H. Dierkes and P. Körner, Food Technol. Europe, 3 (1995) 106.
8.   S. Shimada; M. andou; N. Naito; N. Yamada; M. Osumi and R. Hayashi, Appl. Microbiol. Biotechnol., 40 (1993) 123.
9.   R.G. Earnshaw, Nut. & Food Sci., 2 (1996) 8.
10. Y. P. Pandya, F. F jewett and D. Hoover, J. Food Protect. 58 (1995) 301.
11. H. Ogawa, K. Fukuhisa, Y. Kubo and H. Fukumoto, Agric. Biol. Chem., 54 (1990) 1219.
12. C. Chen and C. W. Tseng, Process Biochem., 32 (1997) 337.
13. E. Palou, A. López-Malo, G. V. Barbosa-Cánovas, J. Welti-Chanes and B. G. Swanson, Letters in Applied Microbiology, 24 (1997) 417.

Trends in High Pressure Bioscience and Biotechnology
R. Hayashi (editor)
© 2002 Elsevier Science B.V. All rights reserved.

The effect of pressure processing on food quality related enzymes: from kinetic information to process engineering

Ludikhuyze, L., Van Loey, A., Indrawati, Denys, S., and Hendrickx, M.

Laboratory of Food Technology, Department of Food and Microbial Technology,
Katholieke Universiteit Leuven, Kasteelpark Arenberg 22, B-3001 Leuven, Belgium

The present paper will give an overview of the combined effects of pressure and temperature on enzymes related to quality of fruits and vegetables, and this from a quantitative point of view. The enzymes considered include polyphenoloxidase (PPO) which is responsible for enzymatic browning, pectinmethylesterase (PME) which induces cloud loss and consistency changes and lipoxygenase (LOX) which is responsible for development of off-flavors, loss of essential fatty acids and color changes. Complete kinetic characterization of pressure-temperature inactivation of these enzymes will be discussed, as well as the use of integrated kinetic information in process engineering.

1. INTRODUCTION

Throughout the last decade high pressure technology has been shown to offer great potential to the food processing and preservation industry in delivering safe and high quality products. Large-scale industrial implementation of this new technology will be facilitated when a scientific basis to assess quantitatively the impact of high pressure processes on food safety (for 'regulatory approval') and quality (for 'consumer acceptance') becomes available. Quantitative data on the effects of pressure and temperature on safety and quality aspects of foods are indispensable for design and evaluation of optimal high pressure processes, *i.e.* processes resulting in maximal quality retention within the constraints of the required reduction of microbial load and enzyme activity. Indeed it has to be stressed that new technologies should deliver, apart from the promised quality improvement, an equivalent or preferably enhanced level of safety.

2. EXPERIMENTAL SETUP

Pressure-temperature inactivation experiments were performed in a multivessel high pressure equipment, consisting of eight individual thermostated pressure vessels (8 ml), each provided with a thermocouple. This equipment allows to combine pressures up to 1000 MPa with temperatures between −20 and 100°C and is especially designed to perform kinetic experiments because the samples can be submitted simultaneously to the same pressure and temperature for different pre-set times.

Validation and uniformity experiments were performed in a single vessel (0.5 l) warm isostatic press (−20-100°C, 0.1-600 MPa), which is equipped with one pressure sensor and four thermocouples positioned at different heights in the pressure vessel.

After treatment, enzyme activities were determined spectrophotometrically (PPO, LOX), titrimetrically (PME) or polarographically (LOX).

518

3. QUANTITATIVE DATA ANALYSIS

All pressure-temperature inactivation experiments were analyzed using a kinetic approach, *i.e.* firstly enzyme activity was evaluated as a function of time to determine a (inactivation) rate model and secondly the variation of the time-dependent model parameters (inactivation rate constants) with temperature, pressure and other intrinsic/extrinsic factors was assessed. Hitherto, three different models have been used to describe pressure-temperature inactivation of food quality related enzymes: (1) a first order model, (2) a biphasic model which supposes the existence of two enzyme fractions with different pressure and/or temperature stability and both inactivating according to a first order model and (3) a fractional conversion model which is a special case of a first order model taking into account a resistant enzyme fraction that is not affected even after prolonged treatment at the pre-set pressure-temperature condition.

$$A = A_0 \, \exp(-kt) \tag{1}$$

$$A = A_l \, \exp(-k_l t) + A_s \, \exp(-k_s t) \tag{2}$$

$$A = A_{rf} + (A_0 - A_{rf}) \exp(-kt) \tag{3}$$

Temperature and pressure dependence of the inactivation rate constants (k) was expressed as the activation energy (E_a) and activation volume (V_a) according to the Arrhenius (4) and Eyring (5) equation respectively.

$$k = k_{refT} \, \exp\left(\frac{-E_a}{R} \left(\frac{1}{T} - \frac{1}{T_{ref}} \right) \right) \tag{4}$$

$$k = k_{refP} \, \exp\left(\frac{-V_a}{RT} (P - P_{ref}) \right) \tag{5}$$

Based on the complete kinetic data sets for enzyme inactivation pressure-temperature contour diagrams were constructed, *i.e.* two-dimensional diagrams representing combinations of constant pressure and temperature resulting in the same inactivation rate constant. Besides, it was tried to develop mathematical models capable of predicting the combined effect of pressure and temperature on the inactivation rate constant.

4. PRESSURE-TEMPERATURE INACTIVATION KINETICS OF FOOD QUALITY RELATED ENZYMES

Complete kinetic characterization of pressure-temperature inactivation in a broad pressure-temperature domain was accomplished for several food quality related enzymes (table 1). Based on these data, pressure-temperature kinetic diagrams were constructed from which possible antagonistic and synergistic effects could be derived (figure 1 and 2). In general it can be concluded from the kinetic studies that pressure stability of food quality related enzymes is largely dependent on many factors such as the enzyme type, the enzyme origin, the pH of the medium, the temperature, the system complexity and the presence of (de)stabilizing additives. Therefore, enzyme stability rankings have to be interpreted with care and are merely relevant when similar environmental conditions are considered.

Table 1. Kinetic characterization of pressure-temperature inactivation of food quality related enzymes.

Enzyme	Source	P/T area studied	Inactivation kinetics	Ref.
LOX	soybean	0.1-650 MPa, 10-64°C	first order	[1,2]
	soybean	0.1-625 MPa, −15-68°C	first order	[3]
	green bean extract	0.1-650 MPa, −10-55°C	first order	[4]
	green bean *in situ*	0.1-550 MPa, −10-55°C	first order	[4]
	pea extract	0.1-625 MPa, −15-70°C	first order	[4]
	pea *in situ*	0.1-500 MPa, −10-70°C	first order	[4]
PME	orange	0.1-900 MPa, 15-65°C	fractional conversion	[5]
PPO	avocado	0.1-900 MPa, 20-77.5°C	first order	[6]

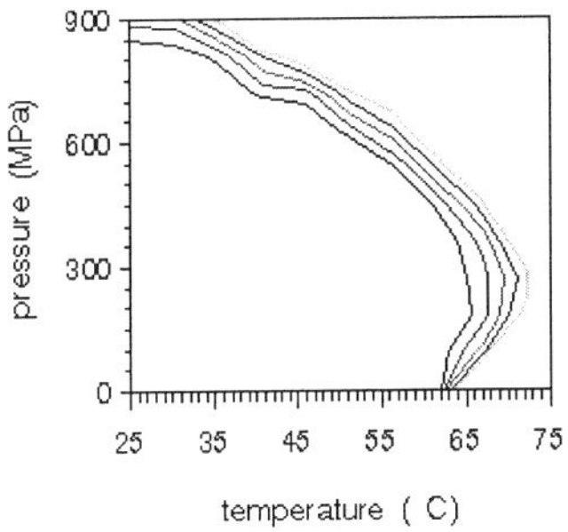

Figure 1. Pressure-temperature kinetic diagram for inactivation of avocado PPO. (inner line: k=0.005 min^{-1}; outer line: k=0.015 min^{-1})

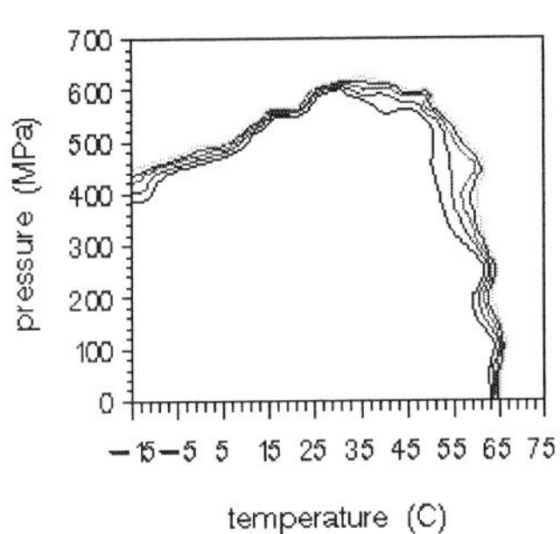

Figure 2. Pressure-temperature kinetic diagram for inactivation of soybean LOX. (inner line: k=0.03 min^{-1}; outer line: k=0.07 min^{-1})

5. MATHEMATICAL MODELING OF PRESSURE-TEMPERATURE INACTIVATION KINETICS

Based on the detailed kinetic data sets it was tried to develop mathematical models capable of predicting the pressure-temperature dependence of the inactivation rate constants. For this purpose, both a thermodynamical strategy starting from a general thermodynamic equation (6) and a pure mathematical strategy starting from the Arrhenius, Eyring or elliptical equation have been evaluated.

$$d(\Delta G) = -\Delta S dT + \Delta V dP \tag{6}$$

In the former case, a modified thermodynamic equation (7) was obtained by taking into account the variation of entropy and volume with temperature and pressure as well as the relation between the Gibbs free energy and the reaction rate constant. In the latter case, pressure or temperature dependent parameters of the Arrhenius (k_{refT}, E_a) and Eyring (k_{refP}, V_a) equation respectively were replaced by empirical equations expressing this dependency. The

520

resulting mathematical models (8-11) are summarized in table 2. The accuracy of these models was evaluated by plotting the experimental k-values versus the k-values calculated according to the model and its concomitant parameters (figure 3).

Table 2. Mathematical models describing pressure-temperature inactivation of food quality related enzymes.

Enzyme	Model		Ref.
All	$$\ln(k) = \ln(k_0) - \frac{\Delta V_0^{\neq}}{RT}(P - P_0) + \frac{\Delta S_0^{\neq}}{RT}(T - T_0) - \frac{1}{2}\frac{\Delta \kappa^{\neq}}{RT}(P - P_0)^2 \\ - \frac{\Delta \zeta^{\neq}}{RT}(P - P_0)(T - T_0) + \frac{\Delta C_p^{\neq}}{RT}\left[T\left(\ln\frac{T}{T_0} - 1\right) + T_0 \right]$$	(7)	[4]
Avocado PPO	$$\ln(k) = (a + bP + cP^2 + dP^3) + \left[\frac{f \exp(-eP)}{R}\left(\frac{1}{T_{ref}} - \frac{1}{T}\right) \right]$$	(8)	[6]
Soybean LOX	$$\ln(k) = (aT^2 + bT + c) + \left[\frac{dT \exp(-eT)}{RT}(P - P_{ref}) \right]$$	(9)	[2]
Green bean LOX	$$\ln(k) = \ln(aT^2 + bT + c) + \left[\frac{d(T - SF)\exp(e(T - SF))}{RT}(P - P_{ref}) \right]$$	(10)	[4]
Orange PME	$$\ln(k) = (a + bP + cP^2 + dP^3) + \left[\frac{f - eP}{R}\left(\frac{1}{T_{ref}} - \frac{1}{T}\right) \right]$$	(11)	[5]

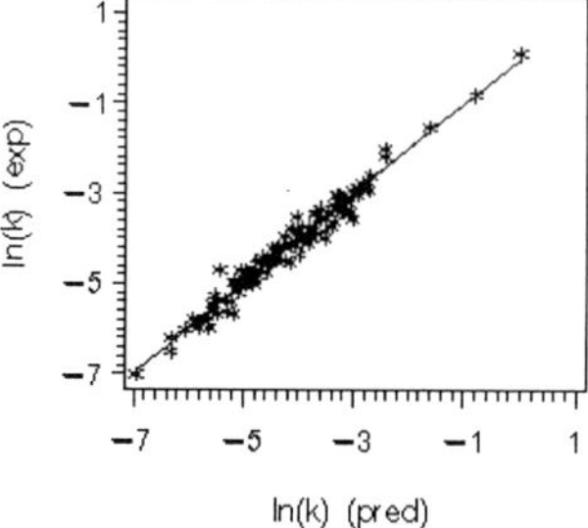

Figure 3. Correlation between experimental k-values for pressure-temperature inactivation of avocado PPO and k-values predicted according to equation (8) and concomitant parameters.

7. FROM KINETIC INFORMATION TO PROCESS ENGINEERING

7.1. Use of pressure-temperature kinetic diagrams for process design and optimization

This paragraph illustrates the usefulness of mathematical models describing the pressure-temperature dependence of the inactivation/degradation rate of food safety/quality attributes in design and optimization of high pressure processes. In combination with the rate model of the attribute considered, it allows to convert the iso-rate diagram in contours of equal degree of inactivation/degradation. As an example, pressure-temperature contours of equal degree of soybean LOX inactivation were simulated by combining a first order inactivation model with the modified thermodynamic equation (7). The pressure-temperature levels needed to achieve different degrees of enzyme activity reduction (*i.e.* 1, 2 or 5 log units) within a fixed process time of 30 minutes are depicted in figure 4. At 500 MPa, the degree of LOX inactivation can be increased from 90% to 99.999% either by decreasing temperature from -1 to -10°C or by increasing it from 59 to 72°C. Similar examples can be elaborated for other processing times and other degrees of inactivation.

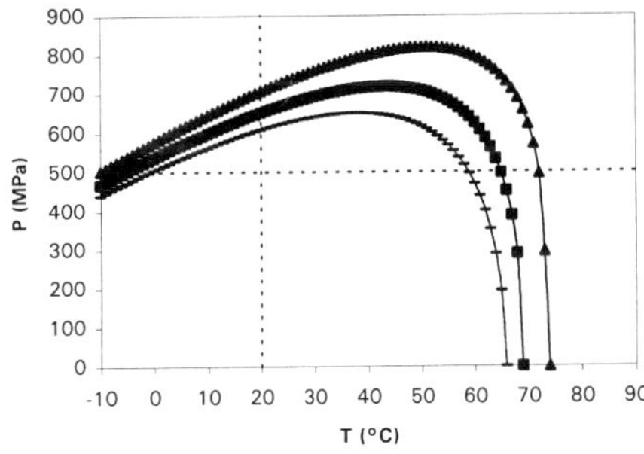

Figure 4. Simulated pressure-temperature combination (equation 7) resulting in different degrees of soybean LOX inactivation in a fixed process time of 30 minutes: 1 (-), 2 (■) and 5 (▲) log unit reductions.

For process optimization, kinetic information on different food safety and quality aspects has to be combined. A theoretical case study on high pressure process optimization was elaborated by combining pressure-temperature kinetic diagrams of the above mentioned quality related enzymes with pressure-temperature kinetic diagrams of microbial inactivation [7-9] and of chlorophyll degradation [10]. For this purpose pressure-temperature combinations resulting in six log-unit reductions of microbial load, in 90% enzyme activity reduction and in 90% loss of total chlorophyll content after a process time of 15 minutes are combined in figure 5. From this figure it can be noticed that for enzymes the shape of the pressure-temperature diagrams does not only depend on the type of enzyme but also on the source. For avocado PPO, orange PME, alkaline phosphatase (ALP) and LOX from peas, a pronounced antagonistic effect of low pressure and high temperature can be observed, *i.e.* application of low pressure counteracts thermal inactivation. Inactivation of soybean LOX on the other hand is characterized by an antagonistic effect of elevated pressure and low temperature (<40°C). Besides it can be seen that the enzyme stability ranking is strongly dependent on the applied pressure and temperature. Thermal stability ranking perceived at atmospheric pressure cannot be extrapolated towards higher pressures. In a similar way, pressure stability ranking largely changes with the process temperature.

In contrast to the enzymes, the shapes of the kinetic diagrams for microbial inactivation are very similar. Besides, it can be seen that in general vegetative microorganisms are less pressure stable as compared to food quality related enzymes, which become, rather than the microorganisms, critical issues in defining optimal high pressure processes.

Moreover it can be seen that at pressure-temperature conditions resulting in sufficient microbial and enzyme inactivation, total chlorophyll content remains almost unchanged. This supports the general statement that nutritional and sensorial quality is only minimally affected by high pressure processing.

522

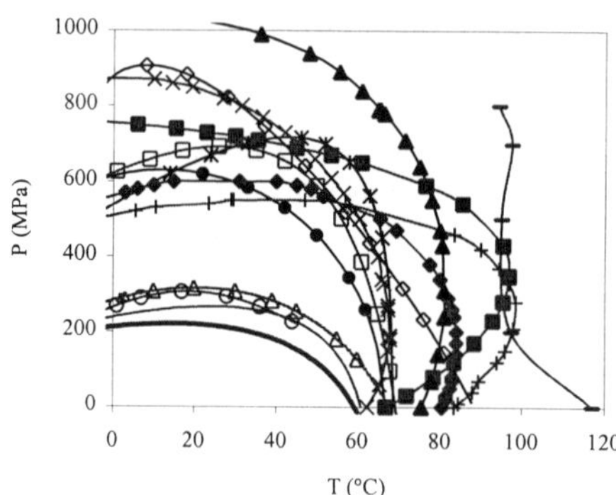

Figure 5. Simulated pressure-temperature combinations (equation 7) resulting in six log-unit reductions of microbial load, 90% reduction of enzyme activity and 90% chlorophyll loss after treatment for 15 minutes: PPO (▲); alkaline phosphatase (■); *Bacillus subtilis* α-amylase (◊); pea LOX *in situ* (+); pea LOX in juice (♦); green bean LOX *in situ* (•); green bean LOX in juice (□); soybean (✳); PME (x), total chlorophyll content (–); yeast (Δ); *Z. bailii* (o), *L. casei* (—), *E. coli* (—).

7.2. Evaluation of process uniformity on the basis of enzyme inactivation kinetics

It is generally admitted that high pressure is set immediately and uniformly throughout the pressure vessel, making the process independent of product size and geometry [11]. However, this statement should be interpreted with care because also heat transfer through the medium and the product should be taken into account. Compression of the workload results in a proportional temperature increase due to adiabatic heating, the extent of which is dependent on thermal and physical properties of the compressed products. After compression, dissipation of heat through the metal wall induces temperature gradients inside the vessel and the product.

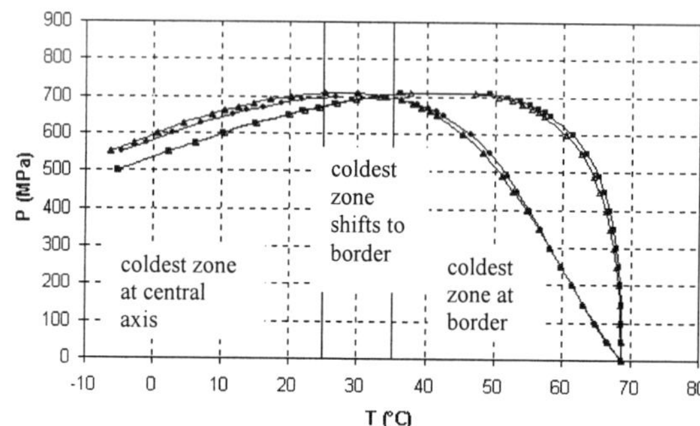

Figure 6. Pressure-temperature combinations resulting in 10% activity retention for soybean LOX in the sample center (closed symbols) and at the sample border (open symbols) after treatment for 15 minutes: (■) isobaric-isothermal; (•) pressurization rate of 50 MPa/min; (▲/Δ) pressurization rate of 200 MPa.

The variation in the pressure-temperature-time profile perceived at different positions in the vessel can result in a spatial distribution of the integrated impact on safety and quality attributes, *i.e.* non-uniform enzyme or microbial inactivation and nutritional or sensorial quality degradation. Design of safe pressure processes therefore has to be based on the pressure-temperature history at the coldest point in the pressure vessel. This coldest zone varies with process conditions (figure 6).

Next to the non-uniformity within one process, non-uniformity between different processes resulting in the same impact on average enzyme activity can be observed [12] and is more or less pronounced depending on the pressure-temperature inactivation kinetics of the attribute that is focused on. Non-uniformity of batch high pressure processing for inactivation of soybean LOX was examined on a theoretical and experimental basis. Hereto, 10 different processes resulting in 50% average activity retention after 25 minutes were simulated and the average activity retention at different positions in the vessel was calculated. A box plot, representing the spatial variation in residual activity, *i.e.* the (non)-uniformity of these processes, is shown in figure 7.

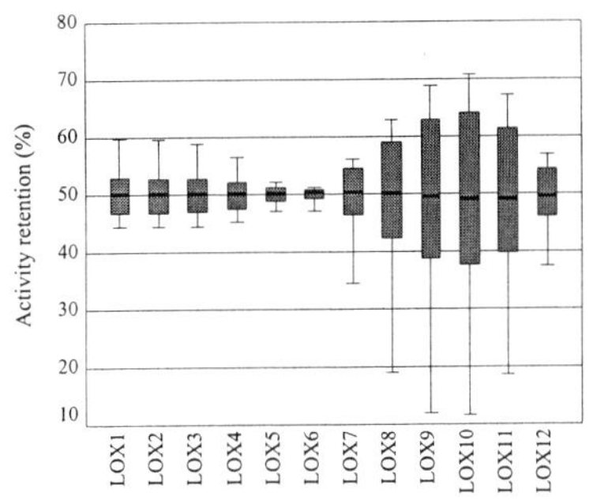

Figure 7. Box-plot representing the (non)-uniformity of LOX inactivation for different processes.

Processes 1 to 6, which were conducted at temperatures below 40°C, are quite uniform whereas processes at higher temperature are characterized by an extreme non-uniformity. This can be explained by the shape of the pressure-temperature contour diagram. At low temperature the inactivation rate constants are almost temperature independent. Processes with holding temperature in this area will consequently be much more uniform with regard to LOX inactivation than processes at higher holding temperature.

Following the theoretical evaluation, some experimental validation studies were carried out by submitting the enzyme system, positioned at different levels in the pressure vessel, to the above mentioned processes and determining the residual enzyme activity [13]. The low correlation between experimentally determined and calculated activity retentions for soybean LOX can be possibly explained by the occurrence of cold denaturation during decompression, a phenomenon that is not accounted for in the inactivation model developed.

CONCLUSIONS

By means of some limited case studies, the potentials of using kinetic models in process design and optimization and in process uniformity studies have been shown. Besides, it should be emphasized that the observed difference in thermostability ranking at atmospheric and elevated pressure, and the difference in pressure stability ranking at different processing temperatures can create possibilities to selectively knock-out certain enzymes by a scientifically based choice of pressure and temperature conditions, and hence can open new and challenging routes for high pressure applications in the food industry.

524

NOTATION

symbols A: enzyme activity; E_a: activation energy (kJ/mol); R: universal gas constant (8.314 J/mol,K); k: inactivation rate constant (min^{-1}); T: temperature (°C or K); P: Pressure (MPa); V_a: activation volume (cm^3/mol); t: time (min); ΔG: Gibbs free energy; ΔV: volume change; ΔS: entropy change; $\Delta\kappa$: compressibility factor (cm^6/J,mol); $\Delta\zeta$: thermal expansibility factor (cm^3/mol,K); ΔCp: heat capacity (J/mol,K); SF: scaling factor; a,b,c,d,e,f: model parameters
sub- and superscripts 0: initial state; l: labile fraction; s: stable fraction; rf: resistant fraction; $refT$: at reference temperature; $refP$: at reference pressure; $\neq$: activated state

REFERENCES

1. Ludikhuyze, L., Indrawati, Van den Broeck, I., Weemaes, C., Hendrickx, M. 1998a. The effect of combined pressure and temperature on soybean lipoxygenase: I. The influence of extrinsic and intrinsic factors on isobaric-isothermal inactivation kinetics. J. Agric. Food Chem. 46: 4074-4080.
2. Ludikhuyze, L., Indrawati, Van den Broeck, I., Weemaes, C., Hendrickx, M. 1998b. The effect of combined pressure and temperature on soybean lipoxygenase: II. Modeling inactivation kinetics under static and dynamic conditions. J. Agric. Food Chem. 46: 4081-4086.
3. Indrawati, Van Loey, A.L., Ludikhuyze, L.R., Hendrickx, M.E. 1999. Soybean lipoxygenase inactivation by pressure at subzero and elevated temperatures. J. Agric. Food Chem. 47: 2468-2474.
4. Indrawati, 2000. Lipoxygenase inactivation by high pressure treatment at subzero and elevated temperatures: a kinetic study. Ph.D. Dissertation, Faculty of Agricultural and Applied Biological Sciences, KULeuven, Leuven, Belgium.
5. Van den Broeck, I., Ludikhuyze, L. R., Van Loey, A. M., Hendrickx, M. E. 2000. Inactivation of orange pectinesterase by combined high pressure and temperature treatments: a kinetic study. J. Agric. Food Chem. 48: 1960-1970.
6. Weemaes, C., Ludikhuyze, L., Van den Broeck, I. and Hendrickx, M. 1998. Kinetics of combined pressure-temperature inactivation of avocado polyphenoloxidase. Biotechnol. Bioeng. 60: 292-300.
7. Hashizume, C., Kimura, K., Hayashi, R. 1995. Kinetic analysis of yeast inactivation by high pressure treatment at low temperatures. Biosci. Biotech. Biochem. 59: 1455-1458.
8. Reyns, K.M., Soontjens, C.C., Cornelis, K., Weemaes, C.A., Hendrickx, M.E., Michiels, C.W. 2000. Kinetic analysis and modeling of combined high pressure-temperature inactivation of the yeast *Zychosaccharomyces bailii*. Int. J. Food Microbiol. 59: 199-210.
9. Sonoike, K., Setoyama, T., Kuma, Y., Kobayashi, S. 1992. Effect of pressure and temperature on the death rates of *Lactobacillus casei* and *Eschericia coli*. In: 'High Pressure and Biotechnology', pp. 297-301. Balny, C., Hayashi, R., Heremans, K. and Masson, P. (Eds), John Libbey Eurotext, Montrouge.
10. Van Loey, A., Ooms, V., Weemaes, C., Van den Broeck, I., Ludikhuyze, L., Indrawati, Denys, S., Hendrickx, M. 1998. Thermal and pressure-temperature degradation of chlorophyll in broccoli (*Brassica oleracea* L. *Italica*) juice: a kinetic study. J. Agric. Food Chem. 46: 5289-5294.
11. Mertens, B. 1995. Hydrostatic pressure treatment of food: equipment and processing. In: 'New methods of food preservation', pp. 135-138. Gould, G.W. (Ed.), Blackie Academic and Professional, Glasgow.
12. Denys, S., Van Loey, A., Hendrickx, M. 2000a. Modeling conductive heat transfer and process uniformity during batch high pressure processing of foods. Biotechnol. Prog. 16: 92-101.
13. Denys, S., Van Loey, A., Hendrickx, M.E. 2000b. A modeling approach for evaluating process uniformity during batch high hydrostatic pressure processing: combination of a numerical heat transfer model and enzyme inactivation kinetics. Innovative Food Science and Emerging Technologies 1: 5-19.

Trends in High Pressure Bioscience and Biotechnology
R. Hayashi (editor)

Effect of pressure, temperature, time and storage on peroxidase and polyphenol oxidase from pineapple

A. Rosenthal[a], D. Ledward[b], A. Defaye[b], S. Gilmour[c] and L. Trinca[d]

[a] Embrapa (Brazilian Agricultural Research Corporation) - Food Technology Center, Av. das Américas 29501 - CEP 23.020-470, Rio de Janeiro-RJ, Brazil

[b] Food Science and Technology Department, University of Reading, Whiteknights, Reading RG6 3AZ, United Kingdom

[c] School of Mathematical Sciences, Queen Mary and Westfield College, London E1 4NS, United Kingdom

[d] Biostatistics Department, IB, University of Sao Paulo State, Botucatu, Brazil.

The individual and combined effect of pressure, temperature, time and post-pressurisation storage on pineapple polyphenol oxidase and peroxidase activity was evaluated in the study. The experiments were carried out according to a 27-basic treatments design using the following levels of the independent variables: temperature: 12, 40 and 60°C; pressure: 0, 100, 300, 500 and 600 MPa; time: 15, 30 and 45 min.; storage: 24 h storage at 5°C or absence of storage after pressure treatment, previously to the activity determination. Enzymatic extract including both peroxidase and polyphenol oxidase was obtained by blending internal part of the fruit with tris buffer (pH=6.5). After being filtrated the extract to be pressurised was divided into small portions in polietilene bags, which were then termosealed. The samples were frozen and kept in a freezer at −18°C and defrosted just before the pressurisation. Polyphenol oxidase activity was determined through an espectofometric method by measuring the absorbance resulted from the residual enzyme activity on catechol, used as substrate. Peroxidase was similarly determined but using in that case hydrogen peroxide as substrate and p-phenylenediamine as acceptor. A second-degree polynomial model was adjusted for each dependent variable polyphenol oxidase activity and peroxidase activity. Due to the restrictions in the aleatorisation procedure, a mixed model was adjusted, i.e., aleatory effects were used for the factor "day in which the experiment was carried out" and fixed effects for the polynomial coefficients. The aleatory effect for the day of treatment relates to the fact that the level of temperature was adjusted and maintained all day long for a set of three treatments carried out in each day of experiments. The statistical analysis for polyphenol oxidase has shown that there was a negative significant effect of storage factor on enzyme activity, i.e., the activity was lower for stored samples. No interaction among the independent variables was detected. Operational time presented a slight negative linear effect i.e., activity dropped linearly with time. Temperature and time did not result in linear and quadratic significant

526

effects on polyphenol oxidase activity. The maximum polyphenol oxidase activity was verified for a temperature of 30.8°C and pressure of 274 MPa. Change in temperature had a higher effect on polyphenol oxidase activity than change in pressure.

In the case of peroxidase, the analysis revealed that both time and storage of enzymatic extract after pressurisation did not affect significantly the enzymatic activity. However, the effect of interaction between temperature and pressure on peroxidase activity was significant. In those terms, for low temperatures the enzyme activity was almost constant with pressure increasing, while for high temperatures the peroxidase activity dropped abruptly with the increase of pressure. Curvature resulting from the quadratic effect was only observed for temperature in the case of peroxidase. The maximum peroxidase activity was verified at the temperature of 36.8°C and absence of operational pressure (0 MPa) inside the pressure vessel. The lowest level of enzymatic activity was achieved at the highest level of quantitative variables for both enzymes. Maximum reduction of enzymatic activities obtained were 60.08% for peroxidase, and 33.17% for polyphenol oxidase, in comparison to the original levels obtained for the enzymatic extract without treatment.

1. INTRODUCTION

High pressure has been regarded as a novel technology for food processing and preservation, which is capable of maintaining nutritional and sensory characteristics of the fresh food. Several applications to food products have been studied and some of them are already being used in food industries mainly in developed countries. Such applications include change in food texture or rheological behaviour and freezing and thawing processing, besides food preservation (1, 2).

Food preservation by high pressure is based on the effect of pressure on spoilage micro-organisms and enzymes. High pressure allows the destruction of most vegetative cells micro-organisms and at higher levels and multiple pressure cycles even the destruction of most bacteria spores (3, 4). In terms of enzymes, high pressure can either activate or inactivate enzymes, depending not only on the enzyme and pressure levels themselves but also on other complementary parameters (pH, temperature, substrate, medium composition) (1, 5).

Polyphenol oxidase (PPO) and peroxidase (POD) are among the main enzymes responsible for enzymatic deterioration of fruit and vegetables products. Both enzymes catalyse oxidation of polyphenol substances naturally present in raw materials, originating products responsible for changing colour, sensory and nutritional characteristics.

Polyphenol oxidases, also known as tyrosinases, are copper containing enzymes which catalyse two types of reactions in a sequential biochemical pathway. The first reaction corresponds to hydroxylation of monophenols to *o*-diphenols and the second to dehydrogenation of *o*-diphenols to *o*-diquinones (6). The resulting *o*-quinones polymerise mutually, or react with proteins or amino acids, originating brown complexes (7).

Peroxidase is considered to be ubiquitous in the plant kingdom. The enzyme has been implicated in a variety of physiological processes (8-11) and is also considered to be related to off-flavours and off-colours in food products (12, 13). Peroxidase is recognised to be one of the most heat stable enzymes in fruit and vegetables. It has a low specificity for hydrogen donor substrate and oxidises aromatic amines, e.g. *o*-dianisidine, and phenols, e.g. guaiacol, as well as other organic compounds, at the expense of hydrogen peroxide (14).

Several studies have focused on the effect of high pressure on polyphenol oxidase and peroxidase activities. However, it seems that so far the reported results have not been completely conclusive (5, 15).

In the case of polyphenol oxidase, although it is considered generally as a thermosensitive enzyme, most reports show it presents a rather high-pressure resistance (7). In this sense, Asaka and Hayashi (16) detected the increase in Bartlett pears polyphenol oxidase under pressure in the range of 200-500 MPa. On the other hand, Knorr (17) found that pressure at 100-400 MPa caused a decrease in the PPO activity. Eshtiaghi et al. (18) have reported a complete inactivation of potato PPO at pH 7 in phosphate buffer at 900 MPa and 45°C for 30 min. Gomes and Ledward (15) verified that while mushroom PPO exhibited a marked increase at a level of 140% in the original activity at 400 MPa for 10 min, potato PPO steadily lost activity with increasing applied pressure. The remaining mushroom and potato PPO activity after 10 min at 800 MPa was 60 % and 40 % of the original non-treated samples, respectively (15).

This study aimed at evaluating the combined effect of main operational parameters – pressure, temperature, time and post-treatment storage – on residual activity of polyphenol oxidase and peroxidase from pineapple.

2. MATERIALS AND METHODS

2.1. High pressure treatment

Pineapple fruits (*Ananas comosus* L.) originated from Ivory Coast were obtained from commercial sources in the United Kingdom. Fruits used in the experiments were selected to be mature and disease free. Enzymatic extract was obtained by mincing and blending 10 g of edible fruit pieces together with 20-25 mL of tris-buffer (pH 6.5). The resulting suspension was filtered and centrifuged (2224 *g* for 10 min at 5°C) and packed in portions of 10 mL in polyethylene bags. The samples were frozen and kept in a freezer at –18°C and defrosted just before the pressurisation. For each experimental design combination three bags containing the enzymatic extracts were inserted in the pressure vessel. High pressure treatments were performed in a lab model Stansted high-pressure rig (Stansted Fluid Power Ltd, Stansted, UK). After completing the treatment, samples were removed from the vessel and kept in the ice until the activity determination.

2.2. Chemical and physical analyses

Moisture, protein, ash, total sugars, reducing sugars, titrable acidity and pH contents were analysed by standard methods according to the AOAC (19).

2.3. Enzymatic activity determination

Peroxidase and polyphenol oxidase activities were assayed spectophotometrically according to Cano et al. (20). In the case of peroxidase aliquots (0.025 mL) of extract and a reaction mixture composed of 2.7 mL 0.05 M sodium phosphate buffer (pH 6.5) with 0.2mL 1% (w/v) p-phenylenediamine as H-donor and 0.1 mL 1.5% (w/v) hydrogen peroxide as oxidant. The oxidation of p-phenylenediamine was measured at 485 nm and 25°C.

Polyphenol oxidase activity was assayed using aliquots (0.075 mL) of extract and 3.0 mL of a solution of 0.07M catechol in 0.05M sodium phosphate buffer (pH 6.5). The reaction was measured with an spectrophotometer at 420 nm at 25°C. The enzyme activities were defined as the change in absorbance/min.

Polyphenol oxidase and peroxidase activities were determined in duplicate for each bag-containing sample. Resulting values from biochemical analyses were average of the results for the three samples obtained from each experimental treatment.

2.4.Experimental design and statistical analysis

The combination of independent variables levels used in the experimental runs are shown in Table 1.

Table 1
Combination of independent variables levels in the 27-run experimental design

Run	Temperature (°C)	Pressure (MPa)	Time (min)	Storage*
1	40	300	30	Yes
2	40	100	45	No
3	40	500	15	No
4	40	300	30	No
5	40	300	45	Yes
6	40	100	15	Yes
7	20	600	30	No
8	20	0	30	No
9	20	300	15	Yes
10	60	300	15	No
11	60	0	30	Yes
12	60	500	45	No
13	20	300	15	No
14	20	100	45	Yes
15	20	600	30	Yes
16	60	300	30	No
17	60	300	15	Yes
18	60	500	45	Yes
19	60	100	15	No
20	60	600	30	Yes
21	60	300	45	No
22	40	300	45	Yes
23	40	600	30	No
24	40	0	30	No
25	20	0	30	Yes
26	20	500	15	Yes
27	20	300	45	No

*24 h storage after treatment

A second-degree polynomial model was adjusted for each dependent variable polyphenol oxidase activity and peroxidase activity. Due to the restrictions in the aleatorisation procedure, a mixed model was adjusted, i.e., aleatory effects were used for the factor "day in which the experiment was carried out" and fixed effects for the polynomial coefficients. The aleatory effect for the day of treatment relates to the fact that the level of temperature was adjusted and maintained all the day long for all consecutive three treatments carried out in each day of experimentation.

3. RESULTS AND DISCUSSION

Results from physico-chemical determination of pineapple fruit are shown in Table 2.

Table 2
Physico-chemical determinations of pineapple fruit

Assay	Mean value
Moisture content (%)	88.80
Total sugars (%)	8.30
Reducing sugars (%)	3.70
Protein (%)	0.07
Titrable acidity (g citric acid/100 g)	0.98
Other constituents (%)*	1.85
pH	3.23

*by difference

Model resulting for PPO was:

$$PPO = K - 0.001351X_1 - 0.0001204X_2 - 0.0008569X_3 - 0.002781X_1^2 - 0.001394X_2^2 \qquad (1)$$

where $K = 0.01405$ for non-stored samples and $K = 0.01199$ for stored samples.

Independent factors presented in the models resulted from mathematical transformation of original variables, as shown in Table 3.

Table 3
Coding for independent variables and mathematical transformation expressions from actual to coded levels used in PPO and POD activities models

MODEL FACTORS	MATHEMATICAL TRANSFORMATION EXPRESSIONS
X1	(Temperature-40)/20
X2	(Pressure-300)/300
X3	(Time-30)/15

530

The effect of pressure and time on polyphenol oxidase activity at a constant temperature of 30°C are shown in Figures 1 and 2, for non-stored and stored samples, respectively.

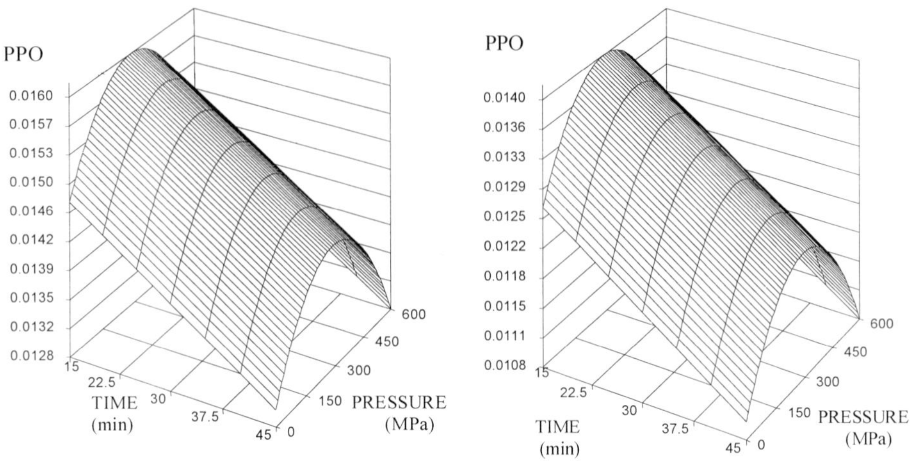

Figure 1. Effect of pressure and time on PPO activity (non-stored sample; 30°C)

Figure 2. Effect of pressure and time on PPO activity (stored sample; 30°C)

The effect of pressure and temperature on polyphenol oxidase activity at a constant time (15 min) is shown in Figures 3 and 4, for non-stored and stored samples, respectively.

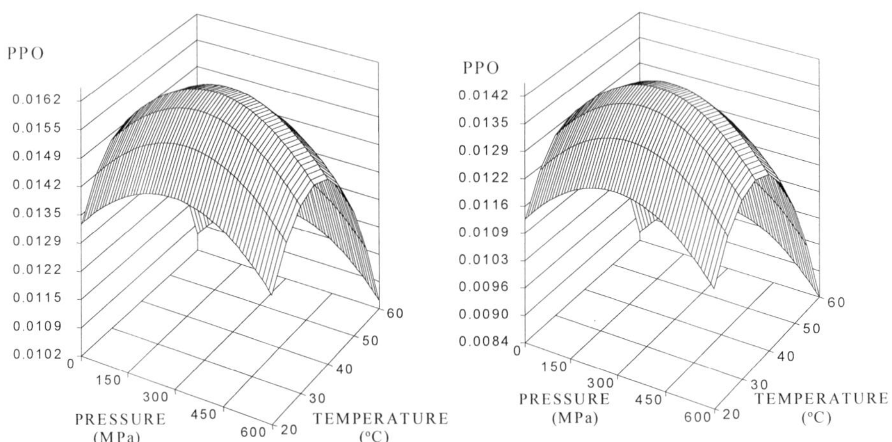

Figure 3. Effect of pressure and temperature on non-stored PPO activity (15 min)

Figure 4. Effect of pressure and temperature on PPO (stored sample; 15 min)

The surface curvatures shown in Figures 3 and 4 indicate the significant quadratic effect of temperature and pressure. However, temperature results in a greater effect on PPO than pressure as it is evident from the Figures.

It is possible to see from PPO model and Figures that although the storage affected the overall PPO activity level, it did not seem to interact with the other factors. Time had a small linear negative effect, i.e. PPO decreased with increasing time. Maximum curvature for PPO was achieved at X1=-0.4858 and X2=-0.08637, i.e. temperature of 30.3 °C and pressure of 274 MPa.

For peroxidase activity the final model fitted was:

$$POD = 0.04290 - 0.004502X_1 - 0.002783X_2 - 0.003605X_1^2 - 0.003343X_1X_2 \qquad (2)$$

Neither storage nor time seemed to have any effect on POD activity. In the same way as verified for PPO, the effect of temperature on POD resulted again as non-linear. On the other hand, the effect of pressure resulted linear for POD activity, but an interaction between pressure with temperature was verified in this case.

Figures 5 presents the combined effect of pressure and temperature on peroxidase activity for non-stored sample at a constant time of 15 min.

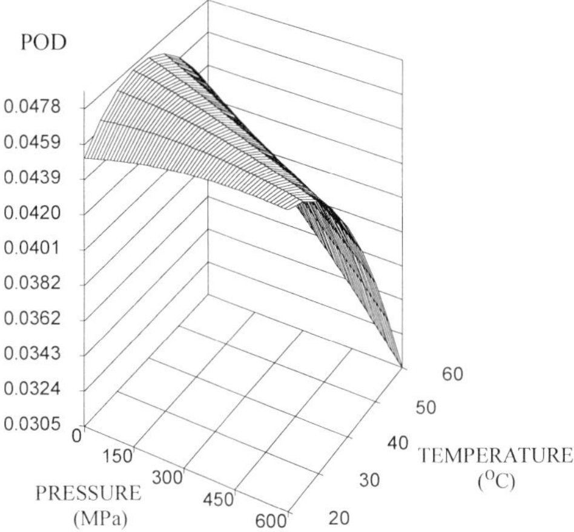

Figure 5. Combined effect of pressure and temperature on pineapple peroxidase activity (non-stored sample; time =15 min.)

The significant interaction between pressure and temperature is evident from Figure 5. In those terms, enzyme activity was almost constant with pressure increasing, while for high temperatures the peroxidase activity dropped abruptly with the increase of pressure. The significant quadratic effect for temperature was also evident from the surface curvature, which was not observed for pressure. Maximum POD activity was verified at X1=-0.1607 and X2=-1, i.e. when the temperature is 36.8 and the pressure is 0.

4. CONCLUSION

The lowest level of enzymatic activity was verified at the highest levels of independent variables for both enzymes. Maximum reduction achieved in enzymatic activities were 60.08% for peroxidase, and 33.17% for polyphenol oxidase, in comparison to the original levels obtained for the enzymatic extract without treatment. Further studies are required to kinetically characterise inactivation of pineapple enzymes and to elucidate the effect of other factors such as pH on the residual enzymatic activity.

REFERENCES

1. J.C. Cheftel, Food Sci. Technol. Int., 1 (1995) 75.
2. W. Messens, J. Van Camp and H. Huyghebaert, Trends Food Sci. Technol., 8 (1997) 107.
3. R. Meyer, K. L. Cooper, D. Knorr and H.L.M. Lelieveld, Food Technol., 54:11 (2000) 67.
4. J.P.P.M. Smelt, Trends Food Sci. Technol., 9 (1998) 152.
5. M. Hendrickx, L. Ludikhuyze, I. Van den Broeck and C. Weemas, Trends Food Sci. Technol., 9 (1998) 197.
6. J.R.Whitaker, Polyphenol oxidase, In: Principles of Enzymology for Food Science, O.R. Fennema (ed.), Marcel Decker, New York, 1972.
7. C. Weemas, L. Ludikhuyze, I. Van den Broeck, S. De Cordt, M. Hendrickx and P. Tobback, In: High Pressure Research in the Biosciences and Biotechnology, K. Heremans (ed.), Leuven University Press, Leuven, Belgium, 1997.
8. F.B. Abeles and C.L. Biles, Plant Physiol., 95 (1991) 269.
9. T.J. Miesle, A. Proctor and L.M. Lagrimini, J. Am. Soc. Hort. Sci., 116 (1991) 827.
10. C. Rothan and J. Nicolas, Hort. Sci., 24 (1989) 340.
11. E. Silva, E.J. Lourenço and V.A. Neves., Phytochem., 29 (1990) 1051.
12. B.S. Chang, K.H. Park and D.B.Lund, J. Food Sci., 38 (1973) 40.
13. L. Vámos-Vigyazo, C.R.C.-Crit. Rev. Food Sci. Nutr., 15 (1981) 49.
14. K. M. McLellan and D.S. Robinson, Food Chem., 23 (1987) 305.
15. M.R.A. Gomes and D.A. Ledward, Food Chem., 56 (1996) 1.
16. M. Asaka and R. Hayashi, Agric. Biol. Chem., 55 (1991) 2439.
17. D. Knorr, Food Technol., 47 (1993) 156.
18. M. N. Esthiaghi, R. Stute and D. Knorr, J. Food Sci., 59 (1994) 1168.
19. A.O.A.C. Official Methods of Analysis, 15th ed. Association of Official Analytical Chemists, Washington D.C, 1990.
20. M. P. Cano, A. Hernandez and B. de Ancos, J. Food Sci., 62 (1997) 85.

Trends in High Pressure Bioscience and Biotechnology
R. Hayashi (editor)
© 2002 Elsevier Science B.V. All rights reserved.

Numerical Simulation of Thermofluiddynamics and Enzyme Inactivation in a Fluid Food System under High Hydrostatic Pressure

Chr. Hartmann and A. Delgado

Technische Universität München, Chair of Fluid Mechanics and Process Automation, Weihenstephaner Steig 23, D-85350 Freising, Germany

In the present contribution a high hydrostatic pressure (HHP) treatment of a fluid food system is analysed by means of numerical simulation. The considered process is subdivided into a phase of pressure increase from ambient pressure to 500 MPa and a phase of constant pressure application. For both phases temperature and fluid velocity distributions are studied. In addition, results on the simulation of an enzyme inactivation are shown where the inactivation is described by a model equation which is coupled with the equations of fluid dynamics. It can be shown that the inactivation is heterogeneous when the viscosity of the matrix fluid is sufficiently high.

1. INTRODUCTION

Depending on the nature of food, the application of high hydrostatic pressure (HHP) to food systems is carried out by volume reduction of either the liquid food itself or of both the pressure fluid and the packed food placed in a pressure chamber.

The volume reduction changes the thermodynamic and fluid-dynamic state of the food system. A fluid flow is generated by forced convection due to the reduction of the volume. Furthermore, the compression generates an increase of temperature. Heat transfer within the fluid and heat exchange with the walls of the fluid volume introduce temperature gradients. As a consequence, density differences occur and lead to free convection of the fluid. The fluid motion generated by forced and free convection strongly influences the temporal and spatial distribution of the temperature as already could be observed by experimental techniques [1].

The consequences of these inhomogeneous process conditions act on both the technological parameters of the HHP-process and the food system itself.

Technologically significant physical properties such as viscosity and specific heat capacity vary under high pressure conditions as functions of pressure and of temperature [2, 3]. This fact has to be taken into account in order to properly

534

describe the fluid-dynamic and thermodynamic processes. The food system is affected by the resulting process inhomogeneity.

Thus, it is necessary to get insight into the phenomena of heat and mass transfer during the HHP treatment of foods in order to enhance product quality (e.g. microbial safety) and to improve the control of the process with the aim of achieving a high degree of process uniformity.

2. OBJECTIVES

In order to analyse the fluid-dynamics and thermodynamics of the HHP treatment of food, the following processes are considered:

A laboratory-scale high-pressure cell is filled with 4 ml pure water representing a model food system. The water is compressed starting at ambient pressure up to a maximum pressure of 500 MPa in 25 seconds. When the maximum pressure is reached the volume is conserved during 200 seconds. The representation and the analysis of the temporal and spatial distribution of the fluid velocity and the temperature within the volume are the principal aims of this investigation. Typical time scales of temperature compensation and hydrodynamic decay are going to be determined additionally.

Another process on a larger geometrical scale (0.8 l volume) is analysed in order to provide results on the coupling between thermo-fluiddynamical processes and a HHP-transformation. The HHP-treatment represents an enzyme inactivation and consists of a compression phase to 500 MPa within 25 seconds and a holding phase at 500 MPa during 20 minutes. The activity retention is analysed for different process parameters.

3. METHODS

The simulations of both processes are based on the numerical solution of the governing equations of fluid-dynamics: the Navier-Stokes equation, the continuity equation and the energy equation. Furthermore, the equation of state for water valid at high pressures is included in the set of equations as well as the pressure- and temperature-dependency of the specific heat capacity. Both functions have been taken from [3]. The dependency of the dynamic viscosity on pressure and temperature has been included based on data from [2] and [4]. An estimation yields a Reynolds number of about ten. Therefore the flow can be assumed to be laminar. Although the considered liquid is compressible, the bulk viscosity is supposed to be zero in the present work.

The enzyme inactivation is taken into account using a passive scalar partial differential equation with a sink term representing a first order kinetics. The kinetic constant as a function of pressure and temperature is taken from [5] for *Bacillus subtilis* α-amylase (BSA).

The geometry of the fluid volume of a laboratory scale high pressure cell is digitized by application of CAD-techniques. It represents a volume (4 ml) of two interpenetrating drill-holes (14 mm diameter) which is connected to a thin inlet vane (1.6 mm diameter, 60 mm length, figure 1). Different computational grids are used for the numerical solution of the governing equations in order to provide a mesh independent solution.

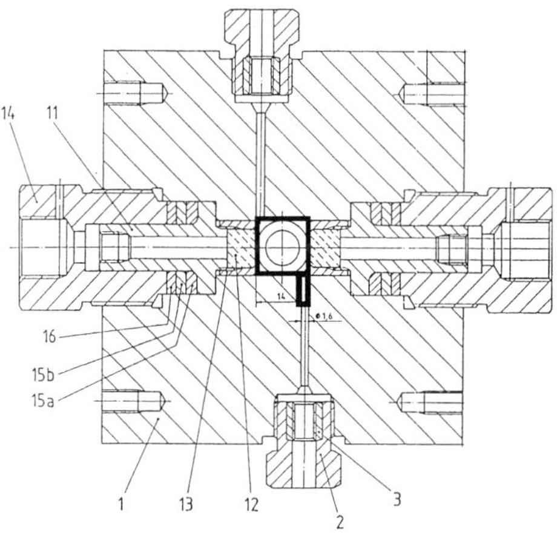

Figure 1: Geometry of the high pressure cell

At the inlet, a normal fluid velocity is prescribed. Depending on the magnitude of this velocity, different pressure ramps can be carried out in the simulation. The temperature of the incoming fluid is supposed to be constant. In practice, this assumption means that the fluid goes through a temper bath before entering the high pressure device.

At the remaining surfaces representing the walls of the steel housing the kinematic boundary condition requires zero fluid velocity, which implies that the fluid is attached to the wall. As thermal boundary condition a constant wall temperature is applied. This setting allows heat transfer across the walls. The steel vessel is considered as an infinite reservoir of heat with an ideal heat conductivity such that the heat is rapidly distributed and temperature gradients in the steel vanish immediately. In reality, the wall temperature is variable in time and might vary over the wall surface. However, a complete analysis of the heat transfer between the pressurized fluid and the surrounding steel vessel is considered as a future work and not subject to the present investigation.

4. RESULTS AND DISCUSSION

A HHP process has been simulated by choosing a specific inlet velocity corresponding to the compressibility of the fluid. The pressure, which is spatially constant up to negligible acoustic variations within the fluid volume is determined numerically during the simulated process. As can be seen from figure

536

2, the ramp represents a slightly non-linear function of time. The pressure level of 505 MPa is reached within 24.5 seconds. The phase of pressure increase is followed by the phase of pressure holding. During this phase no more fluid enters into the high pressure cell. The pressure level is slightly decreasing with time, which is due to the heat removal from the pressurized fluid.

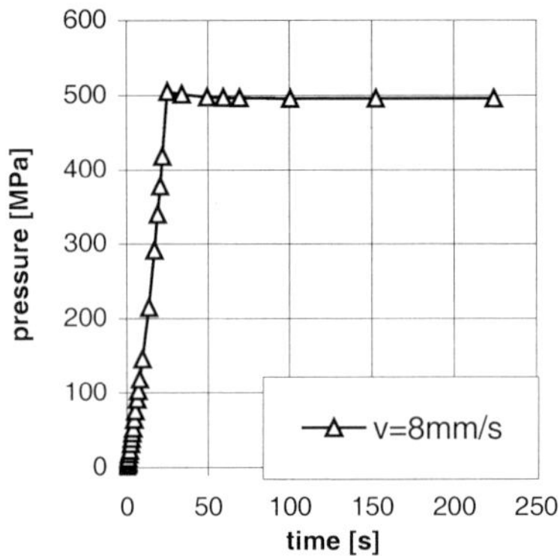

Figure 2: Pressure over time

Both temperature and velocity fields are unsteady during the phase of pressure increase and the phase of pressure holding.

In figure 3, the temperature field and the velocity distribution are shown at t = 24.5 s (left) and t = 30.0 s (right). The presented cross section corresponds to the marked area in figure 1. The shaded distribution corresponds to the distribution of temperature in the given range. The velocity information is shown as a vector plot in a range between 0 mm/s and 15 mm/s.

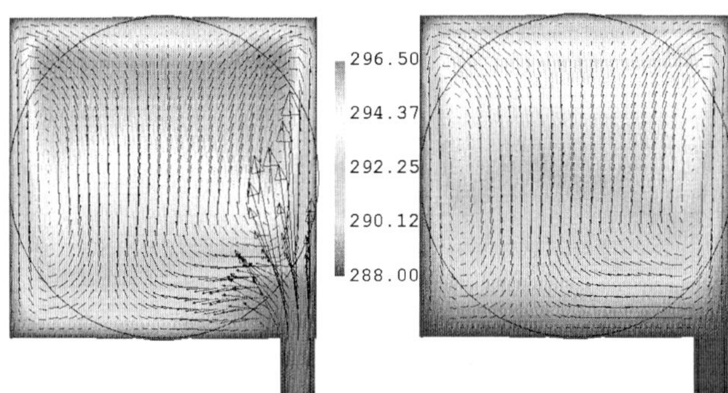

Figure 3: Temperature and velocity distribution in a section through the high pressure cell with 8 mm/s inflow velocity at t = 24.5 s (left) and t = 30 s (right)

At t = 24.5 s, the pressure reaches its maximum of 505 MPa. The temperature covers a range between 288 K and 296.3 K throughout the volume. In the lower part the fluid is heated up due to compression and rises into the upper part of the high pressure cell. There, the maximum temperature is reached. The fluid then moves to the walls. Heat removal at the walls leads to a cooling of the fluid and drives the downward motion which interferes with the motion of the incoming upward directed flow on the right. The maximum velocity at this stage is 14 mm/s. The characteristic velocity of the vortex motion is about 2 mm/s. At this point, it becomes obvious, that the fluid velocity distribution influences strongly the temperature distribution and vice versa.

Micro-organisms or other pressure- and temperature sensitive-substances (e.g. proteins) trapped in this vortical motion are exposed to a periodic thermal treatment where the temperature variation is at a maximum of about $\Delta T = 6$ K.

This instant represents the termination of pressure increase and the begin of the pressure holding phase. Thus, no more fluid is pumped into the high pressure volume. The inlet velocity is set to zero.

At t=30 s (figure3, right), the fluid field is now driven by free convection only. The maximum velocity therefore reduces to 2 mm/s. The vortex motion is of slightly decreasing intensity. The maximum temperature decreases also to about 294.5 K.

The heat removal in the long and thin inlet vane is quite efficient. Therefore, the fluid in the inlet vane remains at an approximately constant temperature of 288 K. A micro-organism transported by the liquid would be exposed to high pressure but constant temperature in the inlet vane until it reaches the HHP-volume. There, it will be exposed to transient thermal conditions. The pressure-temperature history of this individual micro-organisms will clearly differ from that of another micro-organism situated in the centre of the volume from the beginning of the HPP-treatment.

During the phase of pressure holding, the pressure decreases slightly from its maximum of 505 MPa to 495 MPa. The temperature decreases to 288 K due to heat removal. This temperature level corresponds to that of the initial distribution. This stage can be characterized as an equilibrium state from the point of view of thermodynamics and fluid-dynamics. Irreversible pressure and temperature sensitive conversions like the denaturation of proteins or the inactivation of micro-organisms which have taken place during the non-equilibrium phase result in a non-uniformity that subsists throughout the complete HHP-application and represents an uncertainty in the quality of the HHP-process.

In order to investigate, how momentum exchange and heat transfer conditions can influence the result of a HHP-application, a different configuration is analysed. Therefore, the geometry of a cylindrical container with a central, axially oriented inlet vane is considered. The volume of the container is 0.8 litres. The matrix fluid contains BSA which is inactivated by application of HHP. The

538

solution is compressed within 25 seconds to 500 MPa. The pressure is maintained throughout 20 minutes. The initial temperature is T = 313 K. The temperature on the walls of the container is kept constant at T = 313 K throughout the process.

The result of the HHP-treatment is shown in figure 4. The retention of the initial enzyme activity is shown as a shaded distribution in a symmetry plane of the container. The left viewgraph shows the activity field for the case that water is used as matrix fluid. It can be seen, that the activity retention is equally distributed at about 27.5 % throughout the considered volume.

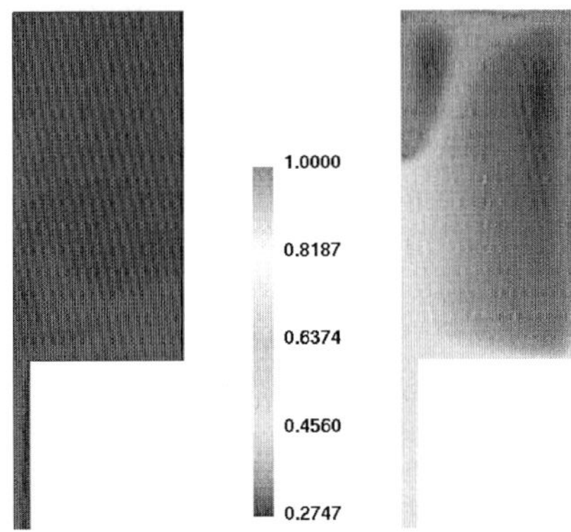

Figure 4: Activity retention of BSA after 20 minutes at 500 MPa in a 0.8 litre-volume; left: $\eta = \eta_{H2O}$, right: $\eta = 100 \times \eta_{H2O}$

The right viewgraph represents the result obtained for a matrix fluid, which of the viscosity is hundred times higher than that of water. Here, the situation is substantially different. The activity retention varies in a range between 28 % and 48 %. The process leads to a non-uniform result. In the volume, the inhomogeneous distribution results from the transport of the dissolved enzyme through non-isothermal conditions due to natural and forced convection. The inlet vane is a region of low temperature due to efficient heat removal from the small volume. As a consequence, the inactivation is less intensive and the activity retention remains on a higher level but is homogeneously distributed.

The viscosity influences the time scales of momentum exchange and, therefore, heat transfer. For the watery solution the hydrodynamic and thermal decay time scales in the considered volume are much lower than the typical inactivation time scale. This means, that temperature gradients are compensated before non-uniform thermal conditions can influence the HHP-transformation.

A higher viscosity, as it can be found in many situations in food technology (e.g. sucrose solution), leads to larger decay time scales for both heat and momentum transfer. In the observed case, they are of the same order of magnitude as the typical inactivation time scale. This is in excellent agreement with the results of Pehl *et al.* shown in the present book. Therefore, the inhomogeneous thermal conditions influence significantly the inactivation. Since thermal and hydrodynamic decay time scales increase with increasing volume size, the macroscopic transport properties will influence microscopic high pressure transformations especially on industrial scale volumes.

5. CONCLUSION AND PERSPECTIVES

In the present study, the temporal evolution of temperature and velocity fields of different food related substances during application of high hydrostatic pressure is analysed by the means of computational fluid-dynamics.

In the first part an analysis on a 4 ml-lab-scale-volume is presented. Water is pressurized up to 500 MPa. For a pressure increase of 500 MPa in 25 seconds, the maximum instantaneous temperature variation in the fluid volume is about 8 K. It can be observed that due to a vortex motion, pressure and temperature sensitive substances or micro-organisms trapped inside the vortex undergo a periodic temperature treatment with a variation of 6 K. The typical time scale for thermal decay is about 70 seconds and for hydrodynamic decay about 200 seconds.

The inactivation of *Bacillus subtilis* α-amylase (BSA) in a (different) 0.8 l-volume has been simulated for two different matrix fluids. A watery solution leads to a uniform process behaviour. The activity retention of about 27.5 % is equally distributed throughout the volume. The use of a matrix fluid which of the viscosity is a hundred times larger than that of water leads to a strong non-uniformity in the process result. The activity retention for this case varies between 28 % and 48 %. This effect is due to the change in the time scales of hydrodynamic and thermal decay. Since for an increasing viscosity the thermal decay time scale is increasing as well, it can reach the order of magnitude of the typical inactivation time scale.

The future work will basically address three subjects. At first, the formulation of the thermal boundary conditions should be improved in order to take into account the local temperature changes of the housing. Furthermore, the inactivation of micro-organisms is going to be modelled by different kinetic models. Finally, the analyses are going to be generalized by means of dimensional analysis.

REFERENCES

[1] Pehl, M., Werner, F., Delgado, A.: First Visualisation of Temperature Fields in Liquids at High Pressure Using Thermochromic Liquid Cristals, Experiments in Fluids, 29/3, 2000

[2] Först, P., Werner, F., Delgado, A.: The viscosity of water at high pressures - especially at subzero degrees centigrade, accepted in Rheologica Acta

[3] Saul, A.; Wagner, W.: A fundamental equation for water covering the range from the melting line to 1273 K at pressures up to 25000 MPa. J. Phys. Chem. Ref. Data 18 (1989), 1537-1564.

[4] Watson JTR, Basu RS, Sengers JV (1980) An improved representative equation for the dynamic viscosity of water substance. J. Phys. Chem. Ref. Data 9:1255-1290.

[5] Denys, S., Van Loey, A. M., Hendrickx, M. E.: A modeling approach for evaluating process uniformity during batch high hydrostatic pressure processing: combination of a numerical heat transfer model and enzyme inactivation kinetics, Innovative Food Science & Emerging Technologies 1 (2000), 5-19

Trends in High Pressure Bioscience and Biotechnology
R. Hayashi (editor)
                                541

EFFECT OF HIGH PRESSURE ON FOOD ENZYME ACTIVITIES: BEHAVIOR OF CATHEPSIN D

S. Jung, N. Chapleau, M. Ghoul & M. de Lamballerie-Anton

ENITIAA GEPEA - BP 82225 - 44322 Nantes Cedex 3 - FRANCE
e-mail : anton@enitiaa-nantes.fr

ABSTRACT

During the ageing (acquisition of beef meat tenderness) several enzymatic systems are involved. Among them, the lysosomal enzymes, *i.e.* cathepsins and specially the cathepsin D, are responsible for the myofibrillar proteins alterations.

Previous results have shown, after high pressure treatment of *post rigor* beef meat, an important increase of the toughness of meat, in spite of an increase of the cathepsin D activity was also observed.

The aim of this work was to determine if the application of a high pressure treatment modified the reaction between cathepsin D and its natural substrate (myofibrils).

The pressurization of cathepsin D does not modify its active site as the electrophoretic patterns and the amount of myofibrillar solubilised proteins were similar to the control. The HP treatment does not perturb the recognition between the enzyme and its natural substrate.

Keywords : Myofibril, high pressure, cathepsin D, meat, ageing, enzymatic activity.

1. INTRODUCTION

The modulation of the enzymatic activities is an important research area for high pressure development in food industries. This process has been presented as a tool to improve the quality of food products from a nutritional, sensorial and microbiological point of view. Furthermore, this process has highlighted the possibilities of inactivating the enzyme responsible for food deterioration or, on the contrary of increasing certain enzymatic activities in order to improve food quality.

During the acquisition of beef meat tenderness (ageing), which requires a refrigerated storage of around one week, several enzymatic systems are involved. Among them, the lysosomal enzymes, *i.e.* cathepsins and specially the cathepsin D, are responsible for the myofibrillar proteins alterations.

The treatment of meat by high pressure makes it possible to increase the lysosomal enzyme activities. This modification could improve meat tenderness or also accelerate ageing. This increase has been related to the breakdown of the lysosomes in which these enzymes are enclosed. On the one hand, as high pressure treatment could activate the enzymatic system, it is also conceivable that the catheptic activities increase. The mechanism occurring during the processing of meat is not yet fully understood and therefore the study of the purified enzyme is useful.

Our previous results (1, 2) have shown after high pressure treatment of *post rigor* beef meat an important increase of the toughness of meat, in spite of an increase of the catheptic activity was also observed.

The aim of this work was to accurately determine the effects of pressurized meat or commercial cathepsin D towards the myofibrillar proteins which are its natural substrate, in order to determine if the treatment could modify the behavior of this enzyme.

2. MATERIALS & METHODS

2.1. Sample source
Bovines muscles (*Biceps femoris*) obtained from a commercial abattoir (SOVIBA, Le Lion d'Angers France) 24 h *post mortem* were stored at + 4°C before the extraction of myofibrils and enzymatic extract.

2.2. Myofibril extraction
Myofibrils were extracted using the method of Busch et al. (3). Minced meat was homogenised in a Waring blendor with six volumes of extraction buffer (20mM Tris-HCl, 100 mM KCl, 5 mM EDTA, pH 7.6,). After centrifugation at 1000 g for 10 min, the pellets were resuspended in the extraction buffer and the same operation was carried out five times. The connective tissue was then eliminated by mean of o filtration trough a 20 mesh nylon net. The final myofibril sample was homogenised in 0.1 M KCl, 1mM NaN3.

2.3. Preparation of enzymatic extract
Commercial cathepsin D was purchased from Sigma (E.C. 3.4.23.5).

Meat cathepsin D extract was prepared according to Homma *et al* (4). Minced meat was homogenised in distilled water (1:2, w:v) with an Ultra Turrax. The extract was centrifuged for 15 min at 10 000 g. The supernatant was filtered and dialysed against distilled water. The extract was then centrifuged at 12 000 g for 20 min and the precipitate was discarded.

2.4. High pressure treatment
Enzymatic sample was pressurised at 10°C in a reactor of 1.5 litres from Alstom Fluides et Mécanique (Nantes, France). The high pressure conditions were : 130 MPa, 260 s.

2.5. Enzymatic reaction
The incubation of myofibrils with enzymatic extract was conducted at 37°C during 6h.

The extract was then centrifuged at 5000 g for 10 min.

The supernatant was submitted to protein content determination and SDS PAGE, and the sedimental fraction to SDS PAGE.

2.6. Protein content and SDS PAGE
The protein concentration was determined following the bicinchoninic acid method (Sigma procedure N° TPRO-562) calibrated with BSA.

The electrophoresis was conducted in a separation gel of 12.5% of acrylamide, with a ratio bisacrylamide : acrylamide of 1:37.

3. RESULTS & DISCUSSION

Table 1
Amount of solubilised proteins at t = 6h (% of initial protein content)

	Purified cathepsin D	Crude meat extract
Control sample	6.0 %	3.7 %
Pressurised sample	5.3 %	9.1 %
	Not statistically different	*Statistically different*

Results show that the amount of solubilised proteins from myofibrils is the same whatever the treatment of purified cathepsin D, because the amounts of solubilised proteins from myofibrils are not statistically different. Then pressurised purified cathepsin D is as active as native cathepsin D.

The table 1 shows that the amount of solubilized proteins from myofibrils is higher with pressurized crude extract than with control crude extract : there is a more important degradation of the myofibrillar structure when the extract has been pressurized.

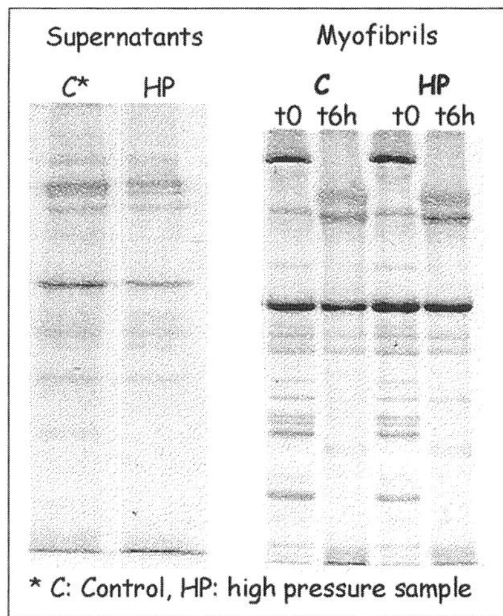

Figure 1. SDS PAGE of supernatants and myofibrils after incubation with commercial cathepsin D

544

SDS PAGE shows on the figure 1 than the pattern of the sedimental fraction of myofibrils is modified by commercial cathepsin D. In the 6h sample, the myosin heavy chain is almost off, the actin band is slightly reduced, and the migration front containing small size proteolytic fragments is enhanced.

There is no difference between the control and the pressurized samples electrophoretic patterns : the pressurised enzyme is able to proteolyse myofibrillar proteins; then the high pressure treatment of the enzyme does not prevent this enzyme to recognize its substrate.

4. CONCLUSION

The pressurization of cathepsin D does not modify its active site as the electrophoretic patterns and the amount of myofibrillar solubilized proteins were similar to the control. The high pressure treatment does not perturb the recognition between the enzyme and its natural substrate.

Further investigations are necessary to understand the complex mechanisms occuring during pressurization of meat : we already wonder why the proteolytic activity is enhanced and toughness is increased, whereas pressurized enzyme is able to act.

ACKNOWLEDGEMENT

This work was supported by the Region Pays de la Loire. We gratefully acknowledge Sylviane Delépine for her technical assistance.

REFERENCES

(1) Jung S., de Lamballerie-Anton M., Taylor R.G. & Ghoul M. (2000a). High pressure effects on lysosome integrity and lysosomal enzyme activity in bovine muscle. *J. Agric. Food Chem.* 48 2467-2471.

(2) Jung S., Ghoul M. & de Lamballerie-Anton M. (2000b). Changes in lysosomal enzyme activities and shear values of high pressure treated meat during ageing. *Meat Sci.* 56 3 239-246.

(3) Busch W.A., Stromer M.H., Goll D.E. & Suzuki A. (1972). Ca^{2+}-specific removal of Z lines from rabbit skeletal muscle. *J. Cell Biol.*, 52, 367-381.

(4) Homma N., Ikeuchi Y. & Suzuki A. (1994). Effects of high pressure treatment on the proteolytic enzymes in meat. *Meat Sci.*, 38, 219-228.

Trends in High Pressure Bioscience and Biotechnology
R. Hayashi (editor)
© 2002 Elsevier Science B.V. All rights reserved.

Scanning Electron Microscopic Study of High Pressure Induced Microstructural Changes of Proteins in Turkey and Pork Meat

M. Scheibenzuber [a], W. Ruß [a], A. Görg [b], R. Meyer-Pittroff [a]

[a] Lehrstuhl für Energie- und Umwelttechnik der Lebensmittelindustrie, 85350 Freising / Weihenstephan, Germany
Phone +49 8161 715256, Fax +49 8161 714415
e-Mail sche@eul.blm.tu-muenchen.de

[b] Fachgebiet Proteomik, 85350 Freising / Weihenstephan, Germany
Phone +49 8161 714265, Fax +49 8161 714264
e-Mail angelika.gorg@lrz.tum.de

Pressure treatment of turkey and pork meat leads to significant alterations in the microstructure. SEM showed that high pressure induces different changes due to denaturing mechanisms. Micrographs of turkey and pork meat depicted denaturative modifications of actin/myosin complexes in the myofibrills and of sarcoplasmic proteins.

1. INTRODUCTION

Meat and meat products are often the subject of food-technological investigations concerning application of high pressure (HP). The sensory attributes of meat show various modifications after treatment with high hydrostatic pressures.
In our study we were able to show that changes already documented, for example solidification of the texture and change of colour, depend very much on the modification of the proteins. In contrast to other techniques, scanning electron microscopy (SEM) of meat samples prepared by the freeze/break method is very suitable for studying the microstructure as well as the HP-induced modifications of meat. SEM provides an excellent method to investigate samples, which by fixation in liquid nitrogen become highly vacuum tolerant in their native structure.

546

2. MATERIALS & METHODS

For SEM examination, turkey and pork meat was treated at 100, 200, 300, 400, 500 and 600 MPa. Untreated meat and meat heated to 100 °C (turkey) served as controls. The pressure preservation time was 10 minutes. Increasing and decreasing rate of pressure amounted to 200 MPa/min, and the processing temperature was 20 °C.
In this study an SEM (type S 360 / 1989, Cambridge / England) with a sputtering device (Polaron) and a cryo preparation chamber (Bio-Rad) was used. Pieces of meat were fixed on a sample plate and immediately chilled in liquid nitrogen to inhibit formation of ice crystals. The samples were then placed in the vacuum chamber of the SEM and kept at − 160 °C. After sublimating the ice and covering with a thin gold layer, a break surface of each sample was examined.
Soluble proteins received by pressing out the meat samples after HP treatment were analysed by SDS-Electrophoresis (12,5 % polyacrylamide gel) after Coomassie blue staining.

3. RESULTS

Fig. 1 shows the inner structure of a turkey meat muscle fibre after breaking it horizontally in its longitudinal direction. Areas separated more or less clearly by gaps are single myofibrills forming the contractile filaments of the muscle. *In vivo* these gaps are filled with sarcoplasm. Due to sublimation during the preparation process, the sarcoplasm is eliminated from the surface and myofibrills are now clearly visible. Myofibrills containing mostly actin and myosin proteins have a diameter in a range from 1,5 µm to 2,5 µm. The two proteins named above form actin/myosin complexes. After HP treatment at 100 MPa, characteristic changes can be found (Fig. 2).

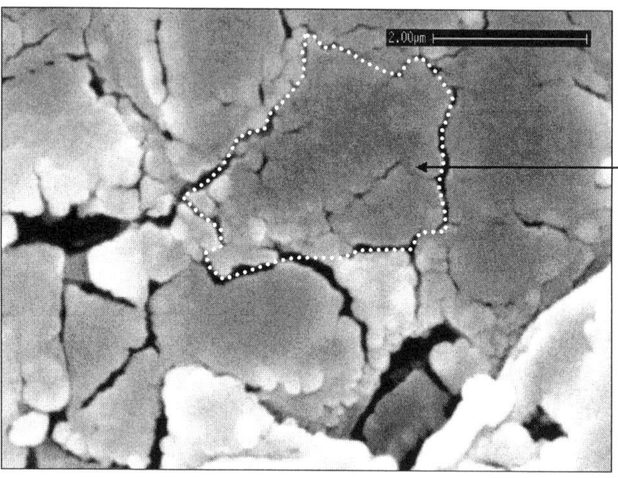

Figure 1. Native turkey meat

Figure 2. Turkey meat 100 MPa

Up to 300 MPa, increasing swelling of acto/myosin leading to spherical, hump-like structures was observed (Fig. 3). These structural changes lead to a moderate solidification of meat texture. Similar results confirming our data have already been published by Yamamoto et al. [1]. They documented an aggregation of pure dissolved actomyosin at 250 MPa.

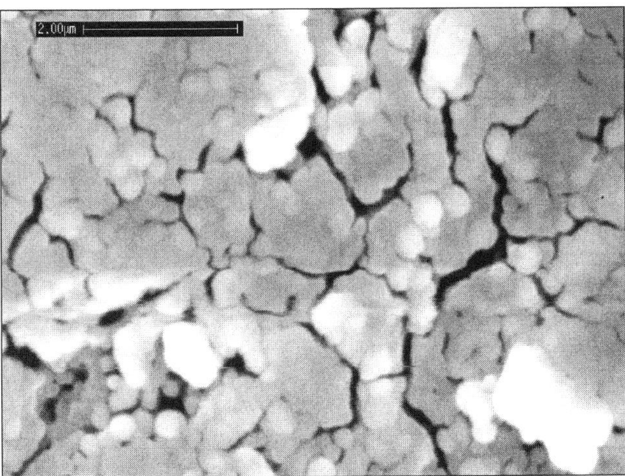

Figure 3. Turkey meat 300 MPa

Fig. 4 shows more drastic changes in a sample treated at 400 MPa. Gaps which can still be found after treatment at 100 to 300 MPa virtually disappear and single myofibrills are no longer observed.

548

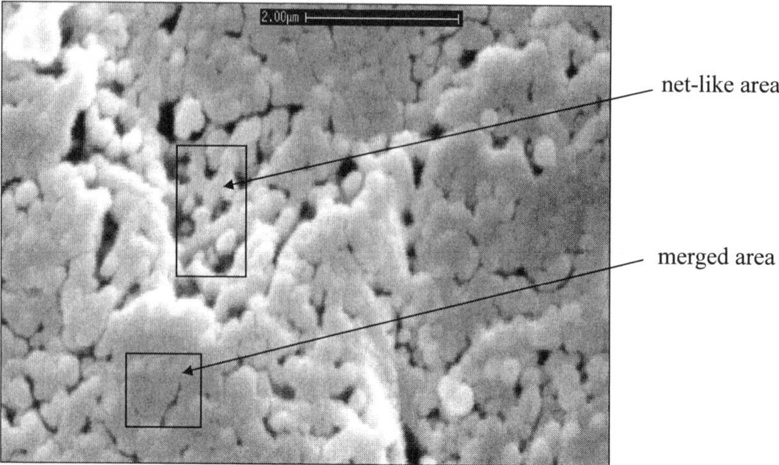

Figure 4. Turkey meat 400 MPa

From a sensory perspective, a clear solidification of texture as well as a change of colour from pink to grey-brown are obvious.

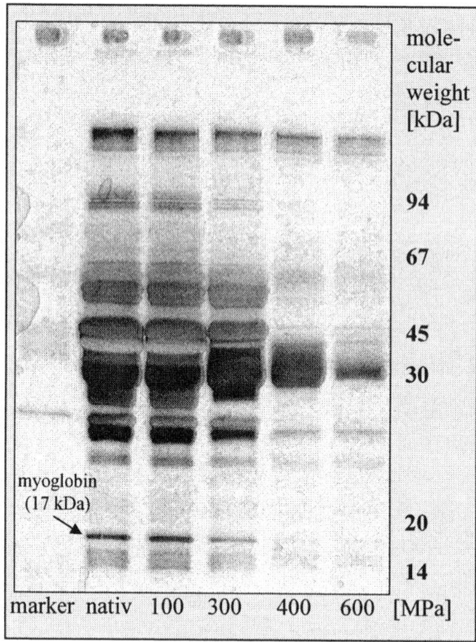

Figure 5. SDS-Electrophoresis of press
extract from HP treated pork

This is can be explained by examinations performed on press extract from HP treated pork meat. Fig. 5 shows that an aggregation of soluble sarcoplasmic proteins especially myoglobin, is responsible for these changes. Myoglobin in its native state causes the red colour of meat. SDS electrophoresis of these press extracts shows a decreasing protein content at 400 MPa. In particular the band corresponding to myoglobin at 17 kDa is nearly invisible at pressures > 300 MPa. These observations can be explained by HP induced denaturation and aggregation of sarcoplasmic proteins. Denatured and aggregated proteins are retained in the sample and cannot be found in the press extract.

Heat treatment of turkey meat leads to a greater solidification of texture. No net-like structures are visible and sarcoplasmic gaps still exist (Fig. 6). Throughout the sarcoplasmic areas, myofibrills are connected by threadlike structures. Probably the myofibrillar protein connectin, which forms the so-called g-filaments, is responsible for these changes. G-filaments contribute greatly to solidity of heat treated meat.

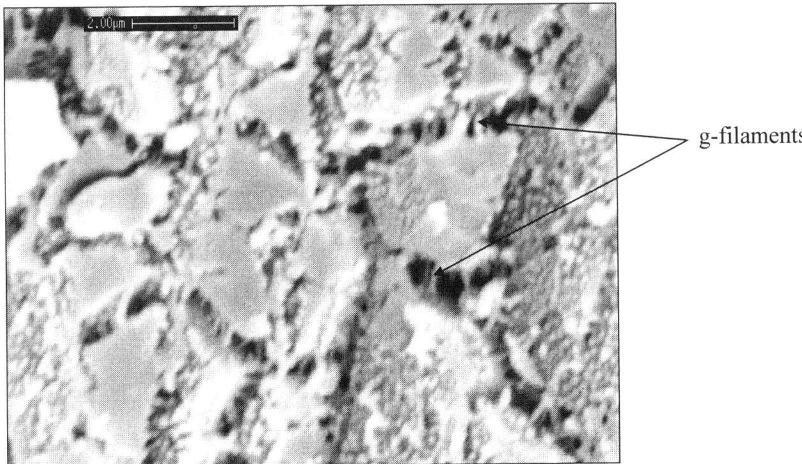

Figure 6. Turkey meat 100 °C

Commercially available pork meat shows no alterations in muscle fibres. Due to longer storage for maturation of meat, myofibrillar proteins are so denatured that HP treatment cannot cause further visible alterations. Sensory same pressure induced changes as in turkey meat can be found. Interestingly swelling of the endomysium, a connective tissue between the myofibres, can be observed after treatment at 400 MPa (Fig. 7 + 8).

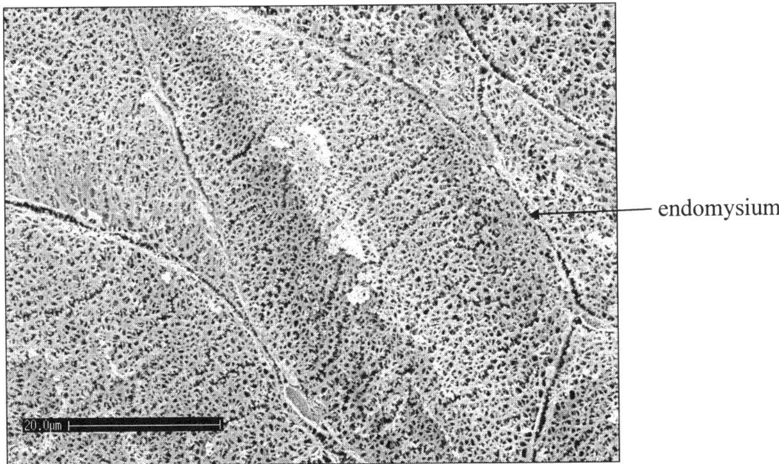

Figure 7. Native pork meat

550

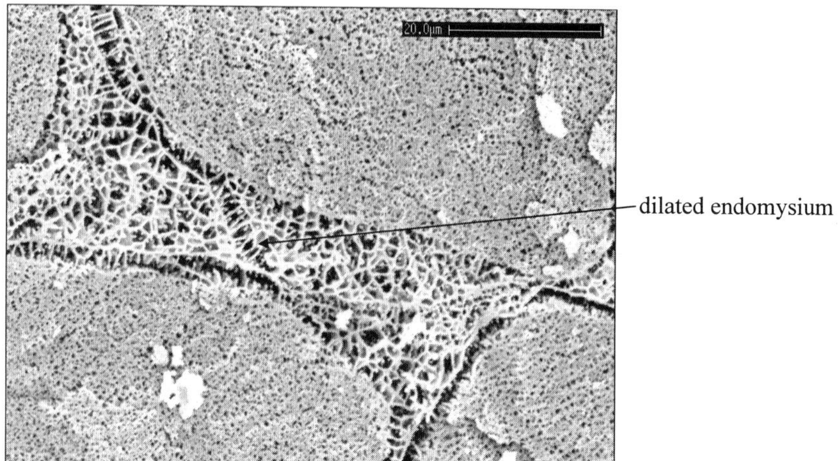

Figure 8. Pork meat 400 MPa

4. CONCLUSION

SEM examination of turkey and pork meat showed that, after application of HP under various conditions, different mechanisms of denaturation lead to sensory alterations of meat samples. At 100 MPa swelling of actin/myosin complexes can be observed. At pressures > 300 MPa denaturation of soluble sarcoplasmic protein starts. In contrast to heat treatment of turkey meat, formation of g-filaments cannot be found after HP treatment.

Aggregation of sarcoplasmic proteins after HP treatment might be a useful application during production of meat and meat products to avoid a weight loss during subsequent thermal processes. Studies by Schöberl [2] have already shown that weight loss during heat treatment can be reduced after HP application.

Swelling of the pork meat endomysium can be also explained by denaturation of fiber proteins. During HP treatment denaturation of myofibrillar and sarcoplasmic protein inhibits expansion of myofibres to their former volume during decompression. For this reason the endomysium is dilated during decompression and fills the wider gaps between muscle fibres.

REFERENCES

1. Yamamoto K., Hayashi S., Yasui T.; Hydrostatic Pressure-induced Aggregation of Myosin Molecules; Bioscience, Biotechnologie, Biochemistry; 1992, 57; p. 383-389
2. Schöberl H.; Physikalisch-chemische und strukturelle Auswirkungen einer hydrostatischen Hochdruckbehandlung auf Lebensmittel; Dissertation; München; Techn. Univ.; 1999

Trends in High Pressure Bioscience and Biotechnology
R. Hayashi (editor)
© 2002 Elsevier Science B.V. All rights reserved.

Pressure-induced denaturation of monomer β-Lactoglobulin-B is partially irreversible

Y. Ikeuchi[a], K. Nakagawa[b], T. Endo[b], A. Suzuki[b], T. Hayashi[a], and T. Ito[a]

[a]Dept. of Bioscience and Biotechnology, Graduate School of Agriculture, Kyushu University, Fukuoka, 812-8581, Japan.
[b]Department of Applied Biological Chemistry, Faculty of Agriculture, Niigata University, Niigata, 950-2181, Japan.

ABSTRACT

This experiment was conducted to assess the effect of high hydrostatic pressure on monomer β-Lactoglobulin-B (BLg) at acidic pH by fluorescence spectroscopy under pressure and by circular dichroism and ^{1}H NMR spectroscopies after release of pressure. The intrinsic (tryptophan) fluorescence measurement and the study on 8-anilinonaphthalene-1-sulfonate (ANS) binding to BLg indicated that at pH 2.0 the recovery of the center of spectral mass or ANS fluorescence was almost complete upon the pressure release. No difference in the ^{1}H NMR spectrum was observed between pressurized and unpressurized BLg. In addition, NMR detection of the H/D exchange of amide protons indicated that the conformation at the vicinity of tryptophan residues can be refolded almost completely after release of pressure. These results confirm that the pressure-induced denaturation of BLg at pH 2.0 is reversible. However, cis-parinaric acid binding ability of pressurized BLg was largely lost although its retinol binding ability was the same as in the unpressurized species. Furthermore, the CD spectra of the far- UV region and the 2D ^{1}H NMR spectra clearly revealed the difference in the molecular conformation between unpressurized and pressurized BLg. These results are interpreted in terms of the existence of a partially fragile structure in the BLg molecule due to high pressure.

1. INTRODUCTION

Whey protein isolate (WPI) is widely utilized in processed foods as a potential food source material due to its excellent functionalities such as gel forming ability, emulsifying activity or foaming capacity in food processing. Recently, it has been reported that when WPI (above 10 %) at neutral pH is subjected to high pressure at 300-400 MPa, it formed a gel, which is weaker, less elastic and more exudative than corresponding thermal gel. This gel, however, does not readily form at acidic pH conditions. These properties of the pressure-induced WPI gel are considered to be mostly dependent on the change in the properties of β-lactoglobulin (BLg), which comprises about half of the WPI, under high pressure [1].

On the other hand, BLg is expected to be useful as a source for the study of the unfolding/refolding mechanism of protein molecules. Pressure techniques have recently been utilized as powerful tools for this kind of study. From several studies, it was suggested that the unfolding of dimeric BLg at neutral pH is irreversible because of misfolding of dimers by the formation of a novel disulfide bond [2-3]. Dufour et al .(1994) also found from the measurements of fluorescence of BLg-retinol   and BLg-cis-parinaric acid complexes that at highly acidic pH the pressure-induced denaturation of monomer BLg is partially irreversible [2]. Their results suggested that the conformation of cis-parinaric acid binding site in the BLg molecule is more susceptible to pressure than retinol binding site in β-sheeted structures of β-barrel [2,4].

The objectives of this study were to elucidate whether the unfolding of BLg induced by high pressure at highly acidic pH (pH=2.0) is reversible or irreversible.

2. MATERIALS AND METHODS

2.1 Materials and High pressure equipment

Crystallized bovine milk β-Lactoglobulin (BLg) was purchased from Sigma (St. Louis, Mo, USA) and further purified by DEAE-cellulose ion exchange   chromatography [5].

Two different types of high pressure devices were used for this study. One was a device consisting of a thermostated high pressure vessel equipped with sapphire windows and a pump capable to elevate pressure up to 400 MPa (Teramecs Co., Ltd., Kyoto). The other was basically the same apparatus as a cold isostatic pressing (CIP) used conventionally for molding of ceramics (Nikkiso Co. Ltd., Tokyo).

2.2 Analytical instruments

Fluorescence measurements were done on a Hitachi F2000 fluorospectrophotometer, in which the high-pressure vessel was placed. The excitation wavelength for the intrinsic fluorescence spectrum was 295 nm which excites tryptophan residues in the BLg molecule. Fluorescence spectra of BLg solutions saturated by 8-Anilinonaphthalene-1-sulfonate (ANS) were recorded between 420 and 560 nm with excitation of 350 nm under pressure. Measurements of retinol and cis-parinaric acid fluorescence were performed at 487 nm (excitation: 354 nm) and 414 nm (excitation: 326 nm) upon binding to the pressurized BLg, respectively.

CD spectra of BLg after pressure treatment were recorded on a JASCO J-725 spectropolarimeter at 20℃ using a quartz cell with a 1 mm light path for far UV (200-250 nm) or a 10 mm light path for near UV (250-340 nm).

1D ^{1}H NMR spectra were recorded on a Bruker DPX-400 spectrometer at 298K using a standard 5 mm ^{1}H probe. 3-(Trimethylsily)-1-propanesulfonic acid (DSS) was used as an internal reference. The presaturation procedure was adopted to suppress the signals of solvent. 1D spectra were recorded with 64K data points, and 512 scans.

β-Lactoglobulin sample solution for NMR detection of the H/D exchange reaction at 600 MPa was prepared at 270 μ M (0.5 wt%) to prevent aggregation after release of pressure [6]. Exchange reactions under high pressure were performed in a CIP type pressure vessel as mentioned above. The BLg samples lyophilized after NMR measurement or release of pressure were dissolved again either in a mixed solvent of 90% H_2O/10% D_2O or 100% D_2O and then the H/D-exchanged protein sample was transferred into an NMR tube (Sigemi Co. Tokyo). Both solvents contained 20 mM H_3PO_4 at pH 2 or pD 2 (as direct pH meter reading), respectively.

3. RESULTS AND DISCUSSION

3.1 Fluorescence measurements

The values of $<v>$ (center mass of the intrinsic fluorescence spectrum) of BLg at pH 2.0 decreased sharply upon increase of pressure from 150 to 300 MPa and reached plateaus upon further rise in pressure. The almost returned to the initial value after release of pressure although the $<v>$ at a specific pressure value during decompression was lower than that at the same pressure value during compression. The fluorescence of ANS has been used to detect the conformational changes of the hydrophobic region in the protein molecule. The ANS emission maximum was at around 470-480 nm which was different from the specific emission maximum (530 nm) of free ANS in aqueous solution and the fluorescence of ANS was enhanced when it bound to the pressure denatured BLg [7]. Fluorescence intensity (at 480 nm) of ANS-BLg at pH 2.0 increased with the increase of pressure intensity and returned to its original value as tracing almost the same curve in the compression direction when the pressure was gradually released. The facts revealed by intrinsic tryptophans (Trp19 and Trp61) and ANS fluorescence imply that the pressure-induced denaturation of BLg at pH 2.0 is reversible.

Dufour et al. (1994) have reported that the dissociation of BLg-retinol complex at pH 3.0 were reversible between 0.1 MPa and 400 MPa, but BLg-cis-parinaric acid complexes dissociated irreversibly under pressure higher than 200 MPa [2]. We also confirmed those facts although data are not shown. However, the hydrophobic probe such as cis-parinaric acid is likely to become insoluble in water (i. e. decrease in solubility) once it dissociates from the protein under pressure. If so, it seems to be difficult to analyze the interaction between protein and ligand under pressure. Alternatively, we examined changes in retinol and cis-parinaric acid binding abilities of BLg after being exposed to a pressure value of 400 MPa at pH 2.0. Pressurized BLg at pH 2.0 had the same rethinol binding ability as the unpressurized species. In the case of cis-parinaric acid the situation was quite different, namely the cis-parinaric acid binding ability of pressurized BLg was fairly lost. Thus, it seems that pressure induced denaturation of BLg at acid pH is not always reversible. Our present data also suggest that at acid pH the conformation of the cis-parinaric acid binding site is much more fragile against pressure than that of the retinol binding site, which may be located at the central hydrophobic pocket formed by the β-barrel [2, 8].

3.2 The circular dichroism (CD) spectra of the pressurized BLg

Pressurization at pH 2.0 affected the CD spectra of the far-UV region to some degree (0.1 MPa: $[\theta]_{217}$=-3004 deg cm^2 d mol^{-1}; 400 MPa: $[\theta]_{215}$=-3917 deg cm^2 d mol^{-1}), indicating the increase in the disordered region. On the other hand, in the near -UV region few changes in the CD spectrum were observed. This fact may indicate that the conformation around the tryptophan residues is well retained and the original structure is recovered after release of pressure. Thus, the cis-parinaric acid binding site of BLg is not present in the vicinity of the tryptophan residues (Trp19 and Trp61). Rather, it may be present at the hydrophobic surface patch located in a groove between the strands and the helix as described above [2, 8].

3.3 ^{1}H NMR spectra of the pressurized BLg

There is a difference in the state of BLg in solution which is dependent on pH conditions. Namely, at neutral pH BLg is present as a dimer, while under strong acidic conditions it is present as a monomer due to electrostatic repulsion. The difference is reflected in the ^{1}H NMR spectrum. Pressurization of BLg (above 300 MPa) at pH 2.0 did not lead to a reduction in the signal intensity and a further broadening of the resonance lines, suggesting no association of the molecules. As far as it can be judged from ^{1}H NMR spectra, the pressure-induced denaturation of BLg at acidic pH condition is seemingly reversible.

The data at acid pH revealed by the near-UV CD, fluorescence and ^{1}H NMR suggested that BLg was unfolded to expose intrinsic tryptophan residues on the surface of the molecule under pressure and was refolded right after release of pressure. Support for this idea will come from the NMR detection of the H/D exchange reaction (Figure 1).

When BLg was pressurized at 600 MPa for 60 min in the presence of 100 % D$_2$O, the signals in the low-filed region (6-10 ppm), which is mainly composed of amide protons and aromatic protons, mostly disappeared. According to the report of Tanaka and Kunugi (1996), the extensive disappearance of these signals of BLg at pH 7.0 occurs at 300 MPa for 3 h [6]. It is noteworthy that the two sharp singlet resonances at 10.424 ppm and 10.087 ppm, which are likely to originate from indole proton of two tryptophan residues (Try19 and Trp61) in the BLg molecule, disappeared completely in the pressurized sample spectrum. This was because the pressure-induced exposure of the indole protons in the tryptophan residues to the solvent was followed by replacement with deuterium atoms. The remaining signals in the range from 7.5 to 9.5 may suggest that the amide protons not freely accessible to the solvent are present even under a high pressure value of 600 MPa as well as the result of the unfolding of BLg induced by urea [9]. Ragona et al. (1997) demonstrated the presence of the buried cluster in BLg which was very stable at high temperature [10].

After lypophilization of BLg pressurized in D_2O, the sample was dissolved in H_2O and then was employed to measure the NMR spectrum (Figure 1C). Interestingly, only the signal at 10.087 ppm appeared. This indicates that the H/D exchange reaction occurs promptly when the indole proton of the tryptophan residue, which is located at the molecular surface and had higher mobility, is exposed to the solvent (H_2O). The other tryptophan residue seems to be buried in the bottom of the calyx.

In order to clarify whether the tryptophan regions refolded reversibly or not, the sample was pressurized again at 600 MPa for 60 min and then was dissolved in H_2O after lyophilization. As can be seen in Figure 1D, the signal at 10.424 ppm was observed with that at 10.087 ppm and a spectrum feature returned to that of the control sample (Figures 1A and 1D). From these results, it seems that at acidic pH conditions the tryptophan regions are refolded almost perfectly after release of pressure. In other words, the conformation in the vicinity of the retinol binding site involving Trp19 is resistant to pressure.

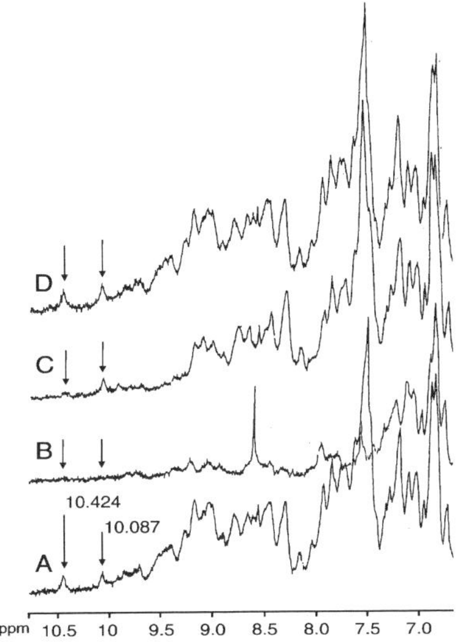

3,Figure 1. The aromatic region of ^1H NMR spectra of hydrogen/deuterium exchanged β-lactoglobulin at pH 2.0 and 20°C. (A) intact; (B) H/D exchanged in D_2O under pressure of 600 MPa for 60 min ; (C) dissolved in H_2O after lyophilizing sample B; (D) H/D exchanged in H_2O under pressure of 600 MPa for 60 min following step C. (A), (C) and (D): 90% H_2O/10% D_2O; (B): 100% D_2O.

4. CONCLUSION

It is clear that monomer BLg is partially irreversible when exposed to high pressure at acidic pH conditions on the basis of the results of the changes in the far-UV CD spectrum and the 2D NMR spectrum (data not described in this text) or incomplete recovery of the cis-parinaric acid binding ability. The reason for this is not completely understood, but one possibility is that the conformation of a certain hydrophobic surface in the BLg molecule, which may be the cis-parinaric acid binding site, collapses partially by misfolding. Further detailed 2D NMR analysis will be needed to detect change in the structure more precisely in the refolding process of the protein after the release of pressure.

5. ACKNOWLEDGEMENT

This study was supported in part by a Grant-in-Aid for Scientific Research from the Ministry of Education, Science, Sports and Culture of Japan (No. 10460118)

REFERENCES

1. Cheftel, J.& Dumay, E. In high pressure bioscience and biotechnology (Hayashi, R. Balny, C. Eds), Elsevier, Amsterdam, **1996,** pp 299-308.
2. Dufour, E., Hui, G. H. B. & Haertle' T. Biochim. Biophys. Acta. **1994**, 1206, 166-172.
3. Tanaka, N., Tsurui, Y., Kobayashi, I.& Kunugi, S. Int. J. Biol. Macromol. **1996**, 19, 63-68.
4. Papiz, m. Z., Sawyer, L., Eliopoulos, E. E., North, A. C. T., Findlay, J. B.C., Sivaprasadarao, R., Jones, T. A., Newcomes, M. E. and Kraulis, P. J. Nature **1986**, 324, 383-385.
5. Piez, K. A., Davie, E. W., Folk, J. E. & Gladner, J. J. Biol. Chem. **1961**, 236, 2912-2916.
6. Tanaka, N. & Kunugi, K. Int. J. Biol. Macromol. **1996**, 18, 33-39
7. Hu, L. Y. & Runa, K. C. Chin. J. Biochem. Sin. **1998**, 14, 567-572.
8. Monaco, H., Zanotti, G., Spadon, P., Bolognesi, M., Sawyer, L. & Eliopoulos, E. E. J. Mol. Biol. **1987**, 197, 695-706.
9. Civera, C., Sevilla, P., Moreno, F. & Churchich, J. E. Biochemistry and Molecular Biology International. **1996**, 38, 773-781.
10. Ragona, L., Pusterla, F., Zetta, L., Monaco, H. L. & Molonari, H. Fold Des. **1997**, 2, 281-290.

Trends in High Pressure Bioscience and Biotechnology
R. Hayashi (editor)
                                             557

FUNCTIONAL PROPERTIES OF SOY PROTEINS AS INFLUENCED BY HIGH PRESSURE: EMULSIFYING ACTIVITY.

E. Molina[*+], A. Papadopoulou, A. Defaye and D. A. Ledward

Department of Food Science and Technology, University of Reading. P.O. BOX 226, Reading RG6 6AP. UNITED KINGDOM.

The different soy proteins studied, soy protein isolate (SPI), 7S globulin and 11S globulin are differently affected by high-pressure (HP). HP unfolds the proteins, exposing hydrophobic sites, leading to improve functional properties of the proteins. In our study, 7S showed the highest emulsifying activity index (EAI) and surface hydrophobicity after treatment at 400 MPa, whereas 11S showed its highest EAI and surface hydrophobicity after treatment at 200 MPa. SPI showed the optimum value of EAI after treatment at 400 MPa although its surface hydrophobicity was low. It is suggested that pressure at 400 MPa dissociated the 7S of the SPI into partially or totally denatured monomers that enhanced the surface activity but at the same time, the unfolding of the polypeptides of the 11S within the hexamer led to aggregation, negatively affecting the surface hydrophobicity of the SPI.

1. INTRODUCTION

Physical (heat or mild alkali treatments), chemical (acylation, phosphorylation and deamidation), and enzymatic methods, (for references see 1-2) have all been used to modify the functional properties of soya proteins, but HP offers a technique by which the functional characteristics of proteins can be modified in a consumer friendly, green way to lead to products of added value. Very limited information is available on the effect of the HP on the emulsifying properties of soy proteins. To our knowledge, only Denda & Hayashi (1992) (3)

---

* Present address of corresponding author: E. Molina, Instituto de Fermentaciones Industriales, CSIC. C/ Juan de la Cierva, n° 3. 28006, Madrid. SPAIN.
+ E. Molina was financially supported by a EU grant, FAIR n° 98-5040.

have demonstrated that HP at certain pH's and temperatures improves the emulsifying stability of SPI, and an improvement in the emulsifying activity of soymilk after HP treatment has also been reported (4). Further experimental work is required to understand the conformational changes or interactions taking place under pressure of the 7S and 11S proteins, which constitute 65-80% of the total seed protein. HP 11S (*Vicia faba*) showed poorer emulsifying properties than the native protein (5). Therefore the objective of this work was to investigate the effect of HP on the emulsifying properties of SPI and its two major globulins, 7S and 11S at different pHs. The hydrophobicity of the proteins which relate to functionality was also determined and conformational changes were followed by differential scanning calorimetry (DSC).

2. MATERIALS AND METHODS

2.1 Extraction of native Soy Protein Isolate (SPI) and 7S and 11S globulins

SPI was prepared from defatted flour by alkaline extraction (pH 8) followed by precipitation at pH 4.5 (6). The crude 7S and 11S were obtained following (7).

2.2 High Pressure Treatments

10% (w/v) protein dispersions in water (pH 6.5 or 7.5) were pressurised for 15 min. at 200, 400 and 600 MPa in a Stansted 'Food lab' HP apparatus (Stansted Fluid Power, Essex, UK). After the HP the samples were stored overnight at 5 °C. A non-treated aliquot was always used as control.

2.3 Emulsifying activity index (EAI)

Emulsifying activity index (EAI) was measured (8). Different concentrations (0.25, 0.5 and 0.75%) of an aqueous protein solution and an oil volumetric fraction (ϕ) of 0.25 were homogenised using a hand-operated piston-type homogeniser at a pressure of 6 MPa. After the homogenisation (MFC Microfluidizer^{TM} Series 5000, MIcrofluidics Corporation, Newton, MA, USA) 25 or 50 µL of the emulsions were pipetted into 50 mL of 0.1% SDS, the final protein dilution being 1/1000 for the SPI and 1/2000 for the 7S and 11S fractions. The absorbance was measured at 500 nm at 0 (A_0) and 10 (A_{10}) minutes after emulsion formation.

2.4 Surface Hydrophobicity

Surface hydrophobicity was determined by the SDS-binding method and represented as µg of SDS bound to 500 µg of protein. The samples were prepared according to (9). The determination of the absorbance of the SDS-methylene blue mixture in the chloroform phase was measured at 655nm (10).

2.5. Differential scanning calorimetry (DSC)

Samples of 20% (w/v) dispersions in distilled water were used. The experiments were conducted against water as reference, at a scanning rate of 5 °C/min from 30-100 °C.

3. RESULTS AND DISCUSSION

The comparative study of the EAI of HP soy proteins was conducted at pHs 7.5 and 6.5 and at 0.50% and 0.75% protein concentrations for the 7S and 11S globulins and 0.25% and 0.50% protein concentrations for SPI. The EAI of the 7S globulin (Figure 1a) and SPI (Figure 1c) showed similar patterns at both pHs and at all concentrations tested. The highest values were reached after treatment at 400 MPa. The 11S fraction followed a different pattern, but the trend was also the same at both pHs and at all the concentrations tested (Figure 1b). The 11S globulin showed the highest values of EAI when treated at 200 MPa, 0.50% protein concentration giving the highest values at both pHs. It is noteworthy that after treatment at 400 and 600 MPa the 11S gelled. Under these conditions the emulsions produced were foamy, and the sample of the emulsion used was not representative (11).

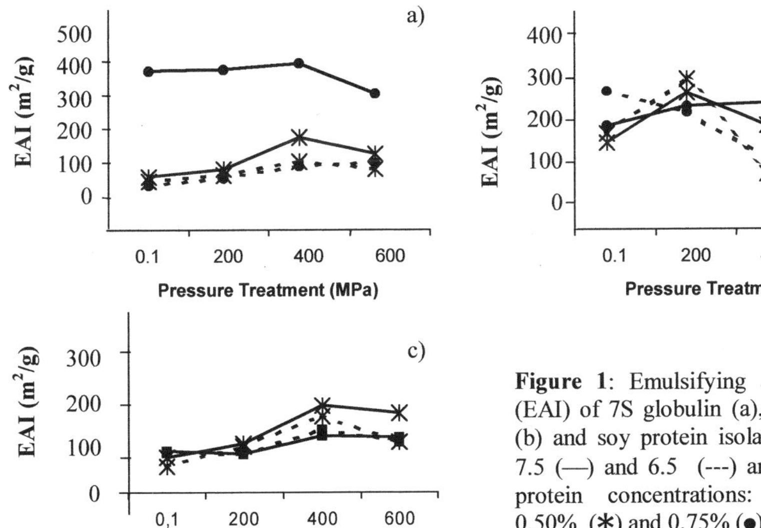

Figure 1: Emulsifying activity index (EAI) of 7S globulin (a), 11S globulin (b) and soy protein isolate (c), at pHs 7.5 (—) and 6.5 (---) and at different protein concentrations: 0.25% (®), 0.50% (✳) and 0.75% (●).

Surface hydrophobicity of SPI and the two globulins, 7S and 11S, at pH 6.5 and 7.5 are shown in Table 1. For both fractions 200 MPa treatment increased the surface hydrophobicity presumably due to partial denaturation, and subsequent exposure of hydrophobic groups (5). But, whereas for the 7S the surface hydrophobicity was increased at 400 MPa, the 11S at this pressure exhibited decreased surface hydrophobicity. At 600 MPa both globulins presented slightly lower or equal surface hydrophobicity than the 400 MPa sample. At 200 MPa the 7S is partially unfolded and dissociation of the subunits and enhancement of surface activity is observed (12). At 400 MPa the maximum unfolding of the subunits occurs thus giving the highest EAI at this pressure. On the other hand, at 200 MPa the 11S is partially unfolded, exposing hydrophobic sites buried inside the native protein, but at 400 MPa the high extent of denaturation of the 11S probably leads to aggregation due to oxidation of the sulphydryl groups leading to disulphide bond formation (5).

Table 1: Surface hydrophobicity (μg SDS / 500 μg of protein) for 7S, 11S and soy protein isolate (SPI) after treatment at 0.1, 200, 400 and 600 MPa, at pHs 7.5 and 6.5. Results are means ± standard deviation of three determinations.

	pH 7.5				pH 6.5		
	7S	**11S**	**SPI**		**7S**	**11S**	**SPI**
0.1 MPa	15.06	39.22[a]	10.68[b]		35.63	38.16[a]	10.81[b]
	(0.68)	(053)	(0.06)		(5.89)	(7.14)	(0.28)
200 MPa	40.89	54.85	13.88[c]		43.39	66.06	11.66
	(2.03)	(0.28)	(0.17)		(1.05)	(4.90)	(0.3)
400 MPa	46.45[e]	51.57[e]	10.81[bf]		72.53	54.89	10.1[f]
	(0.90)	(0.61)	(0.5)		(3.51)	(0.40)	(0.32)
600 MPa	47.66[e]	50.09[e]	14.06[c]		52.42	51.41[e]	12.36
	(2.19)	(0.98)	(0.58)		(0.76)	(0.83)	(0.34)

Means within a column and a row with same superscripts are not significantly different (p< 0.05).

In general, a positive correlation between hydrophobicity and emulsifying ability has been found for the 7S and 11S (except for the 11S at pH 6.5), but not for the SPI. This suggests that the emulsifying ability of HP 7S and 11S can be explained on the basis of protein hydrophobicity. The 7S, a trimer without any disulphide bonds, is more pressure labile than the 11S whose subunits are linked by many disulphide bonds. The different way HP affects the 7S and 11S fractions could well explain the behaviour of SPI, the enhanced emulsifying activity can be attributed to the effect of pressure on the 7S although it does not correlate with surface hydrophobicity (13), which is mainly affected by the behaviour of the 11S.

DSC also revealed the different effects of HP on the two globulins. Significant decreases ($p < 0.05$) in the total calorimetric enthalpy (ΔH) values were found for the 7S treated at 400 and 600 MPa compared to the 0.1 and 200 MPa samples (Table 2), which indicates that the 7S is largely unfolded at 400 MPa. The 11S globulin showed significant decreases in ΔH values at all pressures compared with the native, as well as a shift in the denaturation temperature of the HP treated samples, which indicates a major loss of protein structure. Once again, it is seen that at 200 MPa there was a partial unfolding of the polypeptides within the hexamer (11S consists of two hexamers linked by disulphide bonds), that expose the hydrophobic sites to the surface, but higher pressures leads to aggregation.

Table 2. Denaturation temperature (T_d) (°C) and total calorimetric enthalpy (ΔH) (J/g) values of 7S and 11S globulins in 20% dispersions of SPI, pH 7.5, after treatment at 0.1, 200, 400 and 600 MPa. Results are means ± standard deviation of 3 determinations.

	7S		11S	
	T_d (°C)	ΔH (J/g)	T_d (°C)	ΔH (J/g)
0.1 MPa	78.44 (0.29)	0.277[a] (0.01)	96.88[a] (0.3)	1.370[a] (0.06)
200 MPa	78.95 (0.24)	0.292[a] (0.05)	99.09[b] (0.25)	1.306[b] (0.14)
400 MPa	78.01 (0.33)	0.065[b] (0.25)	99.40[b] (0.13)	0.959[c] (0.05)
600 MPa	76.80 (1.24)	0.043[b] (0.14)	99.54[b] (1.80)	0.117[d] (0)

Means within a column with different superscripts are significantly different ($p < 0.05$)

In conclusion, pressure treatment at neutral pH can improve the emulsifying activity of soy proteins. The 7S reached maximum values of EAI after treatment at 400 MPa, and the 11S after treatment at 200 MPa and these changes correlated well with the changes observed in the surface hydrophobicity. The EAI of the SPI is mostly governed by the emulsifying properties of the 7S, although the 11S is the most abundant globulin on pressure treatment it negatively affects both the hydrophobicity and solubility of the SPI.

REFERENCES

1. Nir, I., Feldman, Y., Aserin, A., and Garti N. (1994). *J. Food Sci.,* 59, 606-610.

2. Qi, M., Hettiarachchy, N. S., and Kalapathy, U. (1997). *J. Food Sci.,* 62, 1110-1115.

3. Denda, A., & Hayashi, R. (1992). In C. Bandy, R. Hayashi, K. Heremans & P. Masson. High Pressure and Biotechnology (Vol. 224) (pp. 333-335). Colloque INSERM/John Libbey Eurotext Ltd. ©.

4. Kajiyama, N., Isobe, S., Uemura, K., & Noguchi, A. (1995). *Int. J. Food Sci. Tech.,* 30, 147-158.

5. Galazka, V. B.; Dickinson, E.; Ledward, D. A. (1999). *Food Hydrocolloids,* 13(5), 425-435.

6. Wolf, W.J., & Sly, D.A. (1967). *Cereal Chem.,* 44, 653-668.

7. Thanh, V.H., & Shibasaki, K. (1976). *J. Agric. Food Chem.,* 24, 1117-1121.

8. Pearce, K.N., & Kinsella, J. E. (1978). *J. Agric. Food Chem.,* 26(3), 716-723.

9. Kato, A., Matsuda, T., Matsudomi, N., & Kobayashi, K. (1984). *J. Agric. Food Chem.,* 32, 284-288.

10. Hayashi, K. (1975). *Anal. Biochem.,* 67, 503-506.

11. Kinsella, J.E. (1979). *JAOC'S,* 56, 242-258.

12. Hermansson, A.M. (1978). *J. Text. Stu.,* 9, 33-58.

13. Were, L., Hettiarachchy, N.S., & Kalapathy, U. (1997). *J. Food Sci.,* 62(4), 821-823, 850.

Trends in High Pressure Bioscience and Biotechnology
R. Hayashi (editor)
                                              563

High Pressure Thawing. Application to selected sea-foods.

A. LeBail[a], D. Mussa[c], J. Rouillé[a], H.S. Ramaswamy[b], N. Chapleau[a], M. Anton[a], M. Hayert[a], L. Boillereaux[a] and D. Chevalier[a]

[a] GEPEA-ENITIAA BP 82225, F-44322 Nantes Cedex 03, France. Tel/Fax : 02.51.78.54.54/02.51.78.54.67. lebail@enitiaa-nantes.fr

[b] McGill University, Dept. of Food Science & Agriculture Chemistry, 21-111, Lakeshore Road, Ste-Anne-de-Bellevue, Québec, H9X-3V9, Canada.

[c] PostDoctoral Fellowship at ENITIAA (Région Pays de Loire-France)

The melting temperature of water is depressed under pressure from 0°C at atmospheric pressure down to -21°C at 207 MPa. This phenomena permits to achieve rapid thawing of foods. Additional gains are related to the possible reduction of the drip volume and of the microbial load. High pressure thawing of aiguillat fish and of salmon was evaluated in comparison to thawing in water at atmospheric pressure from selected parameters (drip, microbial load...). High pressure thawing of aiguillat fish permitted to reduce the drip volume. Farm salmon was inoculated with a suspension of *Listeria innocua*. A significant reduction was observed. The lower the vessel temperature was the higher the reduction was (3 log CFU maximum at 200 MPa and 5°C). This result can also be attributed to the fact that thawing was longer at lower vessel temperature. These results confirmed the interest of this process for food industry.

1. INTRODUCTION

The quality of frozen foods is evaluated after thawing and processing (cooking...). Even though, the initial quality of the product is optimal, the control of the freezing rate, the storage conditions and the thawing conditions might interfere with the final quality. Several studies related to the effect of the freezing conditions on the microstructure of frozen food are available in the literature (Bomben et al. 1982; Chevalier et al. 2000-I; Chevalier et al. 2000-II). On the other hand, research efforts focused on improving thawing process are limited. Some alternative techniques such as plate thawing have also been evaluated (Merts et al. 1999). The main limitation of conventional processes (atmospheric pressure) are due to the fact that minimal ambient temperature must be ensured to minimize microbial growth and other quality loss related phenomena (oxidation ...). High pressure thawing (HPT) offers a new alternative. Sanitation, texturation and phase change which are the three main applications of HP in the food domain can all be combined in high pressure thawing. The use of HPT for biological substances permits to decrease the phase change temperature of pure water (down to -21°C at 207 MPa) which enhances the heat flux rate to the melting zone. The phase change temperature increases with pressure above 207-220 MPa which makes that 207-220 MPa appears as the optimal and maximum pressure of interest in HPT with respect to the heat transfer criterion. A schematic plot of the HPT process is shown in Figure 1 for a case of frozen sample of ice (Le Bail *et al.*, 1997).

HPT is usually realized in a high pressure vessel containing water as a pressurization fluid. Because of the pressure rise (phase #1), the melting temperature is depressed. After thawing (end of phase #2), the temperature of the sample increases up to the ambient temperature of the vessel (phase #3). The sample temperature decreases by a few degrees during the depressurization (phase #4), due to the positive coefficient of thermal expansion of water. Similar temperature and pressure evolution are obtained with real foods which mainly contain water.

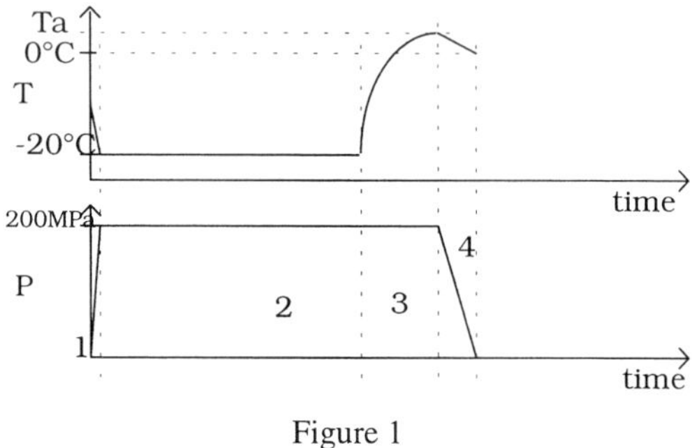

Figure 1

Scheme of food temperature and pressure evolution during HP thawing

The most put forward advantage of HPT is the reduction of the thawing time (Chevalier et al. 1999). Reduction of the microbial load and reduction of the drip volume has been pointed out by some authors (Murakami et al. 1992), (Chevalier et al. 1999). (Taylor 1960) applied HPT to thaw cells of human conjunctiva, and observed that a slow freezing rate combined with a HP thawing at 225 MPa permitted to significantly increase the survival ratio. HPT of beef meat was studied by (Deuchi et al. 1992) and (Zhao et al. 1998). The former authors found 50 MPa as the optimal thawing pressure for beef with respect to drip volume and to color change, where as the latter authors founded that drip loss, cooking loss, color change and penetration force were not significantly affected by HPT (210 MPa) . (Takai et al. 1991) observed color changes of tuna meat and surimi, and attributed them to protein denaturation under pressure. (Murakami et al. 1992) also noticed an increase of the L-value of tuna meat and a reduction of the drip volume. (Chourot 1997) studied HPT of whiting. A minimum drip volume and drip from cooking was obtained at 150 MPa. Pressure level between 100 and 200 MPa is also known as the threshold pressure range over which protein denaturation might occur. This result was confirmed from electrophoretic pattern of sarcoplasmic proteins extracts from flesh (Chevalier et al. 1999), by the reduction of the enthalpy of denaturation (Chevalier et al. 2000), (Iso et al. 1994) or by texture measurement (Ko et al. 1990). (Eshtiaghi et al. 1996) observed improvement of sugar uptake during HPT of strawberries (immersion in a sugar solution) by a pressure induced permeabilization of the cell membrane.

(Chevalier et al. 1999) studied HPT of whiting filets. Among the parameters taken into account, the duration of the pressurization appeared to be related to the minimization of drip volume in comparison to atmospheric thawing (the longer the pressure plateau after thawing, the smaller the drip volume). (Murakami et al. 1992) observed a significant drip reduction for HPT of tuna meat.

The use of HP for sanitation (microbial count reduction) of foods is maybe one of the most investigated domain in the high pressure science. Nevertheless, very few work related to the behavior of microorganisms during HPT can be found in the literature. It is known that bacteria inactivation under high pressure is enhanced either by a relatively elevated temperature (over 50°C) or by a relatively low but positive temperature. (Murakami et al. 1992) observed a slightly constant microbial load during HPT of tuna. (Eshtiaghi et al. 1996) observed a two-log cycle reduction of the vegetative micro-organisms during HPT of strawberries with a sugar solution as pressurization fluid. More recently, (Mussa et al. 2000) presented results obtained with salmon which was inoculated with *Listeria innocua*.

Thus, HPT offers several advantages in comparison with atmospheric thawing : reduction of the thawing time, partial microbial destruction or growth limitation and a possible reduction of the drip volume. The objective of this work is to present some recent results obtained on HPT of sea foods. These results will bring some additional information needed to evaluate the interest of HPT for industrial application.

2. EFFECT OF HPT ON DRIP LOSS OF AIGUILLAT FISH (*Squalus acanthias*)
2.1. Material and methods

Frozen blocks of fish (aiguillats, *Squalus acanthias*) were bought from a fishery institute (Gaspé, Québec). The initial moisture of the fish was 76.5%. Slabs were prepared as parallelepipeds (150*100*30 mm or around 400 g) vacuum packaged, and frozen stored at − 20°c until use. Thawing of fish was done at 4 different pressure levels. Thawing times are presented in Table 1 as a function of pressure and were adjusted from preliminary experiments realized with a thermal sensor installed at center of the sample.

Table 1
Thawing time vs. pressure

Pressure (MPa)	Thawing time (min)
0.1 (atm)	60
100	30
150	25
200	20

Thawing of fish was done either in water or in the high pressure system : for thawing in water, stirred water bath at 12°C ± 2°C was used. In the high pressure system, the temperature of the pressure vessel and immersion medium were maintained at 10°C using a jacketed water circulation system. The internal dimension of the high pressure unit (Model 422, Autoclave

Engineers) were 55cm, length and 10cm, diameter. The pressurization fluid was water mixed with oil to lubricate the pump (2% of mineral oil in water).

2.2. Cooking of the samples

Vacuum packaged thawed fish fillets samples were cooked in a stirred water bath at 80°C for 30 min, then cooled for 5 min in cold water (T ≈ 10°C). This corresponded to a pasteurization value of 95 min at 70°C, with a theoretical decrease in the total microbial population by a factor of 10^{32} (z = 10°C, $D_{70°C}$ = 2.95 min). Thus, the samples were rather overcooked to test their water retention performance.

2.3. Calculation of drip loss

Drip percentage was measured just after the thawing and cooking to avoid any additional reactions or modifications of volume. Fish fillets and corresponding plastic bags, were weighed before HP treatment. After thawing and cooking, fish were taken out from the plastic bags then wiped with a paper towel and weighed again. Drip loss percentage was then calculated from the ratio between the difference in mass (drip) and the initial mass of the sample. Each experimental point is a mean of 3 drip measurements and percentages were expressed on a dry basis in order to prevent any bias introduced by thawing drip losses or cooking drip losses.

2.4. Results

The evolution of the drip ratio as a function of the pressure during thawing is shown in figure 2. One can observe that a significant reduction of the thawing and cooking drip was observed for HPT in comparison with atmospheric thawing.

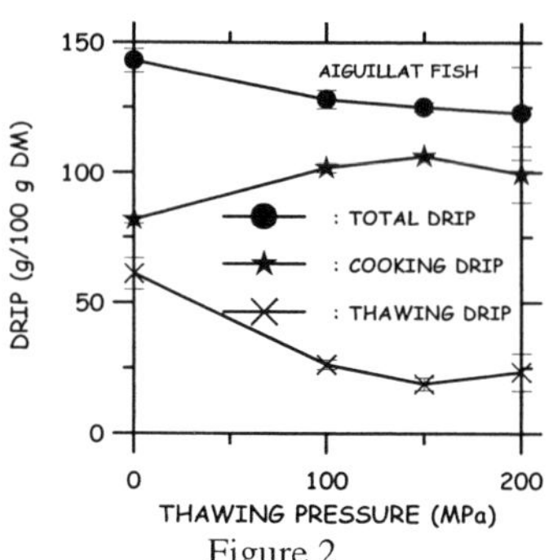

Figure 2

Thawing and cooking drip of *aiguillat* fillets as a function of pressure during thawing
Total drip = thawing drip + cooking drip (80°C-30minutes)

As observed by (Chourot 1997; Chevalier et al. 1999), the drip mass due to thawing seemed to be minimum for pressures around 150 MPa. On the opposite, the drip mass was higher for higher pressure. The thawing drip mass was 15% on a wet basis for atmospheric conditions and 9.5% for HPT at 200 MPa. This results can be explained by the fact that a partial denaturation of proteins was realized during HPT process which would affect the water holding capacity of the flesh. The total drip combining cooking and thawing drip was reduced for HPT in comparison with atmospheric pressure thawing. The higher the pressure, the lower was the drip loss. The cooking done was more than required for pasteurization, and it is likely that a reduced cooking time, would reduce the cooking drip.

3. EFFECT OF HPT ON MICROBIAL QUALITY OF SALMON FISH
3.1. Material and methods

Salmon fish obtained from Norwegian Farm were received in the laboratory 3 days postmortem. Samples were then filleted in close to aseptic condition. Pure cultures of *Listeria innocua* (cow brain isolate) were obtained from the Pasteur Institut (Paris, France) and were maintained on Tryptic Soy Agar (TSA) slants (Pasteur Institut (Paris, France) at 4C. The inoculum was prepared by transferring isolated colonies of *Listeria innocua* from TSA slants

into 5 mL of sterile Tryptic Soy Broth with 0.6% yeast extract and incubating at $37^{\circ}C$ for 24 h, to give a stock suspension of 10^9 CFU/mL. Microorganisms at the stationary stage were used in the experiment because of their high resistance to high pressure destruction. The fish fillet was cut with a cylindrical cuter to fit a 7 cm diameter sample holder. Samples of 100g were inoculated with the culture of *Listeria innocua*. The indigenous microbial load of salmon fish samples were first evaluated and was $\sim 10^4$ CFU/g. The 24 h cultures of *Listeria innocua* were diluted with 0.1% sterile peptone broth to give an inoculum level of 10^8CFU/mL. Exactly 1 mL of this diluted cultures were inoculated onto the 100 g fish to achieve 10^6 CFU/g sample. All inoculated samples and controls were kept at 4°C for two hours to allow the cells recondition to the new environment before freezing. Samples were frozen in a air freezer and were stored for 1 week at –21°C. HPT was realised at pressure levels of 100 and 200 MPa with two thawing temperatures (5 or 20°C). Thawing at atmopsheric pressure was realised in a water bath at 5°C. Duplicate samples were selected randomly and the experiment repeated three times (three replicates). All thawed samples were allowed to rest in the refrigerator (4°C) before analysis to allow resuscitation of injured cells. This procedure has been reported to be effective in resuscitating thermally injured cells. Samples were then added to 0.1 peptone water and homogenised in a stomacher for 1 min. Survival of indigenous microorganisms was evaluated by aerobic standard plate count while *Listeria innocua* were enumerated on oxford medium with a spread plate technique.

Table 2
Thawing time for thawing of salmon samples (7 cm diameter)

PRESSURE (MPa)	TIME (Min)	
	5°C	20°C
0.1	150	53
100	83	32
200	50	28

3.2. Results

The initial microbial load of *Listeria innocua* (10^6 CFU/g) was reduced by 1.5 log-cycles at 200 MPa/20°C and 3 log-cycles at 200 MPa / 5°C. These were accompanied by a 2 and 4 log-cycle reductions in the population of indigenous microorganisms at 200 MPa/20°C and 200 MPa / 5°C, respectively. The results thus demonstrated that a strong sanitation effect is associated with HP thawing even at the relatively low pressure levels employed (200 MPa) in comparison with conventional HP sanitation (not associated with thawing). It seems that the pressure and phase change applied to the microorganisms during HP thawing has a synergistic effect and yield more pronounced inactivation of the microorganisms. (Murakami et al. 1992) are the only researchers who have proposed equivalent data in the available literature. They observed a slight reduction of the total count (indigenous flora) on tuna samples. It is obvious that much more investigations needs to be done in the domain of microbial inactivation during HPT. There is no evidence of the impact of the water to ice phase change on the inactivation mechanisms. It might also be due to a stress of the lipids contained in the membranes, affecting the functionality of the membranes. Also, the improved reduction with lower vessel temperature can be attributed to the fact that thawing was longer at lower vessel temperature.

4. DISCUSSION

These results from these studies, in addition to those available in the literature, confirm that a significant reduction of the drip volume can be achieved with HPT. Moreover, a significant reduction of the microbial load can be observed for both indigenous flora and *Listeria innocua* in salmon. An economic analysis done by (LeBail et al. 1997) showed that an HP equipment designed for thawing can be paid back quickly as a result of reducing the drip loss alone for selected products. Depending on the assumption (tonnage between 150 and 300 tons/year, product cost between 7.7 and 15.3 EU/kg), a 5% drip saving allow a payback after 2 to 3 years on an average. These estimations did not even take into account the fact that a HP thawing is twice faster (even up to 5-times faster in certain situations) than a conventional equivalent atmospheric thawing (i.e. water bath). Thus, one HPT machine is worth between 2 and 5 conventional machines in term of productivity. This is also accompanied by a reduced maintenance and labor costs. Moreover, a significant reduction of the microbial load will have an impact on the final cooking process which can be shorter (i.e. for pasteurization of ready to eat meals).

5. CONCLUSION

Thawing remains a major bottleneck in the food industry. It induces weight and quality losses. The excessive duration of this process can be troublesome. Even though HPT will necessarily be applied to high value products, it brings several major advantages which include a possible reduction in the drip loss and in lowering of microbial load. In our studies, up to 4 log-cycle was observed for indigenous micro flora in salmon (3 log-cycles for *Listeria innocua*). A 5% drip volume reduction was observed for *aiguillat* fish confirming some previous results. Additional investigations should be done on these points with other foods. The mechanisms of microbial inactivation should be deeply investigated to elucidate if the inactivation has to be attributed to the melting of water under pressure or to some other phenomena (i.e. lipids phase change in membranes). The control of the pressurization and depressurization rates which are interacting with the phase change kinetic during HPT are also some important process related parameters. Indeed, a good control of the pressurization rate will allow to control the velocity of the melting zone and by the way might interact with the inactivation kinetic. HPT appears as a very promising process for industry. The relatively low pressure level makes it a relatively affordable process.

ACKNOWLEDGEMENTS: Dr MUSSA was supported by a post doctoral grant from Région Pays de Loire. Experimental facilities were co-funded by Région Pays de Loire, Pôle Agronomique Ouest and ENITIAA. Thanks to O. Rioux (HP facilities at ENITIAA), to S. Delepine and to M. Pasco

REFERENCES

Bomben, J. L. and J. King (1982). "Heat and mass transport in the freezing of apple tissue." Journal of food technology 17: 615-632.

Chevalier, D., A. Bail, J. M. Chourot and P. Chantreau (1999). "High pressure thawing of fish (whiting): influence of the process parameters on drip losses." Lebensmittel Wissenschaft und Technologie 32(1): 25-31.

Chevalier, D., A. LeBail and M. Ghoul (2000). "Effects of high pressure treatment (100-200MPa) at low temperature on turbot (*Scophthalmus Maximus*) muscle." Food Research International accepted.

Chevalier, D., A. LeBail and M. Ghoul (2000-I). "Freezing and ice crystals formed in cylindrical model food. Part I. Freezing at atmospheric pressure." Journal of Food Engineering 46: 277-285.

Chevalier, D., A. LeBail and M. Ghoul (2000-II). "Freezing and ice crystals formed in cylindrical model food. Part II. Comparison between freezing at atmospheric pressure and pressure shift freezing." Journal of Food Engineering 46: 287-293.

Chourot, J. M. (1997). Contribution à l'étude de la décongélation par haute pression. Nantes-FRANCE, University of Nantes.

Deuchi, T. and R. Hayashi (1992). High pressure treatements at subzero temperature : application to preservation, rapid freezing and rapid thawing of foods. High Pressure and biotechnology, colloque INSERM. C. B. al., John Libbey Eurotext Ltd. 224: 353-355.

Eshtiaghi, M. and D. Knorr (1996). "High hydrostatic pressure thawing for the processing of fruit preparations from frozen strawberries." Food Technology 10(2): 143-148.

Iso, S., H. Mizuno, H. Ogawa, Y. Mochizuki, T. Mihori and N. Iso (1994). "Physico-chemical properties of pressurized carp meat." Fisheries Science 60(1): 89-91.

Ko, W. C., M. Tanaka, Y. Nagashima, T. Taguchi and K. Amano (1990). "Effect of high pressure treatment on the thermal gelation of sardine and Alaska pollack meat and myosin pastes." Journal of Japanese Society of Food Science and Technology [Nippon Shokuhin Kogyo Gakkaishi] 37(8): 637-642.

LeBail, A., J. M. Chourot, P. Barillot and J. M. Lebas (1997). "Congélation-décongélation par haute pression." Revue Générale du Froid(972, avril 97): 51-56.

Merts, I. and S. D. Cotter (1999). A comparison of air, water and plate thawing methods for meat. 20th International Conference of the International Institute of Refrigeration, Sydney, Australia, in press.

Murakami, T., I. Kimura, T. Yamagishi and M. Sujimoto (1992). Thawing of frozen fish by hydrostatic pressure. High Pressure Biotechnology, Colloque Inserm. C. B. al., John Libbey Eurotext Ltd. 224: 329-331.

Mussa, D. M. and A. LeBail (2000). High pressure thawing of fish : evaluation of the process impact *on Listeria innocua.* Institute of Food Technologists, Dallas-USA.

Takai, R., T. Kozhima and T. Suzuki (1991). Low temperature thawing by using high Pressure. 17th Congress of the International Institute of Refrigeration, Montréal, Québec, CANADA.
Taylor, A. C. (1960). "The physical state transition in the freezing of living cells." Annals New York Academy of Science 85: 595-609.
Zhao, Y., R. A. Flores and D. G. Olson (1998). "High hydrostatic pressure effects on rapid thawing of frozen beef." Journal of Food Science 63(2): 272-275.

Trends in High Pressure Bioscience and Biotechnology
R. Hayashi (editor)
© 2002 Elsevier Science B.V. All rights reserved.

Effects of high hydrostatic pressure-thawing on pork meat

A.Okamoto[a] and A.Suzuki[b]

[a]Graduate School of Science and Technology, Niigata University, Ikarashi, Niigata 950-2181, Japan

[b]Department of Applied Biological Chemistry, Faculty of Agriculture, Niigata University, Ikarashi, Niigata 950-2181, Japan

ABSTRACT

Frozen pork meat was thawed under high pressure (100-500 MPa) and physicochemical and histological parameters (thawing loss, tenderness, color, drip, ultrastructure, etc.) were compared with those of meat thawed by running water.

Effects on physicochemical properties. The meat drip decreased by high hydrostatic pressure-thawing, and the water holding capacity of the meat improved. The discoloration of the meat induced by high pressure-thawing was not recognized with the naked eyes up to the pressurization of 200 MPa. The meat tenderization was induced during high pressure-thawing. When the frozen pork was pressurized at 200 MPa, the most desirable results were obtained. However, the changes caused by high pressure-thawing over 200 MPa were not favorable.

Effects on ultrastructure and myofibrillar proteins. The regular structure of myofibrils was gradually lost with the increase of the pressure applied, and the remarkable changes were observed in the meat thawed by pressure above 200 MPa. The measurements of the size distribution of the myofibrils prepared from high pressure-thawed meats indicated that frequency of distribution centered on a certain point with the increase of the pressure applied, in different from the progress of the fragmentation as observed in the myofibrils prepared from pressurized meats. The SDS-PAGE analysis showed the progressive solubilisation of α-actinin, troponin T, tropomyosin, and myosin light chain (MLC) in the meat thawed by pressure above 200 MPa.

1. INTRODUCTION

Following two points are important to prevent the deterioration of meat quality during thawing process of the frozen meat: 1) shortening the time for thawing, 2) thawing at low temperature as possible. High hydrostatic pressure-thawing is new technology to resolve these points. The property of water to remain in the liquid state down to about $-22^{\circ}C$ at pressure up to 210 MPa, allows the rapid thawing of frozen foods through pressure applications[1,2].

This paper describes the changes in the physicochemical and histological properties of frozen pork meat thawed under high pressure (100-500 MPa) in comparison with those of meat thawed by running water.

2. MATERIALS AND METHODS

Pork loin meats were molded (70-80×130-150 mm, thickness 15 mm) after the slaughter within 24 hours, and then frozen at -20 ℃ for more than 7 days. Frozen pork meats were vacuum-packaged in polyethylene bags, and them thawed by high pressure at 100, 200, 300, 400, and 500 MPa for 10 min at about 2-4 ℃. The control meat was thawed by running water at about 15-20 ℃ for 60 min. Thawing drip loss was determined by measuring the weight difference between before and after treatment, and in the same manner cooking loss was determined. Total loss was calculated by addition of thawing and cooking losses. The color difference ($\triangle$E) on the surface of the pork meat was evaluated on the basis of C. I. E. values (L*, a*, b*) measured with a spectrocolorimeter ColorMeter ZE 2000 (Nippon Denshoku Co.). Changes in textural properties of thawing and cooking meat were measured by a texturometer RHEONER RE-33005 (Yamaden Co.) with maximum load. An electronmicroscopic observation of muscle strips taken from pressure-thawed pork meats was carried out by the method of Suzuki *et al.* [3]. Myofibrils were prepared from pressure-thawing and running water-thawing meats according to the procedure of Busch *et al.*[4]. SDS-PAGE analysis of the myofibrils solubilized in 0.01 M sodium phosphate buffer (pH 7.0) containing 5 % SDS and 1 % 2-mercaptoethanol was conducted according to the procedure described by Laemmli [5] with a slight modification. The particle size distribution of the myofibrils were determined by the particle size distribution meter LS-2000 (Coulter Co.) of the light scattering type with the laser analysis.

3. RESULTS AND DISCUSSION

3.1. Effects on drip loss and color

The effects of the high pressure-thawing on drip and color difference were shown in Table 1 and 2, respectively. As shown in the Table 1, the high pressure-thawing of pork meat is likely to reduce thawing loss. The thawing loss reduced with the increase of pressure applied, and the highest reduction was obtained on the meat thawed at 400 MPa. The cooking loss increased apparently with high hydrostatic pressure-thawing, but total loss was less than that of the meats thawed by running water.

As shown in the Table 2, the discoloration of meat color as expressed ΔE was a little on the meat thawed at 100 MPa, but progressed with the increase of the pressure applied above 200 MPa. However, the difference in the color was not recognized with the naked eyes between the meats thawed at 100 and 200 MPa. These results are a little different from the results reported by Massaux *et al.* [7].

3.2. Effects on texture

Effects of high pressure-thawing on the hardness of meat were shown in Figure 1. The tenderization of the meat was induced by high pressure-thawing, and accelerated with the increase of the pressure applied. However, further progress was not observed on the meat thawed by high pressure over 200 MPa. In the case of ribloin cooked at 70 ℃, the increase of hardness was observed on the meat thawed by high pressure over 300 MPa. The higher amount of connective tissue than sirloin and the complicated distribution of connective tissue may cause the difference.

Table 1. Thawing and cooking losses of the high pressure-thawed pork meat

Thawing pressure	Thawing loss (%)	Cooking loss (%)	Total loss (%)
Control	4.28±0.52	15.31±2.08	19.30±2.57
100 MPa	4.97±0.46	18.29±2.18	22.70±2.29
200 MPa	2.76±0.39	18.93±0.46	21.40±0.74
300 MPa	1.77±0.24	16.95±1.18	18.39±1.31
400 MPa	1.51±0.45	14.92±0.79	15.60±1.10
500 MPa	2.29±0.25	16.59±0.83	18.48±0.91

Thawing and cooking losses were determined by weight difference before and after treatment. Total loss was calculated by adding thawing and cooking losses.

Table 2. Evaluation of the color difference on the surface of the high pressure-thawed pork meat

Thawing pressure	ΔE	evaluated*
Control	—	—
100 MPa	3.60±0.55	appreciable
200 MPa	14.89±1.19	very much
300 MPa	20.63±1.82	very much
400 MPa	20.49±1.29	very much
500 MPa	20.31±0.89	very much

*Discoloration was evaluated on the basis of NBS units.

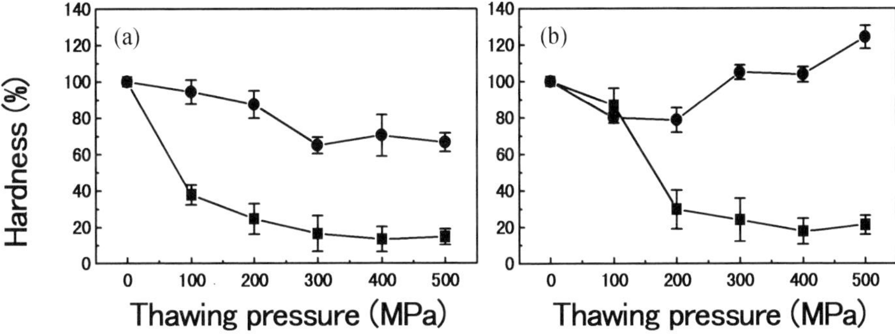

Figure 1. Hardness of the high pressure-thawed pork meat
Relative hardness was expressed as % of that from thawed pork meat by running water (control). The result shown in the figure is representative of those obtained by repeated experiments for five different pork meats.
(a) sirloin; (b) ribloin
■, non-cooked meat; ●, cooked meat at 70 ℃

3.3 Effects on ultrastructure

The ultrastructures of the muscle strips prepared from high pressure-thawed muscles are shown in Figure 2. In the muscle strips thawed at 100 MPa, a contraction of the sarcomere was observed, and the difference in density between the A-band and the I-band became indistinguishable as compared with the control. Marked rupture of the filamentous structure of the I-band and a loss of the M-line materials were observed in the muscle strips thawed at 200 MPa. The Z-line structure seemed to be out of register. In the muscle strips thawed at 300 MPa, the structural continuity of the sarcomere was almost completely lost, with broken A- and I-filament spread over the sarcomere. Complete loss of the M-line and thickening of the Z-line, probably due to a collapse of the I-filament, were also observed. Cleavage of the A-band adding to the many changes already mentioned was observed in the muscle strips thawed at 400 and 500 MPa. These results are almost in agreement with the results reported by Suzuki *et al.* [8].

3.4. Effects on SDS-PAGE profile of myofibrillar proteins

The effects of high pressure-thawing on the protein constituents of the myofibrils are shown in Figure 3. No significant difference in the electrophoretic pattern of myofibrillar proteins was observed between the control and pressure-thawed meats at 100 MPa. But, the progressive solubilisation of α-actinin, troponin T, tropomyosin, and MLC was observed in the meats thawed above 200 MPa. These results are almost in agreement with the results reported by Nishiwaki *et al.* [9].

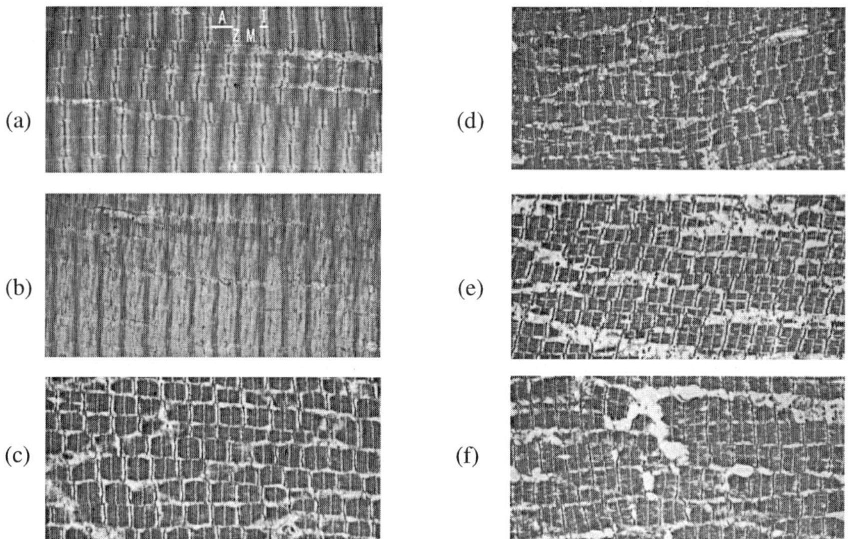

Figure 2. Electron micrographs of muscle strips prepared from the high pressure-thawed pork meat

(a) control (thawed by running water); (b) thawed at 100 MPa; (c) thawed at 200 MPa; (d) thawed at 300 MPa; (e) thawed at 400 MPa; (f) thawed at 500 MPa

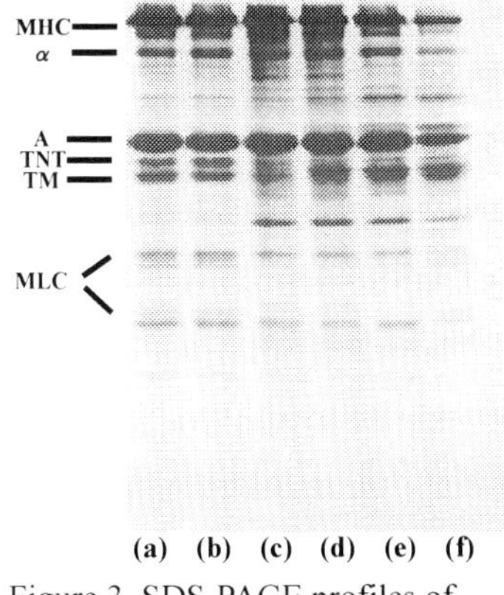

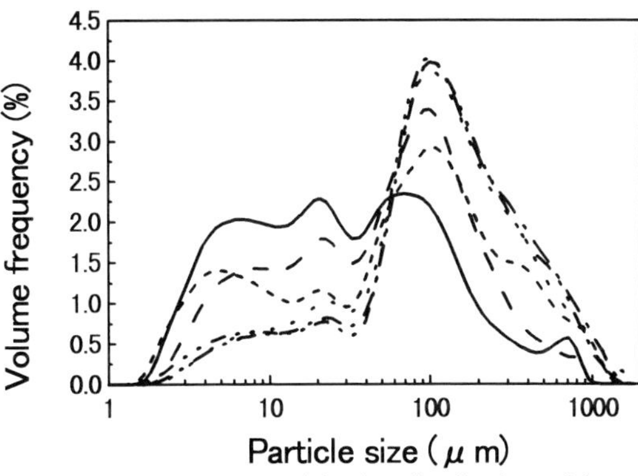

(a) (b) (c) (d) (e) (f)

Figure 3. SDS-PAGE profiles of the myofibrils prepared from the high pressure-thawed pork meats

Figure 4. Particle size distribution of the myofibrils prepared from the high pressure-thawed pork meats

MHC, myosin heavy chain; α, α-actinin; A, actin; TNT, troponin T; TM, tropomyosin; MLC, myosin light chain. (a) control(thawed by running water); (b) thawed at 100 MPa; (c) thawed at 200 MPa; (d) thawed at 300 MPa; (e) thawed at 400 MPa; (f) thawed at 500 MPa.

——— , control; - - - , thawed at 100 MPa; ·········· , thawed at 200 MPa; — · — · , thawed at 300 MPa; — ·· , thawed at 400 MPa; ········· , thawed at 500 MPa

3.5. Effects on particle size distribution of the myofibrils

Effects of high pressure-thawing on the particle size distribution of the myofibrils are shown in Figure 4. The measurements of the size distribution of the myofibrils prepared from high pressure-thawed meats indicated that frequency of distribution centered on a certain point with the increase of the pressure applied, in different from the progress of the fragmentation as observed in the myofibrils prepared from pressurized meats.

4. CONCLUSION

From these results, it was proved that the high hydrostatic pressure-thawing was applicable to the meat processing. The most interesting method of pork meat thawing seems to be process under a pressure of 200 MPa. It should be applied to the meat products requiring no long-time heating, such as cured smoked pork, roast beef, tataki, etc.

ACKNOWLEDGMENTS

We express our thanks to Dr. T. Tsujita and Mr. Sudo of the Ehime University School of Medicine for their assistance in preparing electronmicrograph, and President K. Ozawa of the Nihonshokken Co., Ltd. for his support during the experiments.

REFERENCES

1. H. D. Lüdeman, In High Pressure and Biotechnology, (eds. C. Balny, R. Hayashi, K. Heremans and P. Masson), John Libbey Eurotext, Montrouge, pp.371, 1992.
2. R. M. George, Trends in Food Science and Technology, 4 (1993), 134.
3. A. Suzuki, M. Saito, H. Sato, and Y. Nonami, Agric. Biol. Chem., 42 (1978), 2111.
4. W. A. Busch, M. H. Stromer, D. E. Goll, and A. Suzuki, J. Cell Biol., 52 (1972) 367.
5. U. K. Laemmli, Nature, 227 (1970) 680.
6. C. Massaux, F. Béra, B. Steyer, M. Sindic, and C. Deroanne, In Advance in Pressure Bioscience and Biotechnology, (ed. H. Ludwig), pp. 485, 1999.
7. C. Massaux, F. Béra, B. Steyer, M. Sindic, and C. Deroanne, In Advance in Pressure Bioscience and Biotechnology, (ed. H. Ludwig), pp. 497, 1999.
8. A. Suzuki, M. Watanabe, K. Iwamura, Y. Ikeuchi, and M. Saito, Agric. Biol. Chem., 54 (1990) 3085.
9. T. Nishiwaki, Y. Ikeuchi, and A. Suzuki, Meat Science, 43 (1996) 145.

Trends in High Pressure Bioscience and Biotechnology
R. Hayashi (editor)
                    577

Pressure shift freezing of turbot (*Scophthalmus maximus*) and carp (*Cyprinus carpio*) : effect on ice crystals and drip volumes.

D. Chevalier[1], A. Le Bail[1*], A. Sequeira-Munoz[2], B.K. Simpson[2] & M. Ghoul[3].

[1] GEPEA-ENITIAA, la géraudière BP 82225 F-44322 Nantes cedex 3, France

[2] Dpt Food Science and Agricultural Chemistry, McGill University, Ste Anne de Bellevue, Quebec, Canada H9X3V9

[3] ENSAIA, 2 av. de la forêt de Haye F-54500 Vandoeuvre les Nancy, France
*corresponding author : lebail@enitiaa-nantes.fr

Turbot and carp fillets were frozen either by Pressure Shift Freezing (PSF, 140MPa, -14°C) or Air-Blast Freezing (ABF), and then stored at -20°C for 75 days. PSF resulted in smaller and more regular ice crystals compared to ABF samples whatever the fish species was. Ice crystals area in PSF samples was about ten and seven times smaller than in ABF ones for turbot and carp, respectively. In addition, PSF induced less thawing and cooking drips than ABF. The storage time did not adversely influence the ice crystal size, thawing drip and cooking drip for turbot and carp for PSF samples.

1. INTRODUCTION

Freezing is a convenient preservation process particularly used for fish and other seafood. Statistics show that as much as 60% of all fishes consumed in North America and Europe have been frozen once or possible twice (Morrison, 1993). Nevertheless, freezing usually results in undesirable physical, chemical and structural changes leading to quality losses (Krivchenia & Fennema, 1988). Yet, the effect of freezing on microbial and biochemical processes is now well understood, but physical changes as a result of ice formation remain relatively unclear. Nevertheless the size and location of ice crystals formed during freezing are some of the most put forward factors affecting the fish quality. A slow freezing rate induces a high salt concentration in the extracellular fluid, which in turn draws out moisture from inside the cell by osmosis resulting in the formation of mainly extracellular ice crystals particularly stretching the tissues (Shenouda, 1980). Conversely, a fast freezing rate is supposed to minimise the migration of water into the extracellular spaces, and consequently forms smaller intracellular ice crystals (Shenouda, 1980). In addition, loss of moisture during freezing and ice crystal coarsening during storage are known to be related to the quantity of drip losses (Petrovic, Grujic & Petrovic, 1993). Supercooling (difference between the temperature of sample and the expected solid-liquid equilibrium temperature at a given pressure) is driving force of nucleation (Burke et al., 1975). The freezing temperature of water is depressed under pressure down to -21°C at 207 MPa (Bridgman, 1912). PSF consists in placing a food in a high pressure vessel whose temperature is regulated below the initial freezing temperature of the food at atmospheric pressure. Then, pressure is increased up to a pressure level for which freezing is not possible for the sample. When the temperature of the sample is homogeneously close to the regulated temperature, the pressure is rapidly released. Thus, the water contained in the sample finds itself in supercooling condition and an homogeneous ice nucleation is carried out at core which much-preserved microstructure of sample. The percentage of frozen water after the pressure release is 32% maximum (Sanz et al., 1997). After the pressure release, ice crystals growth is then achieved at atmospheric pressure in a conventional freezer. Several investigators have studied PSF of food models or systems such as tofu (Kanda et al., 1992) and oil-in-water emulsion (Levy et al., 1999). Some others were interested in PSF of vegetables (Koch et al., 1996 ; Fuchigami et al., 1998 ; Otero et al., 1998) or muscle food (Martino et al., 1998 ; Chevalier et al., 2000). All these studies demonstrated much-preserved structure and microstructure for pressure shift frozen samples compared with samples frozen by conventional methods. In addition, Otero et al. (1998) and Chevalier (2000) showed a decline in the drip losses for PSF samples compared with those conventional frozen for thawed eggplant and turbot, respectively.

For this study, farm turbots and carps were chosen to compare the effects of PSF and ABF on the microstructure of flesh of these two fish species. Firstly, the size and location of ice crystals formed in turbot and carp fillets during PSF and ABF were analysed. Secondly, the influence of the ice crystals formation on thawing and cooking drip volumes was evaluated.

2. MATERIALS AND METHODS
2.1. Fish sampling
Two different lots of 4 years old farm turbots (*Scophthalmus maximus*) were obtained from France Turbot Company (Noimoutier, France) in November 1999. Their average length and weight were 35 cm ± 1.5 cm and 824 g ± 90 g, respectively. Fishes were slaughered and transported within 24 h in ice boxes to the food processing laboratory (Nantes, France) where they were cleaned, skinned and filleted in post rigor condition. Live carps (*Cyprinus carpio*) (weight, ca. 1226 ± 221 g and length 40 ± 7 cm) were obtained from an aquaculture farm (Ferme acquacole d'Anjou, Morannes, France) and immediately transported to the pilot plant (Nantes, France). Fishes were slaughtered, cleaned and filleted the next day and processed in post-rigor condition. Each turbot and carp fillet was divided into two halves in order to compare homogeneous batches; one half was subjected to pressure shift freezing and the second one was tested at atmospheric pressure. 100 g from each half of the flesh fillet were placed in moisture-impermeable polyethylene bags (La Bovida, France), vacuum-packaged and kept on ice before freezing treatments.

2.2. Freezing processes
Air-Blast Freezing was carried out in a freezer (Servathin, France) at -20°C using an air speed of 4m/s. The temperatures of the cooling medium and centre of the sample were monitored using a temperature recorder (Model SA 32, AOIP, France) with K-type thermocouples (Omega, Stamford, USA). Pressure Shift Freezing was carried out in 3L capacity high pressure vessel (ALSTOM, Nantes, France). The stainless steel vessel (12 cm internal diameter and 30 cm internal height) and the pressure transmitting medium (50/50 v/v ethanol/water solution) were maintained at -14°C by circulation of ethanol/water solution (-14°C) from an external cryostat in the internal cooling circuit of the high pressure vessel. The

pressure was supplied by a high pressure pump. A K-type thermocouple (0.3 mm diameter, Omega, Stamford, USA) was installed at the centre of samples and another one in the pressurisation medium. A data logger (Model SA 32 AOIP, Evry, France) recorded temperature and pressure data with an acquisition rate of one measurement per second. Samples were placed in the high pressure vessel and the pressure was increased to 140 MPa at a rate of 100 MPa/min. This pressure level was chosen according to a previous study carried out on the effects of high pressure treatment (100-200 MPa) at low temperature (+4°C) on turbot muscle (Chevalier, 2000). This study showed particularly that a high pressure treatment at 140 MPa limited the colour, lipid and protein changes in turbot fillet and allowed to achieve a high level of supercooling. When the temperature of the samples reached -14°C (≈12 min), pressure was released at a rate of 10 MPa/sec resulting in supercooling of the samples. Nucleation was initiated and as freezing was only partial at the end of the pressure release, samples were immersed in a stirred cooling bath of 50/50 (v/v) ethanol water solution at -20°C for 7 min. Once frozen, the fish samples were stored in isothermal boxes at -20°C for about three months. Untreated samples (controls) and thawed samples after 2, 15, 30, 45 and 75 days of storage at –20 °C were analysed. Thawing of the samples was carried out by placing frozen packaged fish samples in a 4°C cold room overnight.

2.3. Microscopy analysis

Three samples of about 1×0.5×0.5 cm were cut in a cold room (-20°C) transversally to the muscle fibres from the surface and core of frozen fillet directly using a blade previously cooled at -20°C. These samples were prepared by an indirect technique called isothermal freeze substitution (Martino & Zaritzky, 1986). Stained sections of samples were finally viewed under an optical microscope (LEICA, Germany) fitted with a CCD RGB camera (MACC-C71, Sony, Japan). The recorded images were treated with an image analysis software (Visilog 5.2, Noesis, France). The surface of about 150 ice crystals was evaluated for each case considered.

2.4. Drip losses

Thawing and cooking drip volumes were determined by weighting samples prior to freezing and after thawing and prior and after cooking, respectively. Fish samples were cooked for 20 min at 65°C. Drip losses results were expressed as mass of drip per g of dry matter of flesh. The dry matter of flesh fish was determined in triplicate by drying 5 g of flesh at 102°C for 24 h.

2.5. Statistical analysis

The effect of frozen storage time was determined by two-way analysis of variance according to Sokal & Rohlf (1981) using Statgraphics plus version 2.1 software (Statistical Graphics Corp., Princeton, NJ, USA). In all cases the significant level used was 5% (α=0.05). To test significant differences between means for the freezing processes at the same time storage, Student T-test was used, whenever the Fisher F-test was not significant (p≥0.05).

3. RESULTS AND DISCUSSION
3.1. Miscroscopy investigations

Figures 1(a) and 1(b) show micrographs of the surface and central zones of the ABF turbot samples after 2 days of storage at -20°C. The differences between the surface and centre locations in terms of ice crystal size (Table 1) showed the effect of the thermal gradient that occurred during the atmospheric freezing process. The surface of the sample was exposed to a higher freezing rate and the ice crystals in this part were significantly (p<0.05) smaller than those formed in the central part for either turbot and carp fillets.

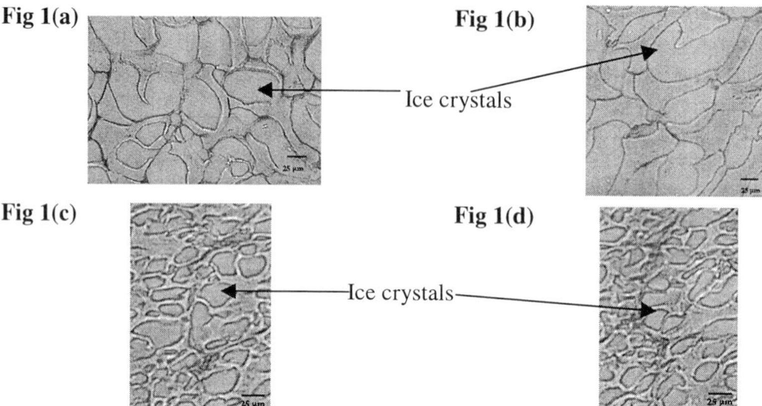

Figure 1 : Micrographs of air-blast frozen (a : surface, b: centre) and pressure shift frozen (c : surface, d: centre) turbot fillets.

Table 1 : Ice crystal surface formed in turbot and carp fillets either by air-blast or by pressure release and stored 2 and 75 days[1].

Storage time (days)	Fish	Air-blast freezing		Pressure shift Freezing	
		Surface (μm^2)	Centre (μm^2)	Surface (μm^2)	Centre (μm^2)
2	**Turbot**	2868±636	4983±540	359±102	332±95
2	**Carp**	3710±824	-	534±115	-
75	**Turbot**	3070±940	5215±639	336±111	364±121
75	**Carp**	4032±631	-	587±145	-

[1]Results are shown as mean value ±standard deviation from about 150 measurements.

Ice crystals formed during ABF freezing process had besides irregular shapes. ABF corresponds to a quite slow freezing process which generally induced important shrinkage of cells and formation of large extracellular ice crystals (Shenouda, 1980). Ice crystals formed in turbot and carp appeared to have statistically similar size at surface for ABF process. Micrographs of turbot samples frozen by PSF process and stored for 2 days at -20°C are presented in Figs. 1(c) and 1 (d) for surface and central zones, respectively. Micrograph from the centre of the PSF samples is very quite similar to the one from the surface. No significant difference appeared in the ice crystal size between the surface and centre locations of these samples (Table 1). That confirms the assumption of an uniform and isotropic ice nucleation ice crystals in PSF samples. Smaller ice crystals than in the previous freezing process (ABF) were moreover observed. PSF reduced about ten times the ice crystal size in turbot fillets and about seven times in carp fillets compared to ABF, as shown in Table 1. In the case of PSF, ice crystals appeared regular with a rounded shape. The fact that smaller ice crystals were created by PSF is attributed to the degree of supercooling reached when pressure was released. The freezing process by pressure release from 140 MPa was theoretically able to achieve a supercooling of around -14°C and so induced a 140-fold increase in nucleation rate according Burke et al. (1975). Nevertheless, in the case of turbot, the supercooling was experimentally estimated at 6.3°C (Chevalier, 2000) that induced the formation of large numbers of small and homogeneous ice crystal nuclei. Martino et al. (1998) obtained similar results on pork meat pieces frozen by pressure release from 200 MPa at -20°C. During the storage time, no gradual accretion of ice crystals was observed in the fillets of the two fish species as shown in Table 1. The relatively short storage time and constant storage temperature had certainly restricted this phenomenon of accretion.

3.2. Thawing and cooking drips

The results concerning the thawing and cooking drips are presented in the case of turbot fillets in Table 2. The thawing drip of the conventionally frozen samples increased significantly after 45 days of storage. This increase is often attributed to the surface dehydration of protein (Petrovic et al., 1993). Conversely no significant difference appeared in the thawing drip of PSF samples. In addition, a significant decrease of the thawing drip was observed concerning PSF samples compared with ABF samples from 15 days of storage.

Table 2 : Thawing and cooking drip volumes of turbot fillets according to the freezing process and the storage time[1].

	ABF		PSF	
Storage time (days)	Thawing drip (g/g)	Cooking drip (g/g)	Thawing drip (g/g)	Cooking drip (g/g)
2	0.21 ± 0.08^a	0.86 ± 0.12^a	0.17 ± 0.07^a	0.80 ± 0.06^a
75	0.33 ± 0.06^b	0.75 ± 0.11^a	0.17 ± 0.03^a	0.47 ± 0.06^b

[1]Results are shown as mean value±standart deviation from 5 measurements.

[a,b]Different letters in the same column indicate significant difference (p<0.05).

The cooking drip of PSF samples showed a tendency to decrease with the storage time. These reduction of drip volumes may be related to the smaller and regular ice crystals created during PSF which were less damaging for cells when compared to ice crystals of ABF samples. Similar observations were carried out in the case of ABF and PSF carp fillets.

4. CONCLUSION

To conclude, it appeared that PSF (140 MPa, -14°C) was effective to produce small and regular ice crystals resulting in reduced drip volumes in comparison to ABF in the case of turbot and carp fillets. This shows the relative advantages and potential use of high pressure technology to freezing and to preserve a good quality of products.

Acknowledgements: The authors wish thank the Ministère de l'Agriculture et de la Pêche, the Association Inter Universitaire France-Québec, P. Cantoni for microscopy analysis and E. Lukomska for help in microscopy techniques.

References

Bridgman, P.W. (1912). Water in the liquid and five solid forms under pressure. *Proceedings of the American Academy of Arts and Sciences*, 47, 441-558.

Burke, M.J., George, M.F., & Bryant, R.G. (1975). Water in plant tissues and frost hardiness. In R.B. Duckworth, *Water relations of foods* (pp. 111-135). London : Academic Press.

Chevalier, D. (2000). Contribution à l'étude de la congélation par détente haute pression. Thèse de Doctorat, Ecole polytechnique de l'université de Nantes, 182p.

Chevalier, D., Sentissi, M., Havet, M., & Le Bail, A. (2000). Comparison of air-blast and pressure shift freezing on Norway lobster quality. *Journal of Food Science*, 65, 329-333.

Fuchigami, M., Kato, N., & Teramoto, A. (1998). High-pressure freezing effects on textural quality of Chinese cabbage. *Journal of Food Science*, 63, 122-125.

Kanda, Y., Aoki, M., & Kosuki, T. (1992). Freezing of tofu (soybean curd) by pressure-shift freezing and its structure. *Japan Shokuhin Kogyo Gakkaishi*, 39, 608-614.

Koch, H., Seyderhelm, I., Wille, P., Kalichevsky, M.T., & Knorr, D. (1996). Pressure-shift freezing and its influence on texture, colour, microstructure and rehydration behaviour of potato cubes. *Nahrung*, 40, 125-131.

Krichvenia, M., & Fennema, O. (1988). Effect of cryoprotectants on frozen Burbot fillets and a comparison with whitefish fillets. *Journal of Food Science*, 53, 1004-1008;1050.

Levy, J., Dumay, E., Kolodziejczyk, E., & Cheftel, J.C. (1999). Freezing kinetics of a model oil-in-water emulsion under high pressure or by pressure release. Impact on ice crystals and oil droplets. *Lebensmittel-Wissenschaft und-Technologie*, 32, 396-405.

Martino, M., & Zaritzky, N. (1986). Fixing conditions in the freeze substitution technique for light microscopy observation of frozen beef tissue. *Food Microstructure*, 35, 19-24.

Martino, M., Otero, L., Sanz, P.D., & Zaritzky, N.E. (1998). Size and location of ice crystals in pork frozen by high-pressure assisted freezing compared to classical methods. *Meat Science*, 50, 303-313.

Morrison, C.R. (1993). *Fish and shellfish*. In C.P. Mallet, Frozen Food Technology. London : Chapman & Hall.

Otero, L., Solas, M.T., Sanz, P.D., de Elvira, C., & Carrasco, J.A. (1998). Contrasting effects of high-pressure assisted freezing and conventional air-freezing on eggplant tissue microstructure. *Zeitschrift-fur-Lebensmittel-Untersuchung-und-Forschung*, 206, 338-342.

Petrovic, L., Grujic, R., & Petrovic, M. (1993). Definition of the optimal freezing rate-2. Investigations of the physico-chemical properties of beef *M. longissimus dorsi* frozen at different freezing rates. *Meat Science*, 33, 319-331.

Sanz, P.D., Otero, L., de Elvira, C., & Carrasco, J.A. (1997). Freezing processes in high pressure domains. International *Journal of Refrigeration*, 20, 301-307.

Shenouda, S.Y. (1980). Theories of protein denaturation during frozen storage of fish flesh. In C.O. Chichester, E.M. Mrak, & G.F. Stewart, *Advances in Food Research* vol.26 (pp. 275-311). London : Academic Press.

Sokal, R.R., & Rohlf, F.J. (1981). *Biometry* (2nd ed.). San Francisco, W.H. Freeman & company.

Trends in High Pressure Bioscience and Biotechnology
R. Hayashi (editor)
© 2002 Elsevier Science B.V. All rights reserved.

Circular Dichroism under High Pressure

R. Hayashi, Y. Kakehi, M. Kato, N. Tanimizu, S. Ozawa, M. Matsumoto*, S. Kawai*, and P. Pudney**

Division of Applied Life Sciences, Graduate School of Agriculture, Kyoto University, Kyoto 606-8502, *Teramecs Company, Kyoto 612-8448, Japan, and **Unilever Research Coloworth Laboratory, Bedford MK4 1LQ, UK

Circular dichroism (CD) cells, equipped with quartz window, are used for measurements of CD in the ultraviolet (UV) region under high pressure. We describe here an improvement in these cells, which are typically distorted by high pressure, thus diminishing the CD signal.   A slit with a small 3 mm diameter hole was placed in the light source-side in contact with the window, thus adjusting the center of the light path.   This method gave reasonable CD spectra of camphorsulfonic acid, poly-glutamic acid, and RNase A at 150 MPa or lower.   For measurements at higher pressure, synthetic diamond was used as windows of the CD cell and reliable CD spectra were obtained in the near-UV and visible light region at pressure of 400 MPa or lower, since the diamond strongly absorbed UV light at 230 nm and shorter wavelengths.   By monitoring the molar ellipticity at 235 nm, it was possible to estimate pressure-induced changes in the secondary structure of poly-amino acids, myoglobin, and RNase A.   The CD spectra of carboxypeptidase Y, lysozyme, and RNase A were also measured at 245-320 nm under high pressure to estimate structural changes which occur at these conditions.   CD spectra of RNase A at 0.1-400 MPa were changed by pressurization at 100 to 250 MPa, showing a reversible two-state transition.   However, this negligible discrepancy may be corrected when diamond windows are used.

1. INTRODUCTION

Absorption, fluorescence, NMR, IR, and Raman spectroscopies have been developed for the *in situ* measurement of protein conformational changes under high pressure.   Some spectroscopies, however, require a large amount of concentrated protein sample, which

584

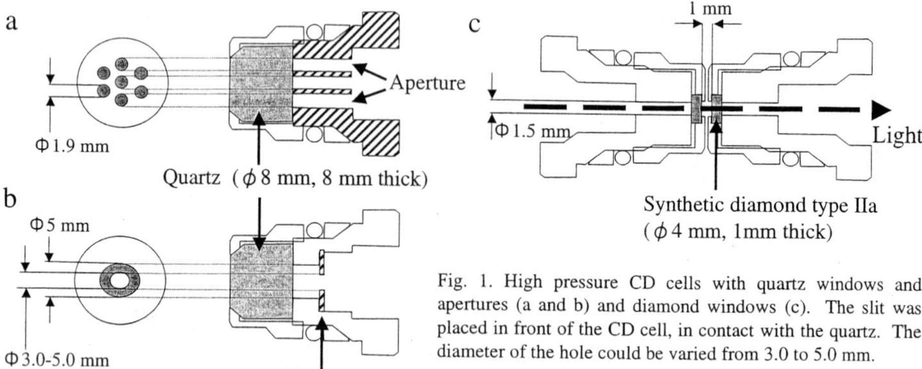

Fig. 1. High pressure CD cells with quartz windows and apertures (a and b) and diamond windows (c). The slit was placed in front of the CD cell, in contact with the quartz. The diameter of the hole could be varied from 3.0 to 5.0 mm.

sometimes induces aggregation or gelation during the high pressure treatment resulting in erroneous interpretations of the spectrometric analysis. This problem can be avoided via the use of circular dichroism (CD) because of low concentration of sample proteins required. However, the *in situ* observation of CD under high pressure, simply referred to as high pressure CD in this report, has been unsuccessful except for measurements at 15 MPa (1).

This report describes the difficulties and the successes encountered in attempts to construct devices for use in high pressure CD: As a material for use in windows for high pressure CD, quartz is not sufficiently resistant to pressure, and sapphire absorbs far ultraviolet (UV) light, but synthetic diamond, which is strongly resistant to pressure, would be expected to result in normal CD signals.

2. MATERIALS AND METHODS

2.1. Materials

Ammonium-D-camphor-10-sulfonate (CSA) (special grade, for CD measurement) was purchased from Katayama Kagaku (Osaka, Japan). Poly-L-glutamic acid sodium salt (MW=8000-8800) (PLG), bovine pancreatic ribonuclease A (RNase A) (Type XII-A) and hen egg lysozyme were purchased from Sigma Chemical (St. Louis, USA) and were used without further purification. Poly-L-lysine sodium salt (PLL) was purchased from the Peptide Institute (Osaka, Japan). Myoglobin and RNase A used for CD measurement in the near-UV were purchased from Nacalai Tesque (Kyoto, Japan). Carboxypeptidase Y (CPY) was purified as described previously (2). Synthetic single crystals of diamond Type IIa (φ 4 mm, 1mm thick), the nitrogen content of which is below 0.1%, were donated by Sumitomo Electric

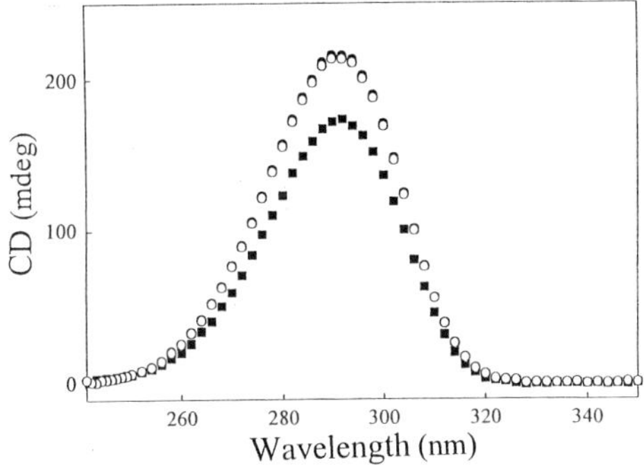

Fig. 2. CD spectra of CSA using quartz windows and a single hole-slit (●, 3.0 mm, ○, 3.5 mm, ■, 4.0 mm) at 200 MPa . CSA conc. was 0.6 % in 0.1 M KCl.

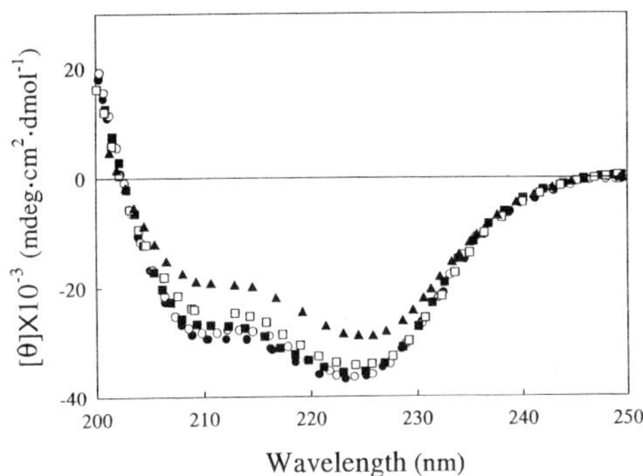

Fig. 3. CD spectra of PLG using quartz windows and a single hole-slit at various pressures (●, 0.1 MPa, ○, 50 MPa, ■, 100 MPa, □, 150 MPa, ▲, 200 MPa). PLG conc. was 0.4 mg/ml in 10 mM sodium acetate, pH 4.2, containing 0.1 M KCl. The diameter of the hole was 3.0 mm

Ind. (Itami, Japan) (3).

2.2. Preparation of sample solutions

CSA was dissolved to a concentration of 0.6 % for CD and 20 % for FTIR measurements, respectively, in 0.1 M KCl; PLG for 0.4 mg/ml in 10 mM sodium acetate, pH 4.2, containing 0.1 M KCl; PLL for 0.4 mg/ml in 0.1 M KCl/NaOH, pH 12; myoglobin for 0.2 mg/ml for CD and 60 mg/ml for FTIR measurements, respectively, in 10 mM MES, pH 6.0, containing 0.1 M KCl; RNase A for 0.2 mg/ml in 10 mM MES, pH 6.0, containing 0.1 M KCl or in 0.1 M KCl/HCl, pH 2.0, for CD and 6 mg/ml in 0.1 M KCl/HCl, pH 2.0, for FTIR measurements, respectively.

2.3. Measurement of CD and FTIR

CD cells (light path, 1 mm) with quartz windows (ϕ 10 mm and 8 mm thick), in conjunction with a high pressure vessel, were constructed in cooperation with Teramecs Co. (Kyoto, Japan). Apertures with small holes were placed in front of the quartz windows and the center of the hole was precisely adjusted in a direct path along the light pass. Two types of apertures were prepared: one with seven holes (ϕ 1.9 mm in size) prepared with SUS630H125 (6.5 mm thick) (Fig. 1a). The other contained one hole (ϕ 3.0, 3.5, 4.0, or 5.0 mm) and was made from SUS304 (1 mm thick) (Fig. 1b).

High pressure was generated by a manual pump of Teramecs Co. (1).

CD spectra were measured at 0.1-400 MPa using a Jasco J700 CD spectropolarimeter

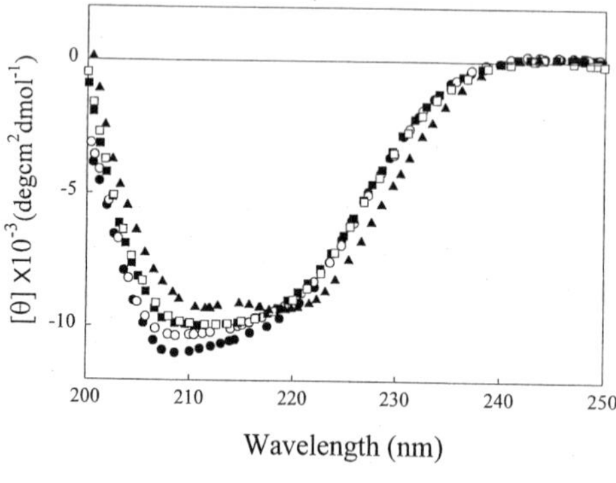

Fig. 4. CD spectra of RNase A using quartz windows and a single hole-slit at various pressures (●, 0.1 MPa, ○, 50 MPa, ■, 100 MPa, □, 150 MPa ▲, 200 MPa). RNase A conc. was 0.2 mg/ml in 0.1 M KCl, pH 3.0. See Fig. 3 for other conditions. Insert shows changes in CD at 291 nm *vs.* pressure.

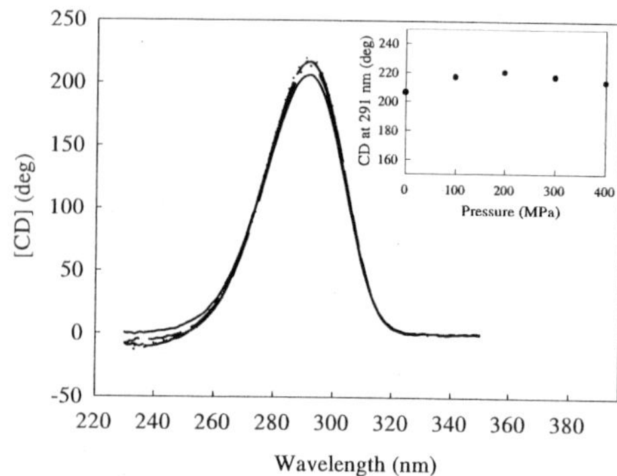

Fig. 5. CD spectra of CSA using diamond windows at various pressures (—, 0.1 MPa, -----, 100 MPa, ·····, 200 MPa, -·-·-, 300 MPa, -··-, 400 MPa). CSA conc. was 0.6 % in 50 mM MES, pH 6.0, containing 0.1 M KCl.

(Nippon Bunko, Tokyo, Japan).

FTIR spectra were obtained on a Bruker IFS 88 model infrared spectrometer using a diamond anvil cell. α-Quartz was used as an internal pressure standard (5). Sample solutions, along with a small amount of powdered α-quartz were placed in a small diameter hole of a 25 mm stainless steel gasket, which was mounted on the diamond anvil cell. A curve-fit software program was used for the analysis.

2.4. CD measurement in the near-UV region

For a CD measurement at 0.1 MPa and high pressure, CD cells with quartz windows (path length, 5.0 mm) and diamond windows (path length, 2.0 mm) were used. The results are expressed as the mean residue ellipticity $[\theta]$ (deg·cm^2·dmol^{-1}), which is defined as $[\theta] = (\theta_{obs}/10)\cdot(MRW/cL)$, where θ_{obs} is the ellipticity in deg; MRW, the mean residue molecular weight; c, the protein concentration in g/ml; L, the path-length of the cell in cm.

In the case of a CD measurement under high pressure, a correction was made for the compressibility of water (6).

3. RESULTS AND DISCUSSION

3.1. Properties of quartz, sapphire, and diamond

Quartz has no absorption at far-UV region, but sapphire and diamond strongly

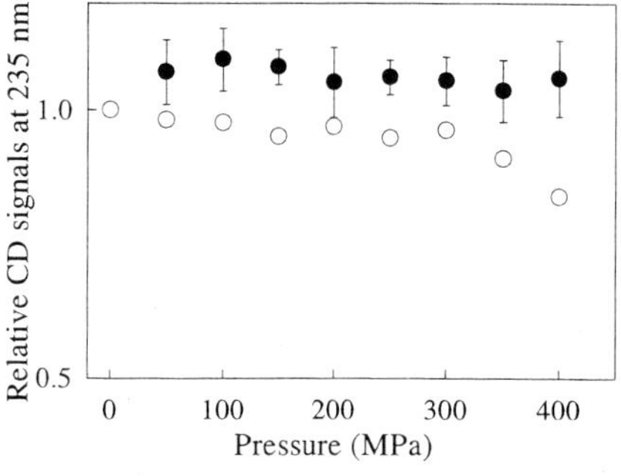

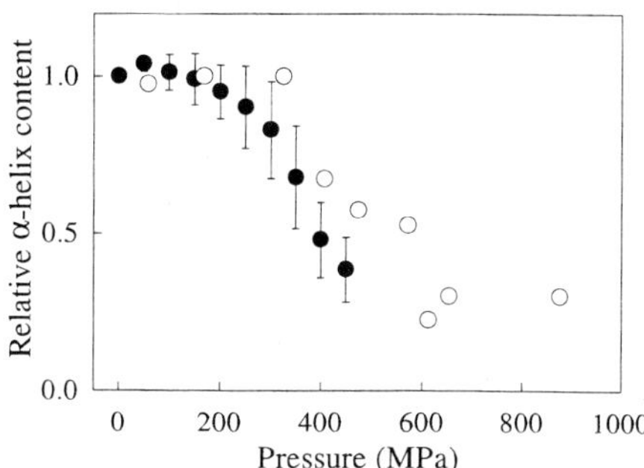

Fig. 6.   Changes in molar ellipticity at 235 nm of PGA (●) and PLL (○) at various pressures.   The ordinate shows the relative molar ellipticity at 235 nm.   PGA and PLL conc. were 0.4 mg/ml in 10 mM sodium acetate, pH 4.2, containing 0.1 M KCl and 0.4 mg/ml in 0.1 M NaOH/KCl, pH 12.0, respectively.   See Fig. 5 for other conditions.

Fig.7. Change in the α-helix content of myoglobin under various pressures measured by CD (●) and FTIR (○) measurements.   See Fig. 5 for other conditions.

absorbed UV light at 250 nm or below and 230 nm or below, respectively.   Quartz and diamond showed no CD signals, but sapphire showed strong signals at 260 nm or below even at 0.1 MPa.   Quartz generally is tolerant of pressure at 200 MPa or lower but sometimes became cloudy when it was repeatedly used at higher pressure, probably due to crystallization. These drawbacks were improved as follows.

3.2. High pressure CD with quartz windows

The CD spectra of PLG and CSA with quartz windows gradually diminished at 210 to 250 nm and 250 to 320 nm, respectively, with an increase in pressure from 0.1 to 250 MPa. However, the FTIR spectrum of CSA, which has a sharp peak at 1730 cm^{-1}, due to C=O vibration, showed no change at 800 MPa or lower.   Thus, the decrease in the CD signal of PLG can be attributed to some distortion of the quartz windows, which is caused by high pressures, and not to conformational changes of the samples.

An attempt to minimize the possible distortion of the quartz windows was unsuccessful in preventing the diminution of CD signals. That is, two supporting metal plates (6.5 mm thick), which contain seven small holes (Φ 1.9 mm), were placed in front of the two quartz windows in tight contact with the quartz, which sacrifices light quantity but would be expected to prevent distortion of the quartz (Fig. 1a).

In the second attempt, one thin metal-plate (1 mm, thick) with a single hole (simply referred to as a single hole-slit), the diameters of which were 3.0, 3.5, 4.0 or 5.0 mm, were placed to contact with the light-side window, thus adjusting the center of the light path (Fig.

588

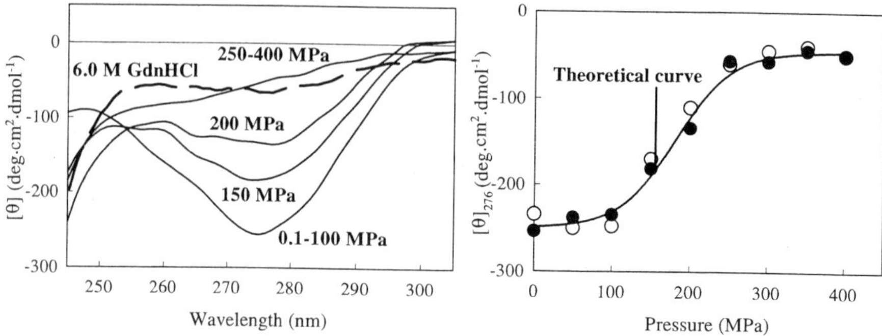

Fig 8. CD spectra of RNase A with diamond windows in the near-UV region under a pressure of 400 MPa or lower, or in 6.0 M guanidine·HCl. The protein conc. was 242 µM in 0.1 M KCl/HCl, pH 2.0.

Fig. 9. Pressure-dependent chages of $[\theta]_{276}$ of RNase A at pH2.0 and 25°C. (●, compression; ○, decompression). See Fig. 8 for details.

1b).   The slit with a single hole of a 3.0 and 3.5 mm diameter permitted the collection of authentic CD spectra of CSA at 150 MPa or lower (Fig. 2), although, the spectrum was reduced at 200 MPa or higher.   Thus, quartz windows equipped with a single hole-slit with a 3.0 mm diameter-hole were applied to measure far-UV CD under pressures at 150 MPa or lower.

High pressure CD spectra of RNase A (Fig. 3) and PLG (Fig.4) measured in this way remained unchanged at 150 MPa or lower, indicating that their secondary structures are stable at pressures of up to 150 MPa.

Although the method using a single hole-slit is useful in measuring CD spectra at 150 MPa or lower, a new material other than quartz needs to be investigated for use as windows at higher pressures.

3.3. High pressure CD with diamond windows

Since diamond strongly absorbs far-UV light, the CD spectra of CSA (Fig. 5), proteins, and poly-amino acid (Fig. 6) using diamond windows were measured at 230 nm to 350 nm under 0.1 MPa or high pressure.   However, the CD spectrum of CSA using the diamond windows remained unchanged within experimental errors by an increase in pressure up to 400 MPa (Fig. 5).

Using the high pressure cell equipped with diamond windows, the stability of the α-helix of poly-amino acids and proteins under high pressure was estimated by monitoring the ellipticity at 235 nm: When the pressure was increased, the α-helix of PLG at pH 4.2 was

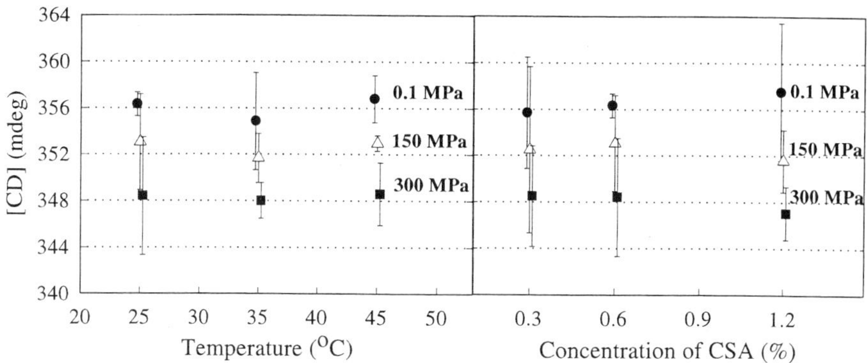

Fig. 10. Effect of temperature and concentration on $[\theta]_{290}$ of CSA at various pressures. See Fig. 8 for details.

stable between 0.1 and 450 MPa (Fig. 6) and, that of PLL at pH 12 began to decrease at 350 MPa (Fig. 6). The α-Helix of myoglobin decreased at 200 MPa and remained, to some extent, at 400 MPa (Fig. 7), consistent with the FTIR measurement: The amide I peak, showing an α-helix started to decrease around 200 MPa and completely disappeared at 600 MPa. The CD signals for RNase A at pH 2.0 was purturbed at 150 MPa and became unchanged at 300 to 400 MPa, indicating that RNase A maintained a partially ordered structure in this pressure region. Since RNase A contains α-helix, β-sheet, and β-turn, measurement of the molar ellipticity at 235 nm was insufficient for characterizing the denaturation of this protein. The main change in RNase A, as the result of pressure at 150 to 300 MPa is estimated to be disappearance of the β-turn structure. CD spectra of RNase A showed no detectable change in the range of 300 to 400 MPa.

3.5. CD spectra with diamond windows in the near-UV

Since diamond strongly absorbs far-UV light of 230 nm or shorter, CD measurements with diamond windows are appropriate for measurements of the tertiary structure of proteins in the near-UV region.

The CD spectra of 61.9 μM CPY at pH 6.0, 207 μM lysozyme at pH 6.0 and 497 μM RNase A at pH 6.0, which were measured in the 245-320 nm region under a pressure of 400 MPa or lower, showed only small changes from those observed at atmospheric pressure.

The CD spectra of RNase A at pH 2 under high pressure were clearly different from those observed at atmospheric pressure (Fig. 8): the CD spectra was decreased with increasing pressure from 100 to 250 MPa and this change stopped at 250 MPa or higher. This spectral change is very similar to that of the denaturation of RNase A which is induced by guanidinium chloride. The CD spectrum was returned reversibly as a result of decompression from 250 to 100 MPa.

The maximum CD change ($[\theta]$ at 276 nm) *vs.* pressure indicated that the pressure-induced denaturation of RNase A is a two-state transition in a completely reversible manner (Fig. 9).

Thus, it is possible to monitor the CD spectrum of a protein under high pressure at 400 MPa or lower in the near-UV region and to detect conformational changes in the protein, which can be attributed to environmental changes in phenylalanine, tyrosine, tryptophan, and/or cystine residues.

3.6. Reliability of CD spectra with diamond windows in the near-UV

The CD of CSA in the near-UV was slightly increased by an increase in applied pressure, even after the application of a correction for water (Fig. 10). This increase is probably due to a small distortion in the diamond windows, because it was completely reversible and independent of both temperature and the concentration of CSA. Although this discrepancy is within 3% at 300 MPa (Fig.10), it is possible to correct this decrease by the following equation.

$$[CD] = (1 + 6.92\ P \times 10^{-5})\ [CD]_{obs}$$

4. ACKNOWLEDGEMENT

We express our sincere gratitude to Dr. P. Lillford, Unilever Research Coloworth Laboratory, for his interest, encouragement, and support in completing this study, and to Dr. S. Sato, Sumitomo Electric Industry, for his gift of the synthetic diamonds, which were used in this study. We are also indebted to Dr. S. Takahashi, Institute for Chemical Research, Kyoto University, for discussions in the initial stage of this study.

REFERENCES

(1) R.D.Harris, M.Jacobs, M.M.Long and D.W.Urry, (1976) *Anal. Biochem.* 73, 363-368.

(2) R.Hayashi, (1976) *Methods in Enzymol.*, 45, 568-586.

(3) H.Sumiya, N.Toda and S.Satoh, (1997) *High Pressure Bioscience and Biotechnology* (eds., A.Suzuki and R.Hayashi), pp.165-170, San-Ei Shuppan Co., Kyoto, Japan.

(4) M.Morishita, T.Tanaka and S.Kawai, (1997) *High Pressure Bioscience and Biotechnology* (eds., A.Suzuki and R.Hayashi), pp.187-194, San- Ei Shuppan Co., Kyoto, Japan.

(5) P.T.T.Wong and K.Heremans, (1988) *Biochim.Biophys.Acta.* 956, 1-9.

(6) R.Lange, J.Frank, J.-L.Saldana and C.Balny, (1996) *Eur.Biophys.J.*, 24, 277-283.

Trends in High Pressure Bioscience and Biotechnology
R. Hayashi (editor)

Microscopic Observation of Biological Substances in Near- and Supercritical Water

Shigeru Deguchi, Kaoru Tsujii, and Koki Horikoshi

The DEEPSTAR Group, Japan Marine Science and Technology Center (JAMSTEC), 2-15 Natsushima-cho, Yokosuka 237-0061, Japan

1. INTRODUCTION

Water at high temperatures and pressures is intriguing from view points of physical chemistry. When the temperature and pressure both exceed the critical point (374 °C, 22.1 MPa), water becomes supercritical water (SCW), where liquid and gas are no longer distinguishable (Fig. 1). In recent years, there has been increasing interest in SCW [1,2]. For instance, SCW has been found to be a very effective medium in waste water treatment, destruction of hazardous organic materials such as PCB and dioxin, decomposition or recycling of waste plastics such as poly(ethylene terephthalate) (PET). Particularly interesting aspect of SCW is that it behaves remarkably differently from water at ambient conditions. Hydrogen bonding network breaks down in SCW, and the static dielectric constant decreases down to 6 at 400 °C and 30 MPa, the value of which is comparable to that of polar organic solvents. Consequently, in SCW, organic substances are highly soluble or completely miscible with water, while inorganic salts precipitate out. In addition, the properties of SCW are tunable over wide range by changing the temperature or pressure.

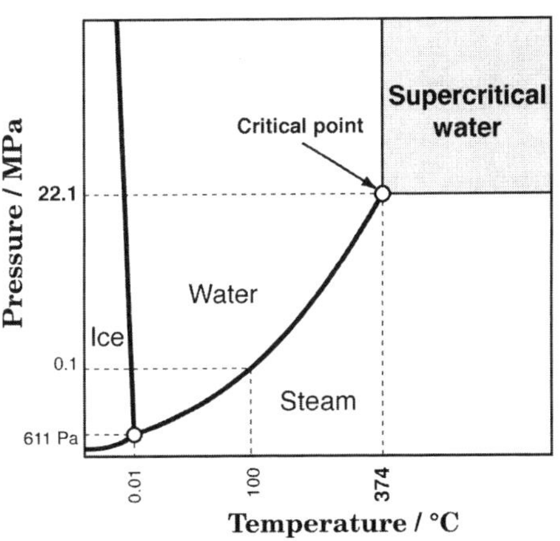

Fig. 1. Phase diagram of water

592

SCW would possibly exist in nature, in the vicinity of a hydrothermal vent or over a sea volcano for example. It has also been hypothesized that life might have originated in hydrothermal vents. At the first look, water under these extreme conditions seems too harsh for life to survive, and the behavior of biological substances in water under such conditions has rarely attracted scientific interest so far. However, due to the recent progress in microbiology, especially the discoveries of extremophiles including hyperthermophiles that grow at high temperatures [3], it is becoming important to understand the behavior of biological substances in water under such conditions.

One of the most fundamental properties of a solvent is to dissolve various substances. In order to survey the behavior of wide variety of biological substances in SCW, direct observation by an optical microscope is the most efficient and effective method. We have designed and developed a new flow type cell that can be used with a conventional inverted microscope to achieve high optical resolution. In this paper, we describe the new flow type cell, and the observations of biological substances under hydrothermal conditions that are made with this cell.

2. EXPERIMENTAL SECTION

2,1. Materials

Microcrystalline cellulose for column chromatography (Trade name, Funacel) was purchased from Funakoshi Co., Ltd. (Tokyo, Japan). Chitin from crab shells was purchased from Nacalai Tesque, Inc. (Kyoto, Japan). The polysaccharides were used as received. An yeast of deep sea origin, *Cryptococcus* sp. N6, was kindly supplied by Dr. Fumiyoshi Abe, JAMSTEC. *Flammulina velutipes* was purchased. Millipore water was used throughout the work (Millipore, Bedford, Massachusetts).

2.2 High Temperature High Pressure Microscope Cell

Cross-sectional view of the high temperature and high pressure cell is shown in Fig. 2. The cell is designed to be mounted on a conventional inverted microscope and allow experiments above the critical point of water (374 °C, 22.1 MPa) up to 400 °C and 40 MPa. Under these conditions, corrosion due to supercritical water is a severe problem [1]. In order to circumvent the problem, the cell body is made of corrosion-resistant Ni-based supermetal alloy, Inconel 600. The dimension of the cell body is $80 \times 40 \times 35$ mm. It has two natural diamond windows for optical access to the sample. The diamond windows are fit to the cell body by titanium washers and fixed by a compression nut. The titanium washer deforms as the compression nut is tightened to make effective seal between the diamond window and the cell body. Two spring washers are placed between the window and the compression nut to accommodate the thermal expansion mismatch between Inconel 600, titanium and diamond. The sample volume in the cell body is approximately 0.3 mL. The cell body is surrounded by heater blocks made of brass, and four 250 W electric cartridge heaters are embedded in the blocks.

Temperature is controlled by a PID controller (DSM5 temperature control unit, Shimaden Co., Ltd., Tokyo, Japan). Temperature of the sample inside the cell is monitored by a chromel-alumel thermocouple (TC2 in Fig. 2), inserted in the cell body and located 10 mm away from the sample. The cell body and the heater blocks are contained in a cooling jacket, and the space between the cell and the jacket is filled with heat insulator for better temperature stability. The dimension of the cooling jacket is 150 × 150 × 62.5 mm. Water, which is kept at 20 °C by an external cooling bath (RTE-110, Neslab Instruments, Inc., Portsmouth NH, USA), is circulated through the slit of the cooling jacket walls. This allows to cool the outer wall of the jacket down to ambient temperature, even when the sample is heated up to 400 °C. On the lower side of the cell, where an objective lens locates, an additional cooling block is placed in order to protect the objective lens from radiative heat. The objective lens is also equipped with a cooling jacket, and cooled by circulating water. The combination of the cooling jacket, the cooling block, and the cooling jacket for an objective lens effectively protect the objective lens and the microscope from the high temperature. The cell was manufactured by AKICO Co., Ltd. (Tokyo, Japan).

The high temperature and high pressure cell is connected to a flow-type pressurizing system, presented schematically in Fig. 3. The system consists of an HPLC pump (PU-1580, JASCO, Hachioji, Japan) , a back pressure regulator (Model 880-81, JASCO, Hachioji, Japan), a pressure transmitter (KH15, Nagano Keiki Co., Ltd., Tokyo, Japan) interfaced to a digital pressure display, and a stop valve. A blow-off valve is also inserted between the cell and the pressure gauge in order to protect the cell from accidental high pressure. The cell is mounted on a bracket, which is built on an Eclipse TE300 inverted optical microscope (Nikon, Tokyo, Japan). The cell position is adjustable in x-y-z directions. Due to the thickness of the cell, a limited range of objective lens can be used. We use two objective lenses for this set up: CF IC EPI Plan SLWD (Nikon, Tokyo, Japan) of ×10 and ×20 magnification. Both lenses have a long working distance (20.3 mm and 20.5 mm, respectively).

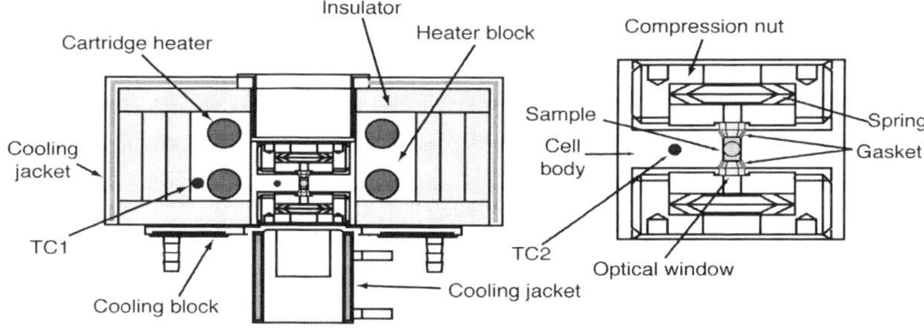

Fig. 2 Cross-sectional view of the high-temperature high-pressure cell for an optical microscope

594

In a typical experiment, the entire system was first filled with pure water. The pipe leading to the cell was disconnected at the tee joint, the stop valve was opened, and the suspension of the specimen (approximately 3 mL) was introduced into the cell by a syringe. The stop valve was then closed, and the pipe was reconnected. The system was pressurized by pumping water into the system at a flow rate of 0.2 mL/min. During pressurization, the specimen sedimented on the bottom

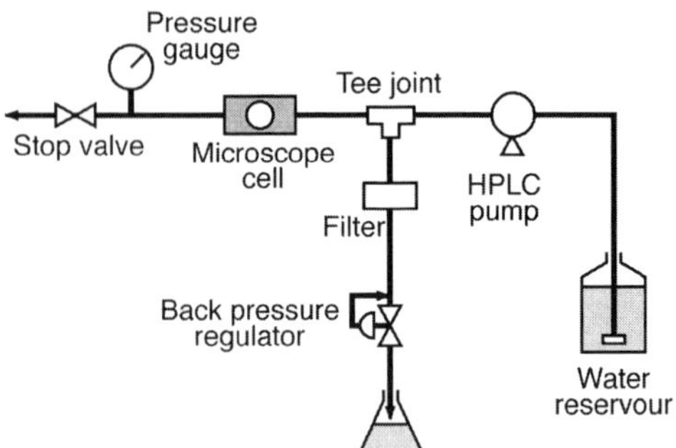

Fig. 3 Schematic diagram of the pressurizing system

diamond window surface. When the pressure reached the desired value, sample was heated. The behavior of the specimen in water during heating at constant pressure was recorded via a color chilled CCD camera (C5810, Hamamatsu Photonics K.K., Hamamatsu, Japan) onto a VCR. The VCR is interfaced to a personal computer, and the recorded images can be transferred digitally to the computer for image analysis or off-line video editing. Constant flow of water at a flow rate of 0.2 mL/min was maintained during the experiment in order to compensate the volume expansion or contraction of water during heating or cooling and to keep the pressure constant. All the experiments reported here were performed at a fixed pressure of 25 MPa.

When the specimen is a large object, such as *Flammulina velutipes*, the specimen was frozen in liquid nitrogen, and ground to fine powder. The specimen in powder form was suspended in water, and was introduced into the cell.

3. RESULTS AND DISCUSSION

3.1. High Temperature and High Pressure Microscope Cell

Supercritical water has been studied by various spectroscopic techniques, and several cell designs with optical windows are already available [4-6]. A diamond anvil cell (DAC) for the observations of minerals under hydrothermal conditions (pressures up to 2.5 GPa, and temperatures between –190 °C and 1200 °C) has also been developed [7]. The DAC has been applied to study the phase behavior of various synthetic polymers in water at high temperatures and pressures [8-10]. However, the DAC is designed to be used with a stereo microscope, and the optical resolution obtained in this set up is not satisfactory for studying biological substances including microorganisms.

Our cell is designed to be set up on a conventional inverted microscope for superior optical resolution. Designing a high temperature and high pressure cell for an inverted optical microscope is not so straightforward as to simply adopt the existing designs, but has its own difficulties. First of all, the cell dimension is restricted by working distance of an objective lens, which is typically less than 10 mm. Secondly, a part of the microscope, especially the objective lens, locates so close to the hot sample that effective cooling system must be implemented to prevent damage to the microscope.

The requirement for the cell thickness was relaxed by employing the objective lens with long working distance. The effective cooling was achieved by circulating cooling

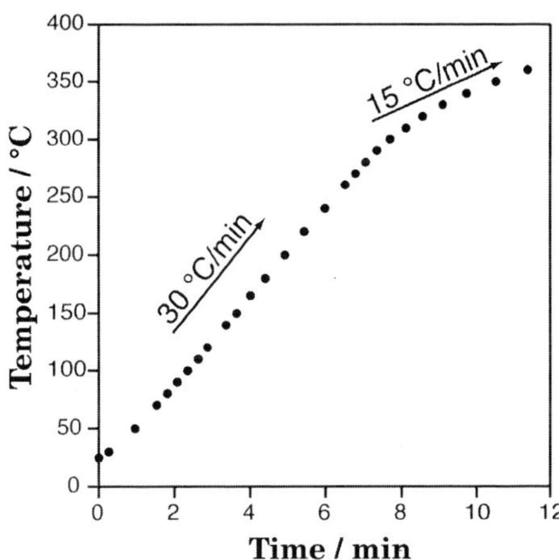

Fig. 4 Typical temperature profile of the sample during a run

water, which is described in experimental section. It should also be mentioned that the gasket material is very important to achieve maintenance-free operation of the cell. The gasket material has to be ductile enough to make effective seal, but also be of mechanical strength. We started with platinum gaskets, but the mechanical strength of platinum was not strong enough and we experienced frequent leakage. The titanium gaskets have performed excellently so far, and the cell has been working without leakage for over an year without gasket replacement or retightening the compression nut.

The heaters incorporated in the cell can heat up the sample up to 400 °C within 15 minutes (Fig. 4). This enables us to heat up the sample from room temperature up to 400 °C, while minimizing the residence time of the sample in superheated water.

Unlike the DAC [7], the flow type design makes it possible to connect the cell to the flow reactor, and monitor the reaction online. Due to the small dimension, the cell could also be mounted on a conventional spectrophotometer with a minimum modification of the apparatus itself.

3.2. Biopolymers in Near- and Supercritical Water

Cellulose and chitin are the most abundant and the second most abundant biomass on Earth. In recent years, these materials have attracted considerable attention in material science, because they are environmentally benign, and are taken from renewable resources. However, both polysaccharides are insoluble in nearly any kinds of solvent, and they have found very limited applications. The behavior of them in water at high temperatures and pressures are also intriguing because they are major components of bacterial cell walls.

596

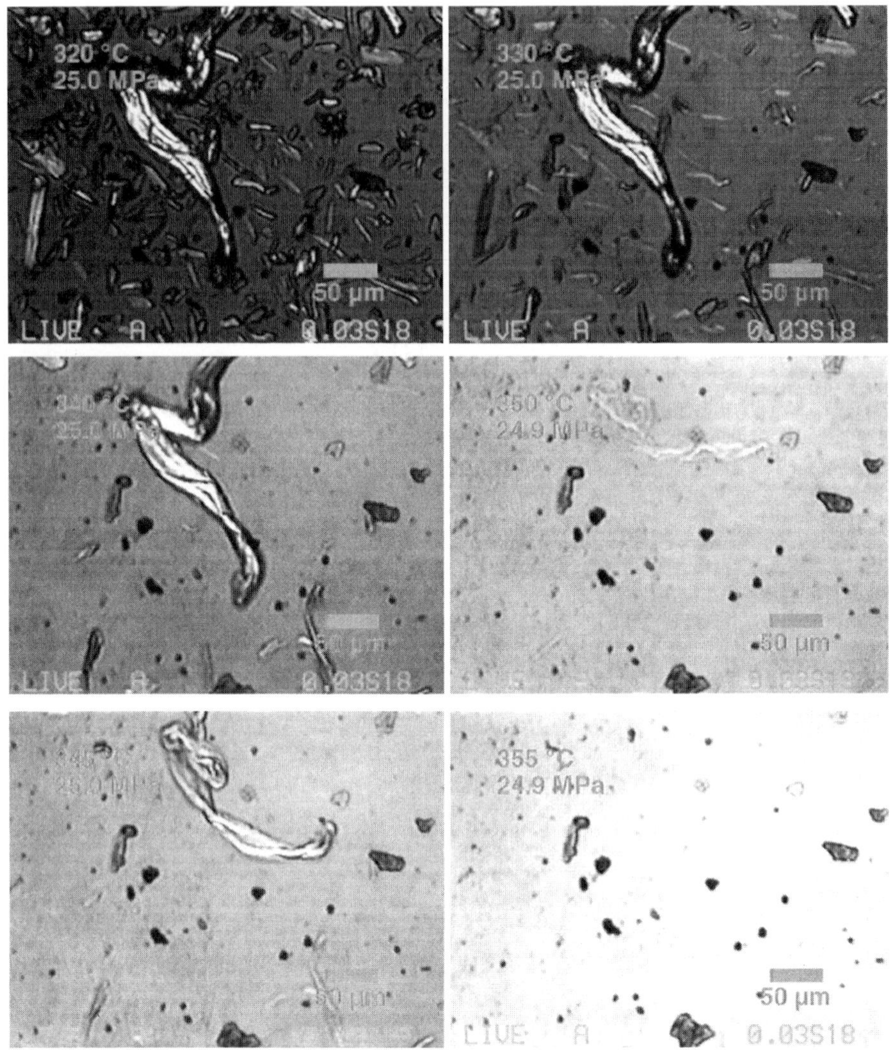

Fig. 5 Polarized microphotographs of crystalline cellulose at 25 MPa and different temperatures.

The insolubility of cellulose in solvents stems from strong interchain hydrogen bonding in the crystal. Figure 5 shows a series of optical micrographs showing the behavior of crystalline cellulose in water at different temperatures and 25 MPa. Upon heating, the crystalline structure of cellulose broke down at 320-350 °C, which was evidenced by disappearance of the color of the crystals under observation with cross-polarizers. Dissolution followed immediately. Noncatalytic hydrolysis of cellulose in near- and supercritical water has been studied [11]. It was found that the hydrolysis rate increased at elevated temperatures. The temperature dependence of the hydrolysis rate would be ascribed to the cellulose dissolution.

Chitin behaves very differently from cellulose. Although chitin is amorphous, it was found to be more stable than cellulose in water at high temperatures, and dissolved slowly at 390 °C. Chitin granule got smaller as the dissolution proceeds, which suggests that the dissolution takes place on the surface of the granule.

3.3. Microorganisms in Near- and Supercritical Water

Microorganisms were also studied with the microscope. The spherical cellular structure of a deep-sea yeast, *Cryptococcus* sp. N6, collapsed at 250 °C to give a residue that remained up to 300 °C. In the case of a mycerial cord of *Flammulina velutipes*, it shrank at 250 − 310 °C along the long axis of the mycerial cord, and then dissolved slowly at 390 °C as chitin did. The shrinkage at lower temperatures would correspond to the dissolution of protein of the cell wall, while the residue that dissolved at 390 °C would be chitin, which is a major component of the cell wall.

4. SUMMARY

We have developed a high temperature and high pressure cell for an optical microscope. Due to the small dimension and the effective cooling system, the cell can be installed on a conventional inverted optical microscope, and the specimens can be observed with high optical resolution. With the cell, we have successfully observed the behavior of biological substances such as polysaccharides and microorganisms in near- and supercritical water. Detailed study of these dissolution phenomena are in progress.

Acknowledgment

Dr. Fumiyoshi Abe, JAMSTEC, is acknowledged for supplying *Cryptococcus* sp. N6.

REFERENCES

1. R.W. Shaw, T.B. Brill, A.A. Clifford, C.A. Eckert and E.U. Franck, Chem. Eng. News, 69 (1991) 26.
2. P.E. Savage, Chem. Rev., 99 (1999) 603.

3. K. Horikoshi and K. Tsujii (eds.), Extremophiles in Deep-Sea Environments, Springer-Verlag, Tokyo, 1999.
4. W. Kohl, H.A. Lindner and E.U. Franck, Ber. Bunsenges. Phys. Chem., 95 (1991) 1586.
5. D.M. Pfund, J.G. Darab, J.L. Fulton and Y. Ma, J. Phys. Chem., 98 (1994) 13102.
6. F.J. Armellini, J.W. Tester and G.T. Hong, J. Supercrit. Fluids, 7 (1994) 147.
7. W.A. Bassett, A.H. Shen, M. Bucknum and I.-M. Chou, Rev. Sci. Instrum., 64 (1993) 2340.
8. Z. Fang, J. Richard L. Smith, H. Inomata and K. Arai, J. Supercrit. Fluids, 15 (1999) 229.
9. Z. Fang, R.L.S. Jr., H. Inomata and K. Arai, J. Supercrit. Fluids, 16 (2000) 207.
10. J. R. L. Smith, Z. Fang, H. Inomata and K. Arai, J. Appl. Polym. Sci., 76 (2000) 1062.
11. T. Adschiri, S. Hirose, R. Malaluan and K. Arai, J. Chem. Eng. Jpn., 26 (1993) 676.

Trends in High Pressure Bioscience and Biotechnology
R. Hayashi (editor)

Differential scanning calorimetry of proteins under high pressure

Kaoru Obuchi[a] and Takashi Yamanobe[b]

[a]Department of Molecular and Cell Biology and [b]Department of Bioresources,

National Institute of Advanced Industrial Science and Technology,

Higashi Tsukuba Ibaraki 305-8566, Japan

Calorimetric performance of a high pressure applicator for MC-2 micro-calorimeter (Microcal Inc. MA) was characterized. Irreversible denaturation under high pressure, up to 100MPa, was investigated for egg white lysozyme, bovine pancreas α-chymotripsin, bovine pancreas ribonucrease A and bovine hemoglobin. The high pressure application increased denaturation temperatures of lysozyme and α-chymotripsin. It did not change the temperature of ribonucrease but decreased it of hemoglobin. An equation is proposed to estimate activation volume of the irreversible denaturation from the difference between the denaturation profiles obtained at different pressures.

1. Introduction

Structural information of protein molecule is available from various spectrometric analyses and the information is specific to the energy level of the electromagnetic wave used. High pressure cells has been developed for *in situ* spectrometric investigations of structural perturbation induced in protein molecules by high pressure application. Differential scanning calorimetry has been applied to investigate protein stability. If the denaturation investigated is a reversible process, then the DSC profile provides thermodynamic parameters describing the equilibrium of unfolding and folding processes of the molecule. Otherwise, DSC profile provides kinetic parameters of the denaturation[1]. Pressure dependence of such parameters are expected to provide information on volume change during protein unfolding. This information would help us to understand protein denaturation mechanism. However, high pressure cell for microcalorimetry of proteins has not been reported so far. Thus, Teramecs Co. (Kyoto) and we collaboratively developed a high pressure applicator for MC-2 micro-calorimeter[2].

Application of this system to whole yeast cells revealed that denaturation of biopolymers in the cells was irreversible, and high pressure decreases stability of cellular components[3]. Membrane lipid from a barophylic microorganisms was also investigated by this system[4]. However, calorimetric performance of this system has not been specified so far. It is expected that comparison of endothermic profiles of irreversible denaturation under different pressures enables to compare denaturation rates at the pressures. Using this high pressure DSC system, irreversible denaturation of proteins was investigated, and the performance was evaluate in this work. Furthermore, we propose a formula to estimate activation volume from denaturation profiles under different pressures.

2. Materials and methods

2.1 proteins

Hen egg white lysozyme (6x crystalized) was purchased from Seikagaku Kogyo Co., Tokyo. Bovine pancreas α-chymotripsin (3x crystalized and lyophilized), bovine blood hemoglobin (type I, 2x crystalized, dialyzed and lyophilized), and bovine pancreas ribonuclease A (type I-A, 5x crystalized, salt fractionated and chromatographycally purified) were purchased from Sigma Chemical Co., MO. Without further purification, all proteins were solubilized in distilled water and subjected to the calorimetry.

2.2 high pressure calorimetry

An MC-2 micro-calorimeter was equipped with high pressure applicator (Teramecs Co. Kyoto) and used to conduct calorimetry. The applicator comprised two high pressure capillaries, capillary holder, a TP200MC high pressure pump driven by a stepping motor, a PG200D digital pressure gauge and SUS tubes connecting among the elements, and was controlled by a desk-top PC. Water was used as pressurizing medium. The capillaries were made of SUS, one end of which was silver blazed. Outer diameter and inner volume of the capillary were 1.0mm and about 100 μl, respectively. Sample protein solution and reference distilled water were loaded in the capillaries and inserted into the sample and reference cells of the calorimeter, respectively.

After pressure of the system reached designated value at 5°C, temperature of the DSC cells were equilibrated at the temperature, and then increased to 95°C at the rate of 60°C/hr. When the temperature reached 95°C, the cells were cooled to 5°C and equilibrated again, and the temperature scan was repeated. Trace of rescan was subtracted from the first scan in order to eliminate the curvature specific to the calorimetric system. For thus obtained profile, cubic polynomial baseline was generated and corrected denaturation profile was obtained by subtraction of the baseline again. Corrected profile was deconvoluted by non-linear least-square curve-fitting (NLSF) using the following equation derived from the kinetic denaturation model as the fitting function:

$$C_p(T) = \frac{2.446a}{w}\exp\left[\frac{2.446(T-T_m)}{w} - \exp\left\{\frac{2.446(T-T_m)}{w}\right\}\right]$$

(1)

where C_p, a, w, T_m denote excess heat capacity, peak area, peak width at the half of maximum height, and peak temperature of the denaturation endotherm. In this formula, scale of temperature is of absolute temperature. However, in the text and figures or tables temperature is expressed with the scale of degree Celsius.

3. Results

3.1 Sensitivity of the system and irreversibility of denaturation   When DSC scans of 10mg/ml solution of lysozyme were conducted using the intrinsic cells of MC-2, the rescan trace included endotherm of comparative but less intensity than that in the initial scan (not shown). Decrease in the intensity by the duplicate scan was of more extent for 100mg/ml solution. Therefore, denaturation in the 10mg/ml and 100mg/ml solutions were almost and partly reversible, respectively. In spite of the increase in the degree of reversibility, the increase in lysozyme concentration did not affect the denaturation temperature.

When the high pressure capillaries were used for the analysis of 100mg/solution, the rescan did not included significant intensity of endotherm(Fig. 1). Thus, the denaturation was almost

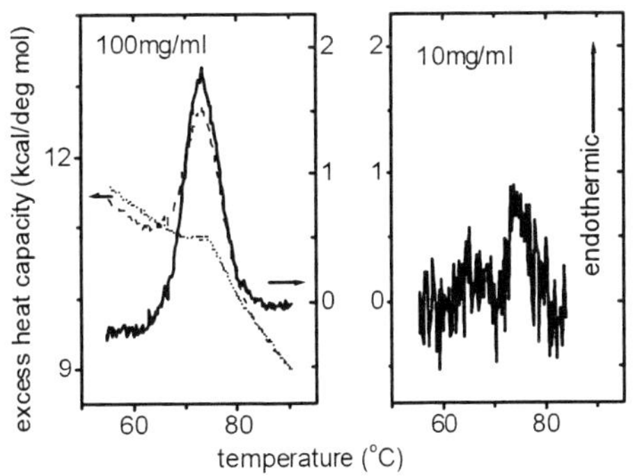

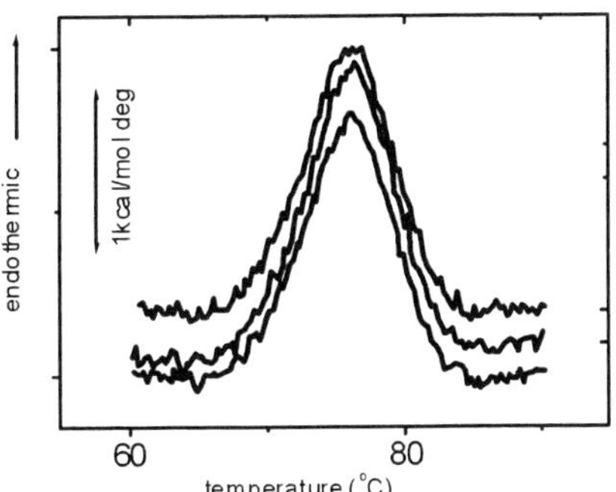

Fig. 1 DSC profiles of lysozyme in high pressure capillary. Dashed, dotted, and solid lines denote traces of scan and rescan and the subtracted profile. Scans were conducted under ambient pressure.

Fig. 2 Reproducibility of corrected profiles. Corrected profiles of 100mg/ml solution of lysozyme were obtained using high pressure capillaries at 100MPa.

Table 1 Peak parameters of lysozyme denaturation.

pressure	peak 1			peak 2		
(MPa)	T_m (°C)	a (kcal/mol)	w (°C)	T_m (°C)	a (kcal/mol)	w (°C)
0.1	72.41±0.11	7.62±0.75	5.98±0.12	75.56±0.48	7.32±0.39	7.25±0.27
100	75.43±0.03	7.13±0.19	5.74±0.30	78.63±0.04	5.61±0.30	6.07±0.25

irreversible. The C_p axes in Fig. 1 were normalized using molar concentration of lysozyme and volumes of the intrinsic cells or capillaries. Height of endothermic profile in the first trace for the 100mg/ml solution in the capillary was approximately a quarter of that for the solution in the intrinsic sample cell. Thus, only a quarter of excess heat capacity was detected during the denaturation in the high pressure application system. This is presumably due to the capillary holder made of SUS that was a heat sink, and loses the adiabatic advantage of MC-2. Due to the reversibility of denaturation and partly to the loss of sensitivity, low intensity of corrected profile was obtained for 10mg/ml solution(Fig. 1). Even though T_m of the endotherm appeared to be the same as that for 100mg/ml solution, poor S/N ratio made it difficult to conduct NLSF of the endotherm. Consequently, this system is suitable to investigate irreversible denaturation of protein in high concentration solution under high pressure. Lysozyme concentration required is at least several tens mg/ml.

Corrected denaturation profiles obtained using the high pressure application system were reproducible both under ambient pressure at 100MPa. The profiles appeared to have one shoulder for each and were deconvoluted by the NLSF. As the fitting parameters (Tab. 1) showed, standard deviations of T_m and w were at most 0.5°C that is comparative to the digital resolution of the C_p data acquisition during the DSC scans, i.e. 0.25°C. High pressure application of 100MPa increased T_m value of each peak components by c.a. 3°C (Tab. 1), implying that high pressure stabilizes lysozyme with respect to denaturation kinetics as described later. The high pressure application did not modify the area of endothermic transition peak 1 significantly. With respect to peak 2, the high pressure application decreased the area significantly. Since integration of C_p with respect to T gives denaturation enthalpy, this decrease suggests high pressure destabilization of protein structure with respect to intramolecular interaction or hydration of protein molecules. The decrease in peak area implies that the high pressure application destabilizes the native conformation. It is notable that the high pressure application affect thermal stability of lysozyme in two contradicting ways: high pressure stabilizes it kinetically and destabilizes it thermodynamically.

Pressure dependence of denaturation profile was investigated for α-chymotripsin over the pressure

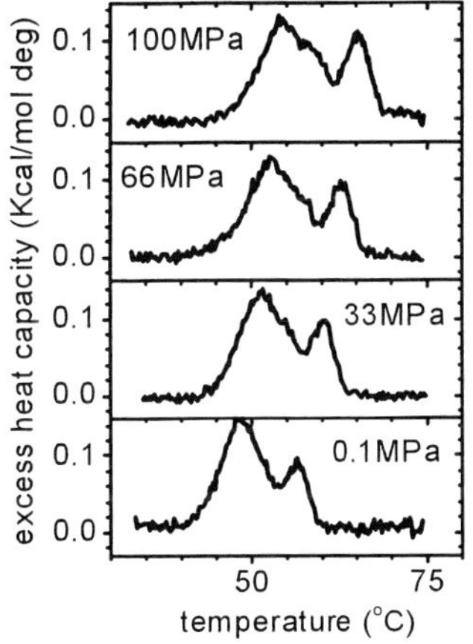

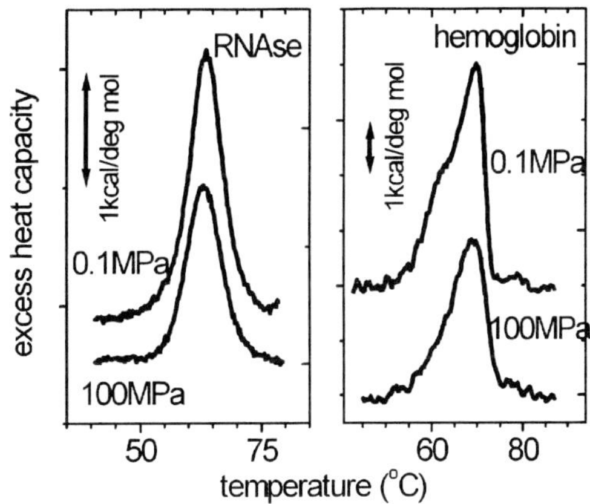

Fig. 3 Pressure dependence of denaturation profiles of α-chymotripsin.

Fig. 4 Corrected profiles of RNAse and hemoglobin

range from 0.1 to 100MPa (Fig. 3). Since the traces of rescan did not include significant intensity of endothermic profile (figure not shown), denaturation of this protein also occurred irreversibly. The denaturation occurred at higher temperature under higher pressure, implying that high pressure application stabilizes α-chymotripsin with respect to the denaturation kinetics over this pressure range.

Since the second trace of 100mg/ml bovine pancreas ribonuclease A did not include significant intensity of endotherm (figure not shown), denaturation of this protein was irreversible. The corrected profile appeared to be without shoulder, and further deconvolution was not conducted. High pressure application did not modify T_m but decreased the intensity of the endotherm (Fig. 4). Therefore, the high pressure application did not modify denaturation kinetics, while it destabilizes the molecular structure.

The intensity of denaturation endotherm of bovine blood hemoglobin was also decreased by the high pressure application (Fig. 4). T_m values were determined for two peak components as 62.90±0.27 and 69.56±0.20°C for the corrected profile at 0.1MPa, and 60.66±1.89 and 69.02±0.28°C for the profile at 100MPa by NLSF using Eq. 1. Thus, the high pressure application destabilizes structure of hemoglobin both thermodynamically and kinetically.

4. Discussion

4.1 Estimation of activation volume from DSC profiles at different pressure   T h e r m a l denaturation occurred irreversibly in the high pressure capillaries for every protein investigated. This is due to the high concentration of sample proteins, (from 10 to) 100mg/ml. The subtraction between the traces of scan and rescan eliminates contribution of reversible transition, if exists. Thus, the corrected profile comprises only the contribution from irreversible transition. Because of this irreversibility, kinetic information is available from the corrected profiles obtained. Although lysozyme was denatured irreversibly, T_m did not appear to depend on protein concentration. Thus unfolding of native conformer is presumably rate determining of the denaturation[5]. DSC profile of this type of denaturation is expressed with $C_p(T)$ that is in proportion to the decrease in native protein concentration, N^1, or:

$$C_p(T) \propto -\frac{dN(T)}{dT} = \frac{1}{v}k(T)N(T)$$

Temperature dependence of k is given by the Arrhenius equation as:

$$k(T) = A\exp\left(-\frac{E}{RT}\right)$$

where A, E and R denote frequency coefficient and activation energy of the denaturation and gas constant, respectively. If unfolding of native conformer that is a unimolecular process is rate determining of the denaturation, then $N(T)$ during DSC scan from initial temperature T_i at the rate of v is given as:

$$N(T) = N_0 \exp\left\{-\frac{1}{v}\int_{T_i}^{T} k(T)dT\right\}$$

Therefore, $C_p(T)$ is given as:

$$C_p(T) \propto \frac{1}{v}\left\{A - \frac{E}{RT} - \frac{1}{v}\int_{T_i}^{T}\exp\left(A - \frac{E}{RT}\right)dT\right\} \qquad (2).$$

This expression with the Arrhenius parameters is transformed to Eq. 1 with peak parameters. This transformation provides relationship between the Arrhenius parameter and the peak parameters as:

$$A = \frac{2.303}{D_m}\exp\left(\frac{2.303\,T_m}{z}\right)$$

$$E = \frac{2.303}{z}RT_m^2$$

where D_m and z denote duration required to reduce k by 1/10 fold at T_m and temperature increase required to increase k by 10 fold, respectively[6].

Over all activation volume of denaturation, ΔV^*, is given as

$$\Delta V^* = -RT\frac{\Delta \ln k}{\Delta P}$$

where $\Delta \ln k$ and ΔP denote difference in natural logarithms of denaturation rate constants at pressure P_1 and P_2 and difference in P_1 and P_2, respectively[7]. From the relationship between the Arrhenius parameter and the peak parameters, ΔV^* is estimated from the peak parameters that are determined from deconvolution of the corrected endotherm as:

$$\Delta V^* = -\frac{2.445RT}{w\Delta P}(\Delta T_m - \frac{\Delta T_m^2}{T}) \qquad (3).$$

Using Eq. 3, ΔV^* of thermal denaturation of lysozyme was estimated to be 36.7 and 33.1 cm^3/mol from peaks 1 and 2, respectively. ΔV^* of the denaturation of hemoglobin was -10.9 and -4.15 cm^3/mol from peaks 1 and 2, respectively.

ΔV^* of thermal denaturation was positive for lysozyme and α-chymotripsin, while it was negative for hemoglobin. It was nearly zero for the ribonuclease. High pressure application of 150MPa increased T_m of peak components of the cellulase system from *Acremonium cellulolyticus* by up to 15°C, and increased optimum temperature of cellulase activity by the corresponding range[8]. The increase in T_m corresponds to 40cm^3/mol of ΔV^*, which increases optimum temperature of the activity. The high pressure application enhances productivity of reducing sugar by almost 8 folds. Optimum temperature of enzyme activity is usually determined only at ambient pressure. This work exposed that some enzymes are stabilized by high pressure application, and reaction conditions of such enzymes should be optimized with respect to both pressure and temperature. If T_m of irreversible thermal denaturation is increased by high pressure application, then optimum temperature at ambient pressure may not be optimum. High pressure DSC of enzyme is useful to examine if pressure should be taken in consideration to determine optimum condition for the catalytic reaction.

High pressure DSC analysis of whole yeast cells showed that high pressure application decreased not only peak temperature but also area of endothermic peak[9]. The latter is presumably due to structural destabilization of cellular components. The extent of the peak area decrease was much more significant than that of the purified proteins, suggesting that some additional destabilizing factors contribute to the stability of cellular components *in vivo*. High pressure DSC may be useful to determine macromolecular stability both *in vivo* and *in vitro*.

REFERENCES

1. K. Obuchi, H. Iwahashi, J.R. Lepock and Y. Komatsu. Yeast 16 (2000), 111.
2. S. Kawai, Namikawa, K. Obuchi and M. Matsumoto. JP Kokai 2000-074863
3. K. Obuchi, H. Iwahashi and Y. Komatsu. in Advances in high pressure bioscience and biotechnology (Ludwig H. ed.), Springer-Verlag, Berlin (1998), pp85-88.
4. H. Kaneko, K. Obuchi, K.. J. Biochem. *(Tokyo)* 128 (2000), 727.
5. J.R. Lepock, K.P. Richie, M.C. Kolios, A.M. Rodahl, KA Heinz, and J Kruuv. Biochemistry. 31 (1992), 12706.
6. P.C. Michels, D. Hei, and D.S. Clark. Advances in Protein Chemistry 48 (1996), 341.
7. C.A. Miles, T.V. Burjanadze, and A.J. Mailey. J. Mol. Biol. 245 (1995), 437.
8. T. Yamanobe and K. Obuchi. This book.
9. K. Obuchi. in (Manghnani MH, Mellis WT, Nicol MF eds) "Science and technology of high pressure" Vol. 1, 301-304. Universities Press (India) Ltd. Jul. 2000.

Trends in High Pressure Bioscience and Biotechnology
R. Hayashi (editor)
© 2002 Elsevier Science B.V. All rights reserved.

High pressure calorimetry; application to phase change under pressure

A. LeBail [a], D. Chevalier [a] and J.M. Chourot [b]

[a] GEPEA-ENITIAA -BP82225-F-44322 NANTES Cedex 3-France
corresponding author: lebail@enitiaa-nantes.fr

[b] CEMAGREF-GPAN-BP44-F92160 ANTONY Cedex-France

Calorimetry is a very versatile technique used to study phase change or reactions of any material. This paper presents a calorimeter which has been designed for studying phase change at elevated pressure. Its originality lies in that a pressure scan is realised at constant temperature. A differential calorimeter head was used with two cells made of stainless steel. Pentane was used as a pressurisation fluid. A high pressure pump driven by a step motor permits to obtain constant pressurisation or depressurisation rates. Data on the latent heat of pure water at –5°C, -10°C and –15°C are proposed. Some statements on the behaviour of the sample of water during the peak are presented and are discussed. Results were within 3% of the existing data from the literature.

Nomenclature

P	Pressure	MPa
Q	Heat	J
T	Temperature	°C
V	Volume	m^3
α	Isothermal compressibility	K^{-1}
σ	Sensibility	$W.V^{-1}$ or mW

1. INTRODUCTION

In a very general matter, calorimetry is related to the measurements of amount of heat. Lavoisier, a French scientist developed a calorimeter during the 18[th][1]. The original drawing of this apparatus is shown in figure 1. The specific heat of given substances was related to the amount of ice that can be thawed by them. Grounded ice was installed in a metallic basket. The melted ice was collected below the system. Ice was taken from frozen ponds and thus experiments were realised exclusively during winter time. These experiments were realised at quasi constant temperature. Modern calorimeters usually work at constant pressure while a temperature scan is used to observe the studied phase change or transition. Differential

608

scanning calorimeters (DSC) are measuring the differential heat flux transmitted to a reference and a sample cell [2]. They usually work with relatively high heating or cooling rates (1-5 °C/min).

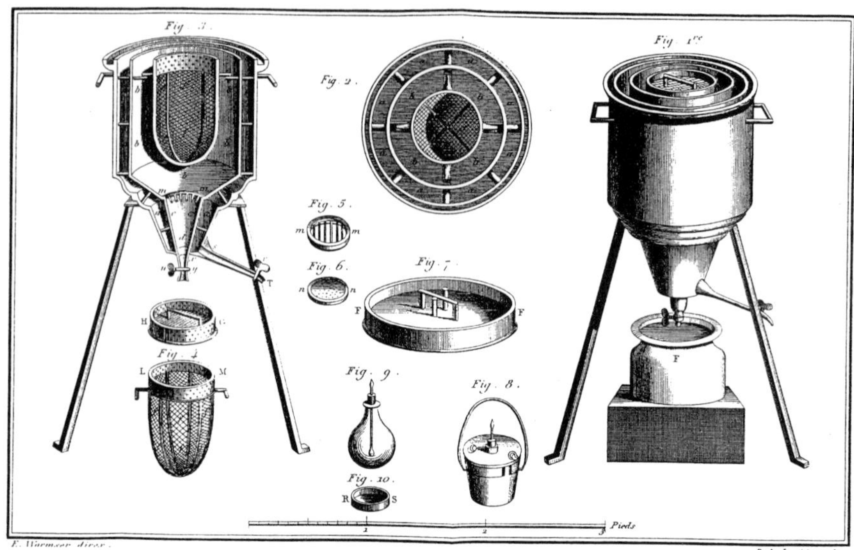

Figure 1
The calorimeter of Lavoisier (1789)

Microcalorimeters are usually more sensitive than DSC and are able to work with low temperature rate (1 °C/min or less) but with larger samples. Constant temperature calorimetry is sometimes realised to study specific reactions (i.e. fermentation). Experiments at constant pressure can be realised [3], [2] to study the influence of pressure on the thermodynamic properties. [4, 5] developed several original HP systems devoted to the measurement of phase change under pressure. He measured the latent heat of water from the Clapeyron equation. The measurement consisted in measuring the evolution of the phase change temperature with pressure and the volume variation due to the phase change. Thanks to the fact that water is a pure substance, it was possible to calculate the latent heat with these data and the Clapeyron equation. Equivalent technique was developed by [6]. They used a novel piezometer based on the measurement of the volume variation to study the phase diagram of a magnesium sulfate aqueous solution. The major drawback for HP calorimetry concerns the heavy cells that must resist to pressure. Their large heat capacity and their thermally resistant walls don't allow accurate temperature scan. An alternative procedure consists in working at constant temperature and to change the pressure. This pressure change can be done by step [7] or at a constant rate [8]. [9-11] measured the thermal expansivity and the isothermal compressibility of selected liquids by using a method based on the measurement of the heat of compression of the sample subjected to a step pressure evolution. Original results were obtained with different ,alkane such as hexadecane, butane-1,4-diol [12], n-alkanes C_{21}, C_{23} and C_{25} [13], Carbone tetrachloride [7]and benzene [7], [14]. For most of these works, the pressurisation fluid was the studied fluid, except for [6] who used mercury. The Heat dissipated during adiabatic compression dP of a substance is given by equation (3).

$$dQ = \alpha \cdot V \cdot T \cdot dP \tag{1}$$

In the case of [9-11] , the coefficient of thermal expansion was determined from the amount of heat which was measured and which corresponded to that given by equation (1). The exact volume of the cell was thus necessary and this kind of approach can be called a "constant volume" technique. It is important to point out that all uncontrolled heat dissipated during pressurisation such as the heat dissipated by the material of the cell due to pressure change has to be subtracted to the calorimetric signal [12]. A second technique consists in installing a sample of a given mass in the cell. This second technique can thus be called "constant mass". Mercury has the advantage of a very small coefficient of thermal expansion. Thus, it is a very convenient pressurisation fluid in the case of constant temperature-constant mass calorimetry. It is usually used for single cell systems. It can also be used for differential systems (2 cells) such as [8]. Nevertheless, differential system allow the use of pressurisation fluids with relatively large coefficient of thermal expansion. Of course, this statement must be referred to the noise to signal ratio. This constant mass technique was used by [15] with a single cell system and by [8] with a differential system. The constant mass approach can be used for solid or liquid materials. [8] used mercury as a pressurisation fluid and installed the material to be studied above the mercury level, below the cell plug (cells were installed with their bottom end facing upward). This has the disadvantage of a poor heat transfer between the sample and the calorimetric cavity, the sample being confined in the bottom of the cell. [16] developed a single cell system which was used to measure the latent heat of pure water and salt aqueous solutions of $MgSO_4$ (w/w) and KCl [15] for pressure until 200 MPa. The sample was installed in a plastic pouch and mercury was used as a pressure transmitting fluid. This experimental apparatus was upgraded by [17] switching from a single cell to a differential system using two cells and pentane as a pressurisation fluid.

This paper is presenting some results obtained with the differential calorimeter developed by [17] which is a high pressure constant mass differential calorimeters working under isothermal condition. Some statements on the behaviour of the sample of water during the peak are presented and are discussed.

2. MATERIAL AND METHODS
2.1. High pressure calorimeter
The differential high pressure calorimeter is schematically shown in Figure 2. The differential system used a calorimetric measurement head (Pass 27, SCERES, Orsay, France) shown in Figure 3. It was 170 mm in diameter and 400 mm in height and the cells were 20 mm diameter and 90 mm high. The calorimetric sensor was composed of multiples thermocouples junctions installed between the two calorimetric cavities. An oven installed around the system was used to control the temperature of the calorimetric head which was kept at subzero temperature by a water glycol circulating bath.

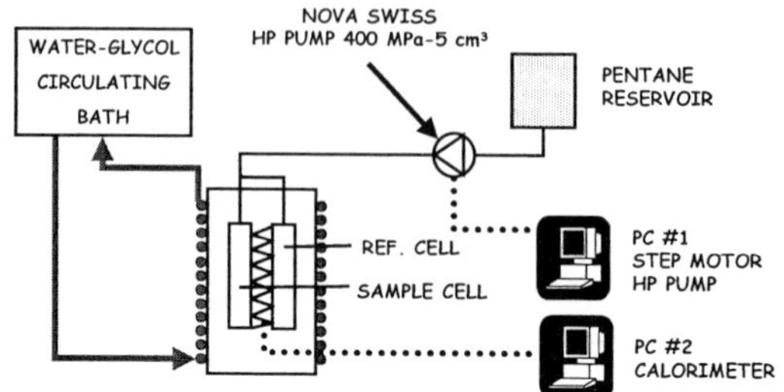

Figure 2. Scheme of the differential high pressure calorimeter

The high pressure cells were placed in the calorimetric cavities. Their volume was 4.6 cm³ (20 mm and 10 mm in ext. and int. diameter and 85 mm in height). They were made in APX steel (Aubert et Duval, France). A metal plug made of APX steel equipped with a nitrile O-ring seal was installed in a slot machined in the plug. A threaded copper-beryllium nut closed the cells and keep the plug in place. The Cells were connected to a high pressure tube (3.2 mm diameter or 1/8 inch) thanks to Harwood miniature fittings (M2 Serie, 100 MPa, Harwood Engineering, Ma, USA) which were used for connections. A high pressure capillary tube (1.6 mm external diameter or 1/16 inch, 0.5 mm internal diameter) was silver brazed to the 3.2 mm high pressure tubing close to the cells to minimise heat losses between the calorimeter head and the ambient.

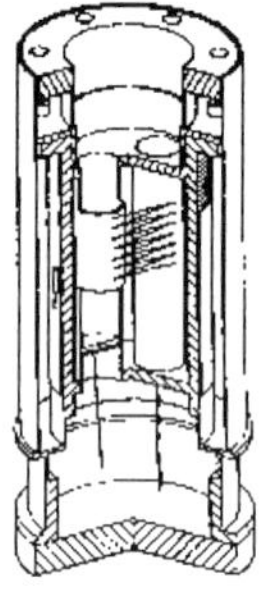

Figure 3
SCERES Calorimetric HEAD (PASS 27). Cell diameter 20 mm (SCERES, Orsay-France)

A T-fitting was used for the differential system to pressurise both cells. Electronics (SCERES A/D converter, Orsay, France) and computer systems were used to record the temperatures and the calorimetric signal. A high pressure pump using pentane (Sigma, St.-Quentin Fallavier, France) as a pressurisation fluid (400 MPa, 5 cm3, Nova-Swiss, Effretikon, CH) was driven by a step motor (MIJNO MO63-LE09). A PID software was developed to drive the motor thanks to a pressure sensor (200 MPa, Asco Instruments, Chateaufort, France). The system was designed for 400 MPa but was currently used in the 0.1-200 MPa domain. Pentane was chosen because of its stability (no phase change) in the pressure-temperature domain of investigation.

2.2. Calibrations

The pressure sensor was calibrated against a reference pressure gauge (Bourdon, France). Temperature of the calorimeter was calibrated against a 0.1 mm diameter K-type thermocouple (Omega, USA) placed in the sample cell at selected temperature between –20°C and +20°C. The calibration of the calorimetric signal was carried out by Joule effect using a 100Ω resistance located in a high pressure cell. A voltage between 1V and 4.2V was applied for duration between 30 and 90 min. The power dissipated by Joule effect was measured with Keithley volt meter and ampere meter. A linear fit of the experimental data yielded eq.(2).

$$\sigma(T) = -3.20 \cdot 10^{-3} \times T + 0.613 \quad (mW),\ T\ in\ °C \tag{2}$$

3. RESULTS AND DISCUSSION

3.1. Latent heat of pure water

Latent heat of pure water was measured for validation of the calorimeter. The water sample had a mass of $\approx$ 1g and was vacuum-packaged in a polyethylene bag. In order to minimise the heat dissipated by pentane in the reference cell for the differential system, the volume of the sample (in the measuring cell) was replaced by a nylon rod (in the reference cell) whose volume was equivalent to that of the water sample placed in the sample cell. Nylon was chosen because of its low compressibility and the absence of phase change in the pressure and temperature domain used. The samples were frozen in the calorimeter. The pressure was then increased at a constant rate which was typically 1 MPa/min from 0.1 to 200 MPa. When the pressure reached the corresponding phase change temperature (temperature of the calorimeter), the samples started melting (point 1 in Figure. 4 and 5).

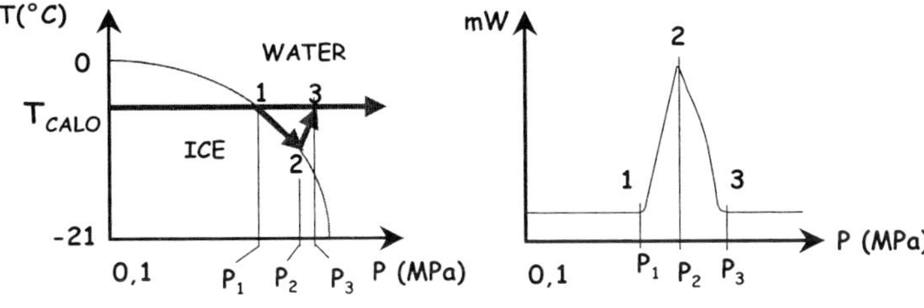

Figure 4	Figure 5
Temperature evolution of the sample of water during pressurisation in the calorimeter	Scheme of the corresponding Calorimetric peak (according to Figure 3)

Figure 4
Temperature evolution of the sample of water
during pressurisation in the calorimeter
1: beginning of the melting
2: end of melting, 3: end of calorimetric peak

Figure 5
Scheme of the corresponding
Calorimetric peak (according to Figure 3)

The melting phenomenon associated to the pressure increase makes that the sample temperature decreases (phase 1-2 in figure 4 and 5). This induced a heat transfer between the cell and the calorimeter through the heat flux sensors (multiple thermocouple junctions). The calorimetric signal increased until complete melting of the sample (point 2) and then decreased at follow to come back to the initial base line (phase 2-3 in figure 4 and 5). An experimental calorimetric plot is given in figure 6.

612

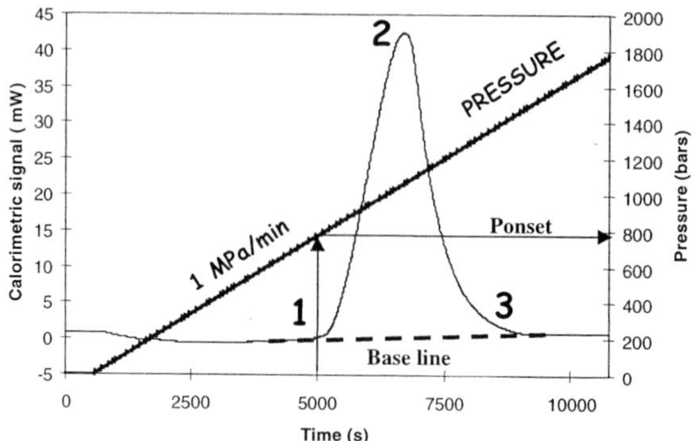

Figure 6. Experimental plot obtained with pure water
Temperature = - 7,3°C, Sample mass ≈ 1g, Pressurisation rate 1 MPa/min
Points 1, 2 and 3 are corresponding to scheme shown in figure 4 and 5

The peak span was 60 MPa. This span can be reduced by using smaller sample mass and slower pressurisation rate. This point is discussed in [18] showing a comparison between the present calorimeter and a previous calorimeter using a single cell with smaller samples [15]. As a comparison, [8] used pressurisation rates between 0.3 and 1.2 MPa .min[-1] with peak span of 70 MPa at 0.3 MPa .min[-1] (melting of benzene at 314.2K). Latent heat of pure water as a function of pressure is shown in figure 7 with data obtained with the present calorimeter.

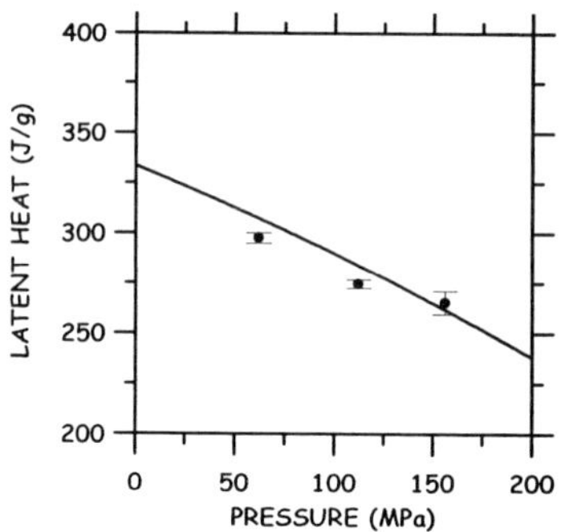

Figure 7. Latent heat of pure water
Line = Bridgman's data [5] - Points = our results (with SD error bars)

It is interesting to point out that the data from the literature were obtained by Bridgman [5] who used the Calpeyron equation. Our results were within 3% from Bridgman's data (experimental error due to calibration, integration of the peak ...). This good agreement allowed us to consider this apparatus reliable to evaluate the latent heat of products under pressure.

4. CONCLUSION

This work allowed us to validate a calorimeter designed to work at high pressure and constant temperature. The latent heat of ice was measured at –5, -10 and –15°C and was within 3% of the existing data from the literature. These apparatus which works isothermally used pentane as pressurisation fluid. A slower pressurisation rate combined to a minimised mass of sample should permit to reduce the span of the peak. Future development will aim to improve the response time of the calorimeter by using smaller cells or more conductive cells. Peak deconvolution has also been evaluated (not presented in this paper) and permitted to reduce the pressure span of the peak.

Acknowledgements

The authors wish to thank ENITIAA, the French Ministère de l'Agriculture et de la Pêche, the Région Pays de Loire and the Pole Agronomique Ouest (PAO) for financial support and L. Guihard, J. Laurenceau and O. Rioux for their technical support.

KEY WORDS: HIGH PRESSURE, CALORIMETRY, WATER

REFERENCES

1. Lavoisier, *Traité Elementaire de Chimie*. 1789.
2. Rollet, A.P. and R. Bouaziz, *L'analyse thermique*, ed. Gauthier-Villars. Vol. Tome 1: Les changements de phase. 1972. 3577.
3. Gutt, W. and A.J. Majumdar, *Differential thermal analysis*. Vol. 2. 1972, London: Academic Press.
4. Bridgman, P.W., *Mercury, liquid and solid under pressure*. Proceedings of the American Academy of Arts and Sciences, 1912-a. **47**: p. 347-438.
5. Bridgman, P.W., *Water in the liquid and five solid forms under pressure*. Proceedings of the American Academy of Arts and Sciences, 1912-b. **47**: p. 411-558.
6. Hogenboom, D.L., *et al.*, *Magnesium sulfate-water to 400 MPa using a novel piezometer : densities, phase equilibria and planetological implications*. Icarus, 1995. **115**: p. 258-277.
7. Pruzan, P., L. TerMinassian, and A. Soulard, *Entropy determined from heat of compression at high pressure for benzene and carbon tetrachloride*. High-pressure science and technology, 1979. **1**: p. 368-378.
8. Randzio, S.L. and J.P. Grollier, *An isothermal scanning calorimeter controlled by linear pressure variations from 0.1 to 400 MPa. Calibration and comparison with the piezothermal technique*. Rev. Sci. Instrum., 1994. **65**(4): p. 960-965.
9. Ter Minassian, L., *et al.*, *Mesures calorimétriques des coefficients de dilatation et de compressibilités absolus*. Journal of Chemistry and Physics, 1970. **67**: p. 265-269.
10. TerMinassian, L. and P. Pruzan, *High-pressure expansivity of materials determined by piezo-thermal analysis*. Journal of Chemistry and Thermodynamics, 1977. **9**: p. 375-390.
11. TerMinassian, L., K. Bouzar, and C. Alba, *Thermodynamic properties of liquid toluene*. Journal of Physical Chemistry, 1988. **92**: p. 487-493.
12. Petit, J.C. and L. TerMinassian, *Measurement of (dV/dP)T , (dV/dT)P and (dH/dT)P by fluxcalorimetry*. Journal of Chemistry and Thermodynamics, 1974. **6**: p. 1139-1152.
13. TerMinassian, L. and F. Milliou, *New developments in the field of piezothermal analysis : an experimental approach to the solid state transitions in the case of n-alkanes C21, C23 and C25*. Journal of thermal analysis, 1992. **38**: p. 181-196.
14. Fuchs, A.H., P. Pruzan, and L. TerMinassian, *Thermal expansion of benzene at high pressure determined by a calorimetric method, its behavior near melting*. Journal of Physics and Chemistry of solids, 1979. **40**: p. 369-374.

15.   Chourot, J.M., A. LeBail, and D. Chevalier, *Phase diagram of aqueous solution at high pressure and low temperature.* High Pressure Research, 2000. **19**(1-6): p. 191-199.

16.   Chourot, J.M., *Contribution à l'étude de la décongélation sous haute pression, étude expérimentale et modélisation, .* 1997, Nantes-France: Nantes. p. 151.

17.   Chevalier, D., *Contribution à l'étude de la congélation par détente haute pression, .* 2000, Nantes-France: Nantes. p. 200.

18.   LeBail, A., D. Chevalier, and J.M. Chourot, *High pressure calorimetry. Comparison of two systems (differential vs. single cell). Application to the phase change of water under pressure.* Journal of Thermal Analysis and Calorimetry, 2001(Submitted dec 2000).

Trends in High Pressure Bioscience and Biotechnology
R. Hayashi (editor)
                                       615

The use of a small-angle X-ray scattering technique with a third-generation synchrotron X-ray source in high-pressure biochemistry

T. Fujisawa[a], Y. Nishikawa[a] and Y. Inoko[b]

[a]Structural Biochemistry, RIKEN Harima Institute/ SPring-8, 1-1-1, Kouto, Mikazuki, Sayo, Hyogo 679-5148 Japan

[b]Division of Biophysical Engineering, Graduate School of Engineering Science, Osaka University, Machikaneyama, Toyonaka, 560-8531 Japan

An improved high-pressure (up to 500 MPa) cell with diamond windows was developed for small-angle X-ray scattering (SAXS) of protein solutions, especially optimized for a third-generation synchrotron X-ray source. The performance of the improved cell was examined by measuring the compressibilities of apo-ferritin and cytochrome-C in solutions. The order of the compressibilities of these proteins was consistent with that shown by ultrasonic measurements.

1. INTRODUCTION

Pressure effects on biological systems have recently been attracting much interest by researchers in both basic and applied science fields. Interest has particularly been shown in pressure effects on protein solutions, which include selective dissociation of protein complexes, unfolding of proteins, inactivation of enzymes, and the recovery of protein from precipitation or an inclusion body [1, 2]. Despite the abundance of available spectroscopic methods, such as Fourier transfer spectroscopy, fluorescence, and NMR, the application of X-ray scattering to protein solutions has long been hampered by instrumental difficulties; small-angle X-ray scattering (SAXS) signal from a protein solution is very weak, while absorption and parasitic scattering from window materials are strong. A major breakthrough in SAXS instrumentation was achieved by Kato & Fujisawa (KF) in the 1990s [3], who used synthetic diamond windows as well as a synchrotron X-ray source. This KF-type high-pressure (HP) cell enabled quantitative measurement of SAXS from protein solutions under high pressure and provided more direct structural information on the hysteresis effect of protein dissociation compared to that obtained in other spectroscopic studies [4]. The cell, however, had some disadvantages; large sample volume (1 ml), difficulty in sample exchange, and leakage problems. Since such deficiencies are not permissible in modern analytical biochemistry, we improved the HP-SAXS cell.

SPring-8 is a third-generation synchrotron facility, having the highest electron energy in the world (8 GeV). In 1997, RIKEN constructed a structural biology beamline I (BL45XU) that included an SAXS branch. The SAXS branch was designed especially for determination of protein structures under hydrostatic pressure: weak interactions between proteins in solution or subunits of fibrous proteins have been studied. To perform SAXS experiments on protein solutions under high pressure, the SAXS beamline must have a small beam size (0.4 mm×0.8 mm at the sample position), high energy (1 Å) and high flux (2×10^{12} photons) with low parasitic scattering [5]. We developed an HP-SAXS cell (up to 500 MPa) with diamond windows for solution X-ray scattering of proteins [6]. The sample volume is 0.4 ml. Together with an efficient detector, an X-ray image intensifier with a CCD (XR-II+CCD), it is now possible to collect highly reproducible data for a dilute protein solution with a protein concentration as low as 1 mg/ml [7]. Typical data collection time is 1 or 2 seconds. Since SAXS reflects the spatial average of the electron density of protein in solution, it is possible to obtain the shape and size of proteins under high pressure. We investigated the performance of this HP cell in the case of direct measurement of isothermal compressibility of a small protein (cytochrome-C) (Fig. 2 A) and a protein complex (apo-ferritin) (Fig. 3 A).

2. METHODS

2.1 HP-SAXS cell

A schematic of the HP-SAXS cell is shown in Fig. 1. A detailed description has been given elsewhere [6]. Basically, the window material is made of synthetic diamond, and the cell incorporates a pressure intensifier (×20). By employing a pressure intensifier, the pressure pump unit can be situated outside the radiation-shielding hatch at the beamline.

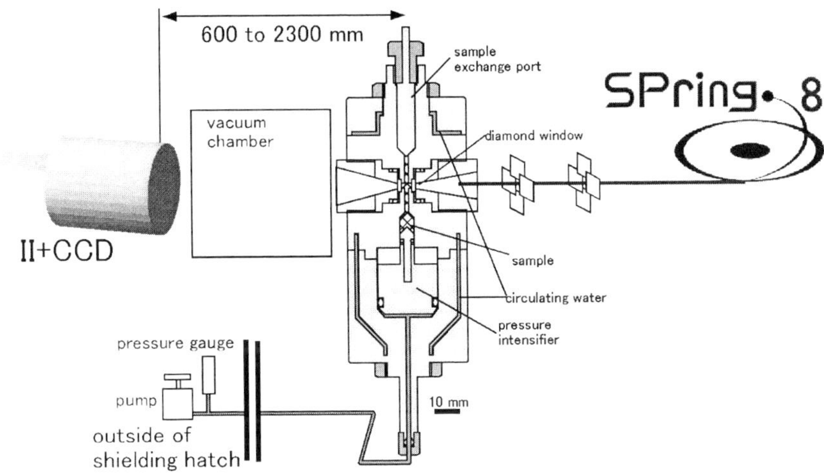

Figure 1. Schematic of the HP-SAXS cell.

2.2 SAXS measurement

HP-SAXS measurement was done at BL45XU in SPring-8.   The basic procedure used was the same as that of standard SAXS measurement except for the use of the HP-SAXS cell: The SAXS of the sample solution and then that of the buffer solution were measured [7].   The sample was inserted into the cell very carefully so as to avoid formation of bubbles, which greatly changes the increasing rate of raising pressure.   Before the SAXS measurement at each pressure, we waited for 10 min in order to equilibrate.   The data collection time was typically 1 sec.   Transmission of X-rays was determined by diffraction intensity of Kapton, which was remotely inserted into the scattering path.

2.3 SAXS data analyses

Two-dimensional scattering images recorded by XR-II+CCD were circular-averaged and scaled by incident X-ray intensity.   The detailed procedure of the data processing has been described elsewhere [7].   The obtained scattering intensity $I(S)$ is a function of the scattering vector $S=2\sin\theta/\lambda$, where 2θ is the scattering angle and λ is the wavelength.   In order to obtain information on the size dependency of protein on high pressure, we employed two types of analyses.   For cytochrome-C (Cyt-C) (Fig. 2 A), we used Guinier analysis for small globular protein [8] (Fig. 2 D).   The intensity of $I(S)$ was fitted by the equation

618

$$I(S) = I(0)\exp(-\frac{4\pi^2}{3}Rg^2S^2) \ ,$$
(1)

where Rg is defined by the second momentum of electron density of the particle, or radius of gyration.   For apo-ferritin, which consists of 24 polypeptides (MW=24×20kDa) (Fig. 3 A), we used classical solid body approximation.   Its structure is well approximated by a hollow sphere. $I(S)$ is fitted by

$$\Phi(t) = 3\frac{\sin t - t\cos t}{t^3} \ \ \text{and} \ \ I(S) = \frac{\left(r_1^3\Phi(2\pi Sr_1) - r_2^3\Phi(2\pi Sr_2)\right)^2}{\left(r_1^3 - r_2^3\right)^2} \ ,$$
(2)

where r_1 and r_2 are outer and inner radii of the hollow sphere, respectively (Fig.3 C) [8].

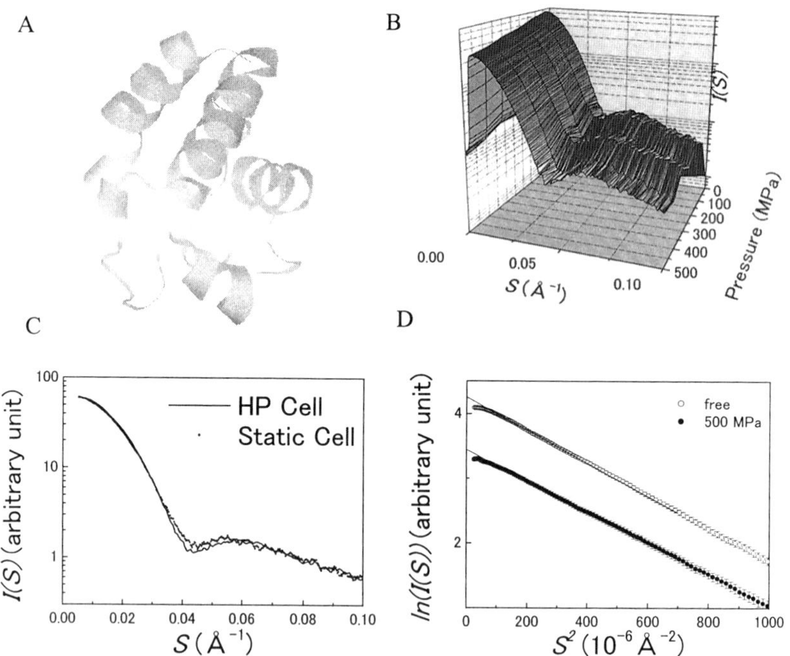

Figure 2. HP-SAXS of cytochrome-C solution.

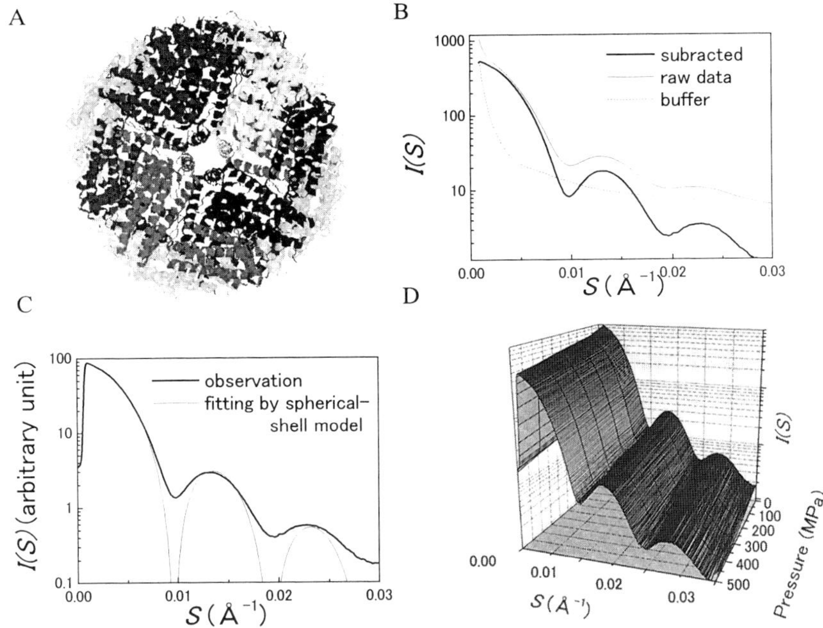

Figure 3. HP-SAXS of apo-ferritin.

3. RESULTS AND DISCUSSION

3.1. High pressure effect on cytochrome-C

First, we compared the Cyt-C profile of the HP-SAXS cell with that of a standard cell. These two curves were identical, showing the reliability of the HP-SAXS cell (Fig. 1 C). The Guinier plot of Cyt-C under pressure showed a slight decrease in Rg with increase in high pressure (by ~2% for 500 MPa), which was not what was expected on basis of the compressibility of protein itself [9]. Rg reflects the structure of protein with its associated water. On the other hand, the change in cavity of proteins has very little effect on Rg. The high-pressure dependence on Rg, therefore, offers isothermal compressibility. The isothermal compressibility of Cyt-C was estimated to be about $\sim 9 \times 10^{-12}$ (cm^2/dyn).

3.2 High pressure effect on apo-ferritin

In a special case such as a protein complex with spherical symmetry, the inner and outer radii of the protein complex can be estimated (Eq. (2) and Fig. 3 A). This analysis is less sensitive to interparticle interference compared with Guinier analysis. Apo-ferritin has very characteristic peaks in the medium angle region, which come from the form factor of this particle (Fig. 3 B). These peak positions are very sensitive to the inner and outer radii (Fig. 3 C). With increasing pressure, these peaks shifted slightly to inner S (Fig. 3 D). The pressure dependence on volume estimated from Eq. (2) showed the isothermal compressibility for apo-ferritin to be $\sim 8 \times 10^{-12}$ (cm^2/dyn).

4. CONCLUSION

The present study showed the possibility of measuring protein compressibility by using an HP-SAXS technique. The obtained isothermal compressibility values ranged in $\sim 8\text{-}9 \times 10^{-12}$ (cm^2/dyn), regardless of the type of SAXS analysis. The values were consistent with those obtained from ultrasonic measurements [10].

This study was supported by Special Coordination Funds of the Ministry of Education, Culture, Sports, Science and Technology, Japan.

REFERENCES

1. P. W. Bridgeman, J. Biol. Chem. 19 (1914), 511.
2. K. Suzuki, Y. Miyosawa, and C. Suzuki, Arch. Biochem. Biophys. 101 (1963) 2225
3. M. Kato and T. Fujisawa, J. Synchrotron Rad. 5 (1998) 1282.
4. T. Fujisawa, M. Kato, and Y. Inoko, Biochemistry 38 (1999) 6411.
5. T. Fujisawa, K. Inoue, T. Oka, H. Iwamoto, T. Uruga, T. Kumasaka, Y. Inoko, N. Yagi, M. Yamamoto and T. Ueki, J.Appl. Cryst. 33 (2000) 797.
6. Y. Nishikawa, T. Fujisawa, Y. Inoko, and M. Moritoki, Nucl.Insturm.Methods. (2001) in press.
7. T. Fujisawa, Y. Inoko and N. Yagi, J,Synchrotron Rad. 6 (1999) 1106.
8. A. Guinier & G. Fournet, Small-Angle X-ray scattering of X-rays, John Wiley, New York, 1955.
9. C.E. Kundrot & F.M. Richards, J.Mol.Biol. 193 (1987) 157.
10. K.Gekko & Y. Hasegawa, Biochemistry 25 (1986) 6563.

Trends in High Pressure Bioscience and Biotechnology
R. Hayashi (editor)

Improvement of a high-pressure vessel for use in a freeze pressure generation method and its application to food sterilization

Kiyoshi Hayakawa[a], Naoto Miyajima[a], Yasushi Fujimoto[a], Kazunari Saho[b], Tetsuya Komai[c], Masamitsu Matsumoto[d] and Rikimaru Hayashi[e]

[a]Kyoto Prefectural Comprehensive Center for Small and Medium Enterprises,
17 Chudoji Minami-machi, Shimogyo-ku, Kyoto 600-8813,
[b]Gunze Co.,Ltd., 1 Zeze Aono-cho, Ayabe-si, Kyoto 623-0011,
[c]Daiichigiken Co.,Ltd., 15-138 Ryogae-cho, Hushimi-ku, Kyoto 612-8082,
[d]Teramecs Co.,Ltd, 97 Takeda higashikoyanouchi-cho, Hushimi-ku, Kyoto 612-8448,
[e]Laboratory of Biomacromolecular Chemistry, Division of Applied Life Sciences, Graduate School of Agriculture, Kyoto University, Sakyo-ku, Kyoto 606-8502

Abstract

A simple method for high-pressure generation and its application for microbial inactivation was studied.   In order to obtain the theoretical pressure, the design of the high-pressure chamber was improved, for use on a large scale.   Two types of high-pressure chambers were constructed: a pin type chamber consisting of a vessel and a cap with a pin stopper, and a screw cap type chamber consisting of a vessel and a screw cap.   Pressure generated in the vessel in the temperature range from 0 to -22°C precisely fitted to the equilibrium curve of water and Type I ice.   The generated pressure increased with decreasing temperature from -22 to -30°C reaching 250 MPa at -30°C.   After storage of food samples in the vessels at -30°C for 24 h, yeasts in *nigori-zake* and beer, and lactic acid bacteria in yoghurt were completely inactivated and microorganisms typically found in milk were partially inactivated.

1. INTRODUCTION

High pressure has recently been used in the sterilization and processing of food [1,2].   In the course of studies on the utilization of high pressure in food science and technology, a simple pressurization method in which high pressure is naturally generated without the need for an oil pump has been reported [3].   This method is based on the well-known phenomenon that water filled in a sealed vessel increases in volume at the freezing temperature and high pressure is then generated in the vessel when it is stored at sub-zero temperatures.   We refer to this technique as "The freeze pressure generation method ".   In a previous experiment [3] a pressure of 140 MPa was thus generated at –15°C in a sealed vessel, but no further increase in pressure beyond 140 MPa was attained as the result of further cooling the chamber to –22°C.   The purpose of this report is to describe the extent of the pressure generated in the improved vessel and the effect of the generated pressure on the inactivation of microorganisms in foods.

2. MATERIALS AND METHODS

2.1. High pressure vessels

To obtain the theoretical pressure [4] , two types of stoppers for the high pressure chamber were constructed: One, a pin type stopper, consisting of a vessel and a cap with a pin stopper (Fig.1). The vessel was sealed with the cap and subsequently stoppered by a pin (diameter 30 mm ϕ). The other was a screw type stopper, consisting of a vessel and a screw cap (Fig.2). The vessel was sealed tightly by the screw cap. Both chambers were equipped with a pressure gauge on the cap. The gauge and connecting pipes were filled with ethanol as the pressure medium, which was separated from the chamber inside by free-pistons. The chamber volumes of pin and screw types were 770 and 1000 ml, respectively.

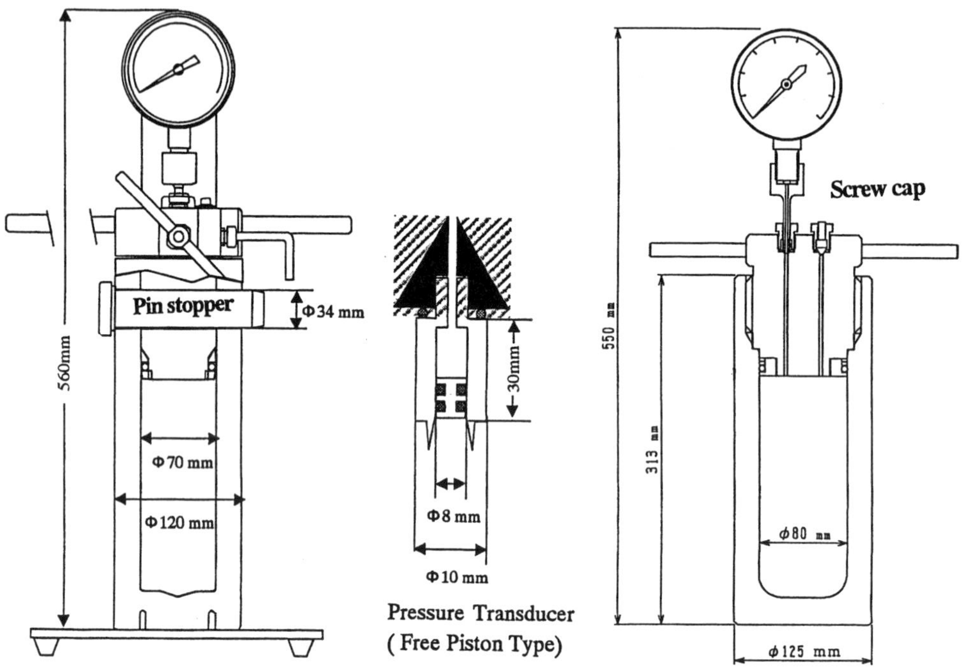

Fig.1 Pin type stopper   **Fig.2 Screw type stopper**

2.2. Measurement of surviving microorganisms

Polyethylene pouches (1 × 5 cm) containing a microbial suspension (approximately 1 ml) were heat-sealed and placed in the pressure chambers. The chambers were filled with distilled water and sealed tightly. The sealed chambers were then stored in the temperature-controlled freezer for 24, 48 or 72 h. After storage, the chambers were removed from the freezer, equilibrated at room temperature, and opened by removing the cap for counting of the surviving cells. Yeasts were cultivated in malt extract agar at 30°C for 2 days and lactic acid bacteria were cultivated in GYP medium at 30°C for 2 days.

3. RESULTS AND DISCUSSION

3.1. Improvement of high pressure vessel

One possible reason for why an ideal pressure was not generated in our previous experiment [3] involves the stopping of the pressure transmission from the chamber to the pressure gauge, because of the large compressibility of ethanol, which was used for the pressure medium.

In order to obtain the theoretical pressure, the design of the high-pressure vessel, the expandability of the vessel and the use of minimum amounts of the pressure-transmission medium (ethanol) have been improved. As a result, two types of high-pressure chamber were constructed, as follows.

One type of chamber with a volume of 770 ml contains pin stopper which consists of a vessel and a cap with a pin stopper (Fig.1). The vessel was sealed with the cap, which is then stoppered by a pin having a diameter of 30 mm. The pin was made from SKD61, which is not easily deformed by pressurization at 250 MPa and can be opened and closed in one action.

The other type of chamber contained a screw cap, consisting of a vessel (inner volume, 1000 ml) and a screw cap (Fig.2). The vessel was covered and sealed tightly by the screw cap.

Both types of vessels were equipped with a pressure gauge on the cap. The gauge and connecting pipes were filled with approximately 1.5 ml of ethanol as the pressure medium, which was separated from the chamber by a free-piston.

3.2. Generation of high-pressures using the improved vessel

The chamber was filled with water, tightly sealed, and stored at sub-zero temperature in a temperature-controlled freezer for 24 h.

The relationship between the pressure generated in the vessel and temperature of the freezer is shown in Fig. 3. The curve precisely fitted the equilibrium curve of water and Type I ice in the range of 0 to -22°C [4]. However, the generated pressure increased with decreasing temperature in the range of -22 to -30°C, reaching 250 MPa at -30°C, indicating that super-cooling had occurred in the range of -22 to -30°C.

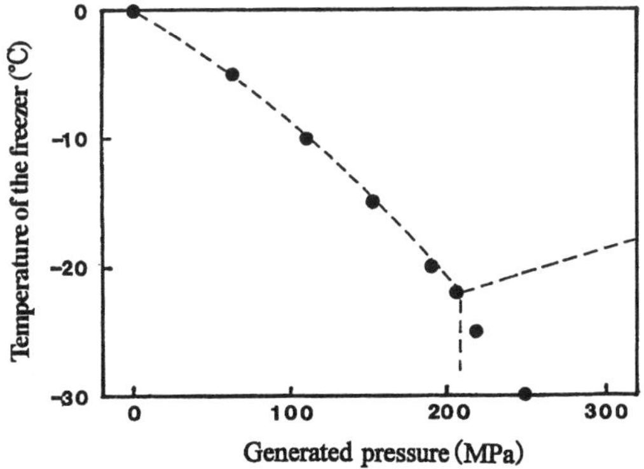

Figure 3. Relationship between pressure generated in the pressure chamber and temperature.

●, experimental value; -----, published values for a water -ice equilibrium curve [4]

3.3. Inactivation of microorganisms in food

The freeze pressure generation method was applied to the inactivation of microorganisms in food (Table 1). Liquid food was placed in a small polyethylene pouch, heat-sealed, placed in the pressure chamber, and kept in the freezer at various temperatures for $1\sim3$ days.

Yeast in *nigori-zake* and beer were completely inactivated by storage at -30°C for 24 h. Lactic acid bacteria in yoghurt were also inactivated by storage at -30°C. However, microorganisms typically found in milk were not completely inactivated.

Table 1. Inactivation of microorganisms in food by the freeze pressure generation method.

Food*	Temperature of the freezer (°C)	Storage time (days)			
		0	1	2	3
Nigori-zake	-10	2.3×10^6	8.0×10	ND	ND
	-20	2.3×10^6	ND	ND	ND
	-30	2.3×10^6	ND	ND	ND
Beer	-10	4.0×10^3	ND	ND	ND
	-20	4.0×10^3	ND	ND	ND
	-30	4.0×10^3	ND	ND	ND
Yoghurt	-10	1.4×10^9	ND	ND	ND
	-20	1.4×10^9	ND	ND	ND
	-30	1.4×10^9	ND	ND	ND
Milk	-10	1.2×10^5	1.0×10^4	9.3×10^3	6.3×10^3
	-20	1.2×10^5	1.0×10^4	5.0×10^3	1.0×10^3
	-30	1.2×10^5	8.9×10^2	2.2×10^2	5.1×10

Nigori-zake is non-pasteurized rice wine. Beer and milk were non-pasteurized one.
ND, not detected.

REFERENCES

1. R. Hayashi, Utilization of pressure in addition to temperature in food science and technology, In "High Pressure and Biotechnology" , eds. C. Balny, R. Hayashi, K. Heremans, P. Masson, Colloques INSERM, vol 224, John Libbey Eurotext Ltd., pp. 185-193 (1992).

2. R. Hayashi, An overview of the use of high pressure in bioscience and biotechnology, In "High Pressure Bioscience and Biotechnology, Progress in Biotechnology 13" , eds. R. Hayashi, C. Balny, Elsevier, pp. 1-6 (1996)

3. K. Hayakawa, Y. Ueno, S. Kawamura, T. Kado and R. Hayashi, Microorganism inactivation using high-pressure generation in sealed vessels under sub-zero temperature, *Appl. Microbiol. Biotechnol.*, **50**, 415-418 (1998).

4. P. W. Bridgman, Water in the liquid and five solid forms under pressure. *Proc. Amer. Acad. Arts Sci.*, **47**, 466 (1912).

Thermal stress of deep-sea dive operations

N. Naraki and M. Mohri

JAMSTEC: Japan Marine Science Technology Center,
2-15 Natsushima, 237-0061 Yokosuka, Japan
narakin@jamstec.go.jp

1. INTRODUCTION

With the progress of technology, the environmental pressure range of human being's activities are extended from 0 Pa (in the space) to 66 MPa (in the deep sea). The methods of penetration in the sea can be classified into two main groups, *1. Environmental pressure diving:* the divers' body is exposed directly to the hydrostatic pressure depending on the dive depth, *2. Atmospheric diving:* the operators stay under atmospheric air environment within the pressure hull (one-atmospheric diving suit or research submersible, Fig.1).

Environmental pressure deep diving, the diver must respire the pressurized gas, He-O_2 or H_2-He-O_2 mixed gas. The sea water temperature goes down to a several degrees Celsius deeper than 200m, then the diver utilize the hot water suit and the inspired gas heater avoiding the respiratory heat loss. Without inspired gas heating system, the diver's body is directly cooled from the body core by their respiration.

Figure 1. Two methods of penetration in the deep sea, the environmental pressure diving and atmospheric diving.

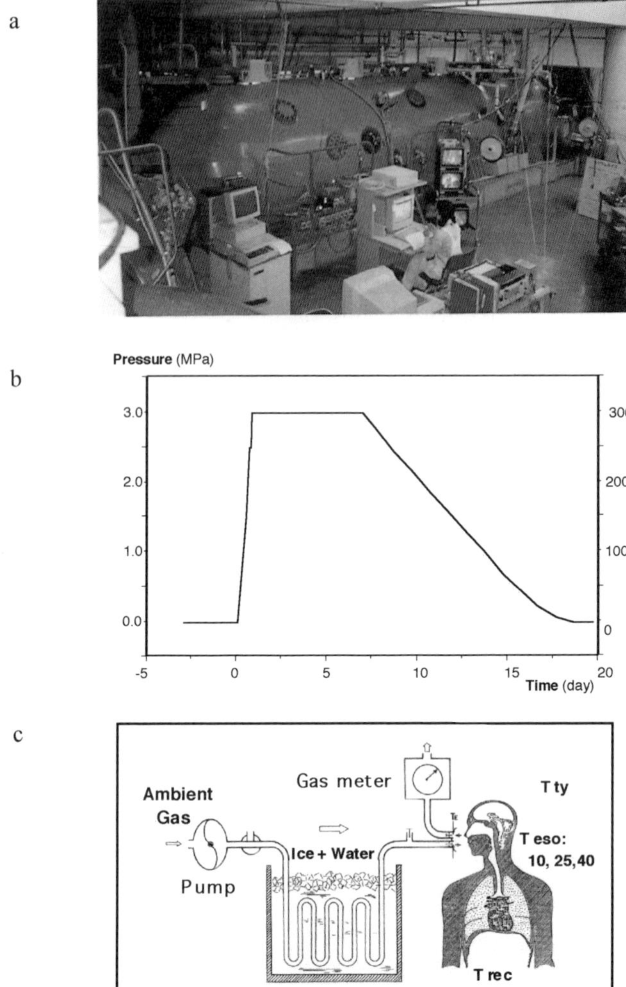

Figure 2. a JAMSTEC Hyperbaric center for the simulated saturation deep diving.
   b Profile of simulated saturation dive tothe depth of 300m with He-O2
   c Cold gas inhalation system

We have a tendency admittedly to image that the atmospheric diving is more comfortable than environmental pressure diving, but let us consider the states of affairs during the deep diving operations from the perspectives of thermal stress.

2. METHODS

Environmental pressure diving, we measured the core body temperatures (tympanic membrane temperature, three esophageal temperatures respectively, 10cm, 25cm and 40cm from the nostril, and rectal temperature) during cold gas inhalation: 5°C (same temperature as the deep sea water) for 20 minutes during He-O$_2$ simulated saturation dive to the 300m (3.1 MPa) under the comfortable ambient temperature 32°C (Fig. 2). This simulated saturation dive was conducted for around four weeks, 4 days pre-dive air condition, a half day compression to 3.1 MPa with helium under 0.04 MPa oxygen partial pressure, 8 days pressure holding at 3.1 MPa, 12 days decompression to atmospheric pressure and 4 days post-dive air. The subjects lived in hyperbaric chamber at 3.1 MPa with nearly pure helium environment (FO$_2$=1.3%, FN$_2$=2.6% and FHe=96.1%). The thermal conductivity of helium is 6 times higher than atmosphric air and its specific heat is 5 times larger than atmospheric air. Then we had to keep the ambient temperature 32°C at 3.1 MPa He-O$_2$ environment with very narrow range ±0.2°C. The subjects inhaled the cooled ambient gas passed through the pipes in the ice water. We measured deep body temperatures for 5 minutes pre-test, 20 minutes cold gas inhalation and 20 minutes recovery rest.

One atmospheric dive apparatus have no air conditioning system. We measured operator's thermal conditions (temperatures and relative humidity) of one-atmospheric diving suit (Newtsuit, 300m class: 3.1 MPa) during the training in swimming pool, in summer and in winter, the water temperatures were respectively 28°C and 12°C.

Also we measured the temperatures in the pressure hull of scientific research submersible Shinkai-6500 (JAMSTEC, 6,500m class: 66 MPa, three crew: main-pilot, sub-pilot and observer, are sitting on the floor mattress about 1.2m diameter, in the pressure hull diameter of 2m) in the real sea operations. The operation of Shinkai-6500 took 30 minutes from the embarkation to beginning of dive, about 3 hours from the sea surface to the bottom of 6,500m depth. After the observation on the bottom, it took also 3 hours from the bottom to the surface and 30 minutes to return to its mother-ship's deck.

3. RESULTS AND DISCUSSION

Environmental pressure diving: The core body temperatures decreased remarkably by several degrees for only 20 minutes cold gas inhalation at 300m He-O$_2$ condition, the esphageal temperature 10cm from the nostril, Teso-10, decreased by 7 or 9 °C comparing from the start of cold gas inhalation, the esophagus temperature of tracheal bifurcation level, Teso-25, decreased by 3 or 5 °C, and its heart level, Teso-40, deceased by 1 °C. The tympanic membrane temperature decreased by around 0.7 °C though, the rectal temperature had delay from others core temperatures and the least change, by only 0.2 or 0.3 °C. The rectal temperature is utilised as an very useful index of deep body temperature, but in these case, the inspired gas heater for deep sea diver has malfunction, the rectal temperature has no more important information for thermal regulation. The more serious hypothermic symptom advanced without conscious (Fig. 3 a & b).

One-atmospheric diving suit: The inside temperature of Newtsuit during dive in the warm waters exceeded 30°C and humidity was higher than 90 % caused by operater's metabolic heat and CO$_2$ absorption heat. During the dive in cold water, the Newtsuit made of aluminum with

628

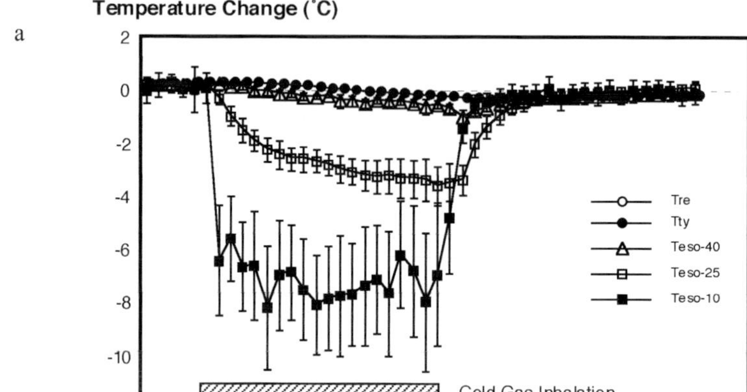

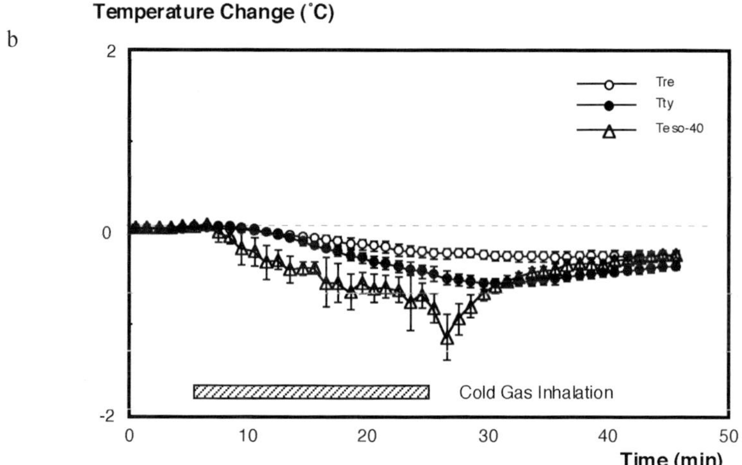

Figure 3. Deep body temperuture changes for 20 minutes cold gas inhalation test,
during the Simulated saturation dive to 300m (3.1MPa with Helium-Oxygen).

a : three esophageal temperatures, respectively distant from the nostril 10, 25, 40cm
b : tympanic membrane temperature: Tty and rectal temperature: Tre
mean and se of 12 data (4 subjects and 3 measurements)

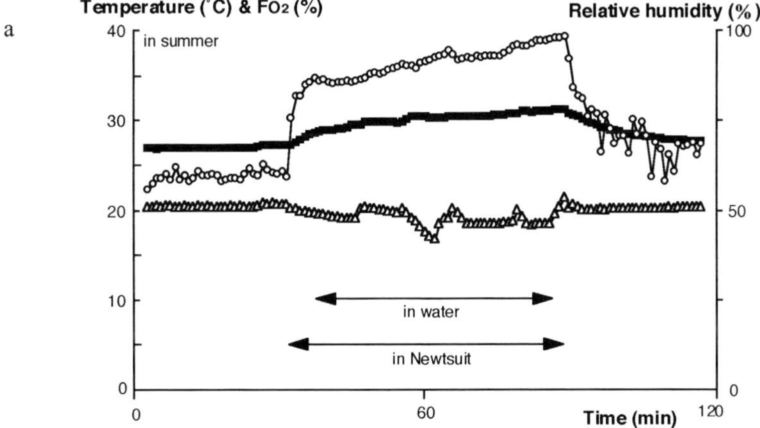

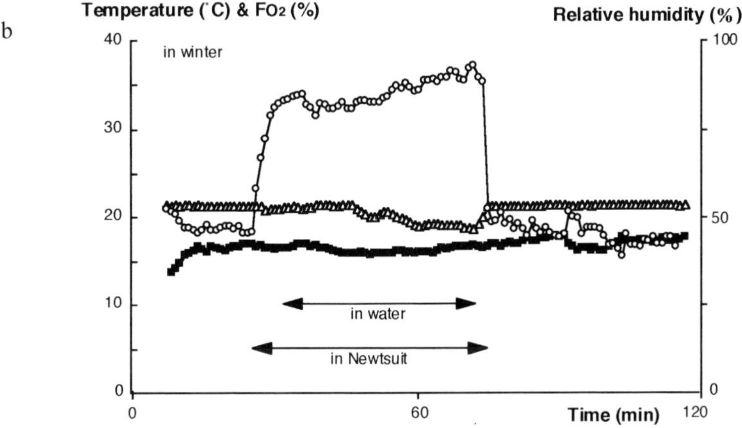

Figure 4. Inside temperature (■), oxygen fraction (△) and relative humidity (○) of Newtsuit during the traning dive in swimming pool.
a : dive in warm water in summer
b : dive in cold water in winter

high thermal conductivity, the operator are relatively well protected by wearing a lot of clothes, but the volume of his or her clothes are limited by small space in Newtsuit (Fig. 4 a & b).

Research submersible: The sea water temperature is around 1°C deeper than 1,000m, the inside air temperature of Shinkai-6500 did not go down below 20°C, but the crew were severely cooled by radiation from the cold wall of pressure hull during the deep dive (Fig. 5). Though the operations of Shinkai-6500 during summer or in the tropics, the pilots and observer must stay in the muggy conditions, like a Newtsuit diving in the warm water, for about 30 minutes, after the entrance hatch was closed on the mother ship til Shinkai-6500 went down to around thousand meters depth. Then they must change their under wears in the small space.

In summary, our research clearly shows that the thermal conditions of the both deep diving methods are very severe for divers, operators and observers. We must not forget that the atmospheric diving is no exception.

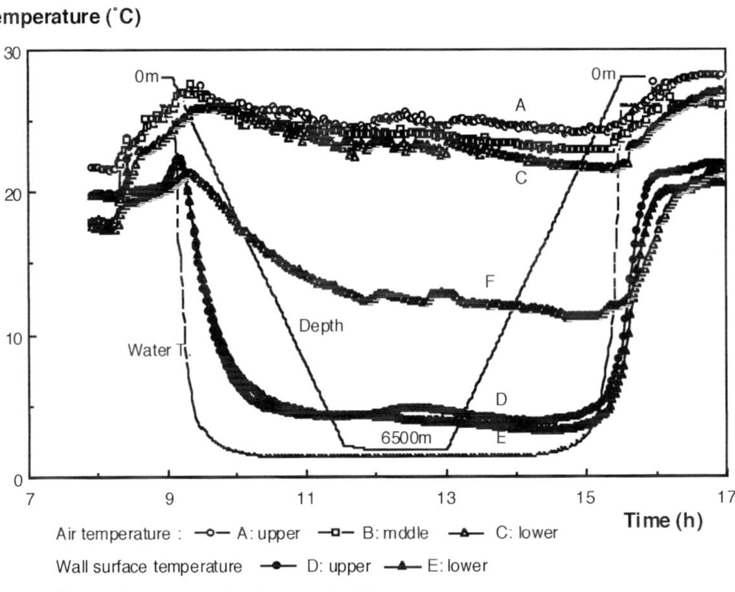

Figure 5. Inside temperatures of Shikai-6500, during dive to 6,500m.
Three air temperatures : upper, middle and lower in puresure hull
Two wall temperatures : upper and lower
Surface temperature of mattres on the floor

Trends in High Pressure Bioscience and Biotechnology
R. Hayashi (editor)
© 2002 Elsevier Science B.V. All rights reserved.

Oligomerization of Glycine in Supercritical Water with Special Attention to the Origin of Life in Deep-sea Hydrothermal System

Dimitar K. Alargov, Shigeru Deguchi, Kaoru Tsujii, Koki Horikoshi

The DEEPSTAR Group, Japan Marine Science and Technology Center, 2–15 Natsushima – cho, Yokosuka 237-0061, Japan

Possibilities of oligopeptide synthesis under conditions simulating the deep-sea hydrothermal system were studied. By injecting a glycine solution to a water at supercritical state (pressure 39.5 MPa and temperature ~400 °C), diglycine, triglycine (trace) and diketopiperazine as well as one unidentified product with high molecular mass (433 Da) were obtained in a flow reactor. It was found that chemical reactions, which lead to the formation of key for the origin of life molecules, take place simultaneously with decomposition. Velocity of flow and the concentration of initial solution as well as temperature and pressure determine the ratios of the products. The results support the hypothesis of chemical evolution of the origin of life in a submarine hydrothermal system.

1. INTRODUCTION

The hypothesis concerning the origin of life in a hydrothermal system was proposed in 1981 [1,2]. Different types of deep-sea hydrothermal vents as a part of submarine volcanism are considered to be continuous flow reactors for abiotic synthesis. Fluids circulate through pores and fractures of oceanic crust, penetrate to the cracking front of magma and then are injected into the deep sea at temperatures up to 350-400 °C [3,4]. Water at these conditions possesses energy high enough to carry out the reactions of oligomerization of amino acids. These amino acids could be the products of terrestrial processes or delivered to the Earth by meteorites or comets [5]. The only problem is that at these conditions processes of decomposition are quite plausible. By heating the mixtures of different amino acids to 240 °C at 26.5 MPa, Bada and Miller found that alkylamines are main products [6,7]. It was found that the main pathway of decomposition of amino acids was decarboxylation. Small oligopeptides were obtained by circulating repeatedly glycine solution through the hot (200-250 °C) and cold (0 °C) regions in a flow reactor at 24.0 MPa, with and without metal catalyst [8]. However, condensation reaction in supercritical water (SCW) has not yet been demonstrated so far. Parameter varieties of SCW provide different behaviors of fluid and its solutions [9,10]. Metastable equilibria may be attained in which important for the origin of life biopolymers to survive [11,12].

Here we report the possibilities of oligopeptide synthesis in the conditions simulating the deep-sea hydrothermal system. For this purpose glycine solution was injected to SCW (pressure 39.5 MPa and temperature 400 °C) in a flow reactor simulating the submarine hydrothermal system.

2. MATERIALS AND METHODS

Glycine was purchased from Nakalai tesque, Inc., Kyoto, Japan. Pure water (Milli-Q, pH 6.95) was used without pH control or added salts. The reactor apparatus consists of two parts - SCW supplying system and sample injecting system (Fig. 1). In both systems, pipes are made of Hasteeluy, a nickel-based alloy. In the SCW supplying system, the water is pressurized by the pump P-1 to 39.5 MPa (controlled by the valve V-1), and heated to about 400 °C by electric heater. After the injecting point, the fluid is cooled externally and leaves the system. In the sample injecting system, 0.6 M solution of glycine is compressed to a pressure slightly higher than that in SCW supplying system to allow injection. The injection is controlled by the valve V-2. The flow rate of a supercritical fluid was controlled by adjusting the capacity of the pump P-1.

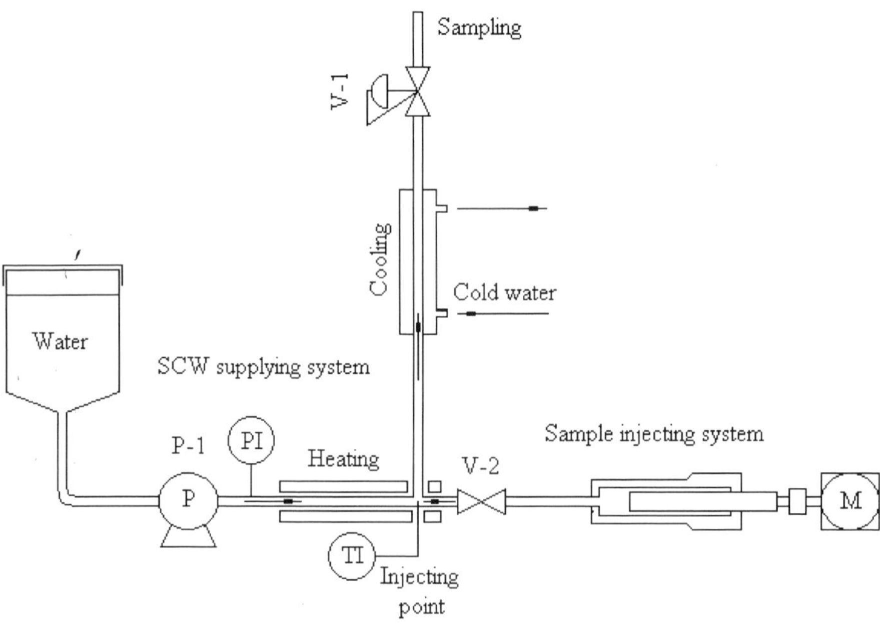

Figure 1. Scheme of the flow reactor simulating a submarine hydrothermal system: P-1– pump; PI – pressure indicator; TI – temperature indicator; V-1 – pressure regulating valve; V-2 – valve.

Table 1
Experimental conditions.

	Flow rate of SCW system, ml/min	Flow rate of inj. system, ml/min	Temp. at inj. point, ^{o}C	Pressure, MPa	Residence time at SCW, sec
Run I	10.0	5.0	386	39.5	14
Run II	20.0	5.0	387	39.5	9
Run III	30.0	5.0	394	39.5	6
Run IV	40.0	5.0	403	39.5	5
Run V	50.0	5.0	408	39.5	4

The residence time of glycine solution at supercritical conditions was calculated taking into account flow rate of the fluid and the length of the tube between the injecting point and the cooler. The experiments at five different flow rates of SCW were performed - 10.0, 20.0, 30.0, 40.0, and 50.0 ml/min (Table 1). The injection rate of the glycine solution was kept constant at 5.0 ml/min. The temperature of SCW was maintained at 400 ^{o}C, and that at the injecting point was 386-408 ^{o}C depending on the flow rate. The reaction mixtures were slightly brownish and slightly turbid. The pH values of the solutions were 8.16-8.66.

Analysis of the reaction products was performed by Shimadzu HPLC apparatus with an auto injector system. Shodex Asahipak column ODP-50 6E (6.0 mm ID, 150 mm L) was used. The mobile phase consisted of 5 mM KH_2PO_4 and 2.5 mM $C_6H_{13}SO_3Na$, and the pH was maintained to 2.7 by H_3PO_4 [13,14]. Flow rate of mobile phase was 0.5 ml/min. The compounds were detected by monitoring the absorbance at 200 nm. As standards, glycine and its oligomers up to hexaglycine were purchased from Sigma. The reaction mixtures were studied also with a thin layer chromatography (TLC). Four solvent systems were used for this purpose: n-butanol - acetic acid - water (4:1:1), phenol - buffer solution (water solution containing 6.3 % sodium citrate and 3.7 % monobasic sodium phosphate) - acetic acid (2000:400:1), n-butanol - methanol –water – ammonia (10:10:5:2), and phenol – methanol - buffer solution – acetic acid (2000:1500:400:1). The compounds were visualized by the appropriate spraying reagents (ninhydrin, chlorine – o-tolidine). To 0.4 ml of sample, 0.4 ml of 37% HCl was added and the mixture was hydrolysed at 110 ^{o}C for 24 hours in a sealed tube.

3. RESULTS AND DISCUSSION

At the HPLC profile (Fig. 2) of a reaction mixture the presence of diglycine and triglycine can be revealed together with diketopiperazine and glycine (raw material). Assignment of glycine oligomers was made by comparison of their retention times with those of their standards and mixture of them. The concentrations of diglycine, triglycine, diketopiperazine, and glycine were determined by referring the area of each corresponding peak to a standard reference of given concentration (Table 2). The linearity between the peak areas and the concentrations was confirmed. The concentration of triglycine is very low and can be calculated only in Run I.

The compounds assigned by HPLC was confirmed also by TLC. After hydrolysis of the reaction mixture, in HPLC chart the peaks of oligoglycines disappeared. On TLC plate only one spot of glycine was detected. Unidentified product **5** possesses high molecular weigh (MS: m/z= 433, 366, 351, 337, 327, 284, 269, 255, 245, 202, 187, 173, 163, 120, 105 (100%). It is ninhydrin and chlorine – o-tolidine positive and disappears after hydrolysis. The reaction mixture is too complex and needs further detailed studies. Small peptides (up to hexaglycine) can be formed [8], but proper confirmation of their structures is difficult. It is possible also some amino acids to be obtained due to the interaction of glycine with the products of decomposition [15].

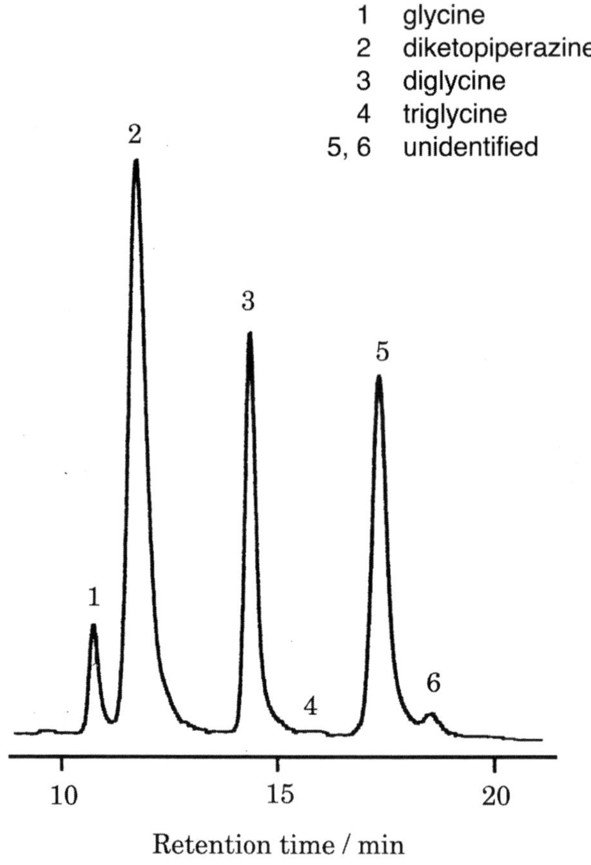

Figure 2. HPLC chromatogram of Run I.

Table 2
The concentrations of oligopeptides obtained at different flow rates.

Compounds	Concentration, mM				
	Run I	Run II	Run III	Run IV	Run V
Gly	37.90	28.62	20.79	13.49	9.67
Gly-Gly	3.67	0.68	0.45	0.21	0.16
Gly-Gly-Gly	0.01	-	-	-	-
Diketopiperazine	3.69	2.86	1.91	0.94	0.59

Our result shows that even at very high temperature (400 °C) almost 20 % of glycine remain unaffected by decomposition. High pressure (39.5 MPa) would suppress the decomposition. At different flow rates, there was no dramatic change in yield of different oligoglycines calculated based on the final concentration of glycine. This means that temperature and pressure determine the specific conditions at which these ratios of products are favored.

4. CONCLUSIONS

Glycine solution was injected to a water at supercritical state (pressure 39.5 MPa and temperature ~400 °C), in a flow reactor simulating submarine hydrothermal system. Diglycine, triglycine and diketopiperazine as well as one unidentified product with high molecular weigh (433Da) were obtained. It was found that chemical reactions, which lead to the formation of key for origin of life molecules, take place simultaneously with decomposition. Velocity of flow and the concentration of initial solution as well as temperature and pressure determine the ratios of the products.

There are three main factors necessary for the abiotic synthesis of peptides- amino acids, water, and heat. All these conditions are a part of global hydroternal system, which is proposed to have existed during the entire length of Earth's history. Together with other conditions like reducing environments, temperature and pressure gradients, presence of different metal catalysts and minerals, and protective environments make a hypothesis of the origin of life in a submarine hydroternal system quite plausible.

REFERENCES

1. J. A. Baross, S. E. Hoffman, Orig. Life, **15** (1985) 327.
2. J. B. Corliss, J. A. Baross, S. E. Hoffman, Oceanol. Acta No SP (1981) 59.
3. T. Gamo, H. Chiba, H. Masuda, H. N. Edmonds, K. Fujioka, Y. Kodama, H. Nanba, Y. Sano, Geophys. Res. Lett. **23** No 23 (1996) 3483.
4   N. J. Holm, Marine Hydrothermal Systems and the Origin of Life, Kluwer Academic Publishers, Dordrecht, 1992.

5. J. L. Bada, Philos. Trans. R. Soc. London, **333** (B) (1991) 349.
6. J. L. Bada, S. L. Miller, M. Zhao, Orig. Life Evol. Biosph., **25**, (1995) 111.
7. S. L. Miller, J. L. Bada, Nature **334** (1988) 609.
8. E. Imai, H. Honda, K. Hatori, A. Brack, K. Matsuno, Science **283** (1999) 831; E. Imai, H. Honda, K. Hatori, K. Matsuno, Orig. Life Evol. Bios. **29** (1999) 249.
9. R. W. Shaw, T. B. Brill, A. A. Clifford, C. A. Ecketr, E. U. Franck, Chem. Eng. News, December 23 (1991) 26.
10. B. R. T. Simoneit, Orig. Life Evol. Biosph. **25** (1995) 119.
11. E. L. Shock, Orig. Life Evol. Biosph. **20** (1990) 331.
12. K. Matsuno, R. Swenson, BioSystems **51** (1999) 53.
13. D. S. Kalonia, S. Musunuri, J. Tanglertpaibul, J. Chromatogr. **475** (1989) 416.
14. I. G. Tucker, M. E. Dowty, M. Veillard, M. A. Longer, J. R. Robinson, Pharml. Res. **6** No 1 (1989) 100.
15. Ch. P. Ivanov, O. Ch. Ivanov, R. A. Simeonova, G. D. Mirkova, Orig. Life, **13** (1983) 97.

Trends in High Pressure Bioscience and Biotechnology
R. Hayashi (editor)
© 2002 Elsevier Science B.V. All rights reserved.

Some Additional Remarks on a High Pressure Conference

J. Claude Cheftel

Unité de Biochimie-Technologie Alimentaires, Université des Sciences et Techniques
F-34095 Montpellier, France

It is always difficult to present "concluding remarks" for a multidisciplinary conference, especially when one's competence is mainly related to one aspect, namely Food Science and Technology.

In the first place, a small **survey of high pressure articles published in 1999 and part of 2000,** based on the Current Contents in Biology, Agriculture and Environmental Sciences, is shown on the following figure. This survey includes Microbiology, Food Science and Technology and some Biotechnology, but not basic Biochemistry or Molecular Biology. Among some 175 papers published in scientific journals, 116 come from European laboratories, and it is true that many European teams are now involved in high pressure research, partly due to recent European projects and contracts. Not shown on the figure is the fact that many studies concerning the effects of high pressure on foods of animal and plant origins have been carried out in Spain. The figure also indicates the distribution of papers among different disciplines and food commodities. Many papers, both in the section "micro-organisms" and in different food sections, deal with microbial inactivation. Studies on food proteins, as modified by high pressure, are also quite frequent, and found in the different sections. In addition to the various food commodities, studies also deal with enzyme inactivation, cold temperature phenomena associated to high pressure, pressure-assisted diffusion, process uniformity, etc.

It would be of interest to consider the sub-distribution of papers, but only the "micro-organisms" and "dairy products" sections will be mentioned here as examples. Each of the following sub-sections contains a significant number of papers:
    - inactivation of pathogenic micro-organisms;
    - a search for more efficient combined processes for inactivation, possibly at lower pressure levels;
    - attempts to understand the structural and metabolic targets of microbial inactivation;
    - effects on the denaturation or gelation of individual dairy proteins; on the casein micelle assembly; on casein coagulation and curd formation;
    - effects on cheese ripening: in some cases, proteases and lipases are released from bacteria and enhance the rate of ripening;
    - effects on cheese-brining, cheese sanitation, yoghurt preparation.

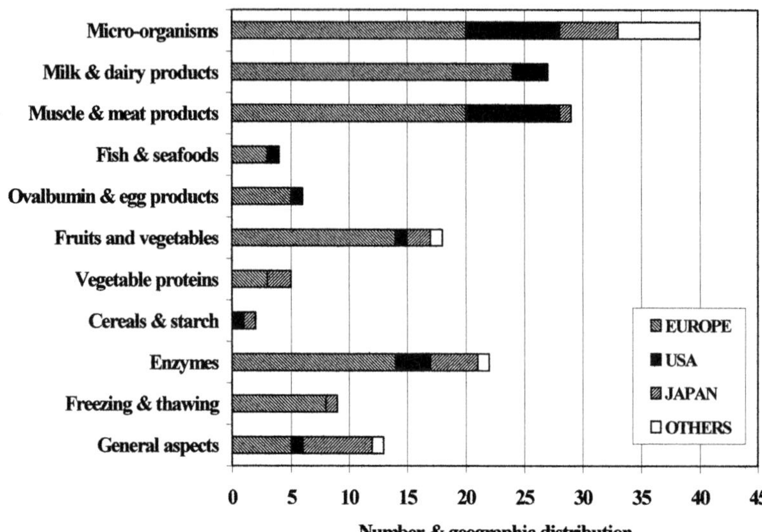

A general trend in food commodity papers is to concern either basic studies on biochemical constituents (e.g. muscle proteins) and micro-organisms or applied studies on product characteristics (e.g. sausages) or process parameters.

The presentations at this conference expanded the knowledge on several of these themes.

Before the conference, Dr C. Balny gave me his comments **on significant recent advances in high pressure Bioscience**. His list appears to be in good agreement with what was presented during the conference:

- high pressure as a tool for studying protein conformation (intermediate steps in protein denaturation and enzymatic catalysis; reversible modifications which can be observed under pressure, by NMR, UV or fluorescence spectroscopy; dissociation of oligomers; aggregation phenomena; protein-ligand interactions; subzero temperature denaturation of proteins in the liquid state; balance between osmotic and hydrostatic solvatation effects; effects on membranes). Impressive such studies and new HP devices for in situ observation were presented here;

- physiology of deep sea organisms and metabolism of micro-organisms under pressure. Thorough studies were certainly presented here on the effects of HP on microbial metabolism;

- potential biomedical applications of high pressure. Dr Balny specially insisted on this point, being surprised that no industrial application had yet occurred, in spite of obvious advantages for the following activities: inactivation of micro-organisms and viruses in pharmaceutical & cosmetic preparations and plasma products; preparation of vaccines; refolding of recombinant proteins; synthesis of given pharmaceuticals in pressure reactors; entrapment of pharmaceuticals in gels or emulsions for protection or slow release; mild freezing processes and subzero storage of biological samples without crystal formation.

Of course the food industry can now boast of several **commercial pressurised food products in Japan, Europe and the USA**, and of impressive **large scale HP machines**. The academic community should not claim credit for these achievements, although it did contribute to them in indirect ways.

The 400 L machine (from ACB Pressure Systems) processing pre-sliced vacuum-packed ham at the Espuna company, near Barcelona, has been shown. ACB has recently installed two such machines in a food plant in north America. Flow International has several large machines operating in the USA on guacamole (Avomex Co.), oysters (Motivated Seafoods Co. near New Orleans), and other products soon to come out. Kobe Steel has machines processing rice at the Echigo Seika Co., as has been reported here. It was also mentioned during the conference that pressurised seafoods were recently commercialised in Japan (kippers by Hokurei Co. and smoked salmon by Kibun Co).

An important point that was not much discussed concerns the **regulatory aspects of introducing pressurised foods on the market**. For a clearer understanding of the problems involved, it may be useful to recall that food stability (microbial, enzymatic, chemical, physical) rests on a complex interplay of several parameters and processes. High pressure application can contribute to microbial (and to a lesser extent enzymatic) stability, but does not play the major role. Also, high pressure alone has little effect on bacterial spores. High pressure must therefore be combined to at least one other parameter (such as an acid pH, the presence of sodium chloride and nitrite, cold temperature storage) or process (such as moderate heating).

The situation of the few commercial pressurised foods with respect to regulations is quite diverse, and it must be realised that the indications given below were not available in published form, and therefore may contain some inaccuracies.

The French orange juice (refrigerated storage, no heat processing, HP processing mentioned on the label) was put on the market before the European Novel Food Directive came out (in May 1997). Permission was requested from the French regulatory administration on the basis of microbial stability for an expanded chilled storage life. A temporary permission was granted and later expanded. This juice could be sold in other E.U. countries, but this is probably not the case because production has remained on a small scale.

The Spanish cooked ham was commercialised after the European Novel Food Directive, without asking for any permission. But it was first (and still is) exported to south America in replacement of a much lower quality product where the slices of cooked ham were cooked (i.e. pasteurised) a second time, after vacuum-packing. HP processing improved the quality and safety, extended the storage life to 1-2 months (depending on the distribution temperature), as necessary for export, and even apparently reduced the packaging plus processing cost. Now the product is also sold, refrigerated, by a Spanish group of supermarkets and department stores. The label of this good quality product mentions that it is pasteurised by high pressure (but it has also been treated with sodium chloride and nitrite, and

640

cooked, as other cooked sliced hams). It seems that no permission whatsoever has been requested or required so far, but the intention of the firm is to prove "Substantial Equivalence" in terms of composition, nutrition and safety, and on this basis to submit a simplified "Notification Procedure" to Brussels. Other E.U. countries may however ask questions and/ or raise objections before the product is allowed on the E.U. market.

A third example is given by the Danone group in France. The group wanted to use pressurised preparations of fresh fruit pieces in a syrup. The products are not yet on the market and the final presentations not yet disclosed (fruit salads, fruit pieces in yoghurts or dairy desserts, or other). The group decided to file a fully documented request under the European Novel Food Directive, and has finally obtained approval, after all E.U. countries were "consulted". However, this procedure took more than two years, and required microbiological, chemical and toxicological tests. This was requested although the product was acid and intended for refrigerated storage and distribution, simply because thermal pasteurisation was being replaced by HP pasteurisation without heating.

The cost and delay of this Novel Food procedure is strongly criticised by some food industries because it partly blocks innovation, but the European public opinion has become very sensitive to food safety issues.

In comparison, the Avomex guacamole product was apparently commercialised in the USA without the need for formal approval by the Food and Drug Administration. It was only necessary to prove that no chemicals leaked from the packaging material into the food as a result of HP processing.

Several **international high pressure conferences** have been held since the first Euro-Japanese Bioscience & Biotechnology meeting in La Grande Motte in 1992. They can be summarised as follows: Biophysics & Enzymology (Madison, 1994); Food Science & Technology (European, Montpellier, 1995); Food Science & Technology (IUFoST, Budapest, 1995); Multidisciplinary (EHPRG + AIRAPT, Warsaw, 1995); Bioscience & Biotechnology (Japan-Europe, Kyoto, 1995); Bioscience & Biotechnology (Euro-Japan and EHPRG 34, Leuven, 1996); Food Science, Bioscience & Chemistry (EHPRG 35, Reading, 1997); Multidisciplinary (AIRAPT 16, Kyoto, 1996); Bioscience & Biotechnology (Euro-Japan, Heidelberg, 1998); Multidisciplinary (EHPRG 37, Montpellier, 1999); Multidisciplinary (AIRAPT 17, Honolulu, 1999); Emerging Food Science & Technology (EFFoST, Tampere, 1999).

The present conference will be followed by two multidisciplinary symposia in 2001 (AIRAPT 18, Beijing and Shanghai; EHPRG 39, Santander).

One should not forget the 11 Japanese HP Bioscience & Food Science meetings already organised by Professor R. Hayashi and his colleagues.

Many of the above mentioned meetings have resulted in published proceedings. To account for the intense research activity in the high pressure field, one should also mention the sessions devoted to minimal food processing during the annual conventions of the American Institute of Food Technologists, the 3 European projects in HP Food Science and Technology

(1992-99), the European COPERNICUS concerted action (1994-97) entitled "Process optimisation and minimal processing of foods", the European COST projects "HP Biotransformations (Enzymes & Biopolymers), the NATO school on HP Molecular Science (Lucca), several meetings on "Extreme conditions", the French HP network (CNRS), etc.

The high proportion of young scientists present at this Bioscience & Biotechnology conference also speaks for the vitality of high pressure biological research.

It is a pleasure to end these remarks with a tribute to the city of Kyoto and to the conference organisers. During our stay in Kyoto, we have crossed several landmarks of Japanese culture and dynamism. We have been within close view of the peace and beauty of the foothill forest and temples. We have enjoyed the friendly and perfect organisation of this conference in the impressive Kyoto International Conference Hall. And we are deeply grateful to Professor R. Hayashi for his long lasting promotion of high pressure studies.

SUBJECT INDEX

AUTHOR INDEX

* Conference participants are underlined.